39th International Symposium on Remote Sensing of Environment (ISRSE 39)

From Human Needs to SDGs

International Archives of the Photogrammetry, Remote Sensing and Spatial Information Sciences XLVIII-M-1

Antalya, Turkey
24 – 28 April 2023

Editors:

Orhan Altan
Filiz Sunar
Doris Klein

ISBN: 978-1-7138-7681-6

Printed from e-media with permission by:

Curran Associates, Inc.
57 Morehouse Lane
Red Hook, NY 12571

Some format issues inherent in the e-media version may also appear in this print version.

International Center for Remote Sensing of the Environment, Inc. (ICRSE)
1955 E. Sixth Street, Suite 115
Tucson, AZ 85719
USA

isrse@symposia.org

Additional copies of this publication are available from:

Curran Associates, Inc.
57 Morehouse Lane
Red Hook, NY 12571 USA
Phone: 845-758-0400
Fax: 845-758-2633
Email: curran@proceedings.com
Web: www.proceedings.com

TABLE OF CONTENTS

Foreword of the Editors

The world is currently undergoing one of the most transformative periods of the past century, brought about by the social, economic, and political impacts of the COVID-19 crisis. Additionally, the effects of climate change and population growth and their impact are becoming more and more apparent. Although we have already seen significant effects, we do not yet fully understand the magnitude of the crisis. As we move forward, we are faced with the challenge of making critical public health, economic, governance, and ecological decisions with limited information, and reversing recently adopted policies, which may lead to significant disruptions in sustainability. In these unprecedented times, it is crucial to fully understand the Earth's system and transform our needs and how we meet them to ensure a sustainable future for all.

Human needs can be defined as the required goods and services necessary for human survival, such as food, clothing, shelter, and healthcare. Throughout history, humans have undergone various stages of evaluation regarding their needs, and it is vital to explore the interrelation between human needs and their evolution over time, as well as future prospects for a sustainable society. Scientific advice and evidence-based decision-making in all matters are essential to meet this goal.

By using Earth Observation, we nowadays have the possibility to observe, quantify and analyze for decades the ongoing changes of our world from space. To understand the current situation and to anticipate the interventions that are needed, we have to understand the impact of human behavior over time on the world in which we live today. Accordingly, we must take responsibility for making positive changes based on these findings and facts.

We hope this collection of scientific and experimental approaches utilizing Earth Observation data will contribute towards the goal of achieving a sustainable future for Earth, by comprehending the current situation and forecasting the needs for a more sustainable management of our world to meet the human needs.

ORHAN ALTAN
Organizing Committee and
Programme Committee Chair
Istanbul Technical University
(ITU), Türkiye

FILIZ SUNAR
Programme Committee Co-chair
Istanbul Technical University
(ITU), Türkiye

DORIS KLEIN
Programme Committee Co-chair
Deutsches Zentrum für Luft-
und Raumfahrt, DLR, Germany

https://doi.org/10.5194/isprs-archives-XLVIII-M-1-2023-1-2023

PREDICTION OF LONG-TERM SENTINEL-1 INSAR TIME SERIES ANALYSIS

S. Abdikan[1]*, S. Coskun[2,3], O.G. Narin[4], C. Bayik[5], F. Calò[6], A. Pepe[6], F. Balik Sanli[7]

[1] HU, Dept. of Geomatics Engineering, 06800 Beytepe Ankara, Turkey – sayginabdikan@hacettepe.edu.tr
[2] YTU, Graduate School of Science and Engineering, Istanbul, Turkey
[3] Ministry of Environment, Urbanization and Climate Change, Bilecik, Turkey - suatcoskun@hotmail.com
[4] AKU, Dept. of Geomatics Engineering, Afyonkarahisar, Turkey - gokberknarin@aku.edu.tr
[5] ZBEU, Dept. of Geomatics Engineering, 67100 Zonguldak, Turkey - caglarbayik@beun.edu.tr
[6] CNR - Institute for the Electromagnetic Sensing of the Environment (IREA), Napoli, Italy - (calo.f, pepe.a)@irea.cnr.it
[7] YTU, Dept. of Geomatic Engineering, Davutpasa Esenler, Istanbul, Turkey - fbalik@yildiz.edu.tr

KEY WORDS: InSAR, time-series, LSTM, landslide, Istanbul, Sentinel-1

ABSTRACT:

This paper presents an initial analysis of predicting time series derived from long-term interferometric Synthetic Aperture Radar (InSAR) data. Time series analysis provides insights into the temporal evolution, variation, and dynamic nature of events. In this study, we focus on the Istanbul region, which is the most populous city in Turkey and spans both Europe and Asia. While the area is prone to seismic risks caused by active tectonic faults, it is also susceptible to other risks due to various phenomena. Therefore, this study investigates landslides triggered by geological structure and human-induced activities, particularly in Tepekent, a landslide-prone area in the town of Buyukcekmece located on the European side. We utilized the StaMPS persistent scatterer InSAR (PSI) method to detect slow movements over time. A total of 157 archive Copernicus Sentinel-1 data, acquired over the region between June 2017 and August 2022, were processed, primarily observing human structures with a maximum displacement amount of approximately 1 cm/year. About 500 persistent scatterer points were identified in the region, and the time series was formed by taking the average of these points. The Long Short-Term Memory (LSTM) neural network method was used to estimate motion. We trained the model with data from the first three years of the time series and used the data from the remaining two years for estimation, while the accuracy analysis was performed with the 5-year time series data. The RMSE values for the training and test data were determined to be 0.725 mm/year and 0.656 mm/year, respectively. Additionally, we estimated the time series for the period from August 2022 to August 2024. The observation and prediction results could be beneficial in developing efficient mitigation risk actions and sustainable urban management strategies.

1. INTRODUCTION

Geohazards represent one of the major concerns in many countries being also exacerbated by the climate changes. Among countries, Turkey is one of the most vulnerable to earthquakes, landslides, fires and floods (Tatar, 2020). Particularly, Istanbul which is the most crowded city of Turkey and Europe is threatened by the North Anatolian Fault that passes through the Marmara Sea representing a relevant source of risk for the city (Ergintav et al., 2011). Meanwhile, there are also small-scale local movements affecting many people and infrastructures of the city. Previously, several studies based on using multi-temporal Interferometric Synthetic Aperture Radar (SAR) techniques, showed the long-term surface displacements triggered by both man-made activities and natural events. Persistent Scatterer Interferometry (PSI) and Small Baseline (SB) techniques have been extensively employed in previous studies to analyse long-term displacement, particularly in large regions and urbanized areas (Ferretti et al., 2001; Berardino et al., 2002; Hooper et al., 2012; Calò et al., 2015; Aslan et al., 2018; Bayik et al., 2021).

Calò et al. (2015) presented one of the first studies over the old peninsula of Istanbul using high resolution TerraSAR-X data. The authors applied SB approach and identify local subsidence areas along the Golden Horn (Halic) where filling and recreation areas are located. The one and half year period of data results indicated displacements up to 4 cm at underground subway stations. The geological data showed that most of the subsidence occurred in the Quaternary layers. Aslan et al. (2018) showed 25 years of movements occurring in Istanbul city using ERS, ENVISAT and Sentinel-1 data processed through PSI method. The authors presented the localized movements on recreational areas, coastal zones, river valley and sliding regions in Istanbul city center. Halicioglu et al. (2021) reported the surface displacements of metro constructions in the Asian side of Istanbul by using Sentinel-1 and PSI approach for the period of 2015 and 2018, and in situ levelling measurements. Yagmur et al. (2022) monitored the site of the

* Corresponding author

https://doi.org/10.5194/isprs-archives-XLVIII-M-1-2023-3-2023

new Istanbul Airport. The authors applied both SB of LiCSAR approach and STAMPS of PSI approach, and showed the slow-developing surface displacement of runway between for the period 2018 and 2021 using Sentinel-1 data.

The analysis of landslide area nearby the case study was carried out by Bayik et al. (2021) using five years (2015-2020) of Sentinel-1 data. The authors investigated the landslide phenomena at the east of Lake Buyukcekmece, over residential areas. They also highlighted the contribution of L-band over C-band data. Previous studies, focused on mapping surface displacements and determining the mean velocity, provided the current situation of the deformation in the area. However, its prediction for future years can play an essential role to prevent/mitigate the associated risks. Recent studies showed the potential of prediction models for InSAR time series. In particular, Long Short-Term Memory (LSTM) neural network method that use time dependency of archived data, is exploited to provide accurate prediction results (Bao et al., 2022). The method has been applied for displacement prediction of the airport in Shanghai (Bao et al., 2022), in Beijing (Chen et al., 2021), of the land subsidence in Iran (Radman et al., 2021) and in the urban area of Wuhan (Ding et al., 2021) using InSAR time-series data.

In this study, the LSTM approach is applied for the prediction of time series related to a landslide-induced deformation. In the analysis, Sentinel-1 data were used, and PSI method was applied to retrieve the slow movements associated with landslides near Buyukcekmece Lake in western Istanbul (Figure 1). The study is aimed at predicting time series of the displacements, to understand the behaviour of the trend and, possibly, the total displacement value.

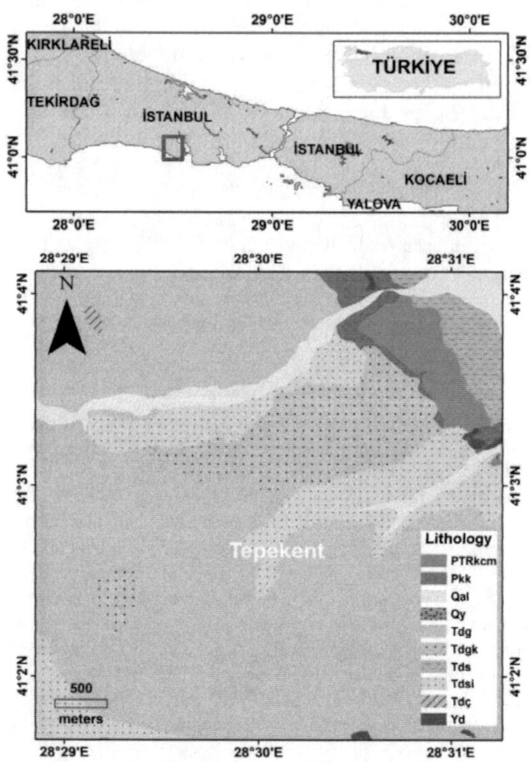

Figure 1. The study area.

2. DATA AND METHOD

2.1 The study Area

The study area is located in the European side of Istanbul, a megacity wherein more than 15 million inhabitants live (Figure 1). In particular, the region under investigation is located in the Buyukcekmece district, prone to two main types of hazards: earthquakes and slow-moving active landslides, threatening thousands of residents. Lake Buyukcekmece is located at the northeast of the study area where a long landslide history exists (Duman et al., 2006). Southern part is bordered with the Marmara Sea.

A large part of the region consists of lithological units such as Oligocene (Tdg, Tdgk, Tds), Miocene (Tdc, Tdsi) (Figure 1). Among the units, Tdsi is composed of ash brown, thick bedded, coarse-grained sandstone-conglomerate, Tdd and Tdgk are greenish, dark gray, purple, mottled, claystone-shale; interbedded with sandstone and claystone, interbedded with thin lignite structures. The rest are alluvium (Qal), artificial filing (Yd), phyllite, garnet schist, graphite schist, calcschist (PTRkcm), and Kızılağaç metagranite formation (Pkk) (Seyyar et al, 2020).

2.2 Sentinel-1 Data Analysis

In order to analyse the landslide kinematics, 157 Copernicus Sentinel-1A data taken along descending orbit between June 2017 and August 2022, were acquired through NASA Alaska Satellite Facility (ASF), as shown in Table 1. The Single Look Complex (SLC) images were pre-processed with SNAP 9.0, and StaMPS (4.1 beta) approach of PSI was utilized for the analysis of deformation time series and mean velocity. Among the images, 03.02.2020 dated one was selected as primary, and 156 interferograms were generated using SNAP software. (SNAP 2022). The Toolbox for Reducing Atmospheric InSAR Noise (TRAIN) was used to reduce the atmospheric effects (Bekaert et al., 2015). The PSI approach identifies persistent scatterers based on their amplitude dispersion and interferometric phase correlated in space. We considered an amplitude dispersion D_A value of 0.4 to eliminate non-PS candidate pixels. The DA is calculated as;

$$D_A = \frac{\sigma_A}{\mu_A} \tag{1}$$

where, σ_A = standard deviation
μ_A = mean of amplitude

Specifications	Description
Sensor	Sentinel-1A
Wavelength	C-band
Orbit	Descending
Acquisitions	06.06.2017-29.08.2022
Image size	157
Pixel spacing	14 m (Rg:4 Az:1)
Polarization	VV

Table 1. Specifications of the Sentinel-1 data used for the study

https://doi.org/10.5194/isprs-archives-XLVIII-M-1-2023-3-2023 | © Author(s) 2023. CC BY 4.0 License.

2.3 The Long Short-Term Memory

To predict time series monitored by InSAR, we exploited the the Long Short-Term Memory (LSTM) approach, which offers advantages, particularly for historical datasets and nonlinear time series. A feedback mechanism has been added to the code of the Recurrent Neural Network (RNN), which is one of the feedforward networks structures, by making additions (Hochreiter and Schmidhuber, 1997). In this way, LSTM can access the input and output information realized in the previous steps when they needed to. The general structure of LSTM comprises layers of forget gate, update gate, Tanh, and output gate layers (Bao et al. 2022). In our study, we utilized the LSTM model developed with the keras package (Url-1), and the analyses were performed using the open-source R programming language (Url-2). As per the code, data normalization is mandatory, and for this purpose, the recipes package was used in the code. Here, the square root is taken while normalizing. The code selected Batch size= 1, Time steps=1, epochs =100. We employed Root Mean Square Error (RMSE) (eq. 2) and %RMSE (eq.3) to assess the accuracy of the test data and determine the accuracy rate, respectively.

$$RMSE = \sqrt{\frac{1}{n}\sum_{i=1}^{n}(y_i^{act} - y_i^{cal})} \tag{2}$$

$$\%RMSE = \sqrt{100 * \frac{1}{n}\sum_{i=1}^{n}(y_i^{act} - y_i^{cal})} \tag{3}$$

where n= the number of samples
y_i^{act} = is the 2-year LOS value
y_i^{act} = is the 2-year estimated LOS value

3. RESULTS AND DISCUSSION

The temporal and geometrical baselines of the 156 interferograms created based on a single image are depicted in Figure 2. The maximum temporal baseline spans approximately 3 years and the temporal baseline is less than 200 m. The findings of PSI analysis revealed that displacement velocity along the Line of Sight (LOS) ranged between -10 to 2 mm/year with a standard deviation of less than 0.5 mm/year, as shown in Figure 3.

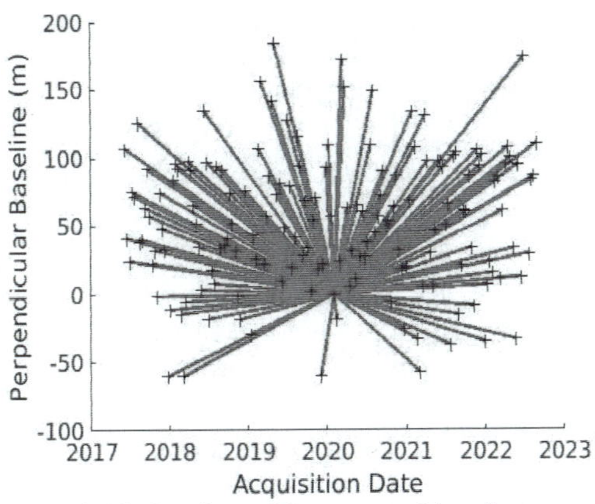

Figure 2. The baselines of the computed interferograms.

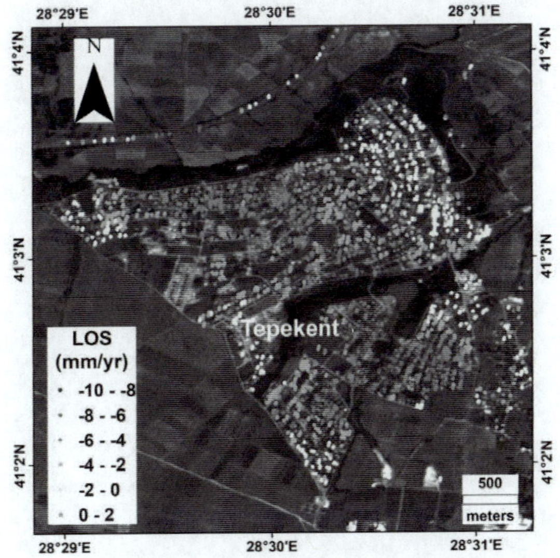

Figure 3. PSI deformation velocity map.

The majority of high movement and PS points were located over the coarse-grained sandstone-conglomerate, claystone, and sandstone structures. PS measure points were predominantly identified on building structures, and along the road line situated in the northern region. As anticipated, no PS points were detected in the agricultural areas surrounding the Tepekent region. Notably, PSs with higher displacements were primarily identified in the valley situated in the central part of the study area. To demonstrate the relationship between the spatial distribution of moving points and the topography, two perpendicular profiles were analysed (Figure 4 and 5). In the A-B profile extending from the north to the southeast, the deformation value increases and decreases in parallel with the increase and decrease in the topography. In the C-D profile, extending from west to east, the movement increased as the elevation decreased. After reaching the minimum at the valley floor, the movement decreased with the height towards the east (Figure 5).

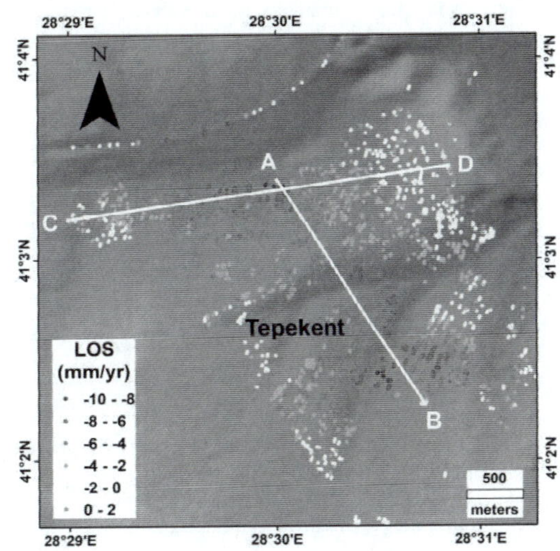

Figure 4. Profiles along the topography and PSI velocity map.

https://doi.org/10.5194/isprs-archives-XLVIII-M-1-2023-3-2023

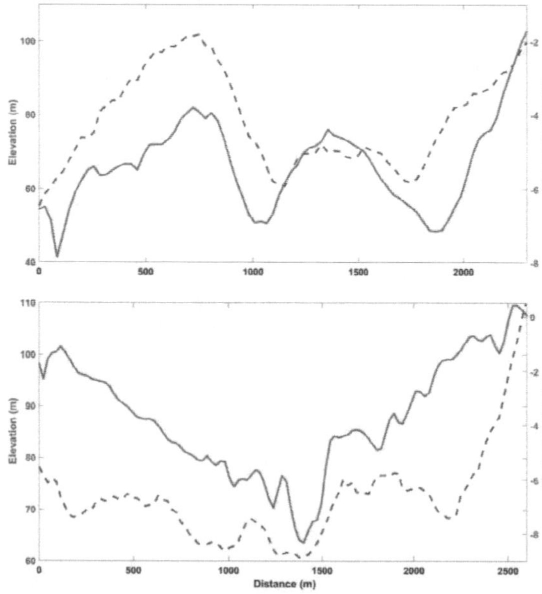

Figure 5. Profile of PSI displacements vs. profile of elevation

A mean time series was constructed by using an average of 513 points in time series analysis. The first three years of the time series (91 images), spanning from June 2017 to August 2020, were used to train the displacement model and predict the last two years (60 images) from August 2020 to August 2022. The predicted values were then compared with the monitored time series, and their accuracy was evaluated using the correlation coefficient (R), Root Mean Square Error (RMSE), and RMSE% evaluation indices. The R values for the train and test data were 0.988 and 0.986, respectively (Table 2). The RMSE values were 0.725 mm/year and 0.656 mm/year for the train and test data, respectively. Although the predicted time series exhibited a velocity for the test data, its RMSE% (8.766 mm/year) was approximately half that of the train data (15.749 mm/year). The results indicate a high level of agreement between the monitored and the predicted time series (Figure 6). Furthermore, a prediction for the next two years of the time series, from August 2022 to August 2024, was also performed (Figure 6).

	LSTM results	
	Train	**Test**
R	0.988	0.986
R^2	0.973	0.973
RMSE (mm/year)	0.725	0.656
%RMSE	15.749	8.766

Table 2. Results of the LSTM method

In a previous study, Chen et al., (2021) compared prediction models for ENVISAT time series data, including LSTM, Multi-Layer Perceptron (MLP) and Recurrent Neural Network (RNN) models. They analysed displacement at Beijing Capital International Airport using images from 2005 to 2010 and found that the cyclic prediction model of LSTM produced the best results, with RMSE of 4.60 mm.

Ding et al. (2021) used Multi-temporal InSAR approach on Sentinel-1 data acquired between 2018 and 2020 to study urban surface subsidence in Wuhan city. They analyzed time series data from eight points and found varying behaviours in the data with RMSE values ranging from 1.89 mm to 3.48 mm.

Bao et al. (2022) conducted an analysis of surface deformation at Shanghai Pudong International Airport using StaMPS PSI technique and Sentinel-1 data. They used data from 2016 to 2021 and applied the LSTM method to predict the time series. The researchers reported that they experimentally combined and determined the parameters for the model. The results showed an RMSE of less than 3.52 mm for the test data.

Radman et al. (2021) conducted research on the environment of Lake Urmia, Iran, using the SB approach to determine surface subsidence, with time series of Sentinel-1 data acquired between 2014 and 2019. In the study, LSTM, multi-layer perceptron (MLP) and convolutional neural network (CNN) methods were compared. In addition to vertical deformation derived from SB analysis, ancillary data such as precipitation, NDVI and groundwater data were included as inputs to the models. Overall, LSTM provided the lowest RMS of 8.1 mm, but CNN showed the lowest RMSE of 8.7 mm in testing results. Among all methods, CNN was considered to be the best approach for the estimation.

Previous studies have examined surface subsidence in various locations, including airports, cities, and natural areas. Two studies comparing different methods have shown that LSTM method provided the best results in one, while CNN was superior in the other. Both PSI and SB methods have been utilized for time series analysis. Considering the studies mentioned above, an RMSE of less than 9 mm was achieved. It should be noted that the differences in how the models are trained can impact their accuracy.

4. CONCLUSIONS AND FUTURE PERSPECTIVES

Due to intense human activities and its geological structure, Istanbul is vulnerable to several anthropogenic and natural hazards. In this study, we focus on investigating landslides that affect some neighbourhoods of the megacity. The time series prediction of InSAR - derived results is investigated through the application of the LSTM method. The approach has shown its potential with RMSE values lower than 1 mm. This proves LSTM method is a promising method for predicting landslide deformation time series, thereby contributing to geo-hazards related risk prevention and reduction.

In the analysis, only the mean of the PS points was used. As further research direction, the study will be extended to time-series of each point to predict spatial distribution of the landslide effects. Furthermore, the time series will be refined by eliminating the observed seasonal effects. Additionally, data obtained with long terrestrial measurements will be incorporated to determine whether the estimation results are more reliable.

ACKNOWLEDGEMENTS

The paper is part of the M.Sc. thesis study of Suat Coskun

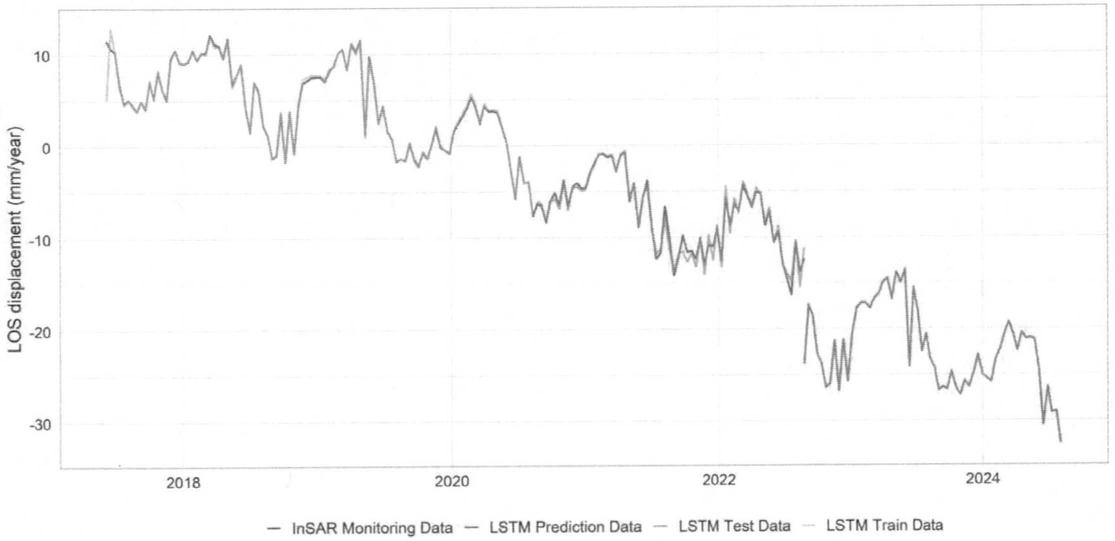

Figure 6. Time series of monitored and predicted displacements. The vertical axes show the LOS displacement velocity (mm/year).

REFERENCES

Aslan, G., Cakır Z., Ergintav S., Lasserre C., Renard F., 2018: Analysis of Secular Ground Motions in Istanbul from a Long-Term InSAR Time-Series (1992–2017). *Remote Sensing* 10(3):408.

Bao, X., Zhang, R., Shama, A., Li, S., Xie, L., Lv, J., Fu, Y., Wu, R., Liu G., 2022: Ground Deformation Pattern Analysis and Evolution Prediction of Shanghai Pudong International Airport Based on PSI Long Time Series Observations. *Remote Sensing*. 14(3):610.

Bekaert, D.P.S., Walters, R.J., Wright, T.J., Hooper, A.J., and Parker, D.J. (2015), Statistical comparison of InSAR tropospheric correction techniques, *Remote Sensing of Environment*, 170, 40-47.

Bayik, C., Abdikan, S., Ozdemir, A., Arıkan, M., Balik Sanli, F., Dogan, U., 2021: Investigation of the landslides in Beylikdüzü-Esenyurt Districts of Istanbul from InSAR and GNSS observations. *Natural Hazards* 109, 1201–1220.

Berardino, P., Fornaro, G., Lanari, R., Sansosti, E., 2002: A new algorithm for surface deformation monitoring based on small baseline differential SAR interferograms. IEEE Trans. Geosci. Remote Sens. 2002, 40, 2375–2383.

Calò F, Abdikan S, Görüm T, Pepe A, Kiliç H, Balik Şanli F. 2015; The Space-Borne SBAS-DInSAR Technique as a Supporting Tool for Sustainable Urban Policies: The Case of Istanbul Megacity, Turkey. *Remote Sensing*. 7(12):16519-16536.

Chen Y., He Y., Zhang L., Chen Y., Pu H., Chen B., Gao L., 2021: Prediction of InSAR deformation time-series using a long short-term memory neural network, *Int. Journal of Remote Sensing*, 42:18, 6919-6942

Ding Q, Shao Z, Huang X, Altan O, Zhuang Q, Hu B, 2021: Monitoring, analyzing and predicting urban surface subsidence: A case study of Wuhan City, China, *Int. J. of Applied Earth Observations and Geoinformation* 102,102422.

Duman, T.Y., Can, T., Gokceoglu, C., Nefeslioglu, H.A., Sonmez, H., 2006: Application of logistic regression for landslide susceptibility zoning of Cekmece Area, Istanbul, Turkey. *Environ Geology*, 51:241–256.

Ergintav, S., Demirbağ, E., Ediger, V., Saatçılar, R., Inan S., Cankurtaranlar A., Dikbaş A., Baş, 2011: Structural framework of onshore and offshore Avcılar, Istanbul under the influence of the North Anatolian fault. *Geophysical Journal International* 185:93–105.

Ferretti, A.; Prati, C.; Rocca, F., 2001: Permanent scatterers in SAR interferometry. IEEE Trans. Geosci. *Remote Sensing*, 39, 8–20.

Halicioglu, K, Erten, E, Rossi, C. 2021: Monitoring deformations of Istanbul metro line stations through Sentinel-1 and levelling observations. *Environ Earth Sci.* 80(9):1–10.

Hochreiter, S.; Schmidhuber, J. Long short-term memory. *Neural Comput.* 1997, 9, 1735–1780.

Hooper, A., Bekaert, D., Spaans, K., Arıkan M., 2012: Recent advances in SAR interferometry time series analysis for measuring crustal deformation, *Tectonophysics*, 514-517, 1-13.

Radman, A., Akhoondzadeh M., Hosseiny B., 2021: Integrating InSAR and deep-learning for modeling and predicting subsidence over the adjacent area of Lake Urmia, Iran, *GIScience & Remote Sensing*, 58:8, 1413-1433.

Seyyar, N., Yildirim, Y.Y., Fitoz E., Yilmaz N., Aksoy, P., Talay, T., Bostancioglu, I., Duran, K., Kahrman T., 2020: İstanbul İli Beylikdüzü-Büyükçekmece-Esenyurt İlçeleri Heyelan Farkindalik Kitapçiği, Deprem Risk Yönetimi ve Kentsel İyileştirme Daire Başkanlığı Deprem ve Zemin İnceleme Müdürlüğü, Istanbul Buyukseir Belediyesi, Istanbul Turkiye,https://depremzemin.ibb.istanbul/wp-content/uploads/2020/07/Beylikduzu-Buyukcekmece-Esenyurt-Heyelan-Farkindalik-Kitapcigi.pdf

SNAP, 2022 - ESA Sentinel Application Platform v9.0 http://step.esa.int/main/

Tatar T., 2020. Review: Understanding of Natural Disaster Risk Management and Where Turkey Stands in the Picture, Academic Platform Journal of Natural Hazards and Disaster Management 1(2), 110-121

Yagmur N., Erten E., Musaoglu N., Safak E., 2022: Assessing spatio-temporal dynamics of large airport's surface stability, *Geocarto International,* https://doi.org/10.1080/10106049.2022.2082554

(Url-1) https://www.r-bloggers.com/2018/04/time-series-deep-learning-forecasting-sunspots-with-keras-stateful-lstm-in-r/

(Url-2) R Core Team (2020) R: A language and environment for statistical computing. R Foundation for Statistical

ANALYSIS OF THE REMOTELY SENSED WATER QUALITY PARAMETERS OF THE INSUBRIC LAKES: METHODS AND RESULTS OF THE INTERREG SIMILE PROJECT

J.F Amieva (1), A. Austoni (1) , M. Bresciani (2), M.A. Brovelli (1,2)

(1) Politecnico di Milano, Dept. of Civil and Environmental Engineering, Lecco Campus,
Via Gaetano Previati 1/c, 23900, Lecco, Italy
maria.brovelli@polimi.it, juanfrancisco.amieva@mail.polimi.it, alessandro.austoni@mail.polimi.it

(2) Istituto per il Rilevamento Elettromagnetico dell'Ambiente,
CNR-IREA, via Bassini 15, 20133 Milano – bresciani.m@irea.cnr.it , maria.brovelli@polimi.it

KEY WORDS: Water Quality Monitoring, Landsat-8 TIRS, Sentinel-3 OLCI, Chlorophyll-a, Total Suspended Matter, Lake Water Surface Temperature, Spatial and Temporal Series.

ABSTRACT:

Lakes are a fundamental component of the environment and the territory and represent a precious source of fresh water for various uses. The area of the Prealps north of the Po valley in Italy is characterized by the presence of lakes which represent almost 80% of the total volume of fresh water in Italy (Rogora et al., 2018). The Insubric lakes (Lugano, Maggiore and Como) have their shared basins between Italy and Switzerland, and they are the objective of the SIMILE project, a cross-border Italian-Swiss project that aims to improve their coordinated management and strengthen stakeholder participation in the processes of knowledge and monitoring of water resources (Brovelli et al., 2019) by analyzing data acquired from in-situ to satellite sensors. The present work refers to data collected by remote sensing methods which offer the possibility to obtain synoptic views of water bodies to monitor water quality parameters (WQPs) such as the chlorophyll-a (Chl-a), the total suspended matter (TSM) and the lake surface water temperature (LSWT) (Giardino et al., 2013). This work presents an extensive evaluation of the space-time trends of the parameters based on the SIMILE remote sensing database.

Dataset	Number of maps	Initial acquisition	Final acquisition
Chl-a	389	15/01/2019	5/11/2022
TSM	392	15/01/2019	5/11/2022
LWST	69	01/01/2019	11/09/2022

Table 1. Summary of available maps per dataset.

1. INTRODUCTION

The paper presents the different analyses that were performed in order to understand the main spatial and temporal trends of some parameters in three insubric lakes, which are part of the SIMILE project: Lake Maggiore, Lake Como and Lake Lugano. The analyses were based on a previously produced set of maps which comprises 3 time-series datasets: chlorophyll-a (Chl-a), Total Suspended Matter (TSM) and Lake Surface Water Temperature (LSWT). Table 1 details the number of maps available in each dataset with the initial and final acquisition dates.

The main objective of the work is to determine if in the analyzed datasets there are trends which can be related to particular moments of the year but also determine if there are present trends in specific areas of the lakes. For the temporal analyses, it was necessary to adopt the seasons defined in the Water Framework Directive (WFD) (Bresciani et al., 2011), according to which are specified the following six time intervals (called, for brevity, "seasons"). The initial and final dates of each of the WFD seasons are specified in the table 2.

WFD Season	Initial date	Final date
Winter	01-Jan	20-March
Spring	01-Apr	15-May
Transit. spring-summer	15-May	15-Jun
Summer	01-Jul	31-Aug
Transit. summer-autumn	01-Sep	01-Oct
Autumn	01-Oct	30-Nov

Table 2. Water Framework Directive seasons' initial and final dates. (Bresciani et al., 2011)

2. MATERIALS AND METHODS

2.1 Analyzed parameters

Remote sensing is one of the possible tools that can help researchers and decision-makers to determine which are the conditions of a lake and, according to them, define the actions to be taken to properly manage its state in the challenging context that nowadays every lake exhibits due to climate change. Among the parameters that could be evaluated according to (Remote sensing for lake research and monitoring – Recent advances, 2016) are the following ones:

- **Transparency:** in (Remote sensing for lake research and monitoring – Recent advances, 2016) it is mentioned the importance of keeping good levels of transparency because it facilitates the penetration of the light in the lakes, allowing "biological, chemical and physical processes such as primary production and formation of macrophytes" (Remote sensing for lake research and monitoring – Recent advances, 2016). In this project, it was used the Total Suspended Matter (TSM) as an indicator for its measurement.

The suspended matter attenuates the lake transparency; it is typically present in lakes due to the contributions of lake tributaries and originates from soil and bedrock erosion or from internal resuspension (Remote sensing for lake research and monitoring – Recent advances, 2016).

- **Chlorophyll-a:** in order to understand the level of nutrient availability which can be a measure of the trophic level of a lake, it is possible to evaluate the amount of phytoplankton which is present in the studied waterbodies. For this purpose, it was adopted the Chloropyll-a indicator, which is a proxy of the level of phytoplankton. In order to determine the value of chlorophyll-a, it is necessary to analyze the absorption in the spectrum that ranges from 440 nm to 560 nm and also at 670nm (Remote sensing for lake research and monitoring – Recent advances, 2016).

- **Temperature:** According to (Remote sensing for lake research and monitoring – Recent advances, 2016), the water temperature is the product of several energy fluxes and it is a factor that affects different aspects of the ecosystem functioning, determining changes in the biodiversity of lake ecosystems and their oxygen concentrations. Specifically, thermal remote sensing instruments provide the capability to derive the lake's surface water temperature.

The analysis of the before-mentioned parameters in time is something that helps to understand which are their potential interactions and it contributes to the determination of the evolution of the global state of the lake's water quality. As was introduced before, this is one of the main objectives of this project.

2.2 Image processing

The data analyzed in the current work was produced in the first part of the SIMILE project and the main aspects of the processing are mentioned in (Toro Herrera et al., 2022). According to (Toro Herrera et al., 2022), in order to obtain the Chl-a and the TSM maps, it was necessary to use images provided by the ESA's Sentinel-3 A/B OLCI instrument which have a daily revisiting time with spatial resolution of 300 m. On the other hand, to produce the LSWT maps it was employed the Landsat-8 TIRS instrument of NASA whose termal infrared sensor has a spatial resolution of 100 m and a revisiting time of 16 days. Regarding the processing techniques, in (Toro Herrera et al., 2022) it is also mentioned that the Case 2 Regional Coast Colour (C2RCC), presented in (Brockmann et al., 2016), was employed to perform radiometric and atmospheric corrections as well as to compute Chl-a and TSM concentrations using Sentinel-3 data by means of a Neural Network model and the LSWT was determined using the Barsi method. The Barsi method is described in (Barsi et al., 2005) In (Toro Herrera et al., 2022) it is also explained that the C2RCC neural net flags were exploited to determine anomalies in the water spectra and also to determine the possible presence of clouds. The pixels with flags were excluded. Finally, (Toro Herrera et al., 2022) describes that in order to remove outliers, the 3 σ rule was applied.

2.3 Data availability

Considering that the different datasets were derived from multispectral imagery, it is possible to anticipate that there are some environmental conditions that may produce the absence of their values or that could also produce undesired values. For this reason, given the different datasets, it was quite important to estimate which are the levels of data availability in each pixel because, according to those levels, it was possible to select which pixels were the proper ones for the later analyses performed in this project. To determine the percentage of available data in each pixel of the different lakes for each of the three studied datasets, it was taken into consideration the total number of maps present for each dataset and then this number was compared with the total count of times that the same pixel did not get a null value.
Finally, were studied the spatial patterns of the data availability by means of a set of maps that analyzed the percentages of available data for all the pixels in each of the datasets by considering the full period of acquisitions but also for each season.

2.4 Statistical analyses

The project has also taken into account statistical analyses to understand the behaviour of the values of each dataset. For this purpose, one of the first considerations was to establish a threshold on the percentage of data availability per pixel: it was selected a minimum requirement of 50% of available data for each pixel to be considered in the analyses. Once defined the group of pixels that was taken into consideration, a histogram representation was prepared to analyze the distribution of the mean values of the pixels of each dataset in each of the different WFD seasons. In the same line, it was created a box-plot representation to complement the analysis of the distribution and to obtain further insights. These statistical analyses were performed for the group of the three lakes but also individually in order to determine if there exist differences in the distributions. In the section 3 section of this document, just the relevant results are described.

2.5 Spatial analysis

In order to understand the spatial distribution of the higher values of TSM, Chlorophylle-a and LSWT and to detect possible recurring patterns, it was chosen to implement a code automated procedure. First, the algorithm allows setting a group of initial parameters, depending on the desired output:

- **Filter_Availability** boolean type, if set to true, the analysis will be performed by also filtering data according to their availability

- **filter_season** boolean type, if set to true, the analysis will be performed individually for each season of the WFD periods, otherwise all periods will be considered together as one

- **lake** string type, to set according to the lake the study needs to focus on (Maggiore, Lugano and Como). If set to "full" the maps will be produced for all the three lakes, otherwise they will be cropped to a specific water basin

- **region** string type, it represents the region in which we want to focus our study (north,center or south). If set to "full" the maps will be fully produced, otherwise they will be cropped to a specific region

After that, the algorithm imports the correct datacube, from a local repository, according to the selected data type (TSM,

LSWT, Chl-a). To let the algorithm properly work, data has to be previously downloaded in a parent repository, in a sub-folder called "tsm", "lswt" and "chl" (depending on the type) inside a folder named "stacks". The datacube has three dimensions:

- dimension 1 and 2: these are the dimensions of a 2D matrix that corresponds to a 2D map of the Insubric lakes. By indexing dimension 1 and 2 it is possible to access a single data pixel from its pixel coordinates
- dimension 3: it corresponds to the time dimension, in particular each maps are stacked so one by one so that each map corresponds to a different day of the year, following a chronological order

For each day in the datacube it is computed the "90th percentile", meaning the pixel value in the 2D map under which 90% of the data falls. Therefore, a single value for each day of the year is extracted from each 2D matrix in the datacube. This "90th percentile" is a fundamental measure, since it will be used as a threshold to find higher pixel values relatively to the pixel values assumed in a specific day of the year. Due to this, greater values will be detected in all WFD periods, and not only in these periods where values are usually higher (for example temperatures will be higher in summer). As already anticipated, each 2D matrix is extracted from the datacube and its pixels are compared to the threshold relative to that day of the year ("90th percentile"). By a assigning a value of 1 if the pixel is greater than the threshold and 0 if not, a new matrix will be created in a form of a binary mask. Later all binary masks (corresponding to each day of the year) are adjusted according to the parameters mentioned at the top of this section and only after that, they are summed up to obtain a single 2D matrix (map), in which each pixel will count the number of days in which the 90th percentile have been exceeded. Finally, the map is normalized with the number of days contained on the datacube, so that each pixel will then contain a percentage value. Thanks to this algorithm, it will be possible to produce a single 2D map (starting from a datacube), where regions in which it is more likely to find high values during the different WFD periods, are characterized by an elevated percentage value.

3. RESULTS AND DISCUSSION

3.1 Data Availability

In the maps that will be presented in the following section, data availability is characterized (as explained in Chapter 2.3) by a value that ranges between 0 (blue) and 1 (yellow), where 0 corresponds to complete unavailability and 1 to the maximum availability. Satellite data provides many benefits to lake monitoring, as it is possible to collect data from lakes without the need to be close to the field. Due to this, sometimes the distance between the sensor and the observed data, may cause some uncertainties in the observations, because of unwanted effects. As a matter of fact, satellite data suffers from the presence of "mixed pixels", shadows or adiacency effects. These effects bring uncertainty in reflectance values. This comes from the fact that these phenomena can generate some unusual responses, which may not be correctly detected from the sensor.

It can be evicted from Figure 1 and Figure 2, representing the data coming from the Sentinel-3 satellite (Chl-a and TSM) for the entire period, that pixels along the coast and in the Lugano

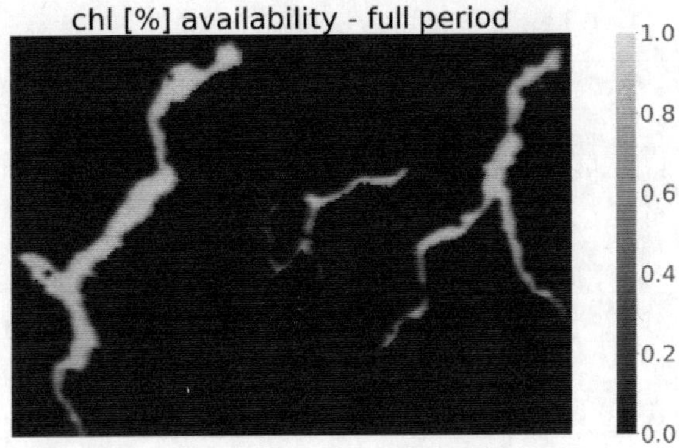

Figure 1. Data availability - Chlorophyll-a, full period.

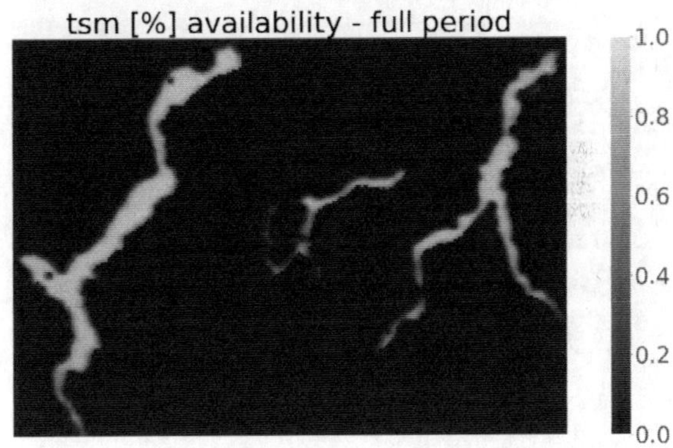

Figure 2. Data availability - TSM, full period.

lake, assume a blue color, which means a level of availability lower than the 50%. Due to this, it is possible to state that these zones are characterized by bigger uncertainty. The reason for that can be attributed to the low resolution of the satellite, that causes a relatively big GSD, which implies that in one single pixel, many different classes of land may be included. Moreover, lake's border are characterized by the presence of both water and coast which may cause mixed pixels and coastal adjacency effects as well. Although this effects are quite evident in the maps, it is also true that borders represents, at least for the two largest lakes, a strong minority of the data. Moreover, TSM and Chl-a data cover almost completely all the zones with an high percentage of availability. For what it concerns the data availability analysis for TSM and Chl-a by period, results are not stating anything different from the full period already mentioned above and therefore the maps are not shown here.

Switching to the temperature, the first remark is that, as immediately evident from Figure 3, the difference in the spatial resolution of Landsat-8 Surface Water Temperature acquisitions, 3 times higher with respect to TSM and Chl-a maps (100m against 300m), makes maps definitely more detailed. Apart from that, data availability is clearly higher with respect to the other two data types, and also the Lugano lake and the coast are characterized by an optimal data coverage. Additionally, in Figure 4 it is possible to appreciate the difference in availability in the various periods, and it can be noticed how the situation is almost optimal in summer-autumn transition where most of the data has 100% availability, but worst in spring and

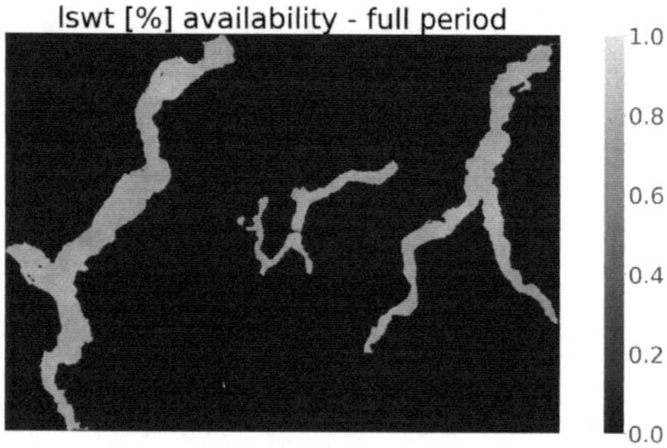

Figure 3. Data availability - LSWT, full period.

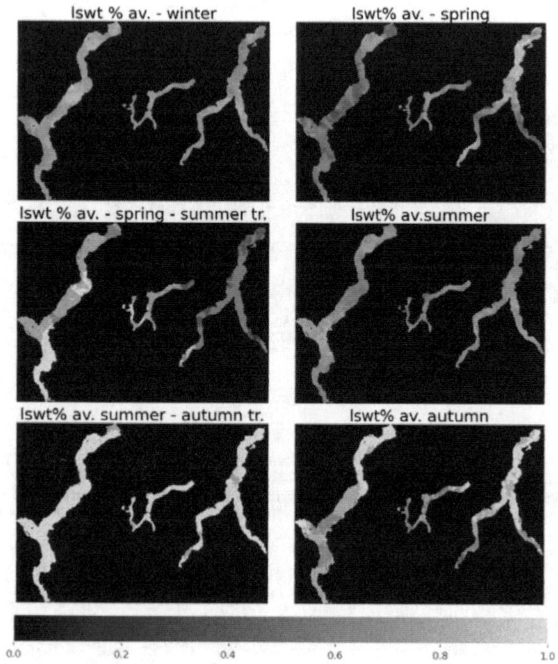

Figure 4. Data availability - LSWT, per season.

spring-summer transition. A further inspection on cloud coverage during this periods can possibly reveal the reason of this phenomenon. Despite that, it can be stated that the situation is almost excellent for all the periods, since generally the availability is never dropping under the 50%.

3.2 Statistical Analysis

This subsection describes the results obtained in the performed statistical analyses. The main objective was to understand the distributions of each of the datasets and to comprehend possible changes in the distributions between seasons and lakes.

TSM

In Figure 5 it is possible to observe the unified distribution of the TSM values for all the three lakes together. Differently from other parameters, the TSM distribution looks almost the same for all the WFD periods. Regarding the boxplots, instead, by

Winter	Spring	Spring-summer	Summer	Summer-aut.	Autumn
5	4	4	4	4	6-7

Table 3. Chl-a peak location per season $\mu g/L$.

comparing the unified plot (three lakes together) in Figure 6 with the one created just with data from the Lugano lake in Figure 7, it is possible to appreciate how the TSM variability for the Lugano lake is increasing during winter period. Despite that, values in winter are generally the lowest between the different periods (as can be seen from the mode in winter season, identified by the light-blue line inside each box in Figure 7). The highest values are, instead, reached during autumn period. As already known, the lake of Lugano is the smallest of the three so it is more likely that the variability of its indexes is higher.

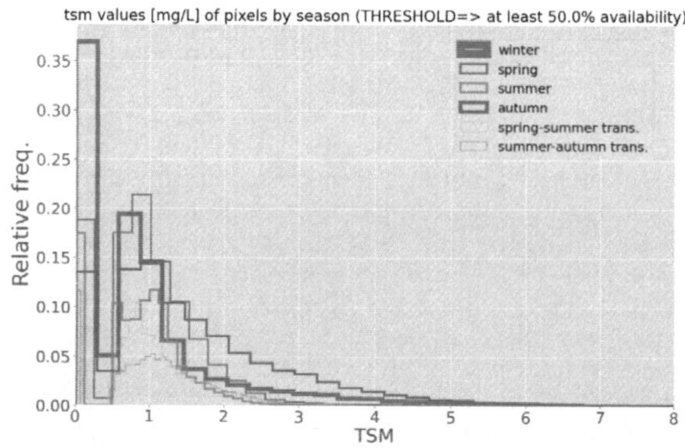

Figure 5. Histograms per season- TSM (three lakes).

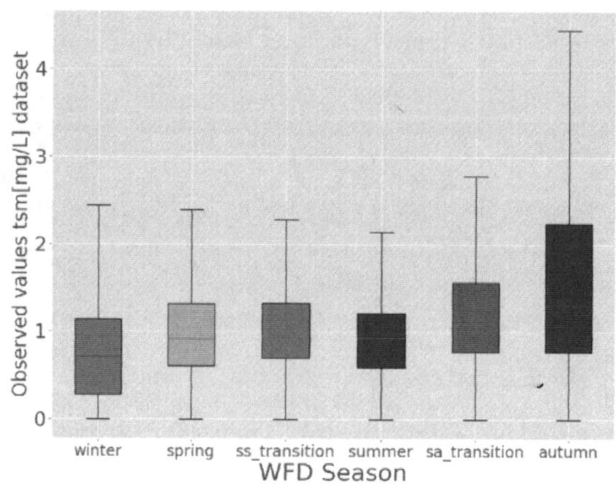

Figure 6. Boxplot per season- TSM (three lakes).

CHLOROPHYLL - A

Regarding Chl-a, the graph in Figure 8 shows how the distributions of the three lakes together, arrange themselves to a normal distribution, by distributing almost homogeneously around a single peak. In particular, for each period the peak is located around a specific value as summarized in table 3.

This trend finds a confirmation also in the boxplot graph (Figure 9), in which it is clearly visible how mean and variance of winter and autumn season are clearly higher with respect to all the other periods.

The reason of this behavior is probably related to cyanobac-

https://doi.org/10.5194/isprs-archives-XLVIII-M-1-2023-9-2023 | © Author(s) 2023. CC BY 4.0 License.

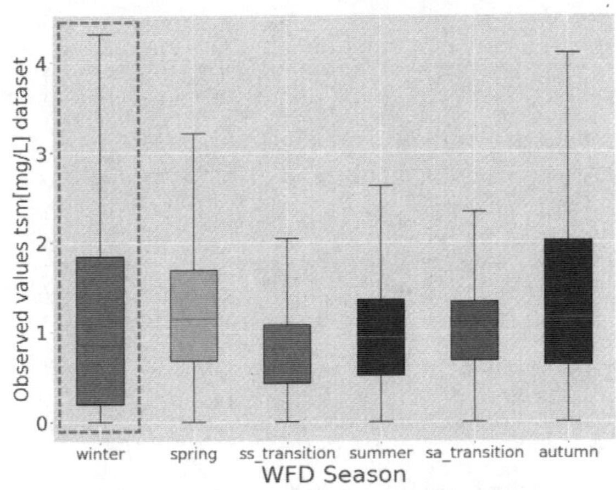

Figure 7. Boxplot per season- TSM (Lake Lugano).

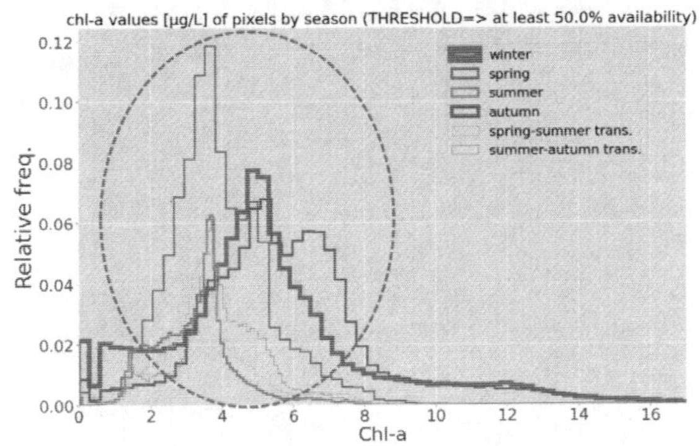

Figure 8. Histograms per season- Chlorophyll-a (three lakes).

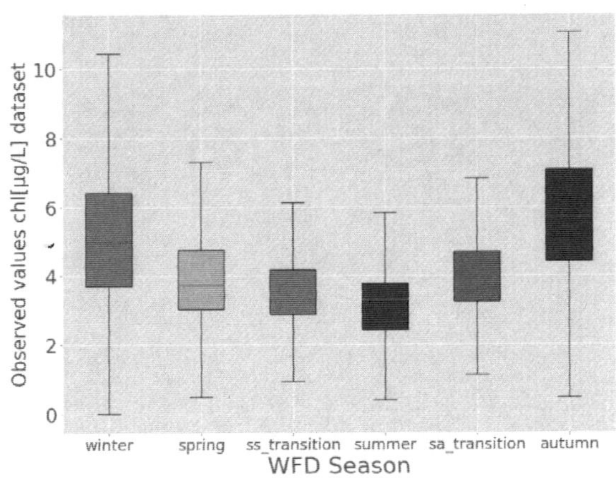

Figure 9. Boxplot per season- Chlorophyll-a (three lakes).

teria's activity: due to the lack of light in these periods, these organisms exploit their buoyancy to reach to the surface in order to get more light from the sun and warmer temperatures, then they duplicate, causing a massive increase in chl-a variability. This phenomenon ususally occur in at the beginning and at the end of winter (Garneau Marie-Ève, 2015) (Yoshihiko Fujita, 2001). It is important to remember that satellites detect only the superficial layers of the water and by doing so acquisitions are mainly detecting surface activity. Moving to single lake analysis, the lake of Lugano shows a very peculiar situation. As can be seen from Figure 10, the distribution of chl-a values for the lake of Lugano tends to assume a bimodal distribution with a double peak particularly evident in winter and summer.

The reason of this is related to the characteristic of the lake,

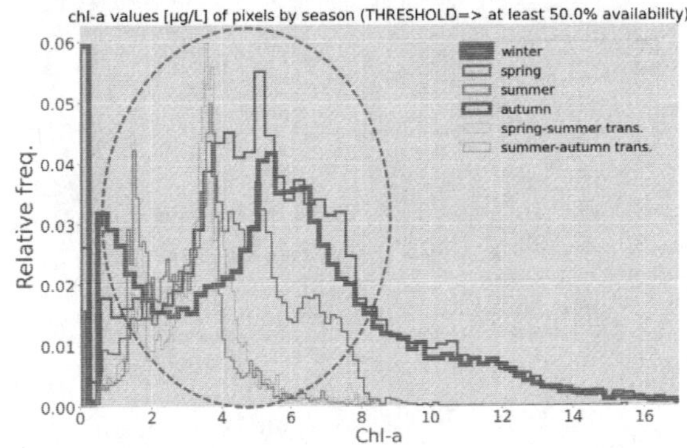

Figure 10. Histograms per season- Chlorophyll-a (Lake Lugano).

which is split in two different basins: the north region and the south one, which tend to have two different behaviours.

Lake Water Surface Temperature

As anticipated in Table 1, the amount of maps for surface temperature is clearly lower with respect to tsm and chl-a (69 maps against 392 and 389 respectively), because of the different satellite's revisit times (16 days for Landsat and 1 day for Sentinel). As a consequence, the distribution in Figure 11 is showing how data has very low variability around a well defined peak. This graph prove that surface temperatures are continuously increasing starting from winter (which has a peak around 7-8°C), up to summer (with a peak around 24°C).

It is fundamental to remember that these temperature are de-

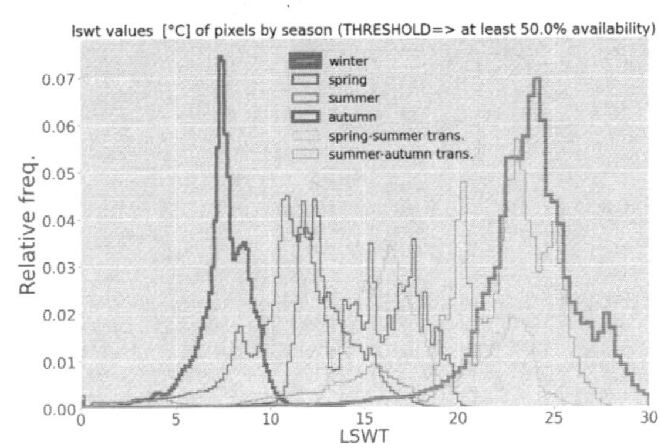

Figure 11. Histograms per season- LSWT (Three lakes).

tected only at the lake surface, due to the nature of the satellite sensor, hence it is usual that temperature may reach such high values. This phenomenon is even more explicit in the boxplot in Figure 12, where it can be appreciated how the temperature are slowly increasing from winter up to summer and then decreasing again going toward autumn, being the two periods of transition and autumn the ones with more variability (as expected, since usually in those periods temperature can vary consistently day by day).

Regarding the lake of Lugano, these normal distributions are

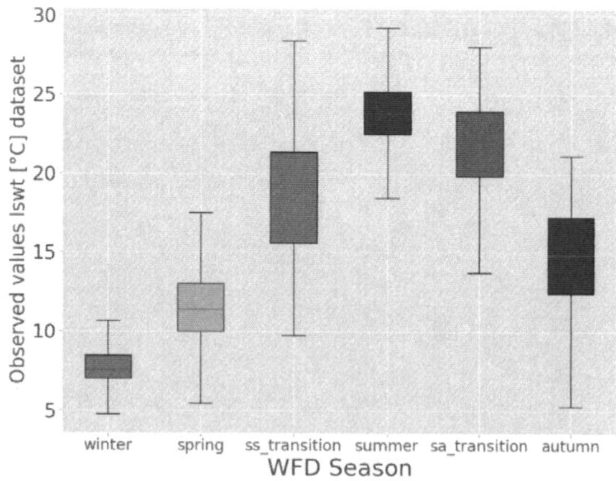

Figure 12. Boxplot per season- LSWT (three lakes).

even more defined, producing a clear and finite number of peaks (especially in autumn, where four bells can be spotted). This phenomenon can be attributed to the even reduced amount of data. Indeed, not only Landsat maps are fewer in number, but the Lugano lake is also the smallest of the three Insubric lakes, hence the amount of pixels contributing to the distribution is even lower. Due to this Lugano's distribution is sparser.

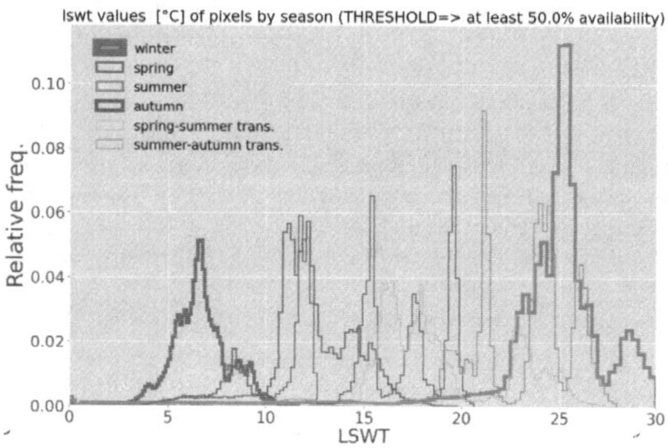

Figure 13. Histograms per season- LSWT (Lake Lugano).

3.3 Spatial patterns Analysis

This subsection presents the results related to the spatial analyses that were implemented in this project. Results associated with the three studied datasets are described in order to emphasize the main spatial patterns which are present for each of them.

TSM

From the point of view of the spatial distribution of the highest values, it is clear how a greater activity belongs to the northern zones of the lakes. Indeed, in figures 14, 15 and 16, especially in the north of the Como lake, the percentage of values exceeding the percentile rises up to almost the 60% in spring, spring-summer transition and summer respectively.

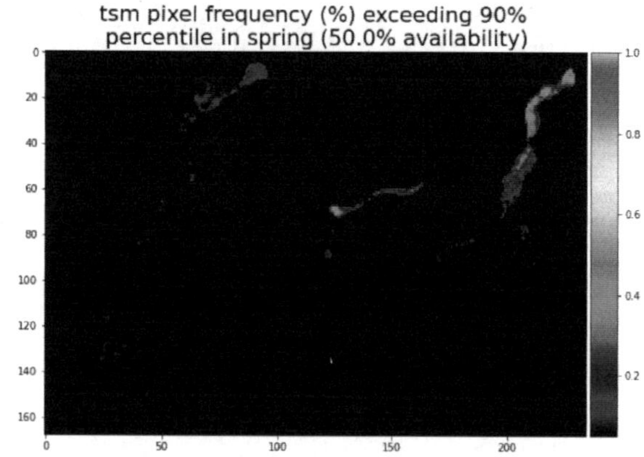

Figure 14. TSM spatial pattern map spring

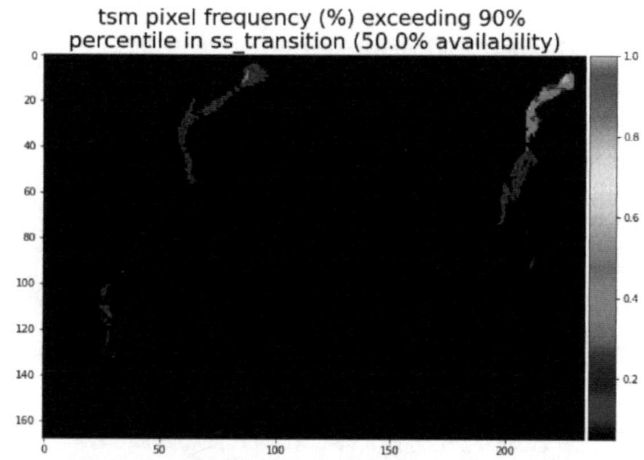

Figure 15. TSM spatial pattern map spring-summer transition

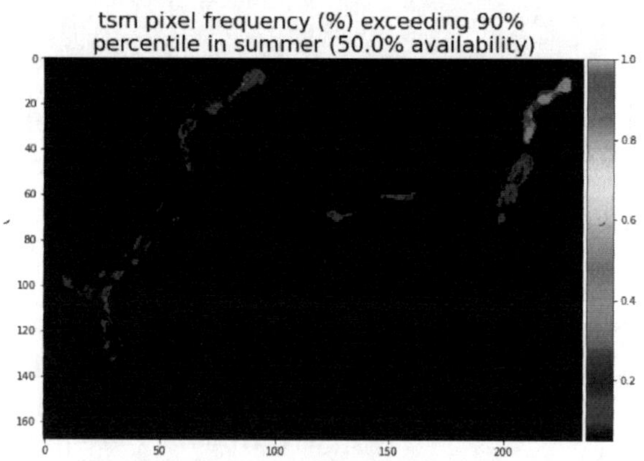

Figure 16. TSM spatial pattern map summer

This spatial effect can be caused by the presence of the main rivers on the northern side of the lakes, which are known to collect suspended matter, which is later brought inside the water basin. In particular the rivers affecting the TSM concentration are the Mera and Adda ones, which are flowing inside the northern side of the Como lake, and the Ticino river which is flowing in the Maggiore lake.

CHLOROPHYLL - A

For chl-a the situation is very close to the TSM one. Indeed, it can be observed from Figure 17, 18, and 19, that in the northern region of Maggiore and Como lake the amount of chlorophyll is increasing up to the 60-70% in the periods of spring-summer transition, summer and summer-autumn transition. Actually, it's not new that chl-a and TMS data are close to one another, as the increase of suspended matter is related to the rivers that flow in the lakes, and knowing that most of that TSM is, indeed, organic matter, it happens that the phytoplankton has more nutrients to reproduce. Due to this, the spatial patterns of chlorophyll-a can be considered as a direct consequence of the TSM increase, hence of the rivers flowing in the lakes.

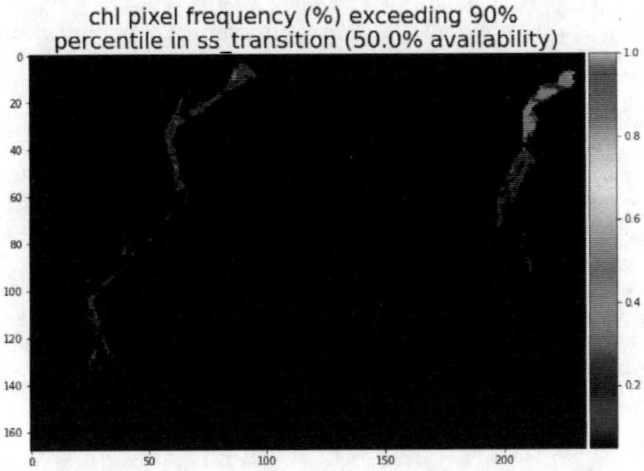

Figure 17. CHLOROPHYLL-A spatial pattern map in spring - summer transition

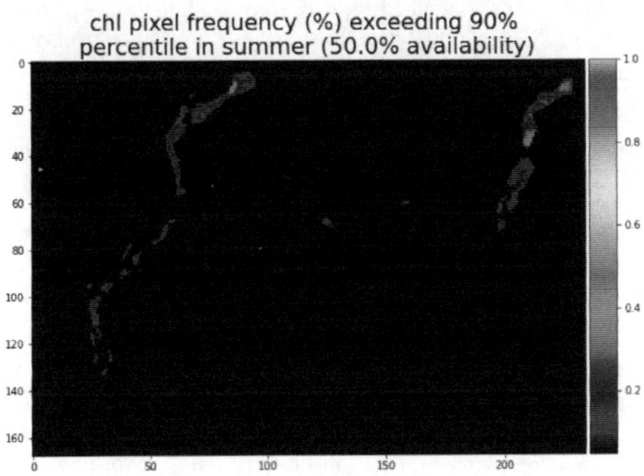

Figure 18. CHLOROPHYLL-A spatial pattern map in summer

Lake Water Surface Temperature

The maps in Figure 20, 21 and 22, corresponding to the two transition periods and summer (which are usually the periods where temperatures are usually the highest), show higher temperatures in the south region of the Maggiore lake and Lugano one. Usually, these temperatures may be affected by light condition and water circulation, since they are detected at the surface. Given that wind may be considered as one of the main parameters that are influencing water circulation, closed environments may "suffer" from higher temperature. Indeed, the southern region of the Maggiore lake, is known to be surrounded by mountains, that may shelter this lake portion from winds.

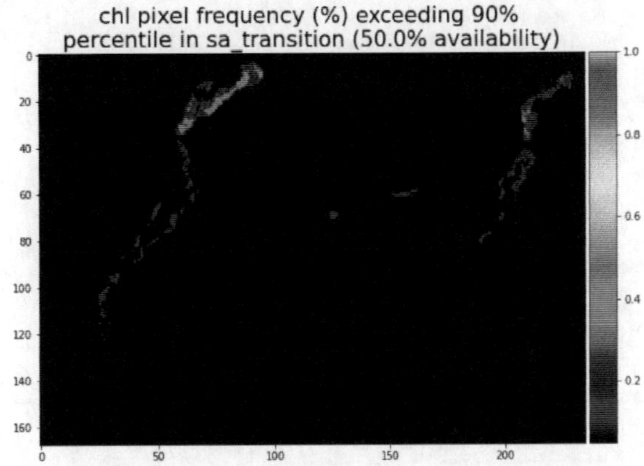

Figure 19. CHLOROPHYLL-A spatial pattern map in summer - autumn transition

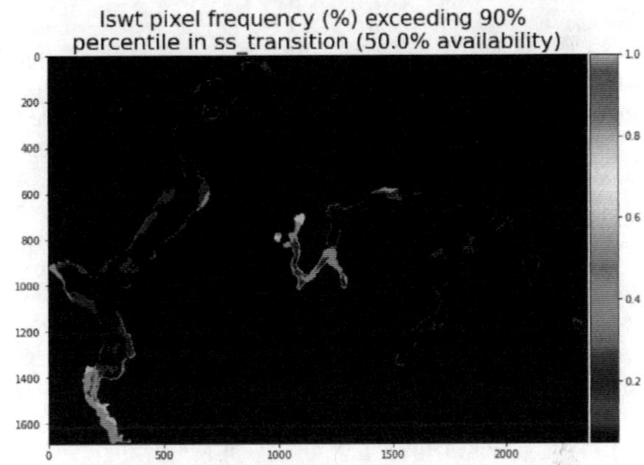

Figure 20. LSWT - spatial pattern map in spring - summer transition

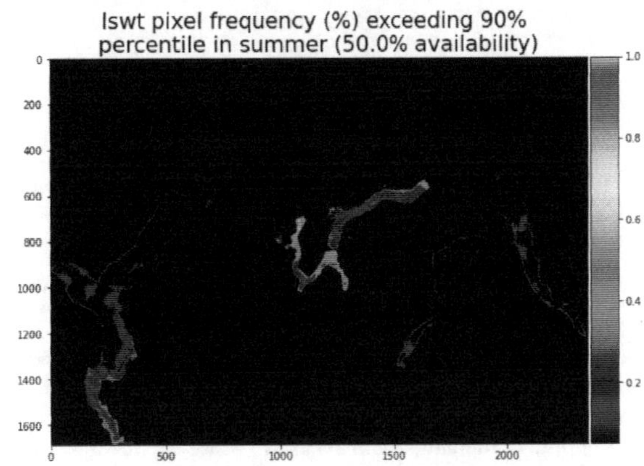

Figure 21. LSWT - spatial pattern map in summer

Regarding the Lugano lake, on the other side, its size may be the main reason of this effect, since by being the smallest and shallowest lake, it may be the most affected by environmental temperatures.

This contribution has been peer-reviewed.
https://doi.org/10.5194/isprs-archives-XLVIII-M-1-2023-9-2023 | © Author(s) 2023. CC BY 4.0 License.

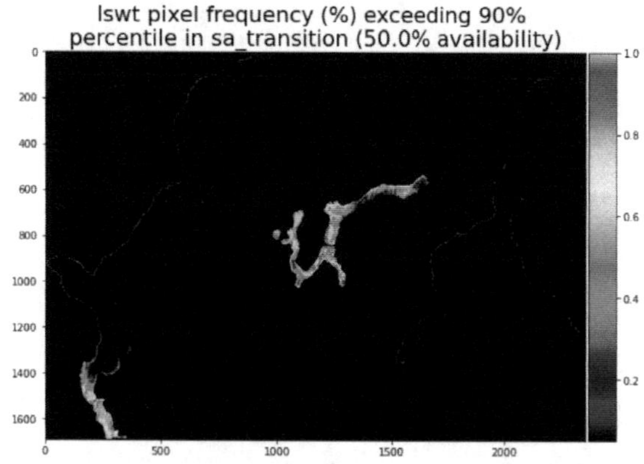

Figure 22. LSWT - spatial pattern map in summer - autumn transition

4. CONCLUSIONS

The lakes which were studied in this project constitute an important source of fresh water for several different activities and for that reason they are a relevant resource for the well-being of the communities and ecosystems. However, their proper management exhibits different challenges; some of which are intrinsic to any water body but also, in this particular case, some of them are related to the fact that Insubric lakes are water basins shared between two countries. Among these challenges, it could be highlighted the need to monitor in an efficient way the main water quality parameters and, according to the analysis of their state during time, define the best actions to be tackled. SIMILE's main objective is the coordination of that management and, in this sense, the current work exhibits a contribution to the understanding of the lake water quality: spatial and temporal patterns have been analyzed by means of statistical techniques that exploited three previously existing datasets (TSM, chl-a and LSWT), developed in the scope of the SIMILE project. This work has shown how satellite data can bring benefits to lake monitoring to help the decision-making processes. The benefits of remote sensing in this field are several, among which:

- sparing costs of expensive field interventions, by collecting data from a far distance
- allows frequent data collection thanks to the short revisit time of newer satellites
- data is freely available on most remote sensing satellite web portals

Of course satellite data analyses are also characterized by minor drawbacks as lower spatial distribution or revisit time for some satellites, and the detection of superficial data only, but their contribution is fundamental if used in synergy with other in-situ technologies, for a complete and accurate lake monitoring campaign.

5. ACKNOWLEDGEMENTS

The work outlined in this paper is part of project SIMILE (Integrated monitoring system for knowledge, protection and valorisation of the subalpine lakes and their ecosystems) and as such it is financed by the European Commission in the frame of Interreg Italy-Switzerland 2014-2021 program.

REFERENCES

Barsi, J. A., Schott, J. R., Palluconi, F. D., Hook, S. J., 2005. Validation of a web-based atmospheric correction tool for single thermal band instruments. J. J. Butler (ed.), *Earth Observing Systems X*, 5882, International Society for Optics and Photonics, SPIE, 58820E.

Bresciani, M., Stroppiana, D., Odermatt, D., Morabito, G., Giardino, C., 2011. Assessing remotely sensed chlorophyll-a for the implementation of the Water Framework Directive in European perialpine lakes. *Science of The Total Environment*, 409(17), 3083-3091.

Brockmann, C., Doerffer, R., Peters, M., Kerstin, S., Embacher, S., Ruescas, A., 2016. Evolution of the C2RCC Neural Network for Sentinel 2 and 3 for the Retrieval of Ocean Colour Products in Normal and Extreme Optically Complex Waters. *Living Planet Symposium*, ESA Special Publication, 740, 54.

Brovelli, M. A., Cannata, M., Rogora, M., 2019. SIMILE, A GEOSPATIAL ENABLER OF THE MONITORING OF SUSTAINABLE DEVELOPMENT GOAL 6 (ENSURE AVAILABILITY AND SUSTAINABILITY OF WATER FOR ALL). *The International Archives of the Photogrammetry, Remote Sensing and Spatial Information Sciences*, XLII-4/W20, 3–10.

Garneau Marie-Ève, Posch Thomas, P. J., 2015. Seasonal patterns of microcystin-producing and non-producing. *Harmful Algae*, 21-31.

Giardino, C., Bresciani, M., Stroppiana, D., Oggioni, A., Morabito, G., 2013. Optical remote sensing of lakes: an overview on Lake Maggiore. *Journal of Limnology*, 73(s1).

Remote sensing for lake research and monitoring – Recent advances, 2016. Remote sensing for lake research and monitoring – Recent advances. *Ecological Indicators*, 64, 105-122.

Rogora, M., Buzzi, F., Dresti, C., Leoni, B., Lepori, F., Mosello, R., Patelli, M., Salmaso, N., 2018. Climatic effects on vertical mixing and deep-water oxygen content in the subalpine lakes in Italy. *Hydrobiologia*, 824.

Toro Herrera, J. F., Carrion, D., Bresciani, M., Bratić, G., 2022. SEMI-AUTOMATED PRODUCTION AND FILTERING OF SATELLITE DERIVED WATER QUALITY PARAMETERS. *The International Archives of the Photogrammetry, Remote Sensing and Spatial Information Sciences*, XLIII-B3-2022, 1019–1026.

Yoshihiko Fujita, Kaori Ohki, A. M., 2001. Acclimation of Photosynthetic Light Energy Conversion to the Light Environments. *Algal Adaptation to Environmental Stresses*, 203-58.

UNDERSTANDING THE BUILT-UP EVOLUTION AND COMPLEXITY OF FAST-GROWING BUILT-UP AGGLOMERATION USING THE CENTROID SHIFT MODEL AND FRACTAL DIMENSION

Keerthikumar Ashritha[1*], Gopal R. Patil[1], Buddhiraju Krishna Mohan[2]

[1] Department of Civil Engineering, Indian Institute of Technology Bombay, Mumbai 400076, India – ashrithasaroja@gmail.com, gpatil@iitb.ac.in

[2] Center of Studies in Resources Engineering, Indian Institute of Technology Bombay, Mumbai 400076, India – bkmohan@iitb.ac.in

KEY WORDS: Land Use Land Cover, Centroid Shift, Growth type, Geometric Complexity, Fractal Dimension.

ABSTRACT

Understanding Built-up evolution and complexity over time is attained to exercise sustainable development and smart implementations in the future. The Spatiotemporal evolution of built-up analysis is carried out using the centroid shift model considering the geometry of the built-up from 1975 to 2019 (1975, 1990, 1999, 2009, and 2019). The centroid shift indicates the trajectory of the growth pattern and the future direction of growth. The study area experienced a unidirectional growth pattern till 2009, considering 1975 as the base year, and by 2019 the growth direction shifted by 3.52 degrees, and the overall centroid shift magnitude was 2.42 km^2. The non-Euclidean geometric nature of the physical world puts forward fractal geometry to understand the complexity of built-up. The mean fractal dimension index shows the growth type in the study area; a directional approach towards the mean fractal dimension provides a critical understanding of the complexity around the built-up Centroid. Among all the eight directions considered, NW_W showed maximum built-up growth with infilling growth type, and the E_SE direction showed a significant jump in an area with external expansion by 2019. This direction indicates the future growth direction. The study concludes that complexity analysis critically enhances the spatiotemporal dynamics of built-up evolution.

1. INTRODUCTION

1.1 Background

Since the Anthropocene Era, Earth's Land Cover has transformed into Land Use unprecedentedly. Currently, 20% of the terrestrial land surface is built-up, and this occupancy is not uniform; it varies from biome to biome. Cold landscapes have experienced fewer human interventions compared to tropical and temperate landscapes. This drastic transformation has led to devasting anthropogenic climate changes, natural resource depletion, and ecosystem imbalance. The factors influencing the transition from Natural Landscape to the built-up landscape are the availability of Natural resources, transportation and communication, industrialization and commercialization, educational and recreational, and economic pull. To curb the actions of humans above biological processes, understanding and quantifying the present built-up landscape is needed, which mitigates the climate crisis and supports biodiversity if managed effectively. A quantitative understanding of the existing built-up and its evolution is complex as the stated factors' influence on built-up growth varies from place to place and over time. But to attain qualitative measures for sustainable planning and scope for future development, understanding the level of complexity and existing development constraints is required.

* Corresponding author

1.2 Literature Studies

Mid-19th Century built-up studies gained importance in the research field (Longley, 2005). Built-up is a terrestrial landscape where human beings interact. As per the United States Geological Survey, Level one scheme of land classification, Urban, Rural, Industrial, commercial, transportation, and other impervious surfaces are categorized as built-ups (Anderson et al., 1976). Sometimes these built-ups are called Urban (National Remote Sensing Centre, India backs this definition). R.LeGates (2001) noted the core twelve domains of Built-up studies. Namely, the evolution of cities; urban culture; urban society; urban politics and governance; urban economics, public finance, and regional science; urban and metropolitan space and form; megacities and global city systems; technology and cities; urban planning, urban design, landscape architecture, and architecture; race, ethnicity, and gender in cities; urban issues and policy; and urban futurism. The domains' evolution of built-up and urban and metropolitan space and form is the scope of our study.

In built-up studies, the topo sheets and past cadastral maps serve as the study documents. These records are updated timely through the survey. But, in inaccessible or remote areas, information update is time-consuming and tedious, hence limiting the studies on the evolution of built-up growth temporally in a particular landscape (Roy et al., 2015). It would have been challenging, if not impossible, to examine the Land Cover (LC) patterns at a

This contribution has been peer-reviewed.
https://doi.org/10.5194/isprs-archives-XLVIII-M-1-2023-17-2023

regional and global scale without the readily available and free satellite imageries. Remote Sensing has been a useful aid in studying the patterns of LC with a decreased cost and less effort. Various global LC data have been produced over the years, such as GLC2000, MCD12, and GLCNMO, GlobCover, IGBP DISCover, MODIS Collection 5 global land cover, Global Land Surface Satellite Global Land Cover (GLASS-GLC), Global Human Settlement Layer, GLC30, GLC10, ESA CCI Land Cover time-series, ESRI 10 m, etc. (Aman et al., 2021).

The foundation of LC is image classification, wherein we give LC classes to a multiband raster image. Pixel-wise classification, sub-pixel-wise classification, and object-based image classification are the three main categories of image classification in remote Sensing. According to the spectral signature and its derivatives, such as vegetation indices, pixel-wise classification categorizes each pixel as pure and assigns it to a specific land use-LC type. This classification is further separated into unsupervised and supervised classification (Li et al., 2014). Numerous research studies have employed unsupervised categorization based on building clusters and assigning classes.

On the other hand, supervised classification uses samples as training data and classifies using them. Land Use Land Cover classification algorithms like Maximum likelihood, Isodata, Artificial Neural Networks, Support Vector Machines, and most advanced algorithms like Convolution Neural Networks, deep learning algorithms, Decision Trees, etc., are used in the preparation of Land Use Land Cover data sets for various spectral and spatial resolution data. Mainly, freely available Landsat data of 30m resolution is widely used to prepare the LULC maps for understanding the built-up growth and dynamics. LULC data products provide information on regions' past and present built-up landscapes. Tracking the movement of the center of change of a particular LC through time is just as important as analysing the current situation of that LC. The varieties of regional land use have seen substantial changes recently due to expanding urbanization. Because of this, they understand the area's development, and providing references for resource management and territorial planning can be accomplished by the study of the central displacement of LC. A change in the Center of Gravity (CoG) of the specific LC class has been researched for this purpose. The geometric center of the LC class is the CoG in terms of space. The CoG model, called the Centroid shift model, helps explain regional shifts along the development axes. (Shi et al., 2015) and (Xiong et al. 2019) studied the built-up growth using the spatial centroid shift method. They also assessed the environmental impact because of built-up evolution over time.

The elemental quantification of Spatio-temporal dynamics is carried out by estimating the landscape growth area. The landscape analysis is categorized into structural, functional, and change characteristics. Landscape structure deals with the spatial relationships among the distinctive ecosystems or elements; Landscape function is the interaction among the spatial features. Landscape change focuses on altering the structure and function of ecological mosaics over time. The landscape structure and complexity measures are carried out using various landscape matrices. These matrices are area/density/edge matrices, shape matrices, core area matrices, isolation/proximity matrices, connectivity matrices, evenness metrics, dominance matrices, and diversity matrices. Among these matrices, the shape matrices deal with the landscape's geometry (Gkyer, 2013). One majorly used shape index is the fractal dimension indices, derived from fractal geometry. Fractal geometry is put forward as the Physical world structures are complex polygons that do not represent any geometric shape of Euclidean (Circle, triangle, etc.); these polygons are measured using fractals. Fractals are complex, hierarchically ordered structures revealing self-similarity across scales (Batty & Longley, 1994). FRAGSTATS interface (MacGarigal, 1995) has a mean fractal dimension index ranging from 1 to 2 in a two-dimensional plane. These values define the complexity of the built-up areas. (1994) presented that fractals depend on the scale and vary from one scale to another. (Ramachandra et al., 2012) (Chen, 2020) studied the fractal dimensions of an urban morphology to understand its structure and ecological functioning.

From past studies, the understanding of evolution is drawn majorly through quantifying the area and its change over time. The studies on the Centroid shift model and the knowledge of geometric complexity are limited. Hence, the study investigates the built-up evolution of a fast-growing city agglomeration using the Euclidean Centroid shift model and the fractal geometry's complexity analysis. The study attempts to understand the built-up's directional complexity, considering the geometric Centroid as the origin over time.

2. STUDY AREA AND DATA

2.1 Study Area

Districts are subdivided into Tehsils as per the Government of India. This study adopted Vasai Tehsil (433 Km^2), located at 19.3919° N, 72.8397° E of Thane District (Thane District is one of the three major units of the Mumbai Metropolitan Region) to analyse the evolution and Complexity of Built-up from 1975 to 2019. The study area is in the Northwest of the Mumbai Metropolitan Region (Figure 3.1), surrounded by Vaitarna in the North, Vasai Creek in the South, Bhiwandi tehsil in the East, and the Arabian Sea in the West. Vasai has a terrain range of up to 663m. This highly elevated area is the Tungareshwar National Forest; this green belt lies in the East of Vasai. Vasai Tehsil remained under the green zone, with small strips of built-up along the coast. With increased urbanization and population explosion in the Mumbai region, the population started growing in Vasai Region; hence in 1983 Maharashtra government converted 8,500 hectares of the green zone into urbanisable land (Sharma et al., 1991). This allocation attracted the human population for settlements resulting in built-up evolution. The availability of good public transport and its connectivity to the neighbouring cities resulted in increased human occupancy in the Vasai region. Vasai road railway station bypasses Mumbai connecting Vadodara to Konkan Railways and Pune Junction towards Bengaluru and Hyderabad. Vasai road railway station bypasses Mumbai connecting Vadodara to Konkan Railways and Pune Junction towards Bengaluru and Hyderabad.

https://doi.org/10.5194/isprs-archives-XLVIII-M-1-2023-17-2023 | © Author(s) 2023. CC BY 4.0 License.

The Municipal Corporation buses connect all the major routes within the town, and State Transport buses provide long-distance travel to and from Vasai. Besides, auto rickshaws are the primary source of transport in the region for shorter trips. The introduction of Local Train Service to MMR in 1867 by Indian Railways enhanced the growth by increasing mobility at a lesser cost; further, the 4 minutes frequency of the local train between Virar and Churchgate made the local train a reliable, time-saving, and cost-effective mode of travel. Noting the transportation and Infrastructure development in Vasai Tehsil, quantifying the complexity of growth concerning the built-up evolution over time is done.

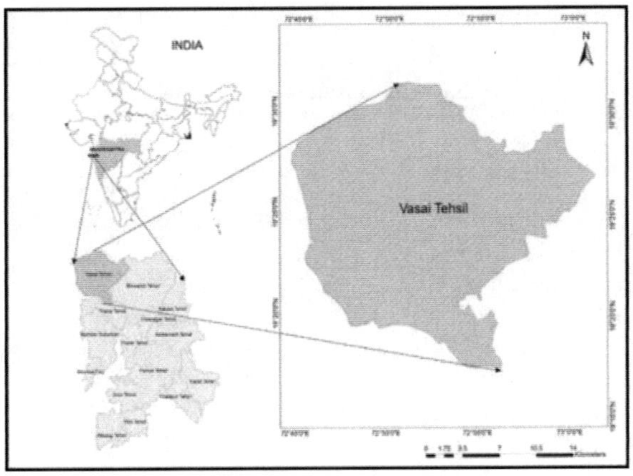

Figure 1. Study Area Vasai Tehsil

2.2 Data

This research required Built-up feature data sets from 1975 to 2019. Cloud-free Landsat Thematic Mapper of 30m spatial resolution for 1999 and 2009, Operational Land Imager + Thermal Infrared Sensor combined Landsat imagery for 2019 of 30m spatial resolution and Acquisition of Global Human Settlement Layer (GHSL) for 1975 and 1990 of 30m spatial resolution. GHSL data set is a Joint Research Centre (JRC) European Commission initiative for providing global time-series geospatial data sets in the public domain. The GHSL data were released in 2016 as an open and free data package (GHS P2016) (Corbane et al., 2018). The GHS P2016 is generated from temporal Landsat imageries, using different Landsat series sensors, namely Multispectral Scanner for 1975 and Thematic Mapper from Landsat 4 & 5 for 1990. The applicability of these data sets has been emphasized and analyzed for Indian conditions proving suitable for estimating Landscape matrices and understanding built-up growth dynamics (Nautiyal & Maithani, 2020). The administrative boundary information of Vasai Tehsil is extracted from the district shapefile of the Mumbai Metropolitan Region obtained from MMRDA (Mumbai Metropolitan Region Development Authority) Development Plan 2016.

3. METHODOLOGY

This research adopts a methodology consisting of three main steps. Step 1 involved the extraction of Built-up features. Step 2 involved computation of the Built-up Centroid and understanding the temporal shift in the Centroid. Step 3 deals with the understanding of complexity in the Built-up geometry directionally.

3.1 Built-up feature extraction

A supervised classification schema is applied to generate land cover maps from satellite images. The main classification features used in this study were the TM and ETM+ multispectral bands. Band composite and mosaic operations are performed to create a False Colour composite considering Red, Green, and NIR bands. The maximum Likelihood algorithm m is adopted for Land Use Land Cover classification, obtaining an overall accuracy and Kappa coefficient of 90.13% and 0.87 for 1999, 91.97% and 0.90 for 2009, and 89.73% and 0.87 for 2019, respectively. The overall accuracy of the land cover maps is >85% for all the years, which was sufficient for extracting the built-up region. Later, the built-up class for 1999, 2009, and 2019 is retrieved from the Land Use Land Cover map. For the years 1975 and 1990, the Built-up class is extracted from the Global Human Settlement Layer. GHSL are prepared based on the symbolic machine learning approach from Landsat images (Pesaresi et al., 2016). It is a time series data for four epochs, 1975, 1990, 2000, and 2014. The built-up area product used for the built-up extraction is 30m spatial resolution data. The data has a 0-6 index, where index six of 1975 and index five of 1990 are considered for the study.

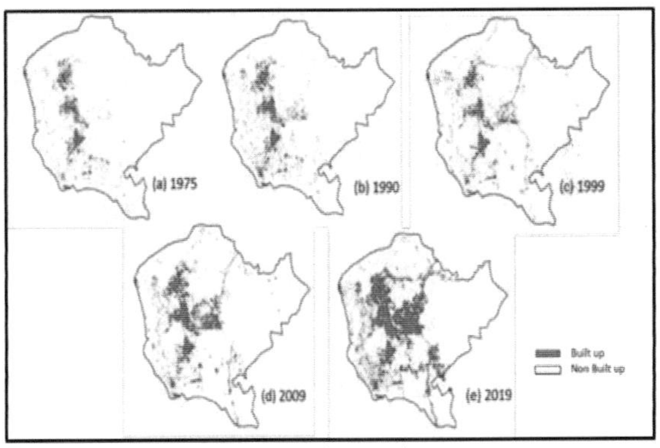

Figure 2. Built-up Areas

3.2 Built-up evolution analysis

A built-up centroid is defined as the geometric center of the centroids of input polygons. For the past four and half decades, analysis of the built-up evolution has been drawn by summation of area-weighted spatial built-up Centroid and tracing the trajectory of changes in the built-up Centroid over time. A quantitative and intuitive description of the spatial change characteristics of built-up is drawn from Euclidean Centroid Shift Model. The model quantifies centroid changes concerning magnitude and direction. Equation 1& 2 spatially locates the built-up Centroid (Xiong et al., 2019) and (Tu et al., 2016). Where X_t and Y_t Indicate the area-weighted abscissa and ordinate of all the built-up land patches in the tth year, respectively; C_{ti} is the area of built-up land patch i in the tth year, n is the total number of patches of built-up land, and X_i and Y_i are the abscissa and ordinate of built-up land patch i, respectively.

$$X_t = \sum_{i=1}^{n}(C_{ti} \times X_i) / \sum_{i=1}^{n} C_{ti} \quad (1)$$

$$Y_t = \sum_{i=1}^{n}(C_{ti} \times Y_i) / \sum_{i=1}^{n} C_{ti} \quad (2)$$

Equation 3 calculates the magnitude with which the built-up centroid movement has occurred over the period of time (1975-1990, 1990-1999, 1999-2009, and 2009-2019). Finally, the overall shift in the Centroid for the past four and half decades is estimated. Equation 4 is the measure of directionality in the centroid movement over a period of time. It is calculated by finding the slope of the line linking the two centroids. Equations 3 and 4 are as follows (Tu et al., 2016):

$$D = \sqrt{(X_{t_2} - X_{t_1})^2 + (Y_{t_2} - Y_{t_1})^2} \quad (3)$$

$$\alpha = \begin{cases} \tan^{-1}\left(\frac{Y_{t_2} - Y_{t_1}}{X_{t_2} - X_{t_1}}\right), X_{t_2} \geq X_{t_1}, \\ \pi - \tan^{-1}\left(\frac{Y_{t_2} - Y_{t_1}}{X_{t_2} - X_{t_1}}\right), X_{t_2} < X_{t_1} \end{cases} \quad (4)$$

Where 'D' is the distance of the centroid shift; 'α' is the centroid shift angle and X_{t_1}, X_{t_2}, and Y_{t_1}, Y_{t_2} is the area-weighted abscissa and ordinate of all the built-up patches in the years of t_1 and t_2, respectively, t_1 is the beginning of the monitoring period, and t_2 is the end of the monitoring period.

3.3 Geometric Complexity of Built-up

Among the various structural landscape matrices, the Fractal dimension is considered to understand the geometric complexity of built-up with time. Fractal dimension matrices define the spatial and structural complexity the study area is experiencing with evolution. The fractal dimension 'D' value is calculated to the whole study area extent, and overall complexity is understood. Later on, the analysis is carried out in eight directions (N, NE, E, SE, S, SW, W, and NW), considering the built-up Centroid for a particular time interval. The study area is divided into eight sectors, and the built-up form lying in each sector is calculated with fractal dimension 'D' value. The equation used for calculation is (MacGarigal, 1995):

$$D = \frac{2 \ln p_{ij}}{\ln a_{ij}} \quad (5)$$

Where p_{ij} is the perimeter and a_{ij} is the area of the built-up patch. The value ranges from 1 to 2. If the value is > 1 for a two-dimensional patch, then it is a departure from Euclidean geometry, and values closer to 2 indicate that the built-up form is highly complex (MacGarigal, 1995). The temporal analysis of built-up complexity also explains the period where the growth was higher and the type of growth. A fundamental characteristic of fractal objects is that their measured metric properties, such as length or area, are a function of the measurement scale. This study adopts a 30 m scale for understanding the complexity.

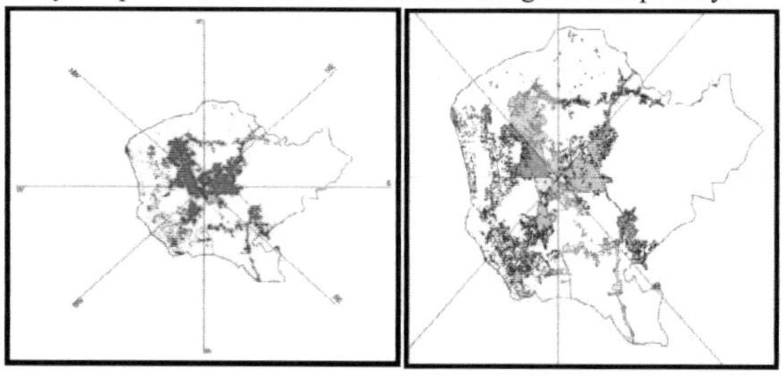

Figure 3. Directional Division of Built-up considering built-up Centroid as the center (2019).

4. RESULTS AND DISCUSSION

4.1 Built-up feature

The built-up area extracted from the GHSL data set for Vasai for 1975 and 1990 is 15.11 km^2 and 19.3 km^2. Similarly, the built-up area extracted from Landsat LULC imagery for 1999, 2009, and 2019 is 22.14 Km^2, 32.42 Km^2, and 65.91 Km^2, respectively. A pictorial representation of the built-up for each year is shown in Figure 2. The quantified area values show that from 1975 to 2019, there was an increasing trend of built-up growth with a varied percentage growth interval. From 1975 to 1990, the growth rate was 27.73%, whereas from 1990 to 1999, the growth percentage was 14.71. From 1999 to 2009, the growth is rapid, with an amount of 45.61%, and from 2009 to 2019, it experienced an increased growth of 104.43%. Vasai's overall built-up class coverage in 1975 was 3.49%, reaching 15.23% by 2019.

4.2 Spatial Centroid Shift

The resulting area trends show that the policy and population explosion in the neighbouring city majorly influenced the spatial pattern of Vasai Tehsil over the past 44 years. Vasai has experienced unprecedented urban growth over the past four decades, with an increased rate of 336% from 1975 to 2019. This amount of growth calls for a deeper understanding of the evolution and complexity of the study area. The city's geometric Centroid is one parameter for quantifying the spatial evolution of built-up. With growth, the built-up geometric Centroid varies spatially, shifting towards the areas where the built-up is more concentrated and has evolved. The geometric centroids of each year is tabulated in table 1.

Year	Longitude (decimal degrees)	Latitude(decimal degrees)
1975	72.818	19.407
1990	72.820	19.408
1999	72.829	19.411
2009	72.833	19.4137
2019	72.840	19.4133

Table 1. Temporal Geometric Built-up Centroid

This contribution has been peer-reviewed.
https://doi.org/10.5194/isprs-archives-XLVIII-M-1-2023-17-2023 | © Author(s) 2023. CC BY 4.0 License.

From 1975 to 2019, the spatial Centroid shifted at a very irregular rate and distance in Vasai. From 1975-1990 the shift distance was 265 meters, whereas, from 1990-1999, the shift was 1 km, indicating a shift rate of 281%. Post-1999, the urban centroid shift from 1999-2009 is 489 meters which are approximately 51% lesser than the 1990-1999 shift rate. From 2009-2019 the Centroid shifted 728 meters, showing a shift rate of 48.7%. The overall shift from 1975 to 2019 is for a distance of 2.42 Km.

Time Interval	Shift Distance(D) in m	Shift Angle(α) in degrees
1975-1990	265	30.3
1990-1999	1011	17.58
1999-2009	489	28.86
2009-2019	728	-3.52
1975-2019	2420	15.73

Table 2. Centroid Shift Distance and Angle

Apart from displacement, directional understanding of centroid shift plays a vital role in understanding the future evolution direction and possibilities. Vasai's overall shift direction between 1975 to 2019 is NE with an amount of 15.73 degrees. Its evolution has been unidirectional till 2009, i.e., in North East Direction. The angular shift between 2009-2019 shows a change in direction from NE to SE with a minimum of 3.5 degrees. The angular shift between 1975-1990, 1990-1999, and 1999 – 2009 is listed in table 2. Figure 4(I) gives the spatial representation of the built-up centroids, and 4(II) shows the shift trajectory.

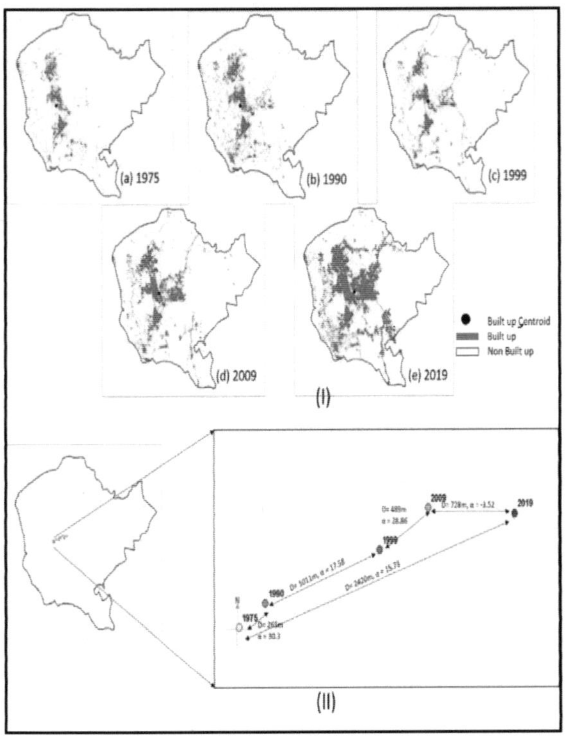

Figure 4 (I) Built-up Area Centroids (II) Centroid Shift Trajectory

4.3 Geometric Complexity of Built-up

Visualizing the spatial pattern of built-up shows that the landscape evolution doesn't follow any Euclidean shapes. Growth is not channelized to one particular geometric shape in the physical world. Vasai Tehsil clearly shows that the built-up evolution is complex. Hence fractal analysis is carried out, where the geometric shapes or boundaries are defined as fractals to understand the complexity involved with the growth. The Mean Fractal Dimension from 1975- 2019 is listed in table 3.

Year	Mean Fractal Dimension (D)	Remarks
1975	1.44	A value higher than 1 indicates an increase in shape complexity.
1990	1.43	Internal Filling
1999	1.42	Internal Filling
2009	1.43	External expansion
2019	1.43	Internal Filling

Table 3. Mean Fractal Dimension of Vasai Tehsil

The Results show that Vasai Tehsil built-up has a complex geometry (Not defining Euclidean Geometry) with an average value of 1.43. The overall mean calculation doesn't critically explain the complexity within the built-up region. Hence the built-up areas are divided into eight directions (Figure 3), considering the built-up Centroid as the reference point. Figure 5 is the wind rose diagram representing the mean fractal values defining the complexity over a time interval in eight different directions. Calculating the directional fractal dimension shows that the Geometric complexity is not the same in every direction. Considering 1975 as the base year for understanding directional geometric complexity, it has been infilling till 2019. In 1975 W_NW had a high Fractal Dimension Value. S_SW has a low Fractal dimension value. In 1990 S_SW experienced an edge filling/external filling indicating a high fractal dimension compared to 1975 S_SW. In all other directions, it is infilling. 1999 shows that in all directions, it has infilling growth.2009 shows an infilling in all directions except the E_SE direction, as the Fractal dimensions indicate that this direction has undergone external growth since 1990. 2019 Fractal dimensions indicate that in N_NE, NE_E, E_SE and SW_W has experienced outward growth from 1990. In the SE_S direction, the geometric complexity was higher than in 1999 and 2009, indicating edge filling. Results also indicate that complexity is not truly increasing with an increasing area over time. It is highly dependent on the type of growth happening in that direction. In 1975 the built-up area in NW_N and SE_S direction was higher, and the mean fractal value in NW_N and NE_E was higher, indicating that SE_S growth was compact in nature, whereas NE_E has a scattered growth increasing the complexity though being a less area coverage. In 1990, NW_N and SE_S has high built-up area coverage but the mean fractal is high in SE_S direction indicating a scattered growth or external filling. In 1999 the trend has a shift showing that N_NE, NE_E has a higher mean

https://doi.org/10.5194/isprs-archives-XLVIII-M-1-2023-17-2023 |

fractal value though the built-up area is high in S_SW, W_NW and NW_N showing an infilling growth in these directions. From 1999 to 2009 a drastic jump in the built-up growth in all directions is seen, where E_SE, S_SW, SW_S and NW_N exhibited higher complexity proving external filling. Further from 2009 to 2019 there is an unprecedented growth in built-up in all the directions on an average of 127.88% from all directions, but directions E_SE and SW_W direction showed some minor external fillings and rest others showed compact and infilling.

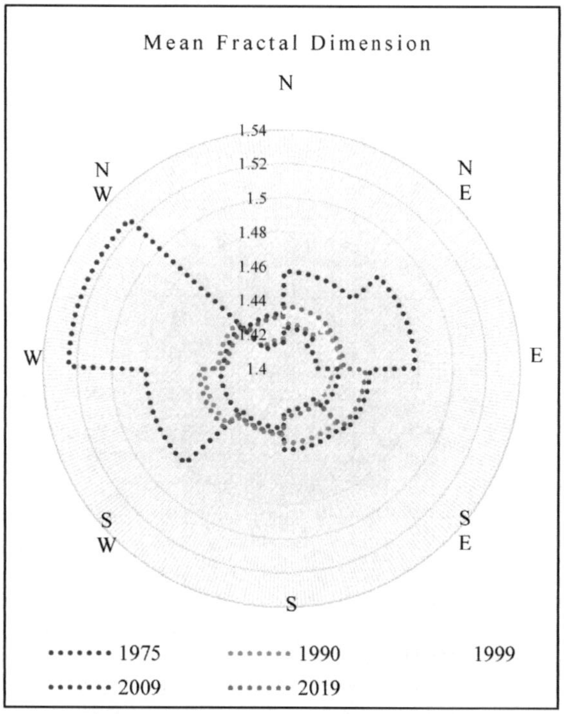

Figure 5. Directional Mean Fractal Dimension

5. CONCLUSION

The Unprecedented built-up growth has put natural Land Cover into Land Use over time, resulting in the need for practicing Sustainable development and smart implementation. Sustainable implementations are needed to curb global warming and rapid climatic changes. Hence, understanding the built-up evolution and its geometrical complexity is studied in the present work for a fast-growing built-up agglomeration. Evolution of built-up for forty-four years from 1975 base year to 2019 is carried out. The Centroid shift model finds the area-weighted geometric Centroid of the built-up temporally and calculates the shift in the Centroid with time. The shift is quantified both in magnitude and direction. The present study area experiences an overall shift of 2.42 Km^2, and the evolution pattern changed from unidirectional (E_NE) to another path (E_SE) by an amplitude of 3.52 degrees. The Centroid shift model yielded the overall growth trajectory of the built-up, and the complexity in the geometry of the built-up was analyzed by fractal dimension Index. Fractal geometry is applied because the built-up in the physical world does not follow. Euclidean geometry of simple shapes. The Mean fractal dimension value for the entire study area remained almost equal to 1.4 showing significant complexity in geometry.

year	Direction															
	NW_N		W_NW		SW_W		S_SW		SE_S		E_SE		NE_E		N_NE	
	D	Area (Km^2)	D	Area (Km^2	D	Area (Km^2)	D	Area (Km^2)	D	Area (Km^2)	D	Area (Km^2)	D	Area (Km^2)	D	Area (Km^2)
1975	1.432	5.21	1.525	0.84	1.48	0.16	1.438	1.59	1.448	4.39	1.451	0.94	1.478	0.58	1.457	1.36
1990	1.417	6.04	1.438	1.42	1.446	0.29	1.439	2.68	1.444	4.32	1.428	1.69	1.433	1.33	1.426	2.1
1999	1.422	5.9	1.419	3.3	1.421	0.95	1.425	3.93	1.417	2.78	1.426	2.34	1.431	1.96	1.433	0.98
2009	1.415	7.69	1.435	4.56	1.437	1.52	1.435	5.77	1.425	3.26	1.432	3.79	1.419	3.95	1.424	1.7
2019	1.43	11.77	1.433	10.11	1.449	3.98	1.438	9.96	1.428	6.21	1.448	8.81	1.435	8.82	1.436	6.25

Table 4. Directional Mean Fractal Dimension (D) and Area of Vasai Tehsil

https://doi.org/10.5194/isprs-archives-XLVIII-M-1-2023-17-2023 | © Author(s) 2023. CC BY 4.0 License.

To draw an in-depth understanding of complexity, the study area's directional division into eight directions considering the built-up Centroid showed that the complexity varies in all directions and is highly dependent on the type of growth than the amount of development happening in the study area. Among all the eight directions considered, NW_W showed maximum built-up growth with infilling growth type, and the E_SE direction showed a major jump in the area with external expansion by 2019. This direction indicates the future growth direction. Hence, the study inferred that complexity analysis critically enhances the spatiotemporal dynamics of built-up evolution.

Fractal Dimension analysis using the Mean fractal index showed that the complexity in geometry is related to the type of growth a city is experiencing. For a comprehensive understanding of geometric complexity, other spatial matrices like the Number of built-up Patches, mean shape index, and built-up patch size also need to be calculated, and statistical correlation with these indices and area needs to be established. The factors like demographic, topographic, socio-economic, transportation, etc., influence the evolution of built-up and play a major role in geometric complexity as well. The results also indicate that the complexity of the urban form needs to be understood on a larger scale. Along with directional complexity analysis using a higher scale, complexity at the proximity of the built-up Centroid need to be assessed to know the variation in complexity within the built-up core.

REFERENCES

Aman, K. C., Wagle, N., Acharya, T. D., 2021. Spatiotemporal analysis of land cover and the effects on ecosystem service values in rupandehi, nepal from 2005 to 2020. *ISPRS International Journal of Geo-Information,* 10(10). doi.org/10.3390/ijgi10100635

Anderson, J. R., Hardy, E. E., Roach, J. T., Witmer, R. E., 1976. A Land Use and Land Cover Classification System for Use with Remote Sensor Data. *Geological Survey Professional Paper, 964*.

Batty, M., Longley, P., 1994. *Fractal cities: a geometry of form and function*. Academic Press.

Chen, Y., 2020. Fractal Modeling and Fractal Dimension Description of Urban Morphology. *Entropy*, 22(9), 1–29. https://doi.org/10.3390/e22090961

Corbane, C., Florczyk, A., Pesaresi, M., Politis, P., Syrris, V., 2018. GHS built-up grid, derived from Landsat, multitemporal (1975-1990-2000-2014), R2018A. doi.org/10.2760/0726

Gkyer, E., 2013. Understanding Landscape Structure Using Landscape Metrics. *Advances in Landscape Architecture*. doi.org/10.5772/55758

Li, M., Zang, S., Zhang, B., Li, S., Wu, C., 2014. A review of remote sensing image classification techniques: The role of Spatio-contextual information. *European Journal of Remote Sensing*, 47(1), 389–411. doi.org/10.5721/EuJRS20144723.

Longley, P. A., 2005. Urban Studies. *Encyclopedia of Social Measurement*, 3, 921–926.

MacGarigal, K., B. J. M., 1995. FRAGSTATS : Spatial Pattern Analysis Program for Quantifying Landscape Structure. European Commission, Joint Research Centre.

Nautiyal, G., Maithani, S., 2020. Study of Urban Growth Dynamics Using Global Human Settlement Layer Data Set : Uttarakhand, India. *Journal of the Indian Society of Remote Sensing*, 48(5), 817–827. doi.org/10.1007/s12524-020-01115-6

Pesaresi, M., Commission, E., Union, E., Property, I., Julea, A., Commission, E., 2016. A New Method for Earth Observation Data Analytics. *Remote Sensing*. doi.org/10.3390/rs8050399

R.LeGates., 2001. Urban Studies: Overview. International Encyclopedia of the Social & Behavioral Sciences.

Ramachandra, T. V., Aithal, B. H., Sreekantha, S., 2012. Spatial Metrics based Landscape Structure and Dynamics Assessment for an emerging Indian Megalopolis. *International Journal of Advanced Research in Artificial Intelligence*, 1(1), 48–57.

Roy, P. S., Roy, A., Joshi, P. K., Kale, M. P., Srivastava, V. K., Srivastava, S. K., Dwevidi, R. S., Joshi, C., Behera, M. D., Meiyappan, P., Sharma, Y., Jain, A. K., Singh, J. S., Palchowdhuri, Y., Ramachandran, R. M., Pinjarla, B., Chakravarthi, V., Babu, N., Gowsalya, M. S., Kushwaha, D., 2015. Development of decadal (1985-1995-2005) land use and land cover database for India. *Remote Sensing*, 7(3), 2401–2430. doi.org/10.3390/rs70302401

Sharma, A. R. N., Development, U., Hinterland, V., 1991. Land Grab , Bombay Style : Urban Development in Vasai-Virar Hinterland of Bombay Land Grab , Bombay Style of Bombay. *Economic & Political Weekly*, 26(8), 413–417.

Shi, L., Liu, F., Zhang, Z., Zhao, X., Liu, B., Xu, J., Wen, Q., Yi, L., Hu, S., 2015: Spatial differences of coastal urban expansion in China from 1970s to 2013. *Chinese Geographical Science*, *25*(4), 389–403. doi.org/10.1007/s11769-015-0765-y

Tu, L., Qin, Z., Li, W., Geng, J., Yang, L., Zhao, S., Zhan, W., Wang, F., 2016. Surface urban heat island effect and its relationship with urban expansion in Nanjing, China. *Journal of Applied Remote Sensing*, 10(2), 026-037. doi.org/10.1117/1.jrs.10.026037

Xiong, Y., Peng, F., & Zou, B., 2019. Spatiotemporal influences of land use/cover changes on the heat island effect in rapid urbanization area. *Frontiers of Earth Science*, 13(3), 614–627. doi.org/10.1007/s11707-018-0747-3

RESEARCH ON NODE NETWORK TRANSMISSION CAPACITY PREDICTION MODEL FOR LARGE SCALE REMOTE SENSING DATA COLLECTION

Lu Bai [1], Xiaogang Liu [2], Man Zhao [3], Zhibao Wang [2, 4 *], Guiying Shi [2]

[1] School of Computing, Ulster University, Belfast BT15 1ED, UK
[2] School of Computer and Information Technology, Northeast Petroleum University, Daqing 163318, China
[3] School of Communication and Electronic Engineering, Qiqihaer University, Qiqihaer 161003, China
[4] Bohai-Rim Energy Research Institute, Northeast Petroleum University, Qinhuangdao 066004, China

KEY WORDS: PSO, BP, prediction model, remote sensing data, network transmission capacity.

ABSTRACT:

In recent years, the use of remote sensing technology has grown exponentially in various industries such as agriculture, forestry, and urban planning. Remote sensing data collection systems rely on a network of nodes to collect and transmit data. The transmission capacity of these node networks is a critical factor in the performance and efficiency of the entire system. However, accurately predicting the transmission capacity of a node network can be a challenging task. To carry out large scale open remote sensing data collection, it is necessary to predict the network transmission capacity of nodes in the face of the difference in the execution speed of each node for various tasks. It is necessary to predict the network transmission capacity of nodes. In this research, we propose a node network transmission capacity prediction model for large scale remote sensing data collection using a combination of Particle Swarm Optimization (PSO) and Backpropagation (BP) algorithms. The proposed PSO-BP model aims to accurately predict the transmission capacity of a node network in a remote sensing data collection system. The model is tested and evaluated using a large-scale dataset and the results show that the proposed model outperforms existing models in terms of prediction accuracy. This work contributes to the field of remote sensing data collection by providing a reliable and efficient method for predicting the transmission capacity of node networks.

1. INTRODUCTION

Space remote sensing technology is entering the era of industrial applications, in which Landsat, Sentinel and other public interest satellite remote sensing data play an important role (Hemati et al., 2021; Phiri et al., 2020; Segarra et al., 2020). With the development of remote sensing technologies and advancement of deep learning-based algorithms, significant progress has been achieved in recent years in many remote sensing tasks, for example, object detection from remote sensing images (Cheng and Han, 2016; Deng et al., 2018; Li et al., 2020; Qian et al., 2020; Song et al., 2020; Wang et al., 2021; Zhang et al., 2019; Zhu et al., 2021), remote sensing change detection (Hecheltjen et al., 2014; Jensen and Im, 2007; Khelifi and Mignotte, 2020; Shi et al., 2022; Q. Wang et al., 2018; Zhang et al., 2021) and remote sensing big data (Chi et al., 2016; Deren et al., 2014; Liu et al., 2018; Ma et al., 2021, 2015; Yu et al., 2021a, 2021b). Remote sensing data user groups such as individuals, small and medium-sized enterprises are gradually increasing. In the process of data application, satellites are generally publicly released on the Internet for global users, which is limited by the service capability of the data source itself and the acquisition capability of data user nodes. Due to the constraints of data source's own service capability and data user node acquisition capability, the problems of low efficiency of large-scale public remote sensing data acquisition and low utilization rate of user acquisition nodes still exist (Lee et al., 2011). The problems such as low efficiency of large-scale public remote sensing data collection and low utilization rate of user collection nodes still exist (Wang et al., 2022).

However, there are few studies on network node transmission capacity prediction. To carry out large scale open remote sensing data collection in crowdsourcing mode, it is necessary to predict the network transmission capacity of nodes considering the difference in the execution speed of each node for various tasks. The traditional Backpropagation (BP) algorithm uses the inertia weights between the input and hidden layers and the inertia weights between the hidden layer and the output layer to establish the prediction model. The inertia weights between the hidden layer and the output layer are randomly initialized, and the learning factors are randomly generated constants. The algorithm has a strong sensitivity to these coefficients during the training process, and it is easy to fall into long iteration time and local extremes. This study analyses the possible factors affecting the transmission speed of the node network at the levels of data source, transmission medium and receiving terminal of open remote sensing data collection. A range of controllable factors were selected to establish a prediction model of network transmission capacity. For the traditional BP neural network algorithm, the inertia weights of network algorithm, it is easy to fall into local minima and other shortcomings. The Particle Swarm Optimization (PSO) algorithm with good robustness and global search capability is combined with the BP neural network algorithm to propose an improved PSO-BP algorithm. The improved algorithm includes the introduction of gradient, momentum factor, and inertia weight adjustment function. To further improve the accuracy of the algorithm in the iteration, the accuracy dynamic adjustment function is introduced, and the learning factor is also introduced into dynamic adjustment function. Finally, a comparison experiment between the

* Corresponding author

https://doi.org/10.5194/isprs-archives-XLVIII-M-1-2023-25-2023

improved PSO-BP algorithm and the traditional BP algorithm is conducted on a real data set collected from Sentinel 2.

2. NODE NETWORK TRANSMISSION CAPACITY PREDICTION MODEL

2.1 Analysis of factors affecting the transmission capacity of node networks

This paper considers the data transmission in large-scale public remote sensing data collection under the crowdsourcing model. During the collection process, data is transmitted from the public remote sensing data source through multiple transmission media to the collection node. The data transmission process mainly involves "two sources and one medium", which are the data source, the transmission medium, and the receiving terminal. The paper analyses the characteristics of the three stages of factors affecting node network transmission capacity, and Figure 1 shows the factors affecting node network transmission capacity.

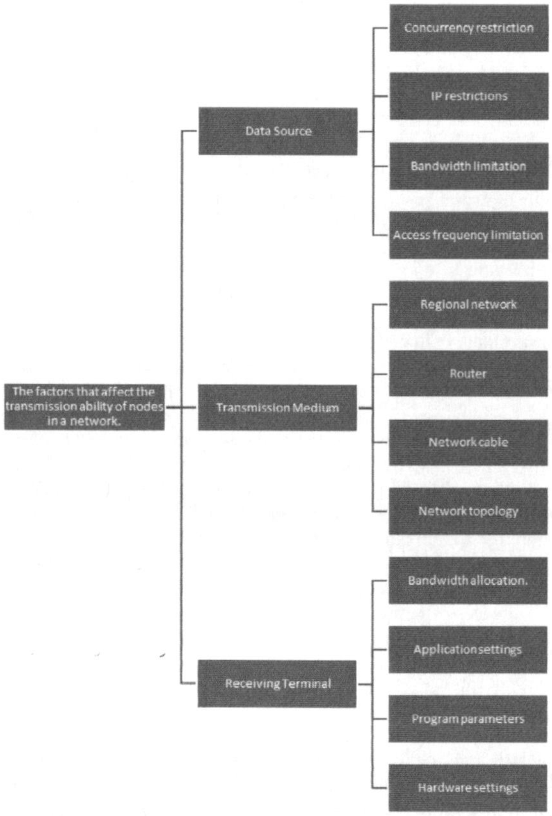

Figure 1. The factors that affect the transmission ability of nodes in a network

2.1.1 Constraint analysis based on the data source side

The main factors that affect the transmission capacity of a node network at the data source are related to the physical configuration and server parameter settings of the data source server.

In order to ensure the long-term stability and security of the data source, the number of concurrent accesses, IPs, access frequencies, access bandwidth speeds, allowed access time periods, and allowed access areas of users are usually restricted based on the physical configuration of the server node.

Through long-term collection, it has been found that when a single user frequently accesses the USGS data source for a long period of time, the user's order production speed is relatively slow, and the data download speed in the morning is much faster than that in the afternoon. The data source is closed for maintenance every Wednesday in Beijing time. User information must be provided when downloading, and the download speed is the fastest in the early morning in Beijing time.

2.1.2 Constraint analysis based on transmission medium

In the long-term practice process, it is found that the acquisition nodes located in different geographical locations also have significant differences. Therefore, it is guessed that when using the network medium for data transmission, the total network sites in different spatial regions sources differ, and the service capacity of configured network bandwidth varies from one spatial region to another. Similarly, in the same spatial region under the same network configuration environment, different physical facilities such as routers, cables, switches, similarly affect the data in transmission speed in the network medium.

To verify the effect of network topology on the transmission speed of data in the network medium, relevant experiments on relay service nodes were conducted. First, taking the public data source as the source end and the relay node as the receiving end, the relay service node host directly collected a certain amount of remote sensing data from the above-mentioned public data source, and stored it in the local physical hard disk of the relay service node. Then, taking the relay node as the source end and the various collection nodes as the receiving end, other collection nodes download data from the relay node to the local. The experimental results show that the speed of using the relay node to collect public remote sensing data is indeed much higher than that of not using the relay node, but due to the high cost of using the relay node, it is not suitable for large-scale long-term public remote sensing data collection.

2.1.3 Constraint analysis based on the acquisition terminal Among the many factors that affect the network transmission capability of the acquisition terminal, the node's own network bandwidth configuration plays a dominant role. Having a higher network bandwidth configuration is more beneficial for network data transmission. During the work process of the collecting terminal, when other applications occupy a large amount of network bandwidth, it seriously affects the node's ability to collect data using the network bandwidth. When the terminal uses third-party software to limit the network bandwidth of the collection client, it also affects the efficiency of data network transmission. The number of threads, task type, start time, and task length set by the terminal collection client also affect the network transmission speed of the node, and these factors are controllable factors of the collection terminal, so these controllable parameters can be used as input for subsequent network transmission ability prediction models. In addition, factors such as disk read and write speed, network card, and other physical hardware facilities on the node also affect the network transmission ability of the node.

2.2 Model parameter feature selection

To establish a model for predicting node network transmission capabilities and accurately predict the network transmission capabilities of crowd-sourced collection nodes, the factors that affect the data in the network transmission environment were analysed in the previous section from the three stages of data source, transmission medium, and receiving terminal. Many of these factors are uncertain and not controlled by the receiving terminal, making it difficult to discover their intrinsic patterns. Therefore, they cannot be considered in the node network transmission capability prediction model. Considering the need for publicly available remote sensing data collection, and from the perspective of being long-term, stable, inexpensive, and practical, many uncontrollable factors are discarded. Without considering many data source-side factors and physical factors, a selection of controllable factors is made from the perspective of dynamic adjustments to the client application parameters of the collection terminal. Using machine learning theory, a model for predicting node network transmission capabilities is established. The model is trained using the node's actual historical collection records, enabling the prediction of the node's network transmission speed. Table 1 shows the selected features that affect network transmission capabilities.

Feature Source	Feature Name
Data source	Number of concurrencies
	Access frequency
	Access time
	Order submission time
	Order capacity
	Whether it is a working day
Transmission medium	Space area
	Networking method
	Router model
	Switch models
	Network cable type
Receiving terminal	Network bandwidth
	Network card type
	Hard drive transfer rate
	Number of task threads
	Task progress
	Task length
	Collection terminal time: hour
	Collection terminal time: minute
	Collection terminal time: second
	Whether it is a working day

Table 1. Selected features affecting network transmission capability.

3. METHODS

3.1 PSO

PSO is an algorithm with advantages of simplicity, fast convergence and easy implementation, and is widely used in scheduling optimization, data mining, model training and other aspects of swarm intelligence optimization algorithms (D. Wang et al., 2018).

The position of the i-th particle in the population in a D-dimensional search space consisting of N particles is denoted as $X_i = (x_{i1}, x_{i2}, \dots, x_{iD}), i = 1,2,\dots,N$ which are the possible solutions of the problem. The velocity of the particle also consists of a D-dimensional vector, denoted as $V_i = (v_{i1}, v_{i2}, \dots, v_{iD}), i = 1,2,\dots,N$. It determines the direction and distance of the particle moving in the population. In addition, the movement of each particle component of the velocity is bounded by the maximum limiting velocity V_m. when $V_{ij} > V_m$, $V_{ij} = V_m$ $(1 \le i \le N, 1 \le j \le D)$. Similarly, the particle s displacement x_{ij} is also bounded by x_m. When $x_{ij} > x_m$, $x_{ij} = x_m$ $(1 \le i \le N, 1 \le j \le D)$, the fitness value of the particle at this time can be calculated according to the objective function. To prevent the blind search of particles, usually $x_{ij} \epsilon (-x_m, x_m)$, $v_{ij} \epsilon (-v_m, v_m)$. The equations of its velocity and displacement update during the particle iterative process are as follows.

$$\begin{cases} v_i^{k+1} = w \cdot v_i^k + c_1 r_1 [p_i - x_i^k] + c_2 r_2 [p_g - x_i^k] \\ x_i^{k+1} = x_i^k + v_i^k, i = 1,2,\dots,N \end{cases} \quad (1)$$

In Equation (1), the number of current iterations of the particle is denoted by k, and the inertia weight coefficient is denoted by w. The larger the value of inertia weight, the stronger the global search ability of the particle. c_1 and c_2 denote the learning factors. c_1 describes the influence of the particle by the individual extremes, so that the particle has global search ability and avoids getting into local solutions. c_2 represents the influence of global optimum on the particle. r_1 and r_2 are random numbers between (0,1). w, c_1 and c_2 are the three constants that jointly determine the spatial search ability of the particle. The position of the optimal fitness value calculated by the particle during the iteration process is represented by the individual extreme value, which are expressed as $p_i = (p_{i1}, p_{i2}, \dots, p_{iD}), i = 1,2,\dots,N$. The position of the optimal fitness value computed by all particles in the population during the iteration is denoted by the global extremum p_g: $p_g = (p_{g1}, p_{g2}, \dots, p_{gD})$.

3.2 PSO-BP Algorithm

In traditional BP neural network algorithms, the weights and thresholds are randomly generated, so the algorithm has randomness and unreliability. This paper proposes the PSO-BP neural network algorithm. First, the particles of PSO algorithm are initialised as the initialization weights of BP neural network algorithm, then iteration of PSO algorithm starts. Each iteration of PSO algorithm is followed by an execution of BP neural network algorithm. The iteration termination condition of BP neural network algorithm is determined by the prediction accuracy, which is dynamically adjusted as the number of iterations of PSO algorithm increases. The training termination condition of BP neural network algorithm becomes stricter in the later stage of PSO algorithm. After each iteration of BP neural network algorithm, the absolute value of the error is used evaluate the fitness of each particle of PSO algorithm. When the PSO algorithm satisfies its iteration termination condition, the PSO-BP algorithm also ends.

The parameters involved in the algorithm are the population size N, the dimensionality of the particles D, the number of iterations of the particles M, the inertia weight w, the learning factors c1 and c2, the maximum displacement x_m , and the maximum velocity v_m. The BP neural network algorithm uses a three-layer network topology. The value of each particle in the PSO algorithm represents the inertia weight in the BP neural network algorithm. The dimensionality of the particles D can be expressed as:

$$D = D_{in} \cdot D_h + D_h \cdot D_{out}, \quad (2)$$

https://doi.org/10.5194/isprs-archives-XLVIII-M-1-2023-25-2023 | © Author(s) 2023. CC BY 4.0 License.

In equation (2), the D_{in} , D_h, and D_{out} denote the number of neuros in the input layer, hidden layer, and output layer of the BP neural network, respectively.

In the process of predicting the transmission capacity of the nodes, six factors including the number of download task threads, task progress, task length, acquisition terminal time (in hours), acquisition terminal time (in minutes), and data source time (in hours) are used as the six neurons in the input layer. The number of neurons in the output layer is 1, and the number of neurons in the hidden layer is 10. The dimension D= 70, which indicates there are 70 inertia weights. For each particle's position $x_i = (x_{i,1}, x_{i,2}, \dots, x_{i,D})$, $x_{i,1}$ to $x_{i,60}$ denote the inertia weights from the input layer to the hidden layer and $x_{i,61}$ to $x_{i,70}$ denote the inertia weights from the hidden layer to the output layer.

3.2.1 Improvement of inertia weights: Inertia weight (w) in Particle Swarm Optimization (PSO) represents the ability of a particle to inherit its current velocity from its previous velocity. A smaller value of w is beneficial for local search, while a larger value is beneficial for global search. To balance the global and local search abilities of particles in PSO, the inertia weight is improved by modifying the inertia weight using the following equation (3)

$$w = \frac{\left[(w_{start} - w_{end}) \cdot \cos\left(\frac{k}{M}\right) + (w_{start} + w_{end})\right]}{2}, \quad (3)$$

In Equation (3), the initial and termination values of inertia weights are denoted by w_{start} and w_{end}, respectively, and $w_{start} < w_{end}$. The current number of particle iterations is k, and the maximum number of iterations is denoted by M. Due to the changing characteristics of the cosine function, the inertia weight (w) decreases slowly during the initial and final iterations. Therefore, the algorithm has a longer global optimization time at the beginning of the iterations, effectively reducing the risk of getting trapped in a local optimum. Additionally, in the later stages of the iterations, The algorithm can perform local search with a finer granularity, allowing for more incremental and precise adjustments.

3.2.2 Improvement of learning factors: In PSO algorithm, the learning factors c1 and c2 are the acceleration factors for the individual and global best values of a particle, respectively. Similarly, to enable the particles to have a good global search ability in the early stages of the iterations and to improve the precision and convergence speed of the particles in the later stages, the values of c1 and c2 are adjusted dynamically. This approach maintains the diversity of the population in the early stages of the search and improves the search performance in the later stages. The formula for calculating c1 and c2 is as follows.

$$c_1 = c_{start} - \frac{(c_{start} - c_{end})(M - k)}{M}$$
$$c_2 = 4 - c_1 \quad (4)$$

In Equation (4), the initial and termination values of the learning factor c1 are denoted by c_{start} and c_{end}, respectively, and $0<c_{start}<c_{end}<4$. M is the maximum number of iterations, and k is the current number of iterations.

3.2.3 Adaptation function: After initializing the neural network, the absolute value of the actual output result and the predicted result of the download force of the node at the next moment as the fitness function of the PSO algorithm:

$$F(X, x_i) = |y_i - y_i'|, i = 1,2, \dots, N \quad (5)$$

In Equation (5), X is the input to the neural network which are the six input metrics mentioned earlier. y_i and y_i' are the node the actual and predicted download speed. In the iterative process, the variation of velocity and displacement of the particles uses equation (1). The inertia weights and learning factors are calculated using in equations (3) and (4), respectively.

3.2.4 Data normalisation: To improve the accuracy of the training and reduce the errors caused by the difference in the magnitude of the input values, the input data is normalized to the range of (0,1). This is done to make the input neurons more sensitive and to reduce the errors caused by the difference in the scale of the factors. The equation for normalization is as follows:

$$X_{i,norm} = \frac{X_i - X_{i,min}}{X_{i,max} - X_{i,min}}, \quad (6)$$

In equation (6), the normalized result of input neuron i is denoted as $X_{i,norm}$. $X_{i,min}$ is the minimum value of the input neuron, and $X_{i,max}$ is the maximum value of the input neuron.

3.2.5 Data inverse normalization: To facilitate the comparison of predicted results with actual results and to make it easier to calculate the fitness value of each particle, it is necessary to reverse the normalization process of the particle's predicted results. The equation for reverse normalization is as follows:

$$Y_{i,norm} = Y_i \cdot (Y_{i,maxV} - Y_{i,minV}) + Y_{i,minV}, \quad (7)$$

In Equation (7), $Y_{i,norm}$ represents the result of the i-th dimension data after de-normalization, Y_i represents the result of the i-th dimension data before de-normalization, that is, the un-normalized result after prediction, $Y_{i,maxV}$ represents the maximum value of the i-th dimension data in the actual results, and $Y_{i,minV}$ represents the minimum value of the i-th dimension data in the actual results.

3.2.6 Accuracy dynamic adjustment function: To achieve the goal of low precision (low expected error) in the early stages of BP neural network training and high precision (high expected error) in the later stages, this paper introduces a precision dynamic adjustment function to dynamically adjust the error precision of the BP algorithm during the training process. The expression for the precision dynamic adjustment function is as follows:

$$accuracy = a - b * k \quad (8)$$

In Equation (8), a is the initial accuracy which usually is 1, b is the accuracy adjustment factor, and k is the current number of iterations.

The execution flow of the PSO-BP algorithm is shown in Figure 2.

This contribution has been peer-reviewed.
https://doi.org/10.5194/isprs-archives-XLVIII-M-1-2023-25-2023 | © Author(s) 2023. CC BY 4.0 License.

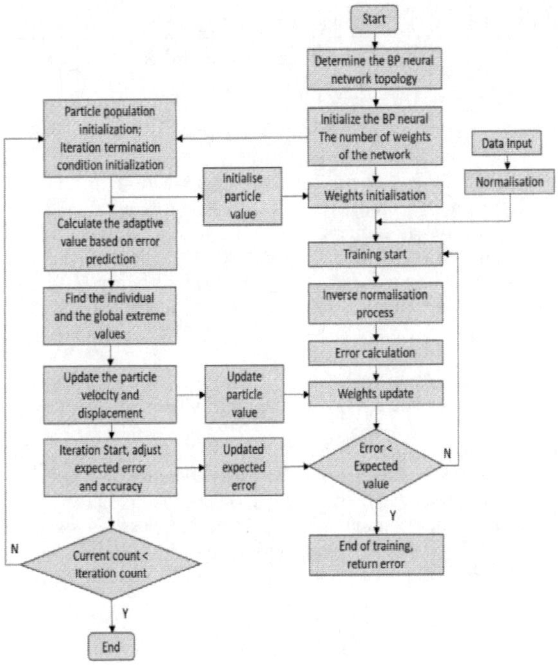

Figure 2. PSO-BP algorithm flow chart

4. EXPERIMENT AND RESULTS

4.1 Experimental parameters setting

To validate the execution results of the above proposed PSO-BP algorithm, BP, and PSO-BP algorithms were evaluated to conduct the experiments with real data sets. In Table 2, the parameters of the above two algorithms are set. The number of neurons in the input layer is 6, the number of neurons in the hidden layer is 10, and the number of neurons in the output layer is 1. The sigmoid function is used for the excitation function of the hidden layer. The inertia weights from the input layer to the hidden layer and from the hidden layer to the output layer are adjusted by combining gradient descent and momentum factor.

Algorithm	Parameters
BP	LearningRate = 0.08, Momentum factor =0.5, AvgTrainError >0.001
PSO-BP	N=50, c_{start}=3.5, c_{end}=2.5, w_{start}=1, w_{end}=0.1, b=0.2, Iter=200, v_{max}=2, $x_{i,D} \in$ [-0.5, 0.5], LearningRate =0.08, Momentum factor = 0.5, Iter=5

Table 2. Algorithm parameters setting

4.2 Experimental results analysis

To validate the execution results of the BP algorithm and the proposed PSO-BP algorithm, experiments have been done on real data sets. After cleaning and filtering the historical dataset, we collected more than 20,000 data files from a collection client of Sentinel 2 data collection. Around 20,000 data files were selected as the training set, and more than 2900 data files were selected as the test set. Table 3 lists the some of the experimental results of the tests using the two algorithms mentioned above.

No	Actual Value	BP		PSO-BP	
		Predicted Value	Error Rate	Predicted Value	Error Rate
1	976	677	0.3064	835	0.1445
2	840	854	0.0167	927	0.1036
3	832	886	0.0649	895	0.0757
4	816	848	0.0392	916	0.1225
5	848	859	0.0130	899	0.0601
6	840	867	0.0321	916	0.0905
7	920	833	0.0946	905	0.0163
8	944	842	0.1081	894	0.0530
9	920	842	0.0848	909	0.0120
10	840	799	0.0488	887	0.0560
11	824	834	0.0121	872	0.0583
12	912	843	0.0757	889	0.0252
13	976	777	0.2039	876	0.1025
14	864	833	0.0359	853	0.0127
15	808	813	0.0062	873	0.0804
16	808	798	0.0124	857	0.0606
17	912	867	0.0493	926	0.0154
18	880	888	0.0091	932	0.0591
19	832	848	0.0192	922	0.1082
20	856	858	0.0023	903	0.0549
21	920	842	0.0848	903	0.0185
22	872	876	0.0046	904	0.0367
23	824	850	0.0316	890	0.0801
24	832	853	0.0252	879	0.0565
25	832	819	0.0156	885	0.0637
26	808	845	0.0458	866	0.0718
27	896	840	0.0625	861	0.0391
28	808	833	0.0309	916	0.1337
29	816	802	0.0172	901	0.1042
30	928	838	0.0970	900	0.0302
31	816	791	0.0306	871	0.0674
32	984	824	0.1626	889	0.0965
33	840	919	0.0940	942	0.1214
34	920	909	0.0120	928	0.0087
35	856	921	0.0759	927	0.0829
36	888	907	0.0214	925	0.0417
37	920	906	0.0152	931	0.0120
38	936	898	0.0406	922	0.0150
39	832	905	0.0877	921	0.1070
40	888	905	0.0191	928	0.0450
41	816	906	0.1103	927	0.1360
42	808	909	0.1250	947	0.1720
43	936	913	0.0246	937	0.0011
44	824	899	0.0910	936	0.1359
45	920	900	0.0217	950	0.0326
46	808	882	0.0916	905	0.1200
47	968	876	0.0950	891	0.0795
48	832	754	0.0938	870	0.0457
49	824	771	0.0643	871	0.0570
50	848	765	0.0979	847	0.0012

Table 3. Selected prediction results

Figure 3 below shows the comparison of the error rates of the two algorithms for the data in Table 3. The prediction errors of the BP algorithm and the PSO-BP algorithm are very different, with the former having a large variance in the prediction results and the latter having a relatively flat prediction error. The errors of the two algorithms are mostly concentrated around 5%.

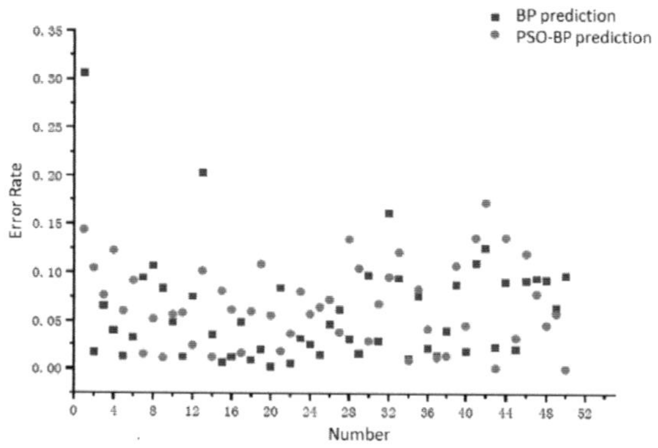

Figure 3. Error rates comparison between BP algorithm and PSO-BP algorithm

In Figure 4, the true value, and the prediction value from the two algorithms are shown. Prediction from PSO-BP algorithm is closer to the true value, while the error of the traditional BP algorithm is larger compared with PSO-BP algorithm.

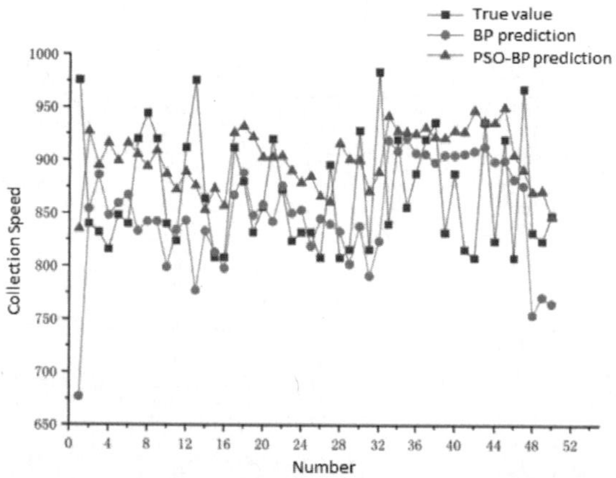

Figure 4. Results comparison between BP algorithm and PSO-BP algorithm

To further evaluate the prediction results, mean relative error and root mean square error are calculated in Table 4.

Algorithm	Mean relative error	Root mean square error
BP	0.0925	0.1163
PSO-BP	0.0595	0.0703

Table 4. Evaluation of algorithm prediction results

From Table 4, the mean relative error and root mean square error of PSO-BP algorithm are smaller than those of BP algorithm, which indicates that the PSO-BP algorithm has better prediction results. Tables 5 show the number of test sets in different error rate ranges for each of the two algorithms. The proposed PSO-BP algorithm achieves better accuracy comparing with traditional BP algorithm.

Algorithm	Error <= 5%	Error <= 15%	Error > 5%
BP	1004	2371	567
PSO-BP	1302	2908	30

Table 5. Error statistics on test set

5. CONCLUSION

This paper proposed an improved PSO-BP algorithm by introducing the accuracy dynamic adjustment function and the learning factor adjustment function. Finally, a comparison experiment between the improved PSO-BP algorithm and the BP algorithm is conducted on the real data set to verify the feasibility of the prediction model and the improved PSO-BP algorithm.

ACKNOWLEDGEMENTS

This work was supported in part by TUOHAI special project 2020 from Bohai Rim Energy Research Institute of Northeast Petroleum University under Grant HBHZX202002 and project of Excellent and Middle-aged Scientific Research Innovation Team of Northeast Petroleum University under Grant KYCXTD201903.

REFERENCES

Cheng, G., Han, J., 2016. A survey on object detection in optical remote sensing images. ISPRS J. Photogramm. Remote Sens. 117, 11–28. https://doi.org/10.1016/j.isprsjprs.2016.03.014

Chi, M., Plaza, A., Benediktsson, J.A., Sun, Z., Shen, J., Zhu, Y., 2016. Big Data for Remote Sensing: Challenges and Opportunities. Proc. IEEE 104, 2207–2219. https://doi.org/10.1109/JPROC.2016.2598228

Deng, Z., Sun, H., Zhou, S., Zhao, J., Lei, L., Zou, H., 2018. Multi-scale object detection in remote sensing imagery with convolutional neural networks. ISPRS J. Photogramm. Remote Sens. 145, 3–22. https://doi.org/10.1016/j.isprsjprs.2018.04.003

Deren, L., Liangpei, Z., Guisong, X., 2014. Automatic analysis and mining of remote sensing big data. Acta Geod. Cartogr. Sin. 43, 1211.

Hecheltjen, A., Thonfeld, F., Menz, G., 2014. Recent advances in remote sensing change detection–a review. Land Use Land Cover Mapp. Eur. Pract. Trends 145–178.

Hemati, M., Hasanlou, M., Mahdianpari, M., Mohammadimanesh, F., 2021. A Systematic Review of Landsat Data for Change Detection Applications: 50 Years of Monitoring the Earth. Remote Sens. 13, 2869. https://doi.org/10.3390/rs13152869

Jensen, J.R., Im, J., 2007. Remote sensing change detection in urban environments. Geo-Spat. Technol. Urban Environ. Policy Pract. Pixels 7–31.

Khelifi, L., Mignotte, M., 2020. Deep Learning for Change Detection in Remote Sensing Images: Comprehensive Review and Meta-Analysis. IEEE Access 8, 126385–126400. https://doi.org/10.1109/ACCESS.2020.3008036

Lee, C.A., Gasster, S.D., Plaza, A., Chang, C.-I., Huang, B., 2011. Recent Developments in High Performance Computing for Remote Sensing: A Review. IEEE J. Sel. Top. Appl. Earth Obs. Remote Sens. 4, 508–527. https://doi.org/10.1109/JSTARS.2011.2162643

Li, K., Wan, G., Cheng, G., Meng, L., Han, J., 2020. Object detection in optical remote sensing images: A survey and a new benchmark. ISPRS J. Photogramm. Remote Sens. 159, 296–307. https://doi.org/10.1016/j.isprsjprs.2019.11.023

Liu, P., Di, L., Du, Q., Wang, L., 2018. Remote sensing big data: Theory, methods and applications. Remote Sens. 10, 711.

Ma, X., Wang, Z., Bai, L., Xu, B., Gao, J., Wen, B., Tao, J., 2021. Implementation of a Federated Large-Scale Remote Sensing Data Sharing Platform, in: 2021 IEEE International Geoscience and Remote Sensing Symposium IGARSS. Presented at the IGARSS 2021 - 2021 IEEE International Geoscience and Remote Sensing Symposium, IEEE, Brussels, Belgium, pp. 5771–5774. https://doi.org/10.1109/IGARSS47720.2021.9554247

Ma, Y., Wu, H., Wang, L., Huang, B., Ranjan, R., Zomaya, A., Jie, W., 2015. Remote sensing big data computing: Challenges and opportunities. Future Gener. Comput. Syst. 51, 47–60. https://doi.org/10.1016/j.future.2014.10.029

Phiri, D., Simwanda, M., Salekin, S., Nyirenda, V.R., Murayama, Y., Ranagalage, M., 2020. Sentinel-2 data for land cover/use mapping: A review. Remote Sens. 12, 2291.

Qian, X., Lin, S., Cheng, G., Yao, X., Ren, H., Wang, W., 2020. Object detection in remote sensing images based on improved bounding box regression and multi-level features fusion. Remote Sens. 12, 143.

Segarra, J., Buchaillot, M.L., Araus, J.L., Kefauver, S.C., 2020. Remote sensing for precision agriculture: Sentinel-2 improved features and applications. Agronomy 10, 641.

Shi, K., Bai, L., Wang, Z., Tong, X., Mulvenna, M.D., Bond, R.R., 2022. Photovoltaic Installations Change Detection from Remote Sensing Images Using Deep Learning, in: IGARSS 2022 - 2022 IEEE International Geoscience and Remote Sensing Symposium. Presented at the IGARSS 2022 - 2022 IEEE International Geoscience and Remote Sensing Symposium, IEEE, Kuala Lumpur, Malaysia, pp. 3231–3234. https://doi.org/10.1109/IGARSS46834.2022.9883738

Song, G., Wang, Z., Bai, L., Zhang, J., Chen, L., 2020. Detection of oil wells based on faster R-CNN in optical satellite remote sensing images, in: Notarnicola, C., Bovenga, F., Bruzzone, L., Bovolo, F., Benediktsson, J.A., Santi, E., Pierdicca, N. (Eds.), Image and Signal Processing for Remote Sensing XXVI. Presented at the Image and Signal Processing for Remote Sensing XXVI, SPIE, Online Only, United Kingdom, p. 17. https://doi.org/10.1117/12.2572996

Wang, D., Tan, D., Liu, L., 2018. Particle swarm optimization algorithm: an overview. Soft Comput. 22, 387–408. https://doi.org/10.1007/s00500-016-2474-6

Wang, Q., Zhang, X., Chen, G., Dai, F., Gong, Y., Zhu, K., 2018. Change detection based on Faster R-CNN for high-resolution remote sensing images. Remote Sens. Lett. 9, 923–932. https://doi.org/10.1080/2150704X.2018.1492172

Wang, Z., Bai, L., Liu, X., Chen, Y., Zhao, M., Tao, J., 2022. Dynamic Task Scheduling in Remote Sensing Data Acquisition from Open-Access Data Using CloudSim. Appl. Sci. 12, 11508. https://doi.org/10.3390/app122211508

Wang, Z., Bai, L., Song, G., Zhang, J., Tao, J., Mulvenna, M.D., Bond, R.R., Chen, L., 2021. An Oil Well Dataset Derived from Satellite-Based Remote Sensing. Remote Sens. 13, 1132. https://doi.org/10.3390/rs13061132

Yu, Z., Wang, Z., Bai, L., Chen, L., Tao, J., 2021a. Remote Sensing Inversion of PM10 Based on Spark Platform, in: 2021 IEEE International Geoscience and Remote Sensing Symposium IGARSS. Presented at the IGARSS 2021 - 2021 IEEE International Geoscience and Remote Sensing Symposium, IEEE, Brussels, Belgium, pp. 1685–1688. https://doi.org/10.1109/IGARSS47720.2021.9554323

Yu, Z., Wang, Z., Bai, L., Chen, L., Tao, J., 2021b. Parameter Optimization on Spark for Particulate Matter Estimation, in: 2021 Workshop on Algorithm and Big Data. Presented at the WABD 2021: 2021 Workshop on Algorithm and Big Data, ACM, Fuzhou China, pp. 9–13. https://doi.org/10.1145/3456389.3456406

Zhang, J., Wang, Z., Bai, L., Song, G., Tao, J., Chen, L., 2021. Deforestation Detection Based on U-Net and LSTM in Optical Satellite Remote Sensing Images, in: 2021 IEEE International Geoscience and Remote Sensing Symposium IGARSS. Presented at the IGARSS 2021 - 2021 IEEE International Geoscience and Remote Sensing Symposium, IEEE, Brussels, Belgium, pp. 3753–3756. https://doi.org/10.1109/IGARSS47720.2021.9554689

Zhang, S., He, G., Chen, H.-B., Jing, N., Wang, Q., 2019. Scale adaptive proposal network for object detection in remote sensing images. IEEE Geosci. Remote Sens. Lett. 16, 864–868.

Zhu, M., Wang, Z., Bai, L., Zhang, J., Tao, J., Chen, L., 2021. Detection of industrial storage tanks at the city-level from optical satellite remote sensing images, in: Bruzzone, L., Bovolo, F., Benediktsson, J.A. (Eds.), Image and Signal Processing for Remote Sensing XXVII. Presented at the Image and Signal Processing for Remote Sensing XXVII, SPIE, Online Only, Spain, p. 33. https://doi.org/10.1117/12.2600008

This contribution has been peer-reviewed.
https://doi.org/10.5194/isprs-archives-XLVIII-M-1-2023-25-2023 | © Author(s) 2023. CC BY 4.0 License.

IRRIGATED AGRICULTURE MAPPING IN A SEMI-ARID REGION IN BRAZIL BASED ON THE USE OF SENTINEL-2 DATA AND RANDOM FOREST ALGORITHM

H. N. Bendini [1], L. M. G. Fonseca [1], C. A. Bertolini [1], R. F. Mariano [1]; A. S. Fernandes Filho [1]; T. H. Fontenelle [2]; D. A. C. Ferreira [2]

[1] National Institute for Space Research (INPE), São José dos Campos, Brazil – (hugo.bendini, leila.fonseca, alexandre.filho@inpe.br), caio.bertolini58@gmail.com, ravimariano@hotmail.com

[2] National Water and Sanitation Agency (ANA), Brasília, Brazil - (thiago.fontenelle, daniel.ferreira)@ana.gov.br

KEY WORDS: irrigated agriculture, phenological metrics, machine learning, time series, earth observation.

ABSTRACT:

Irrigation is important for agricultural production and is often decisive for this, especially in arid and semi-arid areas, where precipitation is insufficient. In Brazil, irrigated agriculture is responsible for 46% of withdrawals from water bodies and 67% of the consumption of the total volume of water collected, representing the highest consumptive use in the country. Remote sensing technologies have great potential for developing methods for monitoring irrigated areas. However, mapping irrigated areas is still a challenge, due to the complexity and diversity of irrigation methods and crops, especially in a country with continental dimensions like Brazil. Remote sensing techniques for mapping irrigated areas in Brazil have been applied mainly in areas with center pivot irrigation in the Cerrado, and with paddy rice in the south of Brazil. But few or no applications, involving mapping of crops irrigated by other irrigation methods, mainly in the semi-arid, have been carried out. The objective of this work was to investigate a method for classifying irrigated agriculture in a semiarid region of Brazil, based on the use of Sentinel 2 imagery and random forest algorithm. We proposed a novel and robust methodology showing with preliminary results that it´s possible to identify irrigated agriculture in this region with a class-f1-score of 74% for complementary irrigation and 95% for center-pivots.

1. INTRODUCTION

Agriculture is the biggest global consumer of water, with irrigated areas constituting 40% of the total area used for agricultural production (FAO, 2014). The rate of increase in irrigated areas was approximately 2.6% per year, going from 95 million hectares (Mha) in the 1940s to 280 Mha in the early 1990s (Van Schilfgaarde 1994, Seckler et al. 2000, Siebert et al., 2005a, b, 2006). FAO estimates that 80% of food needs in 2025 will be covered by irrigated agriculture (Schaldach et al., 2012) and more than 324 million hectares are equipped for irrigation in the world (Dubois, 2011).

In Brazil, irrigated agriculture is responsible for 46% of withdrawals from water bodies and 67% of the consumption of the total volume of water collected, representing the highest consumptive use in the country. It is a dynamic activity that has shown increasing and persistent performance in recent decades, often against the grain of unstable and negative periods in the Brazilian economy. There was an intensification in the most recent period, linked to the greater contribution of credit and private investments, between 2012 and 2019, and growth was around 4% per year in Brazil when around 216 thousand hectares of irrigated fields were incorporated per year. In 2019, the value of irrigated production surpassed the BRL 55 billion mark (ANA, 2021).

In general, the survey carried out by the National Agency for Water and Basic Sanitation (ANA) shows that with the current availability of water, only 36% of the agricultural area and 15% of the pasture area could be converted into irrigated areas in Brazil. However, the potential for expansion of irrigated areas (total and effective) must be observed with caution, as local particularities, infrastructure expansion, and water infrastructure works can change the estimate of additional irrigable area, especially when the water supply is increased with transfers from other basins or reduced with the installation of other uses or with the revision of water supply databases. Furthermore, the potentials were estimated only on current agricultural areas (agriculture and pasture already consolidated). The expansion of the irrigated area in the country has occurred and should continue to occur, according to three main aspects: public perimeters planned by government agencies; joint private initiatives, organized in the form of cooperatives or associations; and individual private initiatives. In this context, it is important to strengthen planning and organize the State's role as a promoter and partner of this development, especially at the federal level, in conjunction with states, municipalities, and the private sector (ANA, 2021).

Irrigation is important for agricultural production and is often decisive for this, especially in arid and semi-arid areas, where precipitation is insufficient. The benefits of irrigation include increased productivity and the dissociation of constraints, ensuring greater resistance to extreme weather events, modifying temperature, humidity, and precipitation regimes on local to regional scales, and evapotranspiration (ET) globally. However, irrigation can also have significant social and environmental impacts, including drainage or maintenance of wetlands, disruption of sedimentation, increased soil salinity, changes in river temperatures, changes in water table depth, decreased water flow, changes in peak discharge and baseflow, and conflicts over water use (Ketchum et al., 2020; Pousa et al., 2019). To mitigate or eliminate negative impacts, the irrigated areas must advance observing the different dimensions of sustainability.

As seen, information on the spatial distribution of irrigated areas is highly relevant for water management and food security, being essential for decision-makers who are facing the transition to a more efficient sustainable agriculture. However, the spatial patterns, extent, and intensity of water use by irrigated agriculture are currently not well understood. In part, this is because statistics and reports are costly to produce and can be biased due to over- or under-reporting of water use (Rufin et al., 2021; Deines et al., 2019; Özdoğan et al., 2006). In addition,

https://doi.org/10.5194/isprs-archives-XLVIII-M-1-2023-33-2023

irrigated area statistics often do not include smaller or informal irrigated areas (e.g., groundwater, small reservoirs, and ponds), due to the difficulty in mapping these smaller areas, as well as due to the inherent underreporting of survey systems by interviews. However, in many countries, these areas are very significant and even exceed the main irrigated areas.

Remote sensing technologies have great potential for developing methods for monitoring irrigated areas (Özdoğan et al., 2010; Mcallister et al., 2015; Deines et al., 2017, 2019; Xie et al., 2019; Vogels et al., 2019; Xiang et al, 2019, 2020; Ketchum et al., 2020; Magidi et al., 2021; Rufin et al., 2021; Zurqani et al., 2021). However, mapping irrigated areas is still a challenge, due to the complexity and diversity of irrigation typologies and crops, especially in a country with continental dimensions like Brazil.

Remote sensing techniques for mapping irrigated areas in Brazil have been applied mainly in areas with center pivot irrigation in the Cerrado (Saraiva et al., 2020; Albuquerque et al., 2021). Recently, some applications involving flood-irrigated rice crop mapping have been carried out (De Bem et al., 2021). But few or no applications, involving mapping of crops irrigated by other irrigation methods, mainly in the semi-arid, have been carried out. There is still a gap in the scientific literature, in terms of concerns the development of semiautomatic methods for mapping irrigated agriculture by remote sensing, which can be transferred to the Brazilian reality.

The objective of this work was to investigate a method for classifying LULC in a semiarid region of Brazil, focusing on irrigated agriculture, based on the use of Sentinel 2 imagery and random forest algorithm. This work is in the context of the project "Irrigated Agriculture Based on Remote Sensing Technologies to Update and Improve ANA's Atlas Irrigation", developed by INPE (National Institute for Space Research) and ANA, which aims at developing a method for automatically mapping irrigated agricultural land and estimating water use in Brazilian irrigated agriculture.

2. MATERIALS AND METHODS

2.1 Study areas

The study area of the western Ceará agriculture hub is located around the municipalities of Guaraciaba do Norte, São Benedito, and Ipu, and is within in the Caatinga biome (Figure 1). Agricultural cultivation areas correspond to small plots, usually close to watercourses. There is no presence of irrigation with center pivots. Irrigation is based on micro spray and dripping systems. In this area, there is a predominance of native vegetation composed of deciduous shrub vegetation, that is, with loss of foliage in the dry season of the year. There is also the presence of evergreen vegetation in the humid areas. The predominant climate is Tropical, Equatorial Zone, hot with an average temperature > 18° C in all months, semi-arid with 7 to 8 dry months. The study area of the Petrolina/Juazeiro agriculture hub is located around the municipalities of Petrolina, in the state of Pernambuco and Juazeiro, in the north of Bahia state, and it is also within in the Caatinga biome (Figure 1). Agricultural cultivation areas are made up of large groups of plots with perennial and annual crops, distributed among a matrix of native vegetation. In this area, there is a predominance of native vegetation composed of deciduous shrub vegetation, with evergreen vegetation on the banks of watercourses. The predominant climate is Tropical, Equatorial Zone, hot with an average temperature > 18° C in all months, semi-arid with 7 to 8 dry months.

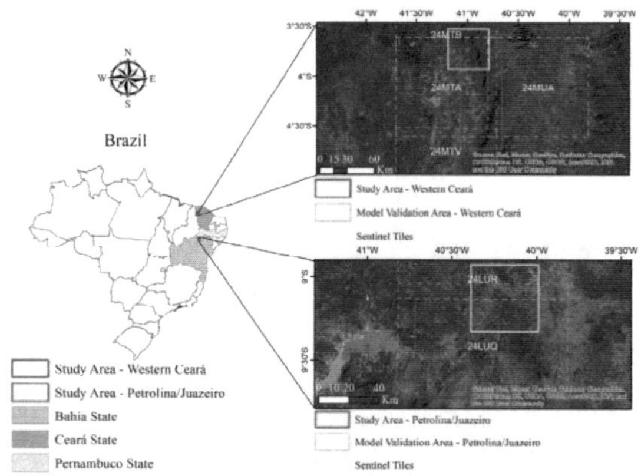

Figure 1. Study areas.

2.2 Sample Generation

The established protocol for sample generation contemplated the segmentation of cloud-free Sentinel 2 image obtained during the dry period (T24MTA_20190810T1 and T24LUQ_20191026T1). Considering bands 2, 3, 4 and 8 (blue, green, red and NIR) we applied the multi-resolution algorithm, with parameters of 80, 0.2 and 0.5 respectively for scale, shape and compactness. Then, with the respective Dynamic World classified image (Brown et al., 2022), we intersected with the segmented images to obtain the major land use and land cover (LULC) class per segment and the probability values of each class. We clipped the segmented images in smaller subsets, considering representative areas of the regions. The segments with the values of the majority classes with the highest probability were assigned to a specialist, whom proceeded the visual interpretation considering the LULC classes showed in Table 1. For the visual interpretation, we assumed that the agricultural areas that showed a stronger color pattern in the infrared during the dry period were irrigated areas. According to information from ANA specialists, in these regions most agriculture has some level of supplementary irrigation, with rainfed areas basically consisting of areas of subsistence agriculture, in small and diffused areas.

Classes	Number of samples
Water bodies	167
Natural Forests	1300
Complementary irrigation	1659
Center-pivot	29
Shrublands	2187
Sugarcane	26
Urban areas	342
Bare soil	347

Table 1. Number of samples.

Finally, for each of the segments, sampled points containing the class information of its respective polygon were generated, and the set of all points generated in the subsets generated in the subsets were used as input data for training.

https://doi.org/10.5194/isprs-archives-XLVIII-M-1-2023-33-2023

2.3 Datasets generation

2.3.1 Environmental data: Terrain data were obtained by the SRTM (Shuttle Radar Topography Mission) (NASA, 2013), from which the elevation and slope images were derived. Precipitation data from CHIRPS (Climate Hazards Group InfraRed Precipitation with Station data) was also used, which is a nearly global rainfall dataset spanning more than 30 years. CHIRPS incorporates 0.05° resolution satellite imagery with in-situ station data to create gridded precipitation time series for trend analysis and seasonal drought monitoring.

2.3.2 Sentinel 2 Analysis Ready Data (ARD) generation: Recent advances in remote sensing technologies offer great opportunities to map land use and land cover over large areas. Several operational sensor systems are currently in orbit and the development of infrastructure for remote sensing data storage and data dissemination allows to derive consistent, analysis-ready images, eliminating the need for pre-processing and storage for the user (Potapov et al. al., 2020, Frantz, 2019). This allows exploring the full potential of integrated data analysis, in which metrics derived from ARD time series (e.g., phenological or spectral-time metrics) are combined with meaningful environmental data for a specific domain. In this work, we used data cubes developed from optical images of multispectral sensors (MSI – Multispectral Instrument) of the Sentinel 2 A and B satellites. Images at surface reflectance level were accessed from the asset "COPERNICUS/S2_SR", in GEE, with atmospheric correction performed with the Sen2cor method (Main-Knorn et al., 2017). The cloud mask was obtained by the CDI (Cloud Displacement Index) algorithm (FRANTZ et al., 2018), which makes use of the three highly correlated near-infrared bands that are observed with different viewing angles. Thus, elevated objects such as clouds are observed under a parallax and can be reliably separated from other objects on the Earth's surface. Dense data cubes were generated, that is, with a temporal resolution of 8 days, for the calculation of phenological metrics, which require a shorter interval between observations, and that they are equally spaced in time. For this, a process was applied to create temporal mosaics of 8 days, based on the highest value, to organize cubes with observations equally spaced in time, using the GEE itself, through the "qualityMosaic" function. For the dense data cubes, in cases where no images were found with a cloud cover percentage lower than 50% within the 8 days, a synthetic image was generated. Subsequently, in a Python programming environment, the pixels contaminated by clouds and cloud shadows found in the data cubes were interpolated using the Radial Basis Funcion method (Schwieder et al., 2016; Bendini et al., 2019). Data cubes with a temporal resolution of 16 days were also generated to calculate the accumulated sums of vegetation indices during the dry period. To identify the dry period, an approach was used in which CHIRPS monthly precipitation data was considered. For each CHIRPS pixel (1 km), the three driest months in 2019 were obtained. The dense data cubes were generated for the following vegetation indices, EVI2 (Two-band Enhanced Vegetation Index) (Jiang et al., 2008), NDWI (Normalized Difference Water Index) (McFeeters, 1996), GI (Greenness Index) (Gitelson, 2003), ARVI2 (Atmospherically Resistant Vegetation Index 2) (Kaufman et al., 1992), and LSWI (Land Surface Water Index) (Jügens, 1997).

2.3.3 Accumulated sum of VI during the dry season: The cumulative sum of the vegetation index values of the Sentinel 2 images was performed by a zonal summation operator on the Sentinel 2 images, according to the grid obtained from the CHIRPS pixels, where the dry period could vary spatially. In this case, as 16 days of temporal resolution would be enough to calculate an accumulated sum, not being necessary to generate a dense cube, which requires more sophisticated interpolation methods not available in GEE, the whole process was carried out in GEE, to reduce computational costs, which would also be amplified by the strategy of using CHIRPS to identify local dry periods. As the presence of null values could impact the calculation of the accumulated sum, an interpolation was also necessary. However, as the process was carried out in the GEE, the interpolation was simpler, being performed by the average of the period values. In cases of negative values, which would also significantly affect the sum, they were replaced by zero. Finally, a smoother based on a moving polynomial function, called Whitaker (Eilers, 2003), was applied to remove false peaks and remaining noise, which could influence the calculation of the accumulated sum. The cumulative sums during the dry period were calculated for the following vegetation indices: EVI2, NDWI, NDVI (Normalized Difference Vegetation Index), NDBI (Normalized Difference Built-up Index) (Zha et al., 2003), GI, GCVI (Green Chlorophyll Vegetation Index) (Gitelson et al., 2003), LSWI, BSI (Bare Soil Index) (Rikimaru, 2003), and ARVI2.

2.3.4 Phenological metrics: We derived the phenological metrics from the Sentinel 2 EVI2 dense data cube described earlier, during the agricultural year 2019 – 2020, i.e., August 2019 to October 2020 (CONAB, 2018). A total of 13 phenological metrics were derived using the Python Stmetrics package (Soares, A; Bendini, H. N.; et al., 2019) (https://github.com/brazil-data-cube/stmetrics, accessed on 20 November 2022). We extracted for the 3 seasonal cycles observed in the EVI2 time series, totaling 39 metrics. These metrics refer, for example, to the beginning and end of the season, the maximum value of station EVI2 or the amplitude and are inspired by the TIMESAT software, explained in detail in Jönsson, Eklundh (2004). Bendini et al. (2019) used them for mapping cropping systems and crop types in Brazil with good accuracies.

2.3.5 Spectral-temporal metrics (STM): Spectro-temporal metrics (STM) (Griffiths et al., 2013, Rufin et al., 2015) were calculated for the dry and rainy seasons of the year 2019, consisting of statistical values (standard deviation, variance, mean, median, and percentiles), obtained from Sentinel 2 surface reflectance images, free of clouds and cloud shadows, of all spectral bands (B, R, G, NIR, Red-edge, and SWIR). Rufin et al. (2018) used them for mapping cropping systems in Turkey with good accuracies.

https://doi.org/10.5194/isprs-archives-XLVIII-M-1-2023-33-2023 | © Author(s) 2023. CC BY 4.0 License.

2.3.6. Neighborhood Green Chlorophyll Vegetation Index (NGCVI): The Neighborhood Green Chlorophyll Vegetation Index (NGCVI) was described by Deines et al. (2019) and consists of a normalization of the Green Chlorophyll Vegetation Index (GCVI) (Gitelson et al., 2005) by a neighborhood index, which is defined as the division of the maximum value of the GCVI divided by the 15th percentile, convoluted by a kernel. Deines et al. (2019) use a 50km radius to map the High Plains aquifer, based on the work of Xu et al. (2019), which explores the autocorrelation of climate variables. Still in Deines et al. (2019), the Landsat image was resampled to 1000 meters, due to the computing limitations of the Google Earth Engine platform, a procedure that was replicated in this work, however, for Sentinel 2 images. Fernandes Filho et al. (2022) used the NGCVI for classifying irrigated agriculture in the Brazilian semi-arid region with promising results. However, the authors suggested incorporating other vegetation indices is necessary to increase the accuracy of the classification.

2.4 Classification

The predictor variables were used together with the reference data to train a Random Forest (RF) classifier. RF is a non-parametric machine learning algorithm based on decision trees. As individual decision trees are error-prone, RF uses a set of many decision trees that have been independently trained with random subsets of the input data to overcome this limitation (Breiman, 2001). The implementation of the algorithm in Python (Python, 2022) also allows evaluating the variable importance of each input variable based on the Gini coefficient. Different models were trained for each study area, and the classification was evaluated using metrics derived from the confusion matrix, such as global accuracy and f1-score. Validation was performed with 30% of the data, in an n-partition scheme, with the remaining 70% used in training. Subsequently, the robustness of these models was evaluated using the same model in an area different from the training area, but in the same irrigation hub. The same sample collection protocol was applied in these test areas, so that the validation was performed with points totally independent of the training set.

3. RESULTS AND DISCUSSIONS

Figure 2 illustrates the mean time series of all pixels corresponding to the training samples used for the global model.

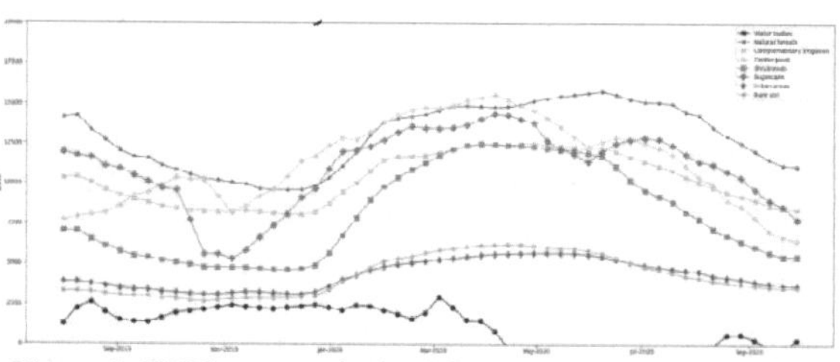

Figure 2. EVI2 curves, obtained for the points of the samples used in the training model.

We can observe that the agricultural areas present a seasonality similar to that of the natural vegetation, however, in the case of center-pivots, there is an indication of seasonality at the beginning of the agricultural year, suggesting the presence of two harvests. This is possibly due to the fact that in this region, there is expressiveness in the presence of perennial agricultural crops, which have a behavior similar to that of natural vegetation.

Figures 3 and 4 illustrate the result of the mapping for respectively the study areas in the Juazeiro/Petrolina, and Western Ceará hubs.

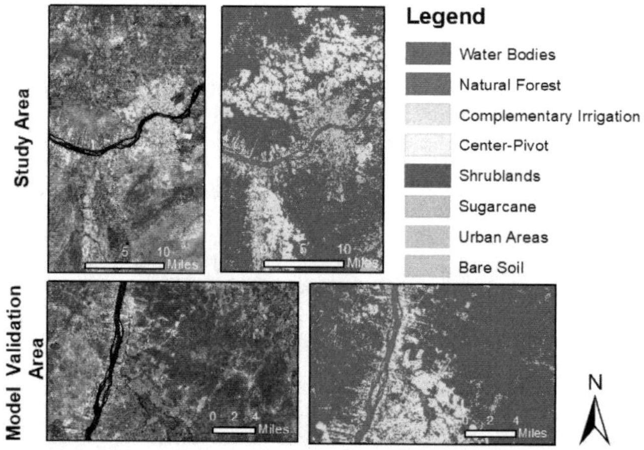

Figure 3. False Color composition (NIR, Red, Green), and result of the classification for training area in the region of Juazeiro/Petrolina hub.

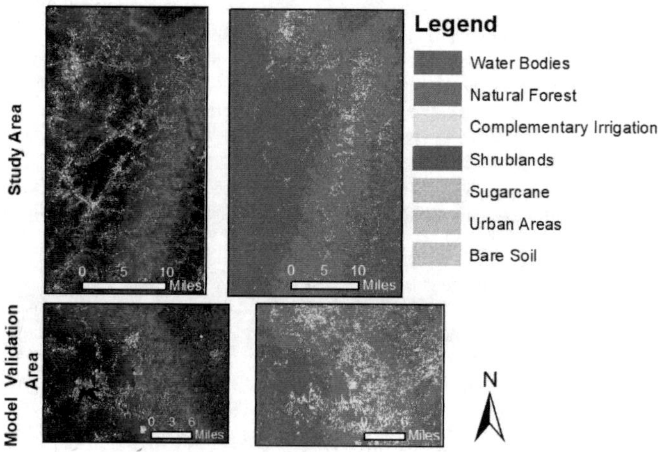

Figure 4. False Color composition (NIR, Red, Green), and result of the classification for training area in the region of Western Ceará hub.

Although the global accuracy of the global model was 72%, the target class, related to irrigated agriculture, achieve a class-f1-score of 74%. We can see that the spatial patterns are very consistent, reflecting the quality of the mapping. Figure 5 shows the confusion matrix of the global model.

https://doi.org/10.5194/isprs-archives-XLVIII-M-1-2023-33-2023

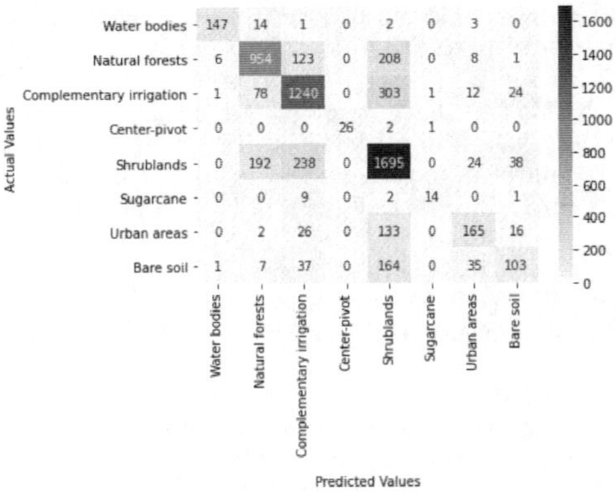

Figure 5. Classification confusion matrix for the global model.

It is possible to verify that there was confusion between agriculture and the classes referring to natural vegetation. We can infer that the more pronounced seasonality of the semiarid vegetation can lead to confusion with agriculture. A more cautious sampling is suggested in this sense, including in the sample collection protocol an evaluation of the EVI time series in these areas, to verify the characteristic pattern of agriculture, as well as an increase in the number of samples, in a more distributed way in the scene.

4. FINAL CONSIDERATIONS AND FUTURE WORKS

We explored a method for classifying irrigated agriculture in the semiarid of Brazil, using Sentinel 2 data and the random forest algorithm. We proposed a global model for classifying irrigated agriculture in this region. Good results were observed for irrigated agriculture, however, there was also a significant inclusion of areas of native vegetation (mainly shrubs). This fact is associated with the high seasonality of this vegetation, as well as its resilience to periods of drought (in the case of native forests areas close to water bodies). In the next steps, a more careful inspection of these areas of inclusion will be carried out, followed by a more representative sampling of these areas.

ACKNOWLEDGEMENTS

The authors acknowledge the CNPq (Process 423959/2021-2) Brazil Data Cube (a project part of the "Environmental Monitoring of Brazilian Biomes", funded by the Amazon Fund through the financial collaboration of the Brazilian Development Bank (BNDES) and the Foundation for Science, Technology and Space Applications (FUNCATE) no. 17.2.0536.1) for the availability of free and open-source software that was useful for this research.

REFERENCES

de Albuquerque, A.O., de Carvalho, O.L.F., Silva, C.R., de Bem, P.P., Gomes, R.A.T., Borges, D.L., Guimarães R.F., Pimental, C.M.M., de Carvalho Júnior, O.A., 2021: Instance segmentation of center pivot irrigation systems using multi-temporal SENTINEL-1 SAR images. *Remote Sensing Applications: Society and Environment*, 23, 100537.

Bendini, H.N., Fonseca, L.M.G., Schwieder, M., Körting, T.S., Rufin, P., Sanches, I.D.A., Leitão P.J., Hostert, P., 2019: Detailed agricultural land classification in the Brazilian Cerrado based on phenological information from dense satellite image time series. *International Journal of Applied Earth Observation and Geoinformation*, 82, 101872.

Breiman, L., 2001: Random forests. *Machine learning*, 45(1), 5–32.

Brown, C.F., Brumby, S.P., Guzder-Williams, B., Birch, T., Hyde, S.B., Mazzariello, J., Tait, A.M., 2022: Dynamic World, Near real-time global 10 m land use land cover mapping. *Scientific Data*, 9(1), 1-17.

de Bem, P.P., de Carvalho Júnior, O.A., de Carvalho, O.L.F., Gomes, R.A.T., Guimarães, R.F., Pimentel, C.M.M., 2021: Irrigated rice crop identification in Southern Brazil using convolutional neural networks and Sentinel-1 time series. *Remote Sensing Applications: Society and Environment,* 24, 100627.

Deines, J.M., Kendall, A.D., Hyndman, D.W., 2017: Annual irrigation dynamics in the US Northern High Plains derived from Landsat satellite data. *Geophysical Research Letters*, 44(18), 9350-9360.

Deines, J.M., Kendall, A.D., Crowley, M.A., Rapp, J., Cardille, J.A., Hyndman, D.W, 2019: Mapping three decades of annual irrigation across the US High Plains Aquifer using Landsat and Google Earth Engine. *Remote Sensing of Environment*, 233, 111400.

Dubois, O., 2011: *The State of the World's Land and Water Resources for Food and Agriculture: Managing Systems at Risk.* Earthscan Publications Ltd., London.

Eilers P.H.C., 2003: "A perfect smoother", *Analytical Chemistry*, 75, 3631-3636.

FAO, 2014: FAOSTAT. http://faostat.fao.org/default.aspx?lang=en (accessed in November 2021).

Fernandes Filho, A.S., Fonseca, L.M.G., Bendini, H., Sanches, I.D. Avaliação de índice de vizinhança para mapeamento de agricultura irrigada. In press. São José dos Campos: INPE; 2023.

Frantz, D., Hass, E., Uhl, A., Stoffels, J., & Hill, J., 2018: Improvement of the Fmask algorithm for Sentinel-2 images: Separating clouds from bright surfaces based on parallax effects. *Remote sensing of environment*, 215, 471-481.

Frantz, D. Force, 2019: Landsat + Sentinel-2 analysis ready data and beyond. *Remote Sens*. 11, 1124.

Gitelson, A.A., Viña, A., Arkebauer, T.J., Rundquist, D.C., Keydan, G., Leavitt, B., 2003: Remote estimation of leaf area index and green leaf biomass in maize canopies. *Geophys. Res. Lett*. 30(5), 1248.

Gitelson, A.A., Viña, A., Ciganda, V., Rundquist, D.C., Arkebauer, T.J., 2005: Remote estimation of canopy chlorophyll content in crops. *Geophysical Research Letters*, 32(8), 1-4).

Griffiths, P., van der Linden, S., Kuemmerle, T., Hostert, P., 2013: A pixel-based Landsat compositing algorithm for large area land cover mapping. *IEEE Journal of Selected Topics in Applied Earth Observations and Remote Sensing*, 6(5), 2088–

2101.

Jiang, Z.; Huete, A.R.; Didan, K.; Miura, T., 2008: Development of a two-band enhanced vegetation index without a blue band. *Remote sensing of Environment*, 112, 3833-3845.

P. Jönsson, L. Eklundh, 2004: Timesata program for analyzing time-series of satellite sensor data. *Comput. Geosci*., 30(8), 833-845.

Jügens, C., 1997: The modified normalization difference vegetation index (mNDVI): a new index to determine frost damages in agriculture based on Landsat TM data. *International Journal of Remote Sensing*, 18, 3583–3594.

Kaufman, Y.J., Tanre, D., 1992: Atmospherically resistant vegetation index (ARVI) for EOS-MODIS. *IEEE transactions on Geoscience and Remote Sensing*, 30(2), 261-270.

Ketchum, D., Jencso, K., Maneta, M. P., Melton, F., Jones, M. O., & Huntington, J., 2020: IrrMapper: A machine learning approach for high resolution mapping of irrigated agriculture across the Western US. *Remote Sensing*, 12(14), 2328.

Magidi, J., Nhamo, L., Mpandeli, S., & Mabhaudhi, T., 2021: Application of the random forest classifier to map irrigated areas using google earth engine. *Remote Sensing*, 13(5), 876.

McAllister, A., Whitfield, D., & Abuzar, M., 2015: Mapping irrigated farmlands using vegetation and thermal thresholds derived from Landsat and ASTER data in an irrigation district of Australia. *Photogrammetric Engineering & Remote Sensing*, 81(3), 229-238.

McFeeters, S. K., 1996: The use of the Normalized Difference Water Index (NDWI) in the delineation of open water features. *International journal of remote sensing*, 17(7), 1425-1432.

Main-knorn, M. et al. Sen2Cor for sentinel-2. In: Image and Signal Processing for Remote Sensing XXIII. SPIE, 2017. p. 37-48.

NASA, LP DAAC, 2013: NASA Shuttle Radar Topography Mission (SRTM) Version 3.0 (SRTM Plus) Product Release. Land Process Distributed Active Archive Center. National Aeronautics and Space Administration.

Ozdogan, M., Yang, Y., Allez, G., & Cervantes, C., 2010: Remote sensing of irrigated agriculture: Opportunities and challenges. *Remote sensing*, 2(9), 2274-2304.

Ozdogan, M., Woodcock, C.E., Salvucci, G.D., & Demir, H., 2006: Changes in summer irrigated crop area and water use in Southeastern Turkey from 1993 to 2002: Implications for current and future water resources. *Water resources management*, 20(3), 467-488.

Potapov, P., Hansen, M.C., Kommareddy, I., Kommareddy, A., Turubanova, S., Pickens, A., Ying, Q., 2020: Landsat analysis ready data for global land cover and land cover change mapping. *Remote Sensing*, 12(3), 426.

,

Pousa, R., Costa, M.H., Pimenta, F.M., Fontes, V.C., Brito, V.F.A.D., & Castro, M., 2019: Climate change and intense irrigation growth in Western Bahia, Brazil: The urgent need for hydroclimatic monitoring. *Water*, 11(5), 933.

Python Software Foundation, 2022. Python Language Reference, version 3.6. http://www.python.org

Rufin, P., Müller, D., Schwieder, M., Pflugmacher, D., & Hostert, P., 2021: Landsat time series reveal simultaneous expansion and intensification of irrigated dry season cropping in Southeastern Turkey. *Journal of Land Use Science*, 16(1), 94-110.

Rikimaru, A., Roy, P. S., & Miyatake, S., 2002: Tropical forest cover density mapping. *Tropical ecology*, 43(1), 39-47.

Saraiva, M., Protas, É., Salgado, M., & Souza Jr, C., 2020: Automatic mapping of center pivot irrigation systems from satellite images using deep learning. *Remote Sensing*, 12(3), 558.

Schaldach, R., Koch, J., der Beek, T. A., Kynast, E., Flörke, M., 2012: Current and future irrigation water requirements in pan-Europe: An integrated analysis of socio-economic and climate scenarios. *Global and Planetary Change,* 94, 33-45.

Schwieder, M., Leitão, P.J., da Cunha Bustamante, M.M., Ferreira, L.G., Rabe, A., Hostert, P., 2016: Mapping Brazilian savanna vegetation gradients with Landsat time series. *International journal of applied earth observation and geoinformation*, 52, 361-370.

Seckler, D., Amarasinghe., U., De Fraiture, C., Keller, A., Molden, D., & Sakthivadivel., R., 2000: *World Water Supply and Demand: 1995 to* 2025. International Water Management Institute, Colombo, Sri Lanka.

Siebert, S., Döll, P., Hoogeveen, J., Faures, J. M., Frenken, K., & Feick, S., 2005: Development and validation of the global map of irrigation areas. *Hydrology and Earth System Sciences*, 9(5), 535-547.

Siebert, S., Feick, S., & Hoogeveen, J., 2005a. A Digital Global Map of Irrigation Areas-an Update for Asia. Frankfurt Hydrology Paper 01. Germany: Frankfurt University, Frankfurt/M. and Rome, Italy: FAO.

Siebert, S., Hoogeveen, J., Frenken, K., 2006. Irrigation in Africa, Europe, and Latin America. Update of the Digital Global Map of Irrigation Areas to Version 4. Frankfurt Hydrology Paper 05. Germany: Frankfurt University (main) and Rome, Italy: FAO.

Soares, A.R., Bendini, H.N., Vaz D.V., Uehara T.D.T., Neves, A.K., Körting, T.S., Fonseca, L.M.G, 2020. Stmetrics: a python package for satellite image time-series feature extraction, IGARSS 2020.

Van Schilfgaarde, J., 1994: Irrigation - A Blessing or a Curse. *Agricultural Water Management*, 25, 203-21.

Vogels, M.F., de Jong, S.M., Sterk, G., & Addink, E.A., 2019: Mapping irrigated agriculture in complex landscapes using SPOT6 imagery and object-based image analysis–A case study in the Central Rift Valley, Ethiopia. *International Journal of Applied Earth Observation and Geoinformation*, 75, 118-129.

Xiang, K., Ma, M., Liu, W., Dong, J., Zhu, X., & Yuan, W., 2019: Mapping irrigated areas of northeast China in comparison to natural vegetation. *Remote Sensing*, 11(7), 825.

This contribution has been peer-reviewed.
https://doi.org/10.5194/isprs-archives-XLVIII-M-1-2023-33-2023

Xie, Y., Lark, T.J., Brown, J.F., Gibbs, H.K., 2019: Mapping irrigated cropland extent across the conterminous United States at 30 m resolution using a semi-automatic training approach on Google Earth Engine. ISPRS *Journal of Photogrammetry and Remote Sensing*, 155, 136-149.

Xu, T., Deines, J. M., Kendall, A. D., Basso, B., Hyndman, D. W., 2019: Addressing challenges for mapping irrigated fields in subhumid temperate regions by integrating remote sensing and hydroclimatic data. *Remote Sensing*, 11(3), 370.

Zha, Y., Gao, J., Ni, S., 2003: Use of normalized difference built-up index in automatically mapping urban areas from TM imagery. *International journal of remote sensing*, 24(3), 583-594.

Zurqani, H.A., Allen, J.S., Post, C.J., Pellett, C.A., Walker, T.C., 2021: Mapping and quantifying agricultural irrigation in heterogeneous landscapes using Google Earth Engine. *Remote Sensing Applications: Society and Environment,* 23, 100590.

This contribution has been peer-reviewed.
https://doi.org/10.5194/isprs-archives-XLVIII-M-1-2023-33-2023 | © Author(s) 2023. CC BY 4.0 License.

WATER RESERVOIRS MONITORING THROUGH GOOGLE EARTH ENGINE: APPLICATION TO SENTINEL AND LANDSAT IMAGERY

F. Bocchino [1,*], R. Ravanelli [1], V. Belloni [1], P. Mazzucchelli [2], M. Crespi [1,3]

[1] Geodesy and Geomatics Division, DICEA, Sapienza University of Rome, Rome, Italy
<filippo.bocchino, roberta.ravanelli, valeria.belloni, mattia.crespi>@uniroma1.it
[2] Aresys S.r.l., Italy – paolo.mazzucchelli@aresys.it
[3] Sapienza School for Advanced Studies, Sapienza University of Rome, Rome, Italy

KEY WORDS: Water reservoir monitoring, Sustainable Development Goals, Google Earth Engine, Sentinel, Landsat.

ABSTRACT:

Water reservoirs are subjected to increasing hydrological stresses, therefore continuous and accurate monitoring of these resources is essential to ensure their sustainable management. This work proposes a methodology to remotely monitor the surface extent of water reservoirs through the analysis of satellite multispectral and Synthetic Aperture Radar (SAR) images. In particular, a segmentation strategy was implemented within Google Earth Engine (GEE) to distinguish water bodies from the surrounding land surface and measure their extension, by applying three different approaches to Sentinel-1, Sentinel-2, and Landsat-8 imagery. The first approach is based on the use of the Automatic Water Extraction Index (AWEI) and the self-adaptive Otsu's thresholding method, the second approach is based on the image conversion from RGB (Red-Green-Blue) to HSV (Hue, Saturation, Value) and the use of a parametric threshold, the third approach is based on the use of SAR imagery and an empirically selected threshold. A "static" validation strategy was developed from scratch and standard segmentation metrics were computed to evaluate the accuracy of the three approaches. The average values of the F1 scores on the Sentinel imagery were equal to 0.95, 0.90, and 0.84 for the three approaches, respectively. The same metric on the Landsat imagery was 0.95 for the first approach and 0.93 for the second approach. The best approach, i.e. the AWEI-based method, was then applied to three water bodies in which the effects of the 2022 drought were particularly significant: Sawa lake (Iraq), Poyang lake (China), and Po river (Italy). The results visually highlighted the good performance of the approach in segmenting the water bodies from the surrounding areas.

1. INTRODUCTION

Natural and artificial water reservoirs are essential freshwater resources for human and animal consumption, agricultural irrigation, and several industrial applications. Currently, water level measurements are generally performed by traditional ground instruments such as gauge stations. Indeed, the data collected by these instruments are used to estimate the area and volume of the reservoirs (and their variations over time) using volume-area-elevation curves derived from bathymetric and topographic information of the reservoir itself (Tong et al., 2016, Hamoudzadeh et al., 2023). However, in-situ monitoring is mainly possible in developed countries, due to the difficulties in installing and maintaining measurement stations in remote areas. In addition, even in developed countries, gauge stations are not installed at small and secondary reservoirs. Conversely, the use of Earth Observation (EO) technologies can remarkably reduce the monitoring costs (independent from the actual extent of the reservoir) and provide frequent and regular data that facilitate the continuous monitoring of water reservoirs, in principle with homogeneous worldwide procedures (Valadão et al., 2021).

The aim of this work is precisely included in this background: the general goal is to lay the foundations for the development of a procedure based on free EO data for the routine monitoring of the extent of whichever water reservoir. In particular, in this preliminary analysis focused on multispectral and SAR satellite images, three different segmentation approaches able to distinguish water bodies from the surrounding land surface and measure their extent were investigated using Sentinel-1, Sentinel-2, and Landsat-8 imagery. Indeed, the continuous computation of 2D metrics, such as perimeter and area, is a first step towards the monitoring of the seasonal and annual variations of water reservoirs, allowing the reconstruction of long-term surface water dynamics. This goal is strictly related to the United Nations Sustainable Development Goals (SDGs) concerning water availability (SDG 6 - Clean water and sanitation) and climate change effects monitoring (SDG 13 - Climate action), and to the Recovery Plan Next Generation EU.

* Corresponding author

2. METHODOLOGY

Three water segmentation approaches were investigated and implemented within Google Earth Engine (GEE). GEE is a cloud computing platform that integrates a wide and freely accessible remote sensing data archive, including Landsat and Sentinel image collections. Through efficient Application Programming Interfaces (APIs), GEE makes it possible to carry out large-scale geospatial analyses, according to the well-known "data-information-knowledge-wisdom" paradigm (Kavvada et al., 2020). In particular, continuing the work started in (Ravanelli et al., 2023), the following three water segmentation approaches were investigated:

- the first (Sengupta et al., 2020, Donchyts et al., 2016) is based on the application of the Automatic Water Extraction Index (AWEI) (Feyisa et al., 2014) and the self-adaptive Otsu's thresholding method (Otsu, 1979) to multispectral imagery;

- the second (Valadão et al., 2021) is based on the use of multispectral imagery and the conversion from the RGB (Red-Green-Blue) to the HSV (Hue, Saturation, Value) color model to which a parametric threshold is applied;
- the third (Geohackweek, 2016) is based on the use of SAR imagery and an empirically selected threshold.

The first two approaches were applied to Sentinel-2 (Google Earth Engine Data Catalog, 2023a) and Landsat-8 (Google Earth Engine Data Catalog, 2023c) multispectral imagery, and the third one to Sentinel-1 SAR (Google Earth Engine Data Catalog, 2023b) imagery. For the approaches based on multispectral imagery, only images with a percentage of cloudy pixels less than 30% were selected and a cloud mask was computed to consider only cloud-free pixels using the QA60 bitmask band of Sentinel-2 (Google Earth Engine Data Catalog, 2022) and the QA_PIXEL and QA_RADSAT bands of Landsat-8 (ESA, 2022). For all the approaches, the median reducer was used to aggregate image data over time (Google Earth Engine Documentation, 2023). The median reducer aggregates all the images of an image collection available in a given period and computes an overall median image by calculating the median value of each pixel – for every band – across the temporal stack of images. The median images were the input to the three segmentation approaches. A period of one year was considered in the analyses, in particular, the images collected in 2019 were analysed into three-month sub-periods. In this way, all the images available in 2019 for the three considered satellites were split into four sub-collections to investigate the four seasons separately. Figure 1 and Figure 2 show a median image for Sentinel-2 (after applying the cloud filters) and Sentinel-1 for the period January-March 2019 over the Marghera dock (Venice, Italy), i.e. the validation area, as described in Section 3.

Figure 1. Sentinel-2 multispectral median image (Marghera dock, January-March 2019).

Each approach returned a water segmentation mask as output, from which the area of the reservoir was computed by counting the pixels classified as water and by multiplying the resulting number by the image resolution. Table 1 shows the resolution (pixel size) of the bands used in the analyses.

The following sections describe each approach in detail.

2.1 AWEI approach

In the AWEI-based approach, the water pixels in the input median image were segmented by implementing the AWEI spectral index and by applying the method of local thresholding based on the Otsu's thresholding and the Canny edge detector algorithms (Donchyts et al., 2022). The AWEI index is presented in Equation 1 (Feyisa et al., 2014):

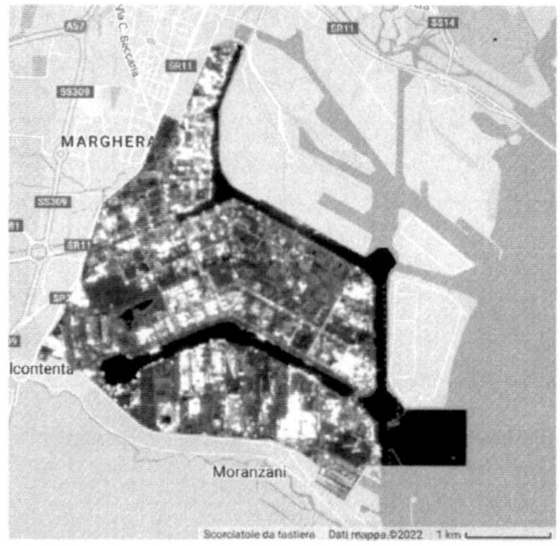

Figure 2. Sentinel-1 median image (Marghera dock, January-March 2019).

Band resolution (pixel size) [m]				
Satellite	RGB	NIR	SWIR1/SWIR2	GRD (IW)
Sentinel-2	10	10	20	-
Landsat-8	30	30	30	-
Sentinel-1	-	-	-	10

Table 1. Band resolution of Sentinel-2, Landsat-8, and Sentinel-1: RGB, Near-infrared (NIR), Short-wave infrared (SWIR), and Ground Range Detected (GRD) Interferometric Wide (IW).

$$AWEI = 4 \cdot (\rho_{GREEN} - \rho_{SWIR1}) - 0.25 \cdot \rho_{NIR} - 2.75 \cdot \rho_{SWIR2} \quad (1)$$

where ρ represents the reflectance for the GREEN (G), SWIR1, SWIR2, and NIR bands, and the coefficients and the arithmetic combinations of the chosen spectral bands were determined by critical examination of the reflectance properties of various land cover types (Feyisa et al., 2014).

The AWEI index allows for maximization of the separability of water and non-water pixels and thereby makes it possible to extract surface water with high accuracy (Feyisa et al., 2014). Unlike the Normalized Difference Water Index (NDWI) or the Modified Normalized Difference Water Index (MNDWI), the use of AWEI index avoids segmentation errors due to many factors such as clouds, snow, and ice. Additionally, the AWEI does not commit errors of commission (false positive segmentation of water) in areas with shadows due to topographic conditions or the presence of cloud shadows (Donchyts et al., 2016). The increase in the contrast between water and other dark surfaces provided by this index compared to the NDWI index is shown in Figure 3.

After the AWEI index computation, the method of local thresholding based on Otsu's thresholding, and the Canny edge detector algorithm were applied to reduce the number of input pixels only to those located near water-land edges (Donchyts et al., 2016). The detection of water pixels was then refined using morphological dilation. Figure 4 shows the water mask generated using the approach applied to Sentinel-2 imagery.

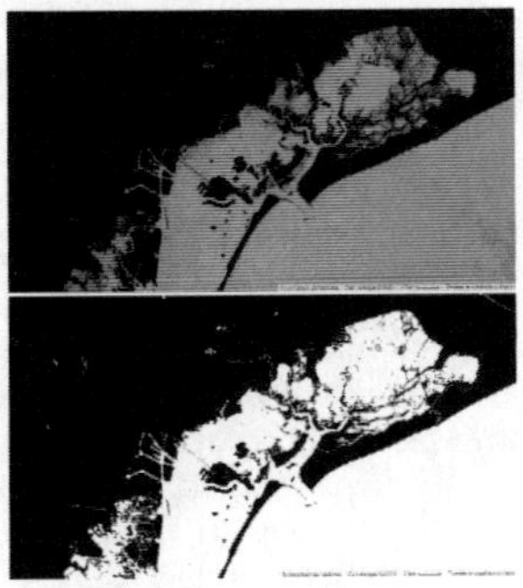

Figure 3. Comparison between NDWI (above) and AWEI (below) resulting images on the same area (Venice lagoon).

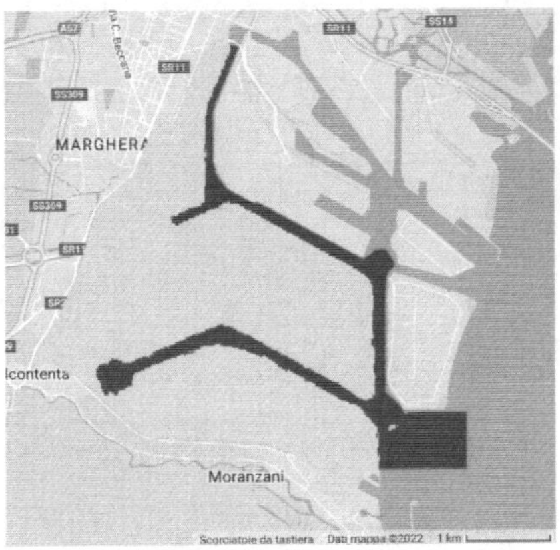

Figure 4. Water segmentation image obtained with the AWEI approach using Sentinel imagery.

2.2 HSV approach

In the HSV-based approach (Valadão et al., 2021), the input median RGB image (see Figure 1) was firstly transformed to HSV. Only the Hue band was selected to facilitate the subsequent identification of the water pixels (Valadão et al., 2021). In this case, the water segmentation was based on a parametric threshold: more specifically, only the pixels in the 0.25-0.95 tonality range were classified as water pixels. The resulting water segmentation mask is shown in Figure 5.

2.3 SAR approach

The Sentinel-1 images were filtered by selecting only the VV polarisation. It is important to underline two issues that affect SAR imagery. Firstly, some surfaces alter the polarisation of the SAR signal, but this is usually not frequent for water body segmentation. Secondly, SAR images are often affected by speckle noise that deteriorates the quality of the image (Geohackweek, 2016). To reduce the speckle, a focal median filter was applied. Then, a simple threshold method was implemented to identify water pixels in the SAR images. Specifically, a threshold of backscatter was chosen and all pixels below that threshold were classified as water. The selected threshold value was -16 dB and it was empirically selected by visually inspecting the histogram representing the number of pixels and the intensity of the backscattered signal (Figure 6).

The water segmentation mask obtained with this approach is shown in Figure 7.

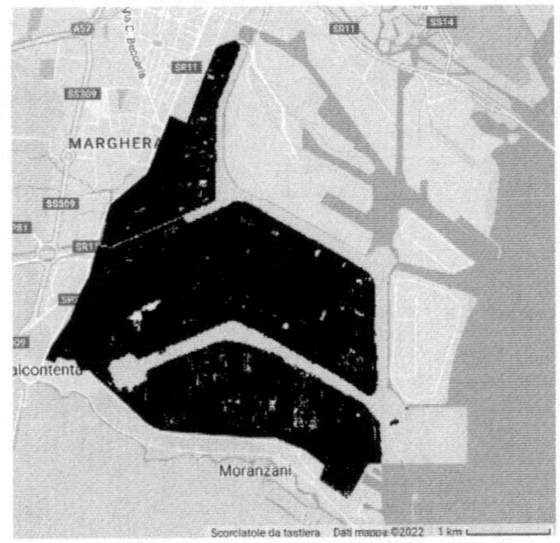

Figure 5. Water segmentation image obtained with the HSV approach using Sentinel imagery.

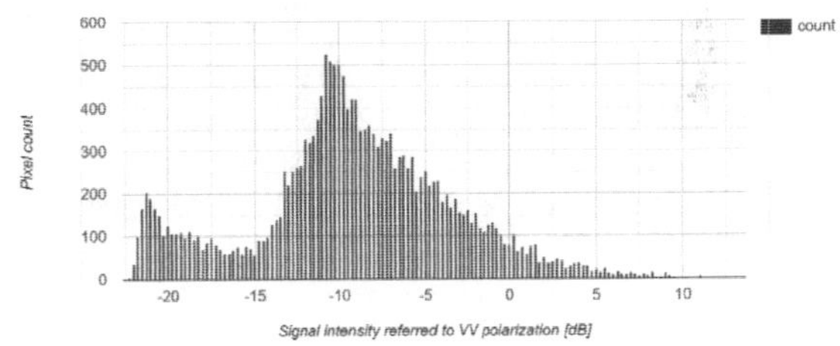

Figure 6. Histogram for SAR threshold selection.

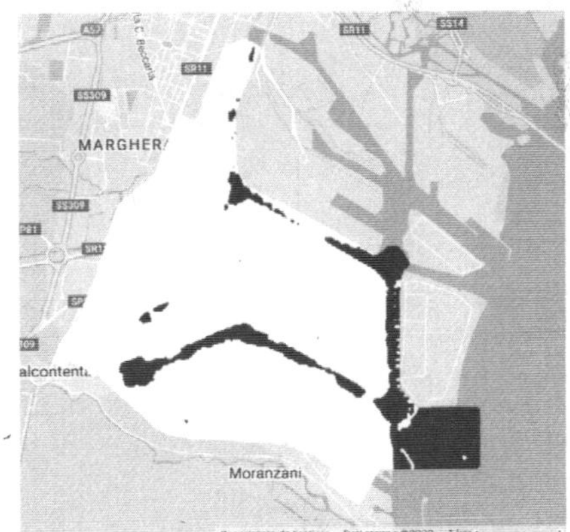

Figure 7. Water segmentation image obtained with the SAR approach using Sentinel imagery.

3. VALIDATION

The three approaches were validated using a "static" validation strategy, i.e, considering a validation area where the water level variations do not impact the water surface extent over time and the geometry of the water borderlines is available as an independent ground truth. Specifically, the different approaches were applied to the harbour area of Marghera considering the 2019 year and the results were compared with the reference shapefile of the Site of National Interest (SIN) of Marghera available on the Ministry of Environment and Energy Secur-

https://doi.org/10.5194/isprs-archives-XLVIII-M-1-2023-41-2023 | © Author(s) 2023. CC BY 4.0 License.

ity website of the Italian Government. In the selected area, the water borderlines are artificial and, therefore, in theory, independent from the water level, allowing to compare the performances of the three approaches using imagery acquired in different epochs. The reference ground truth and the generated water masks are shown in Figure 8.

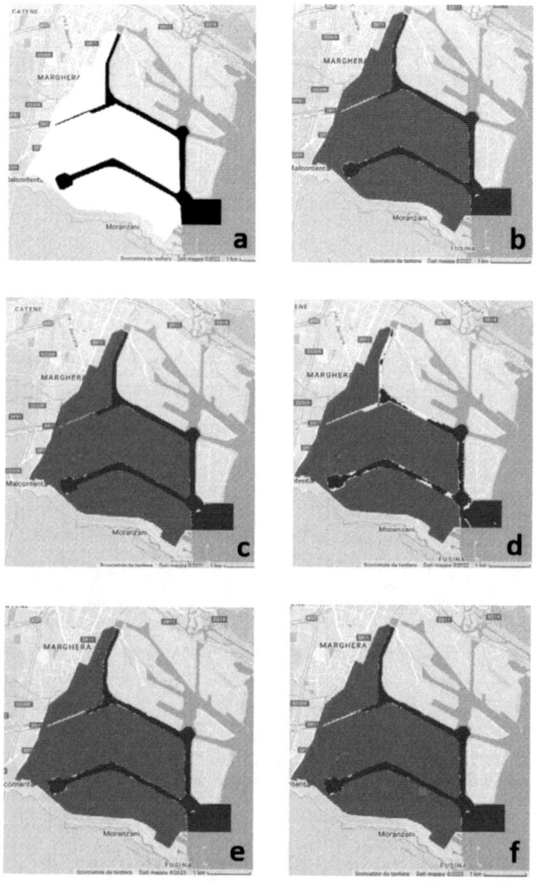

Figure 8. Ground truth mask (a) and water detection masks obtained with the first approach on Sentinel imagery (b) and Landsat imagery (e), the second approach on Sentinel imagery (c) and Landsat imagery (f), and the third approach (d) (Marghera dock, January-March 2019). True positives are in blue, true negatives in green, false positives in red and false negatives in yellow.

Figure 8-b and 8-e visually demonstrate how the AWEI approach performs better than the other ones. Figure 8-c shows that the HSV approach generates a higher number of false positive pixels compared to the AWEI approach. Moreover, the HSV approach not only fails to segment the pixels of land located on the water-land border but also those located inland. The reason is related to the presence of a few roofs in the validation area: their color falls in the Hue range of 0.25-0.95, which was set to classify water in this approach. To further improve the performance of this method, a solution could be to narrow down the Hue range or select a non-urban validation area. Figure 8-d highlights that the SAR approach presents many segmentation issues and provides the worst results. In this case, the number of pixels segmented as false negatives is higher, i.e. a lot of water pixels are segmented as land pixels. Finally, by analyzing Figure 8-d it is possible to notice that the small portion in the hinterland classified as false positive is actually an area where the algorithm works correctly and the reference is not correct. That area is indeed a small wet area that in the reference image was mistakenly considered land. This affects also the second method.

For each approach, the confusion matrix was computed using both Sentinel and Landsat collections to investigate the performance of the investigated approaches. The following standard metrics were computed: Accuracy (A), Intersection over Union (IoU), Precision (P), Recall (R), and F1 score (F1). The results are presented in Tables 2, 3, 4, 5, and 6.

Temporal range	Images	A	IoU	P	R	F1
Q1 2019	12	0.98	0.92	0.97	0.94	0.96
Q2 2019	7	0.98	0.90	0.99	0.91	0.95
Q3 2019	9	0.98	0.91	0.99	0.92	0.95
Q4 2019	8	0.98	0.89	0.95	0.94	0.94

Table 2. Metrics of the AWEI approach with a time range of three months (Sentinel-2).

Temporal range	Images	A	IoU	P	R	F1
Q1 2019	15	0.98	0.92	0.98	0.94	0.96
Q2 2019	13	0.98	0.90	0.97	0.93	0.95
Q3 2019	17	0.98	0.91	0.98	0.93	0.95
Q4 2019	10	0.98	0.89	0.95	0.94	0.94

Table 3. Metrics of the AWEI approach with a time range of three months (Landsat-8).

Temporal range	Images	A	IoU	P	R	F1
Q1 2019	12	0.97	0.86	0.86	0.97	0.92
Q2 2019	7	0.97	0.84	0.86	0.96	0.91
Q3 2019	9	0.96	0.82	0.85	0.97	0.90
Q4 2019	8	0.95	0.78	0.79	0.97	0.87

Table 4. Metrics of the HSV approach with a time range of three months (Sentinel-2).

Temporal range	Images	A	IoU	P	R	F1
Q1 2019	15	0.97	0.87	0.91	0.96	0.93
Q2 2019	13	0.98	0.88	0.95	0.93	0.94
Q3 2019	17	0.98	0.89	0.95	0.93	0.94
Q4 2019	10	0.97	0.84	0.87	0.95	0.91

Table 5. Metrics of HSV approach with a time range of three months (Landsat-8).

Temporal range	Images	A	IoU	P	R	F1
Q1 2019	44	0.95	0.73	0.98	0.74	0.84
Q2 2019	46	0.95	0.71	0.98	0.72	0.83
Q3 2019	47	0.95	0.72	0.98	0.73	0.84
Q4 2019	44	0.95	0.73	0.98	0.73	0.84

Table 6. Metrics of the SAR approach with a time range of three months (Sentinel-1).

Again, Tables 2, 4 and 6 demonstrate that the first approach performs better than the others: all the metrics, except the Recall, are higher than the corresponding ones of the other approaches. In particular, the average F1 scores of the three approaches applied to Sentinel-2 (first and second approach) and Sentinel-1 (third approach) imagery are equal to 0.95, 0.90, and 0.84, respectively. The same metric values on the Landsat imagery are 0.95 for the first approach and 0.93 for the second one. Furthermore, by comparing changes in F1 scores through the subperiods for each approach in Tables 2, 4 and 6, it is possible to notice that the AWEI and SAR methods are less affected by seasonality compared to the HSV method. This behaviour can be related to the different lighting conditions in the median images during winter and summer. The HSV method segments

This contribution has been peer-reviewed.
https://doi.org/10.5194/isprs-archives-XLVIII-M-1-2023-41-2023 | © Author(s) 2023. CC BY 4.0 License.

water pixels based on a Hue range: variations in the lightness conditions cause a change in Hue conditions, and therefore affect the performance of this approach. The differences between the seasons can be visually appreciated in Figure 9, in which black areas highlight the pixels that are misclassified as false positives for the summer (right) period while they are correctly classified in the winter period (left).

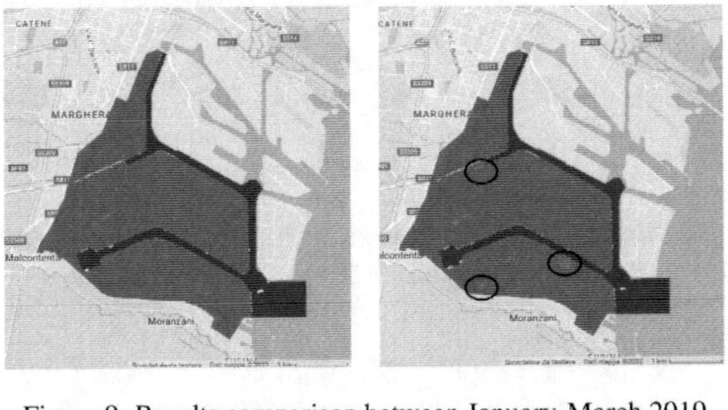

Figure 9. Results comparison between January-March 2019 (left) and July-September 2019 (right) for the second approach.

Finally, Table 6 highlights that the number of SAR images available for each period is higher than the number of multispectral images. This is related to two reasons: firstly the area of Marghera is located at the intersection of several orbits of the Sentinel-1 satellite, both ascending and descending; secondly, SAR images are not affected by clouds, therefore no cloud filter was applied to remove images.

4. CASE STUDIES

The best method (i.e. the AWEI approach) was applied to three case studies in which the effects of the 2022 drought were particularly significant. For sake of brevity, only the results obtained using Sentinel-2 images are shown below.

4.1 Sawa Lake

Sawa Lake is an endorheic lake (4.5 km long and 1.8 km wide) located in southern Iraq in the governorate of Muthanna, near the Euphrates River. The water level in the lake used to fluctuate between dry and wet seasons, but despite being situated in an arid region, in 2022 it did not dry up because of the balance between evaporation and water feed, i.e. based on rainfall and groundwater through cracks and fissure (Service, 2015). Sawa Lake was severely affected by the 2022 water crisis caused by climate change, which led to low rainfall and drought, but also by lack of maintenance of water infrastructure and by the presence of dams in upstream countries that reduced the flow of the Tigris and Euphrates rivers (Radeef and Abdulameer, 2023). Lake Sawa degradation started more than 10 years ago, but in 2022 the lake disappeared for the first time (Copernicus, 2022). This effect can be appreciated in Figure 10 by comparing the median image obtained for the period January-March 2020 with the corresponding one of 2022.

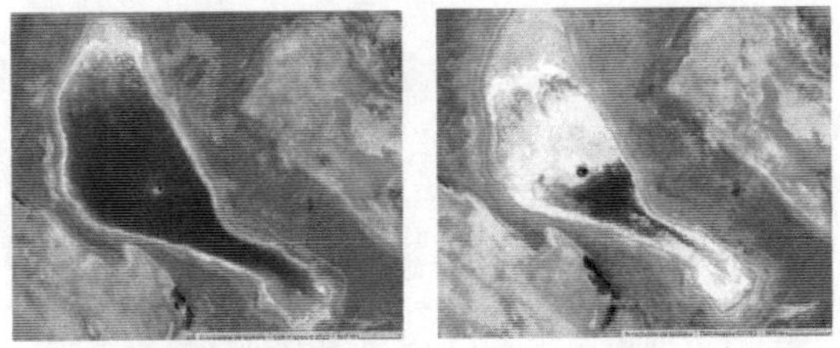

Figure 10. Sentinel-2 median image of Sawa lake on January-March 2020 (left) and on January-March 2022 (right).

The application of the AWEI-based method to this lake returned the water masks shown in Figure 11.

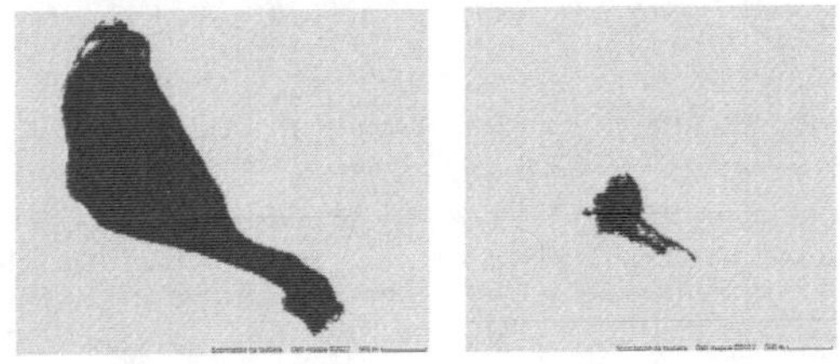

Figure 11. Water detection image on January-March 2020 (left) and January-March 2022 (right).

The water area was obtained from these masks to quantify the change: the area decreased from 3.90 km^2 in 2020 to 0.39 km^2 in 2022.

4.2 Poyang Lake

Poyang Lake is located in Jiujiang, in China's Jiangxi Province and it is the largest freshwater lake in China. The lake is fed by the Gan, Xin, and Xiu rivers which connect to the Yangtze through a narrow channel. Poyang Lake routinely fluctuates in size between the winter and summer seasons: in winter, water levels on the lake are typically low, then summer rains cause it to swell. However, in recent years, the size of the lake has been decreasing overall. In particular, during the summer of 2022, the Yangtze River basin was affected by a prolonged heat wave and drought, and therefore Poyang Lake dried up reaching water levels not seen in decades. On June 23, the Xingzi Station measured the highest water levels of the year on Poyang Lake; after that, the Jiangxi Hydrological Monitoring Center recorded high temperatures and a lack of rain, i.e. the factors that caused the lake to drop rapidly (NASA, 2022). On August 6, water levels had declined to 11.99 meters and the Center identified this date as the start of the lake dry season: water levels have continued to drop, registering 8.96 meters on August 30. In this case, the comparison was made between Spring 2022 (i.e. median image obtained for the period April-June 2022) and Summer 2022 (i.e. median image obtained for the period July-September 2022). The median images depicted in Figure 12 show this trend, and the measures of the areas confirm it: from the masks represented in Figure 13, areas of 2281.48 km^2 and 1054.61 km^2 were obtained for Spring 2022 and for Summer 2022, respectively.

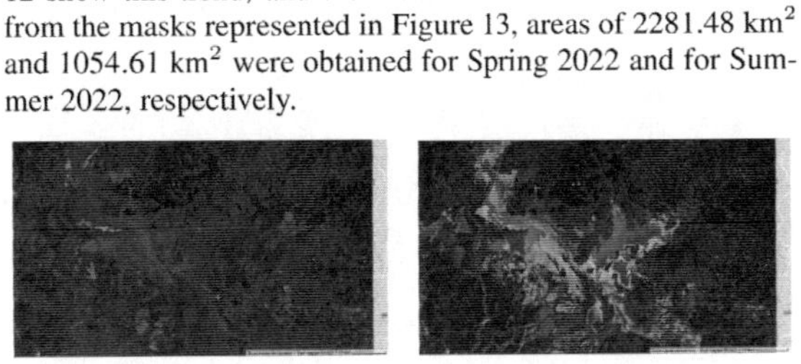

Figure 12. Sentinel-2 median image of Poyang lake on April-June 2022 (left) and on July-September 2022 (right).

4.3 Po river

Po river is the longest river in Italy (652 km). It is located in the North of Italy and flows from east to west crossing many im-

https://doi.org/10.5194/isprs-archives-XLVIII-M-1-2023-41-2023 | © Author(s) 2023. CC BY 4.0 License.

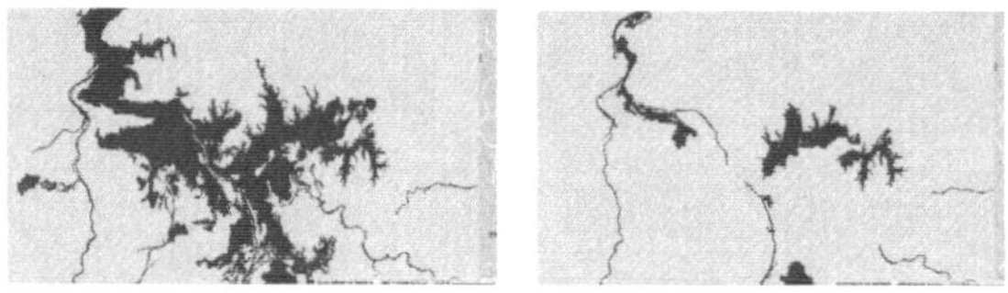

Figure 13. Water detection image on April-June 2022 (left) and on July-September 2022 (right).

portant Italian cities. The vast valley around the Po is called the Po Basin: about 20 million people live there (a third of the Italian population) and it is the main industrial area and the largest agricultural area in the country, accounting for 35% of Italian agricultural production (Borzì et al., 2021). Moreover, the water of the Po river is essential for the energy production of the country. Its flow is controlled by lots of dams that work as hydroelectric power plants and there are several thermoelectric power stations that use the water of the Po basin as coolant. During 2022, due to the intense drought, the water level of the river reached a historic low in the principal sections and the flow touched its all-time low (Po River Basin District Authority, 2022). The analysis was carried out by comparing Spring 2020 and Spring 2022 in a 50 km portion of the Po river near Piacenza. The results are shown in Figure 14 and 15, where it is possible to observe a reduction of the surface from 14.72 km^2 to 11.22 km^2, i.e. about 25% in only two years.

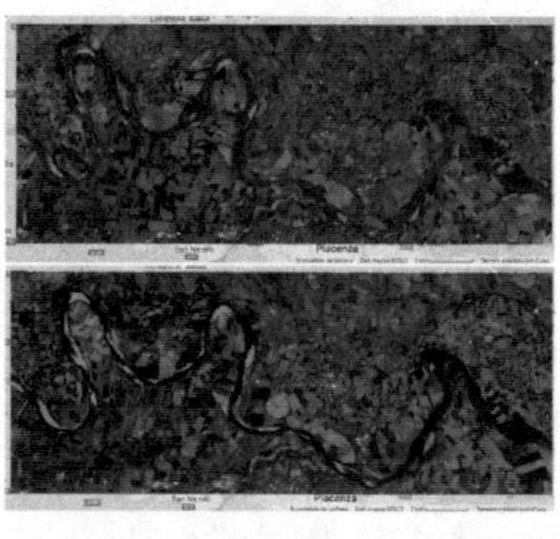

Figure 14. Sentinel-2 median image of Po river on April-June 2020 (above) and on April-June 2022 (below).

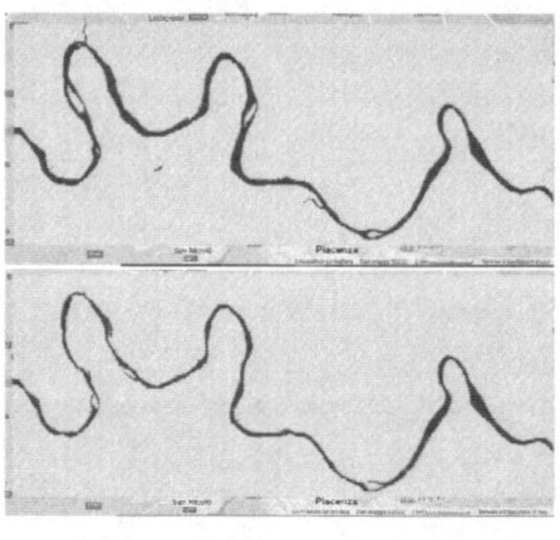

Figure 15. Water detection image on April-June 2020 (above) and on April-June 2022 (below).

5. CONCLUSIONS AND FUTURE PERSPECTIVES

The aim of this work was to lay the foundations for the development of a methodology for the continuous monitoring of water reservoirs through the analysis of multispectral and SAR satellite images. The main objective was to implement a segmentation strategy able to distinguish water from the surrounding land surface and then to measure the planimetric extent of water reservoirs starting from the area occupied by the water pixels identified in the retrieved masks. In particular, three different approaches were implemented within GEE and applied to Sentinel-1, Sentinel-2 and Landsat-8 imagery. The first approach is based on the use of the AWEI and the self-adaptive Otsu's thresholding method; the second approach is based on the image conversion from RGB to HSV and the use of a parametric threshold; the third approach is based on the use of SAR imagery and an empirically selected threshold. A "static" validation strategy was developed from scratch and standard segmentation metrics were computed to evaluate the accuracy of the three approaches. The average values of the F1 scores on the Sentinel imagery were equal to 0.95, 0.90 and 0.84 for the three approaches, respectively. The same metric on the Landsat imagery was 0.95 for the first approach and 0.93 for the second approach. The best approach (the AWEI-based method) was then applied to three water bodies in which the effects of the 2022 drought were particularly significant: Sawa lake (Iraq), Poyang lake (China), and Po river (Italy). The results visually highlighted the good performance of the approach in segmenting the water bodies from the surrounding areas.

Nevertheless, an adequate monitoring of water reservoirs includes not only the estimation of the extent of their surface but also the quantification of their volume. This paper focused on the extent estimation since the continuous computation of 2D metrics is a first step towards the monitoring of the seasonal and annual variations of water reservoirs, allowing the reconstruction of long-term surface water dynamics.

As for future developments, the validation will be extended to other areas. In this work, the validation was carried out on the SIN of Marghera dock, and this choice introduced additional errors in the validation itself due to the presence of nearby buildings and vegetation. Moreover, the dock area was selected because in this area the water surface remains constant over time. The generation of a dataset starting from high-resolution satellite images will allow validation of the methodology on dynamic references too, increasing the generalization of the methodology. Furthermore, the generation of the dataset may lead the way to the development and testing of data-driven approaches, such as deep learning algorithms. Regarding the SAR results, future development will investigate a threshold selection method based on automatic approaches (e.g. Otsu's thresholding), instead of parametric thresholds. Finally, the combination of the three tested approaches will be investigated to improve the results.

ACKNOWLEDGEMENTS

This project has received funding from the European High-Performance Computing Joint Undertaking (JU) under grant agreement No 951745. The JU receives support from the European Union's Horizon 2020 research and innovation programme and Germany, Italy, Slovenia, France, Spain.

REFERENCES

Borzì, I., Monteleone, B., Bonaccorso, B., Martina, M., 2021. Estimating the economic impacts of drought on agriculture through models and surveys in the Po river basin (Northern Italy). *EGU General Assembly Conference Abstracts*, EGU21–12066.

Copernicus, 2022. Al Sawa Lake has completely dried up — Copernicus — copernicus.eu. www.copernicus.eu/en/media/image-day-gallery/al-sawa-lake-has-completely-dried. [Accessed 08-Feb-2023].

Donchyts, G., Schellekens, J., Winsemius, H., Eisemann, E., Van de Giesen, N., 2016. A 30 m resolution surface water mask including estimation of positional and thematic differences using landsat 8, srtm and openstreetmap: a case study in the Murray-Darling Basin, Australia. *Remote Sensing*, 8(5), 386.

Donchyts, G., Winsemius, H., Baart, F., Dahm, R., Schellekens, J., Gorelick, N., Iceland, C., Schmeier, S., 2022. High-resolution surface water dynamics in Earth's small and medium-sized reservoirs. *Scientific reports*, 12(1), 1–13.

ESA, 2022. Cloud Masks. https://sentinel.esa.int/web/sentinel/technical-guides/sentinel-2-msi/level-1c/cloud-masks.

Feyisa, G. L., Meilby, H., Fensholt, R., Proud, S. R., 2014. Automated Water Extraction Index: A new technique for surface water mapping using Landsat imagery. *Remote Sensing of Environment*, 140, 23–35.

Geohackweek, 2016. SAR water body classification. https://mbonnema.github.io/GoogleEarthEngine/07-SAR-Water-Classification/.

Google Earth Engine Data Catalog, 2022. Sentinel-2 MSI: MultiSpectral Instrument, Level-2A. https://developers.google.com/earth-engine/datasets/catalog/COPERNICUS_S2_SR. [Accessed 28-Feb-2023].

Google Earth Engine Data Catalog, 2023a. Harmonized Sentinel-2 MSI: MultiSpectral Instrument, Level-2A — developers.google.com. https://developers.google.com/earth-engine/datasets/catalog/COPERNICUS_S2_SR_HARMONIZED. [Accessed 28-Feb-2023].

Google Earth Engine Data Catalog, 2023b. Sentinel-1 SAR GRD: C-band Synthetic Aperture Radar Ground Range Detected, log scaling — developers.google.com. https://developers.google.com/earth-engine/datasets/catalog/COPERNICUS_S1_GRD. [Accessed 28-Feb-2023].

Google Earth Engine Data Catalog, 2023c. USGS Landsat 8 Level 2, Collection 2, Tier 1 — developers.google.com. https://developers.google.com/earth-engine/datasets/catalog/LANDSAT_LC08_C02_T1_L2#description. [Accessed 28-Feb-2023].

Google Earth Engine Documentation, 2023. Reducer Overview. https://developers.google.com/earth-engine/guides/reducers_intro. Online; accessed February, 2023.

Hamoudzadeh, A., Ravanelli, R., Crespi, M., 2023. GEDI data within Google Earth Engine: preliminary analysis of a resource for inland surface water monitoring. *The International Archives of the Photogrammetry, Remote Sensing and Spatial Information Sciences*. In press.

Kavvada, A., Metternicht, G., Kerblat, F., Mudau, N., Haldorson, M., Laldaparsad, S., Friedl, L., Held, A., Chuvieco, E., 2020. Towards delivering on the Sustainable Development Goals using Earth observations. *Remote Sensing of Environment*, 247.

NASA, 2022. Parched Poyang Lake — earthobservatory.nasa.gov. https://earthobservatory.nasa.gov/images/150285/parched-poyang-lake. [Accessed 09-Feb-2023].

Otsu, N., 1979. A Threshold Selection Method from Gray-Level Histograms. *IEEE Transactions on Systems, Man, and Cybernetics*, 9(1), 62-66.

Po River Basin District Authority, 2022. Report of 22 July 2022 - Po River. *Permanent observatory on water uses in the hydrographic district of the river Po*.

Radeef, H. A., Abdulameer, I. M.-A., 2023. The Water Area of Sawa Lake as Derived from Land Surface Temperature and Remote Sensing Data. *Ibn AL-Haitham Journal For Pure and Applied Sciences*, 36(1), 100–112. https://doi.org/10.30526/36.1.2873.

Ravanelli, R., Mazzucchelli, P., Belloni, V., Bocchino, F., Morselli, L., Fiorino, A., Gerace, F., Crespi, M., 2023. Earth Observation Big Data exploitation for water reservoirs continuous monitoring: the potential of Sentinel-2 data and HPC. C. Ieracitano, N. Mammone, M. Di Clemente, M. Mahmud, R. Furfaro, F. C. Morabito (eds), *The Use of Artificial Intelligence for Space Applications: Workshop at the 2022 International Conference on Applied Intelligence and Informatics*, Studies in Computational Intelligence. In press.

Sengupta, D., Chen, R., Meadows, M. E., Banerjee, A., 2020. Gaining or losing ground? Tracking Asia's hunger for 'new' coastal land in the era of sea level rise. *Science of The Total Environment*, 732, 139290.

Service, R. S. I., 2015. Sawa Lake — Ramsar Sites Information Service — rsis.ramsar.org. https://rsis.ramsar.org/ris/2240. [Accessed 28-Feb-2023].

Tong, X., Pan, H., Xie, H., Xu, X., Li, F., Chen, L., Luo, X., Liu, S., Chen, P., Jin, Y., 2016. Estimating water volume variations in Lake Victoria over the past 22years using multi-mission altimetry and remotely sensed images. *Remote Sensing of Environment*, 187, 400-413.

Valadão, L. V., Cicerelli, R. E., de Almeida, T., Ma, J. B. C., Garnier, J., 2021. Reservoir metrics estimated by remote sensors based on the Google Earth Engine platform. *Remote Sensing Applications: Society and Environment*, 24(100652).

This contribution has been peer-reviewed.
https://doi.org/10.5194/isprs-archives-XLVIII-M-1-2023-41-2023 | © Author(s) 2023. CC BY 4.0 License.

LAND USE/LAND COVER CLASSIFICATION IN A HETEROGENEOUS AGRICULTURAL LANDSCAPE USING PLANETSCOPE DATA

I. T. Bueno [1,3]*, J. F. G. Antunes [1,2], A. P. S. G. D. D. Toro [3], J. P. S. Werner [3], A. C. Coutinho [2], G. K. D. A. Figueiredo [3], R. A. C. Lamparelli [1,3], J. C. D. M. Esquerdo [2,3], P. S. G. Magalhães [1]

[1] Interdisciplinary Center of Energy Planning, University of Campinas, Campinas 13083-896, SP, Brazil – (ibueno, lamparel, graziano)@unicamp.br

[2] Embrapa Digital Agriculture, Brazilian Agricultural Research Corporation, Campinas 13083-886, SP, Brazil – (joao.antunes, alex.coutinho, julio.esquerdo)@embrapa.br

[3] School of Agricultural Engineering, University of Campinas, Campinas 13083-875, SP, Brazil – (a265622, j164880)@dac.unicamp.br, gleyce@unicamp.br.

KEY WORDS: Clustering, Agricultural crops, OBIA, Random Forest, Spectro-temporal signature, Intra-class variability.

ABSTRACT:

Land use and land cover (LULC) classification has long been an essential topic in Earth Observation research and plays a key role in the sustainable development of agriculture. This study evaluated the accuracy of LULC classification based on an initial clustering step in a heterogeneous agricultural landscape using PlanetScope imagery while checking for variability among their Normalized Difference Vegetation Index (NDVI) temporal signatures. We adopt an object-based image analysis to generate image-objects and then extract statistical information of PlanetScope spectral bands and vegetation indices as input information for classification. The exploratory analysis focused on the double crop class and calculated the distance between NDVI temporal signatures of paired land parcels. We applied an unsupervised clustering technique along with Random Forest algorithm based on multiple tests to classify and analyse gains and losses in accuracies produced by these approaches. Our results showed that the initial clustering method outperformed the non-clustered classification of LULC in overall accuracy measures. The exploratory analysis demonstrated that double crops might present high intra-class variability and diverse crop calendars for neighbour land parcels. The accuracies achieved represent promising opportunities for the sufficiently accurate classification of such areas, and the knowledge of the intra-class variability allows the analyst to infer the temporal dynamics of crop fields. We reinforce that further work could assess other types of classifiers, especially in areas with a large number of crop types and distinct management practices.

1. INTRODUCTION

Land use and land cover (LULC) classification is an essential planning tool once it illustrates the spatial distribution of the Earth's surface attributes and plays a pivotal role in the sustainable development of agronomics, environment, and economics (Vali et al., 2020). Over the past few decades, accurate information on crop types and their spatial distribution have become a central issue for food security, policy making, and water and soil resource management (FAO, 2021).

Besides the accurate information, LULC maps for agricultural purposes must present a suitable spatial resolution and be updated frequently to deliver practical benefits. From this perspective, remote sensing data and classification techniques have the unique capability to automate the generation of such maps in a large-scale observation (Bellón et al., 2018) and the potential to discriminate a wide range of crop types (Brinkhoff et al., 2019).

A considerable amount of literature has been published on LULC classification of agricultural areas using remote sensing data (Orynbaikyzy et al., 2019). The increasing availability of public Earth Observation data and the advanced methods in digital image processing leveraged these studies from single imagery (Mathur and Foody, 2008) to time-series analysis (Cai et al., 2018), allowing it to capture the temporal dynamics of crops. In addition, the availability of very high-resolution images has provided fine-grained classifications and more effective target recognition under complex landscapes (Qin and Liu, 2022). For instance, the commercial constellation of Planet CubeSats provides a powerful combination of high spatial (3m pixel size) and temporal (daily) resolution imagery for fine-scale LULC classification and monitoring (Planet Team, 2022).

However, LULC classification of agricultural areas is not a trivial task since it depends on the size of the area and its complexity, for instance, the number of crop classes, parcel sizes, and environmental conditions. In addition, a single crop class can exhibit considerable variability in its within-parcel spatial distribution and spectro-temporal signatures due to different management practices and cropping intensity or even the presence of trees within the parcels. For instance, the growing demand for agricultural commodities has increased the incentive for intensification practices, and multiple-cropping has drawn interest in boosting production without expanding the planting area (Garrett et al., 2018). This system can demonstrate diverse crop calendars for neighbour land parcels, which may lead to similar spectro-temporal signatures to other LULC classes (Azar et al., 2016; Villa et al., 2015).

In order to deal with the high intra-class variability, crop mapping methods would largely benefit from spectral-temporal variability reduction steps prior to classification. Object-based approaches (so-called Object-Based Image Analysis – OBIA) are widely used to reduce spatial variability by dividing an image into groups of pixels that are spatially continuous and spectrally homogeneous (Blaschke, 2010). Moreover, unsupervised classification methods can reduce the spectral variability of crops. In this case, they rely on algorithms for grouping spectro-temporal signatures according to their intrinsic characteristics or similarity and not using a priori labelling. These methods appear as good alternatives for mapping agricultural patterns (Pascucci et al., 2018).

This contribution has been peer-reviewed.
https://doi.org/10.5194/isprs-archives-XLVIII-M-1-2023-49-2023 | © Author(s) 2023. CC BY 4.0 License.

In this context, the objective of our study was to evaluate the accuracy of LULC classification based on an initial clustering step in a heterogeneous agricultural landscape while checking for variability among their Normalized Difference Vegetation Index (NDVI) temporal signatures. This paper first examines the temporal signatures of the double crop class and the intra-class variability, investigating the potential patterns of plant structure, growing behaviour, and management practices. Second, we evaluated the performance of an initial clustering-based classification, investigating the benefit of partitioning the spectro-temporal signatures dataset into initial sub-clusters. We hypothesized that grouping observations with similar spectro-temporal signatures will improve the performance of LULC classification over a non-clustered method because these clusters will reduce the intra-class variability, returning gains in accuracy.

This study addresses the following research questions: (1) How similar are NDVI temporal signatures derived from a dynamic agricultural land use class? (2) How does a LULC classification method based on an initial clustering scheme affect map accuracies in a heterogeneous agricultural landscape?

To answer these research questions, we extracted and classified PS spectro-temporal signatures in the southern region of Mato Grosso (MT) State, Brazil. We applied an unsupervised clustering technique along with RF algorithm based on multiple tests to classify and analyse gains and losses in accuracies produced by these approaches.

2. MATERIAL AND METHODS

2.1 Study area

The study area is the Gravataí farm located in the southern region of MT State, Brazil (Figure 1). Totalling 10,352 hectares, the area is occupied by LULC classes comprising agriculture parcels, pasture, native forest, roads, buildings, and bare lands.

The area presents sixteen land parcels that have been intensively managed as a double crop system based on the rotation of soybean in the summer season and corn during the winter season. Usually, the soybean-sowing period lasts from late October to mid-November according to the onset of the rainy season, which lasts from October to March; and, immediately after the soybean harvest, there is the corn-sowing period for harvest in May and June (called *safrinha* or second season in Brazil).

The crop expansion from the last five decades led MT to be Brazil's major source of agricultural production. Last year, MT produced 26.2% and 36.2% of the national production of soybeans and corn, respectively, making it the most productive state for each of these crops (IBGE, 2021).

The use of intensive practices, such as double cropping, has been widely adopted in MT, which is an additional challenge for accurate crop area mapping. Thus, the need for accurate crop maps and the agricultural attention upon MT reinforce the choice of our study area.

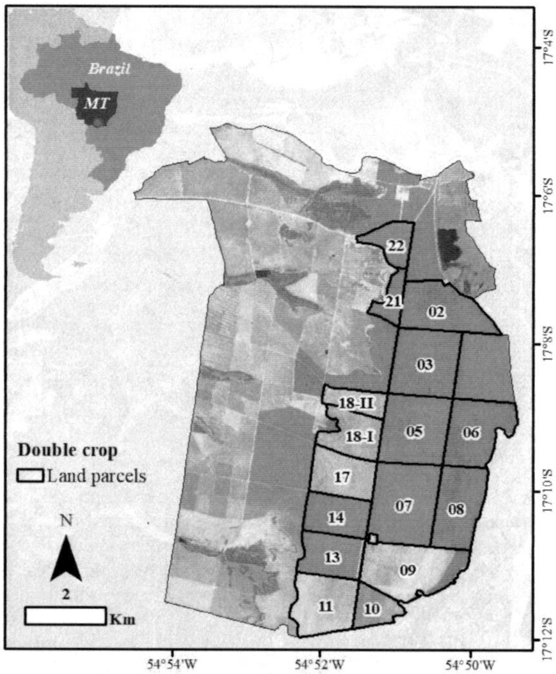

Figure 1. The Gravataí farm and its representation in PS false colour composition (Red = near-infrared, Green = red, Blue = green) from 16/01/2019.

2.2 PlanetScope data processing

We acquired PS images from 1st September 2018, to 31st August 2020, covering the study region. This time interval corresponds to the period of two agricultural years in the region: 2018-2019, and 2019-2020. We selected all surface reflectance images with less than 30% of cloud cover and masked them with the Unusable Data Mask product to generate cloud-free time series for each agricultural year. A total of 717 PS image files were downloaded.

A set of five PS-based spectral indices and four spectral bands was generated to classify the LULC in the study area (Table 1). These spectral variables can indicate vegetation condition, phenological dynamics, plant water, and chlorophyll content by providing information about crops' reflective characteristics at different wavelengths. In addition, these indices have been used in agricultural studies using PS data and have shown the potential to detect important vegetation features for classification and estimation purposes (Reis et al., 2020).

To overcome long image gaps in the time series due to cloud cover, we calculated a 15-day image composition by selecting the median value of each band/index in a 15-day interval. These image composites also provided a consistent time series with equal time intervals. In addition, we linearly interpolated potential remaining gaps in the time series using the nearest valid value before and after the time step.

https://doi.org/10.5194/isprs-archives-XLVIII-M-1-2023-49-2023

Band/ Index	Formula	Ref
B	NA	(Planet Team, 2022)
G	NA	
R	NA	
NIR	NA	
EVI	$2.5((NIR - R)/(NIR + 6R - 7.5B + 1))$	(A. Huete et al., 2002)
GNDVI	$(NIR - G)/(NIR + G)$	(A. Huete et al., 2002)
MSAVI	$\frac{(2NIR + 1) - \sqrt{(2NIR + 1)^2 - 8(NIR - R)}}{2}$	(Qi et al., 1994)
NDVI	$(NIR - R)/(NIR + R)$	(Rouse et al., 1974)
SAVI	$1.5((NIR - R)/(NIR + R + 0.5))$	(A. R. Huete, 1988)

Table 1. Summary of the spectral indices and bands used in this study for LULC classification.

2.3 Image segmentation

Image segmentation is a crucial step in OBIA and divides an image into groups of pixels that are spatially continuous and spectrally homogeneous, also known as image-objects. The segmentation of a remote sensing image minimises the within-object variability compared to the between-object variability (Desclée et al., 2006).

We implemented image segmentation within the Google Earth Engine environment based on the Simple Non-Iterative Clustering (SNIC) algorithm (Achanta and Süsstrunk, 2017) performed on the NDVI images of each agricultural year, so-called multitemporal segmentation. This method created multitemporal image objects by segmenting multiple NDVI images of sequential periods. It incorporated spectral, spatial, and temporal information from NDVI images, which created objects based on the LULC dynamics in time (Bueno et al., 2019).

The multitemporal segmentation was based on setting a grid of seeds to establish a superpixel seed location spacing, influencing the image-objects size. Then, SNIC required setting some main parameters: the "compactness factor" was set to 0.5 and affects the object shape; the "connectivity" was set to 4 and defined the type of contiguity to merge adjacent objects, and a "neighborhoodSize" was set to 256 to avoid tile boundary artifacts. We set these parameters using an essay on trial-and-error to find the appropriate segmentation for the study area, then select the most suitable output based on a visual assessment (Duro et al., 2012).

We generated two sets of multitemporal segmentation, one for each agricultural year.

2.4 Ground Reference data

Field campaigns were conducted in June 2021 to collect terrestrial reference data points. The LULC information was collected based on the current land cover during the field campaigns and by interviewing local farmers to obtain the area's historical LULC. From the reference data, we set six LULC classes: cultivated pasture, double crop (soybean and corn), forest, integrated crop-livestock system (soybean and pasture), and others (roads, buildings, and bare lands).

2.5 Exploratory analysis

We conducted an exploratory analysis of the NDVI temporal signatures of the double crop class to obtain further in-depth information on their time dynamics and intra-class variability. From image-objects of this class, we extracted the mean NDVI and then generated their temporal signatures.

We calculated the mean NDVI temporal signature for each double crop land parcel by filtering their corresponding image-objects. In this case, we calculated the Euclidean distance of NDVI values (q_1 and p_1) for each date (q_2 and p_2), then obtained the Mean Euclidean Distance (MED) between NDVI temporal signatures of land parcels (Equation 1). Euclidean distance was chosen primarily due to its straightforward interpretation.

$$MED = \frac{\sum_{i=1}^{n} \sqrt{\left(p_{1_i} - q_{1_i}\right)^2 + \left(p_{2_i} - q_{2_i}\right)^2}}{n} \quad (1)$$

MED values between all pairs of land parcels generated a matrix of distances, which allowed us to infer their intra-class variability graphically. MED values were also scaled from 0 to 1 based on the absolute minimum and maximum values to relatively monitor their variability.

2.6 LULC classification

From image-objects, we extracted four descriptive attributes: mean, 95th and 5th percentile, and standard deviation based on each band/index value inside the objects, then generated their respective spectro-temporal signatures. The initial pool of variables was screened to limit the potential effects of multicollinearity by calculating correlations between pairs of variables using Pearson's R correlation coefficient. We removed those with R values greater than 0.90.

A clustering step partitioned the double crop image-objects into initial clusters based on their spectro-temporal signatures. We evaluated nine initial clusterings according to their number of clusters (NC), NC = {2, 3, 4, 5, 6, 7, 8, 9, 10}, and a non-clustered test (NC = 0). They were based on the k-means algorithm, a widely used clustering method that partitions the data into k groups such that the within-cluster sum of squares is minimized. For this study, k-means clustering was undertaken using the Hartigan-Wong algorithm in the 'stats' package (R Core Team, 2014).

We used the Random Forest algorithm (RF; Breiman (2001)) to classify the LULC in the study area. In this study, we tuned three RF parameters that control the structure of the algorithm: the number of trees to grow, or *Ntree*; the number of predictors sampled at each tree node, or *Mtry*; and the minimum size of terminal nodes. The control of the node size parameter defines the minimum number of observations in a terminal node. We used the following parameter values in the RF tuning: *Ntree* = {200, 600, 1000}; *Mtry* = $\{\sqrt{(p/4)}), \sqrt{(p/2)}, \sqrt{p}\}$ with p the total number of variables; and node size = {2, 6, 10}.

Ten RF classifications representing each initial clustering step were performed. We balanced the number of observations by sampling 100 image-objects per LULC class when available, then split the data into 70% for training, while 30% was used for the test set to assess the generalization error of the RF model. Observations of training and test sets were sampled from the first agricultural year (2018-2019), while all observations from the

https://doi.org/10.5194/isprs-archives-XLVIII-M-1-2023-49-2023 | © Author(s) 2023. CC BY 4.0 License.

last year (2019-2020) were set for prediction analysis and LULC maps illustration.

We created individual confusion matrices using the test dataset and computed overall accuracy, producer's accuracy of the double crop class (observations characterised as double crop at the reference dataset, but not assigned as such by the model), and user's accuracy (observations detected as a double crop by the model but not identified as such in the reference dataset). Since double crop clusters represent a unique LULC class, we expand the major diagonal of the confusion matrix of this class for clustered tests. This technique merged false negative and false positive observations into the true positive group allowing us to infer accuracies more properly (Congalton and Green, 2009).

Finally, we checked for linear associations between the overall accuracy and the number of clusters by fitting a linear regression model. The model performance was evaluated by the coefficient of determination (R^2).

RF analysis and tunning were undertaken using the *mlr* package (Bischl et al., 2016), while confusion matrices were derived using the *caret* package (Kuhn, 2008).

3. RESULTS

The first set of analyses examined the NDVI temporal signatures of the double crop class and the intra-class variability. This method was particularly useful since it allowed us to create a matrix of distance-based variability among all land parcels and infer the heterogeneity of the agricultural landscape (Figure 2b).

It is apparent from this matrix that some land parcels (e.g., 06, 08, 21, and 22) presented different NDVI temporal signatures compared to others (red tones). The most striking result to emerge from the matrix is the highest MED between parcels 08 and 11, consequently, the most heterogenous land parcel pair (Figure 2a). There was a visible time shift in soybean harvest along with different patterns in NDVI values. We also observed a time shift in corn-sowing and harvest between these land parcels, where parcel 08 started late than parcel 11.

The white tones indicate an intermediary variability between land parcels. Despite showing lower values of MED, those comparisons have a rather interesting outcome since they can present variability from a particular time period. For instance, parcels 14 and 22 returned a MED of 0.53 (Figure 2c). However, they might have adopted a different corn harvest practice or land cover before the soybean sowing while still presenting very similar summer season NDVI temporal signatures.

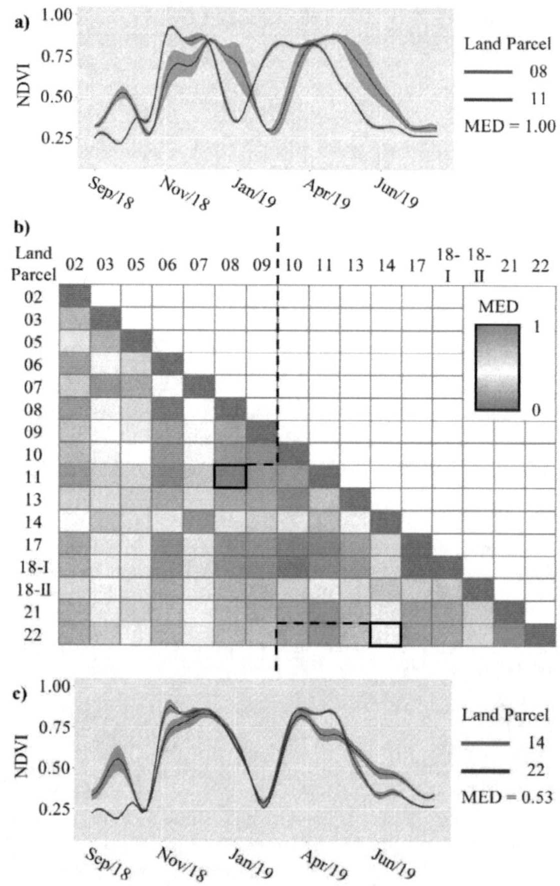

Figure 2. Exploratory analysis of the double crop land parcels' and their paired comparison of NDVI temporal signatures. b) Matrix of distance-based variability among land parcels, illustrating a) the lowest NDVI temporal signature variability between parcels 08 e 11, and (c) a moderate variability between parcels 14 and 22.

Accuracy measures of the test sets are displayed in Table 2. The initial clustering method outperformed the non-clustered classification of LULC in overall accuracy measures.

The most striking observation from this analysis was the classification output with eight initial double crop clusters that returned an overall accuracy of 98.8%, an increase of 3.2% compared to the non-clustered method. This test also returned the highest producer's accuracy of 99.7%, where, considering the 240 double crop observations sampled in the reference dataset, only one was omitted (0.3% of omission error of the double crop class). However, the non-clustered classification returned the highest user's accuracy with an inclusion error of 0.0%.

Although the clustered test with seven and ten clusters completed the top three highest accuracies, we found a weak linear association with R^2 of 0.47 between the overall accuracy and the number of clusters.

NC	OA	Double crop				Rank
		PA	OE	UA	CE	
0	95.6	86.7	13.3	100.0	0.0	10º
2	97.0	98.3	1.7	96.7	3.3	7º
3	96.4	96.7	3.3	96.7	3.3	9º
4	97.4	97.5	2.5	97.5	2.5	4º
5	97.3	98.0	2.0	98.0	2.0	5º
6	96.5	96.7	3.3	98.3	1.7	8º
7	97.8	98.6	1.4	99.0	1.0	2º
8	98.8	99.7	0.3	98.8	1.2	1º
9	97.1	98.9	1.1	97.8	2.2	6º
10	97.8	98.7	1.3	98.3	1.7	3º

Table 2. Accuracy results (in percentage) from cluster-based classification tests. (NC = number of clusters, OA = overall accuracy, PA = producer's accuracy, OE = omission error, UA = user's accuracy, and CE = commission error).

Figure 3 displays the map predictions for each test. Visual interpretation indicates that map predictions were similar and corroborated with the high overall accuracies of the test sets. However, we can observe a few cultivated pasture image-objects included in the double crop class, especially in the tests with 3, 5, and 9 clusters. These commission errors are a rather unexpected result and can not be compared with the user's accuracies of the test sets.

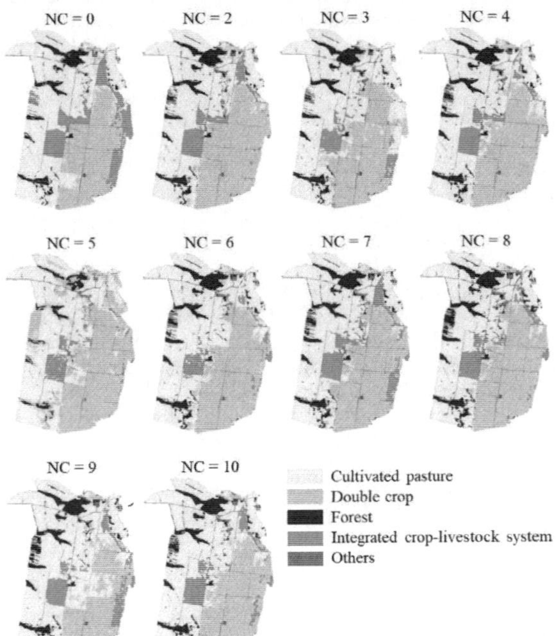

Figure 3. Map predictions of classification tests, where NC: number of clusters.

4. DISCUSSION

In this research, we evaluated the accuracy of LULC classification based on an initial clustering step in a heterogeneous agricultural landscape while checking for variability among their NDVI temporal signatures.

4.1 Effect of an initial clustering step on LULC classifications

Our results showed that an initial clustering method in a heterogenous agricultural class using PS data and RF algorithm produced more accurate results than a non-clustered LULC classification. These results also showed patterns related to plant structure, growing behaviour, and management practice, which allow for highly accurate classifications.

This finding broadly supports the work of other studies in this area, reducing the intra-class spectral-temporal variability of agricultural areas. Bellón et al. (2018) used multiple clustering approaches combining field-level segmentation and landscape stratification to reduce the spectral variability of crops. In that study, the authors demonstrated an improvement in classification accuracies for underrepresented and sparsely distributed cropping systems.

Although we achieved higher overall accuracies based on a higher number of clusters, we found a weak linear association between them. Nevertheless, the present results are significant in one major respect: the very satisfactory accuracy regardless of the clustering step, which indicates the potential of our method.

One unanticipated finding was that accuracy measures of predicted maps, such as the user's accuracy, did not fully mirror those we achieved from the test set. This result may be explained by the fact that we set a different agricultural year for prediction analysis and LULC maps illustration, which was entirely independent of the observations used to train and test the RF model in both time and space. Using yearly independent data sets leads to natural meteorological variations between agricultural years and different phenological responses of vegetation, which may directly affect the map predictions. These results reflect those of Han et al. (2022), who also observed this weather volatility, demonstrating its considerable impact on double crop systems yield variability.

4.2 Intra-class variability of NDVI temporal signatures

Soybeans are recognized as a typical summer crop in double-cropped rotations, particularly in combination with winter corn in the MT state. The exploratory analysis demonstrated that these agricultural land uses may present high intra-class variability and diverse crop calendars for neighbour land parcels. These results reflect those of Arias et al. (2020), who also found considerable variability in crop parcels related to management techniques and cultivars compositions using a crop classification approach based on temporal signatures from Sentinel-1 time series.

In our study, soybean-related NDVI temporal signatures varied little, although we observed a slight time shift between some land parcels during the harvest period. On the other hand, corn-related NDVI temporal signatures varied the most. Paired land parcels presented up to one month of time shift considering the sowing period. Harvest periods also varied depending on the management practice and season length of cultivars, e.g., plant residuals might be left to dry after the harvest or removed mechanically.

Another source of variability was the land cover prior to soybean sowing, where we observed land parcels with both low and high NDVI values. The first indicates the fallow period between rotations, and the latter may indicate a crop emergence of residual seeds left during harvest.

This contribution has been peer-reviewed.
https://doi.org/10.5194/isprs-archives-XLVIII-M-1-2023-49-2023 | © Author(s) 2023. CC BY 4.0 License.

The variability among land parcels may partly be explained by the size of the crop area (approximately 2900 hectares) and the 15-day time interval of PS composites, which makes the time regularity of management practices impracticable.

We reinforce that such high variability should be better addressed by designing parcel-specific strategies and not based only on broad class-related spectro-temporal signatures.

Given the promising results, however, further investigation is needed to provide greater insight into the effects of this methodology on other crop dynamics. Moreover, more robust classifiers, such as deep learning architectures, should be considered to improve model accuracies and to better generalise the predictions to other areas.

5. CONCLUSION

The high intra-class variability of agricultural parcels has demanded reliable approaches for producing accurate LULC classification maps. The present study investigated the accuracy of LULC classification based on an initial clustering step in a heterogeneous agricultural landscape while checking for variability among their NDVI temporal signatures. In a heterogeneous agricultural area located in the state of MT, we examined the hypothesis that grouping observations with similar spectro-temporal signatures would improve the performance of LULC classification over a conventional method.

Results showed that some land parcels presented distinct NDVI temporal signatures.

The initial clustering scheme outperformed the non-clustered classification of LULC in overall accuracy measures regardless of the number of clusters, which addressed the complexity and variability of spectro-temporal signatures. These clusters identified parcel-specific dynamics of management practices and cropping intensity, reducing the uncertainty in classifying the area.

In order for countries to implement more effective and sustainable development of agriculture, there is a need to improve the accuracy of thematic maps produced through the process of LULC classification, especially in heterogeneous agricultural areas. The accuracies achieved with this approach represent promising opportunities toward the sufficiently accurate classification of such areas. In addition, the knowledge of the intra-class variability allows the analyst to classify agricultural areas and infer the temporal dynamics, whether caused by management practices or cropping intensity.

Further work could assess other types of classifiers, such as deep learning architectures, especially in areas with a large number of crop types and distinct management practices.

6. ACKNOWLEDGEMENTS

The authors are thankful for the financial support from FAPESP (Grants: 2021/15001-9 and 2017/50205-9). R. A. C. Lamparelli thanks the National Council for Scientific and Technological Development (CNPq) for the scholarship (Grant: 305271/2020-2)

7. REFERENCES

Achanta, R., Süsstrunk, S., 2017. Superpixels and polygons using simple non-iterative clustering. *30th IEEE Conference on Computer Vision and Pattern Recognition*, 4895–4904. https://doi.org/10.1109/CVPR.2017.520

Arias, M., Campo-Bescós, M. Á., Álvarez-Mozos, J., 2020. Crop Classification Based on Temporal Signatures of Sentinel-1 Observations over Navarre Province, Spain. *Remote Sensing*, *12*(2), 278. https://doi.org/10.3390/rs12020278.

Azar, R., Villa, P., Stroppiana, D., Crema, A., Boschetti, M., Brivio, P. A., 2016. Assessing in-season crop classification performance using satellite data: a test case in Northern Italy. *European Journal of Remote Sensing*, *49*(1), 361–380. https://doi.org/10.5721/EuJRS20164920.

Bellón, B., Bégué, A., Lo Seen, D., Lebourgeois, V., Evangelista, B. A., Simões, M., Demonte Ferraz, R. P., 2018. Improved regional-scale Brazilian cropping systems' mapping based on a semi-automatic object-based clustering approach. *International Journal of Applied Earth Observation and Geoinformation*, *68*, 127–138. https://doi.org/10.1016/j.jag.2018.01.019.

Bischl, B., Lang, M., Kotthoff, L., Schiffner, J., Richter, J., Studerus, E., Casalicchio, G., Jones, Z. M., 2016: Mlr: Machine learning in R. *Journal of Machine Learning Research*, *17*, 1–5. https://github.com/mlr-org/mlr.

Blaschke, T., 2010. Object based image analysis for remote sensing. *ISPRS Journal of Photogrammetry and Remote Sensing*, *65*(1), 2–16. https://doi.org/10.1016/j.isprsjprs.2009.06.004.

Breiman, L., 2001. Random forests. *Machine Learning*, *45*(1), 5–32. https://doi.org/10.1023/A:1010933404324.

Brinkhoff, J., Vardanega, J., Robson, A. J., 2019. Land Cover Classification of Nine Perennial Crops Using Sentinel-1 and -2 Data. *Remote Sensing*, *12*(1), 96. https://doi.org/10.3390/rs12010096.

Bueno, I. T., Acerbi Júnior, F. W., Silveira, E. M. O., Mello, J. M., Carvalho, L. M. T., Gomide, L. R., Withey, K., Scolforo, J. R. S., 2019. Object-Based Change Detection in the Cerrado Biome Using Landsat Time Series. *Remote Sensing*, *11*(5), 570. https://doi.org/10.3390/rs11050570.

Cai, Y., Guan, K., Peng, J., Wang, S., Seifert, C., Wardlow, B., Li, Z., 2018. A high-performance and in-season classification system of field-level crop types using time-series Landsat data and a machine learning approach. *Remote Sensing of Environment*, *210*, 35–47. https://doi.org/10.1016/J.RSE.2018.02.045.

Congalton, R. G., Green, K., 2009. *Assessing the accuracy of remotely sensed data: principles and practices*. CRC Press/Taylor & Francis.

Desclée, B., Bogaert, P., Defourny, P., 2006. Forest change detection by statistical object-based method. *Remote Sensing of Environment*, *102*(1–2), 1–11. https://doi.org/10.1016/j.rse.2006.01.013.

Duro, D. C., Franklin, S. E., Dubé, M. G., 2012. A comparison of pixel-based and object-based image analysis with selected machine learning algorithms for the classification of agricultural landscapes using SPOT-5 HRG imagery. *Remote Sensing of Environment*, *118*, 259–272. https://doi.org/10.1016/j.rse.2011.11.020.

FAO, 2021. FAO Statistical Yearbook 2021: World Food and Agriculture. In *Food and Agriculture Organization of the United Nations (FAO)*. FAO. https://doi.org/10.4060/CB4477EN.

Garrett, R. D., Koh, I., Lambin, E. F., le Polain de Waroux, Y., Kastens, J. H., Brown, J. C., 2018. Intensification in agriculture-forest frontiers: Land use responses to development and conservation policies in Brazil. *Global Environmental Change*, *53*, 233–243. https://doi.org/10.1016/J.GLOENVCHA.2018.09.011.

Han, S.-S., Park, H.-J., Shin, T., Ko, J., Choi, W.-J., Lee, Y.-H., Bae, H.-S., Ahn, S.-H., Youn, J.-T., Kim, H.-Y., 2022. Effects of Tillage System, Sowing Date, and Weather Course on Yield of Double-Crop Soybeans Cultivated in Drained Paddy Fields. *Agronomy*, *12*(8), 1901. https://doi.org/10.3390/agronomy12081901.

Huete, A., Didan, K., Miura, T., Rodriguez, E. ., Gao, X., Ferreira, L., 2002. Overview of the radiometric and biophysical performance of the MODIS vegetation indices. *Remote Sensing of Environment*, *83*(1–2), 195–213. https://doi.org/10.1016/S0034-4257(02)00096-2.

Huete, A. R., 1988. A soil-adjusted vegetation index (SAVI). *Remote Sensing of Environment*, *25*(3), 295–309. https://doi.org/10.1016/0034-4257(88)90106-X.

IBGE., 2021. *Produção Agrícola Municipal*. IBGE. https://www.ibge.gov.br/estatisticas/economicas/agricultura-e-pecuaria/9117-producao-agricola-municipal-culturas-temporarias-e-permanentes.html?=&t=destaques.

Kuhn, M., 2008. Building Predictive Models in R Using the caret Package. *Journal of Statistical Software*, *28*(5), 1–26. https://doi.org/10.18637/jss.v028.i05.

Mathur, A., Foody, G. M., 2008. Crop classification by support vector machine with intelligently selected training data for an operational application. *International Journal of Remote Sensing*, *29*(8), 2227–2240. https://doi.org/10.1080/01431160701395203.

Orynbaikyzy, A., Gessner, U., Conrad, C., 2019. Crop type classification using a combination of optical and radar remote sensing data: a review. *International Journal of Remote Sensing*, *40*(17), 6553–6595. https://doi.org/10.1080/01431161.2019.1569791.

Pascucci, S., Carfora, M., Palombo, A., Pignatti, S., Casa, R., Pepe, M., Castaldi, F., 2018. A Comparison between Standard and Functional Clustering Methodologies: Application to Agricultural Fields for Yield Pattern Assessment. *Remote Sensing*, *10*(4), 585. https://doi.org/10.3390/rs10040585.

Planet Team., 2022. *Planet Imagery Product Specifications*. https://assets.planet.com/docs/Planet_Combined_Imagery_Product_Specs_letter_screen.pdf.

Qi, J., Chehbouni, A., Huete, A. R., Kerr, Y. H., Sorooshian, S., 1994. A Modified Soil Adjusted Vegetation Index. *Remote Sensing of Environment*, *48*, 119–126.

Qin, R., Liu, T., 2022. A Review of Landcover Classification with Very-High Resolution Remotely Sensed Optical Images—Analysis Unit, Model Scalability and Transferability. *Remote Sensing*, *14*(3), 646. https://doi.org/10.3390/rs14030646.

R Core Team., 2014. *A language and environment for statistical computing*.

Reis, A. A., Werner, J. P. S., Silva, B. C., Figueiredo, G. K. D. A., Antunes, J. F. G., Esquerdo, J. C. D. M., Coutinho, A. C., Lamparelli, R. A. C., Rocha, J. V., Magalhães, P. S. G., (2020). Monitoring Pasture Aboveground Biomass and Canopy Height in an Integrated Crop–Livestock System Using Textural Information from PlanetScope Imagery. *Remote Sensing*, *12*(16), 2534. https://doi.org/10.3390/rs12162534.

Rouse, J. W., Haas, R. H., Schell, J. A., Deering, D. W., Haas, R. H., Schell, J. A., Deering, D. W., 1974. Monitoring vegetation systems in the Great Plains with ERTS. *Third Earth Resources Technology Satellite-1 Symposium*, 309–317.

Vali, A., Comai, S., Matteucci, M., 2020. Deep Learning for Land Use and Land Cover Classification Based on Hyperspectral and Multispectral Earth Observation Data: A Review. *Remote Sensing*, *12*(15), 2495. https://doi.org/10.3390/rs12152495.

Villa, P., Stroppiana, D., Fontanelli, G., Azar, R., Brivio, P., 2015. In-Season Mapping of Crop Type with Optical and X-Band SAR Data: A Classification Tree Approach Using Synoptic Seasonal Features. *Remote Sensing*, *7*(10), 12859–12886. https://doi.org/10.3390/rs71012859.

This contribution has been peer-reviewed.
https://doi.org/10.5194/isprs-archives-XLVIII-M-1-2023-49-2023 | © Author(s) 2023. CC BY 4.0 License.

IMPACT OF LEARNING SET AND SAMPLING FOR SNOW AVALANCHE SUSCEPTIBILITY MAPPING WITH RANDOM FOREST

S. Cetinkaya[1,2], S. Kocaman [2*]

[1] Hacettepe University, Graduate School of Science and Engineering, Ankara, Türkiye – sinemcetinkaya@hacettepe.edu.tr
[2] Dept. of Geomatics Engineering, Hacettepe University, 06800 Beytepe Ankara, Türkiye - sultankocaman@hacettepe.edu.tr

KEY WORDS: Snow avalanche susceptibility, snow avalanche inventory, random forest, remote sensing, machine learning

ABSTRACT: Snow avalanche refers to rapid snow mass movement under the influence of gravity and is a frequently observed natural phenomenon in mountainous regions. An area affected by a snow avalanche consists of starting, track (or transition) and runout (or deposition) zones, which have different geomorphologic characteristics that need to be considered in hazard modelling. This study attempted to analyze the impact of the different zones in producing snow avalanche susceptibility maps (ASMs) with data-driven methods. The random forest (RF) method was applied to the data from Gross Spannort Mountain region (Switzerland) for this purpose. Avalanche inventories from two dates were manually delineated to separate the three zones and two RF models were trained with learning datasets; i.e. Inventory-A which includes all snow avalanche zones (original inventory), and Inventory-B with starting and track zones. The conditioning factors were defined based on the literature, data availability and study area characteristics. The trained models (Model-A with Inventory-A and Model-B with Inventory-B) were evaluated with the area under the receiver operating characteristic (ROC), precision, recall and F1 score. The results show that Model-A has AUC of 0.97, precision of 0.89, recall of 0.95 and F1 score of 0.92 and the Model-B has AUC of 0.98, precision of 0.90, recall of 0.94 and F1 score of 0.92 that indicate high prediction performances for both cases. Furthermore, feature importance values were calculated by the Mean Decrease in Impurity (MDI) method, and elevation, aspect and valley depth were found the most influencing conditioning factors.

1. INTRODUCTION

Snow avalanches can be defined as rapid mass movement of snow, ice, rock and soil on snow covered steep slopes. Snow avalanche occurs depending on terrain, meteorological and snowpack properties (Schweizer et al., 2003). A snow avalanche may contain soil, rocks, vegetation, or ice particles, all of which can have profoundly devastating socioeconomic and environmental effects. Snow avalanche susceptibility assessment is crucial for disaster management to identify the spatial probability distribution of hazard-prone areas.

An area affected by a snow avalanche event consists of three sub-regions, such as starting, track and runout zones. An avalanche initiates from the starting zone, where unstable snow originates and then flows through the track to the runout zone. Delineation of avalanche zones, especially starting and runout zones, plays a crucial role for avalanche hazard and risk assessments (Bianchi et al., 2021; Beato Bergua et al., 2019; Chueca Cía et al., 2014; Fazzini et al., 2021; Kim & Park, 2017; Meseşan et al., 2019). The characteristics of starting zone and track can influence the size and speed of avalanche, which in turn affects runout zone. The distribution of snow mass in the runout zone and starting zone can be used for predicting potential damage to buildings and infrastructures (Cappabianca et al., 2018; Pistocchi & Notarnicola, 2013). The geomorphologic characteristics of each zone differ significantly from each other, and they should be considered for accurate avalanche hazard modeling (Choubin et al., 2019).

In recent decades, Geographic Information Systems (GIS) and remote sensing (RS) domains have provided important tools for enabling improved disaster management. Data-driven machine learning (ML) techniques have contributed to numerous natural hazard assessment studies. Recently, several snow avalanche susceptibility studies have been conducted with the ML algorithms such as Support vector machine (SVM), K-nearest neighbours (KNN), Classification and Regression Tree (CART), Multilayer perceptron (MLP), Logistics Regression (LR), Random Forest (RF), and others (Akay, 2021; Rahmati et al., 2019; Tiwari et al., 2021; Wen et al., 2022).

Here, we employed a widely used ML algorithm, the RF, to generate two snow avalanche susceptibility maps (ASMs) for the Gross Spannort Mountain region (Switzerland). A previous study was carried out for producing the ASM for Davos, Switzerland by employing a set of features (partly different from the present study) using the RF and LR methods (Cetinkaya and Kocaman, 2022), which illustrated higher prediction quality of the RF classifier. The present research aimed to provide a better insight into the impact of learning sample selection on snow avalanche susceptibility mapping, by investigating in particular the use of runout zones for this purpose. After a literature survey on the regional characteristics and snow susceptibility assessment studies, a total of 18 features were utilized as conditioning factors that are elevation, slope, land use and land cover (LULC), diurnal anisotropic heating index (DAH), plan curvature, profile curvature, aspect, relative slope position index (RSP), slope length factor (LS), topographic position index (TPI), topographic ruggedness index (TRI), topographic wetness index (TWI), convergence index, wind exposition index (WEI), mid-slope position index (MSP), valley depth (VD), normalized height, and lithology. A snow avalanche inventory produced by (Hafner & Bühler, 2019) was provided as 2D polygons and employed here for learning. In addition, the prediction capabilities of the methods were analyzed using the

* Corresponding author

This contribution has been peer-reviewed.
https://doi.org/10.5194/isprs-archives-XLVIII-M-1-2023-57-2023 | © Author(s) 2023. CC BY 4.0 License.

area under the receiver operating characteristic (ROC) curve (AUC), precision, recall, and F1 score. Furthermore, the influence of the conditioning factors on the models was evaluated with feature importance analysis and presented here.

2. MATERIALS AND METHODS

2.1 Study Area

The study region is located between Erstfeld and Engelberg in Central Switzerland and covers an area of 259 km^2 (Figure 1). Gross Spannort Mountain, part of the Uri Alps, is the highest point of the investigated area with an altitude of 3,198 meters at the peak. Avalanches occur frequently in the region, and the site was selected due to the accurate and reliable inventory.

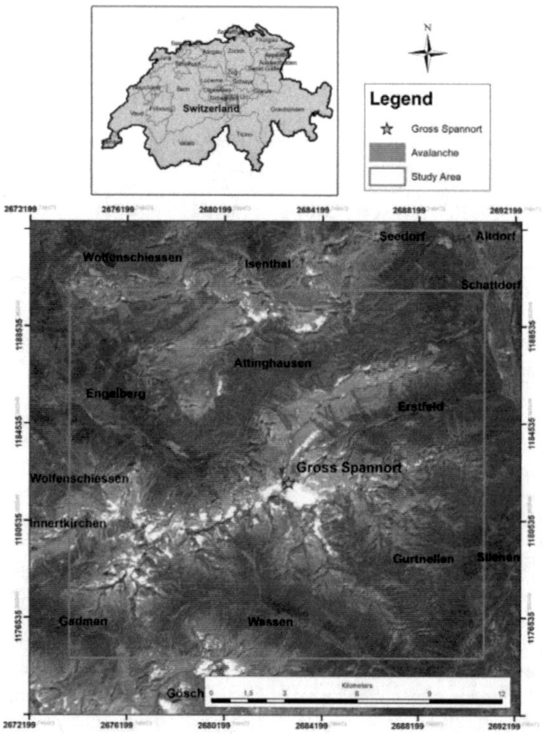

Figure 1. Location of the study area.

2.2 Input Datasets and Features

Here, we aimed to investigate the impact of inventory, in particular the use of different avalanche zones, on ASM production. Initially, the starting, track and runout zones of the avalanche polygons in the inventory data provided by Hafner & Bühler (2019) was manually delineated using a desktop GIS software. Thus, besides the original inventory with all zones (Inventory-A), a second map called Inventory-B involving only starting and track zones was produced. The number of pixels involved in both inventories is given in Table 1. The overall workflow of the study used to produce the ASMs is illustrated in Figure 2.

The digital elevation model (DEM), lithology and LULC of the study area were obtained from the Swiss Federal Office of Topography, Switzerland (Swisstopo, 2021). The conditioning factors, which are depicted in Figure 2 as numerical factors, were generated from DEM in the SAGA GIS environment (Conrad et al., 2015). The statistical summary of numerical factors can be found in Table A1 (Appendix). Both inventory maps and the feature maps representing the conditioning factors (see Figure A1 in Appendix) were converted into raster datasets with a dimension of 1518 x 1713 pixels and 10 m resolution.

	Inventory-A (Starting + Track + Runout Zones)	Inventory-B (Starting + Track Zones)
Avalanche	274,967	207,314
Non - Avalanche	2,325,367	2,393,020

Table 1. The pixel counts of snow avalanche inventories.

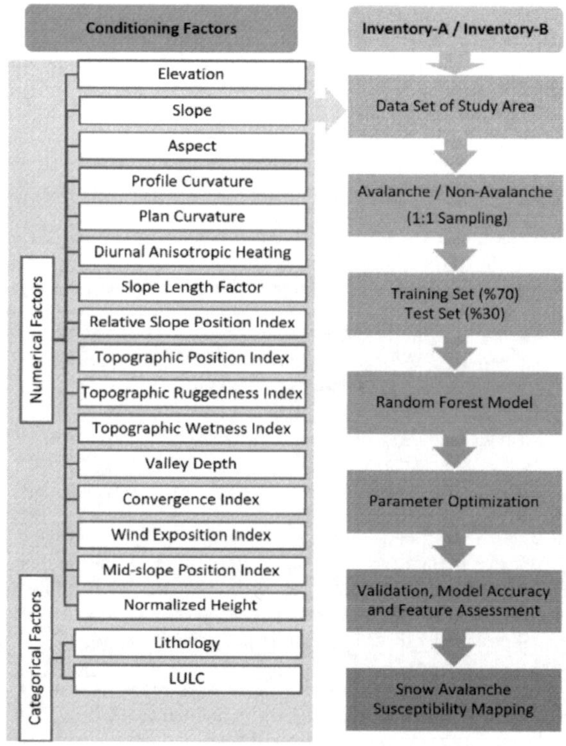

Figure 2. Snow avalanche susceptibility mapping workflow.

2.3 Snow Avalanche Susceptibility Mapping

In the current study, the random undersampling technique of SMOTE (Chawla et al., 2002) was used to handle an imbalanced dataset by random selection relying on avalanche inventories (by using a ratio of 1/1 for avalanche/non-avalanche pixels). The learning dataset was split as training (70%) and validation (30%) for producing the ASMs with Inventory-A and Inventory-B, separately. The parameter optimization strategy was conducted with existing libraries for the RF in a Python environment. The RF proposed by Breiman (2001) is one of the most widely used ML algorithms based on a large number of decision trees. It has the ability to reduce overfitting problems as well as to perform accurate classifications.

Parameter optimization plays a crucial role in the accuracy of ML algorithms. In this study, the HalvingGridSearchCV (Pedregosa et al., 2011) implemented in Scikit-learn library was applied to identify optimal parameters, which are given in Table

https://doi.org/10.5194/isprs-archives-XLVIII-M-1-2023-57-2023 | © Author(s) 2023. CC BY 4.0 License.

2. The area under the receiver operatic characteristic curve (AUC) was used to evaluate the model's predictive abilities. The performance assessment of the models was also investigated using precision, recall and F1-score values.

Parameters	Model-A	Model-B
Number of estimators	500	500
Max depth	32	32
Max feature	32	32
Min sample leaf	1	1
Min sample split	2	2

Table 2. Optimal parameters of RF models.

2.4 Feature Importance

Feature importance enables to compute of the relative importance of each feature in a predictive model. The Mean Decrease in Impurity (MDI) method was selected to measure the effects of each feature on the models. The higher the calculated value, the higher the contribution of the feature to the model is considered and vice versa.

3. RESULTS AND DISCUSSIONS

In this study, two ASMs were produced by using Inventory-A (all zones) and Inventory-B (without runout zone) with 18 conditioning factors by applying the RF model (hereinafter will be referred as Model-A and Model-B, respectively). As summarized in Table 3, the models achieved similarly accurate results in terms of precision, recall and F1 score, which indicate high coherence between the learning datasets and the models. The AUC values obtained from both models (Figure 3 and Figure 4) show that Model-B (AUC = 0.98) performed slightly better than Model-A (AUC = 0.97).

	Class	Precision	Recall	F1-Score
Model-A	Non - Avalanche	0.95	0.88	0.91
	Avalanche	0.89	0.95	0.92
Model-B	Non - Avalanche	0.94	0.90	0.92
	Avalanche	0.90	0.94	0.92

Table 3. Overall precision, recall, and F1-Score of the RF models.

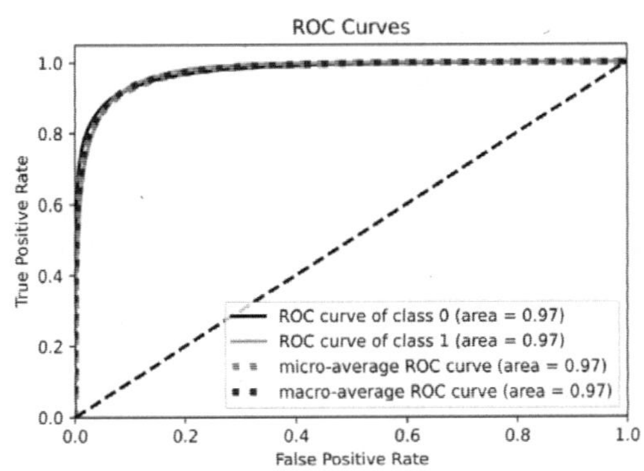

Figure 3. The ROC curve obtained from Model-A.

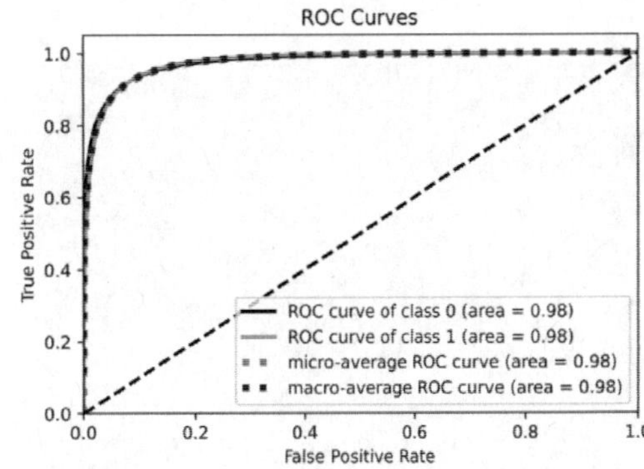

Figure 4. The ROC curve obtained from Model-B.

The feature importance results computed from the models are given in Figure 5. In the figure, the features ranked almost identically according to the importance values of the models. Both models indicated that the elevation is the most important variable, followed by aspect, valley depth, and wind exposition index. Based on the importance values of the elevation feature, Model-B indicated that elevation has even higher importance for modeling when runout zones are not considered. It can also be observed that DEM-derived features contribute to avalanche susceptibility greater than lithology and LULC in the site. Yet, a few lithology types such as serpentinite, talc schist, metaperidotite, and granite, granodiorite were found effective in learning.

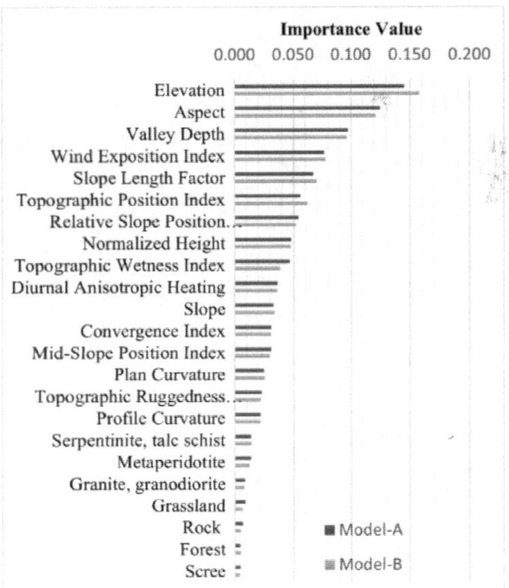

Figure 5. Feature importance of the RF models.

Figure 6 demonstrates the ASMs of the study area produced by the prediction results of the RF models and detailed views from different parts. Although the performance metrics calculated from Model-A and Model-B were quite similar, the spatial distribution of model probabilities differed significantly from each other (see Figure 6). Model-A identified larger areas as snow avalanche-prone than Model-B as expected.

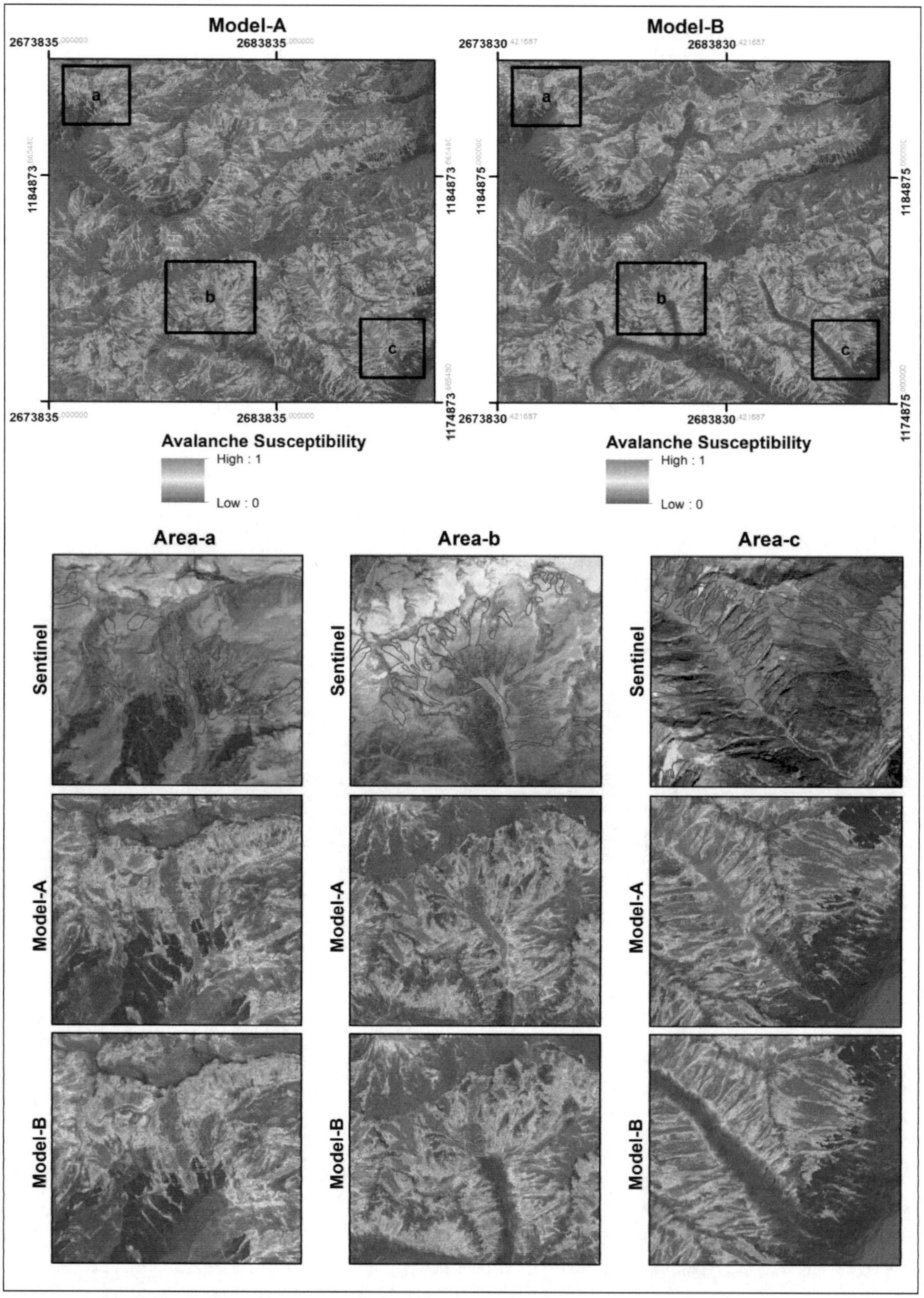

Figure 6. Snow avalanche susceptibility belongs to Model-A, and Model-B.

When the feature statistics given in Table A1 in Appendix is analyzed, it can be seen that removing runout zones from the inventory enabled a differentiation between the avalanche/non-avalanche zones in terms of elevation and its first derivative, the slope. In addition, the slope length factor and mid-slope position index varies between both inventories. Thus, removal of runout zones helped to better classify avalanche-non-avalanche zones. On the other hand, there are other features that keep the mean values even after the removal of runout zones, such as the aspect and wind exposition index, which are highly effective variables in both models. The detailed maps shown in Figure 6 for the three sub-areas as marked in the maps of the whole site indicate that both models can successfully classify the data for which they aim for, i.e. Model-A the avalanches with all zones, and Model-B the starting and the track zones. On the other hand, as mentioned previously, the characteristics of starting zone and track can influence the size and speed of avalanche, which in turn affects runout zone. Thus, the use purposes of both models would be different and an ASM should rather be produced by using the starting and track zones. Yet, the runout zones predicted by Model-A, which can precisely be delineated by differencing ASM-A and ASM-B, would be a good approximation for risk assessment under the given conditions of the avalanche periods mapped by Hafner and Bühler (2019).

4. CONCLUSIONS

Snow avalanches are among destructive natural hazards in cold and snow-covered mountainous regions. Within the scope of this study, the RF model was applied for the avalanche susceptibility mapping of an area in Gross Spannort Mountain region, Switzerland by using two types of avalanche inventory maps. Inventory-A includes avalanche polygons with the starting, track and runout regions delineated as a single zone for each event. Inventory-B includes the starting and runout zones (without runout zones) to assess the influence of the learning data on the produced ASMs. The results show that both models could successfully classify the data for the given inventory types with AUC values of 0.98 (Model-B) and 0.97 (Model-A). The results illustrate high performance of the RF models for snow avalanche hazard and risk assessments. The features employed for this purpose were also found suitable for precise modeling.

On the other hand, the use purpose of an ASM must clearly be defined and they should be produced based on it. In order to assess the avalanche susceptibility, starting and track zones need to be used as the runout zones depend on the characteristics of starting and track zones, which can influence the size and speed of avalanche. Yet, the results of Model-B also show that the RF can classify the runout zones when high quality data for all zones are available that can provide a good approximation for risk assessment under similar conditions.

In previous studies, applying elimination and backward selection of features would be another strategy to improve ASMs. Further research is necessary to demonstrate the ability of transfer learning. Apart from these, the inventory data was manually divided into avalanche zones in the study, which raises the question of whether we can obtain avalanche zones with ML methods.

ACKNOWLEDGEMENTS

The authors thank to Elisabeth Hafner, Dr. Yves Bühler and WSL Institute for Snow and Avalanche Research SLF, Switzerland for the provision of avalanche inventory data. This paper is part of the Ph.D. thesis research of Sinem Cetinkaya and was carried out with the support of Higher Education Council of Turkey (YÖK) within the 100/2000 Programme.

REFERENCES

Akay, H., 2021. Spatial modeling of snow avalanche susceptibility using hybrid and ensemble machine learning techniques. *Catena, 206*. https://doi.org/10.1016/j.catena.2021.105524.

Breiman, L., 2001. Random forests. *Machine Learning, 45*(1), 5–32. https://doi.org/10.1023/A:1010933404324.

Beato Bergua, S., Poblete Piedrabuena, M. Á., & Marino Alfonso, J. L., 2019. Snow avalanches, land use changes, and atmospheric warming in landscape dynamics of the Atlantic mid-mountains (Cantabrian Range, NW Spain). *Applied Geography,* 107, 38-50. https://doi.org/10.1016/j.apgeog.2019.04.007.

Bianchi, F. M., Grahn, J., Eckerstorfer, M., Malnes, E., & Vickers, H., 2021. Snow Avalanche Segmentation in SAR Images with Fully Convolutional Neural Networks. *IEEE Journal of Selected Topics in Applied Earth Observations and Remote Sensing*, 14, 75-82, https://doi.org/10.1109/JSTARS.2020.3036914.

Cappabianca, F., Barbolini, M., & Natale, L., 2008. Snow avalanche risk assessment and mapping: A new method based on a combination of statistical analysis, avalanche dynamics simulation and empirically-based vulnerability relations integrated in a GIS platform. *Cold Regions Science and Technology*, *54*(3), 193–205. https://doi.org/10.1016/j.coldregions.2008.06.005.

Cetinkaya, S., Kocaman, S., 2022. Snow Avalanche Susceptibility Mapping for DAVOS, Switzerland. *The International Archives of Photogrammetry, Remote Sensing and Spatial Information Sciences, 43,* 1083-1090. https://doi.org/10.5194/isprs-archives-XLIII-B3-2022-1083-2022.

Chawla, N.v, Bowyer, K. W., Hall, L. O., & Kegelmeyer, W. P., 2002. SMOTE: Synthetic Minority Over-sampling Technique. In *Journal of Artificial Intelligence Research* (Vol. 16).

Choubin, B., Borji, M., Mosavi, A., Sajedi-Hosseini, F., Singh, V. P., & Shamshirband, S., 2019. Snow avalanche hazard prediction using machine learning methods. *Journal of Hydrology*, *577*(May), 123929. https://doi.org/10.1016/j.jhydrol.2019.123929.

Chueca Cía, J., Andrés, A. J., & Montañés Magallón, A., 2014. A proposal for avalanche susceptibility mapping in the Pyrenees using GIS: the Formigal-Peyreget area (Sheet 145-I; scale 1:25.000). *Journal of Maps, 10(2)*, 203-210. https://doi.org/10.1080/17445647.2013.870501.

Fazzini, M., Cordeschi, M., Carabella, C., Paglia, G., Esposito, G., & Miccadei, E., 2021. Snow Avalanche Assessment in Mass

Movement-Prone Areas: Results from Climate Extremization in Relationship with Environmental Risk Reduction in the Prati di Tivo Area (Gran Sasso Massif, Central Italy). *Land*, *10*(11), 1176. https://doi.org/10.3390/land10111176.

Hafner, E., & Bühler, Y., 2019. *SPOT6 Avalanche outlines 24 January 2018*. EnviDat. https://doi.org/http://dx.doi.org/10.16904/envidat.77.

Kim, S.-M., & Park, H.-D., 2017. Analogy between grid-based modeling of landslide and avalanche using GIS with surface flow analysis. Bulletin of Engineering Geology and the Environment, 78(1), 189-206. https://doi.org/10.1007/s10064-017-1144-y.

Meseşan, F., Man, T. C., Pop, O. T., & Gavrilă, I. G., 2019. Reconstructing snow-avalanche extent using remote sensing and dendrogeomorphology in Parâng Mountains. Cold Regions Science and Technology, 157, 97-109. https://doi.org/10.1016/j.coldregions.2018.10.002.

Pedregosa, F., Grisel, O., Weiss, R., Passos, A., Brucher, M., Varoquax, G., Gramfort, A., Michel, V., Thirion, B., Grisel, O., Blondel, M., Prettenhofer, P., Weiss, R., Dubourg, V., & Brucher, M., 2011. Scikit-learn: Machine Learning in Python. *Journal of Machine Learning Research*, *12*, 2825–2830.

Pistocchi, A., & Notarnicola, C., 2013. Data-driven mapping of avalanche release areas: A case study in South Tyrol, Italy. *Natural Hazards,* 65(3), 1313–1330. https://doi.org/10.1007/s11069-012-0410-3.

Rahmati, O., Ghorbanzadeh, O., Teimurian, T., Mohammadi, F., Tiefenbacher, J. P., Falah, F., Pirasteh, S., Ngo, P. T. T., & Bui, D. T., 2019. Spatial modeling of snow avalanche using machine learning models and geo-environmental factors: Comparison of effectiveness in two mountain regions. *Remote Sensing*, *11*(24). https://doi.org/10.3390/rs11242995.

Schweizer, J., Jamieson, J. B., & Schneebeli, M., 2003. Snow avalanche formation. *Reviews of Geophysics*, *41*(4). https://doi.org/10.1029/2002RG000123.

Swisstopo., 2021. *Online maps & data*. https://www.swisstopo.admin.ch/en/maps-data-online.html.

Tiwari, A., G., A., & Vishwakarma, B. D., 2021. Parameter importance assessment improves efficacy of machine learning methods for predicting snow avalanche sites in Leh-Manali Highway, India. *Science of the Total Environment*, *794*. https://doi.org/10.1016/j.scitotenv.2021.148738.

Wen, H., Wu, X., Liao, X., Wang, D., Huang, K., & Wünnemann, B., 2022. Application of machine learning methods for snow avalanche susceptibility mapping in the Parlung Tsangpo catchment, southeastern Qinghai-Tibet Plateau. *Cold Regions Science and Technology*, *198*. https://doi.org/10.1016/j.coldregions.2022.103535.

APPENDIX

FACTORS			MEAN	STD	MIN	25%	50%	75%	MAX
Elevation	Inventory -A	Non-avalanche	2022.718	505.606	453.632	1684.412	2072.872	2410.474	3237.224
		Avalanche	2076.937	381.084	856.656	1808.086	2084.501	2352.096	3209.036
	Inventory -B	Non-avalanche	2019.035	503.677	453.632	1680.311	2065.898	2406.594	3237.224
		Avalanche	2137.147	349.738	1078.695	1888.180	2136.086	2385.470	3209.036
Slope	Inventory -A	Non-avalanche	33.821	14.979	0.000	23.940	33.590	42.688	86.979
		Avalanche	33.467	12.868	0.000	25.143	33.131	40.300	85.321
	Inventory -B	Non-avalanche	33.508	14.990	0.000	23.453	33.265	42.369	86.979
		Avalanche	36.968	11.481	0.000	29.896	35.591	42.785	85.321
Aspect	Inventory -A	Non-avalanche	183.858	108.067	-1.000	92.180	184.257	281.686	360.000
		Avalanche	153.513	75.167	-1.000	103.181	150.667	197.542	359.994
	Inventory -B	Non-avalanche	182.901	107.500	-1.000	92.224	182.586	279.147	360.000
		Avalanche	154.661	73.836	-1.000	106.980	153.135	196.755	359.994
Profile Curvature	Inventory -A	Non-avalanche	0.002	7.315	-317.300	-2.067	0.034	2.190	325.558
		Avalanche	0.379	5.811	-193.240	-1.720	0.247	2.186	203.003
	Inventory -B	Non-avalanche	0.010	7.235	-317.300	-2.026	0.041	2.160	325.558
		Avalanche	0.413	6.392	-193.240	-2.011	0.271	2.485	203.003
Plan Curvature	Inventory -A	Non-avalanche	0.093	5.056	-281.991	-1.447	-0.002	1.478	294.960
		Avalanche	-0.344	4.267	-173.432	-1.669	-0.199	1.113	188.452
	Inventory -B	Non-avalanche	0.085	5.004	-281.991	-1.430	-0.007	1.447	294.960
		Avalanche	-0.391	4.676	-173.432	-1.967	-0.251	1.301	188.452
Diurnal Anisotropic Heating	Inventory -A	Non-avalanche	-0.018	0.404	-0.974	-0.356	-0.024	0.325	0.973
		Avalanche	0.126	0.367	-0.938	-0.150	0.175	0.426	0.960
	Inventory -B	Non-avalanche	-0.016	0.401	-0.974	-0.350	-0.019	0.321	0.973
		Avalanche	0.151	0.389	-0.938	-0.146	0.239	0.469	0.960
Slope Length Factor	Inventory -A	Non-avalanche	12.148	6.223	0.000	8.274	12.007	15.441	172.068
		Avalanche	13.989	6.301	0.000	10.267	13.196	16.744	104.631
	Inventory -B	Non-avalanche	12.113	6.228	0.000	8.220	11.959	15.405	172.068
		Avalanche	14.988	5.981	0.000	11.360	13.906	17.475	83.772
Relative Slope Position Index	Inventory -A	Non-avalanche	-4.102	668.904	0.000	0.035	0.246	0.705	16080.000
		Avalanche	-4.870	713.527	0.000	0.006	0.070	0.368	51.291
	Inventory -B	Non-avalanche	-4.151	671.935	0.000	0.029	0.229	0.691	16080.000
		Avalanche	-4.549	694.501	0.000	0.021	0.142	0.480	51.291
Topographic Position Index	Inventory -A	Non-avalanche	0.643	14.025	-129.677	-5.978	-0.696	5.462	231.357
		Avalanche	-4.964	11.038	-96.581	-9.522	-4.065	0.099	109.060
	Inventory -B	Non-avalanche	0.503	13.902	-129.677	-6.034	-0.819	5.220	231.357
		Avalanche	-5.178	12.030	-96.581	-10.485	-4.355	0.589	109.060
Topographic Ruggedness Index	Inventory -A	Non-avalanche	5.341	4.548	0.000	2.987	4.397	6.215	191.151
		Avalanche	4.994	3.644	0.000	3.095	4.286	5.684	124.736
	Inventory -B	Non-avalanche	5.276	4.517	0.000	2.921	4.344	6.141	191.151
		Avalanche	5.633	3.760	0.000	3.765	4.712	6.249	124.736
Topographic Wetness Index	Inventory -A	Non-avalanche	5.690	2.056	-0.634	4.277	5.495	6.792	24.869
		Avalanche	6.446	1.944	0.467	5.157	6.344	7.576	18.466
	Inventory -B	Non-avalanche	5.752	2.089	-0.634	4.311	5.549	6.880	24.869
		Avalanche	5.977	1.644	0.467	4.871	5.951	7.010	16.082
Valley Depth	Inventory -A	Non-avalanche	130.806	118.260	-222.702	33.650	93.546	205.858	573.546
		Avalanche	178.456	122.971	-48.971	74.487	159.891	267.324	569.177
	Inventory -B	Non-avalanche	134.620	120.696	-222.702	34.745	96.957	213.554	573.546
		Avalanche	149.982	106.071	-23.466	61.819	134.019	220.390	562.334
Convergence Index	Inventory -A	Non-avalanche	0.186	5.633	-95.105	-1.958	0.003	1.996	93.177
		Avalanche	-1.382	4.365	-88.141	-3.072	-0.908	0.891	67.722
	Inventory -B	Non-avalanche	0.129	5.632	-95.105	-1.995	-0.023	1.961	93.177
		Avalanche	-1.243	4.035	-70.980	-2.986	-0.896	0.900	54.717
Wind Exposition Index	Inventory -A	Non-avalanche	1.057	0.106	0.770	0.983	1.056	1.130	1.342
		Avalanche	1.044	0.083	0.800	0.986	1.045	1.100	1.328
	Inventory -B	Non-avalanche	1.055	0.106	0.770	0.980	1.054	1.128	1.342
		Avalanche	1.062	0.073	0.810	1.011	1.059	1.110	1.328
Mid-Slope Position Index	Inventory -A	Non-avalanche	0.531	0.293	0.000	0.279	0.545	0.794	1.000
		Avalanche	0.535	0.292	0.000	0.284	0.554	0.797	0.999
	Inventory -B	Non-avalanche	0.537	0.294	0.000	0.285	0.554	0.802	1.000
		Avalanche	0.469	0.275	0.000	0.233	0.464	0.701	0.997
Normalized Height	Inventory -A	Non-avalanche	0.475	0.302	0.000	0.199	0.471	0.745	1.000
		Avalanche	0.332	0.254	0.000	0.112	0.276	0.512	0.997
	Inventory -B	Non-avalanche	0.466	0.304	0.000	0.185	0.457	0.737	1.000
		Avalanche	0.392	0.249	0.001	0.181	0.356	0.577	0.997

Table A1. Summary statistics of the numerical factors based on Inventory-A and Inventory-B.

https://doi.org/10.5194/isprs-archives-XLVIII-M-1-2023-57-2023

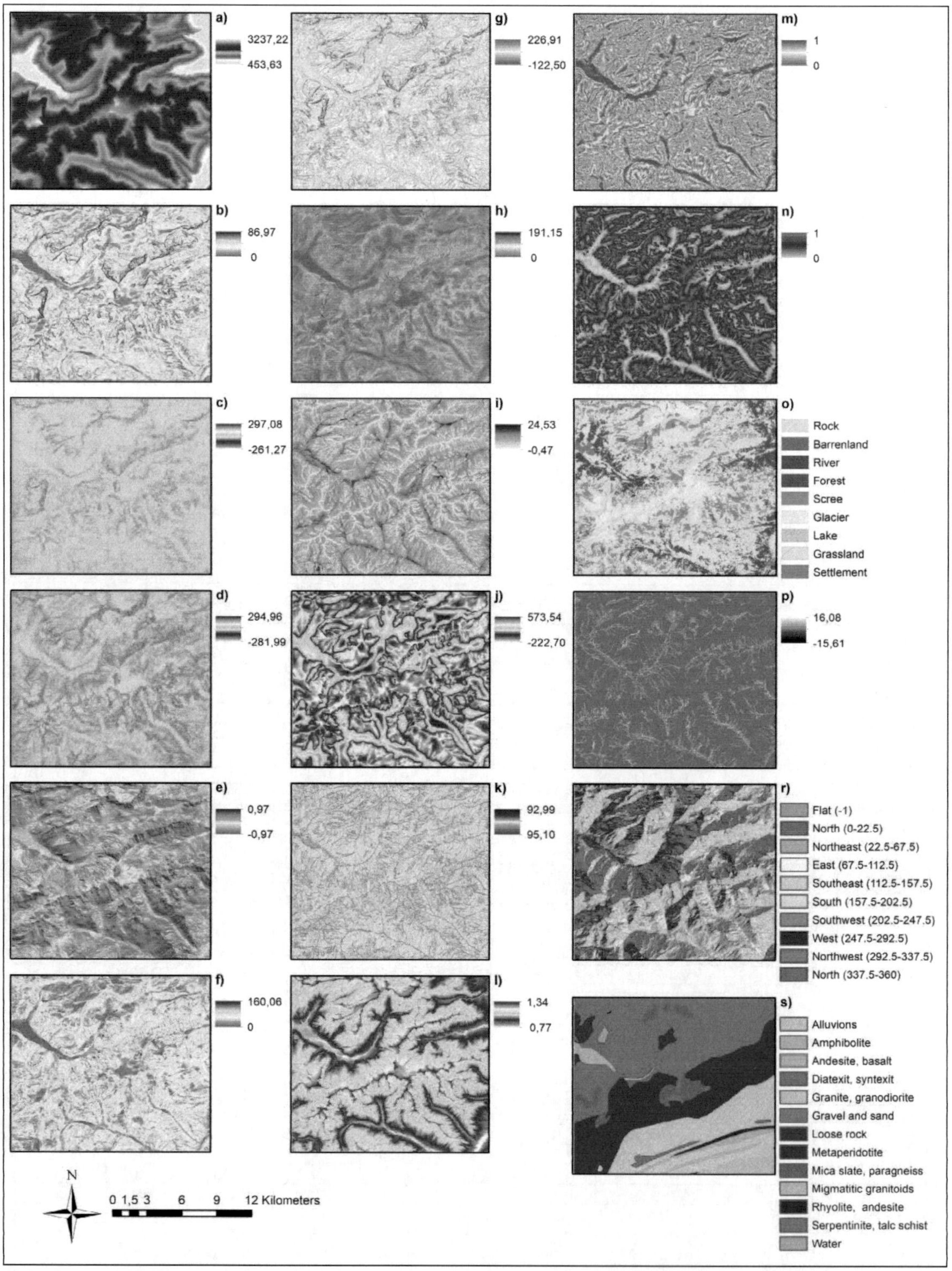

Figure A1. Conditioning factors: a) elevation, b) slope, c) profile curvature, d) plan curvature, e) diurnal anisotropic heating index, f) slope length factor, g) topographic position index, h) topographic ruggedness index, i) topographic wetness index, j) valley depth, k) convergence index, l) wind exposition index, m) mid-slope position index, n) normalized height, o) land use and land cover, p) relative slope position index, r) aspect, s) lithology.

https://doi.org/10.5194/isprs-archives-XLVIII-M-1-2023-57-2023 | © Author(s) 2023. CC BY 4.0 License.

LARGE VECTOR SPATIAL DATA STORAGE AND QUERY PROCESSING USING CLICKHOUSE

Shuaijun Chen [1], Zhibao Wang [1,2 *], Lu Bai [3], Kunyi Liu [1], Juntao Gao [1], Man Zhao [4], Maurice D. Mulvenna [3]

[1] School of Computer and Information Technology, Northeast Petroleum University, Daqing 163318, China
[2] Bohai-Rim Energy Research Institute, Northeast Petroleum University, Qinhuangdao 066004, China
[3] School of Computing, Ulster University, Belfast BT15 1ED, UK
[4] School of Communication and Electronic Engineering, Qiqihaer University, Qiqihaer 161003, China

KEY WORDS: ClickHouse, vector spatial data, query processing, HBase, remote sensing.

ABSTRACT:

The exponential growth of geospatial data resulting from the development of earth observation technology has created significant challenges for traditional relational databases. While NoSQL databases based on distributed file systems can handle massive data storage, they often struggle to cope with real-time query. Column-storage databases, on other hand, are highly effective at both storage and query processing for large-scale datasets. In this paper, we propose a spatial version of ClickHouse that leverages R-Tree indexing to enable efficient storage and real-time analysis of massive remote sensing data. ClickHouse is a column-oriented, open-source database management system designed for handling large-scale datasets. By integrating R-Tree indexing, we have created a highly efficient system for storing and querying geospatial data. To evaluate the performance of our system, we compare it with HBase, a popular distributed, NoSQL database system. Our experimental results show that ClickHouse outperforms HBase in handling spatial data queries, with a response time approximately three times faster than HBase. We attribute this performance gain to the highly efficient R-Tree indexing used in ClickHouse, which allows for fast spatial data query.

1. INTRODUCTION

In recent years, the advancement of Earth observation technology has led to an increasing importance of remote sensing data in various fields such as urban construction, land resource survey, crop disease analysis, smart city, land use and environmental protection (Chi et al., 2016; Kucherov et al., 2017; Li et al., 2020; Shi et al., 2022; Song et al., 2020; Wang et al., 2022; Zhang et al., 2021; Zhu et al., 2021). The exponential increase in remote sensing data volume presents a considerable obstacle to the efficient organization and management of data. Furthermore, it hinders the capacity to analyse and respond promptly to spatial data queries and analyses. Moreover, as a result of the development of global positioning system and mobile intelligent terminals, high accuracy and low latency requirements have emerged for location services including precise positioning, regional query, route planning. In this era of massive spatial data, the key challenges in GIS development is to effectively organize, manage and store large-scale spatial data.

The use of relational databases for spatial data storage dates back to 1970 (Codd, 1970). Today they are widely used for spatial data storage, with popular databases such as Oracle Spatial and PostGIS for PostgreSQL featuring spatial data analysis capabilities. PostGIS, based on PostgreSQL for spatial expansion, has surpassed Oracle Spatial with query speeds that are 300% and 450% faster for spatial data querying and analysis, respectively (Shukla et al., 2016). PostgreSQL provides a simple development path to incorporate new space types, offering numerous features such as reliability, transaction integrity, support for SQL standards, pluggable type extension, and community-oriented development model, along with support for large GIS objects and R-Tree index as a general index structure. PostgreSQL mainly extends the geometry data type for the storage of spatial data, capable of effectively storing point, line, polygon, multipoint, multiline, and multipolygon. Despite its powerful spatial analysis function, traditional relational databases face significant performance bottlenecks. The efficient storage and access of massive data, high scalability, and availability of the database are the major gaps that a relational database cannot resolve. PostgreSQL struggles with decreasing query and processing efficiency as data volume increases. Additionally, deployment and installation of a single node result in limited data storage capacity and low security, and node failures can lead to data loss.

With the continuous growth of data volume and the arrival of the era of big data, data storage has begun to shift from centralized storage to distributed storage on multiple machines. There are three main types of data modes: (1) Tree-structured storage where data is organized in a folder containing file organization. (2) Block storage where data is grouped into blocks of the same size with each block having its own identifier, and (3) Object storage where data is stored in an object unit, and the metadata describing the data is stored in an independent database. In 2003, Google released its own distributed file storage system, Google File System (GFS), which uses a single master and multiple chunkservers to organize data. GFS divides files into chunks, each with a unique 64-bit identifier, and replicates each data block to multiple chunkservers to improve reliability (Ghemawat et al., 2003). Distributed data storage and replication mechanisms are capable of overcoming the limitations of traditional database security and storage scalability. NoSQL databases can be used on distributed clusters, providing strong scalability and storage capabilities, as well as low-latency query services (Wang et al.,

* Corresponding author

https://doi.org/10.5194/isprs-archives-XLVIII-M-1-2023-65-2023

2017). In recent years, there has been extensive exploration of NoSQL database storage for spatial data.

This paper proposes a spatial data storage strategy based on ClickHouse database that utilizes the superior performance of the R-Tree index for spatial data retrieval in traditional relational databases. The R-Tree index is still used as the index of spatial data in ClickHouse. R-Tree is a balanced tree similar to B-Tree but it solves the problem of fast search high-dimensional space, unlike B-Tree, which solves the problem of fast search in one-dimensional space. R-Tree uses the minimum circumscribed rectangle as the boundary of the space geometry, and during the spatial retrieval, it filters out most of the areas that do not overlap with the search rectangle by gradually expanding the boundary (Guttman, 1984). Before storing vector data in ClickHouse, the proposed strategy build the data into an R-Tree, traverses the R-Tree in a hierarchical way during storage, and stores non-leaf nodes in the index table and leaf nodes in the detailed wide table. The data is partitioned based on the root node before storage, so that the non-search areas can be effectively isolated during the query, thereby speeding up the query process.

2. RELATED WORK

In recent years, there has been a surge of interest in using NoSQL databases for managing and storing large-scale datasets. Redis, Elasticsearch, MongoDB and HBase. Compared with these mainstream NoSQL databases, ClickHouse's architecture is highly scalable, fault-tolerant, and distributed, making it an ideal choice for use cases that require high throughput, low latency, and real-time analytics. In this related work section, we provide an overview of the relevant literature that discusses the use of Redis, Elasticsearch, MongoDB, HBase, and ClickHouse, highlighting their advantages and disadvantages in handling Geospatial data.

2.1 Redis

Redis is an open source, memory-based data structure storage system that supports a variety of use cases, including serving as a database, cache, message broker, and streaming engine. It offers support for various data structures such as strings, hashes, lists, sets, sorted sets, bitmaps, hyperloglogs, and geospatial indexs and streams. Redis has built-in replication, Lua scripting, LRU eviction and different levels of persistence. Additionally, it provides high availability through Redis Sentinel and Cluster. When storing geographical locations, Redis uses corresponding key to store the longitude and latitude and the name of the data using Geo Set data structure, which is similar to the Sorted Set data structure. Its implementation leverages Geohash technology to encode the longitude and latitude bits in a staggered manner, and divides the geographical space into grid-shaped buckets with each grid having its own corresponding code (Makris et al., 2019). When querying a spatial range using the georadius method, it transforms the query of two-dimensional space into the comparison of strings of one-dimensional space, which makes it highly efficient for spatial range queries (Liu et al., 2014). (Hao Yu et al., 2012) proposed to use Redis to cache terabytes of geospatial data to cope with highly concurrent queries. However, Redis does not provide data security guarantees and improper design can lead to cache breakdown, cache penetration and cache avalanche.

2.2 Elasticsearch

Elasticsearch is a distributed, real-time search and data analysis engine that is highly scalable. It can perform full-text search, structured search, and analysis, making it a versatile tool for many use cases. Geographical location is represented in Elasticsearch using two different data types: geo_point and geo_shape. The former is used to represent geographical coordinate points with latitude and longitude, while the latter is used for complex geographical shapes using GeoJSON. Geo_points is useful for finding points within a certain range, calculating distances, and aggregating data display on the maps. On the other hand, geo_shape is used to filter the data and analyse whether two shapes intersect, contain or overlap in the geographical space. Elasticsearch uses GeoHash indexing for geographical location queries, ensuring high accuracy. One example of using Elasticsearch is for retrieving geographical location in digital corpus (Bartlett, 2019).

2.3 MongoDB

MongoDB database is a document-based database that organizes data in the form of BJSON. It offers high-performance data persistence, and an API query interface that allows for easy integration with external programs for CRUD operations. MongoDB is highly reliable and offers failover and redundancy mechanisms to prevent data loss as well as horizontal scalability through data partitioning across multiple machines. It supports multiple storage engines, including memory storage engine and WiredTiger storage engine. To efficiently retrieve geospatial coordinates, MongoDB employs two indexing methods: the 2D index for planar geometry and 2dsphere index for spherical geometry. GeoJSON format must be used for organizing geospatial data in MongoDB due to these indexing methods. MongoDB also provides the calculation of geospatial relationships such as geoWithin for inclusion, geoIntersects for intersection, and nearSphere for adjacency. Recent studies have explored different methods of storing remote sensing data in MongoDB. For instance, a remote sensing data management method was proposed based on MongoDB that stores metadata as documents, and image data in GridFS format (Wang et al., 2019) .

2.4 HBase

HBase is a highly scalable NoSQL database that enables the storage of massive amounts of data in a distributed environment. Leveraging the Hadoop Distributed File System (HDFS), HBase uses Namespaces for data space division and Regions for data organization, storing data in a columnar format. Rather than defining specific columns when creating a table, HBase only requires a declaration of the Column Family, providing dynamic field specifications when writing data. This makes Hbase more suitable for scenarios with varying fields, when compared to traditional relational databases. Though HBase does not support geospatial data storage natively, its flexible architecture has led many researchers to explore the use of HBase for this purpose. For example, (Wang et al., 2017) proposed a method of storing vector data in HBase by dividing it into grids using space Z curve filling with the number of space filling curves used as RowKey for storage, and the geometric objects within each grid used as columns (Wang et al.,2017). Similarly, (Wang et al., 2019) proposed an efficient spatial big data storage and query method in HBase, which leverages Hilbert spatial curve filling to convert high-dimensional data into one-dimensional data, allowing for faster query processing.

This contribution has been peer-reviewed.
https://doi.org/10.5194/isprs-archives-XLVIII-M-1-2023-65-2023 | © Author(s) 2023. CC BY 4.0 License.

2.5 ClickHouse

Clickhouse is an open-source column-oriented storage database (DBMS) that specializes in online analytical processing (OLAP) queries, allowing real-time generation of data analysis reports through SQL queries. ClickHouse has the following features: (1) ClickHouse supports various table-level storage engine, with over 20 engines in four categories, namely merge tree, log, interface and others enabling users to choose different storage engines according to specific requirements. (2) ClickHouse has a high throughput capacity and uses a structure similar to LSM tree to achieve sequential write performance, allowing it to write data in a sequential append manner during data import, and merge-sort multiple segments to optimize disk usage. ClickHouse's write performance is superior, with an official open benchmark test showing it can achieve a write throughput of 50MB-200MB/s, equivalent to a write speed of 50W-200W pieces/s (based on an estimated 100 bytes per line). (3) ClickHouse divides the data into multiple partitions and further divides each partition into multiple index granularities, which can then be processed by multiple CPU cores for parallel data processing. This design allows a single query to utilize all the CPU of the entire machine, greatly reduces the query delay. Compared with other databases, ClickHouse has superior performance in various SQL queries, making it an excellent choice for low performance machines and outperforming MySQL which is the most popular open-source database at presently available (Wickramasekara et al., 2020).

3. METHODS

3.1 Storage method of spatial data in ClickHouse

ClickHouse lacks the inherent capability to store geospatial data, but this study builds a logic layer on top of ClickHouse to enable it to do so. This section will focus on the design of this extended logic layer with Figure 1 displaying the comprehensive structure of ClickHouse's storage and retrieval of geospatial data.

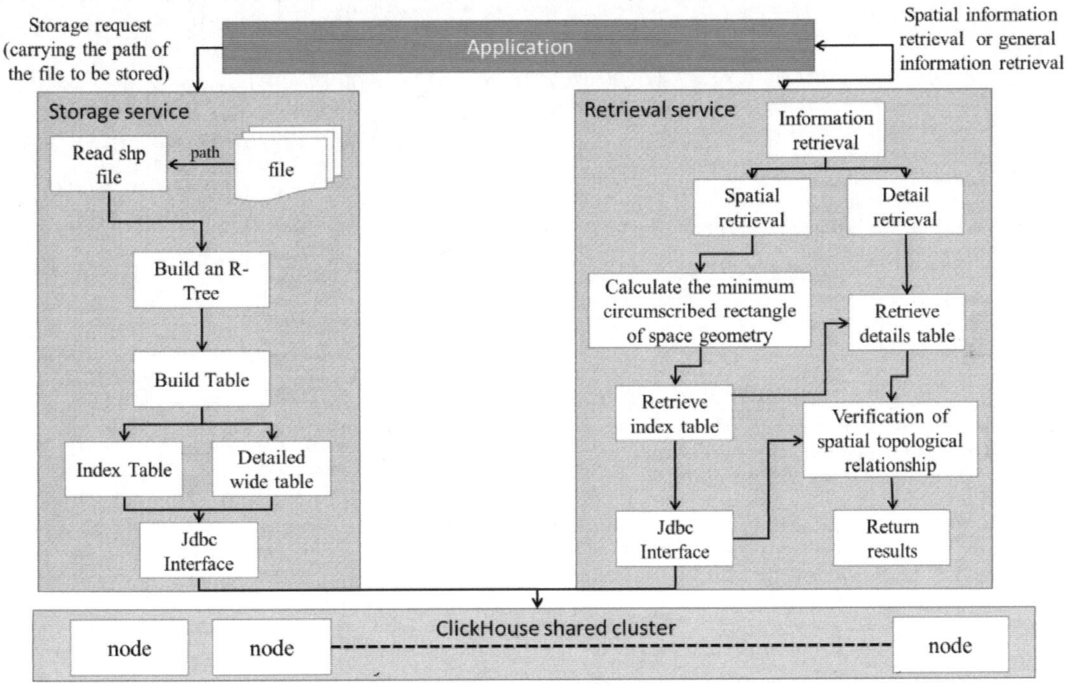

Figure 1. The overall architecture of ClickHouse for storing and retrieving geospatial data.

3.1.1 Storage service: The storage service in the ClickHouse shared cluster for spatial data performs the following operations:

S1: Read the corresponding shapefile file from the corresponding file system based on the path carried in the application request.

S2: Construct an R-Tree using the data from the shapefile file. R-Tree is preferred over the improved R^*-Tree due to its proven ability of R-Tree to retrieve geospatial data has been fully proved in traditional relational databases. The construction time of R^*-Tree is relatively long, and its query efficiency is not optimal when handling small data amounts.

S3: Generate a separate table for each shapefile since they contain unique geographic information. To accelerate geospatial query performance, two tables are created: an index table and a detailed wide table. The index table stores non-leaf node information, while the detailed wide table stores leaf node information in the R-Tree. Table 1 shows the structure of the index table, while Table 2 shows the structure of the detailed wide table.

id	partition_id	level	parent_id	min_x	max_x	min_y	max_y	is_leaf
971074336477675520	0	0	0	563069.6710008697	586413.1604943711	4483095.741016292	4494680.855637121	false

https://doi.org/10.5194/isprs-archives-XLVIII-M-1-2023-65-2023

971074336477675536	0	1	9710743364776755520	563189.8072439035	564977.9116748271	4483095.741016292	4484222.747509402	true
971074336477675537	0	1	9710743364776755520	563069.6710008697	566930.1785916027	4483656.955635705	4484856.83955023	true
971074336477675538	0	1	9710743364776755520	563073.6778494965	568354.4889619449	4484207.59796222	4485157.414893977	true
971074336481869824	0	1	9710743364776755520	563147.7414593172	568336.8224599154	4484451.314492935	4485683.854511977	true
971074336481869825	0	1	9710743364776755520	563875.0872299344	568345.09513854	4484857.718612959	4485934.115992783	true
...	...	...	...	...	...	...	...	...

Table 1. Structure and partial data of index table

id	parent_id	min_x	max_x	min_y	max_y	...	the_geom
971074370996797440	971074336477675536	607240.6270169418	607393.3437808634	4508814.176007433	4508929.788199221	...	MULTIPOLYGON (((607393.3437808634 4508853.954510309, 607263.5223237597 4508814.176007433, 607240.6270169418 4508889.966283751, 607370.2550110875 4508929.788199221, 607393.3437808634 4508853.954510309)))
...	...	...	...	...	...	...	...

Table 2. Structure of detailed wide table and partial data.

ClickHouse's ability to partition data allows for effective isolation of irrelevant data during queries, which is cleverly used when designing the index table. The table is built based on the root node with elements in the root node and the subtree below it stored in the corresponding partition as data of the same partition. The partition number is determined by the left-to-right position of the current element in the root node, while the partition number of the leaf node in the detailed wide table is the ID of the parent node, as shown in Figure 2. The level field records the level of the current node in the entire R-Tree. Since the index data is stored in the database, retrieval needs to be performed recursively with each retrieval taking the ID of the parent node as the filter condition is to isolate unrelated nodes in the same subtree. Fortunately, ClickHouse provides the Prewhere statement, which can filter the data in advance before the selection, and significantly improve efficiency, especially when the number of columns queried is significantly more than that filtered. The minimum bounding rectangle of the current node is defined by the min_x, max_ x, min_y and max_ y values. The is_leaf field identifies whether the current node is the last level of the index node and serves as the exit of the recursive query.

In the detailed wide table, the smallest granularity data of the current shapefile file is stored, with the ID generated by the snowflake algorithm used to identify and sort data. The index of ClickHouse itself is used to speed up retrieval efficiency when querying. The parent_id is used as a partition field, and the data of the same parent node is stored in a partition. The minimum limited rectangle can exclude the data that does not meet the requirements when querying. The_geom field records geometric data, while the remaining fields contain non-spatial attributes of the current data.

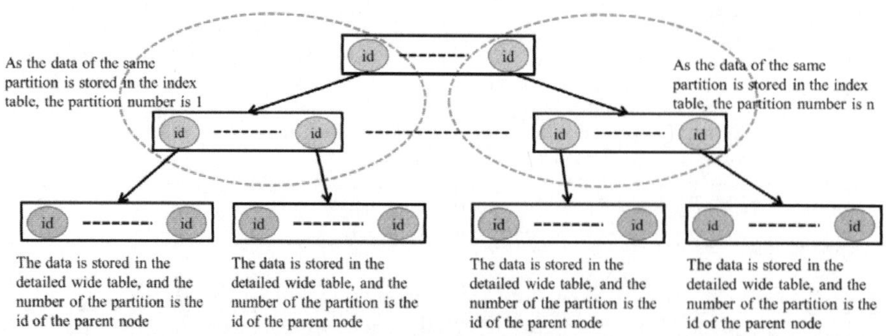

Figure 2. Use R-Tree to build the logic of index table and detail table.

This contribution has been peer-reviewed.
https://doi.org/10.5194/isprs-archives-XLVIII-M-1-2023-65-2023 | © Author(s) 2023. CC BY 4.0 License.

3.1.2 Retrieval service: Data retrieval services can be categorized into geospatial retrieval and conventional retrieval. Geospatial retrieval involves providing spatial geometry information to retrieve data, while conventional retrieval requires general retrieval conditions such as city name, street name, etc. For conventional retrieval, index query is avoided, and the corresponding information is directly queried in the detailed wide table based on the conditions. For example, if the geometric value of a street named "Atlantic commons" is needed, the system will retrieve the relevant information from the detailed wide table and return it to the application. In contrast, geospatial retrieval requires calculating the minimum restricted rectangle of the incoming geometry first. The index table is then retrieved from the root node. When R-Tree is stored in the table, parent_id of the root node is set to 0. Therefore, when searching, the filter condition can be set to parent_id=0 to find the root node. Then, determine the node that intersects with the minimum bounding rectangle of the spatial geometry being queried in the root node. The system then recursively searches the lower level of R-Tree in the index table, taking the id of the parent node as the filter condition at each recursion. When the leaf node of the index is reached, the current node id is used as the filter condition to retrieve the corresponding data in the detailed wide table. During the search in the detailed wide table, the system still uses the intersection of the minimum restricted rectangle as the filter condition. However, the intersection of the minimum restricted rectangle in R-Tree does not necessarily mean that the real geometry is intersected (as shown in Figure 3). Therefore, further judgment is required when the corresponding data is obtained. Finally, when the geometric information is retrieved and the geometric information in the detailed wide table meets the topological requirements of the retrieval, the corresponding results are returned to the upper application.

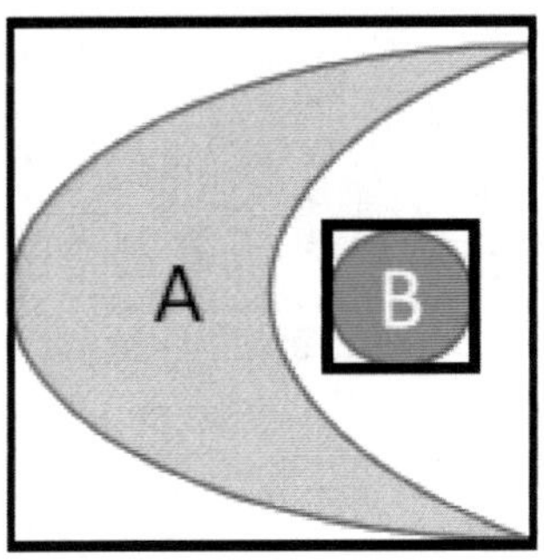

Figure 3. The minimum bounding rectangle of two polygons intersects, but the polygons do not intersect.

3.1.3 ClickHouse Sharded Cluster: ClickHouse provides two mechanisms for distributed data storage: replica and sharded cluster. The replica mechanism ensures the high availability of data by allowing the same data to be obtained from other servers if a ClickHouse node goes down. The write process of the replica mechanism is shown in Figure 4. However, it requires each server to accommodate the full amount of data, limiting horizontal expansion. To overcome this limitation, ClickHouse introduces sharding, which segments a complete piece of data and distributes different shards to different nodes. The Distributed Table Engine acts as a middleware, routing distributed data from multiple nodes with different shards through distributed logical tables. Unlike the replica mechanism, the sharded cluster mechanism allows for horizontal expansion of massive data storage. To ensure both horizontal scalability and high availability, ClickHouse combines sharding and replica mechanisms. In this paper, a sharded cluster architecture is adopted, where three machines are used as the partition cluster. Figure 5 illustrates the design of this architecture. By leveraging sharding and replica mechanisms, ClickHouse provides a powerful solution for distributed data storage and processing.

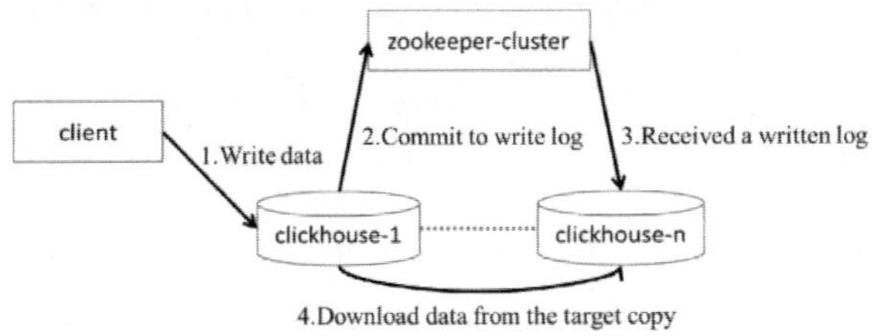

Figure4 . Copy writing process.

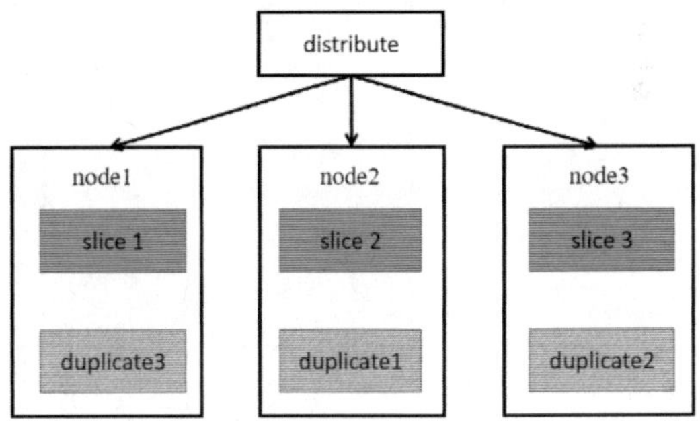

Figure5 . Design architecture of sharded cluster.

3.2 Storage method of spatial data in HBase

To better evaluate the advantages of ClickHouse in querying spatial geographic data, this paper includes a comparative test with Hbase. The experiment expands HBase using R-Tree index while maintaining the independent storage mode for index and detailed wide table to speed up the query. This is achieved by dividing the data into two tables, namely index table and detail table. Table 3 shows the design of the index table while Table 4 shows the design of the detailed wide table.

HBase stores data using a single partition and employs the "parentRowKey_SnowflakesAlgorithm" design rule for Row_Key to ensure its uniqueness. The parameters min_x, max_x, min_y, and max_y represent the smallest circumscribed rectangle, while is_ Leaf identifies whether the current node is a leaf node of the R-Tree index. The Start_Row and Stop_Row parameters specify the range of Row_Key of the child node. The primary purpose of these design features is to accelerate query performance using scan operations. To optimize query performance, HBase stores data for different column families separately. In the design of the detailed wide table, two column families are used: "ordinary" to store non-geometric information and "geometry" to store geometric information. This approach isolates geometric and non-geometric queries, enabling each to be optimized independently. By separating the data into two column families, HBase can accelerate both geometric and non-geometric queries, resulting in faster query performance and better scalability.

https://doi.org/10.5194/isprs-archives-XLVIII-M-1-2023-65-2023 | © Author(s) 2023. CC BY 4.0 License.

	Index_Info						
Row_Key	min_x	max_x	min_y	max_y	is_leaf	Start_Row	Stop_Row
0_973969287091847168	563069.6710008697	586413.1604943711	4483095.741016292	4494680.855637121	false	0973969287091847168_	0973969287091847168|
0_973969287091847169	586214.691145163	593074.0578981118	4488435.770831544	4500914.9872505395	false	0973969287091847169_	0973969287091847169|
0_973969287091847170	567299.8140316624	586522.2945189208	4492040.863678092	4498558.079844725	false	0973969287091847170_	0973969287091847170|
0_973969287091847171	591493.6229326547	598798.5187040224	4489965.561497574	4506856.942924007	false	0973969287091847171_	0973969287091847171|
...	...	...	...	...	...	...	...

Table 3. Design of index table in HBase.

	ordinary	geometry				
rowkey	...	min_x	max_x	min_y	max_y	the_geom
09739692870918471689 73969287091847204_0	...	573351.8780943268	573878.107750165	4487638.46718892	4488463.039496107	MULTIPOLYGON (((572938.2665684235 4486928.674520253, 572933.6947068771 4486936.54728656, 572929.7565177245 4486944.75539434, 572926.4764236218 4486953.247941074, 572923.8747660567 4486961.97226028, 572921.9676791976)))
...	...	...	...	...	...	...

Table 4. Design of Detail Table in HBase.

4. EXPERIMENT AND RESULTS

This section is based on real world data and aims to provide a comprehensive evaluation of the experimental systems. To achieve this, section 4.1 presents the experimental environment used in the study, including details about the hardware and software configurations. Section 4.2 provides a detailed description of the experimental dataset. Finally, in section 4.3, the experimental results are presented and compared to provide insights into the strengths and weaknesses of the systems under test. By using real-world data and a rigorous experimental methodology, this study aims to provide valuable insights into the performance and scalability of the experimental systems, which can be used to inform decisions about their use in real-world applications.

4.1 Experiment setting

The experiments were performed on a virtual machine environment with a 4GB RAM and 2.80GHz Intel(R) Core(TM) i7-7700HQ CPU. ClickHouse was deployed in a single-node storage configuration, while HBase was set up as a three-node cluster with identical hardware and software configurations. Table 5 provides detailed specifications for each of the nodes in the HBase cluster.

CPU type	Intel(R) Core(TM) i7-7700HQ CPU @ 2.80GHz
Processor number	3
RAM	4G
ClickHouse Version	21.7.3.14
Hadoop version	3.1.3
Zookeeper version	3.5.7
HBase version	1.3.1

Table 5. Node configuration for HBase

4.2 Dataset

To evaluate the performance and scalability of the systems under test, a range of real-world datasets were used in the experiments. Specifically, the datasets included community data, census data for New York ("New York Census Data," n.d.), building data, and land data for Papua New Guinea in 2018 ("Papua New Guinea Data," n.d.). The size and characteristics of each dataset are summarized in Table 6, which provides details such as the number of records, file size, and data format. By using real-world datasets, this study aims to provide insights into the systems' ability to handle complex and diverse data types, as well as their performance and scalability under realistic conditions.

Name	Description	Amount of data
nyc_neighborhoods	New York Community Data	129
gis_osm_landuse_a_free_1	Land Data in Papua New Guinea	34700
nyc_census_blocks	New York 2018 Census Data	38700

https://doi.org/10.5194/isprs-archives-XLVIII-M-1-2023-65-2023 | © Author(s) 2023. CC BY 4.0 License.

gis_osm_buildings_a_free_1	Building Data in Papua New Guinea	369254

Table 6. Experimental dataset

4.3 Experiment results

To compare the efficiency of spatial queries between ClickHouse and HBase for different data volumes and query rates, we randomly selected 1% of the data from the tables for cross-checking during the experiments. The results, as shown in Figure 6 indicate that ClickHouse outperforms HBase in terms of query speed, particularly for large amounts of data. When the data volume in a single table data was 129 records, HBase's response time for a data query reached the second level, while ClickHouse returned the query results in milliseconds. When the data volume in a single table increased to 10,000 records, HBase's response time was about three times longer than ClickHouse's. Furthermore, with the increase in the volume of data, HBase's response time increased significantly. These findings demonstrate that ClickHouse is more efficient than HBase is spatial data query, especially for larger datasets.

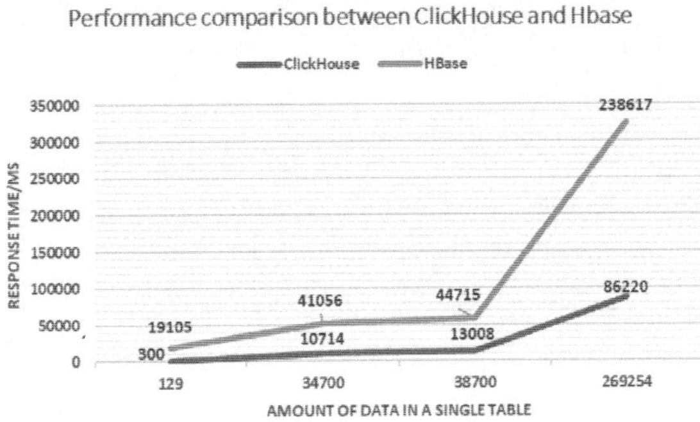

Figure 6. The minimum bounding rectangle of two polygons intersects, but the polygons do not intersect.

5. CONCLUSIONS

This paper proposes a storage model for spatial data using ClickHouse and R-Tree as an index for spatial queries. To evaluate its performance, we compare the geospatial data queries with the HBase database using the same indexes. The results of the experiments demonstrate that ClickHouse outperforms HBase in geospatial data queries, even when using the same indexes.

REFERENCES

Bartlett, R., 2019. Local geographic information storing and querying using Elasticsearch, in: Proceedings of the 13th Workshop on Geographic Information Retrieval. Presented at the GIR'19: 13th Workshop on Geographic Information Retrieval, ACM, Lyon France, pp. 1–4. https://doi.org/10.1145/3371140.3371144

Chi, M., Plaza, A., Benediktsson, J.A., Sun, Z., Shen, J., Zhu, Y., 2016. Big Data for Remote Sensing: Challenges and Opportunities. Proc. IEEE 104, 2207–2219. https://doi.org/10.1109/JPROC.2016.2598228

Codd, E.F., 1970. A relational model of data for large shared data banks. Commun. ACM 13, 377–387. https://doi.org/10.1145/362384.362685

Ghemawat, S., Gobioff, H., Leung, S.-T., 2003. The Google file system, in: Proceedings of the Nineteenth ACM Symposium on Operating Systems Principles. Presented at the SOSP03: ACM Symposium on Operating Systems Principles, ACM, Bolton Landing NY USA, pp. 29–43. https://doi.org/10.1145/945445.945450

Guttman, A., 1984. R-trees: a dynamic index structure for spatial searching. ACM SIGMOD Rec. 14, 47–57. https://doi.org/10.1145/971697.602266

Hao Yu, Yuehu Liu, Chuan Tian, Liang Liu, Mingchao Liu, Yong Gao, 2012. A cache framework for geographical feature store, in: 2012 20th International Conference on Geoinformatics. Presented at the 2012 20th International Conference on Geoinformatics, IEEE, Hong Kong, China, pp. 1–4. https://doi.org/10.1109/Geoinformatics.2012.6270288

Jiajun Liu, Haoran Li, Yong Gao, Hao Yu, Dan Jiang, 2014. A geohash-based index for spatial data management in distributed memory, in: 2014 22nd International Conference on Geoinformatics. Presented at the 2014 22nd International Conference on Geoinformatics, IEEE, Kaohsiung, Taiwan, pp. 1–4. https://doi.org/10.1109/GEOINFORMATICS.2014.6950819

Kucherov, B., Pribyl, O., Artyushenko, V., 2017. Increasing efficiency of getting results of satellite remote sensing for smart cities, in: 2017 Smart City Symposium Prague (SCSP). Presented at the 2017 Smart City Symposium Prague (SCSP), IEEE, Prague, Czech Republic, pp. 1–6. https://doi.org/10.1109/SCSP.2017.7973854

Li, J., Pei, Y., Zhao, S., Xiao, R., Sang, X., Zhang, C., 2020. A Review of Remote Sensing for Environmental Monitoring in China. Remote Sens. 12, 1130. https://doi.org/10.3390/rs12071130

Makris, A., Tserpes, K., Anagnostopoulos, D., Nikolaidou, M., de Macedo, J.A.F., 2019. Database system comparison based on spatiotemporal functionality, in: Proceedings of the 23rd International Database Applications & Engineering Symposium on - IDEAS '19. Presented at the the 23rd International Database Applications & Engineering Symposium, ACM Press, Athens, Greece, pp. 1–7. https://doi.org/10.1145/3331076.3331101

New York Census Data [WWW Document], n.d. URL http://postgis.net/workshops/zh_Hans/postgis-intro/about_data.html

Papua New Guinea Data [WWW Document], n.d. URL http://download.geofabrik.de/australia-oceania/

Shi, K., Bai, L., Wang, Z., Tong, X., Mulvenna, M.D., Bond, R.R., 2022. Photovoltaic Installations Change Detection from Remote Sensing Images Using Deep Learning, in: IGARSS 2022 - 2022 IEEE International Geoscience and Remote Sensing Symposium. Presented at the IGARSS 2022 - 2022 IEEE International Geoscience and Remote Sensing Symposium, IEEE, Kuala Lumpur, Malaysia, pp. 3231–3234. https://doi.org/10.1109/IGARSS46834.2022.9883738

Shukla, D., Shivnani, C., Shah, D., 2016. Comparing oracle spatial and postgres PostGIS. IJCSC 7, 95–100.

This contribution has been peer-reviewed.
https://doi.org/10.5194/isprs-archives-XLVIII-M-1-2023-65-2023 | © Author(s) 2023. CC BY 4.0 License.

Song, G., Wang, Z., Bai, L., Zhang, J., Chen, L., 2020. Detection of oil wells based on faster R-CNN in optical satellite remote sensing images, in: Notarnicola, C., Bovenga, F., Bruzzone, L., Bovolo, F., Benediktsson, J.A., Santi, E., Pierdicca, N. (Eds.), Image and Signal Processing for Remote Sensing XXVI. Presented at the Image and Signal Processing for Remote Sensing XXVI, SPIE, Online Only, United Kingdom, p. 17. https://doi.org/10.1117/12.2572996

Wang, H., Zhu, Y., Wang, J., Han, H., Niu, J., Chen, X., 2022. Modeling of spatial pattern and influencing factors of cultivated land quality in Henan Province based on spatial big data. PLOS ONE 17, e0265613. https://doi.org/10.1371/journal.pone.0265613

Wang, S., Li, G., Yao, X., Zeng, Y., Pang, L., Zhang, L., 2019. A Distributed Storage and Access Approach for Massive Remote Sensing Data in MongoDB. ISPRS Int. J. Geo-Inf. 8, 533. https://doi.org/10.3390/ijgi8120533

Wang, Y., Li, C., Li, M., Liu, Z., 2017. HBase storage schemas for massive spatial vector data. Clust. Comput. 20, 3657–3666. https://doi.org/10.1007/s10586-017-1253-1

Wickramasekara, A., Liyanage, M.P.P., Kumarasinghe, U., 2020. A comparative study between the capabilities of MySQL and ClickHouse in low-performance Linux environment, in: 2020 20th International Conference on Advances in ICT for Emerging Regions (ICTer). Presented at the 2020 20th International Conference on Advances in ICT for Emerging Regions (ICTer), IEEE, Colombo, Sri Lanka, pp. 276–277. https://doi.org/10.1109/ICTer51097.2020.9325483

Zhang, J., Wang, Z., Bai, L., Song, G., Tao, J., Chen, L., 2021. Deforestation Detection Based on U-Net and LSTM in Optical Satellite Remote Sensing Images, in: 2021 IEEE International Geoscience and Remote Sensing Symposium IGARSS. Presented at the IGARSS 2021 - 2021 IEEE International Geoscience and Remote Sensing Symposium, IEEE, Brussels, Belgium, pp. 3753–3756. https://doi.org/10.1109/IGARSS47720.2021.9554689

Zhu, M., Wang, Z., Bai, L., Zhang, J., Tao, J., Chen, L., 2021. Detection of industrial storage tanks at the city-level from optical satellite remote sensing images, in: Bruzzone, L., Bovolo, F., Benediktsson, J.A. (Eds.), Image and Signal Processing for Remote Sensing XXVII. Presented at the Image and Signal Processing for Remote Sensing XXVII, SPIE, Online Only, Spain, p. 33. https://doi.org/10.1117/12.2600008

CANOPY NITROGEN ESTIMATION ON COTTON PLANT USING SATELLITE IMAGERY

B. J. Chew[1] *, W. Wiratama[1], M. H. Goh[1]
[1] ST Engineering Geo-Insights Pte Ltd, Singapore – (chew.boonjin, wahyu.wiratama, minhua.goh)@stengg.com

KEY WORDS: Canopy nitrogen prediction, satellite imagery, remote sensing, Sentinel 2, vegetation indices.

ABSTRACT:

The optimization of nitrogen (N) management is becoming a key challenge to enhance crop yield production while protecting the environment. Analysis of canopy N content in crop plants is used as insights for fertilization management, in which actions can be taken to optimize N fertilizer usage. Traditionally, lab chemical processing is used to measure the crop plant's nutrient content. However, the collection of leaf samples from the field is labour intensive, and it would be costly to increase sampling frequency. Thus, this approach may not be the most optimal for large plantations. Remote sensing applications in agriculture have been widely studied. This study aims to evaluate the potential of using Sentinel 2 imagery to predict canopy N content, as an alternative wide scale method as compared to traditional methods. A cotton plantation with about 50 square km area in the state of Mato Grosso, Brazil, was used as the case study. About 180 samples across the cotton plantation were collected between March and April 2022 and the N contents of the crop plants were measured using lab chemical processes. Sentinel 2 images within 15 days of the sampling dates were retrieved from ESA's Copernicus Open Access Hub. This study proposes a Random Forest (RF) regression algorithm for the generation of an N prediction model. About 52 vegetation indices (VIs) were extracted as the features for model training, such as Normalized Difference Vegetation Index (NDVI) and Enhanced Vegetation Index (EVI). RF model allows easy measurement of the relative importance of each feature with respect to the prediction to achieve a good performance. Validation is done by using mean absolute error (MAE) and mean absolute percentage error (MAPE) to evaluate the prediction accuracy against the ground truth, which resulted to be 3.418 g/kg and 9.29% respectively. Finally, this study analyses the performance of the canopy N prediction model and assesses its ability as an alternative to traditional lab chemical sampling processes.

1. INTRODUCTION

Increasing global demand for agricultural products have resulted in more intensive farming that causes drains nutrients from the soil (Purwanto & Alam, 2019). Application of fertilizer replaces the nutrients for crop uptake which otherwise would have been insufficient. However, over-fertilisation of crop land occurs worldwide and may cause environmental issues due to surface run-off into the natural environment (Ritchie, 2021; Sishodia et al., 2020). Furthermore, application of excess fertiliser contributes to additional costs. This pushes for a need to identify the optimal amount of fertiliser to minimise cost as well as maintaining yield. This optimal amount depends on the plant status and thus the requirement to have an accurate estimate of the plant nutrient status.

Nitrogen is an important element in the plant, and is present in chlorophyll, amino acids and hence protein, nucleic acids, plant tissue, etc (Buchholz, 2022). Traditionally, N level can be obtained via lab chemical processing, however, this process takes time and increases the operational costs (Farella et al., 2022). Furthermore, the sampling results only applies for that set of leaf samples and does not represent the entire crop field. There is a need to consider alternate methods of estimating leaf nutrient that is cheaper and faster.

Satellite-based remote sensing provides an advantage in wide area monitoring and have been used for many different applications in agriculture, such as land use and crop classification, soil health and moisture, and vegetation health (Sishodia et al., 2020). Satellites such as European Space Agency's Sentinel 2 provides global coverage over land once every 5 days and have a wide imaging swath of 290 km (Sentinel-2). Sentinel 2 L2A product has 12 bands ranging from visible to short wave infrared at spectral resolutions ranging from 10m to 60m. It is provided for public access and is atmospherically corrected, hence it is a surface reflectance product. Band combinations – such as band ratios, normalised band differences, or more complex formulae – can be derived for different purposes. There are many indices derived for agricultural purposes, such as Normalised Difference Vegetation Index (NDVI) and Enhanced Vegetation Index (EVI). These vegetation indices (VI) have their own advantages and limitations. One example is the usefulness of NDVI to determine broadly the vigour of a vegetated area, however it faces an oversaturation issue where NDVI value loses sensitivity beyond a vegetation density (Pettorelli et al., 2005).

Machine learning can be used to estimate nitrogen levels on crop leaves using satellite imagery by collecting data, extracting relevant features, training a machine learning model, and validating the model. Marang et al. (2021) proposed hybrid random forest regression, DBSCAN, and PCA to predict N level on cotton crop using hyperspectral UAV and sentinel imagery. Huang et al. (2015) estimated rice nitrogen status based on satellite imagery. R-squared was used to compute relationship between vegetation indices and rice N status. Moreover, Tan et al. (2020) proposed partial least square and remote sensing imagery to predict protein content.

Random forest (RF) regression is one of the common machine learning methods which have been used in estimation of foliar nitrogen levels (Abdel-Rahman et al., 2012; Soltanikazemi et al., 2022). RF is an ensemble method which uses a decision tree as a base estimator. It allows for easy measurement of the relative importance of each feature with respect to the prediction, hence achieving a good performance. In this study, we attempted to use RF regression model to estimate foliar nitrogen concentrations for cotton crops in Brazil. We want to validate the performance of applying RF regression model on selected indices based on Sentinel 2 and determine the limitations of the model.

2. STUDY AREA AND DATA COLLECTION

In this paper, field experiments were conducted from February to April 2022 over a cotton plantation in the state of Mato Grosso, Brazil with about 50 square km area, as shown in Figure 1. Leaf samples were collected during the vegetative and flowering stage of the cotton plants, between 70 to 120 days after emergence. The sampling process involve by first identifying a sampling coordinate (latitude, longitude) within the plantation. Leaf samples are then collected randomly within a 10m radius from the sampling coordinate. This process was iterated across different sampling coordinates and dates. Afterwards, the samples were handled by a professional vendor to measure the nitrogen content using chemical laboratory equipment, carried out with standardized procedure. The measured leaf nitrogen concentration of the leaf samples was then provided as g/kg. In this study, 180 samples were collected and used to generate and validate our model. The statistics of the sampled nitrogen concentration is as shown in Table 1 below.

Figure 1. Cotton Plantation Area

Mean	Standard Deviation	Min	Max
38.3	5.15	23.5	52.6

Table 1. Statistics of Nitrogen samples in g/kg

We collected cloud-free Sentinel-2 L2A data that fall within 15 days from the sample date for each sampling coordinate. L2A products were used for our processing as the atmospheric effects were removed. We then up-sampled all bands to 10m spatial resolution, before extracting the 3x3 context pixels centred on the sampling coordinates. Vegetation indices (VIs) were computed to obtain the percentage of vegetation cover, amount of chlorophyll content, leaf area, and so on (As-Syakur et al. 2012; Brecht 2018; Broge and Leblanc 2001; Chen 1996; Duong et al. 2017; El-Shikha et al. 2008; Frampton et al. 2013; Gitelson et al. 2002; Hiphen-plant 2022; Huang et al. 2012; Main et al. 2011; Metternicht 2003; NDRE index 2023; Pro.arcgis; Rasul et al. 2018; Sentinel Hub; Vincini et al. 2008; Waqar et al. 2012; Xu 2006; Zhao and Chen, 2005). A total of 52 VIs were computed to get handcrafted features, as described in the Appendix. After which, 52 VIs and 12 bands were deployed as the feature inputs to the RF model.

3. METHOD

Figure 2 shows the overall flowchart of the proposed method, including pixel extraction, feature extraction, and RF regression model. Firstly, Sentinel-2 images and sample coordinates were used to retrieve sampled foliar nitrogen values and their respective pixels. The 52 VIs were then computed. Next, the training dataset was generated from the 12 bands and 52 VIs. The best hyperparameters of the RF model were chosen with reference to the training dataset, with the aid of Random Search CV optimization algorithm. Table 2 shows the hyperparameters that were fed into the optimization algorithm. The dataset was then incorporated into the model for training.

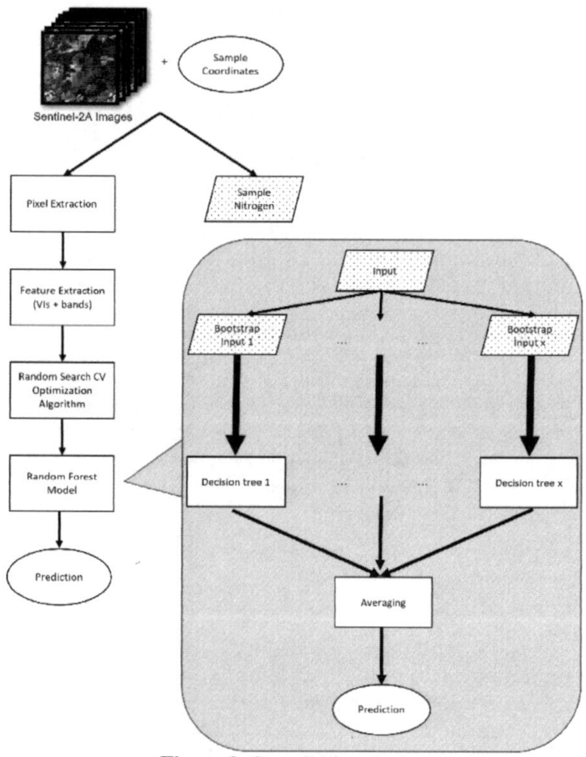

Figure 2. Overall Flowchart

This contribution has been peer-reviewed.
https://doi.org/10.5194/isprs-archives-XLVIII-M-1-2023-73-2023 | © Author(s) 2023. CC BY 4.0 License.

Parameters	Hyperparameters values
Max depth	[10, 20, 30, 40, 50, 60, 70, 80, 90, 100]
Max features	['auto', ''sqrt]
Min sample leaf	[1, 2, 4]
Min sample split	[2, 5, 10]
N estimators	[200, 400, 600, 800, 1000, 1200, 1400, 1600, 1800, 2000]

Table 2. Settings for Hyperparameter Values as Input to the Random Search CV Algorithm

4. PERFORMANCE EVALUATION

Figure 3 shows the histogram of the dataset where we observe most N values ranged between 33 and 47 g/kg, therefore we assumed that values beyond this range were extreme values. In the experiment, we split the dataset randomly into 80% training and 20% testing with respect to the observed value.

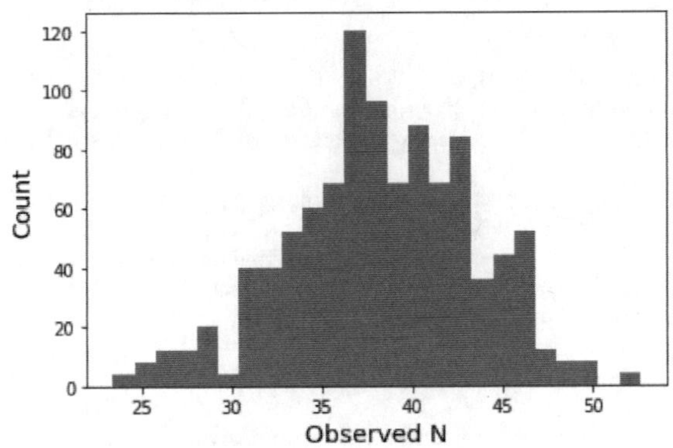

Figure 3. Histogram of Dataset

The optimized RF model was trained using the training dataset, generating the feature importance scores as shown in Figure 4 below. It was shown that the top 15 features were the most important on the N prediction, namely Band 11, Band 1, IRECI, Band 5, MNDWI, BRBA, CCCI_ALT, Band 12, Band 9, NDVI3, MCARI_ALT, BUI, Band 2, Band 3, and NPCI.

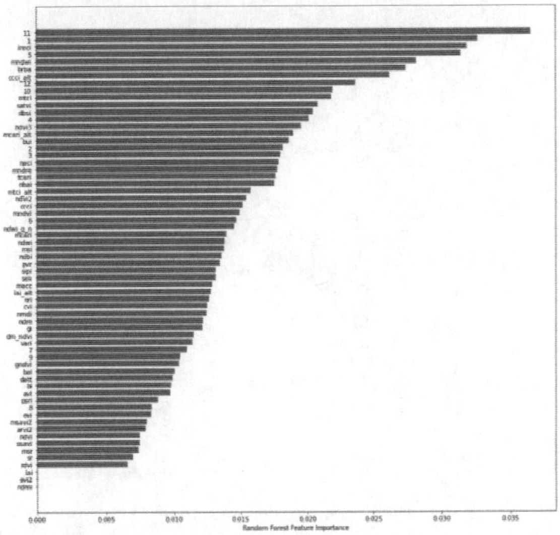

Figure 4. Random Forest Feature Importance

The model was further evaluated using the mean absolute error (MAE) and mean absolute percentage error (MAPE). MAE is a common metric used to evaluate the performance of prediction model by measuring the average absolute difference between the observed and predicted values. MAPE is similar to MAE, except the average absolute difference is being divided by the observed value before summation. The formulas of MAE and MAPE are as shown below:

$$MAE = \frac{1}{n}\sum_{i=1}^{n} |y_{actual} - y_{predicted}| \quad (1)$$

$$MAPE = \frac{100}{n}\sum_{i=1}^{n} \frac{|y_{actual} - y_{predicted}|}{y_{predicted}} \quad (2)$$

where n = number of samples
y_{actual} = observed value
$y_{predicted}$ = predicted value

Figure 5 below depicts the prediction plot using training and testing data that gives a fit prediction for training data. According to our experiment, this study presents MAE of 3.418 g/kg, which is MAPE of 9.29% for the testing dataset.

We observe that the model is able to achieve a better performance for N prediction where the values fall within the middle range highlighted above. Poorer performance for the extreme values could be attributed to lack of training data in that range, and that training with an imbalanced dataset resulted in higher errors. In addition, it is believed that the model loses sensitivity at the extreme ends of the distribution and the dataset is still lacking features with high correlation to N. Increase in variance of N distribution and the size of the dataset can be an alternative solution in the future to overcome the overfitting issue. Moreover, hyperspectral (HS) images can be an option to increase the number of features available for training since they are more sensitive than S2A images, which are multispectral images.

https://doi.org/10.5194/isprs-archives-XLVIII-M-1-2023-73-2023

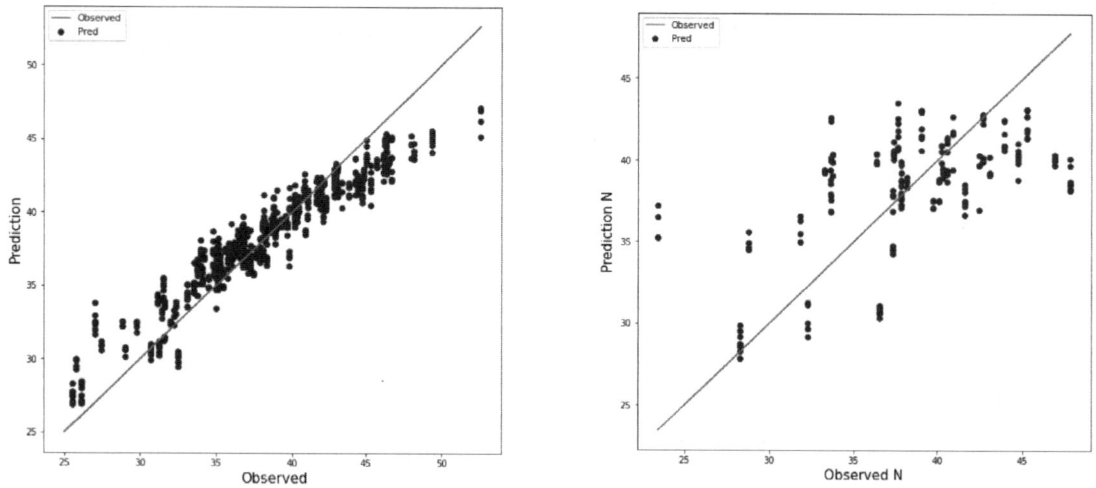

a). Prediction Plot using Training Data

b). Prediction Plot using Testing Data

Figure 5. Prediction Plot a). using Training Data, b). using Testing Data

5. CONCLUSION

This study proposed the use of Sentinel-2 imagery and machine learning method, specifically RF model, to predict the amount of N present in the canopy of the cotton crops, in hopes of replacing the traditional method, with the end goal to save cost and time. RF model was chosen since it could calculate which feature was important to the prediction to achieve a high accuracy, with the top 15 features being the most important on the N prediction to be Band 11, CSI, Band 1, IRECI, Band 5, MNDWI, BRBA, CCCI_ALT, Band 12, Band 9, NDVI3, MCARI_ALT, BUI, Band 2, Band 3. The validation metric used was MAE which resulted to be 3.418 g/kg, which was MAPE of 9.29% for the testing dataset. While MAE is considerably low, more could be done since MAE was mostly contributed by the inaccurate prediction of extreme N values. Improvements such as using hyperspectral imagery instead of Sentinel-2 imagery and using different VIs of different feature importance based on different methods could be done. Even though VIs among different methods could be of different feature importance, stacking them with the consideration of different methods could prevent overfitting and should be used to monitor agricultural fields to improve classification accuracy. In addition, boosting the dataset with more data, especially in the extreme range, would help to resolve the imbalanced dataset issue.

REFERENCES

Abdel-Rahman, E. M., Ahmed, F. B., Ismail, R. (2012). Random Forest regression and spectral band selection for estimating sugarcane leaf nitrogen concentration using EO-1 hyperion Hyperspectral Data. International Journal of Remote Sensing, 34(2), 712–728. https://doi.org/10.1080/01431161.2012.713142

As-Syakur, A.R., Adnyana, I.W.S., Arthana, I.W. and Nuarsa, I.W., 2012. Enhanced built-up and bareness index (EBBI) for mapping built-up and bare land in an urban area. Remote sensing, 4(10), pp.2957-2970. doi.org/10.3390/rs4102957

Brecht, 2018, Remote sensing indices, https://medium.com/regen-network/remote-sensing-indices-389153e3d947, accessed by 19 January 2023.

Broge, N.H., Leblanc, E., 2001, Comparing prediction power and stability of broadband and hyperspectral vegetation indices for estimation of green leaf area index and canopy chlorophyll density. Remote Sensing of Environment. 76. 2. 156-172. doi.org/10.1016/S0034-4257(00)00197-8

Buchholz, D. 2022, March. Nitrogen in the plant. University of Missouri Extension. https://extension.missouri.edu/publications/wq259, accessed by 20 January 2023.

Chen, J.M., 1996. Evaluation of vegetation indices and a modified simple ratio for boreal applications. Canadian Journal of Remote Sensing, 22(3), pp.229-242. doi.org/10.1080/07038992.1996.10855178

Duong, T., Chou, T., Fang, Y., 2017. Integration of GIS and Remote Sensing for Evaluating Forest Canopy Density Index in Thai Nguyen Province, Vietnam. International Journal of Environmental Science and Development. 8. 539-542. doi.org/10.18178/ijesd.2017.8.8.1012.

El-Shikha, D. M, Barnes, E. M, Clarke, T. R, Hunsaker, D. J, Haberland, J. A, Pinter, P. J, Waller, P. M, & Thompson, T. L., 2008. Remote Sensing of Cotton Nitrogen Status Using the Canopy Chlorophyll Content Index (CCCI). Transactions of the ASABE, 51, 73-82. doi.org/10.13031/2013.24228

Farella, M. M., Barnes, M. L., Breshears, D. D., Mitchell, J., van Leeuwen, W. J., & Gallery, R. E., 2022. Evaluation of vegetation indices and imaging spectroscopy to estimate foliar nitrogen across disparate biomes. Ecosphere, 13(3). https://doi.org/10.1002/ecs2.3992

Frampton, W., Dash, J., Watmough, R.G., Milton, E.J., 2013. Evaluating the capabilities of Sentinel-2 for quantitative estimation of biophysical variables in vegetation. ISPRS Journal of Photogrammetry and Remote Sensing. 82. 83–92. doi.org/10.1016/j.isprsjprs.2013.04.007.

Gitelson, A., Kaufman, Y., Stark, R., Rundquist, D., 2002. Novel Algorithms for Remote Estimation of Vegetation Fraction. Remote Sensing of Environment. 80. 76-87. doi.org/10.1016/S0034-4257(01)00289-9.

Hiphen-plant, 2022, Vegetation Indices for Chlorophyll, www.hiphen-plant.com, accessed by 20 January 2023.

Huang, W., Luo, J., Zhang, J., Zhao, J., Zhao, C., Wang, J., Yang, G., Huang, M., Huang, L. and Du, S., 2012. Crop disease and pest monitoring by remote sensing. In Remote Sensing-Applications. IntechOpen. doi.org/10.5772/35204

Huang, S., Miao, Y., Zhao, G., Yuan, F., Ma, X., Tan, C., Yu, W., Gnyp, M.L., Lenz-Wiedemann, V.I., Rascher, U. and Bareth, G., 2015. Satellite remote sensing-based in-season diagnosis of rice nitrogen status in Northeast China. Remote sensing, 7(8), pp.10646-10667. doi.org/10.3390/rs70810646

Main, R., Cho, M.A., Mathieu, R., O'Kennedy, M.M., Ramoelo, A., Koch, S., 2011. An investigation into robust spectral indices for leaf chlorophyll estimation. ISPRS Journal of Photogrammetry and Remote Sensing. 66. 751-761. doi.org/10.1016/j.isprsjprs.2011.08.001

Marang, I.J., Filippi, P., Weaver, T.B., Evans, B.J., Whelan, B.M., Bishop, T.F., Murad, M.O., Al-Shammari, D. and Roth, G., 2021. Machine learning optimised hyperspectral remote sensing retrieves cotton nitrogen status. Remote Sensing, 13(8), p.1428. doi.org/10.3390/rs13081428

Metternicht, G., 2003. Vegetation indices derived from high-resolution airborne videography for precision crop management. Int. J. Remote Sens. 24. doi.org/10.1080/01431160210163074.

NDRE index: Vegetation Analysis in mid and late season. EOS Data Analytics., 2023. https://eos.com/industries/agriculture/ndre/

Pettorelli, N., Vik, J. O., Mysterud, A., Gaillard, J.-M., Tucker, C. J., & Stenseth, N. C., 2005. Using the satellite-derived NDVI to assess ecological responses to environmental change. Trends in Ecology, Evolution, 20(9), 503–510. https://doi.org/10.1016/j.tree.2005.05.011

Pro.arcgis, Indices gallery, https://pro.arcgis.com/en/pro-app/help/data/imagery/indices-gallery.htm, accessed by 20 January 2023.

Purwanto, B. H., & Alam, S., 2019. Impact of intensive agricultural management on carbon and nitrogen dynamics in the Humid Tropics. Soil Science and Plant Nutrition, 66(1), 50–59. https://doi.org/10.1080/00380768.2019.1705182

Rasul, A., Balzter, H., Ibrahim, G.R.F., Hameed, H.M., Wheeler, J., Adamu, B., Ibrahim, S.A. and Najmaddin, P.M., 2018. Applying built-up and bare-soil indices from Landsat 8 to cities in dry climates. Land, 7(3), p.81. doi.org/10.3390/land7030081

Ritchie, H., 2021. Excess fertilizer use: Which countries cause environmental damage by overapplying fertilizers? Our World in Data. Retrieved February 24, 2023, from https://ourworldindata.org/excess-fertilizer

Sentinel Hub, Collection of custom script, https://custom-scripts.sentinel-hub.com/, accessed by 20 January 2023.

Sentinel-2, ESA, https://www.esa.int/Applications/Observing_the_Earth/Copernicus/Sentinel-2, accessed by 21 January 2023

Sishodia, R. P., Ray, R. L., & Singh, S. K., 2020. Applications of remote sensing in Precision Agriculture: A Review. Remote Sensing, 12(19), 3136. https://doi.org/10.3390/rs12193136

Soltanikazemi, M., Minaei, S., Shafizadeh-Moghadam, H., & Mahdavian, A., 2022. Field-scale estimation of sugarcane leaf nitrogen content using vegetation indices and spectral bands of sentinel-2: Application of random forest and support vector regression. Computers and Electronics in Agriculture, 200, 107130. https://doi.org/10.1016/j.compag.2022.107130

Tan, C., Zhou, X., Zhang, P., Wang, Z., Wang, D., Guo, W. and Yun, F., 2020. Predicting grain protein content of field-grown winter wheat with satellite images and partial least square algorithm. PLoS One, 15(3), p.e0228500. doi.org/10.1371/journal.pone.0228500

Vincini, M., Frazzi E., D'Alessio, P., 2008: A broad-band leaf chlorophyll vegetation index at the canopy scale. Precision Agriculture, 9, 303-319. doi.org/10.1007/s11119-008-9075-z

Waqar, M.M., Mirza, J.F., Mumtaz, R. and Hussain, E., 2012. Development of new indices for extraction of built-up area & bare soil from landsat data. Open Access Sci. Rep, 1(1), p.4.

Xu, H., 2006. Modification of Normalized Difference Water Index (NDWI) to Enhance Open Water Features in Remotely Sensed Imagery. International Journal of Remote Sensing. 27. 3025–3033. doi.org/10.1080/01431160600589179.

Zhao, H. and Chen, X., 2005, July. Use of normalized difference bareness index in quickly mapping bare areas from TM/ETM+. In International geoscience and remote sensing symposium (Vol. 3, p. 1666). doi.org/ 10.1109/IGARSS.2005.1526319

APPENDIX

Sentinel-2 L2A bands	Description	Notations
Band 1	Coastal or Aerosol	Co
Band 2	Blue	B
Band 3	Green	G
Band 4	Red	R
Band 5	Vegetation red edge	RE1
Band 6	Vegetation red edge	RE2
Band 7	Vegetation red edge	RE3
Band 8	Near Infrared	NIR
Band 9	Narrow NIR	NNIR
Band 10	Water vapour	WV
Band 11	SWIR	S1
Band 12	SWIR	S2

Table 3. Sentinel-2 L2A bands description

No	Indices	Full name	Formula Based on S2 bands
1	NDVI	Normalised Difference Vegetation Index	$\frac{NNIR - R}{NNIR + R}$
2	MSAVI 2	Modified Soil Adjusted Vegetation Index	$\frac{2\times NNIR+1-\sqrt{(2\times NNIR+1)^2+8(NNIR-R)}}{2}$
3	VARI	Visible Atmospherically Resistant Index	$\frac{G - R}{G + R - B}$
4	MNDWI	Modified Normalized Difference Water Index	$\frac{G - S1}{G + S1}$
5	NDMI	Normalized Difference	$\frac{NNIR - S1}{NNIR + S1}$
6	BAI	Burn Area Index	$\frac{1}{(0.1-R)^2+(0.06-NNIR)^2}$
7	NDBI	Normalized Difference Built-up Index	$\frac{S1 - NNIR}{S1 + NNIR}$
8	EVI	Enhanced Vegetation Index	$2.5 \times \frac{NNIR-R}{NNIR+(6\times R-7.2\times B)+1}$
9	EVI2	Enhanced Vegetation Index	$2.5 \times \frac{NNIR - R}{NNIR + 2.4 \times R + 1}$
10	OSAVI	Optimised Soil Adjusted Vegetation Index	$\frac{NNIR - R}{NNIR + R + 0.16}$
11	SATVI	Soil Adjusted Total Vegetation Index	$\frac{S1 - R}{S1 + R + L} \times (1 + L) + \frac{S2}{2}$
12	BSI	Bare Soil Index	$\frac{R + B - G}{R + B + G}$
13	NDWI	Normalized Difference Water Index	$\frac{NNIR - S1}{NNIR + S1}$
14	SIPI	Structure Insensitive Pigment Index	$\frac{NNIR - CO}{NNIR + R}$
15	MSI	Moisture Stress Index	$\frac{S1}{NNIR}$
16	GNDVI	Green Normalized Difference Vegetation Index	$\frac{NNIR - G}{NNIR + G}$
17	BI	Bare Soil Index	$\frac{(S1 + R) - (NNIR + B)}{(S1 + R) + (NNIR + B)}$
18	DBSI	Dry Bare Soil Index	$\frac{S1 - G}{S1 + G} - NDVI$
19	NBAI	Normalised Built-Up Area Index	$\frac{S2 - S1/G}{S2 + S1/G}$
20	BRBA	Band Ratio for Built-Up Area	$\frac{R}{S1}$

No	Indices	Full name	Formula Based on S2 bands
21	GI	Greenness Index	$\frac{G}{R}$
22	MSR	Modified Simple Ratio	$\frac{SR - 1}{\sqrt{SR - 1}}$
23	SR	Simple Ratio	$NNIR/R$
24	RDVI	Renormalised Difference Vegetation Index	$\frac{NNIR - R}{\sqrt{NNIR + R}}$
25	NRI	Nitrogen Reflectance Index	$\frac{G - R}{G + R}$
26	TCARI	Transformed Chlorophyll Absorption and Reflectance Index	$3\left[(RE1 - R) - 0.2(RE1 - G)\left(\frac{RE1}{R}\right)\right]$
27	PSRI	Plant Senescence Reflectance	$\frac{R - B}{RE2}$
28	NPCI	Normalised Pigment Chlorophyll ratio Index	$\frac{R - Co}{R + Co}$
29	MCARI_alt	Modified Chlorophyll Absorption in Reflectance Index (alternate)	$((RE1 - R) - 0.2 \times (RE1 - G) * (\frac{RE1}{R}))$
30	MTCI	Merris Terrestrial Chlorophyll Index	$\frac{(NNIR - RE2)}{(RE2 + R)}$
31	MCARI	Modified Chlorophyll Absorption in Reflectance Index	$\frac{[(NNIR - RE1) - 0.2(NNIR - G)]}{RE1}$
32	CCCI_alt	Canopy Chlorophyll Content (alternate)	$\frac{\left(\frac{NNIR - RE2}{NNIR + RE2}\right)}{\left(\frac{NNIR - R}{NNIR + R}\right)}$
33	CVI	Chlorophyll Vegetation Index	$NNIR \times \frac{R}{G^2}$
34	AVI	Advanced Vegetation Index	$\sqrt[3]{(NNIR \times (1 - R)) \times (NNIR - R)}$
35	PVR	Photosynthetic Vigour Ratio	$\frac{G - R}{G + R}$
36	IRECI	Inverted Red Edge Chlorophyll Index	$\frac{\rho \times NNIR - \rho \times R}{\frac{\rho\times RE2}{\rho\times RE1}}$
37	ARVI2	Atmospherically Resistant Vegetation Index 2	$-0.18 + 0.17\left(\frac{NNIR - R}{NNIR + R}\right)$
38	mNDVI	Modified NDVI	$\frac{(NNIR - R)}{NNIR - R + 2 \times CO}$

This contribution has been peer-reviewed.
https://doi.org/10.5194/isprs-archives-XLVIII-M-1-2023-73-2023

No	Indices	Full name	Formula Based on S2 bands
39	MNDRE	Modified Normalised Difference Red-edge	$\frac{(RE2-RE1)}{RE2-RE1+2\times CO}$
40	MACC	Maccioni	$\frac{(RE3-RE1)}{RE3+R}$
41	Datt	Datt	$\frac{(NNIR-RE1)}{(NNIR+R)}$
42	NDVI2	NDVI2	$\frac{(RE2-RE1)}{(RE2+RE1)}$
43	NDVI3	NDVI3	$\frac{(R-G)}{(R+G)}$
44	MTCI_at	Meris Terrestrial Chlorophyll Index	$\frac{(RE2-RE1)}{RE1-R}$
45	BUI	Built-up Index	$\left(\frac{S1-NNIR}{S1+NNIR}\right)-\left(\frac{NNIR-R}{NNIR+R}\right)$
46	NDWI_G_N	Normalised Difference Water Index (using G and N bands)	$\frac{G-NNIR}{G+NNIR}$
47	NMDI	Normalised Multiband Drought Index	$\frac{(NNIR-(S1-S2))}{(NNIR+(S1-S2))}$
48	NDRE	Normalised Difference Red Edge	$\frac{NNIR-RE1}{NNIR+RE1}$
49	CCCI	Canopy Content Chlorophyll Index	$\frac{(NNIR-RE2)/(NNIR+RE2)}{(NNIR-R)/(NNIR+R)}$
50	SELI	Sentinel Estimated Leaf-Area Index	$\frac{NNIR-RE1}{NNIR+RE1}$
51	LAI_alt	Leaf-Area Index (alternate)	SELI*5.405-0.114
52	DM_NDVI	Dry Mass using NDVI	NDVI*12.3245-5.70

Table 4. Index description

https://doi.org/10.5194/isprs-archives-XLVIII-M-1-2023-73-2023

The International Archives of the Photogrammetry, Remote Sensing and Spatial Information Sciences, Volume XLVIII-M-1-2023
39th International Symposium on Remote Sensing of Environment (ISRSE-39) "From Human Needs to SDGs", 24–28 April 2023, Antalya, Türkiye

INSAR-BASED INVESTIGATION OF SURFACE DISPLACEMENT OVER THE AREA OF THE COLLAPSED BUILDINGS IN ISTANBUL

Nusret Demir[1,2], Ufuk Yazgan[3]

1-Akdeniz University, Faculty of Science, Department of Space Science and Technologies, 07058, Antalya, Türkiye, nusretdemir@akdeniz.edu.tr

2-Akdeniz University, Institute of Natural and Applied Sciences , Chair of Remote Sensing and GIS, 07058, Antalya, Türkiye

3- Istanbul Technical University, Disaster Management Institute, 34469, Maslak/Istanbul, Türkiye, ufukyazgan@itu.edu.tr

Commission I, WG I/5

KEY WORDS: Interferometry, InSAR, Sentinel 1, Radar, SAR, Buildings

ABSTRACT:

In this research, we investigate the surface displacements in the regions where the buildings collapsed. Istanbul has a surface that is active for earthquakes. As the construction regulations have been changed after the year 2000, it is expected that the buildings built after this year, are much safer due to the strict retime series. But the majority of the buildings are at high risk since they are constructed before the year 2000. Therefore suddenly collapsing buildings are an often case in Istanbul, and we have investigated six buildings that collapsed during the six-year time. The used data are Sentinel 1 Radar images. The methodology is SBAS provided by the COMET-LiCS portal. The study area is Istanbul-Türkiye. six buildings are selected for the investigation. The velocity of the displacements and the time series of the whole displacements on the building locations have been plotted. The neighbouring areas of the collapsed buildings are also investigated and analysed if there are any surface displacement changes between them.

1. INTRODUCTION

Istanbul is a city where the surface is dynamic and active for earthquakes. The urbanization rate is also high, 14% according to the Turkish Statistical Office (Web-1,2022). The regulations of the construction were changed and its enforcement was improved at the beginning of 2000 which resulted in the building being safer than before. At the moment, the majority of the building stock in Istanbul was constructed before 2000 and they possess a significant risk.
In this study, six reported that collapsed buildings are selected for the analysis. InSAR technology provides long-time series displacement measurements with the use of Phase information of the radar datasets.
Sentinel 1 Data are useful to identify the displacement of the surface. SBAS is one of the efficient techniques which use the long time series of the data, with analysis of the multi-interferograms. One of the methods which is called LICSAR and it uses SBAS and pre-produced interferograms are provided on the COMET-LiCS portal (Lazecký et al.,2020). The SBAS method uses interferograms generated from SAR pairs with short base distances that maintain high interferometric consistency. For each pixel, SBAS generates a linear surface displacement ratio map calculated by fitting a straight line to the deformation time series for each binary pair, along with the temporal resolution of the SAR gains and the time series of ground deformation (Berardino et al. 2002).
In this study, we investigate to use of Sentinel 1 data and SBAS analysis for the determination of the surface displacement over the collapsed buildings.

2. STUDY AREA

The study area covers six buildings that collapsed on different dates in different quarters of Istanbul as shown in Figure 1.

Figure 1. The location of the selected collapsed buildings.

The locations and the collapsing dates are shown in Table 1.

Building ID	Lat °	Lon °	Collapsing date
1	28.79782	41.06517	24.08.2021
2	28.85857	41.03716	22.09.2016
3	28.83356	41.01656	21.05.2016
5	28.8983	40.98919	06.05.2021
6	28.9939	41.1768	27.06.2021
7	29.01472	41.02036	16.09.2015

Table 1. The list of collapsed buildings

3. METHOD

For the implementation of the SBAS approach, a variety of tools are available, including GMTSAR (Sandwell et al., 2016), ISCE (https://github.com/isce-framework/), and Mintpy (Yunjun et al., 2019). The LICSAR (Lazeck et al., 2020) package was used in this work. To perform InSAR time series analysis using LiCSAR products (i.e., unwrapped interferograms and coherence), which are freely available on the COMET-LiCS web portal (https://comet.nerc.ac.uk/comet-lics-portal/), use the open-source LiCSBAS package in Python and Bash. The resulting interferograms are published on the COMET-LiCS portal in GeoTIFF format. The relevant data has been processed in a basic manner, and the subsidence maps for the surfaces where each collapsed structure is constructed. All studies are performed with the chosen orbit ascending. We used the Google Colab environment (https://colab.research.google.com) to apply the LICSAR package's stages.

4. RESULTS

The surface displacement has been calculated for the years 2015, 2016, 2017, and 2021 from January to the end of December, covering the dates of each building's collapse. Figure 2 depicts, as an example, the displacement map between January 1, 2016, and January 1, 2017. It was produced in ascending mode.

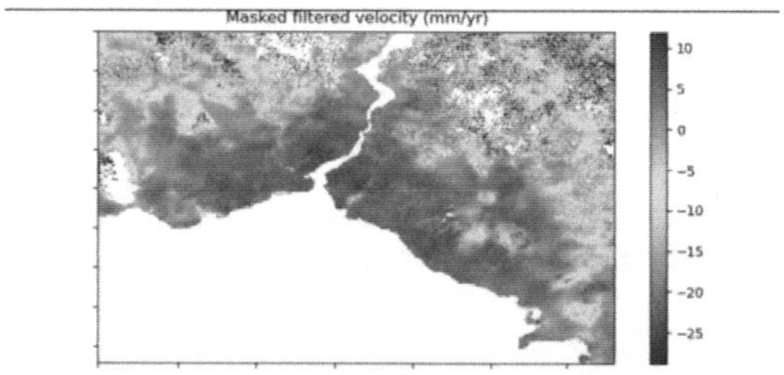

Figure 2. Annual displacement velocity calculated in Ascending mode for Istanbul City (mm/year).

4.1 Analysed Buildings

4.1.1 Building 1

Due to the failure of the retaining wall of the building on the side, the one-story furniture store at Masko Mobilyacilar Sitesi in Ikitelli, Istanbul, collapsed (Web-2). The incident took place on August 23, 2021.

Figure 3 shows the yearly displacement time series. Figure 4 is a photo which shows the collapsing.

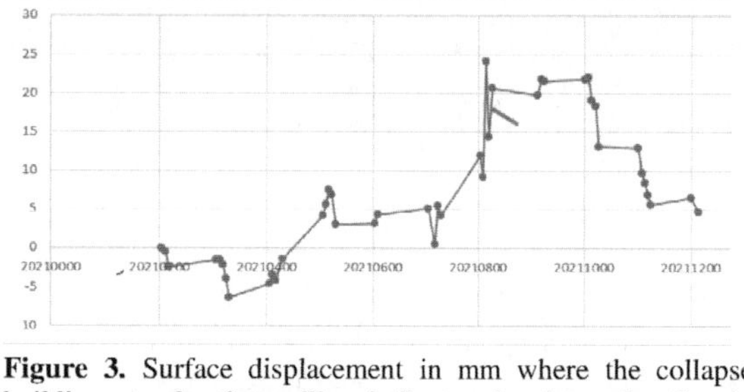

Figure 3. Surface displacement in mm where the collapsed building stands, the redline indicates the date of collapsing 24.08.2021

Figure 4. A picture of te collapsed building (source: DHA, from evrensel.net)

A maximum uplift is depicted in Figure 3 closer to the event's collapse date. The structure collapses shortly after the rising achieves its peak value.

4.1.2 Building 2

Here, we analysed one building which had been collapsed on September 22, 2016 in Bagcilar-Istanbul. The building was a 5-story reinforced concrete frame building (Figure 5a). Dimensions of the building plan area were 21×15 m^2. It was built sometime within the time period from 1980 to 2000 during which many structurally deficient buildings were constructed in Istanbul due to lack of proper seismic design code enforcement. Failure mechanism that was observed after the collapse indicated that building had weak columns and stiff beams which is a typical structural deficiency in substandard buildings (Figure 5b). Moreover, it is seen from the post-collapse photo that the beams along the southern axes of the building spanned in between the 3 columns. As a result, the beams had lengths in the order of 8~9 m which indicates that the axial load ratios in the columns were quite high. Besides the building also had an wedding hall in the basement. Such large open spaces also has the potential for rendering the building unsafe when the material quality is poor and the workmanship is inadequate.

Figure 5. Investigated building: (a) before collapse and, (b) after collapse (Modified from Google, 2022; IHA, 2016)

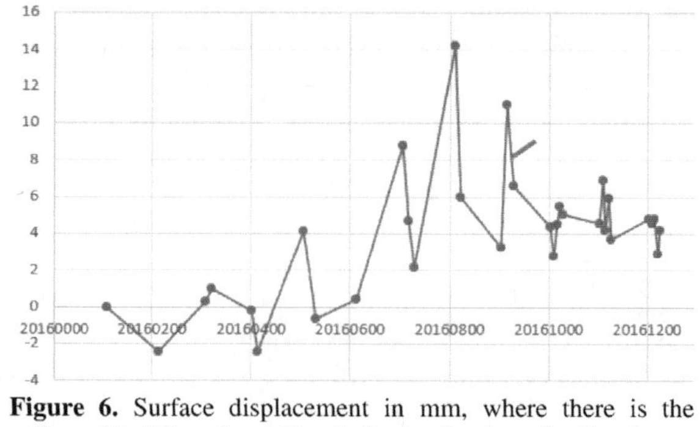

Figure 6. Surface displacement in mm, where there is the collapsed building, the redline indicates the date of collapsing.

Figure 6 demonstrates how the surface acted most uplifted over the year prior to the collapse.

4.1.3 Building 3

Due to the threat of collapsing, a building on Istanbul's Bahçelievler Kocasinan Mahallesi Nurlu Street has already been totally evacuated. The evacuated building eventually collapsed on May 21, 2016. Despite claims that the nearby foundation

https://doi.org/10.5194/isprs-archives-XLVIII-M-1-2023-81-2023

excavation caused the building's destruction, a total of seven nearby structures were evacuated (Web-3).

Figure 7. The collapsed building (Web-3)

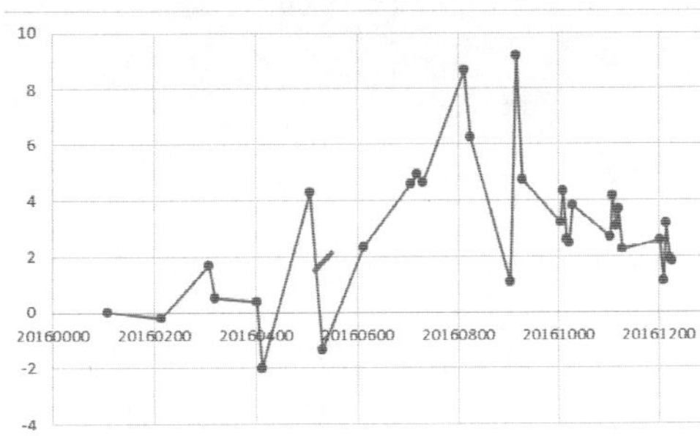

Figure 8. Surface displacement in mm, where there is the collapsed building, the redline indicates the date of collapsing.

The surface displacement relative to the January 1st is shown in Figure. Here, we see again an high uplift surface displacement before the collapsing event.

4.1.4 Building 4

This building was located in Zeytinburnu Sümer Neighborhood. The 5-storey building had been evacuated already because its windows exploded and crackling sounds were heard one day before as stated in the newspaper Hurriyet (Web-4). The Figure 9 shows the collapsed state.

Figure 9. The collapsed building (Web-4)

The graphic depicts an upward movement trend on the surface that reached a height of around 12 cm before the event collapsed.

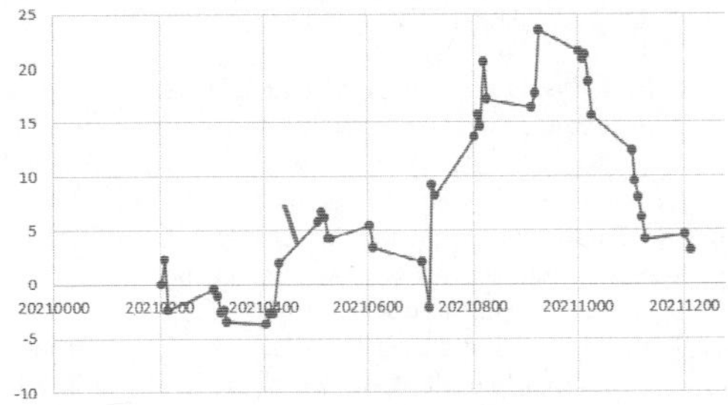

Figure 10. Surface displacement in mm, where there is the collapsed building, the redline indicates the date of collapsing.

4.1.5 Building 5

According to the press, the 10-storey building, located between Bahçeköy Yenimahalle Doğanbey Street and Leylak Street in the town of Sariyer, which was noted to be illegally built, collapsed with a great noise on 27.06.2021. Stating that the building was built gradually without any engineering services (Web-5).

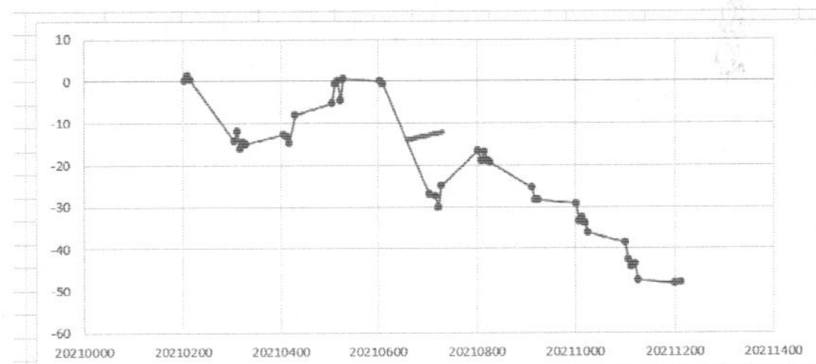

Figure 11. Surface displacement in mm, where there is the collapsed building, the redline indicates the date of collapsing.

This building shows considerable subsidence shortly prior to the incident that caused it to collapse. Subsidence occurs when the surface displacement rises in the same direction. At the conclusion of the year, the overall subsidence was around 5 cm.

4.1.6 Building 6

A 4-storey building on Üsküdar Halk Caddesi was destroyed as a result of the collapse in the construction area behind it, according to the newspaper Cumhuriyet (Web-5) The surface displacement graph demonstrates the subsidence (Figure 12) that occurs before to collapse, as was also highlighted in the news article.

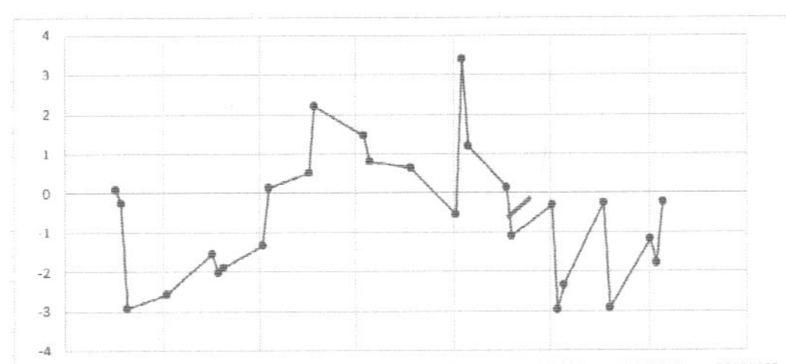

Figure 12. Surface displacement in mm, where there is the collapsed building, the redline indicates the date of collapsing

https://doi.org/10.5194/isprs-archives-XLVIII-M-1-2023-81-2023 | © Author(s) 2023. CC BY 4.0 License.

5. DISCUSSION

The results show that the uplift movement is evident across all situations when measured in relation to the year's beginning. However, it should be emphasized that the measurement's accuracy depends on the wavelength used, which is 5.54 cm from Sentinel 1 datasets. Any movement below 2.27 cm, as is the case in each structure, makes the results in this scenario unreliable. In order to determine whether the primary movement of the buildings had any impact on the plots, we calculated these data for four structures that are next to the grid of the collapsed buildings. The neighbor grids in the east, south, west, and north are examined and plotted appropriately.

The collapse of building 1 had actually occurred on August 24th, 2021. With the exception of the time of observation immediately before the collapse, the grid containing the building in Figure behaves almost exactly like its neighbours. The building's grid reveals a subsidence in comparison to the neighbours (Figure 13).

.

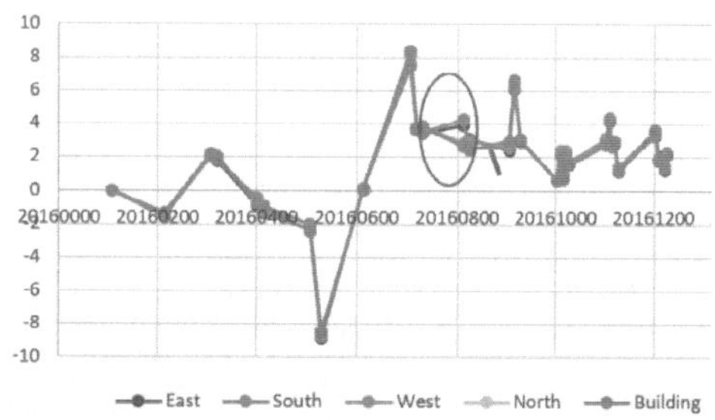

Figure 13. The displacement plots of the building 1 with its neighbours.

The circled area in Figure 14 represents building 2, which has subsidence movement once more compared to its neighbors before the collapse event.

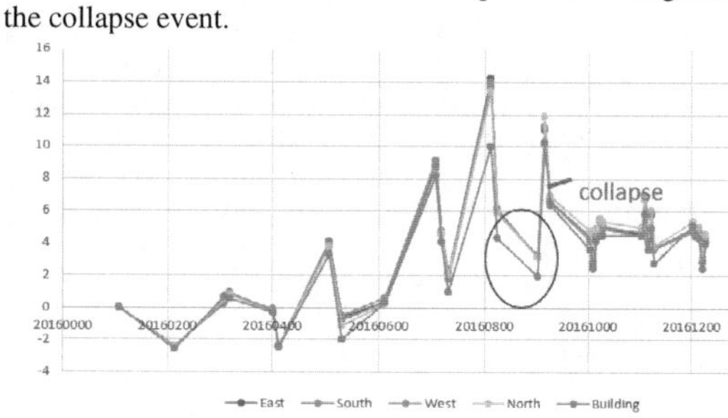

Figure 14. Displacement on Building 2 location, and its neighbours.

Focusing on building 3 reveals a slight variation in movement for the building's grid, once more in the direction of subsidence in Figure 15.

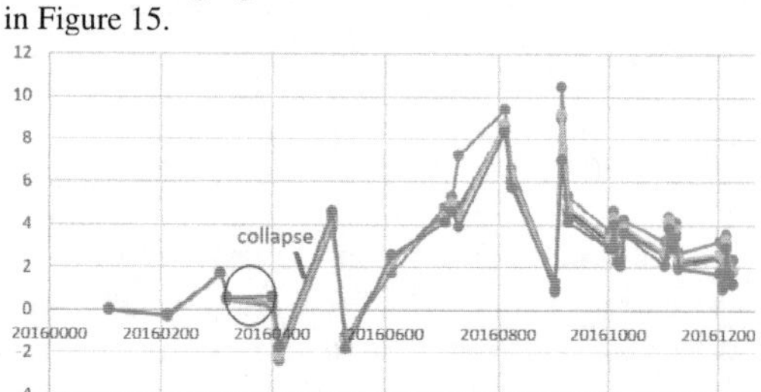

Figure 15. Displacement on Building 3 location, and its neighbours.

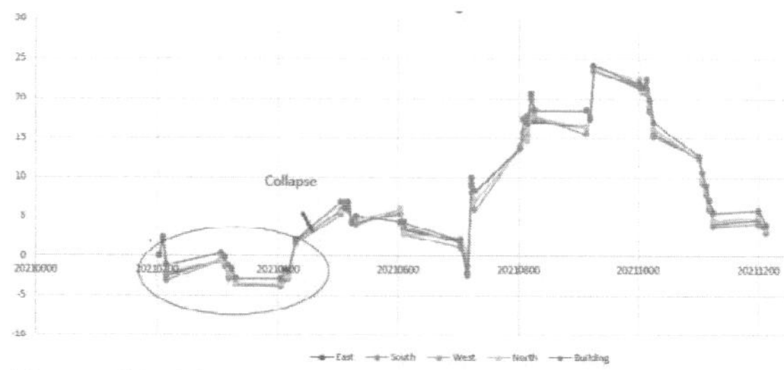

Figure 16. Displacement on Building 4 location, and its neighbours.

Figure 16 shows all neighboring grids have the same surface displacement except the one that has the building.

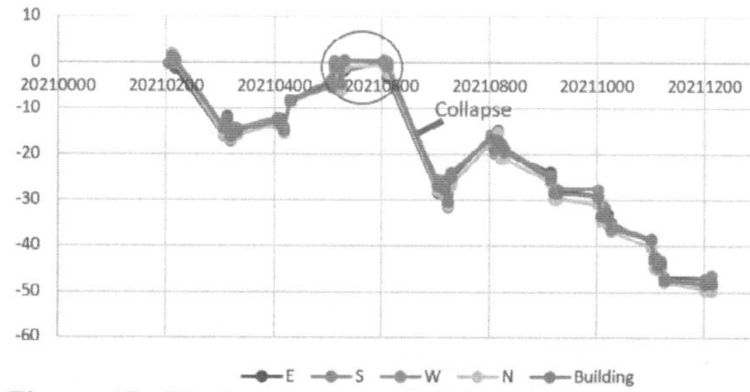

Figure 17. Displacement on Building 5 location, and its neighbours.

The plot (Fig. 17) of the Building 5 does not differ among the neighbors and with the grid with collapsing. Only a minor relative change is shown just before the collapse of the building. The relative difference between the building grid and neighbors reach ca 2 mm in this period.

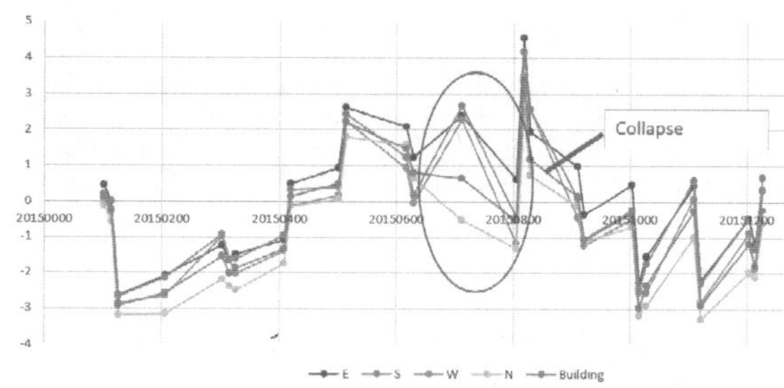

Figure 18. Displacement on Building 6 location, and its neighbours.

The collapse of building 6 took place on September 16, 2015. The plot makes it abundantly clear that the timeline just prior to the event differs noticeably from the other neighboring grids in a way that is not depicted in other timelines.

6. CONCLUSIONS

An effective method for monitoring surface displacement is inSAR technology. The purpose of this work is to demonstrate the capability of risk assessment while looking into the displacement in the areas where the buildings collapsed. The precision of the results is directly influenced by the wavelength and resolution of the produced results. The displacement plots have the potential to monitor risks, but they cannot serve as a toolkit because no tangible result has been produced in this case. It's crucial and significant to pay attention to neighboring displacement. As a result, the results should only be used as a dataset for subsequent, higher resolution investigations. X band datasets for inSAR analysis on a single building scale.The future work will cover to do this analysis with use of different methodologies.

ACKNOWLEDGMENT

This work has been supported by The Scientific and Technological Research Institution of Turkiye with project number 121M706.

REFERENCES

Berardino, P., Fornaro, G., Lanari, R., Sansosti, E. 2002. A new algorithm for surface deformation monitoring based on small baseline differential SAR interferograms. IEEE Transactions on geoscience and remote sensing, 40(11), 2375-2383.

Google, 2022. Google Street View, https://www.google.com/maps/@41.0372561,28.8587687,3a,90y,225.57h,106.78t/data=!3m6!1e1!3m4!1sHfFPYgT7eucYrAuk7qYLFg!2e0!7i13312!8i6656 , accessed on November 20th, 2022

İHA, 2016. A 5 story building collapsed in Bağcılar (in Turkish), İhlas Haber Ajansı (İHA), https://www.iha.com.tr/haber-bagcilarda-5-katli-bina-coktu-588609/, accessed on November 20th, 2022.

Lazecký, Milan, Karsten Spaans, Pablo J. González, Yasser Maghsoudi, Yu Morishita, Fabien Albino, John Elliott, Nicholas Greenall, Emma Hatton, Andrew Hooper, Daniel Juncu, Alistair McDougall, Richard J. Walters, C. Scott Watson, Jonathan R. Weiss, and Tim J. Wright. 2020. LiCSAR: An Automatic InSAR Tool for Measuring and Monitoring Tectonic and Volcanic Activity Remote Sensing 12, no. 15: 2430. https://doi.org/10.3390/rs12152430

Sandwell, D., Mellors, R., Tong, X., Xu, X., Wei, M., Wessel, P. (2016) GMTSAR: An InSAR Processing System Based on Generic Mapping Tools. UC San Diego: Scripps Institution of Oceanography. Retrieved from: http://topex.ucsd.edu/gmtsar/tar/GMTSAR_2ND_TEX.pdf

Yunjun, Z., Fattahi, H., and Amelung, F. 2019, Small baseline InSAR time series analysis: Unwrapping error correction and noise reduction, Computers & Geosciences, 133, 104331, doi:10.1016/j.cageo.2019.104331, arXiv, data, notebooks.

Web-1.2022. Hurriyet Newspaper, https://www.haberturk.com/bina-sayisi-yuzde-14-artti-3354897-ekonomi, accessed on November 16th, 2022.

Web-2. Evrensel Newspaper, https://www.evrensel.net/haber/441062/ikitelli-masko-sanayi-sitesinde-bir-bina-coktu, accessed on November 16th, 2022.

Web-3. Haberler.com internet sayfası, https://www.haberler.com/guncel/bahcelievler-kocasinan-da-bosaltilan-bina-yikildi-8458420-haberi/. accessed on November 16th, 2022.

Web-4. Sözcü Newspaper, https://www.sozcu.com.tr/2021/gundem/zeytinburnunda-bina-coktu-6416252/)., accessed on November 16th, 2022.

Web-5. Sözcü Newspaper, https://www.sozcu.com.tr/2021/gundem/sariyerde-bosaltilan-9-katli-bina-coktu-6508124/, accessed on November 16th, 2022.

Web-6, Cumhuriyet Newspaper, https://www.cumhuriyet.com.tr/haber/uskudarda-5-katli-bina-coktu-1-olu-4-yarali-370495, accessed on November 16th, 2022.

This contribution has been peer-reviewed.
https://doi.org/10.5194/isprs-archives-XLVIII-M-1-2023-81-2023 | © Author(s) 2023. CC BY 4.0 License.

VULNERABILITY OF ECONOMIC SECTORS TO GEOMAGNETIC STORMS AND MITIGATION STRATEGIES

M.B. Demirköz[1], G. Karagöz[2], Y.K. Kadıoğlu[3], H. Sözeri[4]*

[1]M.Bilge Demirköz, Orta Doğu Teknik Üniversitesi, İVMER, 06800, Çankaya, Ankara, demirkoz@metu.edu.tr
[2]Gülce Karagöz, Orta Doğu Teknik Üniversitesi, İVMER, 06800, Çankaya, Ankara, gulce.karagoz@metu.edu.tr
[3]Yusuf Kağan Kadıoğlu, Ankara Üniversitesi, Jeoloji Mühendisliği Bölümü,06830, Gölbaşı, Ankara, kadi@ankara.edu.tr
[4]Hüseyin Sözeri, TÜBİTAK-UME, 41470, Gebze, Kocaeli, huseyin.sozeri@tubitak.gov.tr

KEY WORDS: GNSS, geomagnetic observatories, Kp index, Solar storms, adverse effects on Turkish economy

ABSTRACT:

Solar storms and CMEs increase the amount of radiation that reaches the Earth and can perturb the magnetic shielding of the Earth. K and Kp indices quantify this perturbation by using magnetometers calibrated accordingly to the latitude of the geomagnetic observatory It is possible to contribute this measurement by establishing an observatory specialized in space weather studies. Many basic sectors are vulnerable against a strong solar storm unless necessary mitigation strategies are applied. Moreover, failure in taking an immediate precaution may have long-term effects which include economic destruction, irreversible health problems and weaknesses in national defence. In this work, the requirement such an observatory should have, the most 11 vulnerable sectors and corresponding mitigation strategies are discussed in detail. Turkey's seven geographical regions are examined to find the optimal location for the foundation of such a probable observatory. It is found that Central Anatolia is the ideal region and districts Konya/Karapinar and Konya/Çeltik are the possible candidates for such a location.

1. INTRODUCTION

The activity of the Sun increases and decreases over a period of 11 years. During a solar maximum, the activity of the Sun becomes its highest. In the solar maximum, the number of sunspots and coronal mass ejections (CMEs) gets stronger and more often; therefore, extreme ultraviolet and X-ray emissions occur. Consequently, their impact on the upper atmosphere and surface of the Earth becomes more prominent. Entrance of these high energetic accelerated particles into Earth's atmosphere results in plenty of outcomes. The first one of these is atmospheric heating due to mentioned energy emissions. By the increasing heat, the temperature and the density of the atmosphere also increase in the altitudes that many spacecrafts are confined to. Increasing drag force in Low Earth Orbit (LEO) may result in a substantial decrease in the orbital lifetime of the satellites. Simultaneously, the increase in the number of high energetic accelerated particles boosts the likelihood of an instrument being damaged. The astronauts in these space stations and people who travel in high altitudes and polar routes are also threatened by the radiation emerge from CMEs and solar flares. Even the terrestrial climate of the Earth gets affected by these changes. It is found that the heat on the surface of the Earth maximizes and minimizes in correspondence to Sun's maxima and minima (Hathaway, 2015). Astoundingly, these particles may cause a formation of radiocarbon in the atmosphere of the Earth (Zhang, 2022). Solar storms also affect the total electron content (TEC) in the ionosphere which might result in erroneous positioning of GNSS satellites. These storms may even affect the electronic components and radio control signals of a GNSS satellite (Shprits et al, 2019). Today, electricity is widely used in many sectors therefore economic, health-related and defence sector consequences of such a storm will be disastrous.

The change in the magnetic field of the Earth due to solar storms and CMEs are expressed by K and Kp indices (Bartels, 1957), (Elliot et al, 2013). Kp index is also called the "planetary K index". These indices are calculated from the data collected by geomagnetic observatories all around the world. For a geomagnetic observatory to be able to join this calculation process, it must satisfy the requirements determined by "International Real-time Magnetic Observatory Network (INTERMAGNET)". Any observatory that meets these requirements can be a part of K index calculation; however, since Kp is the planetary K index, observatories in extreme altitudes cannot contribute Kp index calculation. Therefore, any observatory which can collect the data would be used to calculate Kp index must be located in between the limits of sub auroral altitudes (Matzka et al, 2021). In order to obtain the data for K&Kp index calculation, ground-based magnetometers are used. These devices measure the change in the magnetic field of the Earth and record it. In addition to the INTERMAGNET requirements on magnetometers, there are additional scientific requirements, which will be detailed in the following sections, that need to be satisfied for a geomagnetic observatory.

Even though SOHO provides an early-warning system to such a solar storm, it is vital to understand ground-based consequences of a geomagnetic disturbance and analyse it carefully to take ground-based precautions because the mitigation strategies will be different than in the space (Brekke,2005).

2. SOLAR FACTORS AND ITS CONSEQUENCES

2.1 Solar Cycle and Solar Storms

The change in the activity of the Sun causes a disturbance in the magnetosphere-ionosphere of the Earth and this activity is called "geoeffective processes". 11 years and 22 years of solar cycles are the well-known effects of this change. The Earth-Sun system has two main movements, which are called "sideral motion" and "synodic motion", which may cause other repetitions, namely 27-days, 13.5-days and 9-days of periodicities. The incoming flux from the Sun also exhibits

This contribution has been peer-reviewed.
https://doi.org/10.5194/isprs-archives-XLVIII-M-1-2023-87-2023 |

changes depending on these features. Understanding the magnetic field components of such changes increases the chance of foreseeing flux variations. The dynamics of one or multiple coronal holes are thought to be one of the reasons for these periodicities since coronal holes are usually the plasma arcs that connect sunspots. As the number of sunspots increases, the number of coronal holes and, therefore, the activity of the Sun increases, which occurs more often while the Sun is at its maximum in the cycle (Richardson, 2018).
The Sun's temporal variability affects space weather and climate on time spans ranging from millennia to minutes. We refer to the disturbances as "storms" if they are powerful enough. The magnetosphere, in other words, the Earth's magnetic environment, is shaken by these fluctuations in the solar wind. The magnetosphere of the Earth creates a shield against the high energetic charged particles that emerge from the solar wind. The magnetic force deflects these solar wind particles before they reach the surface of the Earth. For a high energetic charged particle to reach a particular altitude, it should exceed a certain threshold which is specified for that particular location. This threshold is called "geomagnetic cut-off" (Chu et al, 2016). The geomagnetic cut-off value decreases towards the poles since the strength of the magnetic field decreases. This allows particles with lower energy being able to penetrate through the magnetosphere and have a higher chance of reaching the surface of the Earth. Therefore, during a potential solar storm, poles are exposed to more radiation than the rest. A solar storm is capable of disturbing this magnetic protection and as a consequence, since the magnetic field becomes weaker than before, particles with lower energy will be able to reach the surface of the Earth. On the ground, there will be an accumulation of charged particles and, therefore, more electromagnetic radiation. Additionally, due to the tilt angle between the Earth's rotation axis and the magnetic field, there is an inhomogeneity which results in South Atlantic Anomaly (SAA). In SAA, the charged particles accumulate, which results in a higher flux than in other locations in the same latitude.

When the ionosphere of the Earth is considered, solar storms also have an impact. Changes in the ionospheric currents can cause rapid time variations on the surface of the Earth, according to Faraday's law. In any conductive system, electric currents due to this electric field will be driven. Moreover, if this electric field penetrates through soil or water, undersea communication cables and gas pipelines will also be affected by this induction current.

SOHO (the Solar and Heliospheric Observatory) is an international project conducted by ESA and NASA to study the dynamics of Sun (Fleck, 1997). It provides unique information about the Sun, its inner and outer dynamics. Since it is in a halo orbit around L1 Lagrangian point, it is positioned outside the magnetosphere of the Earth and therefore is deprived of the magnetic shielding provided by it. SOHO monitors the Sun 24 hours in a day and due to its proximity to the Sun, SOHO has a leading role in the early-warning system for the space weather extremeness (Brekke,2005).

2.2 Consequences of Solar Storms

2.2.1 Space Mission Consequences

As mentioned in section 1, there are several consequences of a solar storm, one of which is atmospheric changes. When a solar storm penetrates through the atmosphere, it heats the environment which results in a density increase that causes drag force to decrease. Engines that cannot overcome this drag force become insufficient to keep the satellite at a particular altitude therefore, it starts falling down, which ends up burning the satellite in the atmosphere of the Earth. Also, solar storms have a large impact on the electronic subsystems of satellites. It may cause single-event effects and cause the computer to reset itself. When the space environment is concerned, most of the failures are due to solar storms (Tafazoli, 2009). In figure 1, it can be seen that the predominant reason for spacecraft failures due to the space environment is solar storms.
Since astronauts cannot benefit from the shielding of the magnetic field, they should protect themselves from solar flares. For example, astronauts protected by the Mir station during the time of the solar storm at the end of 1989 absorbed their yearly radiation dose limit in a matter of hours. During solar storm events, astronauts have reported seeing light flashes in their eyes. They are produced by radioactive particles that enter the bulkhead of the shuttle and generate streaks or flashes in the astronauts' eyes that are comet-like (Marusek, 2007). One can be affected by solar storms even without leaving the atmosphere. Depending on the strength of the storm, air travellers will also be affected by these high energetic charged particles. For example, more than a hundred people were injured in the Qantas plane crash in 2008 (Australian Government, 2008). In the research, it has been revealed that the weather data may be affected by the cosmic ray radiation that the inertial reference unit is exposed to during the flight. As a result of the interaction of cosmic rays with the atmosphere, secondary neutrons formed, which can affect aircraft avionics systems (Australian Government, 2008).

2.2.2 Consequences on GNSS Signals

Solar storms affect the TEC of the atmosphere, which can cause a wrong positioning of GNSS (GPS, GALILEO, GLONASS, BEIDOU, QZSS, IRNSS) satellites. Solar storms may corrupt the radio control signals of a satellite or even its electronic components (Shprits et al, 2019). The change in the ionosphere's total electron content density may change the speed at which the radio waves propagate and therefore, the GNSS signal which will be used to determine the positioning of the satellite will be wrong. Rapid changes in electron density from space weather fluctuations can cause a loop shift in carrier phase tracking. As an example, studies during two successive geomagnetic storms that occurred during the period from 8 to 12 November 2004 clearly demonstrated adverse effects on GPS range latency, as evidenced by Total Electron Content (TEC) measurements from seven dual-frequency GPS chains (Sreeja, 2016). The delay in the ionosphere can be explained by the Fermat principle. It is understood that the sum of the refractive indices along the beam from the satellite to the receiver gives the ionospheric delay in meters. The rapid changes in storm time that occur in the TEC cause a series of phase shifts in the GPS signal compared to those on quiet days. These phase shifts are due to the loss of locking of GPS receivers which results and negatively affects GPS-based navigation (Rao et al, 2009).

2.2.3 Vulnerability and Mitigation Strategies

When we consider the ground-based effects of a solar storm, one can see that the spectrum gets a lot wider. There will be both vital and economic effects which can be either an instant or a long-term one. An example of this situation is the "Carrington solar storm", which occurred during the solar cycle 10 in 1859. It is the strongest solar storm recorded in the history and the probability of such a solar storm repeating again is quite possible. The Carrington solar storm destroyed telegraph lines

This contribution has been peer-reviewed.
https://doi.org/10.5194/isprs-archives-XLVIII-M-1-2023-87-2023 | © Author(s) 2023. CC BY 4.0 License.

and electrical networks at the time of its occurrence. During a solar storm, electronic devices are affected due to the induced current. Today, when electricity is used much more widely, such an event will cause a great disaster. There are 11 fundamental sectors that will be affected enormously by such a solar storm.

i. Civil Aviation Industry: The amount of radiation exposed by an aircraft depends on the solar activity, space weather, geomagnetic latitude and altitude. When the electronic components in spacecrafts are exposed to high radiation, high energetic particle produces an unnatural pulse in the circuit. This pulse may result in temporal or permanent effects in the circuit (Both, 2012). Single Event Upset (SEU) occurs when such a particle hits in a cell in a digital circuit it can flip the logic gate and therefore the stored information. It can take place in a cell or in multiple cells (Multiple Bit Upset). In SEU, by using an error correcting code it is easier to fix a software issue however when MBU is considered, it is not as straight forward as SEU. For full recovery, the system needs to be restarted (Prado, 2013) (Maniatakos, 2012). While SEU and MBU are reversible corruptions, various types of Single Event Effects (SEE) such as Single Event Latch-Up (SEL) and Single Event Burnout (SEB) are irreversible corruptions. Qantas plane incident mentioned in section 2.2.1 is one of the recent examples of an SEE (Australian Transport Safety Bureau, 2008) that is reported. In the report, SEE is given as the most probable reason for such a damage. Reference unit of the Qantas airplane is thought to send a false signal due to a SEE which resulted in flight control failure causing the airplane to sudden ascent and descent with the passengers inside. According to SKYTRAX, Turkish Airlines is the 7th biggest company around the world (World's Top 100 Airlines 2022, SKYTRAX) and it will be greatly affected by such a radiation exposure. During such a solar storm, one way of protecting from the radiation is to decrease the amount of time spent in high radiation zone therefore delaying non-urgent flights until the effect of solar storm passes is a solution. Due to the latitude and altitude effects, a change in flight route is another option. Altitude change might cause more fuel consumption however it can be atoned by adapting the flight speed depending on the altitude (Meier,2020).

ii. International Health Sector: An electromagnetic pulse (EMP) is a temporary electromagnetic disturbance which can also stem from solar storms. A high altitude EMP is capable of producing an electric field strength of 20-50 kV/m (Vandre, 1993). Modern electric systems with highly complex circuits are very sensitive to EMP. Any wire that is attached to the electronic equipment can serve as an antenna, such as power cords, patient leads, or metal chassis, transfers energy from the EMP electric field. An EMP deposits energy that is enough to burnout semiconductors and damage the circuit components. Burnout in the junction in the transistor or an integrated circuit results in permanent damage. Any medical technological equipment facing with this consequence stops at that instant and cannot be made functional again. To prevent damage due to high electromagnetic voltage induced by an EMP, unused equipment should be unplugged, life support units and MR devices in intensive care units should be turned off during a solar storm. Designing medical equipment with a wire as short as possible. Wiring should be close to the ground as much as possible since the higher the wire gets more voltage is generated. Finally, sensitive equipments can be placed to ISO (International Organization for Standardization) shelter (Vandre,1993). These precautions can be taken by Turkish Ministry of Health to prevent damages in hospitals and health sector.

iii. Telecommunication Sector: Ionosphere of the Earth acts as a shielding and protects the Earth from solar storms. However, depending on the strength of the geomagnetic disturbance, this shielding might weaken due to the change in the total electron content (TEC) density in ionosphere (McManus,2021). The increase in the background noise due to the storm will result in a signal loss. Today, most of the communication is done via phones or internet. Maintain the telecommunication sector operational is crucial during a disaster to call for emergency services and organizing purposes.

iv. Defence Sector: In addition to communication sector, ionospheric disturbances also affect radio signals up to 10 GHz. A first-order propagation delay stems from the interaction between radio waves and plasma and this delay is proportional to $1/f^2$. Because of the change in the ionospheric refraction index, satellite mis-positioning might occur which cause an error in the navigation signals. GNSS signal disturbances can also cause communication breakdowns in the defence and military industries. In such a case, for example, it will not be possible to detect a potential air attack. Preventing such a vulnerability is possible with using a dual frequency receiver (Kuusisto, 2013). The Republic of Türkiye and The Republic of Türkiye Ministry of National Defence General Staff may highly benefit from such a precaution.

v. Pharmaceutical and Medical Equipment Sector: Power outages caused by solar storms will also endanger the storage of medicines that must be kept in the cold chain. The effects will be great in areas where extreme care is required during transport such as organ transplantation and tissue samples. In this case, if measures are not taken, the loss that will occur in the pharmaceutical sector will be large-scale economically.

vi. Financial Technologies Sector: When a solar storm hits the Earth, facing regions with the shock will be affected more. While this solar storm may cause electrical and technical problems in one region particularly, others may continue to operate. This will make these regions vulnerable towards to a cyber attack and moreover they will not be able to establish the security back before this problem is solved. All transactions (money transfer, investments etc.) may be disrupted and cyber-attacks may occur due to both internet outages and system shutdowns.

vii. IT Sector: Even though the IT equipment does not affected by a solar storm directly, signal loss, communication interruptions and data transfer & storage will be affected by the electrical outages. Since solar storms may be powerful enough to affect a whole hemisphere and has long-term effects, any sector based on IT will suffer from this corruption. A back-up data or software centre in a different geographical region may make it easier to overcome such a problem(Sergeant, 2012).

viii. Maritime Sector: After a shockwave hits the Earth it disturbs the geomagnetic field followed by a geoelectric field which can penetrate through the conducting interior of the Earth. In grounded lines, a Geomagnetically Induced Current (GIC) is produced as a response. Subsea communication cables are designed to operate at 1A current, in case of a burnout it takes 2-3 years to construct the submarine cables. Due to the loss of signal in the automatic identification system (AIS), which is mentioned in section 2.1, vessel tracking traffic will not be possible, and the probability of loss of route and accident will also increase (Castellanos, 2022).

ix. Livestock Sector: Animals may need to be transported to closed areas depending on the strength of the storm.

x. Beverage and Food Sector: Foods that need to be kept intact in the cold chain may also be wasted, as in medicines. It will have a negative economic impact.

xi. Automotive Sector: The ignition of vehicle batteries is a serious threat. As awareness arises from such a storm, all batteries should be disconnected and especially electric vehicles should not be used for transportation. Additionally, software using GPS signal to determine the route of the car will also be disturbed and therefore may cause the car to move in routes that do not exist or even worse dangerous.

Solar storms affect other industries apart from these ones. However, as a result of the research has been done, the damage in these industries is more than the rest and taking precaution for the damage in these industries is much more important. Otherwise, in the case of a strong solar storm, global possibilities that may progress to major disasters such as loss of life, mass extinctions, and the onset of famine may be triggered.

3. MEASURING THE STREGTH OF SPACE STORMS

As mentioned previously, even though the ground is much more protected by the magnetic shielding of the Earth, the effects of solar storms can be seen from the ground. There are many ways to indicate the strength of a solar storm, all of which are specified according to its purpose. In this work, the "K" and planetary K index "Kp" are focused on. Various ground-based instruments are used to measure different solar storm indices, one of which is magnetometers.

3.1 Solar Strom Indices: K and Kp Index

In order to describe the strength of solar storms, most of the time, sunspot numbers are widely used; however, they do not give any clue for the strength of individual storm; instead, it indicates an extended time range. Solar storms are classified into classes which can be seen in table 1:

A	$10^{-8} - 10^{-7}$ Wm^{-2}
B	$10^{-7} - 10^{-6}$ Wm^{-2}
C	$10^{-6} - 10^{-5}$ Wm^{-2}
M	$10^{-5} - 10^{-4}$ Wm^{-2}
X	$\geq 10^{-4}$ Wm^{-2}

Table 1. Solar X-ray emission classes (Koskinen, 2011).

To characterize solar storms in the magnetosphere, many activity indices have been used to measure the amount of magnetic deviation. For many years, these indices have been obtained by using the Mayaud rules. Indices are specified for features such as latitudes, time scales and decay current. Examples of widely used indices are "Dst" index and "K" index for global storm levels and AE (auroral electrojet) index for auroral regions (Koskinen, 2011).

K and Kp indices are used to indicate the deviations in the horizontal component of the magnetic field of the Earth (Bartels, 1957), (Elliot et al, 2013). Measurement of K and Kp indices is made by 3 hours of data obtained by the geomagnetic observatories. K index can be measured by any geomagnetic observation station at any latitude. Unlike the K index, the Kp index is called the planetary K index; therefore, the geomagnetic observatory which contributes to the measurement should be in the sub-auroral latitude region (Matzka et al, 2021). By doing this, it will be possible to avoid extreme conditions in polar regions since the magnetic field is weaker in that regions and therefore, the optimum measurement can be performed. K and Kp indices have values ranging from 0 to 9 depending on the strength of the solar storm. Values between 0-3 are calm, 4 is perturbed and 5-9 are called geomagnetic storm regions which can be seen in table 2:

K and Kp Indices	Meaning
0	Calm Region
1	
2	
3	
4	Perturbed Region
5	Geomagnetic Storm
6	
7	
8	
9	

Table 2. Values of K&Kp indices and their meaning.

The Kp network is oriented to Western Europe and North America, with seven of the thirteen stations, respectively". There are only two stations in the Southern hemisphere and none in East Europe and Asia. Besides two Europe stations (Brorfelde/Denmark and Uppsala/Sweden), data coming from two stations in the Southern hemisphere (Eyrewell/New Zealand and Canberra/Australia) are combined and involved in the calculation with their mean value. Therefore, the Southern hemisphere contributes to indices as one.

Figure 1. Distribution of Kp observatories (ISGI - International Service of Geomagnetic Indices, n.d.).

The Kp network is oriented to Western Europe and North America, with seven of the thirteen stations, respectively". There are only two stations in the Southern hemisphere and none in East Europe and Asia. Besides two Europe stations (Brorfelde/Denmark and Uppsala/Sweden), data coming from two stations in the Southern hemisphere (Eyrewell/New Zealand and Canberra/Australia) are combined and involved in the calculation with their mean value. Therefore, the Southern hemisphere contributes to indices as one. For many years, the Kp index has been calculated by following morphological rules, so-called Mayaud rules and scaled by hand using analogue magnetometers. All geomagnetic observatories follow these rules while regular hand scaling is performed. With the digitalization of geomagnetic observatories, experienced people who hand scale the data from magnetometers for K index calculations are decreased gradually. Therefore, observatories needed to find substitute computational methods in order to produce the K index. IAGA (International Association of Geomagnetism and Aeronomy) detailed all the electronic, mechanical, temperature-related, data-taking and sharing requirements that such a geomagnetic observatory should have in a report. Today, all observatories use ground-based magnetometers in order to obtain K and Kp indices.

3.2 INTERMAGNET and Magnetometers

INTERMAGNET (International Real-time Magnetic Observatory Network) is a programme that serves as a near-real-time data collector to monitor and predict the response of the Earth to solar activity changes. Magnetic observatories all around the world are encouraged to share real-time data with INTERMAGNET. Its foundation was determined at the "Workshop on Magnetic Observatory Instruments" in Ottawa in August 1986 (Kerridge, 2001). Its purpose is defined as "The INTERMAGNET objective is to establish a global network of co-operating digital magnetic observatories adopting modern standard specifications for measuring and recording equipment in order to facilitate data exchange and the production of geomagnetic data products in close to real-time." (Louis et al, 1999).

The Kp index is measured by ground-based magnetometers in the corresponding geomagnetic observatory. These magnetometers must meet the requirements and standards that are determined by INTERMAGNET. After that, these observatories can be a member of the INTERMAGNET network and can contribute with their data. The standards of a magnetometer that an observatory should have in order to become an INTERMAGNET member are given in table 3:

Magnetometer	Vector	Scalar
Resolution	0.1 nT	0.1 nT
Dynamic Range	8000 nT for Auroral and Equatorial Latitude 6000 nT Mid-latitude	-
Bandpass	Dc 0.1 Hz	-
Sampling Rate	1 Hz (1s)	0.033 Hz (30s)
Thermal Stability	0.25 nT/Celcius	-
Long Term Stability	5 nT/year	-
Accuracy	-	1 nT

Table 3. Minimum magnetometer standards determined by INTERMAGNET

Observatory that meets all the requirements can start to operate after its magnetometer is calibrated according to the latitude that the observatory is founded. After becoming a member of the INTERMAGNET network, it can start to contribute to K index data. Moreover, it can compare its own results with the others. In Turkey, the measurement of the K index was started with the geomagnetic observatory that is located at Kandilli/Istanbul in 1947. This centre became a member of the INTERMAGNET network in 1997 and started to share its own data. However, due to the rapidly increasing city growth, the observatory remained in the city noise, and then the studies continued to be calculated with the data taken from the Iznik Magnetic Observatory affiliated to Boğaziçi University Kandilli Observatory and Earthquake Research Institute since 2005.

It can be concluded as a result of the consultations that have been done, The Scientific and Technological Research Council of Turkey National Metrology Institute (TÜBİTAK UME) is capable of producing such a magnetometer locally. High-sensitive (nT) magnetometer research have been begun approximately ten years ago in TÜBİTAK UME. As a sensor element, a superconductive material is used in the first magnetometer produced and a resolution level of less than 1 nT is reached (Topal, 2013). Since superconducting materials need to be cooled with liquid nitrogen or liquid helium, the usage area of this type of sensor is very limited. However, this study showed us that superconducting materials are also very sensitive to the magnetic field. In later developed devices, soft magnetic alloys were used as sensor material, as in similar commercial products. The resolution level is reached up to pT magnitude (Topal et al, 2021), (Can et al, 2017). At the same time, software and electronic cards that control these sensors are also produced in TÜBİTAK UME. The last device produced at TÜBİTAK UME has been prepared with space qualification and has passed all tests successfully.

4. DETERMINING A SUITABLE LOCATION FOR KP INDEX CALCULATION

A geomagnetic observatory should avoid certain features to be able to take correct measurements. Since a magnetometer detects the anomalies in the magnetic field of the Earth, it should be away from any fault zone to take pure space-based magnetic field effects (Gao et al, 2014), (Johnston, 2002), (Robbes, 2006). Another feature is the DC electrified railways. The observatory should be away from the railways for a similar reason (Pirjola, 2011), (Love, 2009). The soil must be free of elements that can cause geomagnetic disturbance (Girdler, 1971). Due to its conductivity, it should be away from salty and soda water and should be protected from the effect of lightning and strike (Grant, 2006), (Čop, 2014). Requirements that a geomagnetic observatory for Kp index measurements are given in table 4 as a summary:

Requirements	Reasons	Reference
An observatory should be away from fault lines.	Magnetometers in these observatories measure the change in the magnetic field and earthquakes produce magnetic field. This feature should be satisfied in order to avoid false signals.	Gao et al, 2014 Johnston, 2002 Robbes, 2006
An observatory	For observatory to	Pirjola, 2011

https://doi.org/10.5194/isprs-archives-XLVIII-M-1-2023-87-2023 | © Author(s) 2023. CC BY 4.0 License.

should be away from railways.	be far away from the electromagnetic effect of the railways it should be at least 20km away from a DC electrified railways and tram ways.	Love, 2009
Soil must be free of elements that can cause geomagnetic disturbance.	It should be far away from the element sources, such as iron, which have magnetic properties, in order to be protected from magnetism.	Girdler, 1971
An observatory should be away from salty and soda water.	The amount of salt and soda in the water source changes the conductivity of the water. For this reason, it is necessary to stay away from this feature in order to keep the conductivity at a minimum.	Grant, 2006
An observatory should be protected from the effect of lightning and strike.	Weather events such as strikes and lightnings are fast-moving electric currents and therefore they create an electromagnetic field.	Čop, 2014

Table 4. Requirements for a geomagnetic observatory which will be founded with the purpose of space-weather study.

4.1 Potential Locations in Turkey

To determine the location of a potential geomagnetic observatory, elimination method is used. In this research, where all seven geographic regions of Turkey are examined, related maps are studied in order to satisfy the requirements given in table 4.

i. An observatory should be away from fault lines: EM wave emissions due to earthquakes are one of the most important features for a geomagnetic observatory to avoid from. Any geomagnetic observatory that is located near an active fault line would not reach a pure enough space weather sample to analyse because of the frequent EM wave signals coming from the earth crust and its signal would be contaminated by those signals. While this is a necessity for a geomagnetic observatory which searches for earthquakes, it should be avoided for an observatory which is specified to study space weather. The Iznik geomagnetic observatory, which was established for earthquake research, cannot be used for calculating the Kp index due to its proximity to the North Anatolian Fault, and for the same reason, the Marmara and Western Black Sea Regions are not suitable for hosting such an observatory due to the North Anatolian Fault Line. The East Anatolian Fault Line has caused the elimination of a large part of the East and Southeast. The West Anatolian Fault Line, which exhibits a large number of graben and horst structures, has been concluded that the Aegean is not suitable for a geomagnetic observatory for space weather purposes, since it affects the entire Aegean Region. The structure of Turkey was determined by examining the geological and tectonic aspects in order to establish the appropriate observatory for the calculation of the Kp index. According to this, when an area is far from the active earthquake fault line, away from an iron deposit formed as a result of any magmatic intrusion or their influence, and especially away from saline solutions due to the lacustrine sequence and evaporitic rock deposits due to them, the Konya plain was determined as the most suitable region. In case of the start of the project, on-site sampling will be performed for the selection of the exact location and geochemical analyzes will be made, and at the same time, the location of the appropriate observatory will be determined for the calculation of the Kp index by selecting the areas away from the buried faults by taking geophysical measurements.

ii. An observatory should be away from railways: In order not to be affected by electromagnetic fluctuations, geomagnetic observatories need to be located far from railways and should be at least 20 km away from DC electric railways and tram lines. It is seen that the Marmara and Aegean Regions do not meet this requirement, while some of the regions in Central Anatolia and Eastern Anatolia may be away from these effects. The lower parts of the South-eastern Anatolia region do not meet this requirement either.

iii. The soil must be free of elements that can cause geomagnetic disturbance: The geomagnetic observatory should be away from the soil structure where the geological elements with magnetic properties are located. It is seen that the Eastern Black Sea region does not meet the requirement due to magnetic anomalies. Likewise, although not spread in the Central Anatolia region, there is some anomaly.

iv. An observatory should be away from salty and soda water: The change in pH level changes the ionization rate of the soil. The conductivity level of a region closes to salty or soda water will increase due to the ionization rate. Since electrical disturbances may occur due to the increase in conductivity, the observatory should be away from this factor. For this reason, the coastlines of our country, which is surrounded by seas on 3 sides, especially the Eastern Black Sea region, which are far from fault lines, are eliminated. Since most of the lakes in our country (Van, Eğridir, Beyşehir, etc.) are soda, most of the Central Anatolia and Eastern Anatolia regions are eliminated.

v. An observatory should be protected from the effect of lightning and strike: The station should also be far from areas where lightning and strikes are common due to electromagnetic effects. Lightning and strikes are caused by fast-moving charged particles, which cause the electromagnetic effect.

This contribution has been peer-reviewed.
https://doi.org/10.5194/isprs-archives-XLVIII-M-1-2023-87-2023 | © Author(s) 2023. CC BY 4.0 License.

After examining all the requirements in detail, Central Anatolia and the Mediterranean regions seem to be the most appropriate candidates for such an observatory. However, because of the mountain formations and the requirement of being away from salty water, the Mediterranean region is also eliminated. While Central Anatolia meets all the requirements for a potential geomagnetic observatory, Konya province, which has a flat geography and provides an easy-to-reach area, is mainly focused. A magnetic anomaly is spotted; however, as a result of the discussions with the experts, it is seen that avoiding this magnetic anomaly is possible. Districts Konya/Karapınar (37° 42'K, 33° 33'D) and Konya/Çeltik (39° 1′ K, 31° 47′ D) are determined to be the most suitable candidates for a potential location in order to establish such a geomagnetic observatory which will be specified for space weather research.

5. CONCLUSION

The atmosphere and the surface of the Earth are affected by the flux changes caused by our Sun, which has a solar cycle that repeats itself every 11 years. The Sun reaches its maximum and minimum in this interval and when it is at its maximum, solar activity also reaches its maximum, which results in frequent solar flares and CMEs. High energetic charged particles mediated by these solar storms may affect space missions by changing the density of the atmosphere and interfering with the electronics of the satellites. Moreover, by changing the TEC of the ionosphere, they can contaminate GNSS signals which cause delays in communications. On the ground, vital and economic consequences may occur and their negative impact might be long-term. By using K and Kp indices, these changes can be studied and understood in detail and therefore taking precautions becomes easier. Approaching the solar maximum expected in 2024, by becoming a member of INTERMAGNET, Turkey might become a partner in this research and by producing its own magnetometer, even trade of such an instrument would be possible in the future. For these benefits, requirements for such an observatory have been studied in this research and it is decided that the ideal locations are Konya/Karapınar and Konya/Çeltik.

6. ACKNOWLEDGEMENTS

The authors thank associate professor Cengiz Çelik, who shares his knowledge about the studies conducted in the Iznik observatory and Sultan Eylül Öcal for her contributions to this research during the years 2021-2022.

7. REFERENCES

Australian Transport Safety Bureau, Aviation Occurence Investigation, AO-2008-070, "In-flight upset 154 km west of Learmonth, WA, 7 October 2008, VH-QPAAirbus A330-303", 2022

Bartels J. 1957. The technique of scaling indices K and Q of geomagnetic activity. Ann Intern Geophys Year 4: 215– 226.

Both, Thiago & Wirth, Gilson & Pereira, Cicero & Gonçalez, Odair & Vaz, Rafael Galhardo & Pereira, Marlon & Milagres, Diogo. (2012). Analysis of Total Ionizing Dose Effects on a Pseudo-Static Random Access Memory (PSRAM). ECS Transactions. 49. 69-76. 10.1149/04901.0069ecst.

Brekke, P., Chaloupy, M., Fleck, B., Haugan, S. V., van Overbeek, T., & Schweitzer, H. (2005). Space weather effects on SOHO and its space weather warning capabilities. In *Effects of Space Weather on Technology Infrastructure* (pp. 109-122). Springer Netherlands.

Can H., Peter Svec, Jan Bydzovsky, Peter Svec, Hüseyin Sözeri, Uğur Topal, Systematic optimization of the sensing properties of ring-core fluxgate sensors with different core diameters and materials, Sensors and Actuators A: Physical, Volume 255, 2017, Pages 94-103

Castellanos, J. C., Conroy, J., Kamalov, V., & Hölzle, U. (2022). Solar storms and submarine internet cables. *arXiv preprint arXiv:2211.07850.*

Čop, R., Milev, G., Deželjin, D., & Kosmač, J. (2014). Protection against lightning at a geomagnetic observatory. Geoscientific Instrumentation, Methods and Data Systems, 3(2), 135-141

Elliott, H. A., Jahn, J.-M., & McComas, D. J. (2013). The Kp index and solar wind speed relationship: Insights for improving space weather forecasts. Space Weather, 11(6), 339–349. doi:10.1002/swe.20053

Fleck, B. (1997). First results from SOHO. *Astrophysics and space science*, *258*, 57-75.

Gao, Y., Chen, X., Hu, H., Wen, J., Tang, J., & Fang, G. (2014). Induced electromagnetic field by seismic waves in Earth's magnetic field, Journal of Geophysical Research (Solid Earth), 119, 10.1002/2014JB010962

Girdler, R. W. (1971). Notes on Geomagnetic Observatory and Survey Practice K. A. Wienert (UNESCO, Paris, 1970, 217 pp., £2.11s.). Geophysical Journal International, 23(2), 262–262. doi:10.1093/gji/23.2.262

Grant, W. D. (2006). "Alkali ortamlarda ve Ekstremofiller içinde biyolojik çeşitlilik.", 2006, UNESCO / Eolss Yayınları, Oxford, UK

Hathaway, D. H. (2015). The solar cycle. *Living reviews in solar physics*, *12*, 1-87.

ISGI - International Service of Geomagnetic Indices. https://isgi.unistra.fr/indices_kp.php

Johnston, M. (2002). 38 Electromagnetic fields generated by earthquakes. International Geophysics, 81, 621–635. doi:10.1016/S0074-6142(02)802418

Kerridge, D. (2001, December). INTERMAGNET: Worldwide near-real-time geomagnetic observatory data. In Proceedings of the workshop on space weather, ESTEC (Vol. 34).

Koskinen, Hannu. (2011). Physics of Space Storms. 10.1007/978-3-642-00319-6_8.

Kuusisto, I. GNSS–Risks and threats. In *GEONAVPOS: Seminar publications on Geodesy, Navigation and Positioning* (p. 139).

Love, J.J., (2009), Proceedings of the XIIIth IAGA Workshop on geomagnetic observatory instruments, data acquisition, and processing: U.S. Geological Survey Open-File Report 2009–1226, 271 p.

M. Maniatakos, M. K. Michael and Y. Makris, "Vulnerability-based Interleaving for Multi-Bit Upset (MBU) protection in modern microprocessors," *2012 IEEE International Test Conference*, Anaheim, CA, USA, 2012, pp. 1-8, doi: 10.1109/TEST.2012.6401594.

Marusek, J. A. (2007). *Solar storm threat analysis*.

Matzka, J., Stolle, C., Yamazaki, Y., Bronkalla, O., & Morschhauser, A. (2021). The geomagnetic Kp index and derived indices of geomagnetic activity. Space Weather, 19, e2020SW002641. https://doi.org/10.1029/2020SW002641

Meier, M.M.; Copeland, K.; Klöble, K.E.J.; Matthiä, D.; Plettenberg, M.C.; Schennetten, K.; Wirtz, M.; Hellweg, C.E. Radiation in the Atmosphere—A Hazard to Aviation Safety? *Atmosphere* **2020**, *11*, 1358. https://doi.org/10.3390/atmos11121358

Pirjola, R. (2011). Modelling the magnetic field caused by a dc-electrified railway with linearly changing leakage currents. Earth, Planets and Space, 63(9), 991–998. doi:10.5047/eps.2011.06.032

Prado, Adriane C.M., Federico, Claudio A., Pereira Junior, Evaldo C.F., & Goncalez, Odair L. (2013). Effects of cosmic radiation on devices and embedded systems in aircrafts. INAC 2013: international nuclear atlantic conference, Brazil

Rao, P., Seemala, G., Prasad, J., Prasad, S., Prasad, D., & Niranjan, K. (05 2009). Geomagnetic storm effects on GPS based navigation. Annales Geophysicae, 27. doi:10.5194/angeo-27-2101-2009

Richardson, I.G., "Solar Wind Stream Interaction Regions Throughout the Heliosphere," Living Rev Sol Phys, 2018.

Robbes, D. (2006). Highly sensitive magnetometers—a review. Sensors and Actuators A-physical - SENSOR ACTUATOR A-PHYS, 129, 86–93. doi:10.1016/j.sna.2005.11.023

Robert H. Vandre, DC USA, Janis Klebers, BSEE, Frederick M. Tesche, PhD, Janie P. Blanchard, MSEE, Electromagnetic Pulse (EMP), Part II: Field-Expedient Ways to Minimize its Effects on Field Medical Treatment Facilities, *Military Medicine*, Volume 158, Issue 5, May 1993, Pages 285–289, https://doi.org/10.1093/milmed/158.5.285

Robert H. Vandre, Janis Klebers, Frederick M. Tesche, Janie P. Blanchard, Electromagnetic Pulse (EMP), Part I: Effects on Field Medical Equipment, *Military Medicine*, Volume 158, Issue 4, April 1993, Pages 233–236, https://doi.org/10.1093/milmed/158.4.233

Sergeant, A. M., & Weber, J. C. (2012). Sun storm dangers! Protecting IT, business, and society. *Journal of Corporate Accounting & Finance*, *23*(5), 3-11.

Shprits, Y., Vasile, R., & Zhelavskaya, I. (2019). Nowcasting and Predicting the K p Index Using Historical Values and Real-Time Observations. Space Weather, 17. doi:10.1029/2018SW002141

Sreeja, V. Impact and mitigation of space weather effects on GNSS receiver performance. Geosci. Lett. 3, 24 (2016). https://doi.org/10.1186/s40562-016-0057-0

St-Louis, B.J., Sauter, E.A., Trigg, D.F., and Coles, R.L., INTERMAGNET Technical Reference Manual, Version 4, 1999.

Tafazoli, M. (2009). A study of on-orbit spacecraft failures. *Acta Astronautica*, *64*(2-3), 195-205.

Topal, U. 2013 Meas. Sci. Technol. 24 105110

Topal, U. et al., "Optimization of the Temperature Stability of Fluxgate Sensors for Space Applications," in IEEE Sensors Journal, vol. 21, no. 3, pp. 2749-2756, 1 Feb.1, 2021

W.Chu and G.Qin, "The Geomagnetic Cutoff Rigidities at High Latitudes for Different Solar Wind and Geomagnetic Conditions," Annales Geophysicae, 2016.
World's Top 100 Airlines 2022 | SKYTRAX. (n.d.). SKYTRAX. https://www.worldairlineawards.com/worlds-top-100-airlines-2022/

Zhang,Q. (2022), Modelling Cosmic Radiation Events in the Tree-Ring Radiocarbon Record, Proceedings of the Royal Society A, https://doi.org/10.1098/rspa.2022.0497

IN AND END OF SEASON SOYBEAN YIELD PREDICTION WITH HISTOGRAM BASED DEEP LEARNING

E. Erik[1], M. Durmaz[2,*], A. Ö. Ok[3]

[1]HAVELSAN A.Ş., Ankara, Türkiye - eerik@havelsan.com.tr
[2]Dept. of Geomatics Engineering, Hacettepe University, Ankara, Türkiye - muratdurmaz@hacettepe.edu.tr
[3]Dept. of Geomatics Engineering, Hacettepe University, Ankara, Türkiye - ozgunok@hacettepe.edu.tr

KEY WORDS: Crop Yield Prediction, QGIS Plugin, Google Earth Engine, Histogram, Deep Learning

ABSTRACT:

One sector that feels the effects of global warming and climate change on all levels is agriculture. In order to prepare for possible yield loss, as well as market, storage, and import planning challenges brought on by climate change, businesses can utilise agricultural decision support applications. Within the scope of this study, a crop yield prediction module has been developed that can provide in and end of season estimation of crop yields to be obtained from the determined regions. The Python programming language was used in the creation of the module as a QGIS plugin. The area for which crop yield predictions are to be made is covered by retrieving MODIS SR, MODIS LST, and Daymet data from the Google Earth Engine data catalogue. Histograms obtained from remotely sensed images are used as input data to two deep learning methods (CNN-LSTM and HistCNN). As a result, the HistCNN model outperformed CNN-LSTM for in season soybean yield prediction, with an R^2 of 0.72, while the CNN-LSTM model outperformed it for in end of season soybean yield prediction, with an R^2 of 0.67.

1. INTRODUCTION

The Earth's surface temperature is gradually rising as greenhouse gas emissions from the usage of fossil fuels expand, resulting in long-term local, regional, and global climate change. The effects of climate change include the melting of ice, the change of precipitation patterns, and the migration of wildlife to different regions. In particular, agricultural production is the activity most affected by climate change, and due to unexpected climate conditions, agricultural yields can be reduced. Therefore, it must be carefully monitored and managed.

Apparently, the quantity of yield is particularly vital for developing countries and any decrease has also the impact on farmers, government agencies, and agricultural insurance companies (Basso et al., 2013). In addition, governments require yield estimates prior to harvest for purposes such as arranging for imports and exports, estimating market prices, and determining storage requirements (Cunha and Silva, 2020). Furthermore, post-harvest yield estimates are required to determine the agricultural insurance and governmental support.

Predictions of crop yield can be used to generate pre- and post-harvest forecasts. With the help of remote sensing data, we can extract the factors affecting agricultural yield and then use deep learning techniques to analyse the link between these features and the crop yield. As a first step, the data sources that provide the attributes that have the potential to affect crop yield should be identified. For crop yield analysis, it is essential to analyse the crop's growth status indicators and the environmental conditions where the crop is located.

With the advancement of technology, the availability of remote sensing data and the volume of data that may be utilised have considerably grown (Weiss et al., 2020). Besides, the characteristics that determine crop yield can be obtained from various public sources. Nonetheless, the temporal and spatial integrity of the obtained information is critical. On top of all, a substantial quantity of data is required for the development of effective training models in deep learning approaches. All of these point to the necessity of collecting long-term data for remote sensing from a reliable data source. Besides, such a vast quantity of collected data must be stored and efficiently processed. Because the decision support module to be built is an end-user package, it must be capable of responding without storing and processing data on local computers, allowing the application to operate on as few hardware resources as possible. Therefore, Google Earth Engine (GEE) was chosen as the base platform for this study since it met all of the above-given criteria.

GEE is a cloud computing platform for scientific geospatial data analysis and visualisation (Kennedy et al., 2018). In addition, this Google cloud platform service enables the processing of massive amounts of data with high-performance resources. GEE is therefore one of the most popular platforms for processing the largest geodata sets available. It is highly beneficial for mapping and monitoring crops, phenology-based classification, agricultural yield estimation, and other studies (Kennedy et al., 2018). In this study, GEE's services were used both to access remote sensing data and to analyse the acquired information.

Deep learning algorithms, a subfield of machine learning, accelerate analytical learning by simulating human neural networks with artificial neural networks. In crop yield prediction studies, the Convolutional Neural Network (CNN), the Long Short-Term Memory (LSTM), and the Deep Neural Network (DNN) are the most often adopted deep learning algorithms. Hybrid techniques that combine the two algorithms are also popular (van Klompenburg et al., 2020). CNN-LSTM has been utilised in studies on crop yield estimation and forecasting to extract important characteristics from historical information. It is appropriate to use LSTM networks to make a scalar estimate of crop yield from n temporal images of a year (Khaki et al.,

* Corresponding author

This contribution has been peer-reviewed.
https://doi.org/10.5194/isprs-archives-XLVIII-M-1-2023-95-2023 | © Author(s) 2023. CC BY 4.0 License.

2020). It has been also reported that the CNN-LSTM model is more accurate than CNN-only and LSTM-only models (Sun et al., 2019). Besides, in histogram-based CNN (HistCNN), by using a multi-layer CNN, a hierarchical learning model can be created in which close associations with the input occur in the lower layer and distant relations occur in the downstream CNN layers (Gehring et al., 2017).

As is well known, it's possible to browse, edit, print, and analyse geospatial data with Quantum GIS (QGIS), a free and open-source cross-platform desktop GIS application. Therefore, in this work, a QGIS plugin is developed using Python for crop yield prediction. Besides, the GEE Python API is utilised to access and process the remotely sensed data. Meanwhile, CNN-LSTM and HistCNN deep learning models are tested from the utilised data, which is then used to predict the crop yield. We preferred the 15 states in the United States' CONUS region where soybeans are the most widely produced crop for our study.

This paper is organised as follows: We discussed the study area and the datasets utilised in Section 2. Section 3 goes over the methodology and QGIS plugin that were developed. Section 4 discusses the evaluation strategies for the proposed methodologies as well as the achieved accuracy. Section 5 provides a conclusion with some final remarks.

2. STUDY AREA AND DATASET

This study is conducted in the 15 U.S. CONUS states where soybeans are cultivated the most (Figure 1). Ground truth information of soybean yield was obtained from United States Department of Agriculture (USDA) National Agricultural Statistics Service (NASS) ((dataset) USDA National Agricultural Statistics Service (2017)", n.d.a). As is public knowledge, the USDA's National Agricultural Statistics Service (NASS) conducts hundreds of surveys annually and produces reports covering practically every aspect of U.S. agriculture, and this service provides county-level soybean yield information for the years 2006–2021. As required by deep learning models, ground-truth yield data serves as the necessary labelled information.

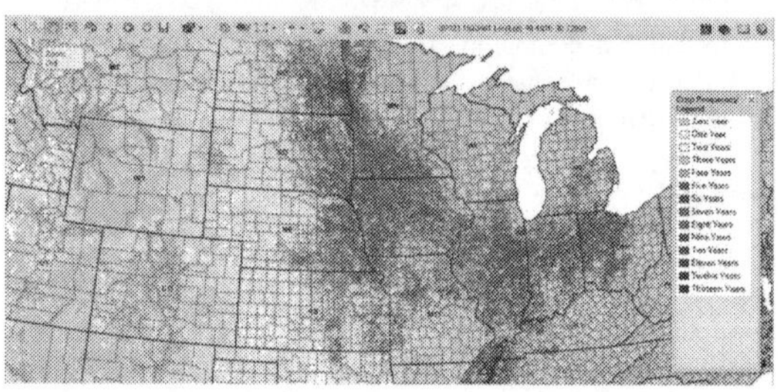

Figure 1. Density map of soybean of the U.S. CONUS region.

In this study, the shapefile vector file can be used to gather information about the borders of the study area. The county boundaries of the United States are collected from the GEE DataCatalog TIGER/2018/Counties dataset. It is used to gather the related geometric information for the region being processed. Moreover, it is used to filter both the input data and the crop data based on the region shape that was derived from the collected information.

Cropland Data Layers (CDL) is a land-cover data layer tailored to crops. Using satellite and agricultural ground truth data, it is produced annually for the United States ((dataset) USDA National Agricultural Statistics Service (2017)", n.d.b). This information is also freely accessible through GEE's USDA/NASS/CDL data catalogue. A crop mask is applied to remotely sensed images to remove areas where the crop is not grown. The code block provided in Appendix 1 demonstrates the generation of a CDL soybean mask using region and year data. Using the produced mask, non-soybean pixels in remote sensing data are filtered out.

The seven band values in the MODIS surface reflectance (SR) data can provide information such as crop growth status indicators. MODIS land surface temperature (LST) data provide daily and nightly ground surface temperature information, whereas Daymet weather data include information on vapour and precipitation (Table 1). These data are analysed to determine the crop's environmental conditions. All of this input data can be accessed using the GEE data catalogue by providing the necessary study area information (Table 2).

Data Name	Catalog Name	Resolution
MODIS SR	MOD09A1	500 m - 8 days
MODIS LST	MOD11A2	1 km - 8 days
DAYMET V4	DAYMET/_V4	1 km -1 day

Table 1. Image dataset utilized in this study.

Data Catalog Name	Bands
MODIS/006/MOD09A1	sur_refl_01- sur_refl_07
MODIS/006/MOD11A2	LST_Day, LST_Night
NASA/ORNL/DAYMET_V4	prcp, vp

Table 2. Data catalog names and the bands utilized.

3. METHODOLOGY

In this section, we detail our strategy in four sub-sections: (i) preparation of the input dataset, (ii) tensor generation, (iii) prediction strategy, and (iv) QGIS plugin development.

3.1 Preparation of the Input Dataset

In the first step, the data from MODIS SR, MODIS LST, and Daymet are required to be put together to generate a total of eleven bands. Note that the theoretical upper and lower bounds for these bands are too large; therefore, in this study, we used the min-max pixel values within the study area as our upper and lower bounds, respectively. In Table 3, the Old Min-Max column shows the real band limits, whereas the New Min-Max column displays the values valid for our study region. During implementation, the related remote sensing data are collected using county geometry and year information. Note that, unlike other data, the cloud removal process is applied to MODIS SR data.

Feature	Old Min-Max	New Min-Max
MOD09A1	-100 - 16000	1 - 5000
MOD11A2	7500 - 65535	12400 - 15600
Precipitation (mm)	0 - 544	0 - 35
Vapor pressure (Pa)	0 - 8230	0 - 3200

Table 3. Theoretical and calculated ranges for input features.

It is necessary to collect historical data features and crop yield information in order to obtain the training data necessary for the deep learning model. In this study, a histogram-based approach is used to collect information on the yield of all band values and to gain more information with fewer data. This approach assumes that meaningful information may be derived from the number of different pixels in images acquired for the study area, disregarding the region's geographical features (You et al., 2017).

3.2 Tensor Generation

Using the histogram method, it is possible to determine the characteristics of annual yield data by combining histograms sequentially. Obtaining histograms of standard sizes and associating them with the annual yield delivers a tensor. Figure 2 illustrates the generation steps for tensors. While creating the tensors, the extent of the study area and the starting and ending dates of the training data are given.

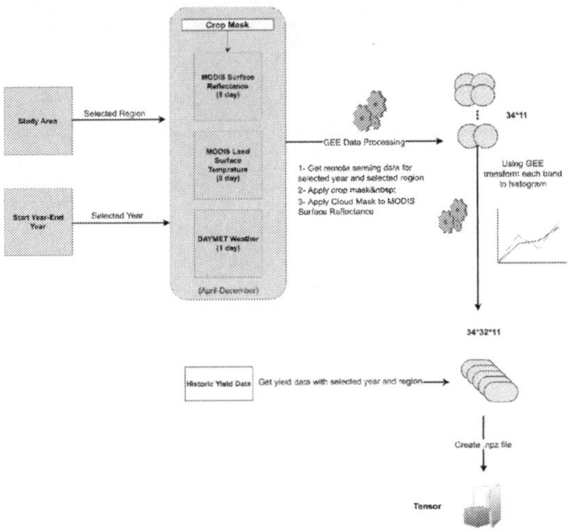

Figure 2. General workflow of tensor generation.

This study follows the soybean plant growth period from April to December. During this period, 34 different remote sensing images are obtained at 8-day intervals. With the help of GEE, we are capable of collecting MODIS SR, MODIS LST, and Daymet weather data for our study area and applying crop masking. To ensure that Daymet weather data has the same periodicity as MODIS data, 8-day average values are calculated. The 34 images that result, each with 11 bands, are saved as a GEE ImageCollection. In this case, filtering, masking, and band merging are all accomplished with the assistance of GEE's capabilities, and the GEE functions. GEE functions used for filtering and masking MODIS LST data are provided in Appendix 2. Notice that all image data is analysed directly on Google Cloud servers, without the datasets being downloaded to the local computer. Tensors are created for each county in the study region for the years 2006–2019. If there is no information regarding soybean production during a year, the tensor for that year is ignored.

In this study, CNN layers are used to extract the properties affecting crop yield using surface reflection values, day-night temperature values and weather data. The 3D sequential histograms obtained in $time \times bin \times bands$ format are used in the 2D CNN to extract important features that affect the crop yield. As mentioned before, the created ImageCollection has 34 images, each with 11 bands. Each image band is divided into 32 bins during the histogram generation step. Hence, an npz-formatted tensor is formed utilising the $34 \times 32 \times 11$ ImageCollection and the yield information.

The CNN-LSTM and HistCNN models are selected as prediction models in this study. In addition, 5-fold cross validation is carried out to achieve an appropriate deep learning model configuration. During implementation, the tensors for the years 2006–2019 are used as training data, while the tensor data for the year 2020 is reserved for testing purposes.

3.3 Prediction Strategy

Crop yield prediction is important both for in season and end of season. In season prediction is important and preferred to supply support for agricultural decisions. Farmers, decision makers and other users can take related precautions and build plans. Besides, end of season prediction is important for assessing farmer statements for agricultural insurance and government agency support payments.

In this study, we performed crop yield prediction for both in season and end of season types. Only data from April through the period determined in the season are used in this work for in season predictions. When predicting the end of season out, all relevant information from April to December is considered.

3.4 Development of QGIS Plugin

QGIS Crop Yield Prediction module has three different actors; user, GEE and QGIS. Figure 3 depicts "The Prepare Training Dataset," the first activity diagram for which a user interface was also developed. This interface requires the user to select the application working directory, study region, crop type, and training years. The application's workspace is called the "Working Directory." It stores study area, trained models, prediction feature, trained features, yield label data, crop data layers. All of the information is kept in different subfolders consisting of StudyAreas, TrainedModels, TrainingData, YieldLabelData and Auxiliary-Data. It should be selected by the user at the start of the application. Additionally, the "Prepare Training Dataset" UI receives input data from GEE. After that when with "Create Training Data" action tensor creation process is started. When training data is ready to use, it saves under the working directory to use when training phase.

"Train Prediction Model" is another UI tab inside the developed plugin. In this step, the plugin expects the user to select the prediction type and training model. The activity diagram of this UI tab is displayed in Figure 4. "Predict Yield" is the final tab in the plugin's interface (Figure 5). Both user and GEE inputs are received. The prediction process begins with the selection of a region and a model. When the prediction is completed, results is shown as temporary shape file on the map. In Figure 6, predictions that in the prepared temporary shape file is shown. Predictions are given in kilogrammes per hectare.

4. RESULTS AND DISCUSSION

This section presents the study's findings and addresses the issues experienced during the data analysis. In total, six different models were created for HistCNN and seven different models

https://doi.org/10.5194/isprs-archives-XLVIII-M-1-2023-95-2023

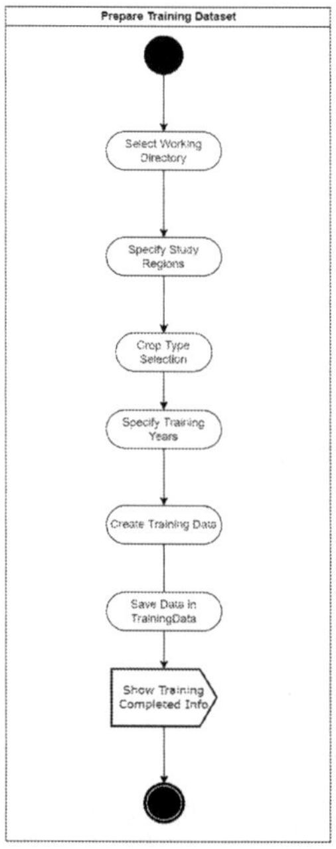

Figure 3. Prepare Training Dataset

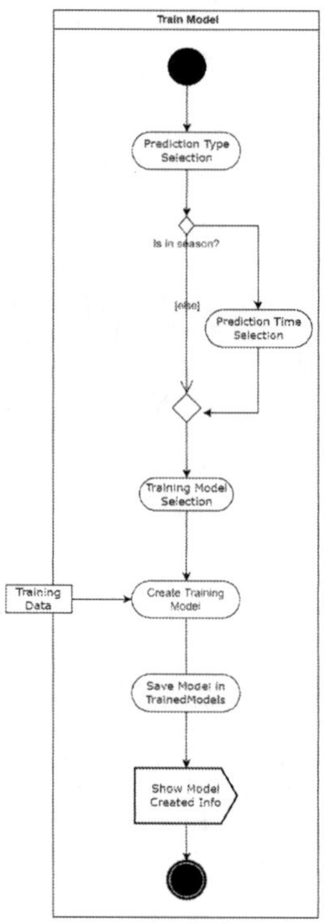

Figure 4. Train model.

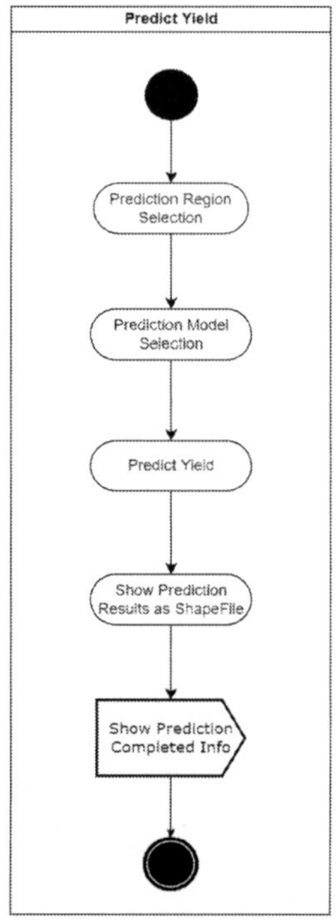

Figure 5. Predict yield.

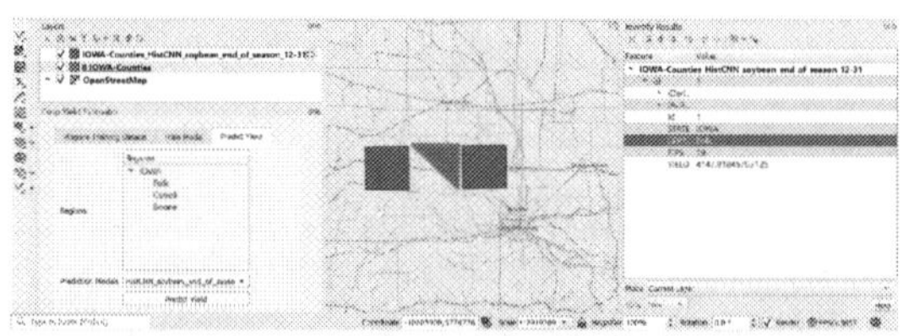

Figure 6. Preparing training dataset is completed interface.

were created for CNN-LSTM. Different model combinations were created by differentiating the number of convolution layers, the number of filters, and the number of dense layer neurons compared to the base model. Each model was separately evaluated for in season and end of season predictions. For in season yield predictions, the 18th time is selected.

The effects of different parameters and their combinations on the accuracy of the crop yield prediction are investigated. The training and test data are split randomly to compare all model configurations. Thus, a 5-fold cross validation is used to compare each model objectively. Finally, model selection is performed using training data. Table 4 and Table 5 show the created model configurations.

Table 6 presents the 5-fold cross validation results of the HistCNN models. The 2th and 4th models provide the worst results for in season and end of season predictions. This result proves that the convolution layer and the filters are important for the performance of the HistCNN model. Table 7 presents the model comparisons of the CNN-LSTM models after the cross validation. Amongst the models, 3th model configuration is the most successful model for both prediction types. 7th model configur-

HistCNN Models	Conv Filter Sizes	Conv Count	Dense Count	Dense Neuron Count
1.Model	128, 256, 512	3	1	64
2.Model	32, 64, 128	3	1	64
3.Model	64, 128, 256, 512	4	1	64
4.Model	128, 256	2	1	64
5.Model	128, 256, 512	3	2	64,32
6.Model	128, 256, 512	3	1	32

Table 4. HistCNN model configuration matrix.

CNN-LSTM Models	Conv. Filters	Conv. Count	LSTM Neurons	Dense Count	Dense Neurons
1.Model	32,64	2	256	1	64
2.Model	64,128	2	256	1	64
3.Model	128,256	2	256	1	64
4.Model	32,64,128	3	256	1	64
5.Model	32,64	2	256	2	64,32
6.Model	32,64	2	256	1	128
7.Model	32,64	2	512	1	64

Table 5. CNN-LSTM model configuration matrix.

ation has the worst error rate among the in season predictions. Therefore, it can be said that the increase in the number of dense layer neurons negatively affected the deep learning output in our case.

Model Name	In Season (18th)		End of Season	
	RMSE	R^2	RMSE	R^2
1.Model	330.807	0.77	309.950	0.80
2.Model	353.455	0.74	331.638	0.77
3.Model	308.418	0.80	317.138	0.79
4.Model	331.427	0.77	336.802	0.76
5.Model	323.097	0.79	333.207	0.77
6.Model	334.020	0.77	324.690	0.78

Table 6. 5-fold cross validation results of HistCNN model.

CNN-LSTM and HistCNN model were tested with test data consisting of 128 tensors that were not included in the training data. According to CNN-LSTM test results, the best end-season yield estimates were obtained with the 2nd model configuration with R^2 of 0.67. The best in season yield estimation results were obtained with the base model with R^2 of 0.56.

In the HistCNN test results, while the 3rd model is the most successful model in predicting soybean yield in season with R^2 of 0.72, the 2nd model is the most successful model in end of season predictions with R^2 of 0.62. Besides, the lowest prediction results were obtained with 4th model configuration for both prediction types with R^2 of 0.3.

The R^2 values derived from the prediction data are lower than those derived from the 5-fold cross validation data. This could be due to the fact that the only data used for testing is for the year 2020. Using 128 tensors from 15 randomly selected states was also incapable of producing reasonable R^2 results. In addition, the prediction results indicate that HistCNN models perform better than CNN-LSTM models for both in season and end of season predictions.

Model Name	In Season (18th)		End of Season	
	RMSE	R^2	RMSE	R^2
1.Model	364.152	0.73	333.333	0.77
2.Model	350.693	0.75	338.191	0.76
3.Model	337.279	0.77	314.482	0.79
4.Model	363.638	0.73	320.531	0.79
5.Model	347.959	0.75	315.476	0.79
6.Model	363.347	0.73	331.090	0.77
7.Model	409.001	0.66	325.191	0.78

Table 7. 5-fold cross validation results of CNN-LSTM model.

The module for predicting crop yields that has been developed relies on surface reflectance data for determining the current state of the crop and on analyses of environmental factors such as temperature, precipitation, and vapour to better better predict crop yields. However, in reality, the prediction module needs to account for farmers' spraying, irrigation, and fertilisation practises. Examining how the soil's properties have changed over time is also crucial for estimating yield.

5. CONCLUSION

This study focuses on developing a QGIS plugin for crop yield prediction. It delivers in season and end of season crop yield predictions by using remote sensing data and deep learning methods. A key component is the use of GEE, a cloud computing platform for the collection and processing of remote sensing data.

During the methodology, images were standardised using a histogram-based strategy so that deep learning models could learn more information with less input. HistCNN and CNN-LSTM models were constructed using the Tensorflow library. Several model configurations were tested, and RMSE, MSE, and R^2 metrics were used to evaluate the results. 5-fold cross validation was applied to determine the optimum deep learning model configuration. As a result, the HistCNN model outperformed CNN-LSTM for in season soybean yield prediction, with an R^2 of 0.72, while the CNN-LSTM model outperformed it for in end of season soybean yield prediction, with an R^2 of 0.67.

In future studies, in addition to remote sensing data, farmers' agricultural practises and soil characteristics can be used to obtain more reliable and accurate yield predictions. In addition, it is thought that the GEE "Batch Environment" method can decrease the training data duration. Moreover, the plugin user interface can be effective when using the batch processing method. However, in future studies, the performance of the plugin developed in this study on different crop types, such as corn and cotton, can be addressed.

ACKNOWLEDGEMENT

This study was carried out as the first author's Master of Science thesis titled "In and End of Season Soybean Yield Prediction with Histogram Based Deep Learning."

REFERENCES

Basso, B., Cammarano, D., Carfagna, E., 2013. Review of Crop Yield Forecasting Methods and Early Warning Systems. *The*

First Meeting of the Scientific Advisory Committee of the Global Strategy to Improve Agricultural and Rural Statistics, 1–56.

Cunha, R. L. D. F., Silva, B., 2020. Estimating Crop Yields with Remote Sensing and Deep Learning. *2020 IEEE Latin American GRSS and ISPRS Remote Sensing Conference, LAGIRS 2020 - Proceedings*, 273–278.

(dataset) USDA National Agricultural Statistics Service (2017)", year=2021, m. n., n.d.a. NASS - Quick Stats. USDA National Agricultural Statistics Service. https://data.nal.usda.gov/dataset/nass-quick-stats.

(dataset) USDA National Agricultural Statistics Service (2017)", year=2021, m. n., n.d.b. Usda national agricultural statistics service cropland data layer. https://nassgeodata.gmu.edu/CropScape.

Gehring, J., Auli, M., Grangier, D., Yarats, D., Dauphin, Y. N., 2017. Convolutional Sequence to Sequence Learning. http://arxiv.org/abs/1705.03122.

Kennedy, R. E., Yang, Z., Gorelick, N., Braaten, J., Cavalcante, L., Cohen, W. B., Healey, S., 2018. Implementation of the LandTrendr algorithm on Google Earth Engine. *Remote Sensing*, 10(5), 2019–2021.

Khaki, S., Wang, L., Archontoulis, S. V., 2020. A CNN-RNN Framework for Crop Yield Prediction. *Frontiers in Plant Science*, 10(January), 1–14.

Sun, J., Di, L., Sun, Z., Shen, Y., Lai, Z., 2019. County-level soybean yield prediction using deep CNN-LSTM model. *Sensors (Switzerland)*, 19(20), 1–21.

van Klompenburg, T., Kassahun, A., Catal, C., 2020. Crop yield prediction using machine learning: A systematic literature review. *Computers and Electronics in Agriculture*, 177(August), 105709. https://doi.org/10.1016/j.compag.2020.105709.

Weiss, M., Jacob, F., Duveiller, G., 2020. Remote sensing for agricultural applications: A meta-review. *Remote Sensing of Environment*, 236(December 2018), 111402. https://doi.org/10.1016/j.rse.2019.111402.

You, J., Li, X., Low, M., Lobell, D., Ermon, S., 2017. Deep gaussian process for crop yield prediction based on remote sensing data.

A. APPENDIX

Appendix 1- The function of generating soybean mask using CDL data.

```
def create_soybean_mask(self):
# Eliminating non soybean pixels
  cdl = ee.ImageCollection('USDA/NASS/CDL')
  .filter(ee.Filter.date
  (self.historic_start_date,
  self.historic_end_date))
  .filterBounds(self.region)

   def createMask(cdl):
     # Soybean mask for selected year
     imageCdl = ee.Image(cdl.toList(1)
     .get(0))
     soybeanMask = createMask(imageCdl)
     return soybeanMask
```

Appendix 2- Function to retrieve MODIS LST data for the selected year and region.

```
def getMODLST(self):
 MOD_lst = self.create_image_collection
 ('MODIS/006/MOD11A2',
 self.start_date,
 self.end_date, self.region)
 MOD_lst = MOD_lst.select(self.LST_bands)

 def updateMasking(image):
  return image.updateMask(self.soybeanMask)

 MOD_lst = MOD_lst.map(updateMasking)
```

CONTRIBUTION OF URBAN CEMETERIES TO ECOSYSTEM SERVICES: EVIDENCE FROM TOURISTIC ANTALYA CITY OF TURKEY

B. Eyileten [1], S. Selim [2]*

1 Akdeniz University, Institute of Science, Department of Remote Sensing and Geographical Information Systems, 07058 Antalya, Turkey - buketeyileten@gmail.com
2 Akdeniz University, Faculty of Science, Department of Space Science and Technologies, 07058 Antalya, Turkey - serdarselim@akdeniz.edu.tr (* corresponding author)

KEY WORDS: Ecosystem Services, Urban Cemeteries, Carbon Sequestration, I-Tree Canopy, Remote Sensing.

ABSTRACT:

This study focuses on concretely determining the contribution of cemeteries to the urban ecosystems in Antalya, one of the most important tourism destinations of Turkey. The study covers three cemeteries (Andızlı, Arapsuyu, and Uncalı Cemeteries) located in the city center of Antalya, with an area of 14.02 ha, 5.12 ha, and 22.96 ha, respectively. These areas are classified into two classes as vegetation and non-vegetation to calculate ecosystem services with open access and user-friendly i-Tree Canopy software to calculate the contribution of green areas in the mentioned cemeteries to the regulatory services of the ecosystem. According to the results, the green cover in the Andızlı Cemetery constitutes 81.64% of the total cemetery area and holds 54.35 tons of carbon per year. In terms of combating air pollution, it has been determined that the cemetery contributes to ecosystem services by removing a total of 1085.60 kilos of CO, NO_2, O_3, SO_2, and PM annually. Similarly, with its 4.88% impermeable buildings, 2.93% other impermeable surfaces, 18.55% soil ground, and 73.63% vegetation, the green cover of Arapsuyu Cemetery captures 357.96 kg of pollutants per year. Lastly, Uncalı Cemetery was shown to remove 1466.01 kg of air pollutants and 73.40 tons of carbon. Cemeteries that are part of the green infrastructure system and its contributions to ecosystem services are emphasized among the proposals to be made in line with the 2050 targets of the European Green Deal. In this context, by making legal and administrative additions to the spatial planning throughout the city, the realization of this intention might be accelerated.

1. INTRODUCTION

Cemeteries are a reflection of the history of the countries they are in and are considered sacred places all over the world (Kowarik et al., 2016). Cemeteries also form protected open green spaces which are elements of the urban green infrastructure system (Sallay et al., 2022). The presence of both natural and cultural elements in cemeteries, particularly in urban areas, gives them significant value for their potential to enhance ecosystem services. In Turkey, cemeteries are considered sacred and are primarily used for worship rather than recreation, much like in other predominantly Muslim countries. However, these are still valuable components of the urban green infrastructure system due to their nature as protected green areas. (Quinton and Duinker, 2019; McClaymont and Sinnet, 2021). The green areas of cemeteries provide important contributions to ecosystem services in terms of improving air, water and soil quality, balancing the microclimate and increasing carbon sequestration (Balzan et al., 2021). The status of the ecosystem services provided by cemeteries, the factors that improve or weaken the conditions of these services and all their possible future effects, and the benefits of these services being and staying high quality can have on people can be increased through creating more effective policies and decision-making mechanisms by regional administrations. This study focuses on concretely determining the contribution of cemeteries to the urban ecosystem in the city center of Antalya, one of the most important tourism destinations of Turkey, whose foreign population has increased significantly in parallel with current international developments.

Evaluation of Urban Ecosystem services contributes significantly to learning about the state of the environment and the sustainable management of natural resources and making improvements (Frélichová et al., 2014). A large number of studies have been and continue to be made in the literature on the conservation of biodiversity, the management of ecosystems, sustainable urban life, the maintenance of natural resources, and ecosystem services for human well-being (Reid et al., 2006; Elmqvist et al., 2015; Schmidt et al., 2016; Fagerholm, et al., 2019; McPhearson et al., 2022). The main objectives of evaluating the contribution of green areas to ecosystem services are: to make the value of nature visible, to integrate the value of ecosystem services into urban decision-making processes, and to enable ecosystem services to be managed effectively by developing tools such as maps that identify and quantify these services (Niemelä et al., 2010; Cilliers et al., 2013; Kabisch, 2015). In addition, in the biodiversity and green infrastructure strategies of the European Green Consensus, it is recommended to develop various strategies in order to identify ecosystem services and convert them into digital data, and to create a carbon-neutral environment in the 2030 - 2050 targets (European Commission, 2019). In this context, determining the contribution of urban green spaces such as cemeteries to ecosystem services can help strengthen green space management and conservation efforts by measuring the environmental benefits of green spaces.
There are many methods for identifying and modeling ecosystem services and their advantages. Especially in the last 10 years, as a result of the developments in remote sensing and geographic information systems, the contributions of ecosystem services can be determined and mapped quickly and with high accuracy with the help of various interfaces such as GIS and similar environments. For example, (i) ESTIMAP models from land cover and land use maps to which it adds other spatial information to map various ecosystem services (Zulian et al., 2014; Stange et al., 2017); (ii) The QUICKScan tool performs a spatial modeling combining expert knowledge with spatial and statistical data, visualizing the results on interactive maps. (Winograd et al., 2014; Dick et al., 2017), (iii) InVEST produces a set of models for mapping and assessing the ecological or economic value of multiple ecosystem services at a local or

https://doi.org/10.5194/isprs-archives-XLVIII-M-1-2023-101-2023

regional scale (Arcidiacono et al., 2015; Benra et al., 2021), and (iv) I-Tree Canopy classifies green areas according to the presence of vegetation and quantifies the contribution of green areas to ecosystem services (Olivatto, 2019).

In this study, I-Tree Canopy tool was used to quantify the role of urban cemeteries in regulating ecosystem services. The I-Tree Tools, developed by the USDA Forest Service and other collaborators, are freely available and can be used to assess ecosystem services online. This tool is widely used globally and offers a practical and efficient way to evaluate the presence and absence of tree cover, from individual trees to large forested areas, and their contributions to ecosystem services, as demonstrated in previous studies (Mills et al., 2015; Nowak et al., 2018; Selim et al., 2022). In addition, this tool allows users to make statistically reliable predictions about tree and other cover types, as well as calculating and correcting uncertainties that may occur in forecasts because Google uses aerial images (Selim et al., 2022). In this study, the 3 largest cemeteries of Antalya were chosen as the study area. The land covers of these cemeteries were classified and following values were calculated: carbon monoxide (CO) removed annually, nitrogen dioxide (NO_2) removed annually, ozone (O_3) removed annually, sulfur dioxide (SO_2) removed annually, particulate matter less than 2.5 microns removed annually ($PM_{2.5}$) and finally, particulate matter greater than 2.5 microns and less than 10 microns (PM_{10}) removed annually. Lastly, suggestions have been developed for central and local governments and policy makers to evaluate cemeteries in the creation of green infrastructure systems.

2. MATERIAL AND METHODS

2.1 Study Area

As of the end of 2021, Antalya is the fifth most populous city in Turkey with its 2,619,832 residents. Especially in the summer months, the active population increases with the number of tourists. The area of the city is 20,177 km2 (Antalya Governorship, 2023). The city is between 29° 20'-32°35' east longitudes and 36° 07'-37° 29' north latitudes in southwestern Turkey. (Figure 1).

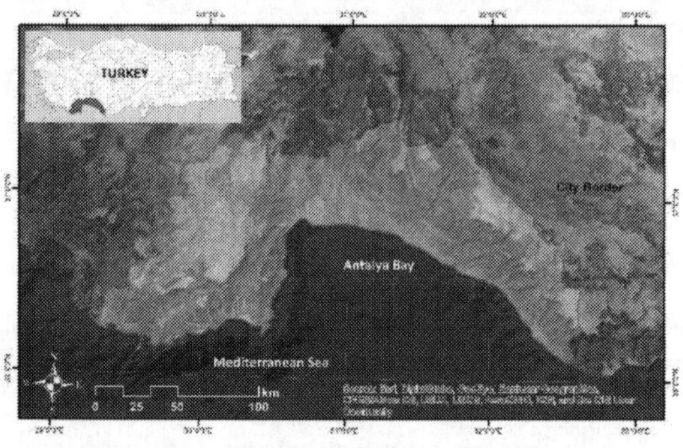

Figure 1. Study area

On average, 77.8% of the province's land is mountainous, 10.2% is plain, and 12% is rough. The climate of Antalya is generally characterized with Mediterranean climate. The average temperature in summer is between 28-36 °C. In January, the average temperature varies between 10-20 °C (Selim et al., 2018; Selim, 2021; Olgun et al., 2022). The cemeteries that are the subject of the study are Uncalı cemetery (22.96 ha), Arapsuyu cemetery (5.12 ha) and Andızlı cemetery (14.02 ha) located in the city center of Antalya (Figure 2, 3, 4).

Figure 2. Uncalı cemetery

Figure 3. Arapsuyu cemetery

Figure 3. Andızlı cemetery

Uncalı cemetery at 36°53'53.57"N and 30°37'29.17"E coordinates, Arapsuyu cemetery at 36°52'49.52"N and 30°39'19.76"E coordinates and Andızlı cemetery at 36°53'37.96"N and 30°42'31.54"E coordinates. The dominant plant species in cemeteries is mostly Cupressus Sempervirens. In addition, Pinus brutia and Pinus pinea populations are also present.

https://doi.org/10.5194/isprs-archives-XLVIII-M-1-2023-101-2023 | © Author(s) 2023. CC BY 4.0 License.

2.2 Method

The study employed the I-Tree canopy tool and a recent Google Earth satellite image as the primary data sources. The region's zoning plans were also utilized to establish the boundaries of the relevant cemeteries, with the process of digitization being conducted via ArcGIS 10.4.1 software (see Figure 4).

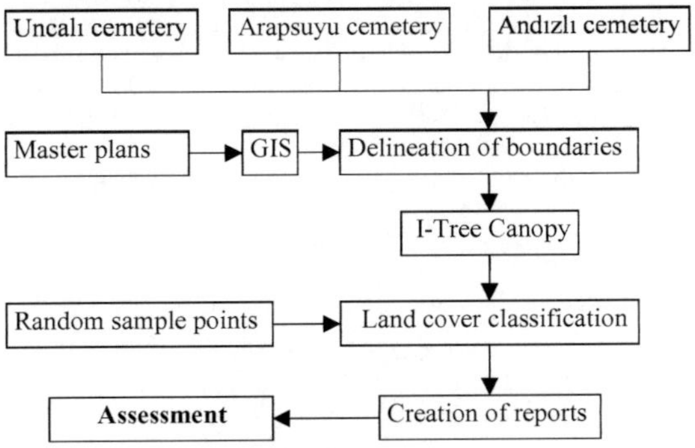

Figure 4. Method flowchart

The study consists of 3 main stages: determining the boundaries of the cemetery, assigning a sufficient number of sampling points, and creating analysis reports. In the first stage, the boundaries determined in the zoning plans were transferred to the I-Tree canopy software. Then, sufficient number of sampling points were assigned for each cemetery to capture the minimum standard error (SE) predicted by the mentioned software. Land cover is divided into 4 classes: buildings, impervious surface, soil ground and tree cover. Buildings are defined as "areas with impervious cover occupied by buildings". Impervious surface is defined as "areas with other impervious cover (e.g., pavements, roads, parking lots and surface areas covered with concrete, key parquet stone coating)". Soil ground is defined as "soil surface" or" bare areas without vegetation". Finally, tree cover was defined as "areas covered with trees and tall bush vegetation". I-Tree Eco software and literature information (Ersoy Tonyaloğlu & Atak, 2021; Selim & Atabey, 2020) were used to calculate the contribution of the land cover per unit area to the regulation services of the ecosystem. After classification, reports of each cemetery were created within the scope of the CO, NO_2, O_3, SO_2, $PM_{2.5}$ and PM_{10} values of the tree cover, and numerical values were obtained based on air pollution and carbon sequestration. Finally, the findings were interpreted by discussing with the literature information within the scope of the contributions of the related cemeteries to the ecosystem services.

3. RESULTS

The carbon holding capacity of the cemeteries, which are an important part of the urban green space system, and the amount of their effects on reducing air pollution have been determined. The cemeteries, located in the city center of Antalya and dominated by vegetation, provide important contributions to the urban ecosystem and urban ecosystem services. Uncalı cemetery (Table 1), 67.29% of which is covered with tree cover, absorbs 73.40 tonnes of carbon annually.

Cover Class	Points	% Cover ± SE	Area (ha) ± SE
Impervious Building	12	0.52 ± 0.15	0.12 ± 0.03
Impervious Other	48	2.09 ± 0.30	0.48 ± 0.07
Soil/Bare Ground	691	30.10 ± 0.96	6.91 ± 0.22
Tree/ Shrub	1545	67.29 ± 0.98	15.45 ± 0.22
Total	2296	100.00	22.96

Table 1. Land cover classes of Uncalı cemetery.

2.09% of Uncalı cemetery is classified as impervious other and 30.10% as soil/bare ground. The effect of the cemetery on reducing air pollution is given in Table 2.

Abbr.	Description	Amount (kg)	±SE
CO	Carbon Monoxide removed annually	22.45	±0.33
NO_2	Nitrogen Dioxide removed annually	104.49	±1.52
O_3	Ozone removed annually	989.36	±14.40
SO_2	Sulfur Dioxide removed annually	20.64	±0.30
$PM_{2.5}$	Particulate Matter less than 2.5 microns removed annually	43.89	±0.64
PM_{10}*	Particulate Matter 2.5<PM<10 microns removed annually	285.18	±4.15
Total		1,466.01	±21.33

Table 2. Contribution of Uncalı cemetery to ecosystem services

It is estimated that Uncalı Cemetery's contribution to ecosystem services that reduce the effects of air pollution is 1,466,01 kg per year.

Arapsuyu cemetery, on the other hand, is a small cemetery with an area of 5.12 ha compared to other cemeteries (Table 3). The annual carbon holding capacity of the vegetation in this cemetery is 17.81-tonnes.

Cover Class	Points	% Cover ± SE	Area (ha) ± SE
Impervious Building	104	5.22 ± 0.50	0.27 ± 0.03
Impervious Other	59	2.96 ± 0.38	0.15 ± 0.02
Soil/Bare Ground	372	18.66± 0.87	0.96 ± 0.04
Tree/ Shrub	1459	73.17 ± 0.99	3.75± 0.05
Total	1994	100.00	5.12

Table 3. Land cover classes of Arapsuyu cemetery.

73.17% of the Arapsuyu cemetery is covered with vegetation and consists of Cupressus sempervirens populations. Arapsuyu cemetery absorbs 5.45 kg of CO, 240.06 kg of O3 and 5.01 kg of SO2 per year (Table 4).

Abbr.	Description	Amount (kg)	±SE

https://doi.org/10.5194/isprs-archives-XLVIII-M-1-2023-101-2023

CO	Carbon Monoxide removed annually	5.45	±0.07
NO_2	Nitrogen Dioxide removed annually	25.35	±0.34
O_3	Ozone removed annually	240.06	±3.26
SO_2	Sulfur Dioxide removed annually	5.01	±0.07
$PM_{2.5}$	Particulate Matter less than 2.5 microns removed annually	10.65	±0.14
PM_{10}*	Particulate Matter 2.5< PM<10 microns removed annually	69.20	±0.94
Total		355.71	±4.82

Table 4. Contribution of Arapsuyu cemetery to ecosystem services

Arapsuyu cemetery is spread over a smaller area compared to other areas in the study and its contribution to urban ecosystem services has been estimated as 355.71 kg in total.

Andızlı cemetery, on the other hand, has the largest green cover with an area of 14.02 ha and 81.64% vegetation (Table 5). Andızlı cemetery cleans 54.35-tonnes of carbon from the city annually.

Cover Class	Points	% Cover ± SE	Area (ha) ± SE
Impervious Building	13	0.93 ± 0.26	0.13 ± 0.04
Impervious Other	52	3.71 ± 0.51	0.52 ± 0.07
Soil/Bare Ground	192	13.71 ± 0.92	1.92 ± 0.13
Tree/ Shrub	1143	81.64 ± 1.03	11.44 ± 0.15
Total	1400	100.00	14.02

Table 5. Land cover classes of Arapsuyu cemetery.

The CO clearance amount of the Andızlı cemetery was estimated to be 16.62 kg, the NO_2 removal amount to 77.38 kg and the O_3 removal amount to 732.63 kg (Table 6).

Abbr.	Description	Amount (kg)	±SE
CO	Carbon Monoxide removed annually	16.62	±0.21
NO_2	Nitrogen Dioxide removed annually	77.38	±0.98
O_3	Ozone removed annually	732.63	±9.28
SO_2	Sulfur Dioxide removed annually	15.29	±0.19
$PM_{2.5}$	Particulate Matter less than 2.5 microns removed annually	32.50	±0.41
PM_{10}*	Particulate Matter 2.5< PM<10 microns removed annually	211.18	±2.68
Total		1,085.60	±13.76

Table 6. Contribution of Andızlı cemetery to ecosystem services

The total contribution of the Andızlı cemetery to ecosystem services within the scope of reducing the effects of air pollution is approximately 1 ton per year.

The results of the study show that, considering their area size and tree cover rate, the cemetery with the highest contribution to the ecosystem services of the city among the three studied cemeteries is the Uncalı cemetery. Uncalı cemetery has an annual sequestration value of 1466.01 kg of CO, NO_2, O_3, SO_2, $PM_{2.5}$ and PM_{10}. This is followed by the Andızlı cemetery with 1085.60 kg and the Arapsuyu cemetery with 355.71 kg, respectively. The annual total carbon sequestration amount of these three cemeteries in the whole city is 145.56 tons. In other words, these cemeteries absorb a significant amount of carbon in the atmosphere. As the area and tree cover rate increases, the contributions provided by the cemeteries also increase. However, as can be seen from the maps produced, these cemeteries are independent of other green areas in the city and their ecological connections have not been established. Hence, they are isolated ecosystems.

4. CONCLUSIONS

Regulating services, provisioning services, cultural services and supporting services of urban ecosystems are vital for the health, sustainability and resilience of urban environments (Evans et al., 2022). The increasing population and the pressure of global climate change on urban environments necessitate the creation and maintenance of green infrastructure systems (Lourdes et al., 2022). Similar to urban parks, cemeteries represent an important part of the urban ecosystem as a semi-natural habitat for many plant and animal species. Although cemeteries have limited use, they are an important part of the urban green infrastructure network (Sallay et al., 2022). Since the 2000s, ecosystem services and green infrastructure approaches have begun to be considered more holistically in spatial planning applications involving urban systems (Ersoy Mirici, 2022). The European Commission posits that ecosystem services serve as a supplementary mechanism to green infrastructure, and it is crucial to ascertain the role of individual components of the green infrastructure system, such as cemeteries, in providing ecosystem services (Klimanova et al., 2018; Marando et al., 2022). Cemeteries constitute an array of ecosystem services and diverse habitat types as constituents of green infrastructure in urban localities (Długoński et al., 2022). Cemeteries play a significant role in augmenting the contribution of ecosystem services towards enhancing the adaptability and resilience of cities to climate change. The European Commission advocates for the integration of distinct green spaces, including cemeteries, into a systematically planned green infrastructure system, along with their ecological connectivity (Lanzas et al., 2019).

The results obtained from this study clearly indicate that urban cemeteries contribute significantly to the regulatory services of the ecosystem. Especially considering the reduction of carbon sequestration and air pollution effects, the vegetation in the cemeteries contributes significantly to the regulatory services of the ecosystem. Regulating the microclimate and creating habitat for urban wildlife can be listed as other roles of these cemeteries. Nonetheless, in the urban setting of Antalya, where population density is high and mobility is frequent, cemeteries are disconnected from the surrounding green spaces and remain confined within the city. It is not linked to other components of the green infrastructure system such as urban parks, rivers, ecological corridors and other green spaces. This limits the potential contribution of related cemeteries to ecosystem services. Because interconnected green spaces have the capacity to provide ecosystem services to a large extent (Phillips et al.,

This contribution has been peer-reviewed.
https://doi.org/10.5194/isprs-archives-XLVIII-M-1-2023-101-2023 | © Author(s) 2023. CC BY 4.0 License.

2020). In accordance with the patch-corridor-matrix paradigm of landscape ecology, cemeteries, viewed as a component of the landscape patch, can serve as an integral part of the green infrastructure system in tandem with ecological corridors. Thus, increases the connectivity possibility between important patches, such as cemeteries, urban parks, roads, urban rivers and other green spaces (Peng et al., 2017). This system, including cemeteries, creates resilient ecosystems that play an important role in supporting the ecological structure, such as integrating landscapes, eliminating the isolated island effect, ensuring the spread and gene flow of species, regulating the microclimate, reducing urban heat island effects, etc. Also, as the European Green Deal states under the heading "Restoring biodiversity and ecosystem services", resilient ecosystems are natural absorbers of atmospheric CO2 and can support adaptation to climate change (European Commission, 2020).
As a result, cemeteries, which have been determined to make significant contributions to ecosystem services based on numerical data, should be integrated into a comprehensive green infrastructure plan. Ecological connections with other components of the urban green infrastructure system should be established. This will increase their current contributions to ecosystem services and facilitate the city's adaptation to climate change.

ACKNOWLEDGEMENTS

We express our gratitude to the Department of Remote Sensing and Geographic Information Systems at Akdeniz University's Institute of Science for providing us with their assistance in terms of data, software, and hardware infrastructure.

REFERENCES

Antalya Governorship, 2023. Antalya Governorship Official website. http://www.antalya.gov.tr/# (accessed date: 06.02.2023).

Arcidiacono, A., Ronchi, S., & Salata, S., 2015. Ecosystem Services assessment using InVEST as a tool to support decision making process: critical issues and opportunities. In Computational Science and Its Applications--ICCSA 2015: 15th International Conference, Banff, AB, Canada, June 22-25, 2015, Proceedings, Part IV 15 (pp. 35-49). Springer International Publishing.

Balzan, M. V., Zulian, G., Maes, J., & Borg, M., 2021. Assessing urban ecosystem services to prioritise nature-based solutions in a high-density urban area. *Nature-Based Solutions*, 1, 100007.

Benra, F., De Frutos, A., Gaglio, M., Álvarez-Garretón, C., Felipe-Lucia, M., & Bonn, A., 2021. Mapping water ecosystem services: Evaluating InVEST model predictions in data scarce regions. *Environmental Modelling & Software,* 138, 104982.

Cilliers, S., Cilliers, J., Lubbe, R., & Siebert, S., 2013. Ecosystem services of urban green spaces in African countries—perspectives and challenges. *Urban Ecosystems,* 16, 681-702.

Dick, J., Verweij, P., Carmen, E., Rodela, R., & Andrews, C., 2017. Testing the ecosystem service cascade framework and QUICKScan software tool in the context of land use planning in Glenlivet Estate Scotland. International Journal of Biodiversity Science, *Ecosystem Services & Management*, 13(2), 12-25.

Długoński, A., Dushkova, D., & Haase, D., 2022. Urban Cemeteries—Places of Multiple Diversity and Challenges. A Case Study from Łódź (Poland) and Leipzig (Germany). *Land*, 11(5), 677.

Elmqvist, T., Setälä, H., Handel, S. N., van der Ploeg, S., Aronson, J., Blignaut, J. N., ... & de Groot, R., 2015. Benefits of restoring ecosystem services in urban areas. *Current opinion in environmental sustainability*, 14, 101-108.

Ersoy Mirici, M., 2022. The ecosystem services and green infrastructure: a systematic review and the gap of economic valuation. *Sustainability*, 14(1), 517.

European Commission, 2019. Report From The Commission To The European Parliament, The Council, The European Economic And Social Committee And The Committee Of The Regions. Brussels, COM(2019) 236 final, 12p.

European Commission, 2020. Restoring biodiversity and ecosystem services. https://converges.eu/wp-content/uploads/2020/11 (Acessed date: 07.02.2023)

Evans, D. L., Falagán, N., Hardman, C. A., Kourmpetli, S., Liu, L., Mead, B. R., & Davies, J. A. C., 2022. Ecosystem service delivery by urban agriculture and green infrastructure–a systematic review. *Ecosystem Services*, 54, 101405.

Fagerholm, N., Torralba, M., Moreno, G., Girardello, M., Herzog, F., Aviron, S., ... & Plieninger, T., 2019. Cross-site analysis of perceived ecosystem service benefits in multifunctional landscapes. *Global Environmental Change*, 56, 134-147.

Frélichová, J., Vačkář, D., Pártl, A., Loučková, B., Harmáčková, Z. V., & Lorencová, E., 2014. Integrated assessment of ecosystem services in the Czech Republic. *Ecosystem Services*, 8, 110-117.

Kabisch, N., 2015. Ecosystem service implementation and governance challenges in urban green space planning—The case of Berlin, Germany. *Land use policy*, 42, 557-567.

Klimanova, O., Kolbowsky, E., & Illarionova, O., 2018. Impacts of urbanization on green infrastructure ecosystem services: The case study of post-soviet Moscow. *Belgeo*. Revue belge de géographie,. https://doi.org/10.4000/belgeo.30889

Kowarik, I., Buchholz, S., von der Lippe, M., & Seitz, B., 2016. Biodiversity functions of urban cemeteries: Evidence from one of the largest Jewish cemeteries in Europe. *Urban Forestry & Urban Greening*, 19, 68-78.

Lanzas, M., Hermoso, V., de-Miguel, S., Bota, G., & Brotons, L., 2019. Designing a network of green infrastructure to enhance the conservation value of protected areas and maintain ecosystem services. *Science of the Total Environment*, 651, 541-550.

Lourdes, K. T., Hamel, P., Gibbins, C. N., Sanusi, R., Azhar, B., & Lechner, A. M., 2022. Planning for green infrastructure using multiple urban ecosystem service models and multicriteria analysis. *Landscape and Urban Planning*, 226, 104500.

Marando, F., Heris, M. P., Zulian, G., Udías, A., Mentaschi, L., Chrysoulakis, N., ... & Maes, J., 2022. Urban heat island mitigation by green infrastructure in European Functional Urban Areas. *Sustainable Cities and Society*, 77, 103564.

McClymont, K., & Sinnett, D., 2021. Planning Cemeteries: Their Potential Contribution to Green Infrastructure and Ecosystem Services. *Frontiers in Sustainable Cities*, 136.

McPhearson, T., Cook, E. M., Berbes-Blazquez, M., Cheng, C., Grimm, N. B., Andersson, E., ... & Troxler, T. G., 2022. A social-ecological-technological systems framework for urban ecosystem services. *One Earth*, 5(5), 505-518.

Mills, G., Anjos, M., Brennan, M., Michael, J., McAleavey, C., & Ningal, T., 2015. The green 'signature'of Irish cities: an examination of the ecosystem services provided by trees using iTree Canopy software. *Irish Geography*, 48, 62-77.

Niemelä, J., Saarela, S. R., Söderman, T., Kopperoinen, L., Yli-Pelkonen, V., Väre, S., & Kotze, D. J., 2010. Using the ecosystem services approach for better planning and conservation of urban green spaces: a Finland case study. *Biodiversity and Conservation*, 19, 3225-3243.

Nowak, D. J., Maco, S., & Binkley, M. J. A. C., 2018. i-Tree: Global tools to assess tree benefits and risks to improve forest management. *Arboricultural Consultant*, 51(4), 10-13.

Olivatto, T. F. (2019). Using i-Tree canopy to estimate and value ecosystem services of air pollutant removal. In *Proceedings of the 3rd Brazilian Technology Symposium: Emerging Trends and Challenges in Technology 3 (pp. 291-299).* Springer International Publishing.

Peng, J., Zhao, H., & Liu, Y., 2017. Urban ecological corridors construction: A review. *Acta Ecologica Sinica*, 37(1), 23-30.
Phillips, B. B., Bullock, J. M., Osborne, J. L., & Gaston, K. J., 2020. Ecosystem service provision by road verges. *Journal of Applied Ecology*, 57(3), 488-501.

Quinton, J. M., & Duinker, P. N., 2019. Beyond burial: Researching and managing cemeteries as urban green spaces, with examples from Canada. *Environmental reviews*, 27(2), 252-262.

Reid, W. V., Mooney, H. A., Capistrano, D., Carpenter, S. R., Chopra, K., Cropper, A., ... & Shidong, Z., 2006. Nature: the many benefits of ecosystem services. *nature*, 443(7113), 749-749.

Sallay, Á., Mikházi, Z., Gecséné Tar, I., & Takács, K., 2022. Cemeteries as a Part of Green Infrastructure and Tourism. *Sustainability*, 14(5), 2918.

Schmidt, K., Sachse, R., & Walz, A., 2016. Current role of social benefits in ecosystem service assessments. *Landscape and Urban Planning*, 149, 49-64.

Selim, C., & Atabey, S. (2020). Kentsel yol ağaçlandırmalarının sağladığı faydaların belirlenmesi: Antalya Atatürk Bulvarı örneği. ***Bursa Uludağ Üniversitesi Ziraat Fakültesi Dergisi**, 34*(Özel Sayı), 235-248.

Selim, S., 2021. Yeşil Mutabakat çerçevesinde kentsel yeşil alanların yeşil altyapı sistemine entegrasyonu: Antalya-Konyaaltı örneği. *Süleyman Demirel Üniversitesi Fen Bilimleri Enstitüsü Dergisi*, 25(3), 636-643.

Selim, S., Dönmez, B., & Kilçik, A., 2022. Determination of the optimum number of sample points to classify land cover types and estimate the contribution of trees on ecosystem services using the I-Tree Canopy tool. *Integrated Environmental Assessment and Management.* https://doi.org/10.1002/ieam.4704

Selim, S., Koc-San, D., Selim, C., & San, B. T., 2018. Site selection for avocado cultivation using GIS and multi-criteria decision analyses: Case study of Antalya, Turkey. *Computers and electronics in agriculture*, 154, 450-459.

Stange, E., Zulian, G., Rusch, G., Barton, D., & Nowell, M., 2017. Ecosystem services mapping for municipal policy: ESTIMAP and zoning for urban beekeeping. *One Ecosystem*, 2, e14014.

Winograd, M., Pérez-Soba, M., & Verweij, P., 2014. QUICKScan: a pragmatic approach for decision support in ecosystem services assessment and management. In Handbook on the Economics of Ecosystem Services and Biodiversity (pp. 208-230). Edward Elgar Publishing.

Zulian, G., Polce, C., & Maes, J., 2014. ESTIMAP: a GIS-based model to map ecosystem services in the European Union. *Annali di Botanica*, 4, 1-7.

INVESTIGATION OF GREEN SPACE COOLING POTENTIAL ON LAND SURFACE TEMPERATURE IN ANTALYA CITY OF TURKEY

S. Selim[1*], B. Eyileten[2], N.Karakuş[2]

1 Akdeniz University, Faculty of Science, Department of Space Science and Technologies, 07058 Antalya, Turkey - serdarselim@akdeniz.edu.tr (* corresponding author)
2 Akdeniz University, Institute of Science, Department of Remote Sensing and Geographical Information Systems, 07058 Antalya, Turkey - buketeyileten@gmail.com; nkarakus@akdeniz.edu.tr

KEY WORDS: Green Spaces, Urban Heat Island, Land Surface Temperature, Vegetation Index, Remote Sensing.

ABSTRACT:

The aim of this study is to investigate the potential cooling effect of urban green spaces on land surface temperature (LST). Two significant green areas and their surroundings in the Konyaaltı district of Antalya province were selected as the sample area. Antalya is one of the most important tourist destinations in Turkey and has experienced rapid population growth along with increasing construction. Because the sample area is located in the Mediterranean Climate Zone, the research datasets were limited to a 6-month time series covering May and October, as recommended by previous studies. In the study, cloudless Landsat 8 satellite images were utilized to obtain vegetation density and land surface temperature. After making atmospheric corrections to the satellite images, the researchers used a series of algorithms to calculate NDVI and LST values from Landsat 8's NIR, Red, and thermal bands. A linear cross-section was assigned from the maps, and statistical analysis was performed by applying the Pearson correlation test to the data obtained from the pixels in contact with the section. The analysis revealed that there is a strong opposite correlation between plant density and land surface temperature, and the land surface temperature decreases as the plant density increases. The cooling effect of green areas range from 0.13 - 1.04 °C, although it varies according to the month. The 1st sample area had an average cooling effect of 0.77 °C, while the 2nd sample area had an average of 0.28 °C cooling effect. Combining the results from both sample areas, the researchers found that urban green spaces provide a cooling effect of 0.52 °C during the hot months of the spring and summer seasons in the Mediterranean climate zone. The results suggest that green areas play an important role in cooling today's cities that have turned into concrete jungles. Therefore, it is suggested that increasing the presence of urban green areas through green infrastructure planning can not only enhance the cooling effect of green spaces in urban areas but also help combat global climate change.

1. INTRODUCTION

The growth of populations and consequent increase in urbanization has led to several undesirable environmental changes. Although urban areas account for a small fraction of global land use, industrial development tends to concentrate in urbanized areas, where half of the world's population is already concentrated. Consequently, it is widely acknowledged that the adverse effects of climate change, one of the most pressing ecological and social problems of the 21st century, will be particularly pronounced in urban areas (Moradi and Görer Tamer, 2017; Moda et al, 2019; Hobbie, and Grimm, 2020). As urbanization progresses, land cover changes gradually and natural surfaces are replaced by urban textures. The expansion of concrete surfaces exerts pressure on natural areas and causes land use changes. Furthermore, urbanization exacerbates temperature increases in cities, leading to the formation of urban heat islands, where atmospheric greenhouse gas concentrations and surface temperatures are higher compared to green spaces (Pielke et al, 2011; Zurqani et al, 2019; Orhan, 2021).

As a result of climate change, factors such as an increase in temperature and direct exposure of concrete surfaces to sunlight, as well as the retention of heat by these surfaces, lead to an increase in Land Surface Temperature (LST) in urban areas, thereby negatively impacting human health. Consequently, various strategies are being developed to reduce urban LST. Scientific studies emphasize the necessity of increasing the current potential of green areas that have both cooling capacity in combating global warming and are included in urban plans (Susca et al, 2011; Koc et al, 2018; Orhan, 2021).

Urban green spaces are anticipated to have a pivotal role in regulating urban climates due to their capacity to mitigate the adverse effects of urban warming. Through various means such as providing shade, regulating air currents, promoting evaporation, and generating favorable micro-climates, urban green spaces have the potential to enhance urban health by acting as a source of cooling. Furthermore, they can reduce land surface temperatures by impeding the sun's rays from reaching reflective surfaces. As a result, urban green spaces are viewed as an effective strategy to counter the negative effects of urban warming and land surface temperature. Consequently, it is imperative to identify urban heat islands and land surface temperatures in regions and cities with high population density and warm climate and to investigate their relationship with green spaces for developing micro-climate strategies. Research conducted by Gill et al. (2007), Bowler et al. (2010), Escobedo et al. (2011), Norton et al. (2015), Yang et al. (2019), and Aydemir et al. (2022) underscores the importance of enhancing the existing potential of green infrastructure in urban planning and development to combat global warming.

The study utilized geographical information systems and remote sensing techniques and focused on two urban green spaces in Konyaaltı, a popular tourist destination in Turkey: Hayat Park and Akdeniz Kent Parkı. The study aimed to evaluate the cooling effect of these green spaces on the urban climate and associate it with green spaces to increase the climatic comfort of individuals, particularly in Antalya. Antalya has experienced significant population growth and construction in recent years, leading to pressure on green areas. Therefore, the city is unable to benefit adequately from the ecosystem services that its green areas contribute to urban cooling. The study's findings could serve as a reference for central and local governments, urban planners, and

This contribution has been peer-reviewed.
https://doi.org/10.5194/isprs-archives-XLVIII-M-1-2023-107-2023 | © Author(s) 2023. CC BY 4.0 License.

open green space planning and evaluation to enhance urban green spaces and reduce the UHI effect in Antalya and other similar cities. By identifying the benefits of urban green spaces and their potential to mitigate the UHI effect, this study could contribute to the development of effective urban planning strategies that prioritize the preservation and creation of green spaces in urban areas.

2. MATERIAL AND METHODS

2.1 Study Area

The research focused on two significant green areas and their adjacent lands in the Konyaaltı district of Antalya province, which is one of the most significant tourist destinations in Turkey and has witnessed increasing construction in tandem with rapid population growth. In order to assess the potential cooling effect of the selected urban green areas, it was essential to collect LST data of the surrounding lands. To accomplish this, the study was performed within a 1x1 km sample area that encompassed the selected green space (Figure 1).

Hayat Park and an urban forest to its north were chosen as the first sample area, which covers an urban green area of 32.01 hectares at coordinates 36°52'40"N and 30°38'00"E. This sample area has a limited amount of green space and a high amount of land and buildings surrounding the urban green area.

The second sample area comprised Akdeniz City Park, situated in the city center near the coast, and the fairground area to the north of the park. The urban green space in this area covers 15.15 hectares and is geographically located at 36°52'50"N and 30°39'30"E. Unlike the first sample area, the surroundings of the urban green space in the second sample area have fewer land and buildings, and a higher proportion of green space.

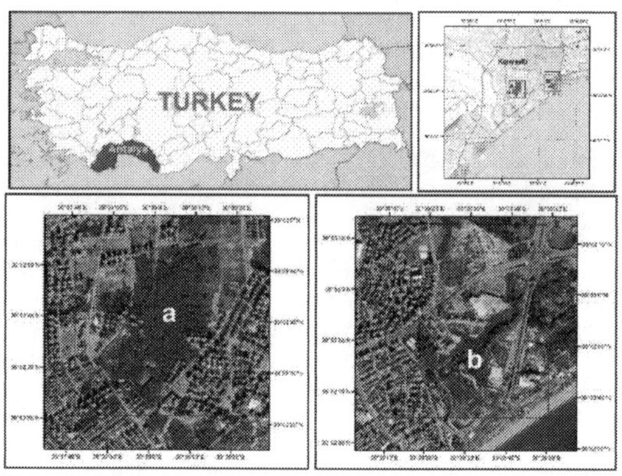

Figure 1. Study area

The decision to select the Konyaaltı district of Antalya province as the study area was influenced by several factors, including the city's population growth, climate, and importance as a prominent tourist destination in Turkey. In terms of selecting the green areas, their central location in the city, areal sizes, vegetation densities, and environmental land uses were taken into consideration.

2.2 Datasets

Olgyay (1973) suggested that thermal comfort conditions deteriorate the most from May to October due to hot stress in the Mediterranean Climate Zone (Karakuş and Selim, 2022; Cinar et al, 2023). Based on this information and previous scientific research, the study focused on the months between May and October. To obtain data on vegetation density and land surface temperature in the study area, the research used cloudless Landsat 8 satellite images, which were downloaded free of charge from the United States Geological Survey (USGS) 'Earth Explorer' website. Landsat 8 provides medium-resolution data ranging from 15 to 100 meters. Landsat 8 satellites dated 14.05.2019, 25.06.2017, 01.07.2019, 15.08.2018, 19.09.2019 and 15.10.2017 were used in the study.

2.3 Method

To investigate the cooling effect of green areas on land surface temperature, open access Landsat 8 satellite imagery was obtained and processed. Atmospheric correction operations were performed in Quantum GIS (QGIS) software, and NDVI (Normalized Difference Vegetation Index) was created using Near Infrared (NIR) and Red bands in ArcGIS software. The Land Surface Temperature was determined using Landsat 8's thermal bands with the help of a series of algorithms. The process consisted of six stages, which are as follows: 1) Calculating the spectral radiance at the top of the atmosphere (TOA), 2) Converting TOA to brightness temperature (BT), 3) Computing the Normalized Difference Vegetation Index (NDVI), 4) Determining vegetation cover (Pv), 5) Calculating emissivity (ε), and 6) Computing Land Surface Temperature (LST). Subsequently, the cooling potential was calculated by subtracting the LST of the areas outside the urban green space from the LST of the entire sample area. In addition, to determine the linear relationship between vegetation and land surface temperature, a linear cross-section was applied to both the NDVI and LST maps. This method involved comparing reflectance values of sample points taken from the images with those from the cross-section lines. Finally, statistical analysis was conducted using the Pearson correlation test on the data obtained from the pixels that intersected with the cross-section lines (Figure 2.).

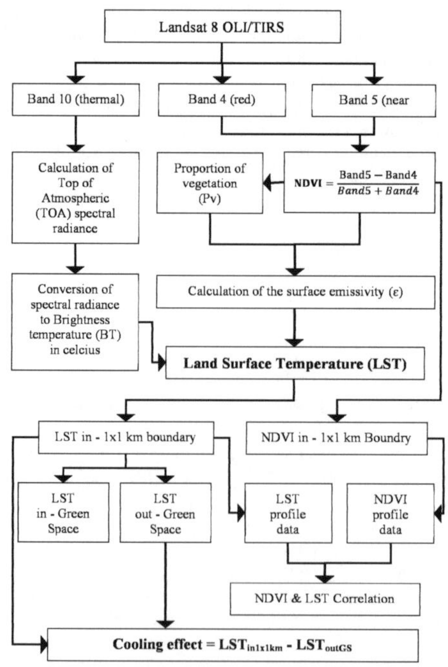

Figure 2. Flowchart

3. RESULT AND DISCUSSION

The main objective of this study was to investigate whether urban green areas have a cooling effect on land surface temperature in specific areas of the Konyaaltı district in Antalya province. To achieve this goal, the study employed Landsat 8 OLI/TIRS satellite images to generate LST and NDVI maps of the green areas and their surrounding areas. Cross-sectional analyses were conducted in a West-East direction to determine the relationship between ground surface temperature and urban green areas. Statistical analysis was then used to determine the relationship between ground surface temperature and the green area using the LST and NDVI values of the pixels in the section plane. The maps and graphics of the first sample area are presented in Figure 3, while NDVI and LST values calculated by month and correlation coefficients are presented in Table 1.

In May, the minimum NDVI value of 0.05 units was detected outside the green area, whereas the maximum NDVI value of 0.36 was observed in both inside and outside the green area in the sample area. The lowest LST value of 33.12 °C was found inside the green area, while the highest LST value of 42.10 °C was detected outside the green area. The difference in average NDVI between the inside and outside of the green area was 0.10, and the average LST difference was 3.18 °C. The results of the Pearson Correlation test applied to the NDVI and LST data in the cross-sectional plane taken in the West-East direction indicated a very strong negative correlation (R=-0.920) between NDVI and LST.

In the month of June, the minimum and maximum NDVI values in the sample area, measuring 0.06 and 0.38, respectively, were discovered to be outside the green area. Within the same area, the minimum LST value of 42.49°C was detected inside the green area, whereas the maximum LST value of 49.52°C was identified outside the green area. The average difference in NDVI between the green and non-green areas was found to be 0.08, whereas the average difference in LST was 2.20°C. The application of the Pearson Correlation test to the NDVI and LST data collected along the West-East cross-sectional plane revealed a strong negative correlation (R=-0.867) between these two variables.

In July, the minimum and maximum NDVI values in the sample area were both detected outside the green area, with values of 0.05 and 0.43, respectively. While the minimum LST value of 39.37°C was found inside the green area, the maximum LST value of 47.82°C was detected outside it. On average, the NDVI difference between inside and outside of the green area was 0.10, and the LST difference was 2.97°C. Results of the Pearson Correlation test on the NDVI and LST data in the West-East cross-sectional plane revealed a highly negative correlation (R=-0.937) between NDVI and LST.

In August, it was observed that the sample area recorded the minimum NDVI value (0.05) and the maximum NDVI value (0.38) outside of the green area. Despite the lowest LST value of 41.10°C being detected inside the green area, the highest LST value of 49.39°C was found outside it. On average, the NDVI difference between inside and outside of the green area was 0.10, while the LST difference was 2.58°C. The Pearson Correlation test conducted on the NDVI and LST data in the West-East cross-sectional plane indicated a very strong negative correlation (R=-0.935) between NDVI and LST.

The sample area in September revealed that the lowest NDVI value (0.05) and the highest NDVI value (0.36) were both detected outside the green area. The minimum LST value (33.79 °C) was found within the green area, whereas the maximum LST value (40.55 °C) was located outside of it. On average, the difference in NDVI between the inside and outside of the green area was 0.08, while the average LST difference was 2.24 °C. The Pearson Correlation test applied to the NDVI and LST data in the West–East cross-sectional plane demonstrated a very strong negative correlation (R=-0.947) between NDVI and LST.

In October, the sample area showed that the lowest NDVI value (0.03) and the highest NDVI value (0.37), along with the minimum LST value (30.24 °C) and the maximum LST value (35.24 °C), were detected outside of the green area. On average, the difference in NDVI between the inside and outside of the green area was 0.04, while the average LST difference was 0.98°C. The Pearson Correlation test, which was conducted on the NDVI and LST data in the cross-sectional plane taken in the West-East direction, indicated a very strong negative correlation (R=-0.915) between NDVI and LST.

Table 1. NDVI and LST values and NDVI & LST Correlation of Hayat Park and surrounding urban forest

Location			Time					
			May	**Jun**	**Jul**	**Aug**	**Sep**	**Oct**
NDVI	In - Green Space Boundry	$NDVI_{min}$	0.10	0.11	0.12	0.10	0.10	0.09
		$NDVI_{max}$	0.36	0.34	0.37	0.34	0.31	0.27
		$NDVI_{ave}$	0.29	0.26	0.28	0.28	0.25	0.23
	Out - Green Space Boundry	$NDVI_{min}$	0.05	0.06	0.05	0.05	0.05	0.03
		$NDVI_{max}$	0.36	0.38	0.43	0.38	0.36	0.37
		$NDVI_{ave}$	0.19	0.18	0.18	0.18	0.17	0.14
	In - 1x1 km Boundry	$NDVI_{min}$	0.05	0.06	0.05	0.05	0.05	0.03
		$NDVI_{max}$	0.36	0.38	0.43	0.38	0.36	0.37
		$NDVI_{ave}$	0.22	0.21	0.22	0.21	0.19	0.17
LST	In - Green Space Boundry	LST_{min}	33.12	42.49	39.37	41.10	33.79	30.71
		LST_{max}	37.47	46.66	43.91	45.55	37.22	33.00
		LST_{ave}	34.56	43.98	41.09	42.55	34.95	31.35
	Out - Green Space Boundry	LST_{min}	33.81	42.89	40.24	41.42	33.92	30.24
		LST_{max}	42.10	49.52	47.82	49.39	40.55	35.24
		LST_{ave}	37.74	46.18	44.06	45.13	37.19	32.33
	In - 1x1 km Boundry	LST_{min}	33.12	42.49	39.37	41.10	33.79	30.24
		LST_{max}	42.10	49.52	47.82	49.39	40.55	35.24
		LST_{ave}	36.70	45.46	43.09	44.28	36.46	32.01
NDVI & LST Correlation		R	**-0.920**	**-0.867**	**-0.937**	**-0.935**	**-0.947**	**-0.915**

In the first study area, the mean variation of NDVI differs depending on the month and is usually 0.09, whereas the average difference in LST is 2.36 °C. The most significant temperature difference of 3.18 °C is observed in May, and the smallest temperature difference of 0.98 °C is recorded in October.

This contribution has been peer-reviewed.
https://doi.org/10.5194/isprs-archives-XLVIII-M-1-2023-107-2023 | © Author(s) 2023. CC BY 4.0 License.

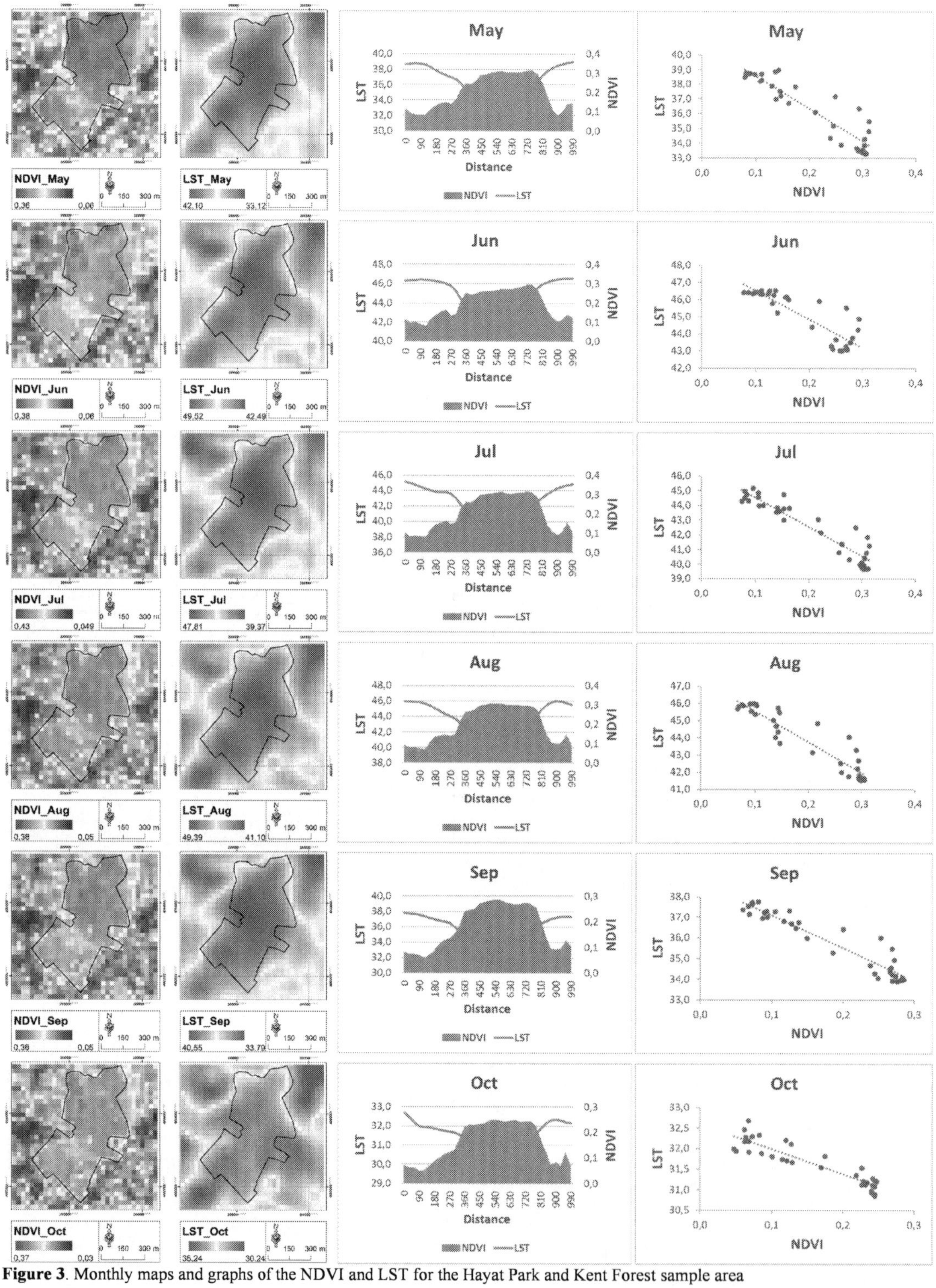

Figure 3. Monthly maps and graphs of the NDVI and LST for the Hayat Park and Kent Forest sample area

Figure 4 displays the maps and graphs for the second research area, and Table 2 demonstrates the NDVI and LST values for each month, including correlation coefficients.

In the month of May, the minimum NDVI value in the sample area was measured as 0.00 outside the green area, while the maximum NDVI value was detected as 0.60 inside the green area. The minimum LST value in the area was recorded as 32.88 °C and the maximum LST value was 47.40 °C outside the green area. The average NDVI difference between the inside and outside of the green area was found to be 0.19, while the average LST difference was 2.06 °C. According to the Pearson Correlation test applied to the NDVI and LST data taken in the West-East section plane, a very strong negative correlation (R=-0.921) was found between NDVI and LST.

In the month of June, the minimum NDVI value outside the green area in the sample area was found to be 0.01, while the maximum NDVI value within the green area was 0.58. The minimum and maximum LST values in the area were recorded as 40.13 °C and 52.41 °C respectively, outside the green area. The average NDVI difference between the inside and outside of the green area was calculated as 0.17, while the average LST difference was found to be 2.11 °C. Based on the Pearson Correlation test conducted on the NDVI and LST data in the West-East direction cross-sectional plane, a strong negative correlation (R=-0.846) was identified between NDVI and LST.

In July, the minimum and maximum NDVI values in the sample area were measured outside the green area, with values of -0.04 and 0.62, respectively. The minimum and maximum LST values in the area were also determined outside the green area, with values of 36.99 °C and 50.59 °C, respectively. The average NDVI difference between the inside and outside of the green area was found to be 0.19, while the average LST difference was 2.53 °C. Analysis of the NDVI and LST data obtained from the West-East direction cross-sectional plane using Pearson Correlation test revealed a strong negative correlation (R=-0.854) between the two variables.

In August, the minimum and maximum NDVI values in the sample area were detected outside the green area, with values of -0.02 and 0.5, respectively. Similarly, the minimum and maximum LST values in the area were also measured outside the green area, with values of 37.64 °C and 49.68 °C, respectively. The average NDVI difference between the inside and outside of the green area was calculated as 0.18, while the average LST difference was 2.16 °C. Analysis of the NDVI and LST data obtained from the West-East direction cross-sectional plane using Pearson Correlation test revealed a strong negative correlation (R=-0.906) between the two variables.

In the month of September, it was observed that the minimum and maximum NDVI values in the sample area were located outside the green area, with values of 0.00 and 0.51, respectively. Likewise, the minimum and maximum LST values in the area were measured outside the green area, with values of 33.28 °C and 42.65 °C, respectively. The average NDVI difference between the inside and outside of the green area was determined as 0.16, while the average LST difference was calculated to be 1.41 °C. The analysis of the NDVI and LST data from the West-East direction cross-sectional plane using Pearson Correlation test showed a very strong negative correlation (R=-0.915) between NDVI and LST.

In October, the minimum NDVI value (-0.01) was found beyond the green area in the sample region, while the highest NDVI value (0.46) was detected inside the green area. The lowest and highest LST values (29.58 °C and 37.14 °C, respectively) were both identified outside the green area. It can be observed that the average NDVI difference between inside and outside of the green area is 0.12, whereas the average LST difference is 0.87 °C. Based on the Pearson Correlation test conducted on the NDVI and LST data in the cross-sectional plane, which was taken in the West-East direction, a strong negative correlation (R=-0.844) was discovered between NDVI and LST.

Table 2. The correlation between the values of NDVI and LST with the NDVI and LST values of the Akdeniz Kent Park and the surrounding Fairground area.

Location			Time					
			May	Jun	Jul	Aug	Sep	Oct
NDVI	In - Green Space Boundry	$NDVI_{min}$	0.12	0.10	0.09	0.08	0.12	0.07
		$NDVI_{max}$	0.60	0.58	0.62	0.54	0.51	0.46
		$NDVI_{ave}$	0.37	0.36	0.38	0.35	0.33	0.28
	Out - Green Space Boundry	$NDVI_{min}$	0.00	0.01	-0.04	-0.02	0.00	-0.01
		$NDVI_{max}$	0.47	0.51	0.54	0.44	0.45	0.40
		$NDVI_{ave}$	0.18	0.19	0.19	0.17	0.17	0.16
	In - 1x1 km Boundry	$NDVI_{min}$	0.00	0.01	-0.04	-0.02	0.00	-0.01
		$NDVI_{max}$	0.60	0.58	0.62	0.54	0.51	0.46
		$NDVI_{ave}$	0.21	0.22	0.22	0.20	0.19	0.17
LST	In - Green Space Boundry	LST_{min}	34.04	41.16	38.76	38.88	34.75	30.62
		LST_{max}	39.86	47.83	45.20	45.16	39.06	34.59
		LST_{ave}	36.04	43.81	40.97	41.68	36.01	31.87
	Out - Green Space Boundry	LST_{min}	32.88	40.13	36.99	37.64	33.28	29.58
		LST_{max}	47.40	52.41	50.59	49.68	42.65	37.14
		LST_{ave}	38.10	45.92	43.50	43.84	37.42	32.74
	In - 1x1 km Boundry	LST_{min}	32.88	40.13	36.99	37.64	33.28	29.58
		LST_{max}	47.40	52.41	50.59	49.68	42.65	37.14
		LST_{ave}	37.79	45.61	43.12	43.52	37.21	32.61
NDVI & LST Correlation		R	**-0.921**	**-0.846**	**-0.854**	**-0.906**	**-0.915**	**-0.844**

The average NDVI difference in the sample area is 0.17, while the average LST difference is 1.86 °C. The maximum temperature difference occurred in July at 2.53 °C, while the least difference was seen in October at 0.87 °C.

In the first sample area, which includes Hayat Park and the urban forest to the north, the amount of buildings and hard ground is high, but the amount of green area around these buildings is low. As a result, the average LST difference between inside and outside the green area (2.36 °C) of the first sample area is higher than the average LST difference between inside and outside the green area in the second sample area. Due to the lower amount of buildings and hard ground and the higher amount of vegetation in the periphery of the second sample area, Akdeniz city park and the fairground area in its north, the average LST difference (1.86 °C) between the inside and outside of the green area is lower than that of the first sample area. The average temperature difference between the inside and outside of the urban green space in both of the sampled areas was calculated to be 2.11 °C.

https://doi.org/10.5194/isprs-archives-XLVIII-M-1-2023-107-2023 | © Author(s) 2023. CC BY 4.0 License.

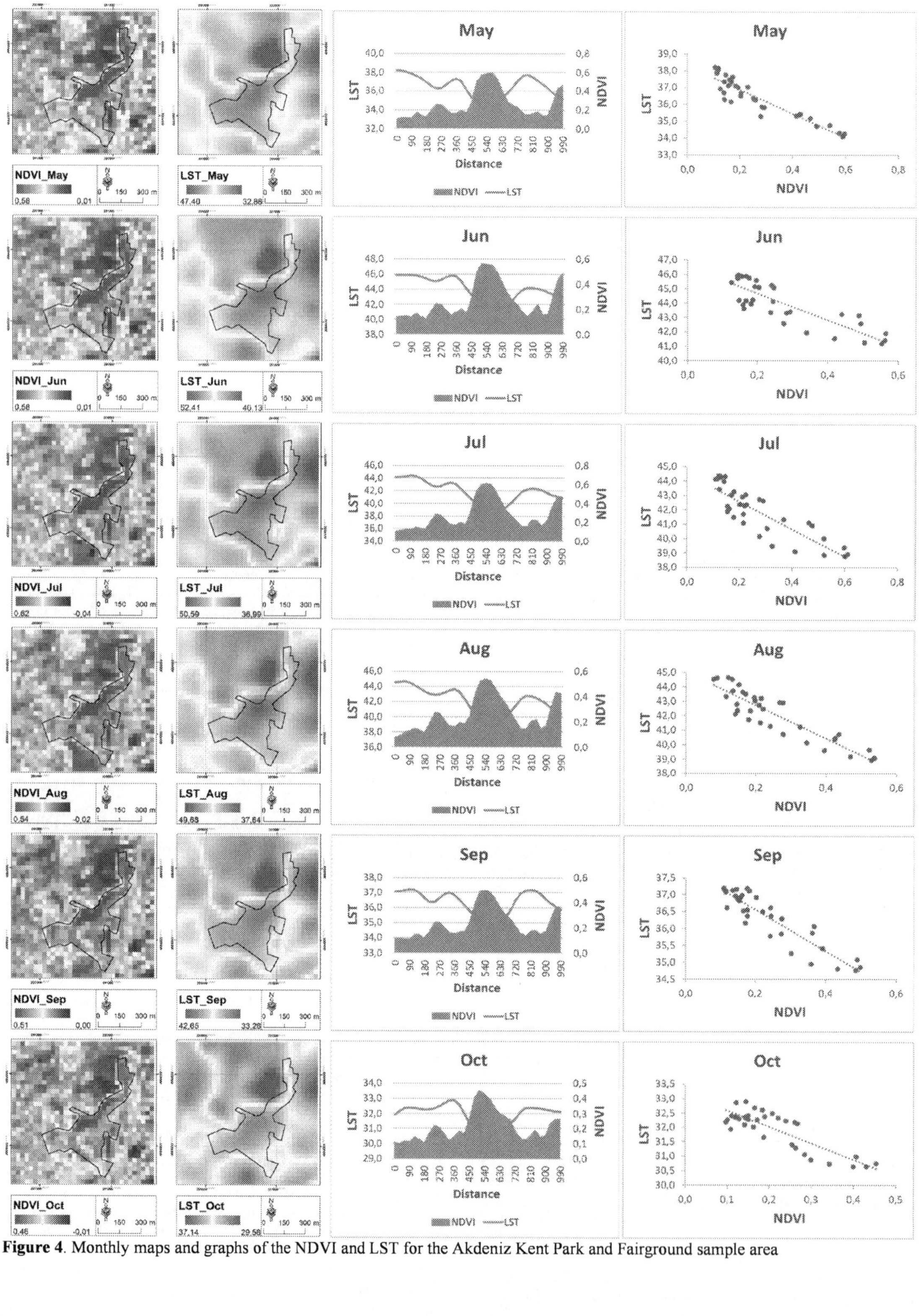

Figure 4. Monthly maps and graphs of the NDVI and LST for the Akdeniz Kent Park and Fairground sample area

This contribution has been peer-reviewed.
https://doi.org/10.5194/isprs-archives-XLVIII-M-1-2023-107-2023 | © Author(s) 2023. CC BY 4.0 License.

In the first sample area, it was found that the plant density and surface temperature differences between green areas and the areas outside the green area border had a positive correlation (R=0.578), based on the results of the Pearson Correlation test. Similarly, a stronger positive correlation (R=0.933) was detected in the second sample area. These findings indicate that the presence of green areas has a cooling effect on land surface temperature in both sample areas. Furthermore, it was observed that as the NDVI difference between the inside and outside of the green area decreases, the LST difference also decreases, while an increase in the NDVI difference leads to an increase in the LST difference (Figure 5).

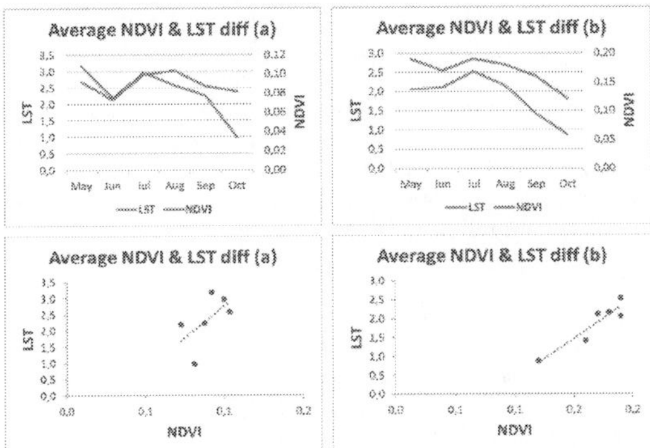

Figure 1. The average difference in NDVI and LST between inside and outside the green area for sample area 1 (a) and sample area 2 (b)

The relationship between vegetation density and land surface temperature has been extensively studied and is supported by the cross-section graphics in Figures 3 and 4 of this study. The literature also confirms that areas with low vegetation density in the land cover have higher land surface temperatures, while areas with high vegetation density have lower land surface temperatures (Leuzinger and Kӧrner, 2007; Koc et al, 2018; Aram et al, 2019). Green areas, especially those with trees, have been shown to effectively reduce the land surface temperature of the surrounding environment (Frumkin and McMichael, 2008; Aram et al, 2019).

The aim of this study was to investigate how green areas affect the temperature in urban environments. To achieve this, temperature changes were measured in two sample areas of 1x1 km. The findings of the analysis are as follows Figure 6:

— In May, sample area 1 exhibited a cooling effect of 1.04 °C, whereas sample area 2 showed a lower cooling effect of 0.31 °C. The average cooling effect of green areas in May was calculated to be 0.67 °C.
— In June, sample area 1 had a cooling effect of 0.72 °C, whereas sample area 2 had a cooling effect of 0.32 °C. The average cooling effect of green areas in June was determined to be 0.52 °C.
— In July, sample area 1 had a higher cooling effect of 0.97 °C compared to sample area 2 which showed a cooling effect of 0.38 °C. The average cooling effect of green areas in July was found to be 0.67 °C.
— In August, sample area 1 had a higher cooling effect of 0.84 °C compared to sample area 2 which showed a cooling effect of 0.32 °C. The average cooling effect of green areas in August was determined as 0.58 °C.
— In September, sample area 1 showed a cooling effect of 0.73 °C, while the second sample area showed a cooling effect of 0.21 °C. The average cooling effect of green areas in September was determined as 0.47 °C.
— In October, sample area 1 showed a cooling effect of 0.32 °C, while the second sample area showed a cooling effect of 0.13 °C. The average cooling effect of green areas in October was determined to be 0.23 °C.

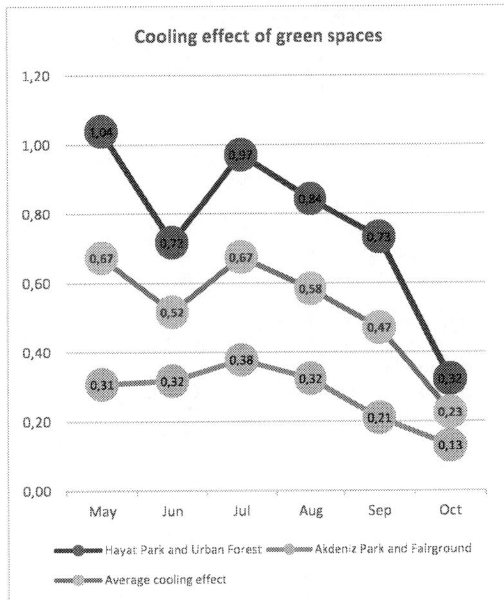

Figure 6. Cooling effect og green spaces in Konyaaltı

The results of the analysis indicated that the first sample area exhibited a greater cooling effect than the second sample area. This discrepancy is primarily attributed to the relatively larger amounts of bare soil and building area surrounding the Hayat park and urban forest, and the consequent lack of vegetation cover between buildings in the first sample area. In contrast, the high plant density in the land cover surrounding the Mediterranean city park and fairground area was the most significant factor contributing to the lower cooling effect observed in the second sample area. Additionally, the cooling effect of different plant species varied, with coniferous plants exhibiting lower surface temperatures than broad-leaved plants, as previously noted by Leuzinger and Kӧrner (2007). Urban green areas have the potential to lower land surface temperatures by up to 5°C compared to other urban land cover types. Notably, the presence of green spaces in residential areas can lead to a cooling effect of up to 4°C on average surface temperature with an increase of 10% in coverage (Gill et al, 2007; Watkins et al, 2007; Frumkin and McMichael, 2008), thus highlighting the significant role that green areas, especially in urban environments, can play in mitigating temperature-related issues.

4. CONCLUSION

The primary objective of this investigation was to explore the cooling effects of urban green areas in the Konyaaltı district of Antalya province, using selected sample areas. The findings indicated significant variations in temperature between the surface temperatures of green areas within the sample areas and those outside the green area boundaries. The temperature discrepancies varied by month, with an average of 7.02 °C in the first sample area and 11.56 °C in the second sample area. The study observed that low plant density led to elevated land surface temperatures, while high plant density resulted in lower land surface temperatures. Additionally, a robust negative correlation between plant density and land surface temperature was established. Moreover, the investigation found that as the Normalized Difference Vegetation Index (NDVI) difference between the inside and outside of the green areas decreased, the land surface temperature difference also decreased. Conversely, the land surface temperature difference increased as the NDVI difference increased.

Through the analyses conducted to ascertain the potential cooling impact of urban green areas on land surface temperature, it has been established that the cooling effects varied between the two sample areas across the months. Upon averaging the six-month cooling effect of the sample areas, it was determined that the first sample area had an average cooling effect of 0.77 °C in its environs, while the second sample area had an average cooling effect of 0.28 °C in its environs. The observed difference is believed to be due to differences in the composition of the land, the density of green spaces within the area, and the plant species present in the area. Collectively, the evaluation of both sample areas indicates that urban green spaces have an average cooling effect of 0.52 °C in the hot months of spring and summer within the Mediterranean climate zone.

The findings of this study reveal that green areas play a crucial role in mitigating the warming effect of urban areas that are dominated by impervious surfaces. Therefore, it is recommended that urban green spaces should be expanded by implementing green infrastructure planning as a means of addressing the pressing challenge of global climate change, which is recognized as a significant ecological and social issue of the current century. This would allow for greater utilization of the cooling capacity of green spaces in urban settings.

ACKNOWLEDGEMENTS

We express our gratitude to the Department of Remote Sensing and Geographic Information Systems at Akdeniz University's Institute of Science for their assistance with the provision of data, software, and hardware infrastructure.

REFERENCES

Aram, F., García, E. H., Solgi, E., & Mansournia, S. (2019). Urban green space cooling effect in cities. *Heliyon*, *5*(4), e01339.

Aydemir, K. P. K., Altınok, G. K., & Ünsal, Ö. (2022). Determining UHI Effect by Remote Sensing Method in Bolu City Centre, Turkey. *ICONARP International Journal of Architecture and Planning*, *10*(2), 735-758.

Bowler, D. E., Buyung-Ali, L., Knight, T. M., & Pullin, A. S., 2010. Urban greening to cool towns and cities: A systematic review of the empirical evidence. Landscape and urban planning, 97(3), 147-155.

Cinar, İ., Karakus, N., & Toy, S. (2023). Analysing daytime summer thermal comfort conditions for Turkey's third largest tourism destination. *Environmental Science and Pollution Research*, 1-11.

Escobedo, F. J., Kroeger, T., & Wagner, J. E., 2011. Urban forests and pollution mitigation: Analyzing ecosystem services and disservices. *Environmental pollution, 159*(8-9), 2078-2087.

Frumkin, H., & McMichael, A. J. (2008). Climate change and public health: thinking, communicating, acting. *American journal of preventive medicine*, *35*(5), 403-410.

Gill, S. E., Handley, J. F., Ennos, A. R., & Pauleit, S. (2007). Adapting cities for climate change: the role of the green infrastructure. *Built environment*, *33*(1), 115-133.

Hobbie, S. E., & Grimm, N. B. (2020). Nature-based approaches to managing climate change impacts in cities. *Philosophical Transactions of the Royal Society B*, *375*(1794), 20190124.

Karakuş, N. and Selim, S. (2022). Temporal and spatial distribution of outdoor thermal comfort conditions: The case of Konyaaltı-Antalya, Turkey. *The Journal of Graduate School of Natural and Applied Sciences of Mehmet Akif Ersoy University*, *13*(2), 259-269.

Koc, C. B., Osmond, P., & Peters, A. (2018). Evaluating the cooling effects of green infrastructure: A systematic review of methods, indicators and data sources. *Solar Energy*, *166*, 486-508.

Leuzinger, S., & Körner, C. (2007). Tree species diversity affects canopy leaf temperatures in a mature temperate forest. *Agricultural and forest meteorology*, *146*(1-2), 29-37.

Moda, H. M., Filho, W. L., & Minhas, A. (2019). Impacts of climate change on outdoor workers and their safety: some research priorities. *International journal of environmental research and public health*, *16*(18), 3458.

Moradi, M., & Görer Tamer, N. (2017). Bursa Örneğinde Kentsel Büyümenin Yerel İklim Değişikliği Üzerine Etkisi. *Planlama* 2017;27(1):26–37.

Norton, B. A., Coutts, A. M., Livesley, S. J., Harris, R. J., Hunter, A. M., & Williams, N. S., 2015. Planning for cooler cities: A framework to prioritise green infrastructure to mitigate high temperatures in urban landscapes. *Landscape and urban planning, 134*, 127-138.

Olgyay, V. (1973). Design with climate, bioclimatic approach to architectural regionalism. Princeton University Press, 190p, Princeton.

Orhan, O. (2021). Mersin ilindeki kentsel büyümenin yer yüzey sıcaklığı üzerine etkisinin araştırılması. *Geomatik*, *6*(1), 69-76.

Pielke Sr, R. A., Pitman, A., Niyogi, D., Mahmood, R., McAlpine, C., Hossain, F., ... & de Noblet, N. (2011). Land use/land cover changes and climate: modeling analysis and observational evidence. *Wiley Interdisciplinary Reviews: Climate Change*, *2*(6), 828-850.

Susca, T., Gaffin, S. R., & Dell'Osso, G. R. (2011). Positive effects of vegetation: Urban heat island and green roofs. *Environmental pollution*, *159*(8-9), 2119-2126.

Watkins, R., Palmer, J., & Kolokotroni, M. (2007). Increased temperature and intensification of the urban heat island: Implications for human comfort and urban design. *Built Environment*, *33*(1), 85-96.

Yang, J., Jin, S., Xiao, X., Jin, C., Xia, J. C., Li, X., & Wang, S. (2019). Local climate zone ventilation and urban land surface temperatures: Towards a performance-based and wind-sensitive planning proposal in megacities. *Sustainable Cities and Society*, *47*, 101487.

Zurqani, H. A., Post, C. J., Mikhailova, E. A., & Allen, J. S. (2019). Mapping urbanization trends in a forested landscape using Google Earth Engine. *Remote Sensing in Earth Systems Sciences*, *2*, 173-182.

This contribution has been peer-reviewed.
https://doi.org/10.5194/isprs-archives-XLVIII-M-1-2023-107-2023 | © Author(s) 2023. CC BY 4.0 License.

RESEARCH ON NATURAL RESOURCES SPATIO-TEMPORAL BIG DATA ANALYSIS PLATFORM FOR HIGH PERFORMANCE COMPUTING

YIN GAO [1], JIANWEI LUI [1], YANG LIU [2], ZHAOKUN ZHAI [1], JIAN CHE [1], HAO LI [1], RU WANG [1], JIANJUN LIU [1] *

[1] National Geomatics Center of China, 100830, Beijing, China
[2] Changjiang Waterway Bureau Survey Center, 430010, Wuhan, China

KEY WORDS: Natural Resources, Cloud Architecture, Spatial-temporal Big Data, Platform Framework, Spatial Analysis, High Performance Computing.

ABSTRACT:

In the era of earth observation big data, a new paradigm for the unified management of natural resource conservation and utilization has been established in China. The current focus of natural resources monitoring is on fully leveraging the value of big data in geospatial and temporal dimensions to support new work on natural resources management. An overall framework for a spatial-temporal big data analysis platform is proposed in this paper, which explores innovative technologies such as heterogeneous cloud collaborative services, business capabilities, data resources, and open sharing. Utilizing private cloud computing resources, the platform establishes collaborative services for storage cloud, computing cloud, and database cloud, demonstrating its ability to perform large-scale online computation of spatial-temporal data. Mainstream spatial data analysis computing methods such as vector computing, grid computing, and other grid computing methods are also established, and an open service model for interdisciplinary applications is explored. The feasibility of the proposed platform is verified using nationwide land cover data of 260 million surface area classifications, indicating its great potential for application.

1. INTRODUCTION

Natural resource monitoring data has the typical characteristics of rich information, authenticity and accuracy, global coverage, high spatial resolution and strong timeliness. It is the authoritative spatio-temporal information data for governments at all levels to find out the resources and carry out planning and decision-making (Ferreira and Vale 2022; Li 2014). It has unique advantages in solving the spatial data support in resource investigation, ecological protection, spatial planning, urban management and macro-decision. After years of continuous monitoring, China has obtained full-coverage geospatial data, identified the status and spatial distribution of natural resources and built a large-scale database (Liu et al. 2019). Additionally, the country has also mapped the distribution of artificial and public service facilities, established a private cloud space computing infrastructure, and achieved efficient management and service of China's spatio-temporal data on geographical conditions. At present, the monitoring of geographical conditions has been carried out regularly. The strategy of "covering the whole country and highlighting key points" is adopted to carry out full-coverage monitoring once a year on the land territory, forming a mass of geographical conditions data achievements with full coverage, seamless, high precision and time sequence, greatly promoting the leap-forward development of the ability to obtain land cover information (Li et al. 2016).

Traditional spatial data management based on single-node relational databases has limitations in terms of data management, concurrent reading and writing, efficient calculation, and scalability (Yang et al. 2011). The emergence of cloud computing has presented both opportunities and challenges for the management and analysis of massive spatial data. Combining cloud computing technology with spatial database technology to achieve elastic management and online analysis of massive spatial data has become a new research field in spatial information technology. In recent years, domestic and international scholars and spatial information enterprises have conducted extensive research on spatial data cloud storage and cloud computing. For example, ESRI, a representative international company in the United States, has developed the Esri Geometry API for JAVA and Spatial Framework for Hadoop to integrate ArcGIS and Hadoop (http://esri.github.io/gis-tools-for-hadoop/) and launched the GeoAnalytics Server distributed spatial computing platform embedded with Spark technology in the newly released ArcGIS Pro software, which has initial distributed analysis and computing capabilities for massive spatial data(Allen and Coffey 2019). Additionally, Oracle Corporation in the United States has launched a package solution (http://www.oracle.com/) that horizontally expands the Exadata database cloud platform, Oracle database, and Oracle Spatial spatial components with OLTP and OLAP for managing and computing massive spatial data. Oracle Spatial's vector and raster data distributed computing capabilities have been significantly enhanced in the latest database version of Oracle 12c (Saygili 2020).

Domestically, Chen et al. (2013) achieved distributed storage of vector and raster data based on the HDFS distributed file system and HBase NoSQL database. Jin et al. (2016) of Zhejiang University conducted in-depth research on the Spark distributed memory computing framework and developed customized spatial computing operators to realize the distributed spatial computing and analysis of large-scale land use vector data. Li (2014) committed to the research and development of the new generation of SuperMap 8C cloud architecture domestic GIS platform software, which has been practically applied in smart city big data and other fields (http://www.supermap.com/). The rapid development of these new technologies provides powerful technical support for the management and analysis of mass result data of geographical national condition monitoring.

The natural resource monitoring data has been extensively utilized at national, provincial, and local levels, with significant positive outcomes observed in China. Specifically, the "multiple planning integration" initiative in Hainan province utilized natural resource data to accurately identify cultivated land, forest land, and natural coastline. This enabled support for overall urban and rural development, optimized industry and facility layouts, and facilitated the realization of a comprehensive plan (Liu et al. 2018). The National Audit Office has leveraged monitoring data of natural resources to quickly ascertain the distribution status of natural resources such as forests, lakes, and wetlands, as well as natural protected

This contribution has been peer-reviewed.
https://doi.org/10.5194/isprs-archives-XLVIII-M-1-2023-115-2023

areas in 11 provinces and cities of the Yangtze River Economic Belt spanning approximately 2.05 million square kilometers. Moreover, the data on heaping and digging surfaces, industrial and mining enterprises, power station, and relevant special topics have facilitated accurate auditing of illegal activities (China 2018).

In many parts of the country, natural resource data has aided comparative analysis of changes in natural resource assets such as land, forest, water conservancy, and minerals before and after taking office during the pilot work of outgoing audit of natural resource assets of leading cadres. This provides objective, accurate, and intuitive data support for the outgoing audit. The national land supervision department employs natural resources data to carry out land supervision in four provinces in southwest China, supporting analysis of suspected problems in the office, and field verification through monitoring data. It effectively promotes the routine land supervision work of the state (Jiao and Wang 2020).

Beijing City applies natural resources data in carrying out "urban physical examinations," realizes scientific and refined urban management, and explores new ideas to solve big city diseases (Guan et al. 2020). Shanghai City uses natural resources data to prepare the Shanghai 2040 Master Plan, providing a solid guarantee for building a happy Shanghai with innovation, humanities, and ecology (Yu et al. 2017). These substantial practical achievements fully demonstrate the great application potential of natural resources monitoring data in critical fields such as the spatial planning system, ecological civilization construction, resources and environment audit, and urban governance and management in the new era.

With the recent round of institutional reforms in China, a new pattern of unified management of natural resources has been established, which imposes new requirements on the management and service of natural resources in the new era. The rapid development of big data technologies, such as Earth observation, satellite remote sensing, and intelligent cities, provides an efficient means of obtaining large-scale, objective, accurate, and real-time surface information throughout the country (Thatcher et al. 2018). Future analysis and decision-making of natural resources will therefore be transformed by big data, characterized by diversification, depth, specialization, and real-time capabilities. As the material basis and basic element of the construction of ecological civilization, geographic information will be actively integrated into the overall situation of natural resources management, exploiting the advantages of massive geographic and spatial data. The goal is to accelerate the emergence of data-driven natural resources planning, management, and scientific decision-making in the era of big data.

The proposed platform in this study is a high-performance computing system for the analysis of spatio-temporal big data related to natural resources. The system is based on a national-level geographic data spatio-temporal database and incorporates big data analysis services for geographic conditions and natural resources. The overall platform structure and technical framework are presented in detail, along with the latest system construction utilizing large-scale cloud computing resources. Additionally, the study performs high-performance computing experiments using national-scale geographic data, providing a valuable reference for supporting spatial natural resources analysis services.

2. GENERAL FRAMEWORK

The linkage between big data and the analysis of geographical conditions is closely tied to cloud computing. As cloud computing technology has evolved, it has transitioned from a single centralized architecture to a distributed application architecture. The most recent advancements in open and transparent cloud architecture technology have been proposed by large internet enterprises based on general cloud resource services and shared services, enabling an open and comprehensive cloud architecture that can accommodate the rapidly growing data and business needs. The traditional centralized system architecture is not able to effectively expand and is inadequate in addressing the computing service requirements of big data in the geographical conditions, particularly in support of natural resources businesses, public welfare industry data services, and other major needs (Jiping et al. 2019).

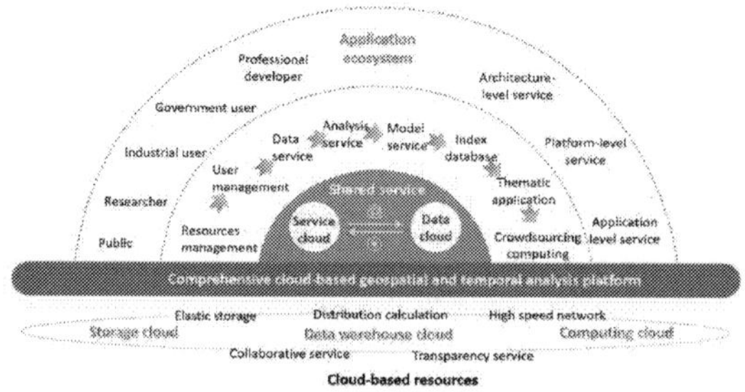

Figure 1. Architecture of the comprehensive cloud-based geospatial and temporal analysis platform

To achieve comprehensive cloud, on-demand access, on-demand response, on-demand service, and open sharing of big data, virtualization is used in the cloud computing environment based on distributed computing and storage, optimized hardware, clustered operation, and maintenance management system. This enables dynamic allocation and deployment of computing, storage, network, and other resources, and allows shared services according to the business center and data center. To establish an agile, scalable, shared, and open geographical big data analysis service framework, a cloud resource collaboration service, a business data sharing service, and a special application service ecology are constructed (Figure 1). This leads to a new design of infrastructure cloud, data platform cloud, business platform cloud, and application model service for geography data analysis service technology. These services provide open, efficient, comprehensive, and personalized solutions for natural resources and industry services.

2.1 Cloud Resource Collaboration Service

The aim of this study is to develop a supportive environment for the geographic national condition service big data by innovatively utilizing the overall design of cloud architecture. This is achieved by offering heterogeneous cloud collaboration and service coupling through the establishment of a cross-platform comprehensive cloud infrastructure that comprises storage cloud, database cloud, and computing cloud. The objective is to accomplish physical isolation and logical coupling of geographical application data, as well as business isolation and application coupling under the support of distributed cloud computing. The service realizes online collaborative and efficient service of distributed storage cloud, database cloud, and distributed computing cloud with service-oriented loose coupling. This approach comprehensively enhances computing power and expands performance, thereby providing comprehensive performance

https://doi.org/10.5194/isprs-archives-XLVIII-M-1-2023-115-2023 | © Author(s) 2023. CC BY 4.0 License.

support for professional, diversified, and in-depth big data analysis services.

2.2 Data Resource Sharing Service

The co-construction, sharing, and open provision of data resources and business applications are necessary for big data in the presence of complex and heterogeneous geographical conditions. This requires the use of state-of-the-art technologies and systematic approaches to reconstruct diverse geographical application data and professional capabilities. Cloud services for data resources and business applications can be realized by coupling them on a unified cloud resource platform and standard standards based on government affairs network, cloud computing, big data, security, and other technologies.

The process of integrated management applications includes resource management, user management, data services, analysis services, model services, index management, subject application, crowdsourcing computing, and other capabilities. These capabilities are designed to address the challenges of the big data era, characterized by large systems, big projects, big science, big research, big management, and big control. The employment of these capabilities can resolve traditional issues such as professional isolation, data isolated islands, and application segmentation.

2.3 Open and Shared Application Ecosystem

A unified authentication and different networks are adopted to provide unified infrastructure-level services, platform-level services, and application-level services for natural resource service subjects, government departments, industry users, scientific research, and the public. Physical computing resources on the physical server hardware platform are abstracted by infrastructure services to provide users with a unified, high-performance, and secure cloud computing environment. Data processing, image interpretation, data analysis, application service management, and other services are offered by platform-level services. Customized services for specific applications are provided by application-level services. Through the construction of a large-scale, data-driven, and application-oriented service ecology, a multi-level, multi-user, multi-theme, flexible, safe, open, and shared geographical analysis service mechanism is established, creating a positive ecosystem for "big demand, big data, big service, and big application."

3. TECHNICAL ROUTE

The advancement of cloud computing and big data technologies has led to the establishment of business cloud service platforms in various professional fields such as telecommunications, petroleum, land, geology, and oceanography. This has facilitated the transition from the traditional centralized application mode to an open and sharing mode(Reddy et al. 2022). This has significantly improved the professional application service capability and provides valuable insights for the development of a geographical analysis platform.

Geographical condition information serves as a public welfare service that caters to both natural resource management and public industry applications, leading to a wider range of service objects, open application fields, and diverse business models. This, in turn, requires higher demands on the flexibility, scalability, and openness of the geographical national condition information analysis platform.

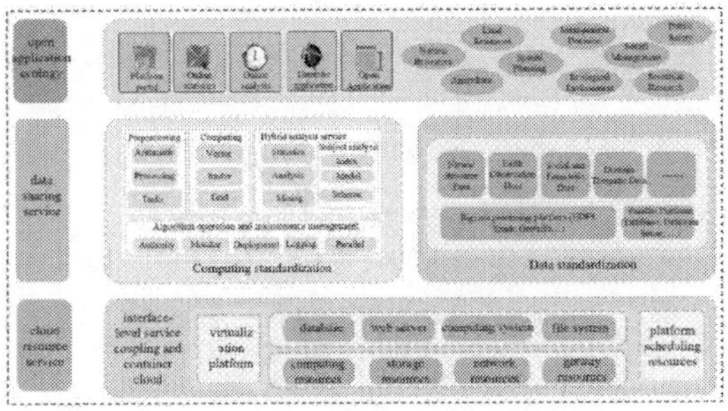

Figure 2. Elastic hybrid cloud computing ecological and distributed spatial analysis framework.

To achieve this goal, it is necessary to adopt new technologies such as heterogeneous cloud collaborative services, business cloud-based construction, and data cloud resource virtualization. These technologies can create a transparent, open, and agile cloud computing support architecture that can provide high-performance and high-availability business service capabilities (Figure 2).

3.1 Resource Layer

At the resource layer, which serves as the infrastructure service layer, a completely heterogeneous cloud resource pool is constructed based on interface-level service coupling, as well as container cloud and other new virtualization technologies. The database server, web server, file server, and distributed computing cluster are logically coupled and coordinated to achieve unified computing power allocation, dynamic memory management, flexible storage maintenance, and service capability migration. This enables the online collaborative and efficient services of the database cloud, distributed storage cloud, and distributed computing cloud, while providing cluster management services to achieve the service capability of "integrated management, transparency, efficiency, on-demand distribution, and on-demand expansion".

To address the traditional problem of data-intensive and computing-intensive spatial big data analysis, a lightweight computing model of micro-service agile design is adopted, with "agile service and elastic expansion" as the primary goal. Computing resources are "fine-grained disassembled, intensive services, and loosely coupled management" to achieve high reliable operation and flexible maintenance of massive supporting computing resources. The computing mode of "centralized service, distributed computing, and flexible expansion" with distributed computing and dynamic load is adopted to alleviate computing pressure, make the user experience agile, improve the robustness and reliability of database computing power, and enhance the overall service user experience.

3.2 Service Layer

At the shared service layer, which serves as the platform service layer, the latest business capabilities and data services are utilized in the design of the middle platform. The traditional server background and application foreground architecture are utilized to move the background business forward and the foreground business backward, thus creating a three-tier service architecture of background, middle, and foreground. This enables the integrated management and sharing of background capability, foreground capability, background data, and foreground data, effectively addressing expansion problems encountered in traditional architecture, such as business model segmentation, service capability segmentation, business data islands, and business migration difficulties (Figure 3).

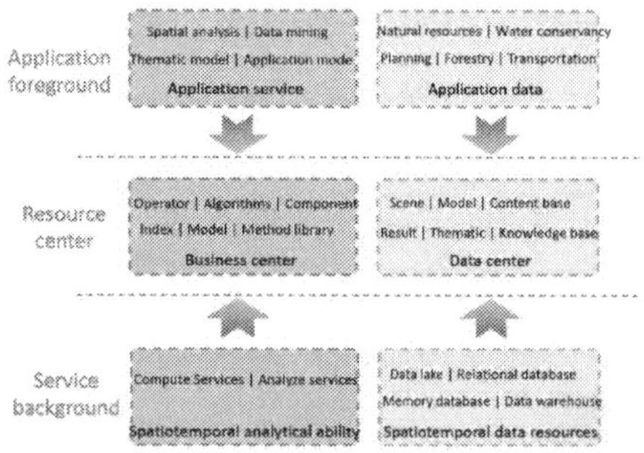

Figure 3. Shared service center in the platform mode.

To enhance the professional support and expansion capabilities of natural resource professional services and industry professional applications, and to ensure a set of cloud platform functions that support diversified application services, the cloud-based businesses adopt a forward sharing approach for computing services, analysis services, and other background capabilities, as well as a backward sharing approach for thematic models, application models, and other business capabilities. This approach maximizes resource sharing, resource reuse, and service co-construction, while promoting and optimizing thematic service capability, and reducing the cost of system maintenance and application migration and development. Moreover, it enhances the iterative ability and reuse vitality of development resources, ultimately improving the application efficiency of the entire service platform.

An open and shared ecosystem for thematic applications is created by opening and sharing the application capabilities of the cluster through data center, forming the index base, model base, method base, and other business capabilities of open services. To manage the challenges of multi-type data related to geographical conditions, a full-link technical approach consisting of data lake, relational database, in-memory database, and data warehouse is adopted for data resource management. Distributed storage architecture supports the unified storage of data with different structures in the data lake, which addresses data integration and distributed computing issues. Structured data is stored in a distributed database to achieve standardized storage, online transactional and analytical service capabilities. Business storage and online analysis of application data are enabled through the use of a distributed data warehouse, facilitating thematic application of geographical conditions and multi-dimensional analysis mining. Centralized management of data resources allows for open sharing of diversified thematic application results, leading to the development of a geographical conditions application result base, special topic database, and knowledge base. The ultimate goal is to establish a self-iterative geography and national conditions application ecology.

3.2 Application Layer

The cloud services provided by the application layer, including portals, visualization services, task monitoring, real-time query, and high-performance analysis services, are investigated in this study. Thematic clusters for natural resources, land, water conservancy, planning, forestry, grassland, and other types are constructed through front-end business topics, supported by the self-resource layer and shared service layer. The aim is to provide active, intelligent, thematic, comprehensive, and personalized services for government agencies, professional departments, scientific research institutions, and the public. Customized models, such as iterative reuse, are available to support stable application patterns and facilitate migration and expansion to new application scenarios or thematic areas. Furthermore, application patterns can be mapped and reconstructed to create a series of reusable service clusters, such as an application scene library, analysis pattern base, and thematic content base. This establishes an interconnected, open, and shared geographical application content ecology, allowing for horizontal promotion and vertical upgrade of application requirements in different industries and regions.

4. SYSTEM EXPERIMENT

This study investigated the construction of an open service analysis system for geographical conditions in a private cloud software and hardware environment based on the national geographic condition monitoring spatio-temporal database and employing large-scale cloud storage facilities, all-in-one database cloud platform, and distributed memory computing cloud facilities. The goal was to address the challenges of ultra-large-scale spatio-temporal data distribution scheduling and online computing core needs, and large-scale data computing experiments were conducted on the scale of 100 million to lay the foundation for the construction of a big data analysis platform for geographical conditions.

4.1 System Structure

A hybrid cloud collaborative model was utilized in the latest system architecture, which integrated storage cloud, computing cloud, database cloud, online thermal coupling, and cloud collaborative load. Key technologies were developed for the interface-level communication and collaborative application of cloud resources. A heterogeneous cloud resource pool, which consisted of elastic storage cloud, distributed memory cloud, and integrated database cloud, was successfully established. Furthermore, the technical framework for very large-scale distributed collaborative parallel computing and cloud collaborative load was implemented. An online analysis system of geographical conditions, based on the collaboration of database cloud and distributed computing cloud, was constructed. This system enabled the parallel transformation of typical spatial operators and model algorithms. The spatial computing was dynamically analyzed and distributed based on bottleneck indexes, such as variable dimension, operator complexity, and data magnitude. As a result, the computing-intensive capability of the database cloud and the data-intensive capability of the distributed computing cloud were fully leveraged. The system achieved large-scale collaborative parallel computing on the magnitude of hundreds of millions of spatial data points, with domestic leading geospatial big data analysis and computing capabilities. The technical level of geospatial data online analysis was significantly enhanced..

4.2 Cloud Resource Pool

A cloud resource pool specifically designed to meet the demands of geographical conditions was established by utilizing advanced software and hardware resources. This pool integrated high-performance storage, computing, and service equipment, ensuring superior performance, availability, and sustainability for big data analysis and calculation in the field. The topology

https://doi.org/10.5194/isprs-archives-XLVIII-M-1-2023-115-2023

and composition of the cloud resource pool were depicted in Figure 4.

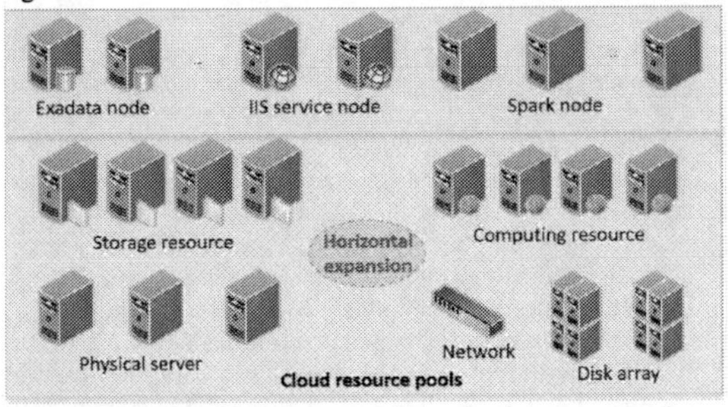

Figure 4. Topology diagram of the cloud resource pool.

4.3 Spatial Calculation Method

Focusing on spatial geographic analysis services, this study constructed three main spatial computing systems of vector computing, grid computing, and point-based computing based on a cloud platform infrastructure and distributed spatial computing environment. The platform enabled large-scale spatial analysis service capabilities and provided third-party development interfaces, as well as customized development capabilities for computing operators and models. The spatial computing architecture of the platform was presented in Table 1.

Computing system	Typical computing content
Vector calculation	The quantitative expression and scientific calculation of spatial attributes, including spatial location, spatial distribution, spatial form, spatial distance, and spatial relationships, are fundamental to supporting complex geographic analysis, such as data validation, extraction, conversion, statistical analysis, modeling, and mining.
Raster computation	Raster data is a crucial representation format for remote sensing images of natural resources, digital elevation models, derived data, and thematic raster data. It provides essential raster computing capabilities for spatial overlay analysis, resource assessment, model computation, and terrain analysis.
Grid computing	Spatial data is recursively encoded through geospatial and scale inclusion relations. With the support of a grid algorithm library and distributed computing engine, traditional spatial relationship and association analyses are transformed into associated queries and recursive statistics of attribute data, enabling accelerated computation of complex spatial calculations and association analyses.

Table 1. Cloud platform spatial computing system.

4.4 Open Service Model

Open service models have been developed in response to diverse analysis scenarios, including geographic analysis service and natural resources. These models are based on storage resource sharing, computing resource sharing, data resource sharing, and model resource sharing. A geographic and national knowledge support base has been preliminarily constructed based on metadata, and an open analysis portal platform has been designed and built to enable open customization and tagged topic management. The open analysis service framework of cluster-user-data-operator-model-index-theme has been realized, effectively supporting the open computing and analysis of geographical data.

4.5 Computational Performance Verification

A computing capability test is conducted on the cloud platform to evaluate the spatial analysis capability of the big data analysis platform, with computing power being considered as a core performance metric. The test employs typical geographic data and spatial computing scenarios to assess the platform's ability to support large-scale data.

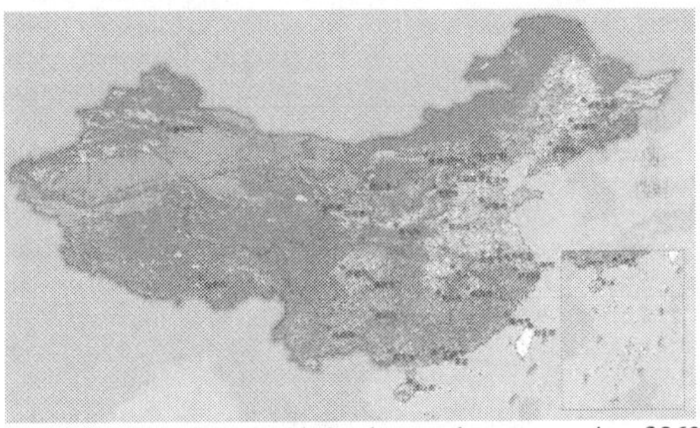

Figure 5. The national wide land cover datasets consist of 260 million vector polygon patches, with a total data amount of 300GB.

The entire land area of the country is covered by a crucial vector data set in geography, known as the land cover data, which comprises 260 million fine vector polygons. These polygons, representing complex polygons such as woodland, grassland, water surface, and road surface, consist of up to 100,000 single-element nodes. As a result, the calculation, analysis, and application of geographical information face bottlenecks due to the complexity of this datasets. To address this issue, the experimental data utilized in this study is the national land cover survey (Figure 5), and the experimental scene is the township area statistics. The data was analyzed and calculated through performance tests. The surface cover of each township spatial unit was determined by using the spatial unit data of township administrative division and summarizing the value of the third-class area of surface cover.

To simplify calculations, the authors converted the national land cover data into WKT format using a 0.2-degree grid and stored it in a distributed file system on the cloud platform. Table 2 presents statistical data on various aspects of the data. Custom-developed space cutting and area statistics operators were utilized on a Spark computing cluster for distributed computing experiments leveraging the existing cloud resource pool. The computing nodes used for the experiment are listed in Table 3.

Data type	Data format	Number of patches	Data volume/GB
Land cover (original)	FileGDB	260 million	/
Township Unit (original)	FileGDB	55,362	/
Land cover	WKT	260 million	1228

This contribution has been peer-reviewed.
https://doi.org/10.5194/isprs-archives-XLVIII-M-1-2023-115-2023 | © Author(s) 2023. CC BY 4.0 License.

(text)			
Township Unit (text)	WKT	55,362	2.78

Table 2. Land cover and township data statistics

Request test memory(GB)	Parallel	Nodes	Threads per node	Time
210	24	8	12	32min

Table 3. Results of land cover intersection of township unit

Excellent computing power is demonstrated by our analysis and calculations, with a total application memory of 256GB, which is highly advantageous for the Spark memory computing framework. A significant reduction in statistical time for national land cover analysis to an hourly level is achieved through our experiments, providing superior computational efficiency and time cost compared to the traditional spatial computing model of "data split, standalone calculation, result collection". Furthermore, an increase in the number of computing nodes can further improve the computational efficiency.

Our findings suggest that cloud-based big data analysis and calculation offers significant advantages and is well-suited for the demands of large-scale geographical analysis. The number of computing nodes in the cloud resource pool can be increased on a large scale as application demand and user concurrency increase, considering the horizontal expansion mechanism of distributed clusters. This approach enhances the computing performance and load capacity of the system, making it better suited to practical applications. Furthermore, for more complex computing-intensive scenarios such as buffer computing, overlay analysis, and exponential computing, the all-in-one cloud platform's rich interfaces and computing power can provide more flexible and efficient computing services.

5. CONCLUSIONS

This paper proposes a big data analysis service platform specifically designed for natural resource management in the contemporary era, which faces new challenges such as the increasing diversification, depth, and real-time requirements of data applications. The proposed platform is built on a new architecture that incorporates features such as comprehensive cloud services, on-demand access, on-demand response, on-demand services, and open data sharing. To realize this architecture, we introduce innovative technologies including heterogeneous cloud collaborative services, business capability, and virtualization of data resources. With the application of geographical conditions as a starting point, we emphasize the role of geographical services in supporting natural resource management and industry applications. This platform offers a broader vision for geographical services, promotes innovative thinking, and presents an opportunity for exploration and support of new work in the field of natural resources.

The analysis of big spatial-oriented data is characterized by its data-intensive and computing-intensive nature, which poses significant challenges in terms of complexity and computational requirements. To overcome these challenges, the authors developed a scalable solution leveraging extensive software and hardware resources. Specifically, they established an elastic cloud-based resource pool comprising storage, computing, and database services, which collectively enable online computational capabilities for large-scale geospatial data analysis.

The conventional methods for spatial data analysis and computation, namely vector computation, grid computing, and lattice computing, have been developed while exploring open service modes, such as resource sharing of storage, computation, data, and model. To assess the performance of the open analysis service framework of cluster-user-data-operator-model-index-theme, a performance test was conducted using 260 million national land cover vector data. The findings suggest that cloud-based big data analysis and computation of geographical conditions offer significant advantages, such as agility, scalability, sharing, and openness, for big data analysis of geographical conditions.

In the forthcoming phase, it is planned to enhance the expansion of the cloud resource pool, collaborative scheduling of cloud resources, upgrade of the business platform, refinement of data extraction, enhancement of the spatial computing interface, development of advanced spatial analysis models, deployment of thematic application models, and provision of multi-user concurrent support. The focus will be on enhancing the design and technology, in tandem with persistent advancement and construction of theme applications, such as ecological value assessment and resource ring environmental audit. The ultimate objective is to establish a robust and effective big data analysis service platform for geographical conditions.

ACKNOWLEDGEMENTS

The authors would like to express their gratitude to the National Geographic Conditions Monitoring Project for providing strong support for this study.

REFERENCES

Allen, D.W., & Coffey, J.M., 2019: *Focus on Geodatabases in ArcGIS Pro*. Esri Press.

Chen, C., Lin, J., Wu, X., Wu, J., & Lian, H., 2013. Massive Geo-spatial Data Cloud Storage and Services Based on NoSQL Database Technique. *Journal of Geo-Information Science* 15(2), 166-174.

China, N.A.O.o.t.P.s.R.o.,2018. Zhonghua Renmin Gongheguo Shenjishu Shenji Jieguo Gonggao. In. Beijing: National Audit Office of the People's Republic of China.

Ferreira, D., & Vale, M., 2022. Geography in the big data age: an overview of the historical resonance of current debates. *Geographical Review* 112(2), 250-266.

Guan, L., Ding, Y., Liu, H., Feng, X., Chen, P., & Ma, H., 2020. Construction and application of health examination evaluation system in new smart city. *Science of Surveying and Mapping* 45(3), 135-142.

Jiao, S., & Wang, Y.,2020. The Aggregation Effect of "Inspector + Mapping". In, *China Natural Resources News* (pp. 7-8). Beijing

Jin, F., Zhang, F., Du, Z., Liu, R., & Li, R., 2016. Spatial overlay analysis of land use vector data based on Spark. *Journal of Zhejiang University(Science Edition)* 43(1), 40-44.

Jiping, L., Chun, D., Xiaochen, K., Shike, Q., Rong, Z., Bin, L., & Lijian, S., 2019. National Geographical Conditions Statistical

Analysis in the Era of Big Data. *Geomatics and Information Science of Wuhan University* 44(1), 68-76.

Li, D., Ding, L., & Shao, Z., 2016. Reflections on Issues in National Geographical Conditions Monitoring. *Geomatics and Information Science of Wuhan University* 41(2), 143-147.

Li, S.,2014. Research on Key technologies of data management and parallel processing under cloud GIS framework. In, *College of Earth and Planetary Sciences*: University of Chinese Academy of Sciences.

Liu, J., Dong, C., Kang, X., Qiu, S., Zhao, R., Li, B., & Sun, L., 2019. National Geographical Conditions Statistical Analysis in the Era of Big Data. *Geomatics and Information Science of Wuhan University* 44(1), 68-76, 83.

Liu, Y., Wang, P., & Huang, C., 2018. A Practical Exploration of the Multi-plan Integration Technical Method in Hainan Province. *NATURAL RESOURCE ECONOMICS OF CHINA* 5(1), 30-34.

Reddy, G.O., Raval, M.S., Adinarayana, J., & Chaudhary, S., 2022: *Data Science in Agriculture and Natural Resource Management*. Springer.

Saygili, O.Y., 2020: *The Introduction to Private Cloud using Oracle Exadata and Oracle Database*. CRC Press.

Thatcher, J., Shears, A., & Eckert, J., 2018: *Thinking Big data in geography: New regimes, new research*. University of Nebraska Press.

Yang, C., Goodchild, M., Huang, Q., Nebert, D., Raskin, R., Xu, Y., Bambacus, M., & Fay, D., 2011. Spatial cloud computing: how can the geospatial sciences use and help shape cloud computing? *International Journal of Digital Earth* 4(4), 305-329.

Yu, J., Chen, Y., & Gu, X., 2017. Spatial Analysis and Management Guideline in the General Urban Design of Shanghai. *Urban Planning Forum* 7(1), 101-107.

This contribution has been peer-reviewed.
https://doi.org/10.5194/isprs-archives-XLVIII-M-1-2023-115-2023

SPATIOTEMPORALLY IMPROVING THE RURAL ACCESS INDEX – A REMOTE SENSING BASED APPROACH

M. Halbgewachs[1]*, A. Schneibel[1], M. Gähler[1]

[1] German Remote Sensing Data Center (DFD), German Aerospace Center (DLR), 82234 Wessling, Germany – magdalena.halbgewachs@dlr.de

KEY WORDS: RAI, infrastructure, state capacity, SDG, Landsat, Sentinel, OpenStreetMap, Lake Chad Basin.

ABSTRACT:

Many countries, especially in the global south still lack the ability to effectively pursue basic policies, which can lead, in the worst case, to armed conflicts. Access to markets is a key factor for economic growth and an important component in reducing poverty. The SDG 9.1.1 addresses the proportion of the rural population who live within 2km of an all-season road, which can be mapped by the Rural Access Index (RAI), introduced by the World Bank in 2006. This requires the road network of so-called all-season roads, population distribution and rural areas. We developed a fully automated approach, using remote sensing and other open source data to calculate the RAI on an annual basis between 2013 and 2020 for the Lake Chad region. We achieved an overall accuracy between 97.0% and 97.5% in detecting all-season roads using a Random Forest classification. Our method shows similar results to those published by the World Bank. However, our approach provides a higher spatial and temporal resolution measuring the RAI compared to previous studies and is independent of field studies.

1. INTRODUCTION

Countries, which lack the ability to effectively pursue domestic or international policies are often at risk of becoming what is known as a 'fragile state'. To assess this risk, scientists use the concept of state capacity, a measure that is closely connected to the fragile state concept (Besley and Persson, 2010). It measures the ability of a state to e.g. ensure political stability, charge taxes and provide essential health or educational supplies to its population (Woodward, 2004, Müller-Crepon, 2021).

One of these key factors, access to markets, is crucial to economic growth and an important component in reducing poverty. Especially the rural population is often insufficiently connected to a well-developed infrastructure. The Sustainable Development Goal (SDG) 9.1 also accounts for this factor and includes the development of "[...] reliable, sustainable and resilient infrastructure, including regional and transborder infrastructure, to support economic development and human well-being, with a focus on affordable and equitable access for all" (United Nations, 2022). The connection of rural population to infrastructure has been translated into the Rural Access Index (RAI) by Roberts et al., 2006 and estimates the proportion of the rural population with adequate access to the transport system, thus directly mapping SDG indicator 9.1.1 (World Bank Group, 2016). Precisely, the RAI is defined as the proportion of the rural population living within two kilometers ($\sim$ 20-25 minutes walking time) of an all-season road. All-season roads are defined as being navigable throughout the year and are only temporarily impassable due to weather conditions. The initial approach to calculate the RAI was based on household surveys (Roberts et al., 2006). In 2015, the World Bank designed a new GIS-based method to provide more accurate and cost-effective results by using specific datasets, i.e. the population distribution, road infrastructure and the condition of these roads (World Bank Group, 2016). According to the World Bank Group, 2016, roads are all-season roads if they are either paved (with an International Roughness Index (IRI) of less than 6 meters/km or in excellent, good or fair condition) or unpaved (with an IRI of less than 13 meters/km or in excellent or good condition). Such an assessment requires an on-site survey of the roads or most often a subjective assessment based on high-resolution imagery, if available. For the RAI calculation, the World Bank proposes the use of different data sets, which are partly publicly available, others restricted, or, depending on the study area, proprietary. The RAI is made partially available for various years at country level, in some cases Administrative Level 1 (admin-1 level) (Workman and McPherson, 2019, World Bank Group, 2019). Other studies calculated the RAI for only one point in time (Akin and Demirel, 2019, World Bank Group, 2016), a specific subregion (Akin and Demirel, 2019), only partly using remote sensing data (Akin and Demirel, 2019, World Bank Group, 2016, Workman and McPherson, 2019), or in a GIS environment (Workman and McPherson, 2021, World Bank Group, 2016).

Our approach consists of using a standardized objective data basis, including remote sensing data, in order to be able to calculate the RAI consistently across countries with a higher temporal and spatial resolution than provided to date. Only open source data is used as part of our framework, which allows researchers and analysts to gather and analyze large amounts of data without incurring significant costs. The OpenStreetMap (OSM) database is used as basis for the road network. Since in 2013 about 72% and in 2020 still about 40% of the roads within our study area do not contain information about the road surface, a Random Forest classification was carried out on Landsat 8 and Sentinel-2 data for each year to identify paved pixels. For this paper, we focus on an annual calculation of the RAI between 2013 and 2020 in the Lake Chad region, which includes parts of Niger, Chad, Cameroon, and Nigeria. As spatial units we use both, admin-1 regions and PRIO-GRID cells therein, a global vector matrix commonly used in social sciences (Tollefsen et al., 2012). In summary, the study specifically addresses the following research questions: (I) Is the differentiation between paved and unpaved road pixels possible by conducting a land cover classification? (II) Can adequate

* Corresponding author

This contribution has been peer-reviewed.
https://doi.org/10.5194/isprs-archives-XLVIII-M-1-2023-123-2023

RAI results be obtained using remote sensing and other open source data? (III) Is it possible to provide a higher temporal and spatial resolution of the RAI than previous studies show?

2. MATERIALS AND METHOD

2.1 Study Area

For this paper, we focus on the Lake Chad region, which covers the cross-border region of Niger, Chad, Cameroon, and Nigeria. A total of eleven admin-1 regions are considered, one in Niger, two in Nigeria, two in Cameroon and six in Chad (Figure 1). All of these countries have been affected by the violent terrorism of Boko Haram, which has built its success on dysfunctional socioeconomic and political gaps (United Nations Development Programme, 2022). The conflict has resulted in more than 2.9 million people being displaced and according to UN OCHA, about 5.6 million people are directly facing food insecurity (UN OCHA, 2022). Insufficient basic social services and the already low availability of natural resources are increasingly deteriorating the situation. Due to years of dominating violence and insecurity, thousands of children lack education, estimations made by UN OCHA account for more than 1,000 closed schools (UN OCHA, 2022). The population derives its income mainly from agriculture and fishing (Masaki and Rodríguez-Castelán, 2021). In some cases, the poverty rate in the region is up to three times higher than in the rest of the neighbouring countries reaching values between 31% and 72% (Masaki and Rodríguez-Castelán, 2021).

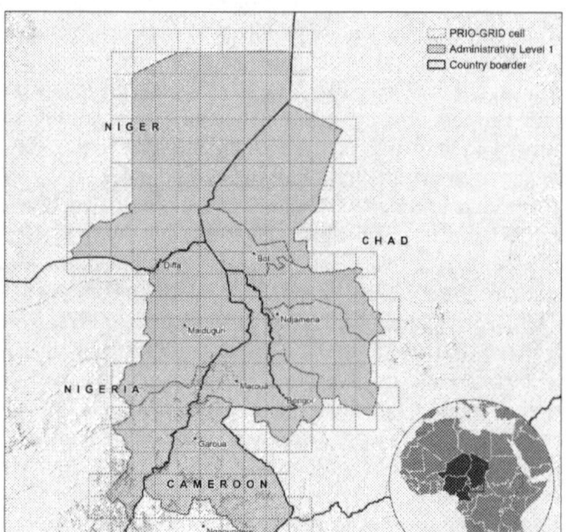

Figure 1. Study Area (data source: Global Administrative Areas, 2012, Tollefsen et al., 2012,OpenStreetMap contributors, 2022; Service Layer Credits: Airbus,USGS,NGA,NASA,CGIAR, NCEAS,NLS,OS,NMA,Geodatastyrelsen,GSA,GSI and the GIS User Community; Esri,HERE,Garmin,FAO,NOAA,USGS)

2.2 Data

Only open source data is used as part of our framework.
The gridded population count dataset from WorldPop, available annually since 2000, is taken as population distribution data (WorldPop and CIESIN, 2018). The used mapping approach is a Random Forest-based dasymetric redistribution (Stevens et al., 2015). This involves matching census-based population counts with the corresponding administrative units and disaggregating them to 3 arc (∼100x100m) grid cells using machine learning techniques that exploit relationships between population densities and a set of spatial covariates (Lloyd et al., 2019). As a second dataset, the Global Human Settlement Layer of the Joint Research Centre (JRC) from 2015 with a spatial resolution of 1km, is used to identify rural regions which is based on satellite and population data (Pesaresi et al., 2016). The dataset from 2015 was chosen because it is the closest to our time period. Only pixels that are not classified as urban centers or urban clusters according to the DEGURBA (Degree of Urbanisation) classification scheme and thus have a population density of less than 300 inhabitants per km^2 are considered in our study (OECD et al., 2021)
The road network is obtained via OpenStreetMap (OSM) and queried for the corresponding, retrospective years through the ohsome API. This API is a data analysis tool, provided by the Heidelberg Institute for Geoinformation Technology (HeiGIT), for working with OSM historical data on a global scale. It allows to query and aggregate the recorded vector data and get insights into the spatial and temporal evolution and associated annotated attributes (Raifer et al., 2019).
Atmospherically corrected surface reflectance from Sentinel-2 MSI and Landsat 8 OLI/TIRS time series data is used for road classification. The data is provided along with a quality assessment (QA) band that indicates the presence of clouds (USGS, 2019, ESA, 2015). In addition, the ESA WorldCover 10m classification and the JRC Global Surface Water Mapping Layer are used to mask water and wetland areas from the satellite mosaics. The ESA WorldCover 10m provides a global land cover classification of 2020, based on Sentinel-1 and Sentinel-2 data. The classification consists of eleven land cover classes, one of which designates permanent water bodies (Zanaga et al., 2021). The JRC Global Surface Water Mapping Layer provides information about spatial and temporal distribution of surface water with a spatial resolution of 30 meters between 1984 and 2021, based on Landsat 5, 7 and 8 data. Both permanent and seasonal water surfaces are recorded (Pekel et al., 2016).
The calculations are performed both on admin-1 level (Global Administrative Areas, 2012), the largest subnational administrative unit of a country, and on PRIO-GRID, which is a global vector matrix of 0.5 x 0.5 decimal degrees spatial resolution and often used in social science research being insensitive of political boundaries (Tollefsen et al., 2012).

2.3 Methods

2.3.1 Random Forest and Road Classification Within our approach we defined all season roads as roads that are paved to calculate the RAI and determine the condition of the roads. Even though the literature refers to paved roads as all-weather roads rather than all-season roads which is rated to a higher standard (World Bank Group, 2016, Workman and McPherson, 2019), our approach does not rely on time-consuming in situ data that must be collected annually. To acquire this information about the road surface, a yearly Random Forest classification was carried out on Landsat 8 (between 2013 and 2018) and Sentinel-2 (for 2019 and 2020) data, where paved pixels were identified. Figure 2A shows a schematic overview of the workflow.

As first pre-processing step, annual multitemporal composite mosaics are created based on the available imagery. Secondly, cloud masking is performed. The provided QA band contains

This contribution has been peer-reviewed.
https://doi.org/10.5194/isprs-archives-XLVIII-M-1-2023-123-2023 | © Author(s) 2023. CC BY 4.0 License.

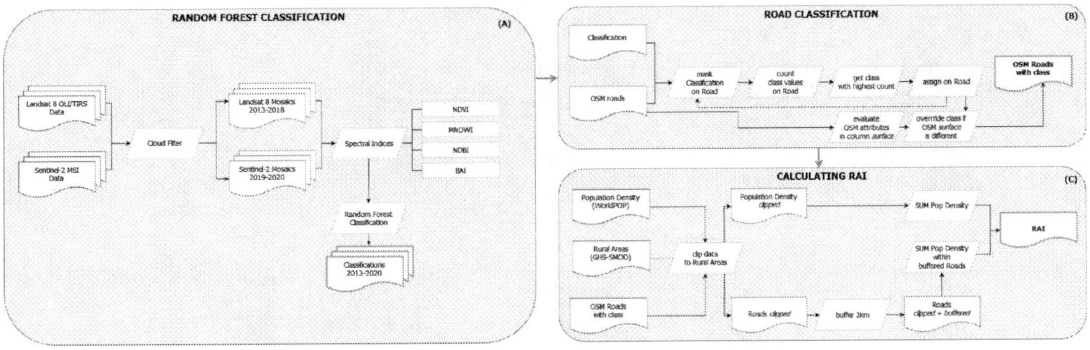

Figure 2. Schematic workflow

bitmasks for clouds and cloud shadows. All pixels containing the corresponding bit value are removed. To avoid misclassification, permanent and seasonal water as well as wetland areas, derived from the ESA WorldCover 10m and Global Surface Water Mapping Layer, are excluded.

In total, 1,648 training samples were collected on the mosaic of the year 2021 for the five classes paved roads, open ground, shrub, vegetation and building. The spectral bands Red, Green, Blue, NIR and SWIR are used for the classification. Additionally, four indices are added to enhance the detection of paved roads and the distinction between the different classes:

The Normalized Difference Vegetation Index (NDVI) is used to highlight the signal from vegetation, which highly reflects in the NIR band (Kamal et al., 2015). The Modified Normalized Difference Water Index (MNDWI) helps to delineate water and artificial surfaces (e.g. asphalt roads). Both surfaces have similar spectral properties and reflect especially in the green wavelength range, the MIR band used improves the distinction between the two (Xu, 2007). The Normalized Difference Built-up Index (NDBI) is particularly useful in detecting built-up areas (Zha et al., 2003). The Burned Area Index (BAI) is additionally used to identify burn scars (Chuvieco et al., 2002). As paved roads also have a lower spectral reflectance in the Red band than vegetation, this index also helps to distinguish asphalt from other land covers (Kamal et al., 2015). During the development of the method it has come to attention that burned areas were often classified as paved roads, which is prevented by incorporating the BAI. The indices are calculated as follows:

$$NDVI = \frac{NIR - RED}{NIR + RED} \tag{1}$$

$$MNDWI = \frac{GREEN - MIR}{GREEN + MIR} \tag{2}$$

$$NDBI = \frac{MIR - NIR}{NIR + NIR} \tag{3}$$

$$BAI = \frac{1}{(0.1 - RED)^2 + (0.06 - NIR)^2} \tag{4}$$

After randomizing, the training collection is split into 70% training and 30% testing points. A 10-tree Random Forest classifier is then carried out, resulting in an overall training accuracy of 95.80% and overall testing accuracy of 69.44%.

The classification is used as input for assigning the road surface on the annual OSM road network. A road was classified as being paved by a winner-takes-all approach, i.e. if most of the road pixels were classified as such. However, if a definite road surface attribute was given by an OSM user (e.g. asphalt, paved, concrete, bitumé, unpaved, sand, earth, dirt, gravel, pebblestone), this was taken, accounting for their expert-based knowledge (Figure 2B).

2.3.2 Calculating the Rural Access Index The calculation of the RAI follows the approach suggested by the World Bank Group, 2019. The RAI represents the percentage of the rural population (RP) living within 2km of an all-season road (RP_{road}):

$$RAI = \frac{RP_{road}}{RP} \tag{5}$$

This is done by determining the sum of the rural population within the Area of Interest (AOI). Subsequently, all roads marked as all-season roads within the rural regions are provided with a buffer of 2km and the population is then summarized within these areas (Figure 2C).

2.3.3 Acccuracy Assessment For the accuracy assessment, stratified random sampling was used to generate reference points for 2017 and 2019, evaluating the recorded all-season roads. These two years were chosen to include both Landsat 8 and Sentinel-2 classifications. All-season roads are defined here as those captured by the Random Forest classification and related OSM attributes, as described in chapter 2.3.1. As paved roads represent a rather small proportion in the study site compared to other land surfaces, the stratified random sampling method allows us to increase the proportion of the underrepresented paved class (Olofsson et al., 2014). Our requirement was to have at least 100 samples in each stratum, which could be ensured by a proportional redistribution of the sample size. In total 4,125 samples were distributed randomly throughout the whole study area for the two points in time, representing the strata paved and unpaved. The points were assigned expert-based to be either paved or unpaved using Google Earth high resolution imagery of the respective year.

3. RESULTS

3.1 Rural Access Index

The results are shown in Figure 3 and Figure 5 on admin-1 levels and in Figure 4 on PRIO-GRID cells. On average, the

https://doi.org/10.5194/isprs-archives-XLVIII-M-1-2023-123-2023

Figure 3. RAI between 2013 and 2020 on admin-1 level (data source: Global Administrative Areas, 2012)

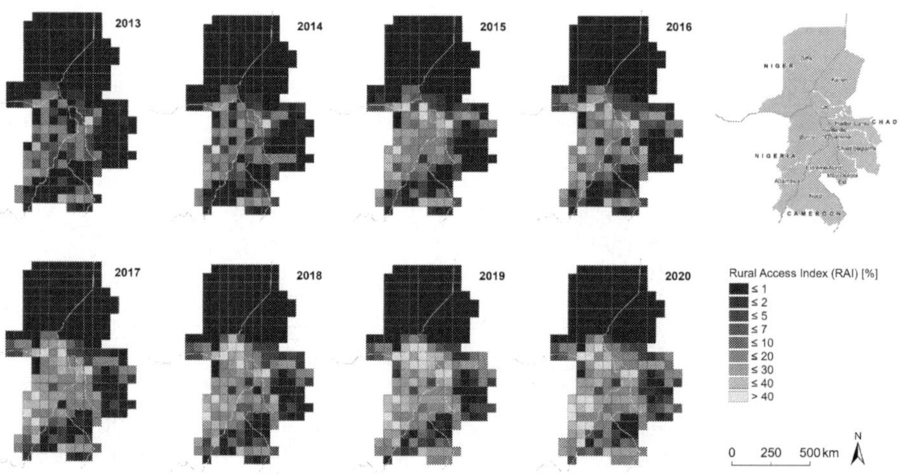

Figure 4. RAI between 2013 and 2020 on PRIO-GRID cell level (data source: Tollefsen et al., 2012,Global Administrative Areas, 2012)

RAI has increased over time and a larger proportion of rural population has gained access to all-season roads. Admin-1 areas with larger towns or cities (such as in Ville de N'Djamena in Chad or Borno in Nigeria) show higher RAI values throughout the whole time series (RAI $>10\%$) than unpopulated regions (RAI $<10\%$). Figure 5 also shows that the RAI in Ville de N'Djamena increased from 35.66% to 81.01% between 2013 and 2020. This admin-1 region has the highest RAI values of all processed admin-1 regions throughout the entire analysis period. This contrasts with the RAI values in Kanem (Chad), where population density is very low and the smallest increase was observed compared to other regions, from 0.00% to 1.53%. This becomes even more noticeable at PRIO-GRID levels where grid cells containing larger settlements (such as Biu in south-western Borno, Nigeria, Guélengdeng in central north Mayo-Kebbi Est, Chad or N'Djamena in Ville de N'Djamena, Chad) stand out clearly (RAI between 28% and 82% over time). The same applies to grid cells holding main road axes to larger nearby cities (such as PRIO-GRID cells in the southern part of Nord, Cameroon where Ngaoundéré is a nearby city). Similarly, it is notable that higher RAI values at admin-1 level often result from single higher values and a large number of unpopulated areas in PRIO-GRID cells contained therein (e.g. Diffa in Niger). In general, the RAI never exceeds 35% during the analysis period, with the exception of Ville de N'Djamena (Chad).

Statistics can be viewed in more detail in Table 2: A positive trend of 2.33% can be observed for the whole region regarding the rural populations' walkable access to all-season roads. For the entire AOI, the value of RAI is 17.17%. Regions in Cameroon rank the lowest, with a mean value of 11.33%, while nigerian regions rank the highest, with 22.79%. Chad regions show the widest range of RAI values, from 0.00% to 81.01%, with the high values being attributed to Ville de N'Djamena.

https://doi.org/10.5194/isprs-archives-XLVIII-M-1-2023-123-2023 | © Author(s) 2023. CC BY 4.0 License.

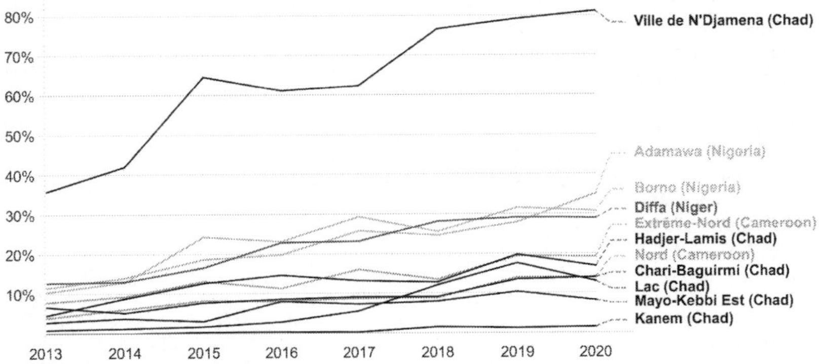

Figure 5. Lineplot of RAI on admin-1 level between 2013 and 2020, colour-coded by country: Chad (blue), Nigeria (turquoise), Niger (red), Cameroon (yellow).

Country (considered regions)	Minimum [%]	Maximum [%]	Mean [%]
Chad (6)	0.00	81.01	16.47
Cameroon (2)	3.91	19.32	11.33
Niger (1)	12.68	29.09	21.79
Nigeria (2)	10.41	35.02	22.79
Study area	**0.00**	**81.01**	**17.17**

Table 2. Country statistics of calculated RAI results.

3.2 Overall Classification Performance

The accuracy assessment in 2017 results in an overall accuracy of 97.0%, and the 2019 verification shows an overall accuracy of 97.5%. The Cohen's Kappa value is 61.6% for the year 2017 and 66.3% for 2019, with values between 61.0% and 80.0% reflecting high agreement (Landis and Koch, 1977). For the paved class, the precision is 54.0% in both years, which means a slight overestimation on paved roads. The recall shows accuracies of 76.1% and 90.0% for the years 2017 and 2019. The class unpaved roads has a precision of 99.1% and 99.7% for 2017 and 2019 and a recall of 97.7% in both years (see Table 3).
By comparing our findings with the World Bank's results, which have been published for parts of our study area, we find an overall strong agreement. For 2014, we obtain a RAI of 15.5% for the admin-1 level of Borno (Nigeria), and 15.9% for Adamawa (Nigeria), the World Bank values of 14.9% and 12.8% correspond respectively (World Bank Group, 2019).

		2017	**2019**
Precision	paved	0.540	0.540
	unpaved	0.991	0.997
Recall	paved	0.761	0.900
	unpaved	0.977	0.977
Overall Accuracy		0.970	0.975
Cohen's Kappa		0.616	0.663

Table 3. Accuracy assessment values for street classifications for 2017 and 2019.

4. DISCUSSION

The main objective of this study was the application of an automated workflow using freely available remote sensing and geo data to calculate the Rural Access Index for admin-1 areas and PRIO-GRID cells for the Lake Chad region between 2013 and 2020. The comparison between the two spatially differing admin-1 level and PRIO-GRID regions has shown that the higher spatial resolution of the PRIO-GRID provides more accurate results with respect to the RAI. Thus, changes can be observed and analyzed more accurately and may serve as valuable input for decision makers.
Based on Landsat 8 and Sentinel-2 mosaics as well as the additional use of four spectral indices over the time period considered, a Random Forest classification was performed to detect paved pixels. Compared to in situ surveys, the remote sensing data used has the advantage of being available free of charge over a long period of time and of allowing an automated classification procedure. The method used is reproducible over different areas of interest, only the training samples would have to be adjusted, depending on the land use of the region. In general, the distinction between paved roads and other land uses (e.g. roofs, water) is difficult because the spectral properties are observed to be often similar (Xu, 2007, Kavzoglu et al., 2009). This could be solved by the initial masking of water areas and bringing in historical OSM road data serving as road database. Due to local wind drifts, asphalt road sections may be covered with sand or vegetation and identified as such. In order to minimize such misclassifications, all OSM roads were assigned the attribute paved, either if the majority of pixels on a road were detected as asphalt or if a corresponding tag was provided in the OSM data itself. Given that OSM is an open user-generated content platform, the quality of the data provided is dependent on the OSM community and their know-how (Neis and Zielstra, 2014). Since the OSM roads serve as data basis for calculating the RAI, a possible mis- or non-editing within the road dataset is inevitable. There can also be large differences in data quality between countries, as well as between rural and urban regions (Neis and Zielstra, 2014). Nevertheless, it is an open source data portal that has seen a strong increase in contributions in recent years (Neis and Zielstra, 2014). It keeps being continuously updated and has been further improved for many years now, thus offering great advantages compared to time-consuming and expensive in situ data collection or surveys, which are often rarely available (Workman and McPherson,

This contribution has been peer-reviewed.
https://doi.org/10.5194/isprs-archives-XLVIII-M-1-2023-123-2023 | © Author(s) 2023. CC BY 4.0 License.

2021). Partial decreases and increases of the RAI within the admin-1 regions can result from different reasons. One factor could be misclassifications resulting from the Random Forest approach, but also due to the used OSM road data. It can happen, for example, that some roads were added incorrectly by OSM users the one year and then deleted in the following year. The RAI calculations depend not only on the input street network, but also on rural areas and population data used. In this study we included rural areas from 2015, derived from the Global Human Settlement Layer of the JRC and annual WorldPop population densities.These data sets represent an approximation to reality by statistical techniques and thus serve as a valuable data basis for the RAI calculations, independent of irregular official government data collections.
Research by the World Bank has shown that on average 57% of the rural population of the International Development Association (IDA) countries in 2003, including the four countries of our study area, have access to the transport network (Roberts et al., 2006). Our results show that the admin-1 regions we analyzed do not fall within this national average, with the exception of Ville de N'Djamena (Chad). These results strengthen the assumptions of the World Development Report that population in poor countries and regions especially have a longer travel time to reach basic services (World Bank Group, 2003). The report also states that children from the poorest 20% of the population from rural areas in Nigeria have to travel more than five times farther to the nearest elementary school than children from the richest 20% and even more than seven times further to the next health facility. In Chad, 80% of the poorest 20% of the population need to travel more than one hour to reach the nearest health facility (World Bank Group, 2003).

5. CONCLUSION AND OUTLOOK

This study analyzes temporal and spatial changes in the rural population in the Lake Chad region between 2013 and 2020 in terms of access to all-season roads. Our approach shows that the calculation of the RAI based on open source data in combination with remote sensing data yields very good results. The availability of time series data from Landsat 8, Sentinel-2 and OSM (in this case road networks) provides a cost-effective and solid basis for detecting asphalt roads using a Random Forest classification. By using the WorldPop population data set, an annual calculation of the RAI is possible. Furthermore, the approach can be fully automatized making it transferrable to other world regions, as long as the ground materials show the same spectral responses as our study area. Otherwise new training samples must be collected. Our approach can effectively contribute to the global calculation of SDG 9.1.1, an important variable for many political, social and natural scientists. The methodology of the RAI can also be extended to other research questions, such as the accessibility of educational institutions or health sites, which allows drawing conclusions about the need for political action.

REFERENCES

Akin, O., Demirel, H., 2019. Measuring Rural Access Index via Spatial Information Science. *2019 9th International Conference on Recent Advances in Space Technologies (RAST)*, IEEE, 461–463.

Besley, T., Persson, T., 2010. State Capacity, Conflict, and Development. *Econometrica*, 78(1), 1–34.

Chuvieco, E., Martín, M. P., Palacios, A., 2002. Assessment of different spectral indices in the red-near-infrared spectral domain for burned land discrimination. *International Journal of Remote Sensing*, 23(23), 5103–5110.

ESA, 2015. *Sentinel-2 User Handbook*.

Global Administrative Areas, 2012. GADM database of Global Administrative Areas, version 2.0. www.gadm.org.

Kamal, M., Phinn, S., Johansen, K., 2015. Object-Based Approach for Multi-Scale Mangrove Composition Mapping Using Multi-Resolution Image Datasets. *Remote Sensing*, 7(4), 4753–4783.

Kavzoglu, T., Sen, Y. E., Cetin, M., 2009. Mapping urban road infrastructure using remotely sensed images. *International Journal of Remote Sensing*, 30(7), 1759–1769.

Landis, J. R., Koch, G. G., 1977. The Measurement of Observer Agreement for Categorical Data. *Biometrics*, 33(1), 159.

Lloyd, C. T., Chamberlain, H., Kerr, D., Yetman, G., Pistolesi, L., Stevens, F. R., Gaughan, A. E., Nieves, J. J., Hornby, G., MacManus, K., Sinha, P., Bondarenko, M., Sorichetta, A., Tatem, A. J., 2019. Global spatio-temporally harmonised datasets for producing high-resolution gridded population distribution datasets. *Big earth data*, 3(2), 108–139.

Masaki, T., Rodríguez-Castelán, C., 2021. *Lake Chad Regional Economic Memorandum*. World Bank.

Müller-Crepon, C., 2021. State reach and development in Africa since the 1960s: new data and analysis. *Political Science Research and Methods*, 1–10.

Neis, P., Zielstra, D., 2014. Recent Developments and Future Trends in Volunteered Geographic Information Research: The Case of OpenStreetMap. *Future Internet*, 6(1), 76–106.

OECD, The European Commission, United Nations Human Settlements Programme, The World Bank, 2021. *Applying the Degree of Urbanisation: A Methodological Manual to Define Cities, Towns and Rural Areas for International Comparisons*. Manuals and guidelines / Eurostat, 2021 edition edn, OECD Publishing, Paris.

Olofsson, P., Foody, G. M., Herold, M., Stehman, S. V., Woodcock, C. E., Wulder, M. A., 2014. Good practices for estimating area and assessing accuracy of land change. *Remote Sensing of Environment*, 148, 42–57.

Pekel, J.-F., Cottam, A., Gorelick, N., Belward, A. S., 2016. High-resolution mapping of global surface water and its long-term changes. *Nature*, 540(7633), 418–422.

Pesaresi, M., Ehrlich, D., Florczyk, A., Freire, S., Julea, A., Kemper, T., Soille, P., Syrris, V., 2016. GHS-BUILT R2015B: GHS built-up grid, derived from Landsat, multitemporal (1975, 1990, 2000, 2014).

Raifer, M., Troilo, R., Kowatsch, F., Auer, M., Loos, L., Marx, S., Przybill, K., Fendrich, S., Mocnik, F.-B., Zipf, A., 2019. OSHDB: a framework for spatio-temporal analysis of OpenStreetMap history data. *Open Geospatial Data, Software and Standards*, 4(1).

This contribution has been peer-reviewed.
https://doi.org/10.5194/isprs-archives-XLVIII-M-1-2023-123-2023 | © Author(s) 2023. CC BY 4.0 License.

Roberts, P., KC, S., Rastogi, C., 2006. *Rural Access Index: A Key Development Indicator*. Transport Papers, 10, World Bank, Washington, DC.

Stevens, F. R., Gaughan, A. E., Linard, C., Tatem, A. J., 2015. Disaggregating census data for population mapping using random forests with remotely-sensed and ancillary data. *PloS one*, 10(2), e0107042.

Tollefsen, A. F., Strand, H., Buhaug, H., 2012. PRIO-GRID: A unified spatial data structure. *Journal of Peace Research*, 49(2), 363–374.

UN OCHA, 2022. Lake Chad Basin: Humanitarian Snapshot: As of 08 December 2022. https://reliefweb.int/report/chad/lake-chad-basin-humanitarian-snapshot-08-december-2022.

United Nations, 2022. Goal 9: Build resilient infrastructure, promote inclusive and sustainable industrialization and foster innovation. https://sdgs.un.org/goals/goal9.

United Nations Development Programme, 2022. Conflict Analysis in the Lake Chad Basin 2020-2021: Trends, developments and implications for peace and stability. https://www.undp.org/africa/publications/conflict-analysis-lake-chad-basin.

USGS, 2019. *Landsat 8 (L8): Data Users Handbook*. Reston, VA, USA.

Woodward, S. L., 2004. Fragile States: Exploring the Concept. *Peace and Social Justice Meeting*.

Workman, R., McPherson, K., 2019. *Measuring Rural Access Using New Technologies: Supplemental Guidlines*. ReCAP for DFID, London.

Workman, R., McPherson, K., 2021. Measuring rural access for SDG 9.1.1. *Transactions in GIS*, 25(2), 721–734.

World Bank Group, 2003. *World Bank Report: Making services work for poor people*. World development report, 26.2004, Oxford Univ. Press, Oxford.

World Bank Group, 2016. *Measuring Rural Access: Using new technologies*. Washington, DC.

World Bank Group, 2019. *Measuring Rural Access : Update 2017/18*. Washington, DC.

WorldPop, CIESIN, 2018. Global High Resolution Population Denominators Project: Global 100m Population. https://hub.worldpop.org/geodata/listing?id=29.

Xu, H., 2007. Extraction of Urban Built-up Land Features from Landsat Imagery Using a Thematicoriented Index Combination Technique. *Photogrammetric Engineering & Remote Sensing*, 73(12), 1381–1391.

Zanaga, D., van de Kerchove, R., de Keersmaecker, W., Souverijns, N., Brockmann, C., Quast, R., Wevers, J., Grosu, A., Paccini, A., Vergnaud, S., Cartus, O., Santoro, M., Fritz, S., Georgieva, I., Lesiv, M., Carter, S., Herold, M., Li, L., Tsendbazar, N.-E., Ramoino, F., Arino, O., 2021. ESA WorldCover 10 m 2020 v100.

Zha, Y., Gao, J., Ni, S., 2003. Use of normalized difference built-up index in automatically mapping urban areas from TM imagery. *International Journal of Remote Sensing*, 24(3), 583–594.

GEDI DATA WITHIN GOOGLE EARTH ENGINE: PRELIMINARY ANALYSIS OF A RESOURCE FOR INLAND SURFACE WATER MONITORING

Alireza Hamoudzadeh [1,]* Roberta Ravanelli [1], Mattia Crespi [1,2]

[1] Sapienza University of Rome, DICEA, Geodesy and Geomatics Division, Italy
[2] Sapienza School for Advanced Studies, Sapienza University of Rome, Italy

<alireza.hamoudzadeh,roberta.ravanelli,mattia.crespi>@uniroma1.it

KEY WORDS: Inland Surface Water Monitoring, Sustainable Development Goals, Google Earth Engine, GEDI

ABSTRACT:

Freshwater is one the most important renewable water resources of the planet but, due to climate change, surface freshwater available in the form of lakes, rivers, reservoirs, snow, and glaciers is becoming significantly threatened. As a result, surface water level monitoring is fundamental for understanding climatic changes and their impact on humans and biodiversity.
This study evaluates the accuracy of the Global Ecosystem Dynamics Investigation (GEDI) LiDAR (Light Detection And Ranging) instrument for monitoring inland water levels. Four lakes in northern Italy were selected for comparison with gauge station measurements. To evaluate the accuracy of GEDI altimetric data, two steps of outlier removal are proposed. The first stage employs GEDI metadata to filter out footprints with very low accuracy. Then, a robust version of the standard 3σ test using a 3NMAD (Normalized Median Absolute Deviation) test is iteratively applied.
After the outlier removal, which led to the elimination of between 80% to 87% of the data, the remaining footprints show an average standard deviation of 0.36 m, a mean NMAD of 0.38 m, and a Root Mean Square Error (RMSE) of 0.44 m, proving the promising potentialities of GEDI L2A altimetric data for inland water monitoring.

1. INTRODUCTION

Inland surface water is the source for about two-thirds of the freshwater for human and animal consumption, agricultural irrigation, and several industrial applications, and a key component of the hydrological cycle. The monitoring of inland surface water is therefore fundamental to understand the effects of climate change on this key resource and to prevent water stresses (Bocchino et al., 2023). This kind of monitoring is strictly related to United Nations (UN) Sustainable Development Goals (SDGs) concerning water availability (SDG 6 - Clean water and sanitation) and climate change effects monitoring (SDG 13 - Climate action) and with the Recovery Plan Next Generation EU (Kavvada et al., 2020, United Nations General Assembly, 2015, Sinha et al., 2020).

Traditionally, water level measurements are obtained using ground-based instruments like gauge stations. However, such in-situ monitoring techniques are typically only feasible in developed countries, since installing and maintaining measurement stations in remote areas can be challenging (Hamoudzadeh et al., 2023). Conversely, the use of Earth observation technologies and methods can remarkably reduce the monitoring costs (independent from the actual extent of the reservoir) and provide frequent and regular data, that facilitate the continuous monitoring of water reservoirs, in principle with homogeneous procedures worldwide.

Access to new global datasets is critical to improving this type of monitoring on a global scale, provided that their accuracy is thoroughly assessed. By taking advantage of Earth observation technologies, such as RADAR (Radio Detection And Ranging) and LiDAR (LIght Detection And Ranging), it is possible to obtain high-quality data with global coverage, enabling a more comprehensive and consistent understanding of water reservoir dynamics worldwide. However, it is important to ensure that the accuracy and reliability of these new data sources are carefully evaluated to enable effective and trustworthy monitoring on a global scale.

In this respect, Google Earth Engine (GEE) is known as a reliable, and real-time cloud-based computation platform, capable of integrating a high variety of up-to-date geospatial datasets with powerful analysis tools (Cardille et al., 2022). GEE provides a High-Performance Computing (HPC) infrastructure that enables fast and convenient access to more than forty years of publicly available data archives, comprising scientific datasets and historical imagery (Ravanelli et al., 2018a, Ravanelli et al., 2018b, Nascetti et al., 2017), making it possible to develop remote sensing applications on a global and large scale. GEE has recently added the Global Ecosystem Dynamics Investigation (GEDI) (University of Maryland, 2022) dataset to its already wide archive.

GEDI was originally developed as an experimental mission onboard the International Space Station (ISS) to enable radically improved quantification and understanding of the Earth's carbon cycle and biodiversity; only lately have GEDI potentialities been investigated for inland surface water level monitoring (Fayad et al., 2022), despite many research has evaluated the accuracy and usability of different LiDAR, RADAR, and SAR (Synthetic Aperture Radar) altimetric data for this purpose (Lee et al., 2021).

The available literature highlights that the quality of GEDI data is variable and impacted by several factors (e.g., latitude, time

* Corresponding author

This contribution has been peer-reviewed.
https://doi.org/10.5194/isprs-archives-XLVIII-M-1-2023-131-2023 | © Author(s) 2023. CC BY 4.0 License.

of observation). Our preliminary analysis is therefore focused on the accuracy assessment of the data collected by GEDI, at first addressing the problem of outliers detection and removal within the GEE platform, and secondly comparing the water levels measured by GEDI with some reference ground truth.

2. DATA

2.1 Gauge measurements

In order to evaluate the accuracy of GEDI, four lakes located in northern Italy (Como, Garda, Iseo, Maggiore (Figure 1)) were chosen as the study area. These lakes have existing gauge stations that provide level measurements for comparison after the outlier removal procedure, and for this objective, the daily level measurements were gathered from the website of (Enti Regolatori dei Grandi Laghi, 2022).

2.2 GEDI Altimetric data

The GEDI (Dubayah et al., 2021) instrument is a geodetic-class, light detection and ranging (LiDAR) laser system consisting of 8 parallel observation tracks (beams) generated from 3 lasers (Liu et al., 2021), with the main objective of precise measurements of forest canopy height, canopy vertical structure, and surface elevation. Each laser emits 242 pulses per second, illuminating a 25 m spot (a footprint) on the surface over which 3D structure is measured. Each footprint is separated by a distance of 60 m along-track, with an across-track distance of about 600 m between each of the 8 tracks. The measurements are made over the Earth's surface nominally between the latitudes of 51.6° and -51.6°.

The standard GEDI data products have some limitations for practical applications, as the footprint-level products provide only a point sample of a limited part of the land surface, leaving most of it without observations. However, for areas with homogeneous surfaces like water bodies, this limitation does not pose a significant issue as much as it would for targeting rare forest change events (Qi et al., 2019). For this study, GEDI L2A (Version 2) data was utilized. This data has 140 different bands and is available in both forms of Image Collection and Vectors within the GEE data catalog (Earth Engine Data Catalog — Google Developers, 2022).

Although GEDI data has been extensively studied for forest monitoring and estimating canopy height, relatively few studies are dedicated to water level monitoring using this altimetric sensor. The potential of GEDI for inland water level monitoring can be recognized due to its high spatial resolution and frequent revisit time, which can provide valuable data for monitoring changes in water levels over time. However, there are still challenges in using GEDI data for water level monitoring, such as the limited coverage of water bodies by GEDI due to lower latitude coverage compared to its rivals like ICESat-2 and the need for accurate calibration and validation against ground-based measurements. Further research is needed to fully explore the potential of GEDI for water level monitoring and to develop effective methodologies for using GEDI data in water resource management, and here we try to preliminarily evaluate these potentialities.

3. METHODOLOGY

The methodology proposed for the outlier removal required a specific procedure to be implemented within the GEE environment. However, some challenges due to the large amount of data to be processed were addressed in several pre-processing steps. These challenges are further discussed before delving into the detailed description of the methodology.

3.1 Pre-processing and footprint selection

The delineation of water bodies is an essential aspect of remote sensing due to the crucial role that EO imagery plays in managing water resources. The ability to accurately extract water boundaries from satellite imagery provides valuable information for various applications, such as water quality assessments, flood risk mapping, and hydrological modeling (Kaplan and Avdan, 2017).

The JRC Global Surface Water Mapping Layers, v1.4 is a widely used dataset that provides detailed information about water bodies worldwide, including their extent and shoreline characteristics (Pekel et al., 2016). By utilizing this dataset, we were able to extract the lake boundaries in a vector form to select the appropriate GEDI footprints within each lake for the analysis (Figure 2).

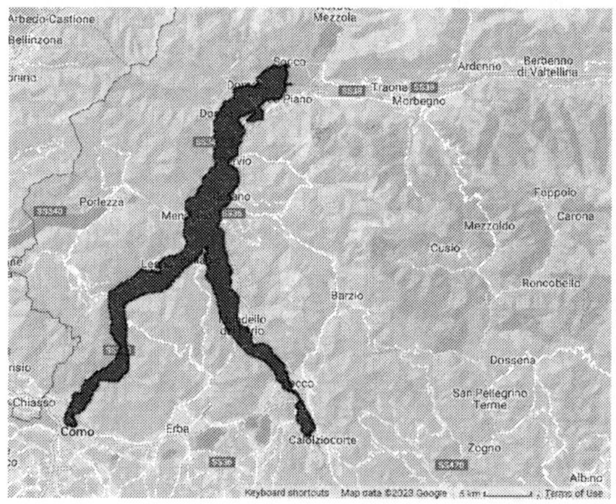

Figure 2. Lake Como extracted boundary from JRC Global Surface Water Mapping Layers and the GEDI footprints within the lake

To enable a proper comparison between the GEDI elevation data and the gauge station measurements, we transformed the GEDI elevation data from ellipsoidal heights referred to WGS84 datum to orthometric heights according to the EGM2008 geoid model. This transformation was necessary to ensure that both the GEDI and gauge station data were referenced to the same elevation model.

3.2 Outlier removal

The proposed outlier detection procedure consists of two different steps for every GEDI passage over each considered water surface. The resulting dataset is more reliable and accurate, which is crucial for an effective inland water resource monitoring.

https://doi.org/10.5194/isprs-archives-XLVIII-M-1-2023-131-2023

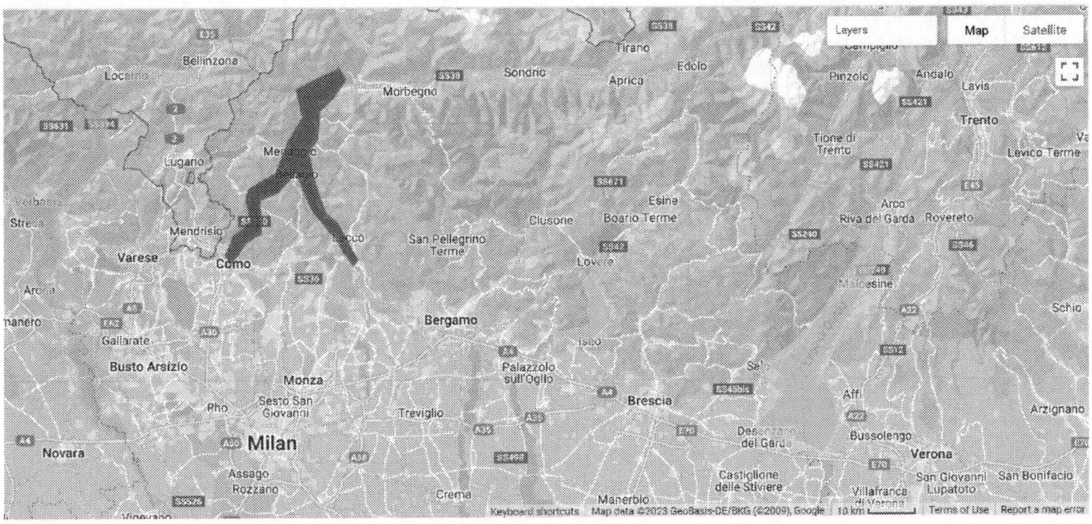

Figure 1. The location of four selected lakes (lake Maggiore highlighted as yellow, Como as dark blue, Iseo as light blue, and Garda as green) on the Google Earth Engine platform.

3.2.1 Quality and degrade flags The first step is based on two flags supplied together within GEDI data. Specifically, the "quality_flag" indicates if the considered footprint has a valid or invalid waveform, due to anomalies in the energy, sensitivity, and amplitude of the signal. The L2A quality_flag uses a conservative sensitivity threshold of 0.9 over land (0.5 over the ocean), but under some conditions (e.g. dense forest) the user may benefit from selecting a higher threshold. In this research, we have only used the footprints with quality_flags equal to one, indicating the laser shot meets all the criteria based on energy, sensitivity, amplitude, real-time surface tracking quality, and difference to Digital Elevation Model (DEM).

The "degrade_flag" indicates the degraded state of pointing (the saturation intensity of the returned photons might reduce the accuracy of the measurement) and/or positioning information (GPS data gap, GPS receiver clock drift) of the GEDI signal. In this study, only footprints with degrade_flags equal to 00 were selected.

3.2.2 3NMAD-test The second step of the outlier removal procedure relies on the robust version of the standard 3-sigma test, here implemented considering the NMAD (Normalized Median Absolute Deviation (Equation 1)):

$$NMAD = Median(H_i - Median(H)) \times 1.4826 \quad (1)$$

where the elevation of each GEDI footprint was denoted as H_i, while H referred to the median water elevation of all the footprints within the considered lake and epoch.

This step was carried out on each lake separately to remove the footprints that had sufficient technical quality (with no anomalies in energy in the returning beam, or GPS drifts), but inconsistent values with the rest of the dataset.

In this level of outlier removal, every GEDI water level measurement (footprint) that is not within the threshold of $\mp$ 3*NMAD from the median of the water level throughout the lake is removed as an outlier.

4. RESULTS

The accuracy of the GEDI elevation data was evaluated by comparing them with gauge measurements, in order to determine GEDI reliability for monitoring lake water levels. The comparison is ongoing over the period from the activation of GEDI in April 2019 until June 2022, for about 3 years.

The quality flag and degraded flag were able to detect only 40% of outliers. The 3NMAD iterative test removed a significant portion of the data, ranging from 80% to 87%, improving their accuracy (Table 1). The GEDI data was quite extensive and rich in numbers of footprints, which resulted in over 4600 footprints for Lake Iseo, despite its relatively small area and after removing 87% of the original data. Although a significant amount of GEDI data was removed, the remaining GEDI data still provided valuable insights into water level monitoring for Lake Iseo and the three other lakes.

To evaluate the accuracy of GEDI data after the outlier removal, the standard deviation (SD), mean, median, NMAD, and root mean squared error (RMSE) were calculated based on the median of each GEDI epoch and the most contemporary gauge measurement (Figure 3). The comparison between the GEDI elevation data and the gauge measurements resulted in a mean standard deviation of 0.36 m and an average NMAD of 0.38 m. Additionally, the RMSE has a value of 0.44 cm. These results indicate that the GEDI data, after the outlier removal, can be used for lake elevation monitoring applications. Figure 3 shows the difference between the median value of each epoch and the water level measured from gauges.

The standard deviation values range from 0.26 to 0.44 m, indicating that the elevation measurements are relatively consistent across footprints as well as agreeing with the NMAD values

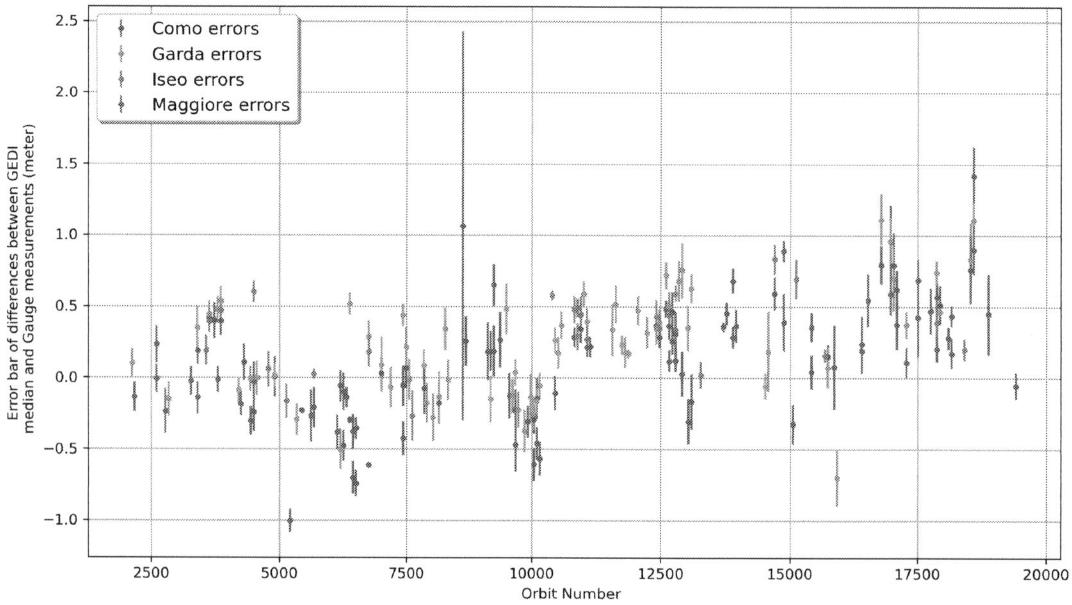

Figure 3. Water level differences between gauge measurements and medians of GEDI measurements after outliers removal over the period 2019/06 – 2022/06 for 4 lakes of Northern Italy (Como, Garda, Maggiore, and Iseo), based on the ISS orbit number; the error bars show the ∓ 3*NMAD intervals centered in the respective difference values

Lake	Como	Garda	Iseo	Maggiore
SD (m)	0.39	0.36	0.26	0.44
Mean (m)	0.27	0.24	0.31	0.04
Median (m)	0.28	0.29	0.37	0.04
NMAD (m)	0.43	0.37	0.25	0.49
RMSE (m)	0.47	0.43	0.41	0.45
Count of footprints before flags (-)	71686	210202	33692	96430
Count of footprint after flags (-)	14672	49326	7793	30195
Count of footprint after 3NMAD (-)	9565	34268	4673	19366
Removal percentage (%)	87	84	87	80

Table 1. Overall statistics for the water level differences between gauge measurements and medians of GEDI measurements (standard deviation (SD), mean, median, and root mean squared error (RMSE) for each lake.) after outliers removal over the whole period of 2019/06 – 2022/06 for 3 lakes of Northern Italy (Como, Garda, and Iseo)

ranging from 0.25 to 0.49 m. The RMSE showed values less than 0.47 m, indicating that the GEDI data correlates well with the gauge station measurements and Figure 4 shows the same pattern and ability of GEDI in following the patterns of the lakes.

The analysis of the GEDI data showed that the error or NMAD was generally below 0.20 m. However, the precision decreases and the NMAD values started to increase and scatter in the last orbits, particularly after orbit number 12500 (Figure 5), for unknown reasons. Also, the same pattern occurs in the differences between gauge measurements and the GEDI median of epoch/orbit right around the same period (Figure 3). This fact indicates the need for further research to understand better and address these limitations.

5. CONCLUSIONS

In conclusion, this study demonstrated the potential of GEDI L2A data for lake water level monitoring, despite the challenges of outlier removal and the need for proper data calibration. To assess the implemented outlier detection procedure and to preliminarily evaluate the accuracy of the GEDI data, we compared the water levels inferred from the median of GEDI measurements after outliers removal with the (as much as possible) contemporary water levels measured by the hydrometric stations at four major lakes (Como, Garda, Iseo, Maggiore) in Northern Italy.

The results showed that GEDI L2A footprints, after the removal of outliers, can provide accurate and reliable measurements of lake surface elevation, in comparison to traditional in-situ gauge measurements. The low values of NMAD (lower than 0.20 m in most cases) showed a high precision in epochs/orbits and the potential for being calibrated to obtain better accuracy.

The use of global EO technologies like GEDI can provide regular and frequent monitoring of water reservoirs worldwide, making it a cost-effective and efficient way to monitor these crucial natural resources. However, further research is needed to assess the accuracy of GEDI L2A data in different types of water bodies and environmental conditions.

REFERENCES

Bocchino, F., Ravanelli, R., Belloni, V., Mazzucchelli, P., Crespi, M., 2023. Water reservoirs monitoring through Google Earth Engine: application to Sentinel and Landsat imagery. *The*

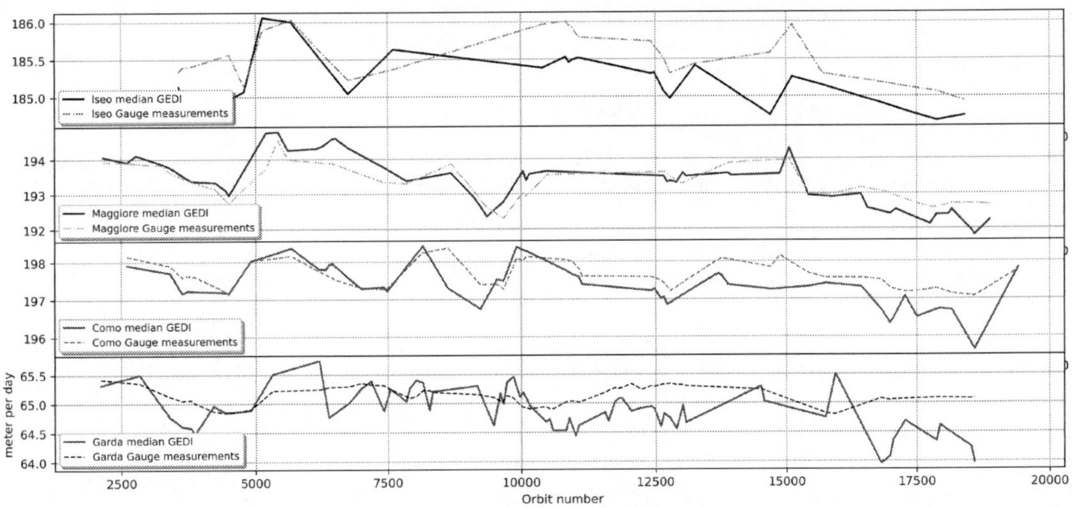

Figure 4. The Four lakes GEDI Median and the gauge measurement changes over the course of orbit number from the period 2019/06 – 2022/06

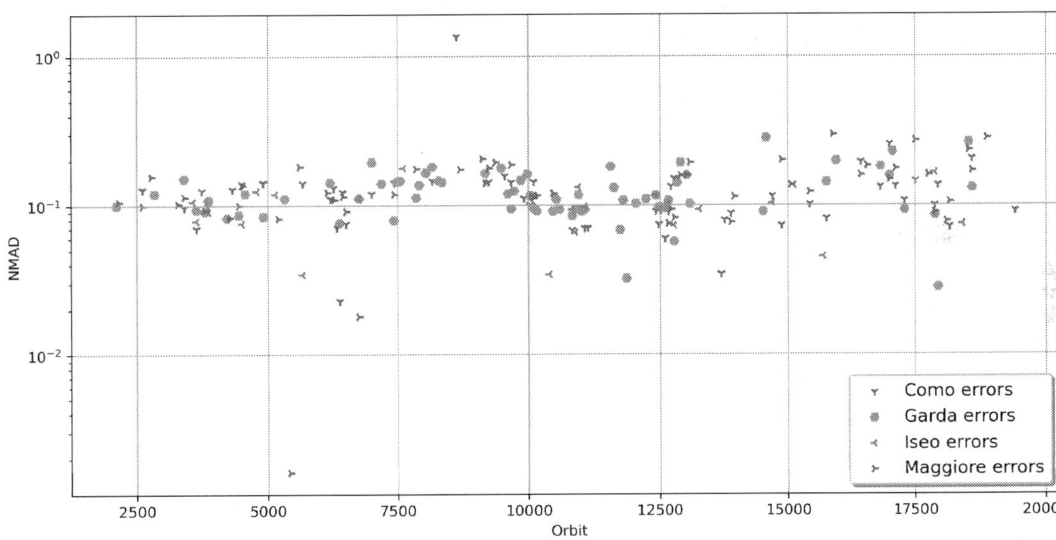

Figure 5. Scatter plot for NMAD (Error) of each epoch/orbit of each of the four lakes based on ISS orbits for the whole period June 2019 to June 2022

International Archives of the Photogrammetry, Remote Sensing and Spatial Information Sciences. In press.

Cardille, J. A., Clinton, N., Crowley, M. A., Saah, D., 2022. *Cloud-Based Remote Sensing with Google Earth Engine*. www.eefabook.org.

Dubayah, R., Luthcke, S., Sabaka, T., Nicholas, J., Preaux, S., Hofton, M., 2021. GEDI L3 gridded land surface metrics, version 1. *ORNL DAAC*.

Earth Engine Data Catalog — Google Developers, 2022. GEDI L2A raster canopy top height (version 2). https://developers.google.com/earth-engine/datasets/catalog/LARSE_GEDI_GEDI02_A_002_MONTHLY.

Enti Regolatori dei Grandi Laghi, 2022. Home Page - Laghi. www.laghi.net.

Fayad, I., Baghdadi, N., Bailly, J.-S., Frappart, F., Pantaleoni Reluy, N., 2022. Correcting GEDI Water Level Estimates for Inland Waterbodies Using Machine Learning. *Remote Sensing*, 14(10), 2361.

Hamoudzadeh, A., Ravanelli, R., Crespi, M., 2023. GEDI data within Google Earth Engine: Potentials and analysis for inland surface water monitoring. EGU, Copernicus Meetings.

Kaplan, G., Avdan, U., 2017. Object-based water body extraction model using Sentinel-2 satellite imagery. *European Journal of Remote Sensing*, 50(1), 137–143.

Kavvada, A., Metternicht, G., Kerblat, F., Mudau, N., Haldorson, M., Laldaparsad, S., Friedl, L., Held, A., Chuvieco, E., 2020. Towards delivering on the sustainable development goals using earth observations. *Remote Sensing of Environment*, 247, 111930.

Lee, Y.-K., Hong, S.-H., Kim, S.-W., 2021. Monitoring of Water Level Change in a Dam from High-Resolution SAR Data. *Remote Sensing*, 13(18).

Liu, A., Cheng, X., Chen, Z., 2021. Performance evaluation of GEDI and ICESat-2 laser altimeter data for terrain and canopy height retrievals. *Remote Sensing of Environment*, 264, 112571.

Nascetti, A., Di Rita, M., Ravanelli, R., Amicuzi, M., Esposito, S., Crespi, M., 2017. Free global dsm assessment on large scale areas exploiting the potentialities of the innovative google earth engine platform. *The International Archives of the Photogrammetry, Remote Sensing and Spatial Information Sciences*, XLII-1/W1, 627–633.

Pekel, J.-F., Cottam, A., Gorelick, N., Belward, A. S., 2016. High-resolution mapping of global surface water and its long-term changes. *Nature*, 540(7633), 418–422.

Qi, W., Lee, S.-K., Hancock, S., Luthcke, S., Tang, H., Armston, J., Dubayah, R., 2019. Improved forest height estimation by fusion of simulated GEDI Lidar data and TanDEM-X InSAR data. *Remote Sensing of Environment*, 221, 621-634.

Ravanelli, R., Cirigliano, R., Di Rico, C., Monti, P., Crespi, M., 2018a. Monitoring Urban Heat Island through Google Earth Engine: potentialities and difficulties in different cities of the United States. *The International Archives of the Photogrammetry, Remote Sensing and Spatial Information Sciences*, XLII-3, 1467–1472.

Ravanelli, R., Nascetti, A., Cirigliano, R. V., Di Rico, C., Leuzzi, G., Monti, P., Crespi, M., 2018b. Monitoring the impact of land cover change on surface urban heat island through Google Earth Engine: Proposal of a global methodology, first applications and problems. *Remote Sensing*, 10(9), 1488.

Sinha, A., Sengupta, T., Alvarado, R., 2020. Interplay between technological innovation and environmental quality: formulating the SDG policies for next 11 economies. *Journal of Cleaner Production*, 242, 118549.

United Nations General Assembly, 2015. Transforming our world: The 2030 agenda for sustainable development.

University of Maryland, 2022. GEDI ecosystem lidar. `www.gedi.umd.edu`.

This contribution has been peer-reviewed.
https://doi.org/10.5194/isprs-archives-XLVIII-M-1-2023-131-2023

A REPLICABLE OPEN-SOURCE MULTI-CAMERA SYSTEM FOR LOW-COST 4D GLACIER MONITORING

F. Ioli [1]*, E. Bruno [2], D. Calzolari [3], M. Galbiati [4], A. Mannocchi [5], P. Manzoni [6], M. Martini [7] A. Bianchi [1], A. Cina [8], C. De Michele [1], L. Pinto [1]

[1] Politecnico di Milano, Department of Civil and Environmental Engineering, Milan, Italy
[2] Politecnico di Torino, Department of Mechanical and Aerospace Engineering, Torino, Italy
[3] CERN, Geneva, Switzerland
[4] Politecnico di Milano, Department of Energy, Milan, Italy
[5] Politecnico di Milano, Department of Aerospace Science and Technology, Milan, Italy
[6] Politecnico di Milano, Department of Mathematics, Milan, Italy
[7] Politecnico di Torino, Department of Electronics and Telecommunications, Torino, Italy
[8] Politecnico di Torino, Department of Environment, Land and Infrastructure Engineering, Torino, Italy

KEY WORDS: Stereo photogrammetry, Time-lapse, Glaciers, Structure-from-Motion, Embedded systems, Rasperry Pi, Arduino

ABSTRACT:

Image-based monitoring has emerged as a prevalent technique for sensing mountain environments. Monoscopic time-lapse cameras, which rely on digital image correlation to quantify glacier motion, have limitations due to the need for a Digital Elevation Model for deriving 3D flow velocity fields. Multi-camera systems overcome this limitation, as they allow for a 3D reconstruction of the scene. This paper presents a replicable low-cost stereoscopic system designed for 4D glacier monitoring. The system consists of independent and autonomous units, built from off-the-shelves components, such as a DSLR camera, an Arduino microcontroller, and a Raspberry Pi Zero, reducing costs compared to pre-built time-lapse cameras. The units are energetically self-sufficient and resistant to harsh alpine conditions. The system was successfully tested for more than a year to monitor the northwest terminus of the Belvedere Glacier (Italian Alps). Daily stereo-pairs acquired were processed with Structure-from-Motion to derive 3D point clouds of the glacier terminus and estimate glacier retreat and ice volume loss. By combining the information about ice volume loss with ablation estimates and ice flow velocity information, e.g., derived from monoscopic-camera time series, a multi-camera system enables a comprehensive understanding of sub-seasonal glacier dynamics.

1. INTRODUCTION

Glacier monitoring in high alpine environments is an essential task for understanding climate change and its effects on natural systems (Barry, 2006, Zemp et al., 2006). Recently, photogrammetry and Structure-from-Motion (SfM) have become increasingly popular tools for 3D reconstruction and monitoring of remote and inaccessible areas, such as high alpine glaciers (Piermattei et al., 2016). Although UAV-based photogrammetry is a cost-effective solution for acquiring high-resolution and accurate 3D models of mountain environments (Chudley et al., 2019, Ioli et al., 2021), it still requires in-situ human intervention, which can be troublesome for long-term monitoring activities. In contrast, fixed time-lapse cameras are designed to operate autonomously and provide images with high temporal frequency (e.g., daily) for long periods, requiring minimal maintenance. These cameras can serve as webcams for monitoring and collecting both qualitative and quantitative data on dynamic scenes, especially in remote alpine areas.

In combination with computer vision and image processing techniques, low-cost time-lapse cameras have been effectively employed for various applications, including monitoring glaciers and ice cliffs (Giordan et al., 2016, Schwalbe and Maas, 2017, Kneib et al., 2022), observing calving glaciers (Vallot et al., 2019), estimating snow accumulation and melting (Ide and Oguma, 2013), measuring streams' and rivers' discharge (Young et al., 2015), monitoring landslide (Roncella et al., 2014), and geomorphological sensing (Eltner et al., 2017). For what concerns glacier monitoring, several software packages which utilize Digital Image Correlation (DIC) and feature tracking to estimate glacier surface velocity from terrestrial time-lapse cameras have been developed. For instance, ImGraft (Messerli and Grinsted, 2015) and Pointcatcher (James et al., 2016) are well-known MATLAB packages for feature tracking and geo-referencing velocity vectors onto Digital Elevation Models (DEMs). More recently, PyTrx (How et al., 2020), a Python toolbox for calculating real-world measurements from oblique time-lapse images, and LAMMA (Dematteis et al., 2022), a local adaptive multiscale image matching algorithm, were released.

A large part of the existing literature exclusively considers single-camera systems for image acquisition. While this option may seem practical due to ease of installation and cost-effectiveness, it ultimately hinders the employment of standard photogrammetric techniques for computing 3D models and deriving 3D velocity fields of moving objects. Indeed, when a single camera is used to derive surface velocities, an external DEM must be provided in order to georeference the pixel-wise displacements obtained from the images by DIC. Roncella et al. (Roncella et al., 2014) first developed a low-cost and handmade stereoscopic system for continuous monitoring of a landslide. They proposed using a fully photogrammetric approach to derive a 3D model of the scene during every image acquisition, eliminating the need for a priori DEMs. More recently, stereoscopic (Marsy et al., 2020) or multi-camera (Kneib et al.,

* Corresponding author

This contribution has been peer-reviewed.
https://doi.org/10.5194/isprs-archives-XLVIII-M-1-2023-137-2023

2022) systems have been tested in a remote glacial environment. However, these studies focused more on the image processing methods and results than on the implementation of the system, providing minimal or no information on the hardware used.

An alpine glacier is a hazardous and challenging environment for the installation of monitoring systems. Considering logistical factors, human intervention should be kept to its bare minimum. Any monitoring system should be capable of withstanding extreme cold temperatures and acquiring data autonomously. In addition, it must be designed to be self-sufficient in terms of power usage: low energy consumption has to be a primary requirement in the design of a monitoring system, both in terms of hardware and data acquisition algorithm.

The aim of this work is to provide a general framework for the realization of low-cost-and-low-power multi-camera monitoring systems. The complete architecture of a low-cost stereoscopic system based on two time-lapse cameras for continuously monitoring an alpine glacier is presented. The setup was successfully tested for over a year on the northwest terminus of the debris-covered Belvedere Glacier in the Italian Alps. Overall, it proved to be accurate and robust in performance, allowing for the continuous data collection necessary for high-frequency monitoring of the remote glacier. Additionally, the system's open-source design enables easy replication for further research studies conducted in other hazardous environments.

2. SYSTEM DESCRIPTION

The low-cost stereoscopic system for continuous monitoring is composed of two independent units, each housing an off-the-shelves DSLR camera. Each unit consists in an autonomous monitoring station that provides power supply, internet connectivity, timing and scheduling of image acquisition, and protection from the harsh environment. The station was designed and built adapting to the alpine environment an existing open-source model developed by Greig Sheridan and published in the GitHub repository `https://github.com/greiginsydney` (Sheridan, 2021). A key aspect of the project was to ensure easy assembly and realization of the system to guarantee future replication or improvement by non-experts. In the following, we describe in detail the services and the nominal performances provided by our system that make it worth consideration for applications in different monitoring contexts. Refer to Figure 1 for a descriptive schematic of the whole system showing the components selected for the Belvedere Glacier case study. The choice of the camera and optics is discussed in Section 3, as these components are dependent on the domain of use of the stereoscopic system.

2.1 Power supply

Each monitoring unit has its autonomous power supply line (yellow arrows in Figure 1) provided by a solar panel combined with a sealed lead-acid battery. An MPPT (Maximum Power Point Tracker) charge regulator directly connects with the unit's internal circuit, providing a regular current supply to the battery and the load and exploiting all the power generated by the panel. This regulator prevents any excess current that may damage the connected device, thus increasing the reliability of the system. Its battery life-saving algorithm modulates the load disconnection level so that a nearly 100% recharge is achieved about once every week and, in case of battery discharging, the whole system is switched off until 100% recharge of

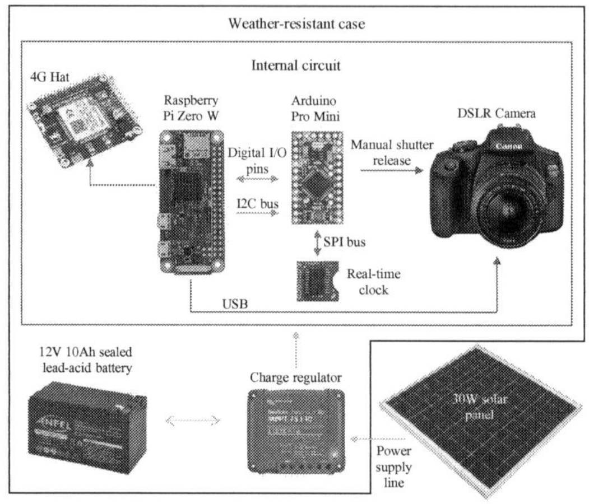

Figure 1. Scheme of the proposed acquisition system configuration for a single monitoring station. Arrows indicate the direction of signal initiation. Image adapted from Greig Sheridan (Sheridan, 2021).

the battery is achieved. The whole monitoring system is designed to minimise power consumption, e.g., by powering off devices responsible for the highest power consumption when not needed. An estimation of the system's energy consumption guided the choice of the components of the power supply line. Specifically, the battery and the associated panel must be accurately dimensioned for the system's purpose. To this end, the Photovoltaic Geographical Information System (PVGIS) (EU Science Hub, 2022), a tool provided by the European Commission, was used. PVGIS can estimate the performance of off-grid photovoltaic systems according to the installation site and it is supported by a database and algorithms for calculating solar radiation. Tuning the parameters for the Belvedere Glacier location and the system consumption, we ended up with the specifications of solar panel power of 30 W and battery voltage of 12 V and capacity of 10 Ah. Some compromises were made, accepting that the system would not be sufficiently powered in the months with lowest solar radiation (November, December, January, and February). In Section 3, further details on PVGIS simulations for the Belvedere Glacier case study are given.

2.2 System control and scheduling

The internal electronic circuit (red box in Figure 1) of the monitoring unit is the only load connected to the power supply and is responsible for all the system's control and scheduling functionalities. Its components are a real-time clock (number 3 in Figure 2) with a 12 mm coin-cell backup battery, an Arduino microcontroller (Pro Mini 328 3.3V 8MHz, number 4 in Figure 2), a Raspberry Pi Zero W with 128 GB SD memory card (number 6 in Figure 2), and minor elements for connections, isolation (capacitors and optoisolators numbers 1 and 5 in Figure 2), and voltage regulations (numbers 2 in Figure 2). The circuit was realized manually by soldering wires and components on a stripboard (see Figure 2).

The Arduino communicates through an SPI (Serial Peripheral Interface) bus with an accurate real-time clock to schedule its actions: waking the camera, firing the shutter to take a photo, and turning on/off the Raspberry. The Raspberry is able to access the camera images and transfer them to a remote server

https://doi.org/10.5194/isprs-archives-XLVIII-M-1-2023-137-2023 | © Author(s) 2023. CC BY 4.0 License.

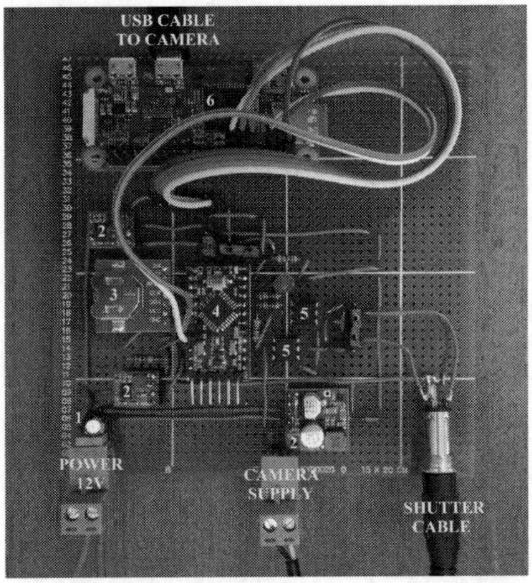

Figure 2. Stripboard on which are soldered the main components of the internal circuit: capacitors (1), voltage regulators (2), real-time clock with coin-cell battery (3), Arduino Pro Mini (4), optoisolators (5), and Raspberry Pi Zero W (6).

via internet connectivity. A large storage memory is added to the unit to save photos before remote transferring. The Raspberry also provides a web-based, user-friendly interface (Figure 3) to configure and monitor the system remotely (Sheridan, 2021). An I2C (Inter Integrated Circuit) bus and digital pin connections enable higher-level communication between these two boards. Among the minor components, we mention the role of the three voltage regulators. They allow each unit to be powered at the appropriate voltage and current: starting from the 12 V in input, one delivers 3.3 V and 500 mA to the Arduino, one 5 V and 2.5 A to the Raspberry and its hat, and the other 7.5 V and 2.4 A to the camera. The Arduino controls the 5V regulator of the Raspberry to reduce quiescent power consumption when the Raspberry board is off.

Acquisition of a defined number of images, camera triggering and timing, and sending of images to a remote server can be scheduled thanks to adequate programming of a cyclic executive program in C++ for Arduino and of services executing automatically Python scripts for the Raspberry. All the monitoring station activities can be remotely scheduled from the terminal or a user-friendly web interface (Sheridan, 2021). This useful feature allows one to change the settings and check the behaviour of the system, avoiding the need for manual intervention. The Raspberry uses the gPhoto2, a Python-based protocol developed to specifically interact with popular cameras' firmware, to communicate with the camera and access the new photos. The web-based interface service is based on NGINX and Gunicorn, open-source software for web servers. The interface (Figure 3) consists of a web page accessible with credentials and provides a summary of the system state and the exact time of the last operations (time of the last shot and the last upload of images), temperatures of the components, previews of the last images, buttons to wake up the camera, take a preview photo and schedule the system routine (time and number of photo acquisition, wake-up time of the Raspberry) and some camera settings.

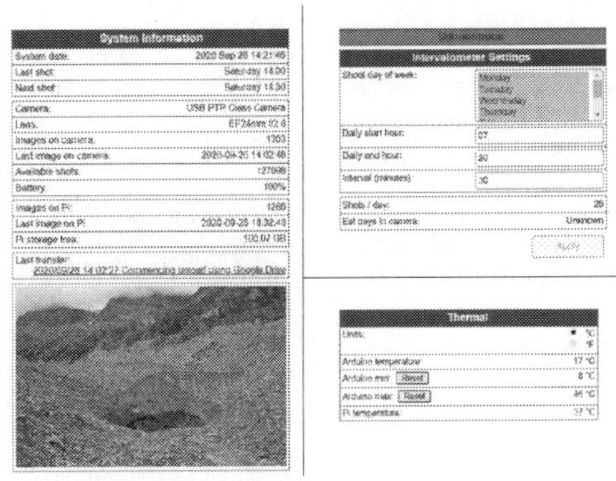

Figure 3. Example of some of the pages of the web-based interface to remotely control the monitoring units.

All the components were chosen because of their low power consumption and their inter-compatibility. The circuit is robust against power losses due to battery discharging and is able to auto-recover when power is regained. A low energy consumption policy is also followed in selecting the scheduling. The camera is switched on only during the data capture activity. The Raspberry is switched on only once per day when the connection with the remote server and image transfer is scheduled.

2.3 Connectivity

Internet connectivity is crucial to transfer images and to control the units remotely. To this end, the Raspberry is equipped with the SIM7600E-H 4G hat by Waveshare. The module supports 4G/3G/2G communication via a SIM card. Raspberry and hat operations are the most expensive in terms of power consumption. For this reason, the board is switched on only once a day for a limited period of time, which is kept adequately low to be sustainable for the system power supply but still sufficient for transferring new images to the server. We adopted IoT (Internet of Things) SIM cards provided by the multi-operator service Emnify (Emnify, 2023), as it offers the ability to connect to the best-quality cellular network available and includes flexible and customizable data plans. Remote access to the devices is allowed by a VPN (Virtual Private Network). The service costs €33 per month for a data plan with 6 GB.

2.4 Case and protection

The majority of the components need to be protected. Therefore, a solid and compact case that allows keeping all the parts (with the exception of the solar panel) in the same location is desirable for the functioning of the system and for convenient transportation, installation and maintenance. The case should also satisfy the thermal insulation requirements according to each component's working temperature range. The case should also be waterproof and robust against different and harsh weather conditions but, at the same time, provide access for connecting the solar panel and taking pictures. Our choice was an HPRC 2250 lightweight, waterproof resin case with internal insulating foam. The foam was found to be useful in enhancing the temperature seal and stabilising the location of some components, preventing unwanted movements. Holes were drilled in the case: one sealed with a UV filter of 82 mm to allow the camera to take shots, two others to give access to the solar panel wires, and others to fix the system to a tripod. The internal case

This contribution has been peer-reviewed.
https://doi.org/10.5194/isprs-archives-XLVIII-M-1-2023-137-2023 | © Author(s) 2023. CC BY 4.0 License.

dimensions (236x182x155 mm) posed a significant challenge in accommodating all the components inside. In particular, the camera lens dimensions were constrained by space availability.

2.5 General performances

The system was subject to several tests before being used in its final application at the Belvedere glacier site. During these tests, the system proved to be autonomous, robust to sudden shutdowns, cold-resistant, water-resistant and self-sufficient for a prolonged period of time with no direct sunlight on the solar panels. Specifically, in the absence of sunlight, the battery is able to power the system for about nine days before it is completely discharged. Battery voltage and temperatures of components were monitored, simulating the absence of solar radiation and low environmental temperatures. See Figure 4 for the performances of the system during these tests.

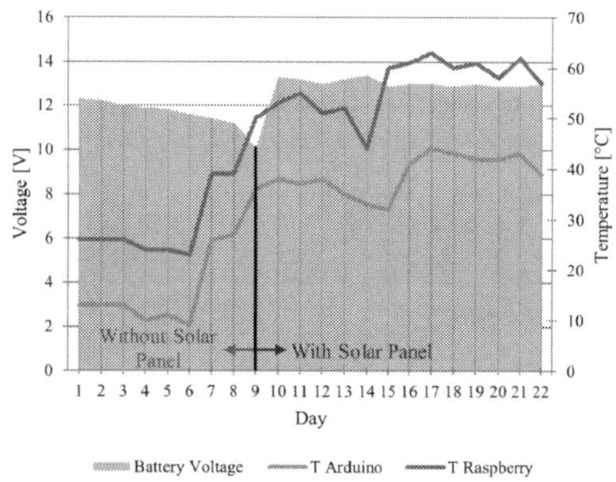

Figure 4. Battery voltage and temperature of boards with and without the solar panel during a period of tests. Starting from a fully charged condition, the system was kept without solar panel at around 2 °C for the first six days. After nine days, the battery was fully discharged and the system was reconnected to the panel and exposed to solar radiation.

It is worth mentioning the performances of the system in terms of costs: our monitoring system can be fully reproduced with an average cost of €1600 per station, including camera (€400), lens (€500) and material for the on-site installation (€150). In comparison, commercial time-lapse cameras (e.g., `https://www.harbortronics.com/Products/TimeLapsePackage`) are expensive ($2,375.00 only for the board) and not adaptable to modifications. The final prototype of the monitoring station is shown in Figure 5.

3. THE BELVEDERE GLACIER CASE STUDY

The system has been tested and it is currently operating (February 2023) at the Belvedere Glacier north-west terminal ice cliff. The Belvedere Glacier is a debris-covered glacier located in Valle Anzasca (Italy), on the east side of the Monte Rosa Massif (located at approximately N 45° 58', E 7° 55') (Figure 6). The choice of the Belvedere Glacier was driven by the fact that a long-term monitoring project has been carried out jointly by Politecnico di Milano, Politecnico di Torino and Alta Scuola Politecnica since 2015. The project is aimed at deriving

Figure 5. (a) Picture of the monitoring unit prototype installed for tests at the Belvedere Glacier site. (b) Picture of the final monumentation of the camera installed on the stream-wise right moraine.

ice flow velocity, volume variations and estimation of the glacier retreat by combining UAV-based photogrammetry, in-situ GNSS measurements (Ioli et al., 2021) and historical aerial photogrammetry (De Gaetani et al., 2021). However, sub-seasonal information on glacier evolution was missing. In this context, a permanent stereoscopic camera system fills this gap.

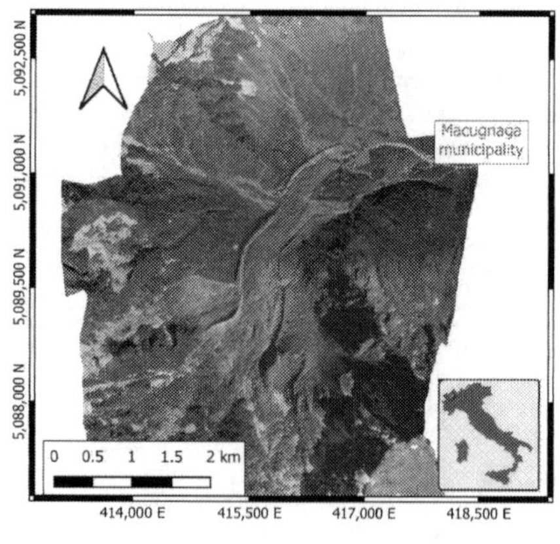

Figure 6. Location of the Belvedere Glacier. The position of the two cameras is marked with red camera symbols.

3.1 Camera and lenses

The choice of the DSLR camera and of the optics can be considered application-dependent. Therefore specifications regarding these components are reported in reference to the Belvedere glacier case study. Each monitoring station was equipped with a DSLR camera Canon EOS 2000D, with 24.1 MP CMOS APS-C sensor. This model was chosen because of the limited costs, the high image resolution and its compactness, as it must fit inside the case. Additionally, another fundamental requirement was camera compatibility with the gPhoto2 software. In the installation inside the system case, the traditional rechargeable battery was replaced by a special battery ("fake battery") equipped with a power supply cable, which provides a voltage of 7.5 V from the circuit. The camera is attached to a sliding plate able to vary, with a limited range, and then fix the position of the camera on the longitudinal axis. Screws fix the camera to the plate and the plate to the case.

The two stations were installed respectively at $\sim$180 m and $\sim$340 m distance to the glacier terminus (see Figure 7), avoiding the steep moraines of the glacier, often prone to collapses and landslides. Therefore, a wide baseline of $\sim$260 m occurs between the two sensors. The final positioning of the two cameras was decisive in the choice of optics. In fact, to ensure comparable Ground Sample Distance (GSD) of the images ($\sim$3 cm px^{-1}), lenses with different focal lengths were employed. In particular, a Canon EF 24 mm f/2.8 IS USM and a Canon EF 35 mm f/2 IS USM were used respectively for Camera 1 (stream-wise right) and Camera 2 (stream-wise left).

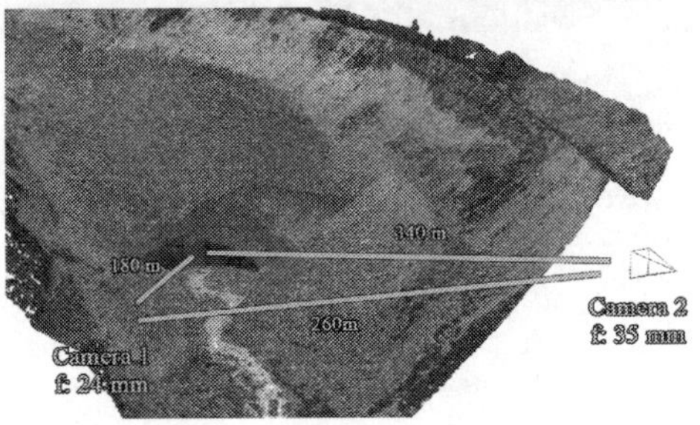

Figure 7. Configuration of cameras installed on the Belvedere glacier. The focal length of each camera is reported next to each camera.

3.2 Monumentation

Monumentation of the two cameras at the Belvedere Glacier site was achieved by choosing two large stable rocks along the moraines. Each camera is supported by an aluminium topographic tripod system anchored with steel dowels and cables to the rocks (Figure 5b). A central steel tie rod keeps the system in a fixed position. The choice of an agile monumentation was aimed at having a system that can be rather easily assembled on site (and also disassembled, if needed), also in a harsh environment, with limited costs and time. To install the cameras, in fact, all the equipment was carried in backpacks along the moraines, sometimes without a marked walking path (e.g., in the case of the stream-wise left camera). As a drawback, the analysis of the acquired photos revealed a non-optimal stabilization of the two cameras. In fact, small rotations, especially around the vertical axis, and vibrations, mostly induced by the wind, were experienced. Though, a perfectly stable monumentation in a mountain environment is hardly achievable. Therefore, it is not possible to assume the camera orientation as stable a priori, but camera orientation must be estimated based on some Ground Control Points (GCPs) located on the ground in stable areas.

3.3 Acquisition performances

The system was installed in August 2021 and is still operating properly (February 2023). Images of the glacier terminus have been collected daily, twice a day, around noon, for more than one year, resulting in $\sim$ 1000 images collected from each camera. Figure 8 reports an example of a stereo-pair. The two monitoring stations operated almost continuously during this period.

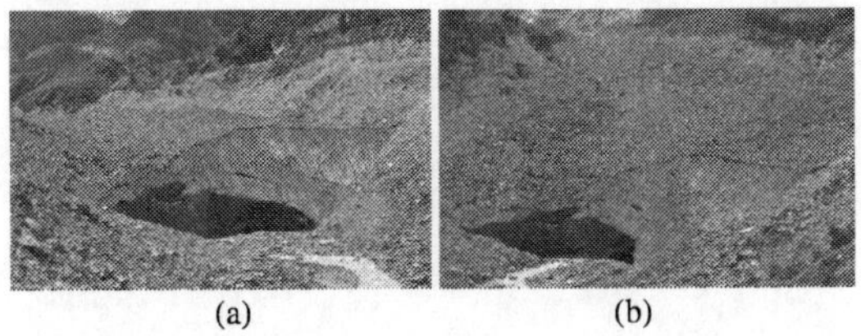

Figure 8. Example of a stereo-pair of images acquired by the two cameras on 28/07/22 at 14:00. (a) Image taken from Camera 1 (located on the stream-wise right moraine); (b) Image from Camera 2 (stream-wise left moraine).

Figure 9a reports the number of shots per day over a one-year period. The number of shots becomes zero when the monitoring station is inactive due to battery discharging. Figure 9b reports the results of the PVGIS simulation considering a 120 Wh solar panel and daily consumption of 40 Wh, which overestimates the actual daily consumption of the system (Raspberry up-time of hours, several shots per day). The simulation predicts that the battery is fully discharged in 16.87% days of the year. Actual performances revealed that Camera 2 was active on 100% of days and Camera 1 on 90% of days. This is because Camera 2 is situated in a more open area than Camera 1. In fact, Camera 1 is located north of the Belvedere hill which can obstruct the sun's radiation during the winter months when the sun's elevation is low.

3.4 Image processing

The daily stereo-pairs are processed by a photogrammetric approach to build 3D models of the glacier terminal ice cliff. The wide baseline between the two cameras, which is typical of complex mountain environments, made it challenging to find corresponding points across different viewpoints (Yao et al., 2021). Indeed, commercial SfM software packages such as Agisoft Metashape (Agisoft Metashape, 2023), as well as an open-source solution such as COLMAP (Schönberger and Frahm, 2016) or MICMAC (Rupnik et al., 2017), which use SIFT (Lowe, 2004) for feature matching, failed to find enough and well-distributed matches. On the other hand, state-of-the-art deep learning-based algorithms for wide-baseline matching, such as SuperGlue (Sarlin et al., 2020), outperformed traditional feature matching. Therefore, a Python routine was developed for finding matches by SuperGlue and performing traditional SfM processing (i.e., relative orientation of the image pair and triangulation of corresponding points). Then, a Bundle Adjustment (BA) was performed with Agisoft Metashape Python API by importing the approximate solution obtained by relative/absolute orientation, as well as a pre-calibrated camera interior orientation. Additionally, the world coordinates of the two cameras, measured by a topographic-grade GNSS receiver, and the coordinates of at least three GCPs (i.e., plastic targets materialized in stable areas along the moraines and measured by GNSS) were included in the BA. Moreover, a dense reconstruction was performed with Agisoft Metashape Python API to build a dense point cloud of the glacier terminus (Figure 10), with $\sim 6 \times 10^6$ pts. Thanks to Agisoft Metashape Python API, Bundle Adjustment and dense reconstruction were fully integrated and automated within the processing workflow.

3.5 Validation with UAV-based Ground Truth

To validate the 3D dense point cloud obtained from the stereo-cameras, Ground Truth (GT) 3D data of the glacier terminus

https://doi.org/10.5194/isprs-archives-XLVIII-M-1-2023-137-2023 | © Author(s) 2023. CC BY 4.0 License.

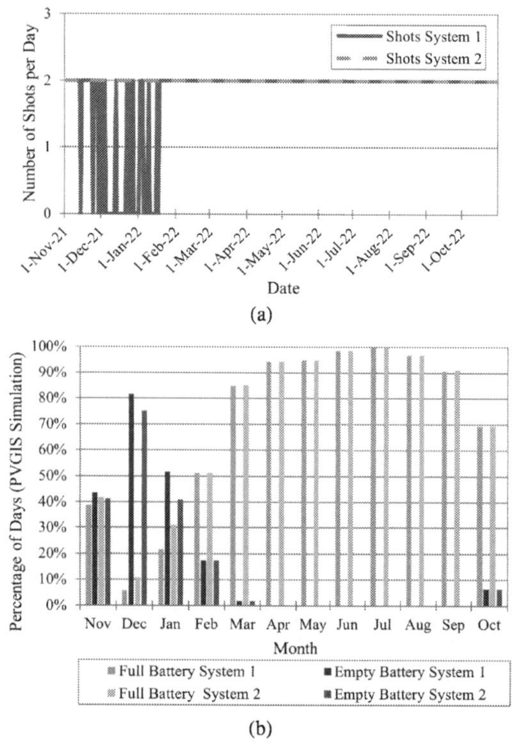

Figure 9. Actual and simulated performances of the system over a year. (a) Number of shots per day of the two systems. (b) Percentage of days in a month when the battery became full or empty according to PVGIS simulations at the location of the two monitoring systems.

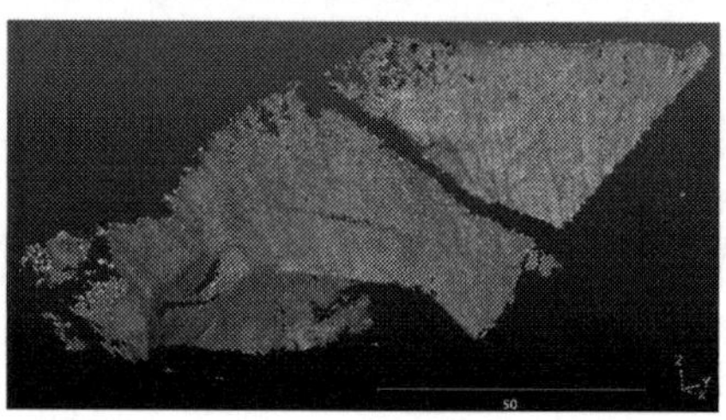

Figure 10. Dense point cloud built from the stereo-pair of 28/07/22. The point cloud was composed of $\sim 6 \times 10^6$ points, with an average point density of $\sim 1000\,\mathrm{pts\,m^{-2}}$.

was acquired on 28/07/2022 by a UAV-based photogrammetric flight, together with 10 GCPs, measured by a GNSS and a total station along the moraines and on the glacier. The images were processed by Agisoft Metashape and the accuracy of the model was evaluated with 5 Check Points (CPs) as an RMSE of $\sim 0.04\,\mathrm{m}$. The dense cloud obtained by the stereo cameras on 28/07/2022 was compared with the concurrent acquired UAV-GT by evaluating the cloud-to-cloud distance along local normals with the M3C2 algorithm (Lague et al., 2013). The resulting differences between the two point clouds, presented in Figure 11, have a mean of 0.01 m and a standard deviation of 0.08 m, highlighting that there are no systematic errors in the stereo-point cloud and its noise is between 2 and 3 times the average image GSD (i.e., 0.03 m). Considering that the forward speed of the Belvedere Glacier terminus has an average annual velocity of $\sim 0.04\,\mathrm{m\,d^{-1}}$ (Ioli et al., 2021), a time-lag of 3 days may be enough to have a rather good signal-to-noise ratio when comparing two consecutive stereo-point clouds.

Figure 11. Cloud-to-cloud distances, computed with M3C2 algorithm, between the UAV-based ground truth point cloud (displayed with RGB colors) and the stereo-point cloud (displayed with a color scale). The colors represent the average distance (i.e., error) between each point of the stereo-point cloud (seed point) and the points of the UAV cloud that falls within a cylinder oriented as the local normal computed around the seed point. Most of the points have an error smaller than ± 10 cm, with the highest values concentrated within the cave.

3.6 Volume variations

From a pair of stereo-points clouds built on different days, the glacier terminus retreat was obtained by computing cloud-to-cloud distances with M3C2. As an example, Figure 12 presents the distances between the stereo-point cloud obtained on 28/07/22 and that obtained one month later on 30/08/22, resulting in an average retreat of 2.7 m. Integrating the displacement over the area in which both point clouds contain points and normalizing the result by the whole area covered by the glacier ice cliff, it was possible to estimate the ice volume variation. Between 28/07/22 and 30/08/22, $\sim 14 \times 10^3\,\mathrm{m}^3$ of ice volume was lost. The same procedure can be applied to the whole time series of data by choosing a proper time step between point cloud pairs, i.e., small enough to appreciate sub-weekly variations but maintaining a good signal-to-noise ratio when computing the displacements.

4. DISCUSSIONS AND FUTURE PERSPECTIVES

The low-cost stereoscopic system proposed in this study demonstrates accurate and robust performance, as it has been collecting daily data from the Belvedere Glacier for more than a year (August 2021-February 2023). The camera system can stand the extreme mountain environment, being weather- and cold-resistant. Its low power consumption sustained by solar panels allowed an autonomous functioning for the whole acquiring period. And these performances were achieved with easy assembly and low-cost components. Thanks to its characteristics, the system performed better than expected. The percentage of days in which the collection of data was prevented by battery discharging was at least 5% lower than the PVGIS simulation prediction. However, the system can be improved in various ways. One is the optimization of the camera's stabilization and the positioning of the tripod. Moreover, the system

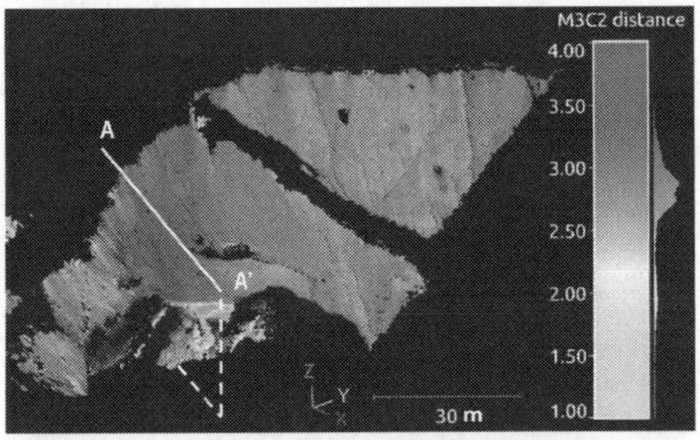

(a)

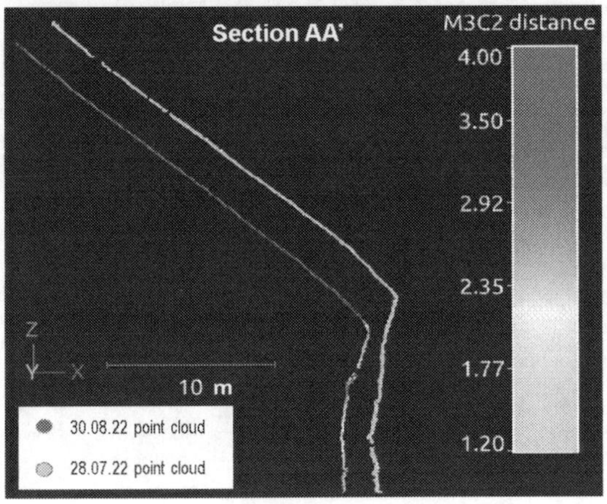

(b)

Figure 12. (a) Glacier terminus retreat between 28/07/22 and 30/08/22, computed as M3C2 cloud-to-cloud distances between stereo-point clouds. The colors represent the distances between each point of the 30/08/22 point cloud (seed point) and the points of 28/07/22 that fall within a cylinder oriented as the local normal computed around the seed point. Distances are expressed in meters. (b) Computed distances extracted along the transversal cross-section AA'. The 28/07/22 point cloud is represented with a gray color as a reference.

could be made even more autonomous, making it independent of mobile networks, which coverage can be unreliable and often missing in alpine environments.

A multi-camera system allows for computing glacier retreat and volume variations in a 3D world, which is not achievable with monoscopic camera systems. Information about the ice volume lost can be combined with ablation (estimated as a reduction of the height of the glacier surface from 3D point clouds) and ice flow velocity information, e.g., derived from single camera time-series by DIC (Messerli and Grinsted, 2015, Schwalbe and Maas, 2017, How et al., 2020), to gather a complete understanding of the glacier dynamics. Furthermore, multi-camera systems do not need any a priori information for reconstructing 3D scenes, overcoming the limit of monoscopic systems that require DEMs to translate 2D pixel displacements computed by DIC to 3D displacement vectors. If a complete DEM of the glacier surface can be reconstructed at each epoch by stereo-cameras, it can be used to derive 3D metric displacements, as proposed by Marsy et al. (Marsy et al., 2020). Otherwise, if the camera viewshed is limited (as for the Belvedere Glacier case study), information about the glacier ablation at the terminal ice cliff can be employed to model the evolution of a glacier DEM (e.g., acquired by UAV) over time.

Although the results of the stereoscopic system are promising, the routine for processing stereo-pairs and extracting relevant information (e.g., volume losses, flow velocities) from point cloud series is still under active development. In particular, alternative open-source solutions for performing BA and dense reconstructions are under study to replace the commercial software Agisoft Metashape. Moreover, change detection algorithms such as M3C2 (Lague et al., 2013), or its improved variations such as CDPB-M3C2 (Zahs et al., 2022), for increasing the signal-to-noise ratio when changes have magnitude similar to the uncertainty, are being tested. The focus of this paper, however, is on the hardware implementation of a low-cost multi-camera monitoring system rather than on image processing methods and results obtained on the Belvedere Glacier. Therefore, a thorough discussion of the processing methods and analysis of the results will be addressed in a future paper, and all the Python routines will be published in an open-source repository. Additionally, in the case study of the Belvedere Glacier, only two cameras were employed. However, the system can be easily extended to include multiple cameras for increasing viewing coverage, which may allow for deriving a full DEM of the area of interest, without limiting the reconstruction to the terminal ice cliff only.

5. CONCLUSIONS

This paper presents a low-cost implementation of an image-based stereoscopic system for continuous glacier monitoring. The system is composed of two autonomous monitoring units that are designed and built with low-cost and off-the-shelves components and can be easily replicated (and possibly also improved) by non-experts. Each monitoring unit includes a DSLR camera, an Arduino Pro Mini microcontroller for camera triggering and waking up the system and a Raspberry Pi Zero W board for storing the images taken by the camera, providing remote access to the system by internet connection and automatically sending the images to a remote server. The system has its autonomous power supply, ensured by a lead-acid battery and a solar panel, and it is enclosed in a waterproof case. The total cost of a single monitoring unit, including a middle-edge DSRL camera with a fixed-focal-length lens, was between €1500 and €2000. This, together with the easy implementation, makes the system affordable and reproducible for several remote glacier sites.

From daily stereo-pairs, 3D dense point clouds of the glacier terminus can be built with decimetric accuracy and used for estimating glacier retreat over time, ice volume loss and applying change detection algorithms on point clouds, with sub-weekly frequency. Especially if combined with forward glacier surface velocity, information regarding glacier retreat and ice volume losses over time is extremely valuable information for glaciologists who are aimed at understanding sub-seasonal glacier dynamics and computing mass balances.

6. ACKNOWLEDGMENTS

The stereoscopic system was designed, realized and deployed at Belvedere Glacier within the Alta Scuola Politecnica (ASP), a joint venture between Politecnico di Milano and Politecnico di Torino, through the multidisciplinary project titled "Kuoleva Jaatikko" (`https://www.asp-poli.it/kuoleva-jäätikkö/`). Many thanks are owed to Iosif Horea Bendea for the realization of the monumentation system and his work on the cases of the two monitoring units. We thank Paolo Maschio and Fabio Giulio Tonolo for their fruitful suggestion in the design phase and their help in the installation. We finally thank the Belvedere ropeway team and all the people of Macugnaga municipality who made with their help possible the transportation and installation of the monitoring systems.

REFERENCES

Agisoft Metashape, 2023. [Online; visited on 22/2/2023].

Barry, R. G., 2006. The status of research on glaciers and global glacier recession: a review. *PPG: Earth and Environment*, 30(3), 285-306.

Chudley, T. R., Christoffersen, P., Doyle, S. H., Abellan, A., Snooke, N., 2019. High-accuracy UAV photogrammetry of ice sheet dynamics with no ground control. *Cryosphere*, 13(3), 955–968.

De Gaetani, C. I., Ioli, F., Pinto, L., 2021. Aerial and UAV Images for Photogrammetric Analysis of Belvedere Glacier Evolution in the Period 1977–2019. *Remote Sens.*, 13(18), 3787.

Dematteis, N., Giordan, D., Crippa, B., Monserrat, O., 2022. Fast local adaptive multiscale image matching algorithm for remote sensing image correlation. *Computers & Geosciences*, 159, 104988.

Eltner, A., Kaiser, A., Abellan, A., Schindewolf, M., 2017. Time lapse structure-from-motion photogrammetry for continuous geomorphic monitoring. *Earth Surf. Proc. Land.*, 42(14), 2240-2253.

Emnify, 2023. [Online; visited on 22/2/2023].

EU Science Hub, 2022. Photovoltaic Geographical Information System (PVGIS). https://joint-research-centre.ec.europa.eu/pvgis-online-tool_en.

Giordan, D., Allasia, P., Dematteis, N., Dell'Anese, F., Vagliasindi, M., Motta, E., 2016. A low-cost optical remote sensing application for glacier deformation monitoring in an alpine environment. *Sensors*, 16(10), 1750.

How, P., Hulton, N. R. J., Buie, L., Benn, D. I., 2020. PyTrx: A Python-Based Monoscopic Terrestrial Photogrammetry Toolset for Glaciology. *Frontiers Earth Scie.*, 8.

Ide, R., Oguma, H., 2013. A cost-effective monitoring method using digital time-lapse cameras for detecting temporal and spatial variations of snowmelt and vegetation phenology in alpine ecosystems. *Ecological Informatics*, 16, 25-34.

Ioli, F., Bianchi, A., Cina, A., De Michele, C., Maschio, P., Passoni, D., Pinto, L., 2021. Mid-Term Monitoring of Glacier's Variations with UAVs: The Example of the Belvedere Glacier. *Remote Sens.*, 14(1), 28.

James, M. R., How, P., Wynn, P. M., 2016. Pointcatcher software: analysis of glacial time-lapse photography and integration with multitemporal digital elevation models. *J. Glaciol.*, 62(231), 159–169.

Kneib, M., Miles, E. S., Buri, P., Fugger, S., McCarthy, M., Shaw, T. E., Chuanxi, Z., Truffer, M., Westoby, M. J., Yang, W., Pellicciotti, F., 2022. Sub-seasonal variability of supraglacial ice cliff melt rates and associated processes from time-lapse photogrammetry. *The Cryosphere*, 16(11), 4701–4725.

Lague, D., Brodu, N., Leroux, J., 2013. Accurate 3D comparison of complex topography with terrestrial laser scanner: Application to the Rangitikei canyon (NZ). *ISPRS J. Photogramm. Remote Sens.*, 82, 10–26.

Lowe, D. G., 2004. Distinctive image features from scale-invariant keypoints. *Int. J. Comput. Vision*, 60(2), 91–110.

Marsy, G., Vernier, F., Bodin, X., Cusicanqui, D., Castaings, W., Trouvé, E., 2020. Monitoring mountain cryosphere dynamics by time lapse stereo photogrammetry. *ISPRS Ann. Photogramm. Remote Sens. Spat. Inf. Sci.*, V-2.

Messerli, A., Grinsted, A., 2015. Image georectification and feature tracking toolbox: ImGRAFT. *Geosci. Intrum. Meth.*, 4(1), 23–34.

Piermattei, L., Carturan, L., De Blasi, F., Tarolli, P., Dalla Fontana, G., Vettore, A., Pfeifer, N., 2016. Suitability of ground-based SfM-MVS for monitoring glacial and periglacial processes. *Earth Surf. Dynam*, 4, 425–443.

Roncella, R., Forlani, G., Fornari, M., Diotri, F., 2014. Landslide monitoring by fixed-base terrestrial stereo-photogrammetry. *ISPRS Ann. Photogramm. Remote Sens. Spat. Inf. Sci.*, 2(5), 297–304.

Rupnik, E., Daakir, M., Pierrot Deseilligny, M., 2017. MicMac–a free, open-source solution for photogrammetry. *Open Geospat. Data, Softw. Stand.*, 2(1), 1–9.

Sarlin, P.-E., DeTone, D., Malisiewicz, T., Rabinovich, A., 2020. SuperGlue: Learning feature matching with graph neural networks. *Proc. CVPR IEEE.*

Schönberger, J. L., Frahm, J.-M., 2016. Structure-from-motion revisited. *Proc. CVPR IEEE.*

Schwalbe, E., Maas, H.-G., 2017. The determination of high-resolution spatio-temporal glacier motion fields from time-lapse sequences. *Earth Surf. Dynam.*, 5(4), 861–879.

Sheridan, G., 2021. Intervalometerator Version 4.4.3. https://github.com/greiginsydney/Intervalometerator.

Vallot, D., Adinugroho, S., Strand, R., How, P., Pettersson, R., Benn, D. I., Hulton, N. R. J., 2019. Automatic detection of calving events from time-lapse imagery at Tunabreen, Svalbard. *Geosci. Intrum. Meth.*, 8(1), 113–127.

Yao, G., Yilmaz, A., Meng, F., Zhang, L., 2021. Review of Wide-Baseline Stereo Image Matching Based on Deep Learning. *Remote Sens.*, 13(16), 3247.

Young, D. S., Hart, J. K., Martinez, K., 2015. Image analysis techniques to estimate river discharge using time-lapse cameras in remote locations. *Computers & Geosciences*, 76, 1–10.

Zahs, V., Winiwarter, L., Anders, K., Williams, J. G., Rutzinger, M., Höfle, B., 2022. Correspondence-driven plane-based M3C2 for lower uncertainty in 3D topographic change quantification. *ISPRS J. Photogramm. Remote Sens.*, 183, 541–559.

Zemp, M., Haeberli, W., Hoelzle, M., Paul, F., 2006. Alpine glaciers to disappear within decades? *Geophys. Res. Lett.*, 33(13).

MODELLING AND VISUALIZING URBAN GROWTH TRAJECTORY IN ABU DHABI USING TIME SERIES SATELLITE IMAGERY

S. Issa[1], M. Al Ameeri[1], H. Abouhaligah[2], N. Alkaabi[1], M. Sultan[1] and B. Dahy[3]*

[1] Department of Geosciences, College of Science, United Arab Emirates University, Al Ain, Abu Dhabi, UAE.

[2] Department of Physics, College of Science, United Arab Emirates, Al Ain, Abu Dhabi, UAE

[3] CITIES Center, New York University - Abu Dhabi, PO Box 129188, Abu Dhabi, UAE – bsd9396@nyu.edu

KEY WORDS: LULC, Change Analysis, Remote Sensing, Urbanization, Landsat, SDGs

ABSTRACT:

Anthropogenic activities are leading to massive changes in the natural environment. An assessment of these changes is crucial for sustainable development. The present study aims to characterize and geo-visualize the spatial and temporal changes in the capital city of Abu Dhabi (AD) in the United Arab Emirates (UAE) and to assess and identify the driving forces behind this process. The temporal assessment of urban growth in AD was carried out using eight Landsat images for the period between 1990 and 2022. Initially the images were classified into seven land use land cover (LULC) classes namely: Urban, Sparse vegetation, Dense vegetation, Shallow water, Deep Water, Sand, and Sabkha. The 7 classes were then collapsed and merged to create a binary mask showing urban/non-urban areas. The post-classification approach was implemented to track the dynamic changes and to quantify the urbanization process of the capital city during the study period. The study showed a rise of 21.4% in urban area at the expense of a significant fall in the sand and sabkha classes. It was also found that, unlike most world urban centers, AD city is not developing at the cost of green areas. To the contrary, vegetated areas (sparse and dense vegetation classes) of AD have doubled more than 60 times over the years 1972 to 2021. We concluded that urbanization in the UAE is supporting the sustainable environmental development in the region and is responding positively to the United Nations (UN)' Sustainable Development Goals (SDGs).

1. INTRODUCTION

Urban areas play a vital role in social and economic development globally. Despite their socio-economic benefits, urban centers are also prone to multiple hazards and risks. Land use conversion for urban development is leading to higher carbon emissions and in turn causing environmental pollution. Urban expansion is a global trend and comes with its pros and cons. It has been reported that carbon footprint and resource consumption in cities is much higher than rural (Churkina, 2016). Therefore, urbanization is posing threat on natural resources. Consequently, accurate estimation and monitoring of urban growth is useful for planning services as well as water and land resources management (Das & Angadi, 2022, Aljaddani et al., 2022).

Urban development and growth patterns need to be monitored regularly to analyze the impact of development activities on natural environments. Hence assessment of LULC is prerequisite for sustainable urban development and planning. Traditional surveying and mapping methods are costly and time-consuming (Al Ahbabi, 2014). Consequently, studies are increasingly relying on remote sensing (RS) and GIS for monitoring urban growth (Ji et al., 2001, Maktav et al., 2005, Jat et al., 2008, Aljaddani et al., 2022, Shah et al., 2022). With the development of RS technology and the launch of new satellites in recent years, satellite or airborne spatial images and data have grown massively over the previous decades (Yu et al., 2014). High resolution data is not easily accessible. However, medium resolution satellites like Landsat series are freely available and provide archives of data (Wang et al., 2009; Sundarakumar et al., 2012; Vishwakarma et al., 2016). Furthermore, satellite images have been found highly effective in assessing the spatio-temporal changes over the earth's surface (Das & Angadi, 2022, Aljaddani et al., 2022). Precise and updated information on variation in expansion trends and rate of LULC change are critical to ensure sustainable development. Over the years, advancements have been made in data acquisition, processing, and analysis (Issa et al., 2021). Change analysis has been successfully used in wide range of applications such as natural resource management, water quality monitoring, and urban planning (Asokan & Anitha, 2019).

Among several change detection techniques, the image difference algorithm (IDA) and the post- classification methods seem to be most advantageous in the case of monitoring and modelling urbanization (Issa, S. M., & Al Shuwaihi, A., 2011; Dahy et al., 2021). The IDA is one of the most effective change detection methods due to its simplicity and wide usage. However, since IDA measures binary change only, it can only compare two images at once, and therefore can be used to compare the earliest and the most recent images to illustrate dramatic growth. On the other hand, the post classification method can be implemented to compare multiple images at once and adjust for variations in atmospheric conditions (as it uses classified maps rather than raw images). The trade-off is that it great expertise, is dependent of the classified image, and is time consuming (Dahy et al., 2021; Dahy et al., 2022).

United Arab Emirates (UAE) is amongst the top tourist and immigrant destinations countries in the world. The large number of immigrants is in turn leading to urbanization (Elessawy & Zaidan, 2014). According to the 2020 Human Development Index Report published by the United Nations Development Program (UNDP), the UAE clinched the top spot in the Arab region, taking a leap of 4 positions from the previous year and ranking 31st globally out of a total of 189 countries (United Arab

This contribution has been peer-reviewed.
https://doi.org/10.5194/isprs-archives-XLVIII-M-1-2023-145-2023 | © Author(s) 2023. CC BY 4.0 License.

Emirates Ministry of Foreign Affairs & International Cooperation, 2020). In the course of urbanization, LULC patterns are significantly altered. Changes in urban growth typically affect adversely water resources and vegetation, as well as land degradation (Aljaddani et al., 2022); however, studies indicate that this is not the case for the UAE (Dahy et al., 2022). Despite urban growth, the UAE has actually become greener over the past 50 years (Bardsley, 2022; Dahy et al., 2022).

Abu Dhabi (AD), the capital of UAE, has shown significant development in the late 20th century due to high rise in revenue from oil and gas exports. During the early 1970s, residents of AD city were depending on fishing and trade as a source of living. As the revenue from oil and gas exports has increased, AD turned into one of the fastest growing cities. Its population increased by approximately 1000% between 1975 and 2018 (Elessawy, 2021.). The immigrant workers form the major part of the population growth of the city, as they constitute around 90% of the population. Fueled by the population rise and economic growth, urbanization has increased drastically. Urban growth was so rapid that it is also termed as instant city development (Sharaf et al., 2018). For the residents of AD, a new era began with Sheikh Zayed bin Sultan Al Nahyan's succession. He initiated multiple projects to pave the path for AD's growth as a metropolis. Sheikh Khalifa continued the development trend, his vision included economic progress, urban development, and environmental stability.

The purpose of this study is to characterize and monitor the spatial and temporal changes in AD through focusing on the dynamics of the urban/non-urban classes in the study area between 1990 and 2022. Using animated maps, we illustrate and geo-visualize the changing aspects of urban growth and sprawl in AD city for the last 32 years. Specific objectives are: 1) To create a time-series of the seven LULC maps for the years: 1990, 1995, 2001, 2008, 2013, 2015, 2020 and 2022 with emphasis on urban/non-urban classes of the study area, 2) To create a From-To-Change map of the AD city over the study period and; 3) To create and publish an animated map showing dynamics of urban growth and urban sprawl of AD city for the past 32 years.

2. MATERIALS AND METHODS

2.1 Study Area

The study area lies between 54°16′ to 54°40′ E longitudes and 24°16′ to 24°41′ N latitudes. The Emirate of Abu Dhabi is located in the southern shore of the Arabian Gulf, within the UAE. The Emirate is bordered to the east by Oman, to the south by Saudi Arabia, and to the northeast by Dubai (Figure 1). With an area of 67,340 km2, the Emirate of Abu Dhabi covers over 75% of the UAE. Among the emirate's many oil fields are those both on land and in the Arabian Gulf. Salt marshes cover much of Abu Dhabi's coast, making it relatively arid. Around 200 islands fall under Abu Dhabi territory; some of the major ones include Al Hudairiyat, Saadiyat, and Yas Island. The Emirate of Abu Dhabi has a warm desert climate. Furthermore, summers are generally hot, humid, and temperature averages above 40°C from June to September. Additionally, frequent sandstorms can occur during this time, reducing visibility. From October to May, temperatures are relatively mild. The weather is coolest between January to February, during which time dense fog frequently forms.

The study area comprises the City of AD which is the capital of the Emirate of Abu Dhabi. Located in the Arabian Gulf, the city is situated on a long, narrow island. While most of the city is on the island of Abu Dhabi, residential projects (informally termed "cities" but typically considered part of the larger City of Abu Dhabi) have also been constructed on the mainland that connects the entire emirate, including Al Maqta City, Musaffah City, Mohamed Bin Zayed City, Khalifa City, Masdar City, Al Raha Beach, Shakhbout City and Bani Yas City (Elessawy, 2021). On nearby islands, such as Saadiyat and Al Reem Islands in the northeast, new projects are constantly being developed. The east comprises the Yas, Samaliyah, and Sas Al Nakheel islands. To the northwest of Abu Dhabi Island is Al Ras Al Akhdar, and to the west is Hudariyat island (Elessawy, 2021).

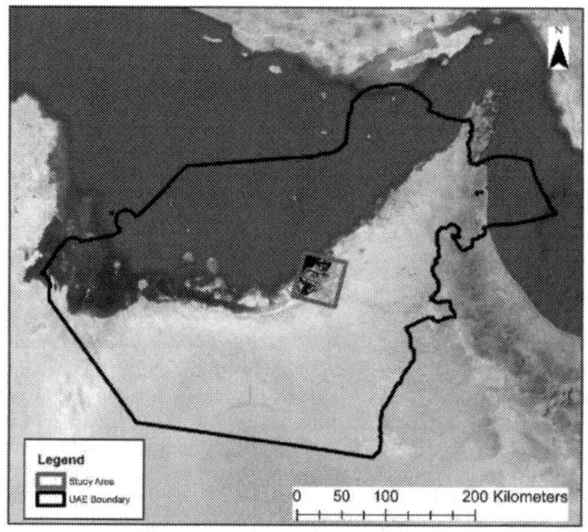

Figure 1. Location map of the study area.

2.2 Data Acquisition

Based on the data availability, image quality, land clouds cover (less than 10%), eight different dates of Landsat scene no 160/43 (path/row) were selected to study and analyze the urban change dynamics of AD city during last 32 years. The Collection 2 and Level 2 (C2, L2) product of OLI, ETM+, and TM images were downloaded from USGS Website. Additionally, panchromatic bands (Level 1) of the same scenes of Landsat-8 OLI and Landsat-7 ETM+ were downloaded and later used in the pan-sharpening process. Table 1 summarizes the data and sources used in this study.

Landsat Sensors	Date	Bands Used
TM	16/05/1990 30/05/1995	B1 (Blue) B2 (Green) B3 (Red) B4 (NIR) B7 (SWIR-2)
ETM+	22/05/2001 17/05/2008	B1 (Blue) B2 (Green) B3 (Red) B4 (NIR) B7 (SWIR-2) B8 (PAN)
OLI	16/06/2013 21/05/2015 18/05/2020 24/05/2022	B2 (Blue) B3 (Green) B4 (Red) B5 (NIR) B7 (SWIR-2) B8 (PAN)

Table 1. Summary of data used in the study.

Land cover classification and mapping are important tools for assessment of the landscape characteristics. Figure 2 displays a flowchart covering steps that were carried out in this study. Main steps include (i) image processing (ii) classification (iii)

This contribution has been peer-reviewed.
https://doi.org/10.5194/isprs-archives-XLVIII-M-1-2023-145-2023 | © Author(s) 2023. CC BY 4.0 License.

urban/non-urban mask production (iv) change detection analysis (v) geovisualization and map animation.

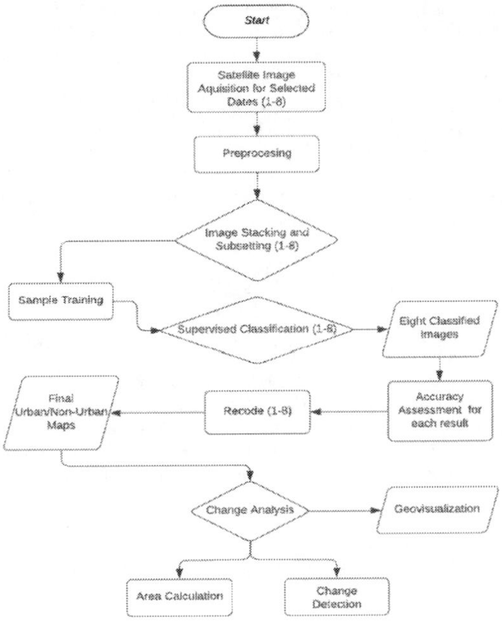

Figure 2. Flowchart showing the methodology of research of the study area.

2.3 Methods

2.3.1 Image Processing: Landsat images covering the study area were acquired from the USGS website. The first step was stacking the bands together to create a multiband image for each study year. The blue, green, red, near-infrared (NIR), and shortwave infrared 2 (SWIR2) bands, in addition to the panchromatic band (for ETM+ and OLI sensors), were used. This task was achieved using ERDAS Imagine 2020 software package. Then the images were pansharpened using the Hyperspherical Colour Sharpening (HCS) resolution method (Padwick et al., 2010) to benefit from the enhanced pixel size of 15 m. The pansharpening task was conducted only for ETM+ and OLI sensors scene (the year 2001 onward), as TM sensor scene has no panchromatic band. For the older dates, the pixel size was resampled to 15 m resolution for compatibility with the other images.

2.3.2 Classification: This paper uses the integrated classification approach, combining both supervised and unsupervised classification while the change detection was carried using the union matrix tool of ERDAS Imagine 2020. The first step was to attempt an unsupervised classification. Twenty clusters (classes) were explored initially. After investigation, seven final classes were produced namely: 1. urban, 2. sparse vegetation, 3. dense vegetation, 4. shallow water, 5. deep water, 6. sand, and 7. sabkha. To produce the final land cover maps of the study area for each of the eight dates (1990, 1995, 2001, 2008, 2013, 2015, 2020, and 2022), a supervised classification was conducted through selecting a number of training sets and by collecting and creating the set of spectral signatures for each class, using the previously produced 8 clusters.

2.3.3 Urban/Non-Urban Mask Production: The seven-class LULC data was simplified using a binary mask to clearly distinguish between urban and non-urban areas. The binary mask generated eight maps that only represented two classes: Urban and Non-Urban. The urban class comprises roads infrastructure in addition to built-up areas. The non-urban class includes all other classes including, sparse vegetation, dense vegetation, shallow water, deep, Water, sand, and sabkha desert.

2.3.4 Change Detection Analysis: Post-classification change detection was applied using the before and after thematic images through union matrix tool in ERDAS Imagine 2020. Union matrix tool is a powerful geospatial analysis tool in comparative studies. The tool helps in comparing and analyzing datasets and it compares pixel values in each layer and produces a map and attribute table which helps is assessing change detection. Through this tool we calculated the change in area of each land use type relative to the other land use classes and produced a map showing change between non-urban and urban class.

2.3.5 Geovisualizing and Map Animation: Geovisualization is the visual representation of geographic data. Geovisualization acts as a strong tool in studies that involve the analysis and understanding of spatial and temporals trends. It has been effectively used in various fields such as geography, statistics, visual analytics, scientific visualization, computer science, art, and cognitive science (Çöltekin et al., 2018). This study has also attempted to display the LULCC in the form a video containing maps of study area. The video shows conversion on landcover classes and urban growth during the period of 32 years i.e., 1990-2022.

3. RESULTS AND DISCUSSION

3.1 Multi-temporal Classification and Mapping of AD LULC Classes with Emphasis on the Urban Change

LULC changes are common worldwide due to natural and human causes, with urbanization playing a major role. The UN predicts global urban population will reach 60% by 2030, and the UAE's urban population is expected to increase from 79% in 1990 to 89% in 2030 (UN & DESA, 2019). Understanding the trends of urban growth over time is crucial for sustainable urban development. Urbanization has both positive and negative impacts, such as improving the economy and degrading groundwater (Sohl, 1999). Hence, mapping and assessment of urban dynamics is crucial for sustainable city development (Elessawy, 2020).

Multiple techniques have been employed for detecting changes in LULC. In areas where comprehensive cartographic information is scarce, geospatial technologies can play a vital role in supplying a foundation for land utilization planning, management, and preservation efforts (Issa et al., 2021). Landsat program including Multispectral Scanner System (MSS), Thematic Mapper (TM), Enhanced Thematic Mapper (ETM+), and Operational Land Imager (OLI) provided moderate-resolution multispectral images with long history repetitive coverage on a regional scale. The data provided a unique baseline for analyzing land-cover changes over the past 50 years. LULC change have become a high priority area for research. Furthermore, analyzing LULC change has become one of the most critical research topics. LULC maps and datasets are essential for understanding and modeling complex interactions and impacts between human and natural systems (Yagoub & Kolan, 2006).

https://doi.org/10.5194/isprs-archives-XLVIII-M-1-2023-145-2023

The study used Landsat data to create 8 maps of LULC from 1990 to 2022. 20 clusters (classes) were first made through unsupervised classification. Later, these clusters where improved with supervised classification to improve the quality of results. In this step, seven classes were created that provide an accurate illustration of LULC in the the study area. Figure 3 displays LULC maps of study area for the 8 selected years. The maps show 7 major LULC classes: urban, sparse vegetation, dense vegetation, shallow water, deep water, sand, and sabkha. The urban area increased from 3831.48 ha in 1990 to 56968.9 ha in 2022.

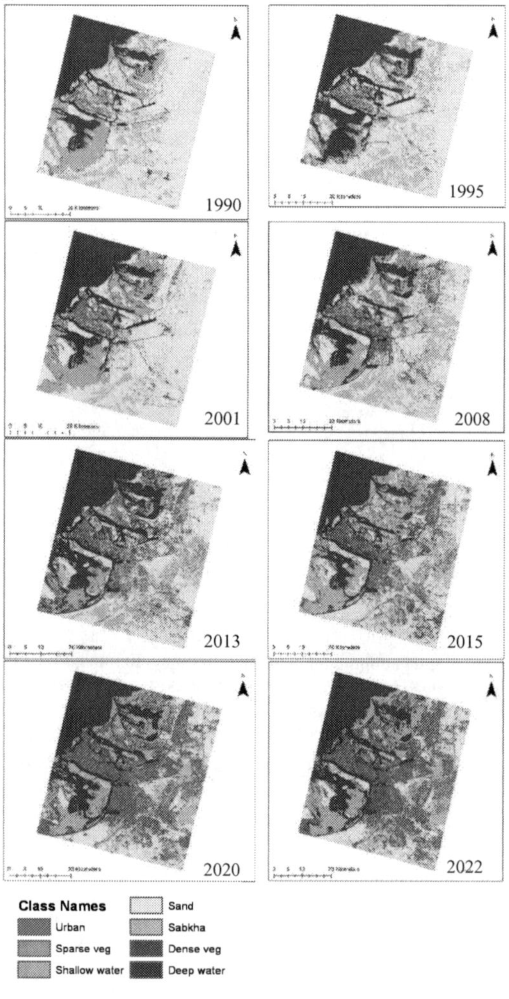

Figure 3. LULC maps of the study area for the years 1990, 1995, 2001, 2008, 2013, 2015, 2020, and 2022.

Urban land in the study area is increasing while sand and sabkha are decreasing. Numerous factors contribute to the urban sprawl of AD city. The prominent factors include population growth, economic development, and government policies (Taileb et al., 2008). Multitemporal analyses of LULC provide insight into long-term trends and help identify future determinants of change. The UAE is a rich oil-producing country that has experienced rapid urbanization and modernization in recent decades. The capital city of Abu Dhabi has grown significantly and become a global economic and financial center (Elessawy, 2021). A variety of developments were approved to be built in the 1990s including schools, hotels, restaurants, and offices. Hence the urban area, which was covering only 1.55% in 1990, became 1.6% in 1995 and since then the pace of urban growth is rising. By 1995, the emirate of Abu Dhabi had 33 public parks (Abu Dhabi Municipality, 1996). The emirate also has a rising green cover, with vegetation increasing from 2.1% in 1990 to 3.5% in 1995, due to policies that consider both urban development and green diversity. Throughout the island's urbanization process, old buildings were converted to newer, more modern structures. By 2001, reclamation and dredging had increased the landmass of the island to 96 km2, with the urban area growing from 3831.48 ha in 1990 to 8217 ha in 2001.

After the passing away of HH Sheikh Zayed and his succession by HH Sheikh Khalifa in 2004, some changes to boost urban development were introduced. Urban development remained the focus of this era too, however, attention was also given to sustainable development. Sustainable development considered peripheral development along with the development of urban centre which is focus of attention for investors (Kyriazis et al., 2017). In particular, new laws allowing Emirati nationals to sell land granted by the government and introducing and legalizing foreign ownership. Moreover, the road network was expanded, and new landmarks such as the Emirates Palace Hotel and Sheikh Zayed Grand Mosque were built in 2005. Despite the global economic crisis of 2008, Abu Dhabi managed to remain stable by building Masdar City, the first "carbon-free" city in the world. Urban cover of AD became 7% in 2005 and reached 10.5% in 2013. Built area kept increasing and was found to be 15.6% in 2015, 21.7% in 2020 and reached near 23% in the year 2022.

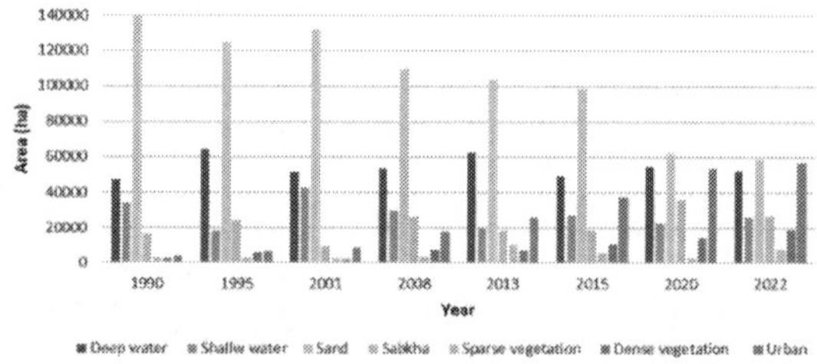

Figure 4: LULC maps in Abu Dhabi from 1990 to 2022.

Figure 4 demonstrates the gradual change in land cover from 1990 to 2022, with a decrease in sand coverage and an increase in urban coverage. The scope of this paper is to find the urban growth in the study area, hence, to enhance the clarity of change between urban and non-urban class binary mask was applied. The use of a binary urban/non-urban mask provided a simple, yet useful, representation of the urban-rural dichotomy. Figure 5 presents the binary maps with urban and non-urban class between 1990 and 2022.

The population of Abu Dhabi is rising due to economic growth, job opportunities, and high rates of immigration. The demand for housing and infrastructure also increases with the rise in population. The statistical data shows a rising trend in population, and urbanization follows the same trend as displayed in Figure 6. Urban sprawl and population growth of AD are showing significant rise between the years 1990 and 2022.

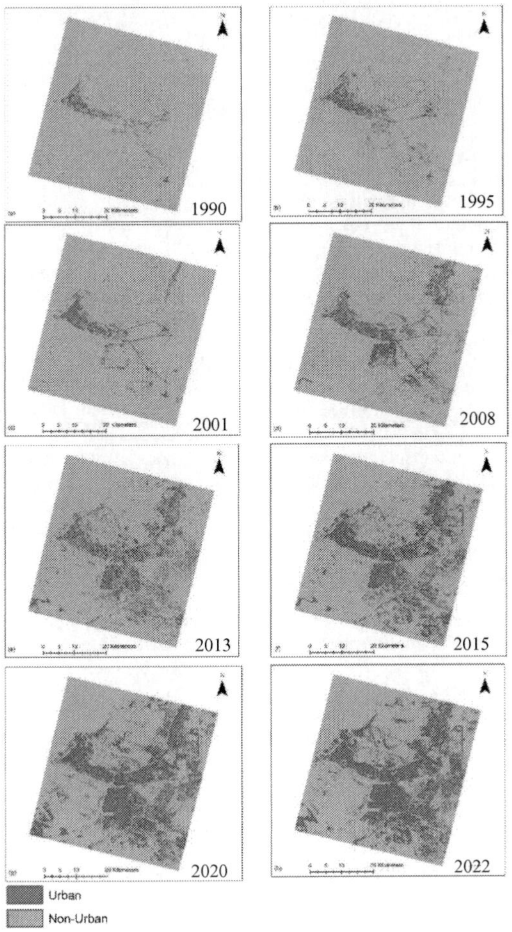

Figure 5: Urban/Non-Urban maps in Abu Dhabi from 1990 to 2022.

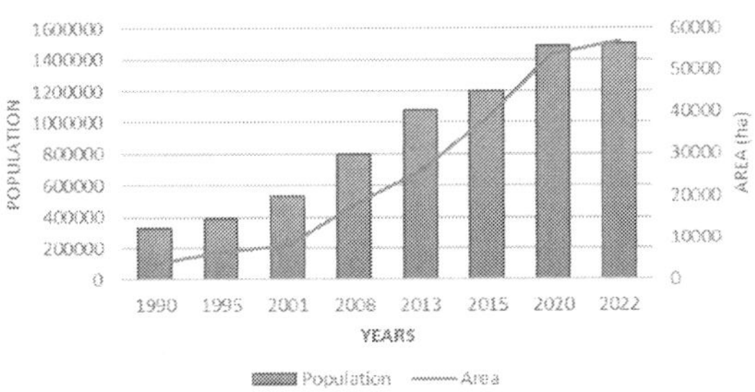

Figure 6: The relationship between population growth and urban sprawl in Abu Dhabi from 1990 to 2022

3.2 Change Detection Analysis

There are various techniques for change detection analysis. The method that was used in this study was post-classification change detection with respect to area as it is the most relevant to the objectives of the study. The method was applied to detect the changes across 7 different classes over the study period.

Change in class	Area (ha)
Deep water to Urban	511.987
Shallow water to Urban	1066.07
Sand to Urban	50216.7
Sabkha to Urban	6428.09
Sparse veg to Urban	66.3075
Dense veg to Urban	85.3425

Table 2. Change in Non-Urban areas to Urban areas in hectares (ha) between 1990 and 2022 in the study area.

The results in (Table 2) and (Figure 7.) indicate that the class that was most affected by urbanization in AD was the sand class; with an overall change of 6564 ha. A total of approximately 19 hectares of the vegetation classes were urbanized between 1990 and 2022, indicating that the vegetation cover was the least affected by the urbanization sprawl.

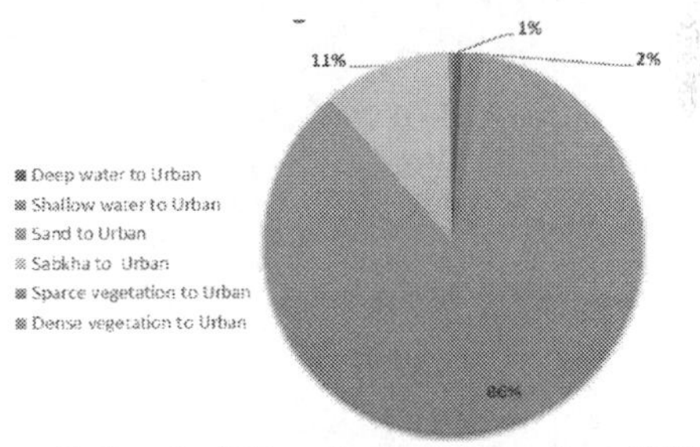

Figure 7: Growth of Urban class in study area from 1990 to 2022 with respect to other classes

3.3 Urban Cover Change

3.3.1 Quantification: To accurately measure and represent the change in the urban class in AD City, a quantification method was applied in which the urban and non-urban areas were measured in hectares and percentages. Table 3. and Figure 8. confirm that the urban cover in the study area has increased significantly between 1990 and 2022. An overall 21.4% of the total area has changed into the urban cover between 1990 and 2022; subsequently decreasing the non-urban area by 21.4%.

Year	Urban		Non-Urban	
	Area (ha)	%	Area (ha)	%
1990	3831.48	1.55	243491	98.45
1995	6579.34	2.66	240870	97.34
2001	8216.91	3.31	240339	96.69
2008	17864.7	7.17	231171	92.83
2013	26093.3	10.52	221865	89.48
2015	38571.6	15.60	208723	84.40
2020	53746.9	21.70	194013	78.30
2022	56968.9	22.95	191338	77.05

Table 3. Urban and Non-Urban areas (ha) and their percentages in the study area from 1990 to 2022.

https://doi.org/10.5194/isprs-archives-XLVIII-M-1-2023-145-2023

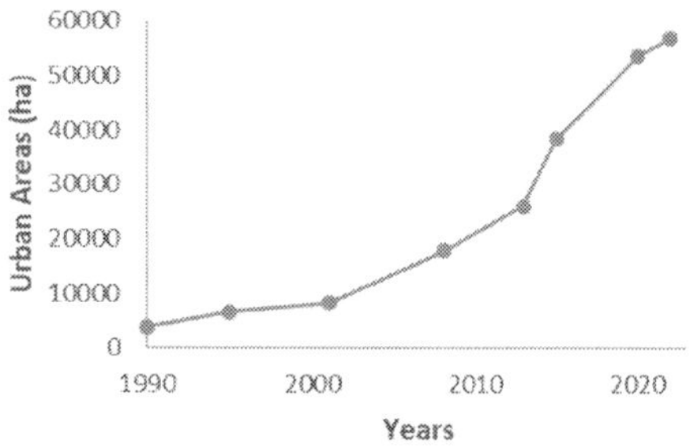

Figure 8: Growth of Urban in Abu Dhabi from 1990 to 2022.

All The study area has experienced significant conversion from non-urban to urban class during the 32 years between 1990 and 2022. The change detection map created using the union matrix tool in ERDAS Imagine displays the conversions between non-urban and urban classes figure 9.

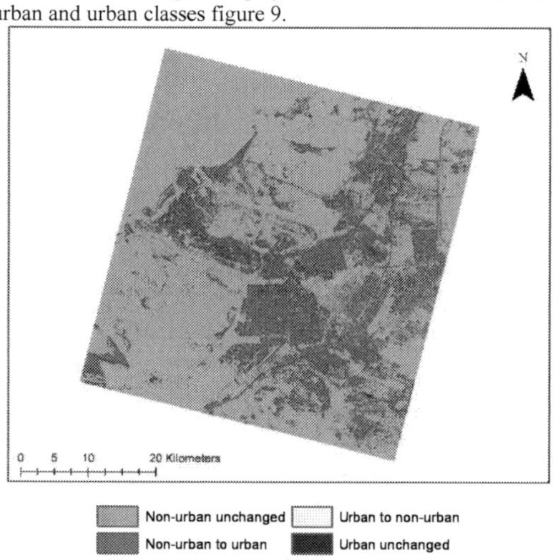

Figure 9: Change map between Non-Urban and Urban classes in the study area (1990-2022).

The conversion from non-urban to urban is very high, while urban to non-urban is very small and the areas of change are on the roadsides which signal the creation of green belts along the roads.

The urbanization of AD has been rising, and major driving factors are population growth, economic growth, and government policies. Urbanization has its positive and negative impacts on the sustainable development. It is therefore crucial to continuously monitor urban growth in order to ensure that growth is in accordance with sustainable development goals. In this regard, current study provides view of LULC changes and urban sprawl trends in AD over a period of 32 years. The study helps tracking trends in urban development, such as conversion of one landcover type to another. The prominent land cover class declining in the study area is sand while urban area is increasing. On one side this is sustainable and environmentally safe approach but on the other hand increased urbanization might lead to groundwater pollution. Therefore, continued monitoring of LULC change and assessment of its impacts is crucial to achieve sustainable development.

4. CONCLUSION AND RECOMMENDATIONS

In terms of urban planning and maintaining sustainable development, it is very critical for decision-makers and planners to be able to analyze the urban cover of the area. The spectral variability in desert environments like AD, caused by the presence of highly reflective surfaces like sand dunes and low reflectance areas like shadows or areas with sparse vegetation, can pose difficulties in accurately mapping and identifying different land covers through remote sensing techniques. Therefore, approaches like integrated classification can be effective. As shown by the results obtained from the study area (AD city) using RS technologies, as well as the results acquired from the study area, it is evident that the urban area's extent has been growing at a constant rate for almost three decades, increasing by approximately 1486.86% from only 3,831.48 ha in 1990 to 56,968.9 ha by 2022 as illustrated by trajectory maps produced earlier. The class most affected by this drastic change was the sand class, which makes this development trend closer to the sustainable development goals. Urban growth in AD is mainly resulting from economic and population growth. Urbanized development is not being carried out at the cost of sustainable development. Along with urban areas, vegetation cover is also increasing, while sabkha and sand are gradually decreasing. These trends are encouraging as desert is being converted into urban land, yet continued monitoring is encouraged. It has been observed that urbanization in arid areas leads to groundwater scarcity and degradation. Future studies may focus on this aspect of urban growth. Temporal change detection and mapping is beneficial for planners and policymakers to continue planned development.

REFERENCES

Al Ahbabi, A. (2014). Vegetation Cover Monitoring in the Abu Dhabi Region Using GIS and Remote Sensing, 1973–2010. The Arab World Geographer / Le Géographe Du Monde Arabe, 150–165.

Abu Dhabi Municipality (ADM) (1996). Abu Dhabi: Dana of the Gulf, Planning and Urban Development (Studies and Research Section). Dubai: Abu Dhabi Municipality & Town Planning Department.

Aljaddani, A. H., Song, X.-P., & Zhu, Z. (2022). Characterizing the patterns and trends of urban growth in Saudi Arabia's 13 Capital Cities using a Landsat Time Series. *Remote Sensing, 14*(10), 2382. https://doi.org/10.3390/rs14102382

Asokan, A., & Anitha, J. J. E. S. I. (2019). Change detection techniques for remote sensing applications: a survey. *Earth Science Informatics, 12*(2), 143-160.

Bardsley, D. (2022, June 18). *UAE desert has grown greener amid urban rise, 50-year study shows*. The National. Retrieved July 2, 2022, from https://www.thenationalnews.com/uae/environment/2022/06/18/uae-desert-has-grown-green-amid- urban-rise-50-year-study-shows/

Churkina, G. (2016). The role of urbanization in the global carbon cycle. *Frontiers in Ecology Evolution, 3*, 144.
Çöltekin, A., Janetzko, H., & Fabrikant, S. (2018). Geovisualization. *Geographic Information Science & Technology Body of Knowledge, 2018*(Q2). https://doi.org/10.22224/gistbok/2018.2.6

Dahy, B., Issa, S., & Saleous, N. (2022). Random forest for classifying and monitoring 50 years of vegetation dynamics in three desert cities of the UAE. *The International Archives of the Photogrammetry, Remote Sensing and Spatial Information Sciences, XLIII-B3-2022*, 69–76. https://doi.org/10.5194/isprs-archives-XLIII-B3-2022-69-2022

Dahy, B., Issa, D., & Saleous, N. (2021). A Review Of Land Change Modelling Techniques Using Remote Sensing and GIS.
Das, S., & Angadi, D. P. (2022). Land use land cover change detection and monitoring of urban growth using remote sensing and GIS techniques: a micro-level study. *GeoJournal, 87*(3), 2101-2123.

Enderle, D. I., & Weih Jr, R. C. (2005). Integrating supervised and unsupervised classification methods to develop a more accurate land cover classification. *Journal of the Arkansas Academy of Science, 59*(1), 65-73.

Elessawy, F. M. (2020). The consequences of population growth on the demographic characteristics of Abu Dhabi City in the United Arab Emirates. *Advances in Economics Business, 8*(5), 268-276.

Elessawy, F. M. (2021). The abnormal population growth and urban sprawl of an Arabian Gulf City: The case of Abu Dhabi City. *Open Journal of Social Sciences, 09*(02), 245–269. https://doi.org/10.4236/jss.2021.92017

Elessawy, F., & Zaidan, E. (2014). Living in the move: Impact of guest workers on population characteristics of the United Arab Emirates (UAE). *The Arab World Geographer, 17*(1), 2-23.

Issa, S. M., & Al Shuwaihi, A. (2011, July). Analysis of LULC changes and urban expansion of the resort city of Al Ain using remote sensing and GIS. In *2011 6th International Workshop on the Analysis of Multi-temporal Remote Sensing Images (Multi-Temp)* (pp. 245-248). IEEE.

Issa, S., Dahy, B., Saleous, N., & Shamsi, M. A. L. (2021). A remote sensing based time-series land use land cover change (LULCC) analysis: A case study from the United Arab Emirates (UAE). *Sixth International Conference on Engineering Geophysics, Virtual, 25–28 October 2021.* https://doi.org/10.1190/iceg2021-029.1

Jat, M. K., Garg, P. K., & Khare, D. (2008). Monitoring and modelling of urban sprawl using remote sensing and GIS techniques. *International journal of Applied earth Observation Geoinformation, 10*(1), 26-43.

Ji, C., Liu, Q., Sun, D., Wang, S., Lin, P., & Li, X. (2001). Monitoring urban expansion with remote sensing in China. *International Journal of Remote Sensing*, 22(8), 1441-1455.
Maktav, D., Erbek, F., & Jürgens, C. (2005). Remote sensing of urban areas. *International Journal of Remote Sensing, 26*(4), 655-659.

Kyriazis, A., Almehairi, A., & Al Seri, I. (2017). The Morphology of Abu Dhabi's Superblocks: A Modernist Vision.

Padwick, C., Deskevich, M., Pacifici, F., & Smallwood, S. (2010). WorldView-2 pan-sharpening. Paper presented at the Proceedings of the ASPRS 2010 Annual Conference, San Diego, CA, USA.

Shah, A., Ali, K., & Nizami, S. (2022). Spatio-temporal analysis of urban sprawl in Islamabad, Pakistan during 1979–2019, using remote sensing. *GeoJournal, 87*(4), 2935-2948.

Sharaf, S. A., Serra, P., & Saurí, D. (2018). A district and sector land-use and landscape analysis of urban sprawl in Al Ain municipality (United Arab Emirates): Just a quick conversion from sand to a built-up environment? *Applied Geography, 95*, 88-100.

Sohl, T. (1999). Change detection in the United Arab Emirates: An investigation of techniques. *Photogrammetric Engineering and Remote Sensing*, 475–484. https://www.researchgate.net/publication/259340978_Change_detection_in_the_United_Arab_Emirates_An_investigation_of_techniques.

Sundarakumar, K., Harika, M., Begum, S. A., Yamini, S., & Balakrishna, K. (2012). Land use and land cover change detection and urban sprawl analysis of Vijayawada city using multitemporal landsat data. *International Journal of Engineering Science Technology, 4*(01), 170-178.

Taileb, A., Arbaoui, A., & Boudiaf, B. (2008). Understanding Urban systems and sprawl in the UAE: Case studies from Ajman, Sharjah, Dubai and Abu Dhabi. *Instant Cities: Emergent Trends in Architecture Urbanism in the Arab World, The Center for the Study of Architecture in the Arab Region, School of Architecture Design, American University of Sharjah, Sharjah*, 39-48.

UN, DESA. *World Urbanization Prospects 2018* (UN, 2019).

United Arab Emirates Ministry of Foreign Affairs & International Cooperation. (2020, December 16). UAE ranks first in Arab World in Human Development Report 2020. Retrieved June 13, 2022, from https://www.mofaic.gov.ae/en/mediahub/news/2020/12/16/16-12-2020-uae-human-development#:~:text=The%20UAE%20ranked%20first%20in,a%20total%20of%20189%20countries.

Vishwakarma, C. A., Thakur, S., Rai, P. K., Kamal, M., & Mukherjee, S. (2016). Changing land trajectories: a case study from India using a remote sensing-based approach. *Eur J Geogr, 7*(2), 61-71.

Yagoub, M. M., & Kolan, G. R. (2006). Monitoring coastal zone land use and land cover changes of Abu Dhabi using remote sensing. *Journal of the Indian Society of Remote Sensing, 34*(1), 57–68. https://doi.org/10.1007/bf02990747

Yu, L., Liang, L., Wang, J., Zhao, Y., Cheng, Q., Hu, L., . . . Zhu, P. (2014). Meta-discoveries from a synthesis of satellite-based land-cover mapping research. *International Journal of Remote Sensing, 35*(13), 4573-4588.

Wang, L., Chen, J., Gong, P., Shimazaki, H., & Tamura, M. (2009). Land cover change detection with a cross-correlogram spectral matching algorithm. *International Journal of Remote Sensing, 30*(12), 3259-3273.

APPENDIX

1. VIDEO SHOWING THE LULC CHANGE OF STUDY AREA BETWEEN 1990 AND 2022

2. VIDEO SHOWING URBAN/NON-URBAN CHANGE OF STUDY AREA BETWEEN 1990 AND 2022

This contribution has been peer-reviewed.
https://doi.org/10.5194/isprs-archives-XLVIII-M-1-2023-145-2023

MATHEMATICAL AND PHYSICAL APPROACHES TO INFER ABSOLUTE ZENITH WET DELAYS FROM DOUBLE DIFFERENTIAL INTERFEROMETRIC OBSERVATIONS USING ERA5 ATMOSPHERIC REANALYSIS

B. Kamm[1], A. Schenk[1], P. Yuan[2], S. Hinz[1]

[1] Institute of Photogrammetry and Remote Sensing (IPF), Karlsruhe Institute of Technology, Karlsruhe, Germany - (bettina.kamm, andreas.schenk, stefan.hinz)@kit.edu
[2] Geodetic Institute Karlsruhe (GIK), Karlsruhe Institute of Technology, Karlsruhe, Germany - peng.yuan@kit.edu

KEY WORDS: PS-InSAR, Water Vapor, Wet Delay, ERA5 atmospheric reanalysis

ABSTRACT:

Atmospheric water vapor (WV) is one of the driving constituents of the atmosphere. The modelling and forecasting of WV and derived quantities like precipitable water is reliable on regional scales but challenging on small scales because of its high spatial and temporal variation. Interferometric synthetic aperture radar (InSAR) can be exploited to retrieve integrated atmospheric water vapor (IWV) from path delay observations along the radar line of sight. InSAR-derived IWV maps feature a very high spatial resolution but the double-differential interferometric observations only provide changes of IWV between acquisition times and with respect to a certain spatial reference. In this study we present a method to derive the absolute IWV by combining ERA5 numerical weather model data with differential path delay observations from InSAR time series. We propose different functional approaches to merge the regional trend of WV from ERA5 with the high resolution IWV signal from InSAR. We apply this to a Sentinel-1 Persistent Scatterer InSAR time series in the Upper Rhine Graben and validate against IWV observations at GNSS stations of the Upper Rhine Graben Network.

1. INTRODUCTION

SAR interferometry (InSAR) is susceptible for changes of atmospheric water vapor (IWV) which can be derived from path delay observations along the radar line of sight. Persistent Scatterer Interferometry (PSI) is an enhanced approach to process long InSAR time series where only those image pixels are chosen, which show high coherence (low phase variance) throughout the time series. PSI enables – apart from height and motion determination - the estimation of differential slant delays due to varying water vapor in the atmosphere. Moreover, single observations of zenith delays (ZD) can be estimated from a PSI stack under certain assumptions and with respect to an arbitrary constant, which appears as a rank deficiency in the estimation procedure. In particular for modern SAR satellite imagery, ZD maps are characterized by a high spatial resolution that can be further used e.g., as input to water vapor tomography.

The high spatial resolution of the InSAR data compared to other meteorological space-borne instruments makes them attractive to address meteorological questions (Hanssen, 2001). Space-borne C-band SAR instruments as, e.g., on the European Remote Sensing satellites (ERS-1/2) and onboard the Envisat satellite are successfully used to construct water vapor maps (Hanssen, 2001, Alshawaf et al., 2015, Heublein et al., 2015). The higher sensitivity and resolution of X-band sensors offers an even more detailed analysis as shown, e.g., by Qin (2013) for an urban example. In turn, the wide footprint of the C-band sensors on the current Sentinel-1 (A/B) satellite missions enables large areas to be investigated, still with a moderately high resolution (Mateus et al., 2017).

2. PROBLEM DEFINITION

A common approach to overcome the rank deficiency of this estimation process is the so called 'zero-mean-assumption', which constraints a zero average delay over all acquisition times for each PS-point. However, this approach neglects local variations, for example height dependent variations due to stratospheric layering or specific regional variations of weather characteristics. Even a conventional DEM-dependent stratification approach can only accomplish a rough approximation of the true situation. In addition, the resulting delays are still relative with respect to an arbitrary constant but for the usage in weather models, absolute values need to be determined. Spatial phase gradients with long wavelengths, so called phase ramps in PSI stacks can be caused by several influences, such as orbital ramps, processing errors, solid earth tides and ocean loading effects or atmospheric signals otherwise. The distinguishment of those contributions is only possible with further knowledge, while a full estimation and reduction from the observables seems hardly realizable in the case of Sentinel-1.

Alshawaf et al. (2015a) derive absolute IWV maps with a sequential approach too. First, a least squares inversion is applied to PSI delays under the constraint of zero temporal mean, whereby the PSI delays have been low-pass filtered before to suppress long-wave signal. Subsequently, these maps are combined with complementary maps from GNSS which provide more reliable absolute and long-wave IWV but at low spatial resolution. Comparing the final combined IWV maps with IWV maps from the MEdium Resolution Imaging Spectrometer (MERIS) onboard the Envisat satellite, shows good spatial correlation of up to 92%.

This contribution has been peer-reviewed.
https://doi.org/10.5194/isprs-archives-XLVIII-M-1-2023-153-2023 | © Author(s) 2023. CC BY 4.0 License.

3. DATA AND PREPROCESSING

In this work, we focus on the processing of the PSI phase observations, which are already unwrapped and corrected. The database is a stack of 169 scenes of Sentinel-1A/B data, acquired in Interferometric Wide Swath Mode (IW) at an altitude of around 690 km and a swath width of 250 km. The ground resolution is about 5 x 20 m. All data are recorded along ascending orbit 88 between March 2015 and July 2019 with the acquisition time of 17:26 UTC. In this study, we used the VV polarized data and the SRTM-1 as reference digital elevation model during preprocessing. The orbit correction was performed with the provided precise orbit files. Those resulting double differential slant delays are corrected for any linear deformations. The data is available online (Fersch et al., 2021).

ERA5 is the fifth generation atmospheric reanalysis provided by ECMWF (Hersbach et al., 2020). With more input data and advanced algorithms in assimilation, ERA5 represents the state-of-the-art atmospheric reanalysis. In this work, we employed the pressure level ERA5 product for the calculation of auxiliary meteorological data. We downloaded the humidity, temperature, and geopotential data with a spatiotemporal resolution of $0.25^\circ \times 0.25^\circ$, 37 vertical levels, and 1-hourly.

As the ERA5 data are provided as spatial grids, spatial interpolation or extrapolation is needed in the calculation of meteorological variables at the PS positions (see Appendix C of (Yuan et al., 2023) and the references therein). In addition, the ERA5 data are provided at integral hours, the values at two nearest hours were linearly interpolated to the PS epoch.

GNSS observations from the GURN (GNSS Upper Rhine Graben Network) were collected and processed by using GAMIT software. The tropospheric delays were corrected with Vienna Mapping Function 1 (VMF1) and associated gridded map products of a priori ZHD and ZWD. The first-order of ionospheric delay was removed by using dual-frequency observations and the second-order delay was modelled. The observations are weighted according to elevation and the observations with their elevations lower than 10° were discarded. The tropospheric delay data are available online (Yuan and Kutterer, 2021). In this study, we used the 25 GNSS stations included in the area covered by the PSI dataset. Additionally 5 stations are chosen as validation stations for specific analysis. More on this separation of the stations can be read in (Fersch et al., 2021).

4. METHODOLOGY

In this study, we consider different approaches to infer additional information from weather models such as ERA5 to resolve the rank deficiency by (temporal) fixing of absolute ZWDs and including statistics on (spatial) variances of average delays. The workflow is presented in Figure 1.

The data basis of this study are the PSI phase observations. Firstly, the observations are mapped from slant to zenith direction using a simple cosine function. Afterwards, an inversion is performed to obtain the atmospheric phase screens for each scene from the interferometric observations. Traditionally the zero-mean assumption is used as restriction of this underdetermined inversion problem. To introduce this condition means that the same long-term mean value of 0 is assumed for each PS point. This neglects the fact, that regional characteristics such as height differences and specific meteorological factors

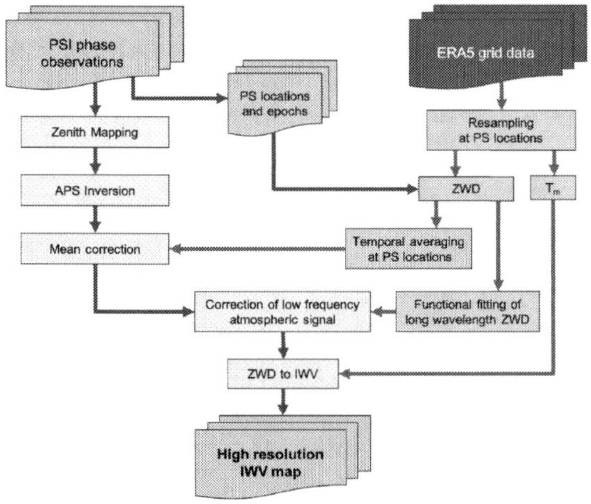

Figure 1. Workflow from PSI phase observations to High resolution IWV maps

influence the true mean value. We therefore performed studies on the ERA5 data to determine the validity of the zero mean assumption. Those tests are performed on the ZWD values obtained by conversion of the phase values to delays in mm. The main influence on the long-term mean is due to the different heights. But even after subtracting this effect, systematic deviations from the zero-mean assumption can be seen for our stack as shown in figure 2. Further tests also showed that those deviations may vary for different years or epochs included in the calculations. This leads to the first correction introduced in our study, the mean correction which is applied onto the APS values. Using statistics of the ERA5 model at the specific PS-points and at the specific SAR acquisition times, ERA5 mean values of the Zenith Wet Delay (ZWD) can be determined for each PS and introduced using the following equation:

$$\overrightarrow{zwd}_{mc} = \overrightarrow{zwd}_{raw} + era_m \tag{1}$$

with

$\overrightarrow{zwd}_{mc}$ vector of all mean corrected InSAR ZWDs

$\overrightarrow{zwd}_{raw}$ vector of all raw InSAR ZWDs

era_m mean ERA5 value over whole InSAR time series

Such mean values are more realistic, as they include regional variabilities and systematic height dependent differences.

As the ERA5 model describes the atmospheric delays at long-wavelengths reliably, it is a suitable dataset to prune the long-wavelength components of the InSAR signal leaving only the atmospheric contribution. Different approaches are presented to introduce the information from the ERA5 model to the PSI dataset. Firstly, simple mathematical models can be used to adjust the difference of the long wavelength component between the mean-corrected PSI ZWD and the nominal ERA5 model. Using low-order polynomial functions assures that only long wavelength components are corrected. We use bivariate polynomials of degree 0, 1, 2 and a thin-plate spline approximation. The latter is based on the MATLAB function *tpaps* (The MathWorks Inc., 2022). In the case of the polynomial approx-

https://doi.org/10.5194/isprs-archives-XLVIII-M-1-2023-153-2023 | © Author(s) 2023. CC BY 4.0 License.

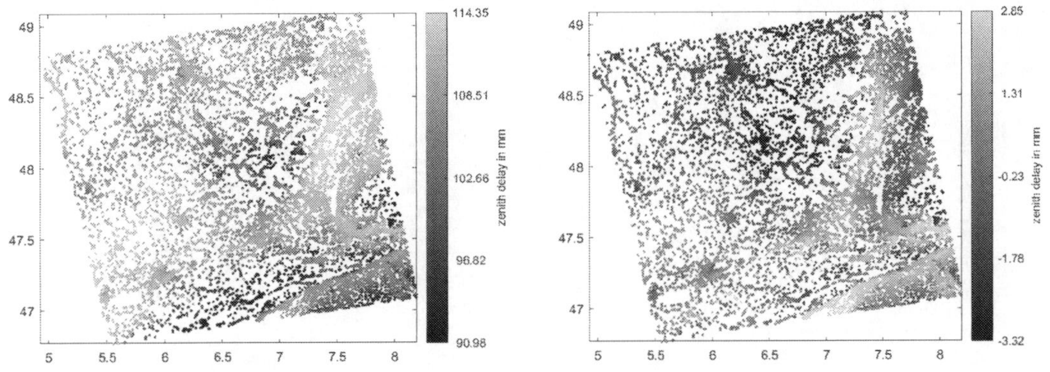

(a) Calculated mean

(b) Remaining signal

Figure 2. Calculated mean for the stack from ERA5 data and the remaining difference after subtracting the height dependent signal.

imations, the functional description is given with the following equations:

$$f_0(\phi,\lambda) = c \tag{2}$$

$$f_1(\phi,\lambda) = a \cdot \phi + b \cdot \lambda + c \tag{3}$$

$$f_2(\phi,\lambda) = a_1 \cdot \phi^2 + b_1 \cdot \lambda^2 + a_2 \cdot \phi + b_2 \cdot \lambda + c \cdot \phi \cdot \lambda + d \tag{4}$$

with

ϕ, λ geographic coordinates

$a, b, c.d$ parameters

As we have over 4 million PS points we use a bootstrapping approach to calculate the ideal parameter estimation for the polynomial approximations. In a loop every 100th point is used, changing the starting index in each run. The mean values of all those 100 parameter sets is then used as final estimation set. Taking into account, that the spline estimation takes for one loop more time then the 100 loops for the other functions, we only use 1 estimation for this approach. Every 1000th point is used for the thin-plate spline interpolation.

An example of the different correction functions and their results using the dataset from 2017/11/24 17:24 UTC is shown in Figure 3. In subfigures c, e, g the correction functions are displayed illustrating the amount of detail captured by the functions.

Another approach is via the spectral description of the long wavelength signal components. After resampling on an equidistant grid of 2 km resolution, the datasets are Fourier transformed and splitted at a certain cutoff wavelength, which is determined using the grid resolution of the ERA5 dataset. While the short wavelength component of the observed signal is maintained, the components with longer wavelengths of the PSI ZWD signal are replaced by the ERA5 model. As the weather model data is originally on a coarse grid of 0.25° resolution, only low frequency components can be used. At the same time, the high spatial variability from the PSI ZWD is preserved. The mask used for filtering the components of the ERA5 data is a gaussian filtering mask. The high frequency parts of the SAR ZWD are then determined by the inversed mask. The result of the filtering can be visualised in power spectra of the datasets. The radially averaged power spectra of the PSI, ERA5 and final ZWD are shown in Figure 5. The final dataset first follows the curve of the ERA5 and then smoothly changes to follow the SAR dataset for shorter wavelengths.

In Figure 4 the different parts are visualized using again the dataset of 2016/02/03 17:24 UTC: (a) shows the low frequency part of the ERA5 data set, (b) the low frequency part of the SAR data which is removed by the filter, (c) displays the high frequency part of the SAR data, which is used for the final product and (d) is the final product using (a) and (c).

The applied corrections influence the absolute values in the regional context and in the long-wavelength range, but do not lower the advantage of the spatially high-resolution InSAR-data and thus, allows the analysis of local variations with absolute ZWD. The proposed upgrading of the zero-mean assumption does not require any commercial data, as both the Sentinel-1 data and weather model data from ERA5 are freely available online.

The final product are high resolution IWV maps, which are generated through transformation of the ZWD using mean temperatures from ERA5 data. For details of this transformation, see (Fersch et al., 2021).

The validation of the different methods is performed on the final IWV products using the GNSS data. For the polynomial approaches the 10 nearest neighbors are chosen for each GNSS station. All values from the PS points are then corrected for the height differences between itself and the GNSS station. For the spectral approach the two nearest grid points are chosen and height corrected. The mean value of the neighboring PS or grid points is then used for time series analyses. Those are computed using the Kling-Gupta-efficiency metric KGE and its constituents: correlation r, the relative variability α (DMEAN) and the bias ratio β (DSTD) (Gupta et al., 2009) as described in equation 5.

$$KGE = 1 - \sqrt{(r-1)^2 + (\alpha-1)^2 + (\beta-1)^2} \tag{5}$$

https://doi.org/10.5194/isprs-archives-XLVIII-M-1-2023-153-2023

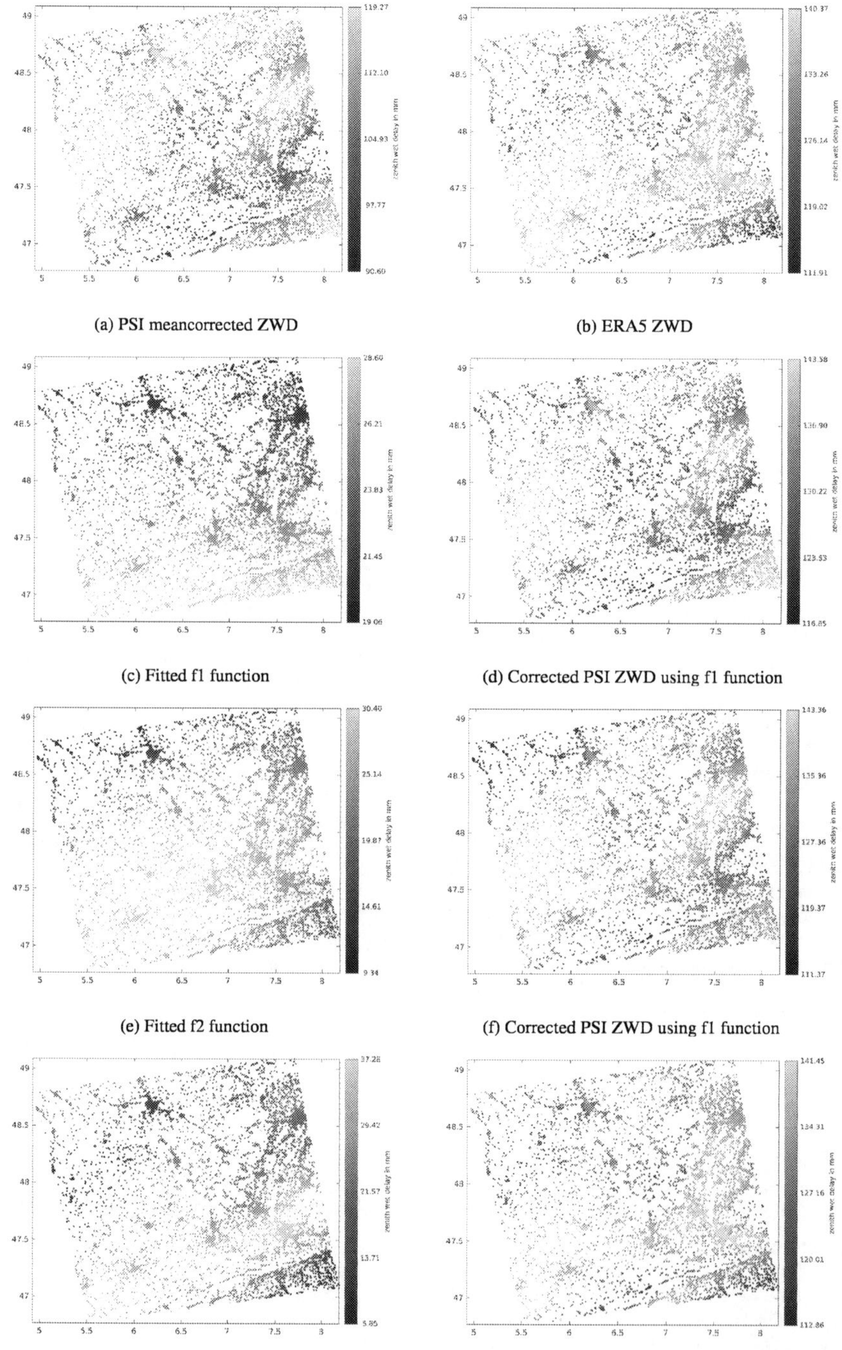

(a) PSI meancorrected ZWD

(b) ERA5 ZWD

(c) Fitted f1 function

(d) Corrected PSI ZWD using f1 function

(e) Fitted f2 function

(f) Corrected PSI ZWD using f1 function

(g) Fitted thin-plate spine function

(h) Corrected PSI ZWD using spline function

Figure 3. The example shows the dataset of 2017/11/24 17:24 UTC. The top row displays the meancorrected PSI ZWD and the ERA5 ZWD. c,e,g show different correction function calculated next to the resulting corrected PSI ZWDs.

This contribution has been peer-reviewed.
https://doi.org/10.5194/isprs-archives-XLVIII-M-1-2023-153-2023 | © Author(s) 2023. CC BY 4.0 License.

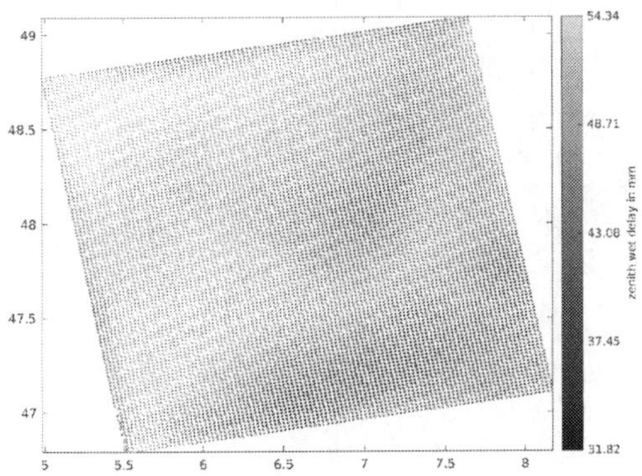

(a) ERA ZWD low frequency part

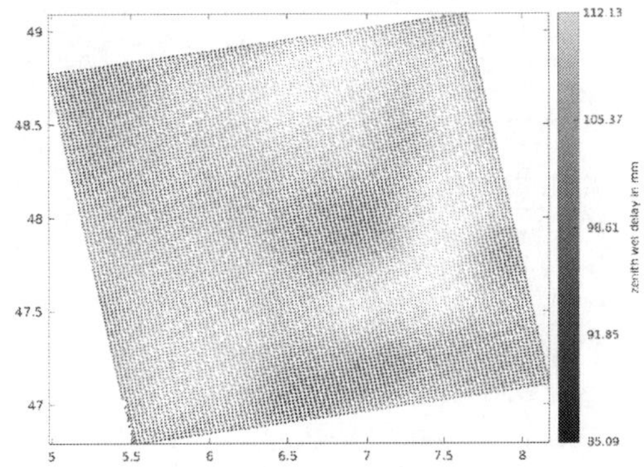

(b) PSI ZWD low frequency part

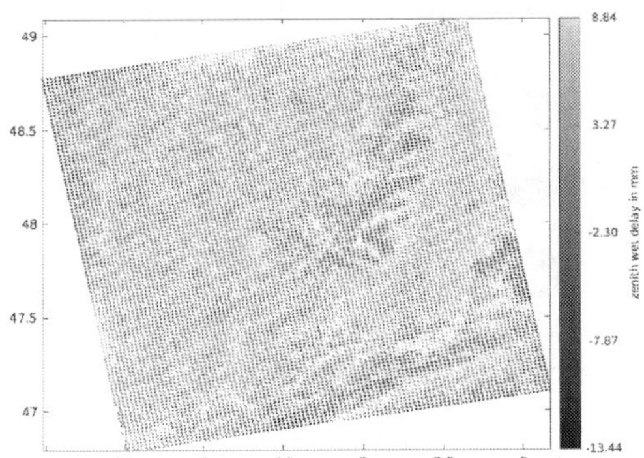

(c) PSI ZWD high frequency part

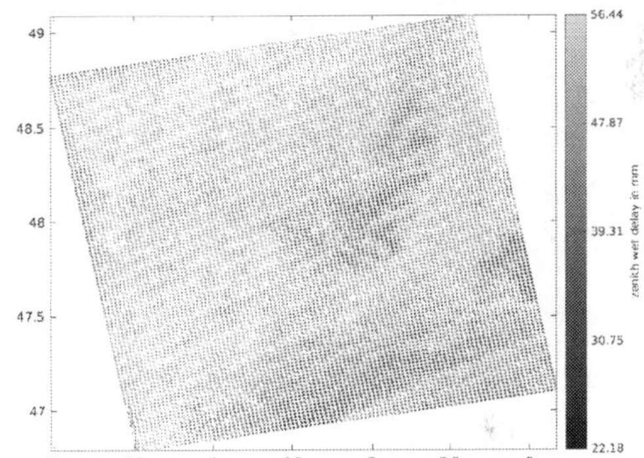

(d) Final ZWD after spectral correction

Figure 4. Different frequency parts of the dataset of 2016/02/03 17:24 UTC are shown: (a) and (b) show the low frequency part of the ERA5 and SAR data, (c) displays the high frequency part of the PSI data and (d) is the final product using (a) and (c).

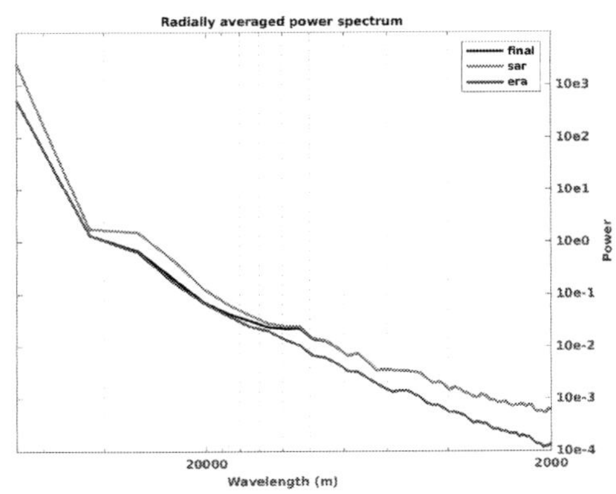

Figure 5. Radially averaged power spectra of meancorrected PSI, ERA5 and final ZWD.

5. RESULTS

We show the results of the upgrading of the ZWD estimation and discuss the plausibility and suitability of the different approaches. The validation of the resulting absolute IWV is performed using GNSS data, including local characteristics at the GNSS sites, which might be contained in the InSAR dataset, but not in the ERA5 model. A first impression of the fitness of the approaches is given by Figure 6b for the GNSS station MNBL (Montbéliard) and the spline correction. Each datapoint represents one epoch included in the SAR and GNSS data. Directly visible are the seasonal dependencies of the IWV amount, which is significantly higher and more variable in summer. That can be explained due to the higher temperatures and accordingly higher water vapour content in the atmosphere as well as the higher turbulence.

The overall evaluation of the different methods is performed on all 25 GNSS stations as shown in Figure 6a. Although more complex functions result in better results, the overall improvement is slight. The best performance is achieved with the thin-plate spline interpolation, followed by the spectral correction in the frequency domain. Notable is the improvement of the

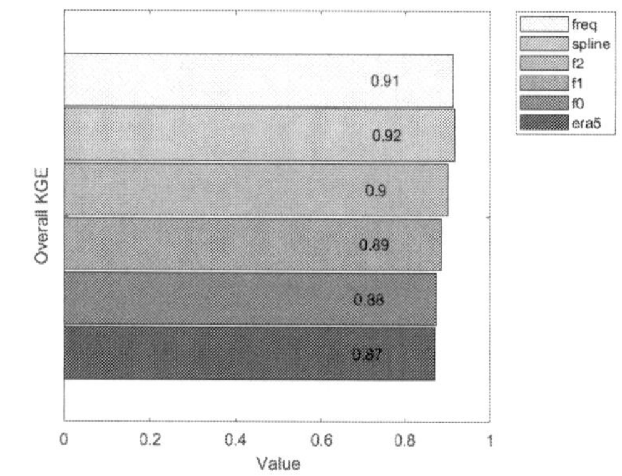

(a) KGE of all methods presented for all GNSS stations and including all epochs.

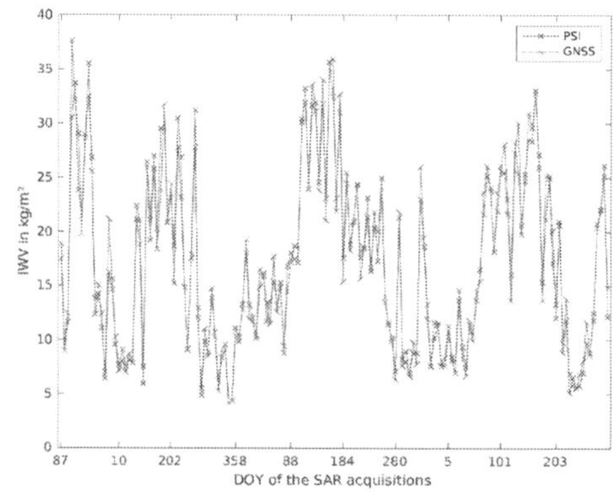

(b) Time series at station MNBL for the spline corrected data.

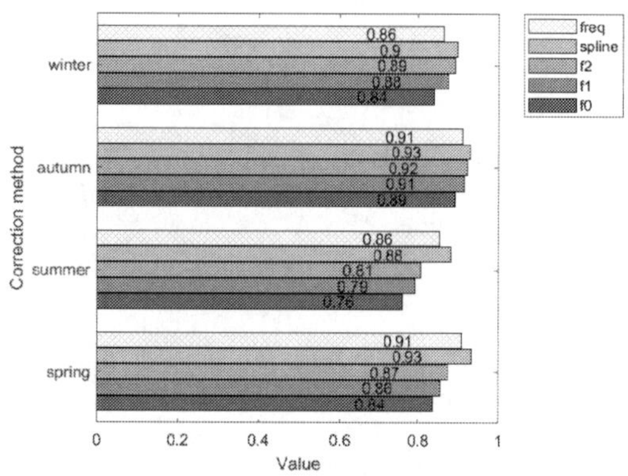

(c) KGE at the four seasons separately for all presented methods.

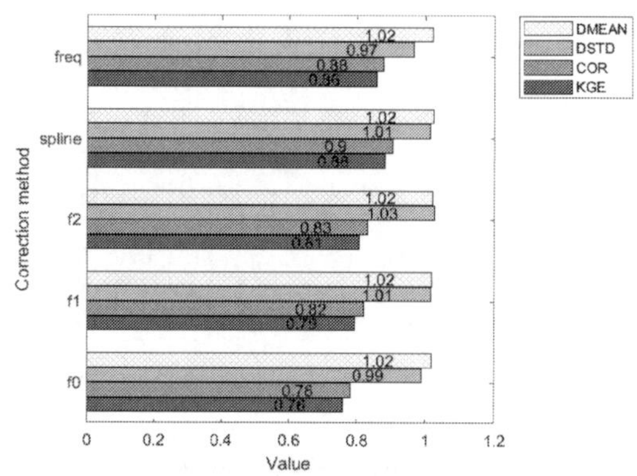

(d) KGE and its constituents for all methods in summer.

Figure 6. Validation results for the presented methods.

validation results from using only ERA5 data to the fused data sets.

A more detailed look at different seasons of the year shows Figure 6c. Clearly the best performance for all approaches is achieved in autumn, which is known for stable weather conditions. In contrast to that, the summer is the most difficult to estimate. The biggest differences between the methods are consequently in summer, followed by spring.

To further go into detail on the reasons of the different performances, we take a look at the different components of the KGE analysis in summer for all the methods. Figure 6d displays the components of KGE. Clearly the results are very similar for the relative bias and variability (DMEAN and DSTD), the biggest differences are visible in the correlation and causing the differences in the KGE.

6. CONCLUSIONS

We presented a workflow to insert mean-correction of the APS and multiple approaches to infer the long wavelength component from ERA5 into the PSI ZWD products. This enables the production of high resolution IWV maps with absolute values derived from PS In-SAR observations. All methods show a significant improvement from the simple zero-mean assumption and should be selected according to the respective requirements. Although the best results were achieved using the spline interpolation, this method comes with high computational costs and needs to be adjusted to avoid overfitting. Physically straight forward is the spectral filtering approach in the frequency domain, which on the other hand leads to a reduced horizontal resolution of the product. For simple applications it might be sufficient to have a polynomial approach, which can be run significantly faster than the spline estimation.

REFERENCES

Alshawaf, F., Fersch, B., Hinz, S., Kunstmann, H., Mayer, M., Meyer, F. J., 2015. Water vapor mapping by fusing InSAR and GNSS remote sensing data and atmospheric simulations. *Hydrol. Earth Syst. Sci.*, 19(12), 4747–4764.

Fersch, B., Kamm, B., Shehaj, E., Wagner, A., Yuan, P., Möller, G., Schenk, A., Geiger, A., Hinz, S., Kutterer, H., Kunstmann, H., 2021. A comprehensive high resolution data collection for tropospheric water vapor assessment for the Upper Rhine Graben, Germany.

Gupta, H., Kling, H., Yilmaz, K., Martinez, G., 2009. Decomposition of the Mean Squared Error and NSE Performance Criteria: Implications for Improving Hydrological Modelling. *Journal of Hydrology*, 377, 80-91.

Hanssen, R. F., 2001. Radar interferometry - Data Interpretation and Error Analysis. Remote Sensing and Digital Image Processing, 2, Springer Netherlands, Dordrecht.

Hersbach, H., Bell, B., Berrisford, P., Hirahara, S., Horányi, A., Muñoz-Sabater, J., Nicolas, J., Peubey, C., Radu, R., Schepers, D. et al., 2020. The ERA5 global reanalysis. *Quarterly Journal of the Royal Meteorological Society*, 146(730), 1999–2049.

Heublein, M., Zhu, X. X., Alshawaf, F., Mayer, M., Bamler, R., Hinz, S., 2015. Compressive sensing for neutrospheric water vapor tomography using GNSS and InSAR observations. *2015 IEEE International Geoscience and Remote Sensing Symposium (IGARSS)*, IEEE, 5268–5271.

Mateus, P., Catalao, J., Nico, G., 2017. Sentinel-1 Interferometric SAR Mapping of Precipitable Water Vapor Over a Country-Spanning Area. *IEEE Transactions on Geoscience and Remote Sensing*, 55(5), 2993–2999. http://ieeexplore.ieee.org/document/7855708/.

The MathWorks Inc., 2022. Curve fitting toolbox. https://www.mathworks.com/help/curvefit/index.html.

Yuan, P., Blewitt, G., Kreemer, C., Hammond, W. C., Argus, D., Yin, X., Van Malderen, R., Mayer, M., Jiang, W., Awange, J. et al., 2023. An enhanced integrated water vapour dataset from more than 10 000 global ground-based GPS stations in 2020. *Earth System Science Data*, 15(2), 723–743.

Yuan, P., Kutterer, H., 2021. Point-scale iwv and zenith total delay (ztd) derived for 66 stations of the global navigation satellite system (gnss) upper rhine graben network (gurn).

This contribution has been peer-reviewed.
https://doi.org/10.5194/isprs-archives-XLVIII-M-1-2023-153-2023

FOREST FIRE SUSCEPTIBILITY ASSESSMENT WITH MACHINE LEARNING METHODS IN NORTH-EAST TURKIYE

O. Kantarcioglu [1,2], K. Schindler [3], S. Kocaman [2*]

[1] Hacettepe University, Graduate School of Science and Engineering, Ankara, Türkiye – omerkantarcioglu@hacettepe.edu.tr
[2] Dept. of Geomatics Engineering, Hacettepe University, 06800 Beytepe Ankara, Türkiye - sultankocaman@hacettepe.edu.tr
[3] ETH Zurich, Photogrammetry and Remote Sensing, 8092 Zurich, Switzerland – schindler@ethz.ch

KEY WORDS: Forest fire susceptibility, forest inventory, random forest, artificial neural network, spatial probability distribution

ABSTRACT:

Forest fires have devastating effects on biodiversity, climate, and humans. Producing detailed and reliable forest fire susceptibility maps is crucial for disaster management. Data-driven machine learning methods can be applied for forest fire susceptibility mapping, and learning data required for this purpose can be obtained from high-resolution satellite imagery along with a fire inventory. In this study, we assessed the performances of Random Forest (RF) and artificial neural network (ANN) classifiers for producing forest fire susceptibility maps of a region in north-east Türkiye covering Trabzon, Gümüşhane, Rize, and Bayburt provinces using freely available Earth observation data and forest inventory provided by the regional directorate. Forest type, EU-DEM v1.1 (25 m), and tree cover density were retrieved from Copernicus Land Monitoring Service. Sentinel-2 images were utilized for calculating spectral indices such as normalized difference vegetation index and modified normalized difference water index to assess surface water and vegetation characteristics. Thus, a total of twelve variables including topographic, anthropogenic, hydrologic, vegetation and land use data were used as input. The RF and ANN illustrated similar prediction performances based on receiver operating characteristics (ROC) area under the curve (AUC) values, which were 0.89 and 0.88, respectively. The RF performed better in terms of overall accuracy and F-1 score. The susceptibility maps with 25 m resolution were also investigated visually. The ANN results predicted higher susceptibility levels and larger areas were found prone to wildfire. Leave-one-out analysis results indicated that elevation was the most influential factor based on the achieved OA.

1. INTRODUCTION

Forest fires have devastating effects on a variety of environmental, social, and economic resources (Gill et al. 2013) Additionally, forest fires cause air pollution and diminish its quality, destroy forest ecosystems, and lead to global warming (Leuenberger et al. 2018). Therefore, the development of an efficient disaster management system to decrease the severity of wildfires and support mitigation efforts is essential (Adab et al. 2013). Although a variety of methods has been utilized in the literature to assess forest fire susceptibility, and the number of studies has increased, further investigations are still needed since the model accuracy depends on the method applied and data availability/quality. More than 30 methods were used for this purpose between 2001 and 2021 (Chicas and Østergaard Nielsen 2022). The most frequently used data-driven machine learning (ML) models have been the random forest (RF) model (Cao et al. 2017) proposed by Breiman (2001), logistic regression (LR) (Rodrigues and de la Riva 2014), support vector machines (SVM) (Sachdeva et al. 2018), and artificial neural networks (ANN) (Zhang et al. 2021; Kantarcioglu et al., 2023). Comprehensive reviews on wildfire susceptibility and hazard mapping can be found in the literature (Jain et al., 2020; Chicas & Østergaard Nielsen, 2022).

The number and severity of forest fires have increased drastically in Turkiye over the last few years. Akinci and Akinci (2023) assessed the forest fire susceptibility for Manavgat region of Antalya province, Turkiye using the ignition inventory of local forest authority and the XGBoost algorithm, based on boosting decision trees. Another recent work by Kantarcioglu et al. (2023) employed an ANN to produce susceptibility map of Thrace region, Turkiye and compared with the RF. An extensive forest database with tree types, tree ages, forest roads, and ignition locations of 1506 fires over a period of eight years was utilized in the study, and the forest roads was found the most predictive variable.

In this study, two widely used ML classifiers, RF and ANN, were again implemented for producing forest fire susceptibility maps for the north-east side of Türkiye covering Trabzon, Gümüşhane, Rize, and Bayburt provinces. The ignition locations were obtained from Trabzon Regional Directorate of Forestry (TRFD). The inventory includes 368 fire events recorded between the years 2013 and 2022. Thus, this study aims to identify areas prone to forest fires (ignition) by predicting their spatial probability using various input variables in combination with RF and ANN. Based on the literature survey, analysis of regional characteristics, and data availability, a total of 12 features were identified, including elevation, aspect, slope, tree cover density, forest type, settlement proximity, settlement density, water proximity, power line proximity, normalized vegetation difference index (NDVI), modified normalized water difference index (MNDWI), and land use land cover (LULC). The susceptibility maps obtained from both models were inspected visually and validated by using receiver operating characteristics (ROC) area under the curve (AUC) values, overall accuracy (OA), and F-1 score. In addition, feature importance analysis based on leave-one-out approach is presented and discussed.

* Corresponding author

This contribution has been peer-reviewed.
https://doi.org/10.5194/isprs-archives-XLVIII-M-1-2023-161-2023 | © Author(s) 2023. CC BY 4.0 License.

2. MATERIALS AND METHODS

2.1 Study Area and Fire Characteristics

The study site is located in the eastern part of Black Sea region, Turkiye, and covers an area of 18,877 km^2, Figure 1. The area is rather mountainous, and the northern parts are located at higher altitudes as can be seen from the EU-DEM v1.1 (2023) data, Figure 2. Trabzon and Rize are more vegetated compared to Gümüşhane and Bayburt, according to European Space Agency (ESA) WorldCover (2023) data, Figure 3. The Pontic mountains located in the north of the region differentiate the climate and vegetation characteristics of the provinces.

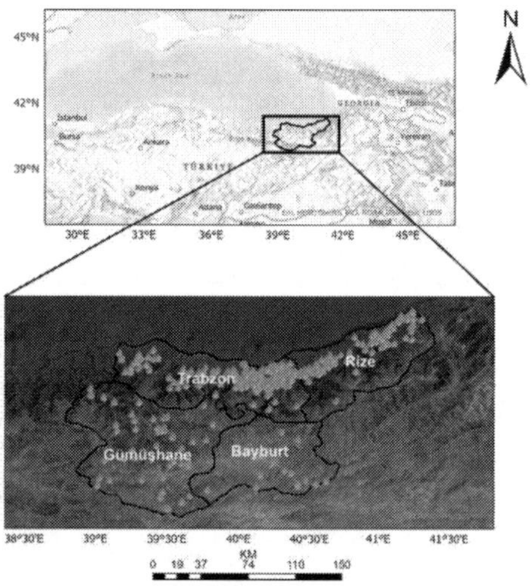

Figure 1. Location map of the study area, and ground truth fire ignition points.

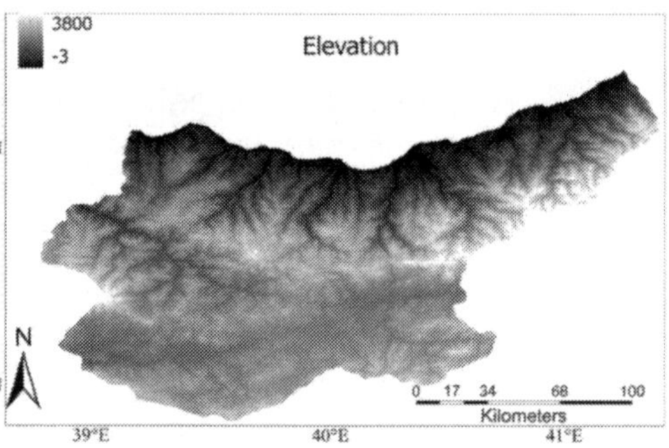

Figure 2. Altitudes within the study area, from EU-DEM v1.1 (2023).

The fire inventory used here consists of locations and dates of ignition events that occurred between 2013 and 2022 (see Figure 1). Further attributes involve risk level description conducted by the emergency response teams and the size of the burned areas, which can be used for hazard assessments in the future. A total of 368 fires served as the basis for forest fire susceptibility mapping. Half of these fires resulted in burnt areas < 1 hectare. The largest burnt area is approximately 130 hectares.

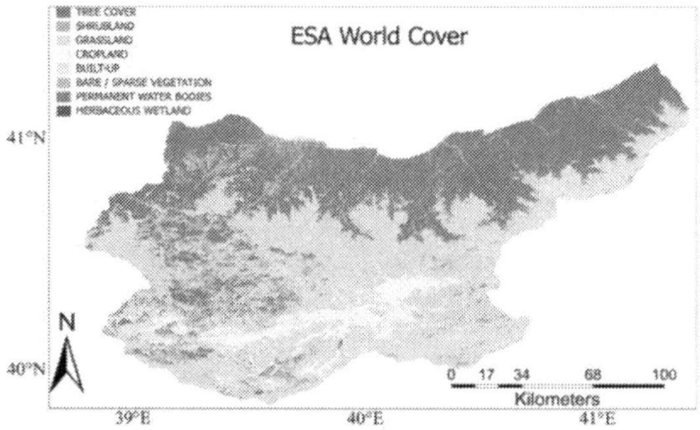

Figure 3. LULC map of the study area, from ESA WorldCover data (ESA WorldCover, 2023).

2.2 Methodology

The overall workflow of the study is presented in Figure 4. It consists of the preparation of the input features, selection of training and test samples, application and validation of the ML models (RF or ANN), and the production of susceptibility maps by applying the learned models. Further details are provided below.

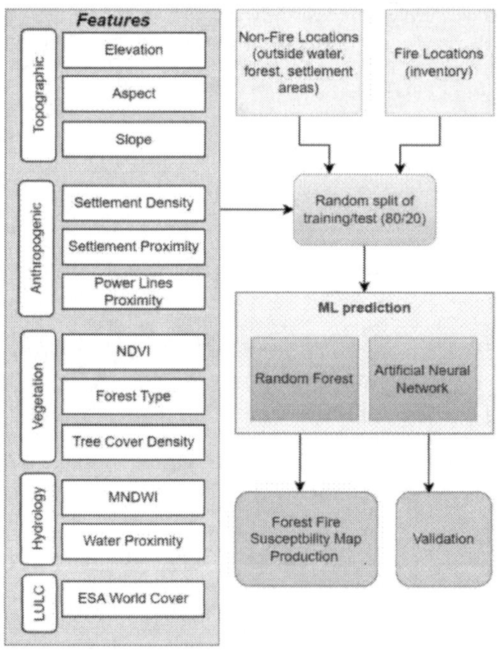

Figure 4. The workflow of the study.

2.2.1 Input Datasets and Pre-processing: A total of six images obtained from Sentinel-2 satellites of ESA were mosaicked and clipped for the site to produce maps of NDVI and MNDWI. All Sentinel-2 images were acquired on the 29th September 2020. As digital elevation model (DEM), EU-DEM v1.1 (2023) with 25 m resolution was employed, additionally slope and the aspect maps were produced using this dataset. Tree cover density and forest type data were obtained from the Copernicus Land Monitoring Service (CLMS, 2023). In

https://doi.org/10.5194/isprs-archives-XLVIII-M-1-2023-161-2023

addition, power line data were extracted from World Bank global electricity transmission and distribution lines (World Bank, 2023). Buildings in the study area were extracted from OpenStreetMap (OSM) Project repository (OSM, 2023). Likewise, water surfaces were extracted from the EU-Hydro database provided by EU Copernicus Programme (EU-Hydro, 2023). A brief summary of the datasets is given in Table 1.

Input Data	Source	Format	SR	Date
DEM	EU EEA[1]	Raster	25 m	2011
Forest Type	EU EEA[1]	Raster	10 m	2018
Tree Cover Density	EU EEA[1]	Raster	10 m	2018
Settlements	OSM	Vector	-	2022
LULC	ESA World Cover[1]	Raster	10 m	2021
Power Lines	Worldbank[2]	Geopackage	-	
Water	EU Hydro[1]			
Sentinel-2 imagery	ESA[1]	Raster	10 m	2020
Fire Inventory	TRDF[3]	Vector	-	2013-2022
Buildings	OSM	Vector	-	2022

[1] via EU Copernicus Service (CLMS, 2023); [2] (Arderne et al., 2020); [3]Trabzon Regional Directorate of Forestry (2022)

Table 1. Input data characteristics.

The data pre-processing was carried out using ArcGIS Pro software by Environmental Systems Research Institute (ESRI). The datasets were clipped according to the study area boundaries. The vector datasets (settlements, buildings, power lines, river network) were rasterized w.r.t. the grid interval of EU-DEM. Density maps were estimated using the kernel density method. Proximity features were derived according to the Euclidean distance. MNDWI and NDVI features were computed from the Sentinel-2 imagery to assess the vegetation and water areas in the study area using Equations 1 and 2.

$$NDVI = (NIR - Red) / (NIR + Red), \quad (1)$$

$$MNDWI = (Green - SWIR) / (Green + SWIR). \quad (2)$$

The negative (non-fire) reference samples were randomly selected. Their number was set 50% greater than fire samples for a ratio of 1:1.5, to avoid any serious class imbalance. The samples were randomly split as training (80%) and validation (20%) sets (Figure 5). Existing Python libraries were utilized for the implementation of the RF and ANN, with manual hyperparameter optimization as explained next.

2.2.2 ANN: An ANN consists of layers of neurons that each implement a linear combination followed by a non-linear activation function (Negnevitsky, 2002). The hyperparameters of the ANN model trained here were set manually, based on the susceptibility map and the AUC value of the result. Keras and Tensorflow (Abadi et al. 2016) libraries were used to implement the ANN method. In ANN, two hidden layers were used. These hidden layers were activated with the "relu" function. The output of these hidden layers was activated with the "sigmoid" function to calculate the predictions. Lastly, for compiling the ANN, the "adam" optimizer parameter was used.

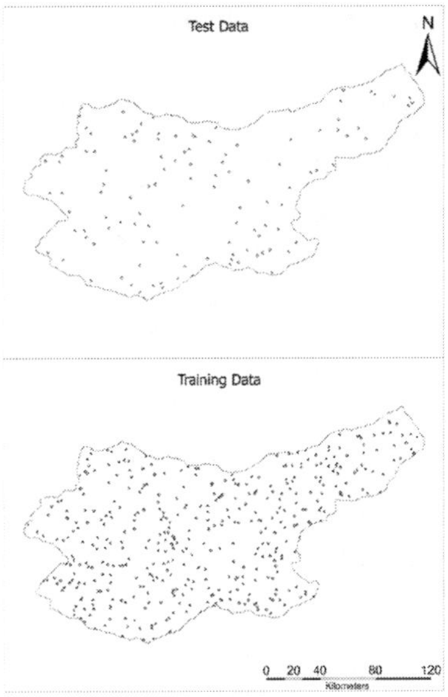

Figure 5. The distribution of training and test samples.

2.2.3 RF: The RF (Breiman 2001) is a supervised classification method consisting of an ensemble of decision trees. Each tree classifies the input according to a sequence of thresholds on the input features, learned from the training data. The RF was implemented using Sci-kit Python libraries (Pedregosa et al. 2011). Parameter selection has been performed via manual testing based on the accuracy results. In RF, the maximum depth of the tree is set as sixteen, the minimum number of samples required to be at a leaf node is four and the minimum number of samples at the leaves is twelve.

2.2.4 Validation: The test data was used for performance assessment via ROC curve, which contrasts recall or sensitivity (Eq. 3) and specificity (Eq. 4). F-1 score (Eq. 5) and OA value (Eq. 6) were also computed. Visual inspection of the predicted maps was carried out by the authors as a further assessment. The importance of different features was analyzed using leave-one-out approach, which is based on removal of one feature from the full set at the learning stage. As a further analysis, the importance of the features was also calculated using

$$Recall = \frac{TP}{TP + FN} \quad (3)$$

$$Specificity = \frac{TN}{TN + FP} \quad (4)$$

$$F1 = \frac{2xTP}{2xTP + FP + FN} \quad (5)$$

$$OA = \frac{TP + TN}{TP + FP + TN + FN} \quad (6)$$

where TP is true positive, FN is false negative, TN is true negative, and FP is false positive.

This contribution has been peer-reviewed.
https://doi.org/10.5194/isprs-archives-XLVIII-M-1-2023-161-2023 | © Author(s) 2023. CC BY 4.0 License.

3. RESULTS AND DISCUSSIONS

In the following, the input feature maps obtained from the pre-processing, the susceptibility maps produced with the RF and the ANN, the validation results, and the feature importance analysis are presented and discussed.

3.1 Input Features: The input features obtained from data pre-processing were categorized as topographic, anthropogenic, LULC, vegetation and hydrological (Table 2). Feature maps are shown in Figure A1 in the Appendix.

Topographic	Anthropogenic	LULC	Vegetation	Hydrology
Elevation	Settlement Density	ESA Word Cover	NDVI	Water Proximity
Aspect	Settlement Proximity		Forest Type	MNDWI
Slope	Power Lines Proximity		Tree Cover Density	

Table 2. Input features implemented in the study.

3.2 Forest Fire Susceptibility Maps: The maps obtained from both the ANN and the RF methods are shown in Figure 6. The RF predicted higher susceptibility values (up to 0.96). Both maps show similar spatial distributions overall. However, there are local differences in different land cover types (water, forest, roads, settlements), as illustrated in Figure 7. The map details shown in the figure also indicate that the ANN predicts larger areas as fire-prone.

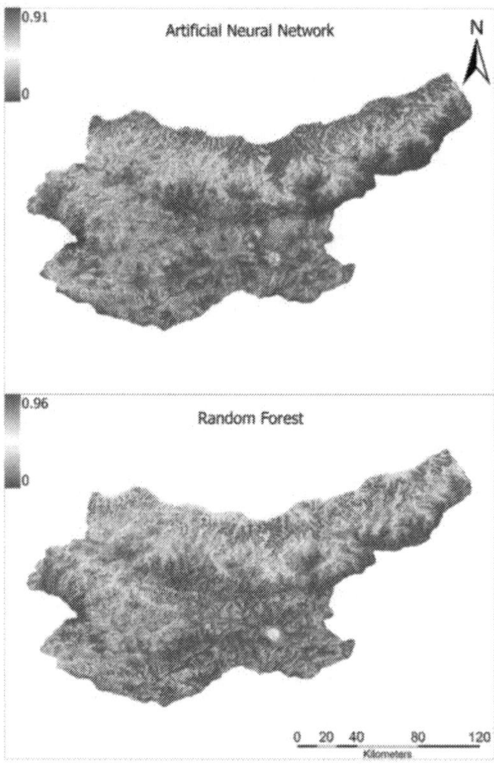

Figure 6. Forest fire susceptibility maps obtained from the ANN (above) and the RF (below).

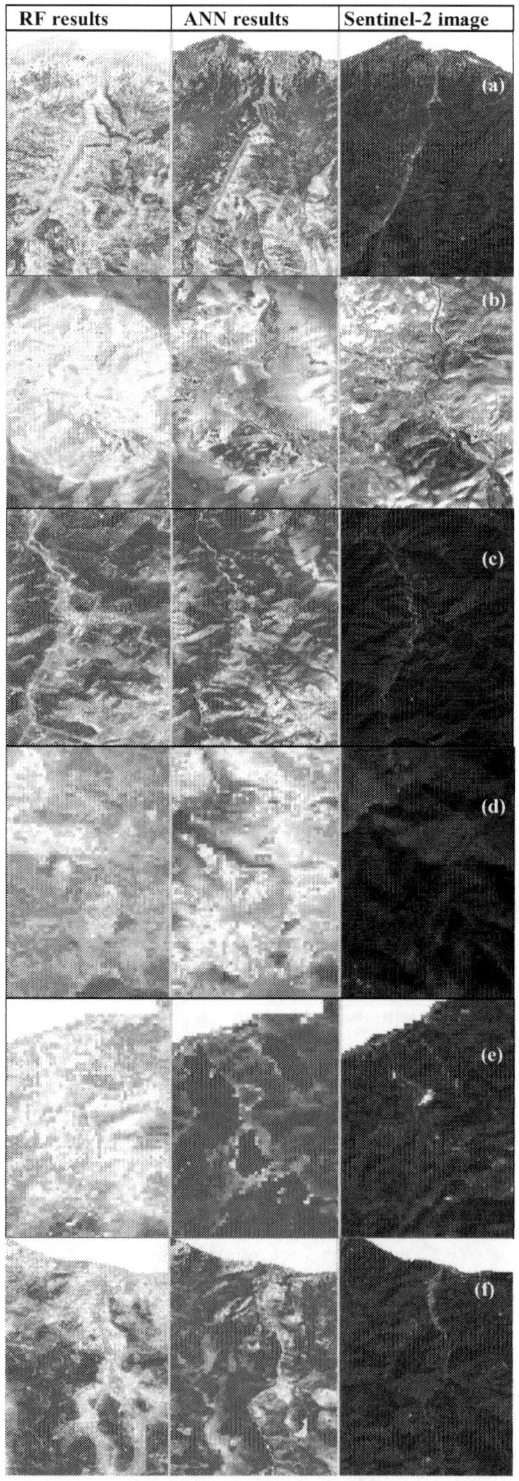

Figure 7. Local details of the maps shown in Figure 6. (a) Trabzon city (b) Bayburt city; (c) Hemşin, Rize; (d) Kızılot, Gümüşhane; (e) Isıkli Village, Ardeşen; (f) Sürmene, Trabzon.

3.3 Model Validation Results: The AUC values obtained from RF and ANN were 0.89 and 0.88, respectively, Figure 8. The corresponding OA were 0.86 and 0.80, and the F1 scores were 0.81 and 0.75. The RF exhibited better prediction performance based on F-1 score and the OA, which indicates a better fit of the model to the learning data. On the other hand, the ANN found larger areas as fire-prone (see above), which needs to be investigated in detail.

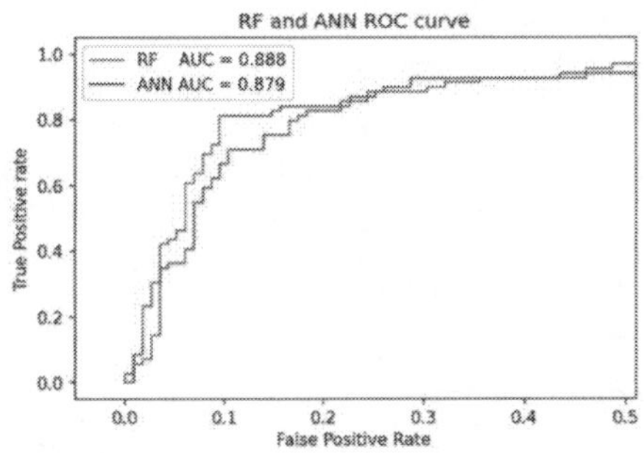

Figure 8. ROC curves obtained for the RF and for the ANN.

3.4 Feature Importance: Based on the feature importance analysis with leave-one-out approach, elevation was found significant for both models (Figure 9, Tables A1 and A2 in Appendix). However, ANN found forest type as the most important based on the F-1 score.

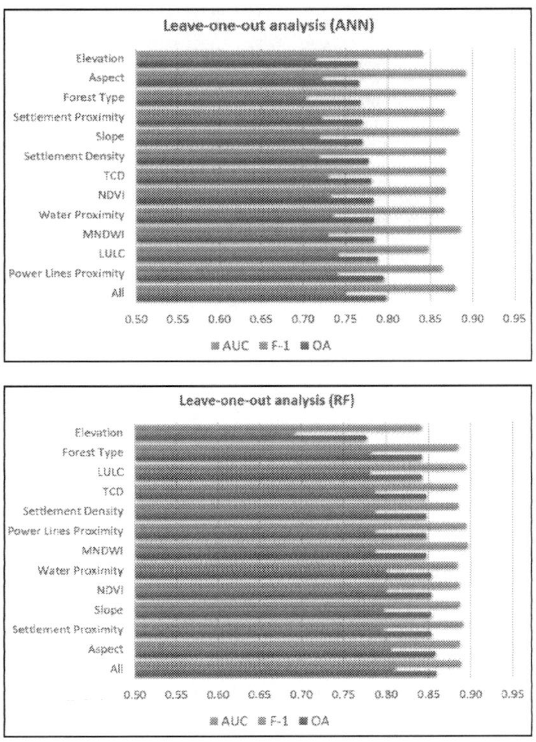

Figure 9. Leave-one-out analysis results ordered by OA values.

When the results presented here are compared with the study of Kantarcioglu et al. (2023) carried out in Thrace Region, Türkiye, the accuracy was better in the latter one (AUC of ANN = 0.94 and RF = 0.95) possibly due to the use of extensive local database including forest roads, and the use of a larger sample size (1506). In addition, although the coverages of both sites are comparable, the environmental and anthropogenic characteristics of both regions are also different.

4. CONCLUSIONS AND FUTURE WORK

In this study, forest fire susceptibility was asessed over a region comprising four provinces in north-eastern Turkiye, using two different ML methods, namely RF and the ANN. The forest fire inventory obtained from the regional forest authority was used to identify reference ignition locations (368). The features such as elevation, land-cover, hydrology, forest, power lines were obtained from openly available datasets. The tested methods achieved comparable performances, with AUC values of 0.89 and 0.88, respectively. Yet, the ANN yielded larger areas as fire prone.

Based on leave-one-out analysis, the elevation was found to be the most influential factor. Forest type, aspect and LULC were also found highly significant, which demonstrates that vegetation characteristics, topography and anthropogenic factors need to be considered for producing accurate forest fire susceptibility maps. Yet, the results presented here are preliminary and require further investigations especially w.r.t. the different LULC types.

ACKNOWLEDGEMENTS

The authors are grateful to Trabzon Regional Directorate of Forestry, Forest Fire Fighting Division, for providing the fire inventory. This study is part of the PhD thesis research of Omer Kantarcioglu.

REFERENCES

Abadi, M., Barham, P., Chen, J.M., Chen, Z.F., Davis, A., Dean, J., Devin, M., Ghemawat, S., Irving, G., Isard, M., Kudlur, M., Levenberg, J., Monga, R., Moore, S., Murray, D.G., Steiner, B., Tucker, P., Vasudevan, V., Warden, P., Wicke, M., Yu, Y., Zheng, X.Q., Assoc, U., 2016. TensorFlow: A system for large-scale machine learning. In, *12th USENIX Symposium on Operating Systems Design and Implementation (OSDI)* (pp. 265-283). Savannah, GA

Adab, H., Kanniah, K.D., Solaimani, K., 2013. Modeling forest fire risk in the northeast of Iran using remote sensing and GIS techniques. *Natural Hazards, 65*, 1723-1743

Akinci, H.A., Akinci, H., 2023. Machine learning based forest fire susceptibility assessment of Manavgat district (Antalya), Turkey. Earth Science Informatics, 16, 397-414

Arderne, C., Zorn, C., Nicolas, C., Koks, E.E., 2020. Predictive mapping of the global power system using open data. *Scientific Data, 7*, 19

Breiman, L., 2001. Random forests. *Machine learning, 45*, 5-32

Cao, Y.X., Wang, M., Liu, K., 2017. Wildfire Susceptibility Assessment in Southern China: A Comparison of Multiple Methods. *International Journal of Disaster Risk Science, 8*, 164-181

https://doi.org/10.5194/isprs-archives-XLVIII-M-1-2023-161-2023

Chicas, S.D., Østergaard Nielsen, J., 2022. Who are the actors and what are the factors that are used in models to map forest fire susceptibility? A systematic review. *Natural Hazards, 114*, 2417-2434

CLMS, 2023. European Union (EU) European Environment Agency (EEA) Copernicus Land Monitoring Service Forests. https://land.copernicus.eu/pan-european/high-resolution-layers/forests (22 February 2023)

EU-DEM v1.1, 2023. https://land.copernicus.eu/imagery-in-situ/eu-dem/eu-dem-v1.1 (22 February 2023).

EU-Hydro, 2023. EU-Hydro - River Network Database. https://land.copernicus.eu/imagery-in-situ/eu-hydro/eu-hydro-river-network-database (22 February 2023).

ESA WorldCover, 2023. ESA WorldCover 10 m 2020 v100. https://worldcover2020.esa.int/ (22 February 2023).

Gill, A.M., Stephens, S.L., Cary, G.J., 2013. The worldwide "wildfire" problem. *Ecological Applications, 23*, 438-454

Jain, P., Coogan, S.C.P., Subramanian, S.G., et al., 2020. A review of machine learning applications in wildfire science and management. *Environmental Reviews*, 28, 478-505, doi:10.1139/er-2020-0019.

Kantarcioglu, O., Kocaman, S., Schindler, K., 2023. Artificial Neural Networks for Assessing Forest Fire Susceptibility in Türkiye. *Ecological Informatics*, 2023. https://doi.org/10.1016/j.ecoinf.2023.102034

Leuenberger, M., Parente, J., Tonini, M., Pereira, M.G., Kanevski, M., 2018. Wildfire susceptibility mapping: Deterministic vs. stochastic approaches. *Environmental Modelling & Software, 101*, 194-203

Negnevitsky, M., 2002. Artificial Intelligence–A Guide to Intelligent Systems, Addison–Wesley Co.: Boston, MA, USA, p. 394.

OSM, 2023. OpenStreetMap Project. https://wiki.openstreetmap.org/wiki/Main_Page (22 Feb 2023)

Pedregosa, F., Varoquaux, G., Gramfort, A., Michel, V., Thirion, B., Grisel, O., Blondel, M., Prettenhofer, P., Weiss, R., Dubourg, V., Vanderplas, J., Passos, A., Cournapeau, D., Brucher, M., Perrot, M., Duchesnay, E., 2011. Scikit-learn: Machine Learning in Python. *Journal of Machine Learning Research, 12*, 2825-2830

Rodrigues, M., de la Riva, J., 2014. An insight into machine-learning algorithms to model human-caused wildfire occurrence. *Environmental Modelling & Software, 57*, 192-201

Sachdeva, S., Bhatia, T., Verma, A.K., 2018. GIS-based evolutionary optimized Gradient Boosted Decision Trees for forest fire susceptibility mapping. *Natural Hazards, 92*, 1399-1418

World Bank, 2023. Derived Map Of Global Electricity Transmission And Distribution Lines. https://datacatalog.worldbank.org/search/dataset/0038055 (22 February 2023)

Zhang, G.L., Wang, M., Liu, K., 2021. Deep neural networks for global wildfire susceptibility modelling. *Ecological Indicators, 127*

APPENDIX

Feature Name	OA	F-1	AUC
All	0.80	0.75	0.88
LULC	0.79	0.74	0.85
Power Lines Proximity	0.80	0.74	0.86
Water Proximity	0.78	0.74	0.87
NDVI	0.78	0.73	0.87
TCD	0.78	0.73	0.87
MNDWI	0.78	0.73	0.89
Settlement Proximity	0.77	0.72	0.87
Aspect	0.77	0.72	0.89
Slope	0.77	0.72	0.88
Settlement Density	0.78	0.72	0.87
Elevation	0.77	0.72	0.84
Forest Type	0.77	0.70	0.88

Table A1. Leave-one-out results of the ANN.

Feature Name	OA	F-1	AUC
All	0.86	0.81	0.89
Aspect	0.86	0.81	0.89
NDVI	0.85	0.80	0.89
Water Proximity	0.85	0.80	0.88
Settlement Proximity	0.85	0.80	0.89
Slope	0.85	0.80	0.89
MNDWI	0.85	0.79	0.90
Power Lines Proximity	0.85	0.79	0.89
Settlement Density	0.85	0.79	0.89
TCD	0.85	0.79	0.88
Forest Type	0.84	0.78	0.89
LULC	0.84	0.78	0.89
Elevation	0.78	0.69	0.84

Table A2. Leave-one-out results of the RF.

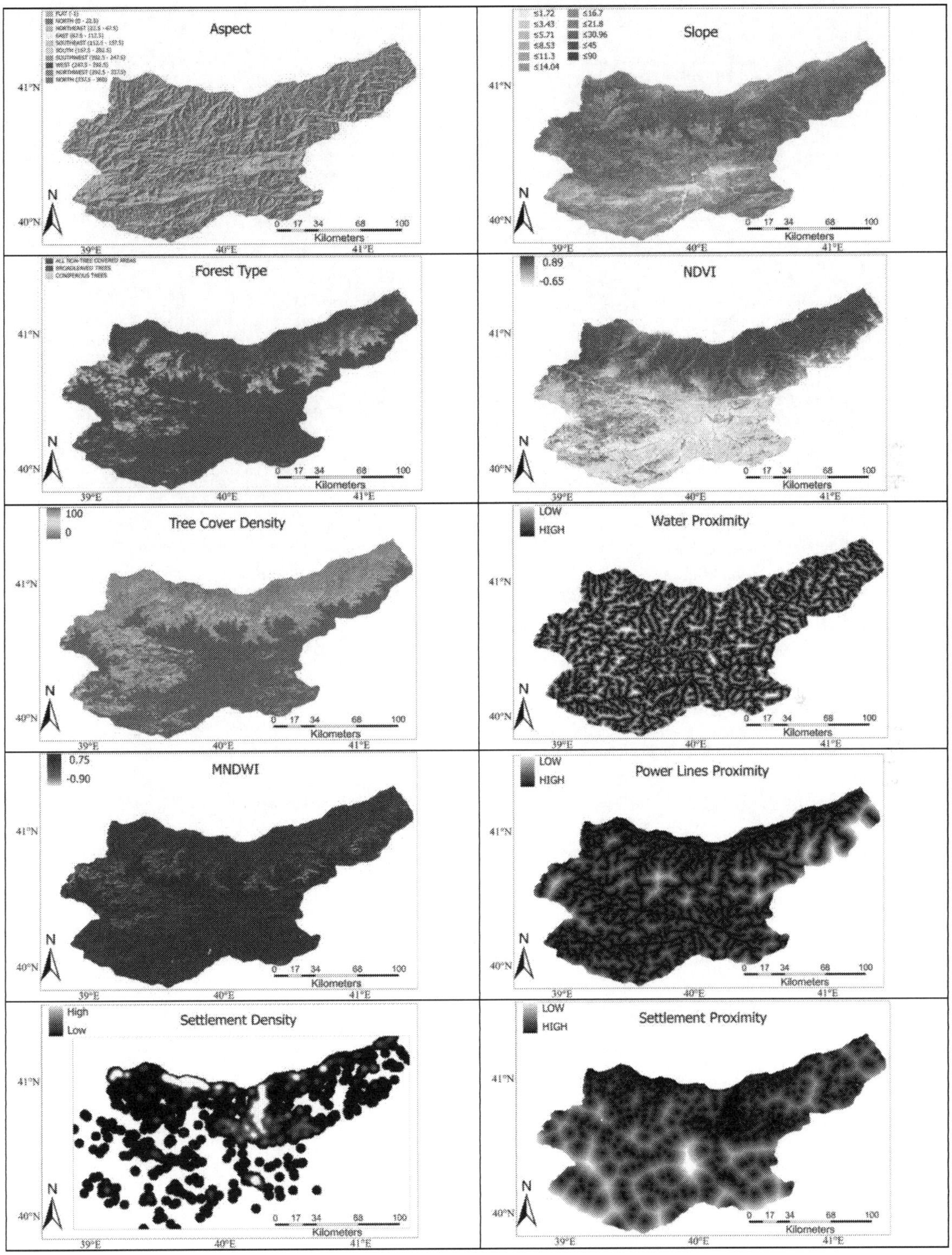

Figure A1. Feature maps used in the modelling.

https://doi.org/10.5194/isprs-archives-XLVIII-M-1-2023-161-2023 | © Author(s) 2023. CC BY 4.0 License.

DERIVING INTERNAL WAVE PHASE SPEED IN THE ARCTIC OCEAN FROM SEQUENTIAL SPACEBORNE SAR OBSERVATIONS

I. E. Kozlov [1]*, T. V. Mikhaylichenko [1]

[1] Marine Hydrophysical Institute of RAS, 299011 Sevastopol, Russia - ik@mhi-ras.ru

KEY WORDS: short-period internal waves, phase speed, wave kinematics, Sentinel-1, SAR imaging, Yermak Plateau, Arctic Ocean.

ABSTRACT:

Here we present direct estimates of phase speeds of short-period internal waves (SIWs) in the Arctic Ocean obtained from the analysis of sequential Sentinel-1 A/B SAR images. Analysis of the data near Svalbard in June-September 2018 has shown a peak in SIW observations in August. Three key regions of SIW generation are identified: deep Fram Strait, southern and central parts of Yermak Plateau and shelf regions northwest of Svalbard. Maximal SIW phase speed values are found over the Yermak Plateau and attain 0.84 m/s. Over Fram Strait and on the Svalbard shelf, phase speed values are similar with mean value about 0.2-0.3±0,03 m/s. We note that obtained SIW phase speeds are higher than maximal tidal currents over all three SIWs' observation sites, suggesting their tidal generation and free propagation at subcritical Froude numbers. Comparison of the satellite-based phase speed estimates with theoretical values obtained using a two-layer model and actual hydrology shows a good correspondence for the cases when the time lag between the satellite and the field data is not exceeding one day.

1. INTRODUCTION

Short-period internal waves (SIWs) are an important element of the dynamic structure of the ocean and affect its "average" state through vertical mixing, horizontal and vertical transfer of momentum, energy, tracers and sediments (Zimin et al., 2016; Fer et al., 2020a; Marchenko et al., 2021; Kozlov et al., 2022). They can influence propagation of acoustic signals and underwater operations. An importance of studying SIW field, its spatial characteristics and kinematics in the Arctic Ocean is also related to their effect on sea ice properties and formation of polynyas (Czipott et al., 1991; Morozov and Pisarev, 2004; Kozlov et al., 2017; Carr et al., 2019; Morozov et al., 2019).

Spaceborne SARs are known to be one of the most effective satellite remote sensing devices capable of observing sea surface manifestations of SIWs and so quantify their horizontal 2-D properties, energetics and infer peculiarities of their generation in the coastal ocean regions (da Silva et al., 2007; Lavrova and Mityagina, 2017; Kozlov et al., 2015a).

Phase speed is an important dynamic property of SIW propagation. Traditionally, it is defined my measuring the distance between coherent sequential SIW trains seen in single SAR images assuming that periodicity of the generation of each wave train is equal to the semidiurnal tidal period M_2 dominant over the most of the World Ocean (da Silva et al., 2015). Another approach is based on utilizing close-in-time satellite observations with a short time gap between sequential measurements. The latter was efficiently used for the analysis of SIW kinematics from visible band and passive microwave data in low latitudes (Liu et al., 2014; Hong et al., 2015; Tensubam et al., 2020).

Yet, for the polar ocean regions such estimates were not available up to now. Due to frequent cloudiness and complex illumination conditions, only spaceborne SARs could be effectively used for this task in the Arctic Ocean (Kozlov et al., 2015b, 2022; Morozov et al., 2017). In this work we attempt to close this gap and explore the potential of Sentinel-1 A/B SAR data to retrieve phase speeds of SIWs in the Arctic Ocean. The obtained results are then compared to SIW phase speed estimates made using a two-layer model and actual *in situ* hydrography data.

2. DATA AND METHODS

Here we use sequential Sentinel-1 A/B SAR data covering the area of Fram Strait, the Yermak Plateau and coastal regions of Svalbard from June to September 2018. Due to their polar orbits, Sentinel-1A and Sentinel-1B have high measurement frequency over the high-latitude regions. The latter is ensured by the operation of Sentinel-1 tandem and the mutual geometry of identical polar orbits with a phase shift of 180°. In this case, one satellite is on the day side and the other is on the night side of the orbit. This fact provides the interval between successive Sentinel-1A/B measurements equal to half of the orbital period, T = 98.6 min. As a result, a distinct set of sequential SAR observations with a time lag of about 50 min is available within the latitude band 70–85° N (Kozlov et al., 2020).

Such a short time lag between consecutive measurements and the high spatial resolution of SAR data, O(100 m), provide a unique opportunity to trace the horizontal movement of surface signatures of SIWs and retrieve their linear propagation speed associated with the phase speed of SIWs, C_{sat}.

The general sequence of satellite data processing has the following steps: 1) selection of consecutive Sentinel-1A/B SAR images in the Copernicus Open Access Hub at https://scihub.copernicus.eu; 2) detection of the SIW signatures on the intersecting parts of the paired SAR images in the Sentinels Application Platform (SNAP) program; 3) calculation and mapping of derived SIW phase speeds in Mathworks ©Matlab.

Dual-polarized HH-band data of gridded Level 1 extra-wide-swath (EW) mode medium-resolution (~90 m) products covering an area of about 400 km × 400 km are used for processing and analysis. To estimate the SIW phase speeds from a pair of sequential SAR images, surface signatures of SIWs in the overlapping fragments of both images are identified. Then, curves matching the positions of the leading waves of all SIW trains detected in both images are drawn and their coordinates are saved. The latter allows to estimate the horizontal shifts in the leading waves' positions during a known time interval from

* Corresponding author

This contribution has been peer-reviewed.
https://doi.org/10.5194/isprs-archives-XLVIII-M-1-2023-169-2023

which the phase speed of SIWs is determined, as standardly made in the literature (e.g., Liu et al, 2014; Hong et al., 2015).

The analysis of satellite data was carried out as follows. At the first step, for each wave packet observed in both radar images, the position of the leading wave was defined in the form of a curve passing along the region of maximum values of the radar signal due to intense backscattering from breaking wind waves in surface current convergence zones formed under the action of SIWs (Kudryavtsev et al., 2014). Further, on the basis of three perpendicular sections from the initial position of the leading wave on the first radar image to its final position on the second radar image, the average horizontal displacement of the leading wave was determined. The latter was used to calculate the average linear speed of the displacement of the SIW signatures associated with the average phase speed of SIWs.

Assuming that over a short time interval between successive SAR measurements (<1 h) the intensity of the background currents and the SIW speed don't change much, the measurement error of the horizontal shifts of SIW signatures is entirely determined by the spatial resolution of the SAR data. Given the spatial resolution of the SAR images is 88 m × 87 m in the range and azimuth directions, and the time lag between sequential images equals to ~49 min, the phase speed retrieval error (and the lower limit) would be ±90 m/49.3 min, i.e., ±0.03 m/s.

Altogether 40 pairs of sequential Sentinel-1A/B images made from June to September 2018 were analyzed. Figure 1 shows an example of two overlapping SAR images acquired on 1 August 2018. The area of their intersection is about 40% of the total area of a single image allowing to trace the movement of various upper-ocean dynamic processes over the area of nearly 64 000 km^2.

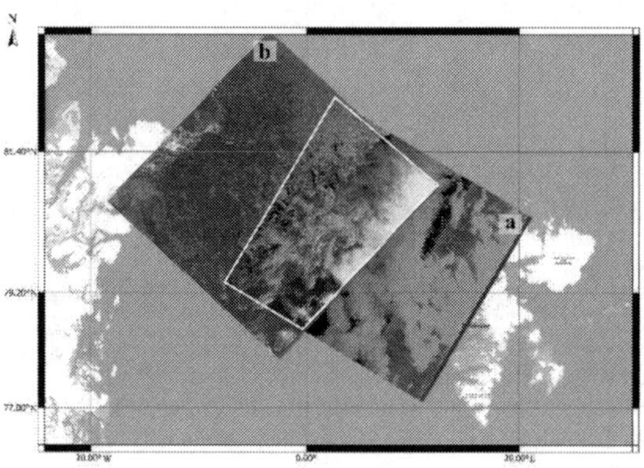

Figure 1. Position of two sequential Sentinel-1A/B images acquired over Fram Strait on 1 August 2018 at a) 07:11 UTC and b) 08:01 UTC. Red rectangle marks the region enlarged in Figure 2.

Figure 2 shows enlarged fragments of both SAR images made on 1 August 2018 exhibiting clear surface signatures of several SIW trains over the southern slope of the Yermak Plateau. The locations, propagation directions and horizontal shifts of leading waves of the four most distinct SIW trains A-D are annotated in Figure 2b. All wave trains travel west-south-west. The mean horizontal shift for the wave train A is about 2 km, for B — 1.6 km, for C — 2 km, for D — 1 km. The mean phase speeds, C_{sat}, for the wave trains A–D are 0.3–0.7 ± 0.03 m/s. The phase speeds of all wave trains considered in the analysis were derived in the same way.

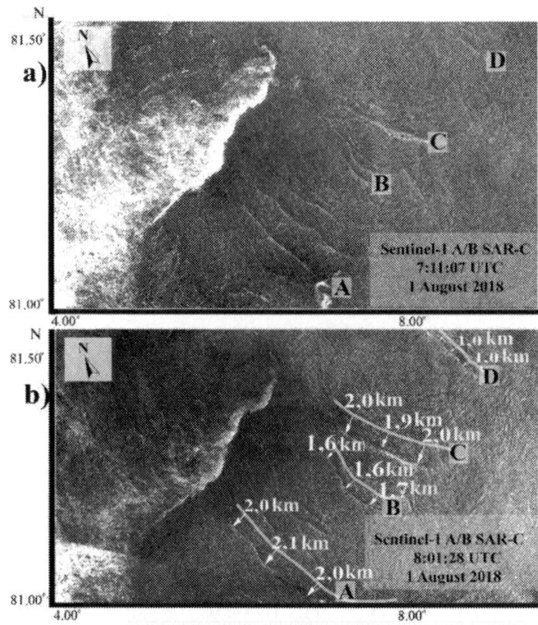

Figure 2. Manifestation of SIW trains in sequential Sentinel-1A/B SAR images acquired on 1 August 2018 over the southern slope of the Yermak Plateau. White lines denote leading waves of four wave trains A-D, arrows show the horizontal shift and direction of SIWs.

3. RESULTS

Analysis of 40 pairs of sequential Sentinel-1 A/B SAR images made from June to September 2018 enabled to detect 231 distinct SIW trains visible in these paired images. It should be noted that a significantly larger number of SIW manifestations (750 SIW signatures) was observed in the initial data. However, due to the difference in the imaging geometry between the Sentinel-1A and the Sentinel-1B satellites (compare a) and b) in Figure 2), SAR signatures of SIWs in one of the paired images could have low radar contrasts for their reliable identification.

The maximum number of SIW signatures (51%) was registered in August, 27% – in July, 12% – in June, and the minimum – in September (10%). Figure 3 presents a generalized map of the spatial distribution of leading waves of all SIW trains detected in the study area in sequential SAR data. Based on Figure 3 it is possible to distinguish three key regions of SIW observations: the deep-sea part of Fram Strait with depths >2000 m (area a), the southern and central parts of the Yermak Plateau with depths of 500–1500 m (area b), and the shelf area northwest of Svalbard with depths <500 m (region c).

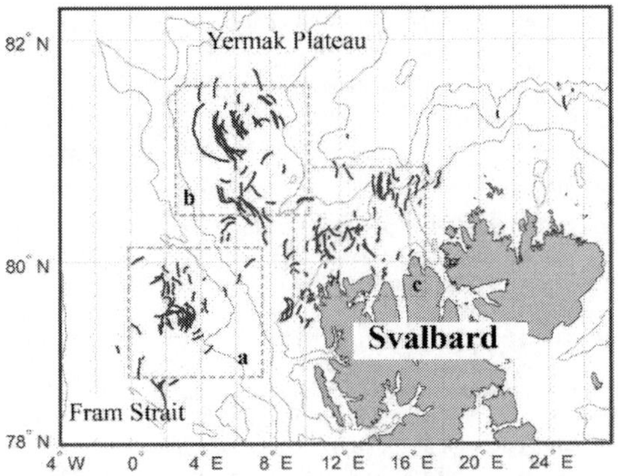

Figure 3. Spatial distribution of leading waves of SIW trains detected in sequential Sentinel-1 A/B images over the study region in June-September 2018.

Figure 4a shows a map of the spatial distribution of the obtained average values of the SIW phase speeds. The range of observed C_{sat} values is 0.08–0.84 ± 0.03 m/s. For the obtained interval of C_{sat} values with an average value of 0.33 m/s, the error in their determination is 4–38% (about 10% for the average value).

As seen, over the deep water of Fram Strait C_{sat} values ranged from 0.2 to 0.4 m/s with an average value of about 0.3 m/s. Significantly higher values of the SIW phase velocity were recorded over the Yermak Plateau. Here they amounted to 0.3–0.8 m/s, including a maximum of 0.84 m/s. On the Svalbard shelf, the range of C_{sat} values was rather high, from 0.1 to 0.8 m/s, but the mean values are also about 0.2-0.3 m/s.

It is interesting to note that the observed difference in the average values of the SIW phase speed between these regions is in general agreement with the spatial inhomogeneity in the tidal current velocity field (see, for example, Figure 1 in (Fer et al., 2020a)). According to Arc5km2018 tidal model (Erofeeva, Egbert, 2020), the maximum velocity of tidal currents in the area of the Yermak Plateau is 0.3–0.6 m/s; near Svalbard - 0.2-0.3 m/s, and over the deep part of Fram Strait - 0.1-0.2 m/s (Fer et al., 2020a).

Regular observations of SIW trains over the same locations suggests they could be generated via the well-known and commonly accepted lee wave mechanism during the interaction of tidal currents with inhomogeneous bottom topography (Vlasenko et al., 2003; Jackson et al., 2012; Fer et al., 2020a). In this case, their kinematic characteristics could be related to the velocity of the tidal currents through the Froude number, $Fr = u/c_0$, where u is the total current velocity; c_0 is the phase velocity of the first-mode IWs. The propagation of internal waves from generation areas is realized at subcritical values of the Froude number, when $Fr < 1$, i.e., when the phase speed of IWs exceeds the velocity of background currents (Jackson et al., 2012). As shown above, the values of the SIW phase speed obtained directly from SAR observations exceed the values of the maximum velocities of tidal currents in all three regions of their observation specified above, which qualitatively confirms the tidal mechanism of their generation and propagation at subcritical values of the Froude number.

The histogram distribution of the average values of SIW phase speed in Figure 4b clearly demonstrates the high probability (about 50% of all observations) of phase speed values in the range of 0.1–0.3 m/s with a pronounced peak for a value of 0.25 m/s. The second peak of observations falls on the phase speed values of about 0.4–0.5 m/s. Values of phase speed exceeding 0.5 m/s occur only in 15% of cases.

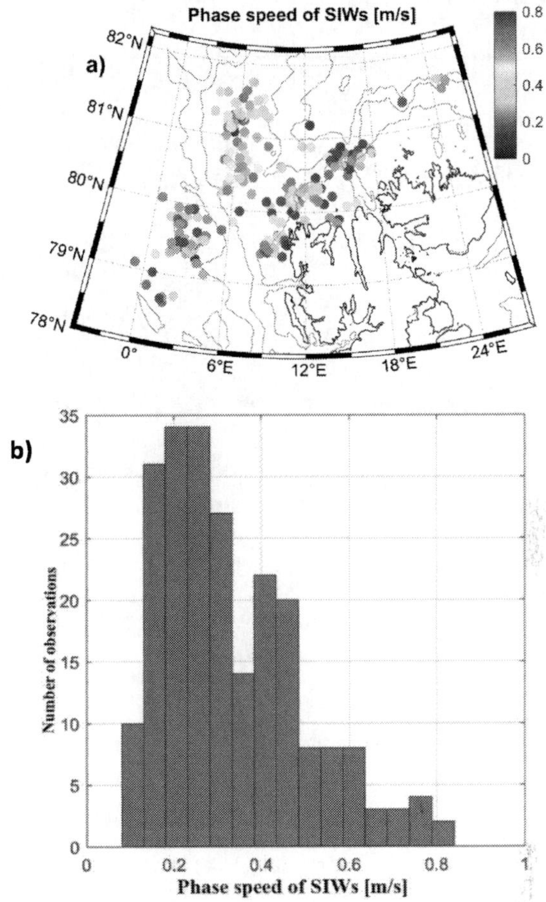

Figure 4. Spatial map (a) and histogram distribution (b) of SIW phase speed values derived from sequential SAR observations in June-September 2018.

The analysis of the distribution histograms for each of the regions a–c shows that they all have several (at least two) main peaks (not shown). The main peak has a minimum value of 0.15 m/s near Svalbard, followed by 0.2 m/s in Fram Strait and 0.3 m/s over the Yermak Plateau. The presence of several peaks in the distribution histograms, apparently, is explained by the seasonal variability of the vertical stratification of waters and the spring-neap cycle of the intensity of tidal currents, leading to the generation of SIWs.

It is important to note that the obtained values of SIW phase speed determined directly from the analysis of satellite data, are, in fact, the sum of the absolute phase speed of the SIWs (C) and the velocity of the background currents (u), including tidal currents. In this regard, it is interesting to compare the observed estimates of the SIW phase velocity (C_{sat}) with the theoretical estimates (C) made on the basis of actual hydrological data. For this task, the results of shipborne CTD measurements made by scientists from the University of Bergen from June 28 to July 10, 2018, near Svalbard are used. The measurement results are described in detail in (Fer et al., 2020b).

Figure 5a shows the location of stations near Svalbard and the locations of SIW manifestations identified in the satellite data during the period of expeditionary work. Figures 5b–5e show vertical profiles of potential density for several stations taken within sections A, D, T.

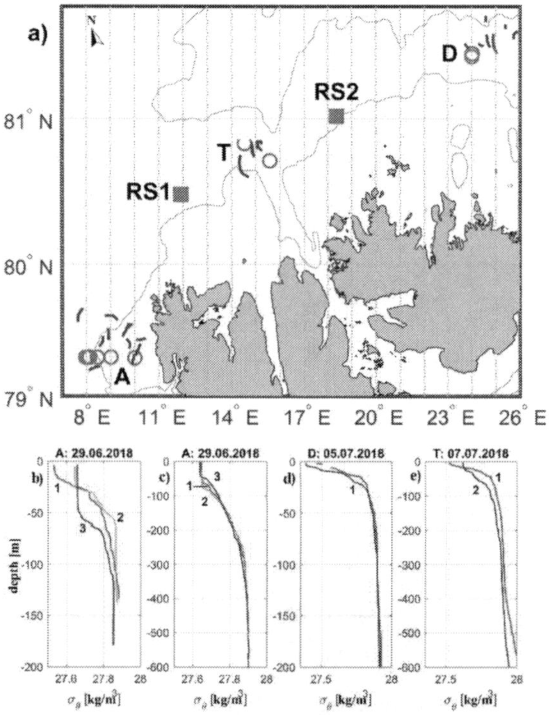

Figure 5. Comparison of satellite SIW observations with *in situ* measurements made near Svalbard in June-July 2018: a) locations of stations along sections A-T (red circles) and leading waves of SIW trains observed nearby (blue lines). b)-d) vertical profiles of potential density at selected stations made along sections A-T.

As can be seen from density profiles shown in Figures 5b–5e, a two-layer model can be satisfactorily used to estimate the phase velocity of the SIWs, as similarly applied e.g., in (Petrusevich et al., 2018; Kozlov et al., 2022). In this case, the phase speed of linear internal waves of the lowest internal mode can be defined as (Phillips, 1977):

$$c_0 = \left(\frac{\delta\rho}{\rho_0}\frac{g}{k}\right)^{1/2}\{\coth kd + \coth k(D-d)\}^{-1/2}, \quad (1)$$

where $\delta\rho$ = density difference between the upper and lower layers

ρ_0 = average density below pycnocline
d = mean pycnocline depth
D = full depth
k = wave number defined as $k = 2\pi/\lambda$, where wavelength λ is taken as a wavelength of the leading waves in ISW trains from SAR observations.

In case of nonlinear internal solitary waves phase speed would be equal to:

$$C = c_0 + \alpha\eta_0/3, \quad (2)$$

where η_0 = amplitude of SIWs

α = coefficient of quadratic nonlinearity defined as:

$$\alpha = {}^3/_2\, c_0(h_1 - h_2)/h_1h_2, \quad (3)$$

where h_1, h_2 = upper- and lower-layer depths, respectively
D = full depth

We will use equations (1)–(3) to calculate the phase speed of the SIWs. According to the field data (Fer et al., 2020a), the amplitude of SIWs recorded at daily stations RS1 and RS2 (see Figure 5a) ranged from 5 to 50 m. The rest of the parameters will be determined from the vertical density profiles. The results of calculations of the absolute values of the SIW phase speed C and their comparison with the data of satellite measurements C_{sat} are presented in Table 1, where for each station the range of C values is determined by the range of SIW amplitudes η_0 = 5–50 m, and the range of C_{sat} values is obtained in the case of observation of several wave trains in the vicinity of a given station or reflects the spatial variability of the phase speed along the front of the leading wave.

St. #	Date in situ	Date SAR	D [m]	d [m]	C [m/s]	C_{sat} [m/s]
A1	29.06	30.06	134	31	**0.26-0.38**	**0,25-0,34**
A1	29.06	05.07	134	31	**0,26-0,38**	**0,36-0,42**
A2	29.06	29.06	149	42	**0,25-0,32**	**0,27-0,33**
A3	29.06	29.06	188	61	**0,24-0,28**	**0,27-0,33**
A3	29.06	05.07	188	61	0,24-0,29	0,34-0,66
A3	29.06	05.07	188	61	0,26-0,30	0,74-0,8
A4	29.06	29.06	397	68	**0,25-0,32**	**0,27-0,33**
A5	29.06	29.06	485	83	**0,25-0,30**	**0,27-0,33**
A6	29.06	05.07	615	49	**0,23-0,32**	**0,3-0,35**
D1	05.07	12.07	281	14	**0,28-0,65**	**0,25-0,39**
T1	07.07	10.07	956	26	0,30-0,53	0,12
T1	07.07	10.07	956	26	0,33-0,58	0,16
T1	07.07	10.07	956	26	0,33-0,59	0,23
A3	07.07	10.07	942	64	0,34-0,42	0,18

Table 1. Comparison of SIW phase speed values obtained north of Svalbard from *in situ* and spaceborne SAR measurements

We note the main features of the results obtained. It can be seen from Table 1 that the best agreement between the observed and absolute values of the SIW phase velocity is obtained for cases where the time difference between satellite and *in situ* measurements is minimal, for example, for surveys made on June 29, 2018 at stations A1–A5 (highlighted in bold in Table 1). In this case, the observed values of the SIW phase speed C_{sat} are, as a rule, somewhat higher than the absolute values of C, but the difference between them does not exceed the error of satellite measurements (±0.03 m/s). On this basis it is difficult to conclude

that there is a systematic addition to the C_{sat} values associated with the background field of surface currents.
It is interesting to note that a good agreement between the SIW phase speed values is also observed for stations A1, A6, D1, despite the fact that satellite measurements were made a week after *in situ* measurements. Yet, the background tidal current velocities at the moment of shipborne and satellite observations were nearly similar, about 6–8 cm/s. The maximum difference in the phase speed values is observed for the T1–T2 stations performed on July 7, 2018. For all the observed SIW trains, the C_{sat} values are two or more times less than predicted by the model. In this case, shipborne measurements were made two days before the neap tide (July 9, 2018), and satellite ones - one day after it. On both of these dates, the amplitude of tidal currents did not exceed 3–4 cm/s, and the waves themselves propagated along the isobaths to the west and southwest. According to the data of ship and satellite measurements, at the time of the satellite imagery, the prevailing wind was about 8–10 m/s of the southwestern direction, i.e., the direction of wind waves and drift currents was actually opposite to the direction of SIW propagation. We, therefore, suggest that this fact can partly explain the noted discrepancy between the model and the satellite estimates of the SIW phase speed.

4. CONCLUSIONS

In this work, we have collected and processed a data set of sequential Sentinel-1A/B images acquired in summer-autumn period of 2018 to directly estimate the phase speed of short-period internal waves in the Arctic Ocean. The analysis of satellite data showed that the main peak of SIW observations in the study region falls on August. Three key regions of the SIW generation are also revealed: the deep-water region of Fram Strait, the southern and central parts of the Yermak Plateau, and the northern shelf of Svalbard.
According to observations, the maximum (average) values of the SIW phase speed are observed over the Yermak Plateau and reach here 0.8 ± 0.03 m/s (0.5–0.6 ± 0.03 m/s). In the other two regions, the phase speed values are close and average at 0.3 ± 0.03 m/s. The observed difference in the average phase speed values between these regions is in good agreement with the spatial inhomogeneity of the tidal currents here.
The observed SIW trains have a regular character and, most likely, are formed during the interaction of tidal currents with inhomogeneous bottom topography. In addition, the observed values of the SIW phase speed usually exceed the values of the maximum velocities of tidal currents in all three regions of their observation, which indirectly confirms their tidal generation via the lee wave mechanism.
Comparison of satellite estimates of the SIW phase speed with theoretical estimates obtained from a two-layer model using quasi-simultaneous hydrological measurements showed the best agreement for cases with a time difference between satellite and *in situ* measurements of no more than one day.
Thus, the analysis of currently available sequential satellite SAR measurements allows not only to determine the sources of internal wave generation and their spatial structure, but also to study their kinematic characteristics and evolution in the Arctic Ocean.

ACKNOWLEDGEMENTS

The study was supported by the Russian Science Foundation grant no. 21-17-00278, https://rscf.ru/project/21-17-00278. The Sentinel-1A/B satellite data was obtained from the archives of the European Maritime Centers system Copernicus Open Access Hub forecasts (https://scihub.copernicus.eu).

REFERENCES

Carr, M., Sutherland P., Haase, A., Evers, K.U., Fer, I., Jensen, A., Kalisch, H., Berntsen, J., Parau, E., Thiem, O., Davies, P.A., 2019. Laboratory experiments on internal solitary waves in ice-covered waters. *Geophysical Research Letters*, 46 (21), 12230-12238. doi.org/10.1029/2019GL084710.

Czipott, P.V., Levine, M.D., Paulson, C.A., Menemenlis, D., Farmer, D.M., Williams, R.G., 1991. Ice flexure forced by internal wave packets in the Arctic Ocean. *Science*, 254 (5033), 832–835.

da Silva, J.C, New, A.L., Azevedo, A., 2007. On the role of SAR for observing "local generation" of internal solitary waves off the Iberian Peninsula. *Canadian Journal of Remote Sensing*, 33(5), 388-403. doi.org/10.5589/m07-041.

da Silva, J.C.B., Buijsman, M.C., Magalhaes, J.M. 2015. Internal waves on the upstream side of a large sill of the Mascarene Ridge: a comprehensive view of their generation mechanisms and evolution. *Deep Sea Research Part I: Oceanographic Research Papers*, 99, 87-104. doi.org/10.1016/j.dsr.2015.01.002.

Erofeeva, S., Egbert, G., 2020: Arc5km2018: Arctic Ocean Inverse Tide Model on a 5-kilometer grid, 2018. doi.org/10.18739/A21R6N14K.

Fer, I., Koenig, Z., Kozlov, I.E., Ostrowski, M., Rippeth, T.P., Padman, L., Bosse, A., Kolas, E., 2020a. Tidally forced lee waves drive turbulent mixing along the Arctic Ocean margins. *Geophysical Research Letters*, 47(16), e2020GL088083. doi.org/10.1029/2020GL088083.

Fer, I., Koenig, Z., Bosse, A., Falck, E., Kolås, E., Nilsen, F., 2020b: Physical oceanography data from the cruise KB2018616 with R.V. Kristine Bonnevie. Dataset. doi.org/10.21335/NMDC-2047975397.

Hong, D.B., Yang, C.S., Ouchi, K., 2015. Estimation of internal wave velocity in the shallow South China Sea using single and multiple satellite images. *Remote Sensing Letters*, 6 (6), 448-457. doi.org/10.1080/2150704X.2015.1034884.

Jackson, C.R.; da Silva, J.C.B.; Jeans, G., 2012. The generation of nonlinear internal waves. *Oceanography*, 25, 108–123.

Kozlov, I., Kudryavtsev, V., Zubkova, E., Atadzhanova, O., Zimin, A., Romanenkov, D., Myasoedov, A., Chapron, B., 2015a. SAR observations of internal waves in the Russian Arctic seas. *IEEE International Geoscience and Remote Sensing Symposium (IGARSS)*, 947-949.

Kozlov, I.E., Kudryavtsev, V.N., Zubkova, E.V., Zimin, A.V., Chapron, B., 2015b. Characteristics of short-period internal waves in the Kara Sea inferred from satellite SAR data. *Izvestiya, Atmospheric and Oceanic Physics*, 51(9), 1073-1087, doi.org/10.1134/S0001433815090121

Kozlov, I.E., Zubkova, E.V., Kudryavtsev, V.N., 2017. Internal solitary waves in the Laptev Sea: first results of spaceborne SAR observations. *IEEE Geoscience and Remote Sensing Letters*, 14(11), 2047-2051. doi.org/10.1109/LGRS.2017.2749681.

Kozlov, I.E., Plotnikov, E.V., Manucharyan, G. E., 2020. Brief Communication: Mesoscale and submesoscale dynamics in the marginal ice zone from sequential synthetic aperture radar observations. *The Cryosphere*, 14, 2941-2947. doi.org/10.5194/tc-14-2941-2020.

Kozlov, I.E., Atadzhanova, O.A., Zimin, A.V., 2022. Internal solitary waves in the White Sea: hot-spots, structure, and kinematics from multi-sensor observations. *Remote Sensing*, 14, 4948. doi:10.3390/rs14194948

Kudryavtsev, V., Kozlov I., Chapron, B., Johannessen J.A., 2014. Quad-polarization SAR features of ocean currents. *Journal of Geophysical Research: Oceans*, 119 (9), 6046-6065, doi.org/10.1002/2014JC010173.

Lavrova, O., Mityagina, M., 2017. Satellite survey of internal waves in the Black and Caspian Seas. *Remote Sensing*. 9 (9), 892. doi.org/10.3390/rs9090892

Liu, B., Yang, H., Ding, X., Li, X., 2014. Tracking the internal waves in the South China Sea with environmental satellite sun glint images. *Remote Sensing Letters*, 5(7), 609-618, doi.org/10.1080/2150704X.2014.949365.

Marchenko, A.V., Morozov, E.G., Kozlov, I.E., Frey D.I., 2021. High-amplitude internal waves southeast of Spitsbergen. *Continental Shelf Research*, 227, 104523, doi.org/10.1016/j.csr.2021.104523.

Morozov, E.G., Kozlov, I.E., Shchuka, S.A., Frey, D.I., 2017: Internal tide in the Kara Gates Strait. *Oceanology*, 57, 8-18. doi.org/10.1134/S0001437017010106.

Morozov, E. G., Marchenko, A. V., Filchuk, K. V., Kowalik, Z., Marchenko, N.A., Ryzhov, I.V., 2019. Sea ice evolution and internal wave generation due to a tidal jet in a frozen sea. *Applied Ocean Research*, 87, 179-191, doi.org/10.1016/j.apor.2019.03.024.

Morozov, E.G., Pisarev, S.V., 2004. Internal waves and polynya formation in Laptev Sea. *Doklady Earth Sciences*, 398(7), 983-986.

Petrusevich V.Y., Dmitrenko, I.A., Kozlov, I.E., Kirillov, S.A., Kuzyk, Z.Z., Komarov, A.S., Heath, J.P., Barber, D.G., Ehn, J.K., 2018. Tidally-generated internal waves in Southeast Hudson Bay. *Continental Shelf Research*, 167, p.65-76, doi.org/10.1016/j.csr.2018.08.002.

Phillips, O.M., 1977: *The dynamics of the upper ocean*. Cambridge Univ. Press, Cambridge.

Tensubam C.M, Raju, N.J., Dash M.K., Barskar H., 2021. Estimation of internal solitary wave propagation speed in the Andaman Sea using multi–satellite images. *Remote Sensing of Environment*, 252, 112123. doi.org/10.1016/j.rse.2020.112123.

Vlasenko, V., Stashchuk, N., Hutter, K., Sabinin, K. 2003. Nonlinear internal waves forced by tides near the critical latitude. *Deep Sea Research Part I: Oceanographic Research Papers*, 50, 3, 317-338.

Zimin, A.V., Kozlov, I.E., Atadzhanova, O.A., Chapron, B., 2016. Monitoring short-period internal waves in the White Sea. *Izvestiya, Atmospheric and Oceanic Physics*, 52(9), 951–960. doi.org/10.1134/S0001433816090309.

MAPPING AREAS IMPACTED BY VOLCANIC FLOWS DURING AN ERUPTION USING SYNTHETIC APERTURE RADAR AND OPTICAL IMAGERY

E. Macorps[1,*], M. Jo[1,2], B. Osmanoglu[1], R. A. Albayrak[1,2]
[1] NASA Goddard Space Flight Center, Greenbelt, MD, USA – (elodie.macorps, minjeong.jo, batuhan.osmanoglu)@nasa.gov
[2] University of Maryland, Baltimore County, MD, USA – albayrak@umbc.edu

KEY WORDS: SAR, NDVI, Disasters Mapping, Volcanic Flows, Surface Roughness

ABSTRACT:

During volcanic disasters, the remoteness of the terrains combined with potentially incapacitated lifelines (e.g., disturbed transportation network) prevent ground-based surveys for timely assessment of damage extents. To that effect, we worked to combine satellite optical and Synthetic Aperture Radar (SAR) data to rapidly delineate the areas impacted by fast-moving volcanic flows during an eruption (e.g., Pyroclastic Density Currents (PDCs), lahars), which can in turn be used to target and organize the response efforts. We used the 2015 eruptions of Volcán de Colima (Mexico) and Volcán Calbuco (Chile) volcanoes to calibrate detection thresholds of different types of volcanic flows, from optical and SAR imagery. Optical imagery is used to calculate temporal changes of Normalized Difference Vegetation Index (NDVI) associated with the presence of erupted materials on the surface. SAR amplitude images are used to detect changes in surface roughness (sigma0) attributed to the emplacement of new volcanic flows. Classification of the respective NDVI and SAR amplitude signal changes for different types of volcanic flows is done using very-high-resolution imagery and ground-based data obtained during field work. Linear rescaling of minimal and maximal threshold signals is used to create probability maps of volcanic flow deposits extent, and then combined into a joint probability map to maximize the accuracy of the deposit extents. We tested our ability to generate volcanic flow extent maps during the April 2021 eruption of Soufrière St Vincent, using this detection method and the calibrated threshold values for PDCs and lahars.

1. INTRODUCTION

The cost of a disaster, both in terms of economic loss and fatalities, is dependent upon the rapidity and efficacy of the event response. Recent volcanic crises like the 2018 eruptions at Fuego (Guatemala) and Anak Krakatau (Indonesia), which were both responsible for over 430 dead and missing persons according to authorities, are tragic examples of how the lack of monitoring capabilities can have dramatic consequences (Auker et al., 2013; Agencias Guatemalteca de Noticias, 2019). Emergency managers can most benefit from satellite remote sensing tools at active volcanoes where ground-based monitoring and field work are dangerous, and sometimes impossible, due to the remote characteristics of the location, natural hazards (eruptive activity, dense vegetation, and rugged relief) or socio-political reasons (Head et al., 2012). Pre-eruptive monitoring can help improve early warnings, followed by syn-eruptive monitoring to provide critical information for hazard mitigation, and post-eruptive monitoring to support disaster response (Poland, 2015).
Multiple remote sensing sensors, platforms and techniques exist in the framework of emergency mapping, and the choice depends on the type and size of disasters being mapped and details required (Boccardo and Giulio Tonolo, 2014). In this research, we focus on medium-resolution (30-m) mapping of fast-moving volcanic flows (e.g., PDCs and lahars) to support emergency responders with rapid delineation of impacted areas. This is important to establish evacuation routes and target rescue efforts. We contrast this to slow-moving lava flows, which can destroy as many infrastructures and lifelines as the other fast-moving volcanic flows but are easily escapable and rarely directly threaten human lives (Harris, 2015).
Multi-sensor analysis strategically increases the range of detectable surface changes during or shortly after an eruption by providing different temporal and spatial-resolution datasets, all the while alleviating the limitations of each sensor types. For instance, while optical (visible and near-infrared) data can provide direct observations of the areas covered by volcanic flows, they require cloud-free skies, which is often restricted during an eruption due to heavy ash clouds or rain. SAR data, on the other hand, compensate for these limitations with all-weather and all-day imaging capabilities. SAR and optical data have been applied to fast-moving volcanic flows in other studies, either individually (Wadge et al., 2002, 2011; Terunuma et al., 2005; Thouret et al., 2010; Solikhin et al., 2012; Bignami et al., 2013; Arnold et al., 2018; Krippner et al., 2018; Pallister et al., 2019; Valade et al., 2019; Dualeh et al., 2021) or in combination (e.g., Solikhin et al., 2015; McAlpin et al., 2013), but the context of emergency mapping has rarely been a focus.

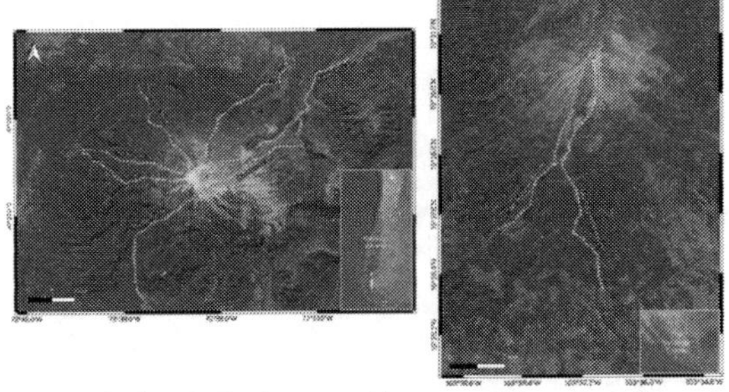

Figure 1. Google Earth maps with topographic contour lines at Volcán Calbuco (left) and Volcán de Colima (right). The valleys in which PDCs were deposited during the 2015 eruptions are highlighted in yellow dashed lines. The inset maps shows the location of both volcanoes, in South and Central America respectively.

* Corresponding author

In this paper, we use the case studies of the 2015 eruptions from Volcán de Colima in Mexico and Volcán Calbuco in Chile (Figure 1) to calibrate detection thresholds of different types of fast-moving volcanic flows from optical and SAR imagery. Surface changes caused by the deposition of PDCs and lahars are measured from differential information between pre- and post-eruption images. Normal Difference Vegetation Index (NDVI) from Landsat-8 OLI images and SAR backscattering intensity changes from ALOS-2 PALSAR images are used in tandem to retrieve the extent of PDCs and lahars. Very high-resolution optical imagery and field-based data are used for validation. The change detection thresholds from NDVI and SAR backscattering intensity are then used in joint probability mapping to refine the delineation of areas impacted by these flows. Finally, we applied the method during a recent volcanic crisis at Soufriere St Vincent.

2. MATERIALS AND METHODS

2.1 Study Sites

2.1.1 Volcán Calbuco (1974 m a.s.l.) located in the Southern Andes of Chile (Figure 1) experienced an abrupt sub-Plinian eruption on the evening of April 22, 2015, after 43 years of quiescence. A first pulsatory ash-rich column, sustained for 90-min, rose up to 15 km (height above crater) in the atmosphere and formed an umbrella cloud expanding and dispersing tephra to the NE. After a 5h30min-break, in the night of April 23, 2015, the second and more violent phase of the eruption began and generated a 6 hour-long sustained eruption plume that reached a maximum height of 28 km, with an umbrella cloud dispersing tephra again to the NE. The partial and eventually complete collapses of the two successive eruption plumes resulted in multiple PDCs that descended into the main drainage pathways around the volcano.

Concurrently, the melting of the summit glacier contributed to the generation of syn- and post-eruption lahars, that were bulking from remobilization of both PDC and tephra deposits. Such secondary lahars can be extremely devastating because they occur after the eruption has ended and can run twice the length of the flows they are eroding. The lack of warning signs and the high intensity of the eruption resulted in significant issues with critical lifelines (blocked roads and destroyed bridges) and evacuations being hindered. Many infrastructures were destroyed by the PDCs, thick accumulation of tephra fall and lahars.

2.1.2 Volcán de Colima (3860 m a.s.l), located at the western end of the Trans-Mexican Volcanic Belt (Figure 1), is one of the most historically active volcanoes in North America. On July 10-11, 2015, Volcán de Colima experienced a significant dome-collapse and partial flank collapse event that sent a series of PDCs onto the southern flank of the volcano into just two main ravines (Macorps et al., 2018). In the following months after the dome-collapse, the rainy season was marked by Hurricane Patricia in October 2015, which contributed to the generation of multiple stream-flows and lahars, cutting into the existing deposits and reworking them to long runout distances. Although the total volume of PDCs at Colima was 10 times lower than that of Calbuco, it was the longest recorded runout distance at Volcán de Colima for a dome-collapse event. This raised the necessity of updating hazard maps accordingly.

2.2 NDVI Change Detection

Optical images from the Operational Land Imager (OLI) sensor of Landsat-8 satellite are used to compute Normalized Difference Vegetation Index (NDVI) values for pre- and post-eruption scenes at Volcán de Colima and Volcán Calbuco using Equation 1:

$$NDVI = \frac{\rho_{NIR} - \rho_R}{\rho_{NIR} + \rho_R}, \tag{1}$$

where: ρ_{NIR} = surface reflectance of band 5 (near-infrared)
ρ_R = surface reflectance of band 4 (red)

Radiometric calibration and atmospheric correction of each Landsat-8 scene is done with the ARCSI Software (Bunting and Clewley, 2018) to remove differences in land surface and atmospheric conditions, as well as solar position and sensor geometry during the different image acquisitions (Lu et al., 2004; Pu et al., 2017). Normalization of pairs of pre- and post-eruption NDVI images is then performed using linear regression of NDVI values for polygons of unchanged land-cover/land-use areas (Pu et al., 2017; Macorps, 2021). This allows to remove seasonal changes of vegetation (i.e., plant phenological stages). Finally, the normalized NDVI image pairs can be used to create absolute NDVI ratio images, hereafter referred to as $|Q_{NDVI}|$, where surface changes caused by the eruption can be detected. The workflow for processing the Landsat-8 OLI bands and producing time-series of normalized pre- over post-eruption NDVI ratio images is presented in Figure 2.

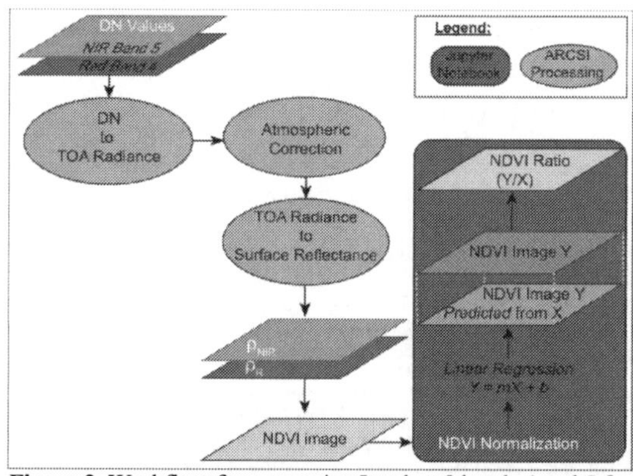

Figure 2. Workflow for processing Landsat-8 bands 4 and 5 for generating pre- over post-eruption $|Q_{NDVI}|$ images

2.3 SAR Backscattering Intensity Change Detection

The SAR dataset consists of ALOS-2 PALSAR images acquired in Fine Beam mode, with dual-polarization (HH and HV) in both ascending and descending orbits for Volcán de Colima, and only in ascending orbit for Volcán Calbuco. The full resolution geocoded backscatter images are produced with the GAMMA software package (Werner and Strozzi, 2000) Finally, the intensity images are converted from linear values to decibels using Equation 2, with ±0.5 dB radiometric accuracy (Motohka et al., 2018):

$$dB = 10 \times log10\ (pwr), \tag{2}$$

where: pwr = linear intensity values
dB = decidbels

SAR backscatter intensity is influenced by the local terrain, including local slope, surface roughness and terrain composition. A smoother surface will produce a weaker backscatter intensity (given the same local slope), thus resulting in a weaker signal (darker pixels), and vice versa. The water content and surface temperature will also result in different backscatter intensity, particularly for crop fields and growing vegetation. Changes in SAR backscattering intensity, hereafter referred to as ΔdB, are computed from differences between pairs of pre- and post-eruption intensity images, thus showing changes in surface properties.

2.4 Reference Data for Signal Change Classification

The final outlines of the extent of PDCs and lahars for validation is obtained from very high-resolution images and fieldwork data (Macorps et al., 2018; Macorps, 2021). These are used to classify the $|Q_{NDVI}|$ and ΔdB signals corresponding to the emplacement of new volcanic flows during the two eruptions. We use histogram thresholding to compute statistics and classify the detection signals for individual types of volcanic flows.

3. MAPPING OF VOLCANIC FLOWS

3.1 $|Q_{NDVI}|$ and ΔdB Detection Signals

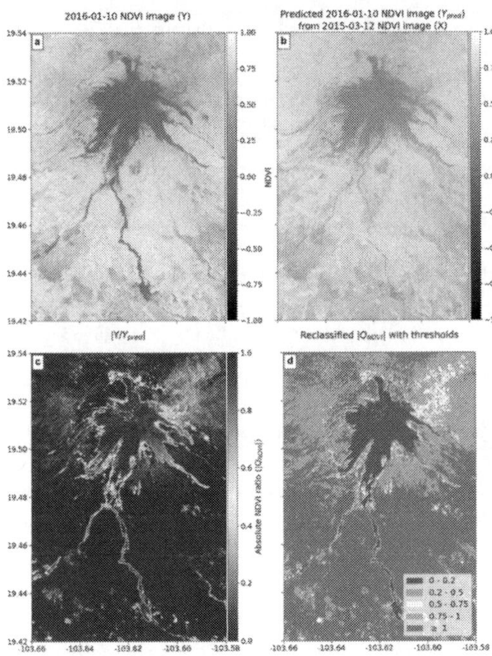

Figure 3. **a)** Post-eruption and **b)** predicted pre-eruption Landsat-8 NDVI images of the southern flank of Volcán de Colima. **c)** $|Q_{NDVI}|$ image obtained from NDVI image (a) divided by (b). **d)** Reclassified $|Q_{NDVI}|$ image with threshold values that emphasize the deposition of new volcanic materials.

High NDVI values represent vegetated surfaces and contrast with bare ground and/or volcanic deposits represented with low NDVI values (Figures 3a,b and 4a,b). Therefore, low $|Q_{NDVI}|$ values can be used to track vegetation removal and deposition of new volcanic flows on the flanks of the volcanoes (Figures 3c and 4c). By reclassifying the $|Q_{NDVI}|$ images using threshold values, we found that the PDCs and lahar deposits have $|Q_{NDVI}| \leq 0.2$ (Figures 3d and 4d). Near the summit, however, where the vegetation is scarce to begin with, it is more difficult to distinguish between different types of volcanic products (e.g., PDCs vs. tephra fall) and to delineate new vs. older materials.

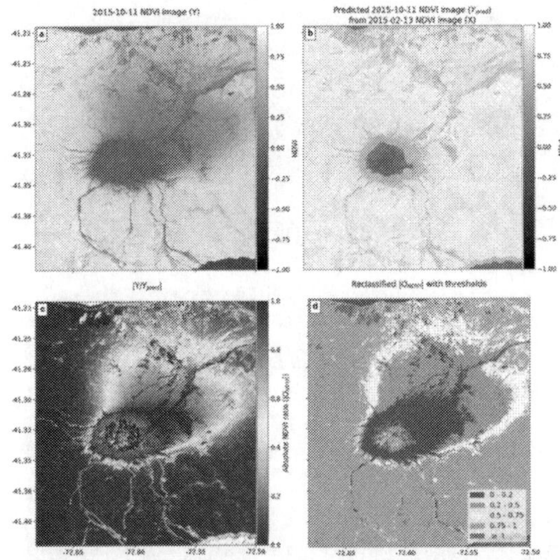

Figure 4. **a)** Post-eruption and **b)** predicted pre-eruption Landsat-8 NDVI images of the southern flank of Volcán Calbuco. **c)** $|Q_{NDVI}|$ image obtained from NDVI image (a) divided by (b). **d)** Reclassified $|Q_{NDVI}|$ image with threshold values that emphasize the deposition of new volcanic materials.

For the detection of SAR backscattering intensity changes, a viewing-friendly method for highlighting the surface changes consists of combining pre-eruption, post-eruption and the ratio of post/pre-eruption in the RGB channels to create a false-color composite image (Figures 5 and 6). The changes in SAR backscattering intensity are associated with changes in surface roughness from the emplacement of new volcanic flows and tephra fall deposits.

On ΔdB images (Figures 7 and 8), $\Delta dB < 0$ values (dark pixels) are detected where PDCs and lahars are deposited because the infilling of the valleys makes the ground smoother on the scale of the radar, thus decreasing the backscattering intensity. On the valley interfluves, the lower magnitude $\Delta dB < 0$ is related to the blanketed topography by thin ash-cloud surge and overbank deposits (i.e., flows that overtopped the valley walls, Macorps et al., 2018), and tephra fall deposits, which often strips the vegetation, making the ground smoother (i.e., removes the contribution of volumetric scattering, hence decreasing the backscatter). The $\Delta dB > 0$ values (brighter pixels) incising through the PDCs deposits are interpreted as erosion and remobilization of deposits by lahars and runoff (Figures 6, 7 and 8). The reworked PDCs and their erosion generate increased surface roughness and can be used to track the occurrence of lahars shortly after the eruption. At Volcán de Colima, the ΔdB>0 values near the summit are caused by increased surface roughness from the emplacement of the post dome-collapse lava flows (Figures 5 and 7). At Volcán Calbuco, the $\Delta dB > 0$ values in the crater/summit area can be explained by increased roughness from crater excavation and by the removal of snow/ice cover (Figure 8).

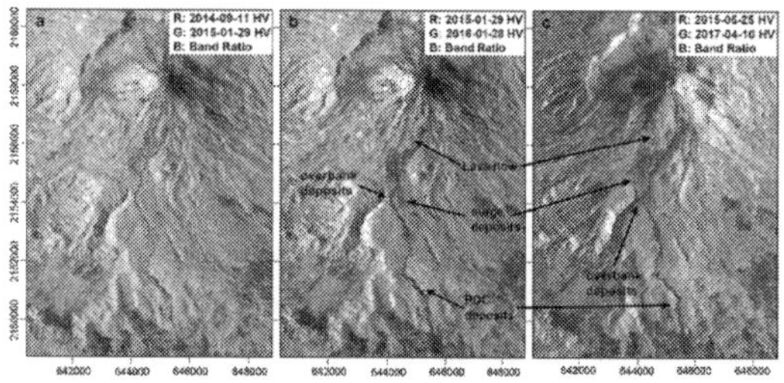

Figure 5. Georeferenced ALOS-2 false-color composite intensity ratio images of the south flank of Volcán de Colima (R: Earlier intensity image, G: Later intensity image, B: Result of the later intensity image divided by the earlier intensity image) obtained from **a)** a pre-eruption image pair and **b) and c)** pre- and post-eruption image pairs. The PDC deposits appear in red (decreased surface roughness) while the lava flows at the summit appear in blue (increased surface roughness).

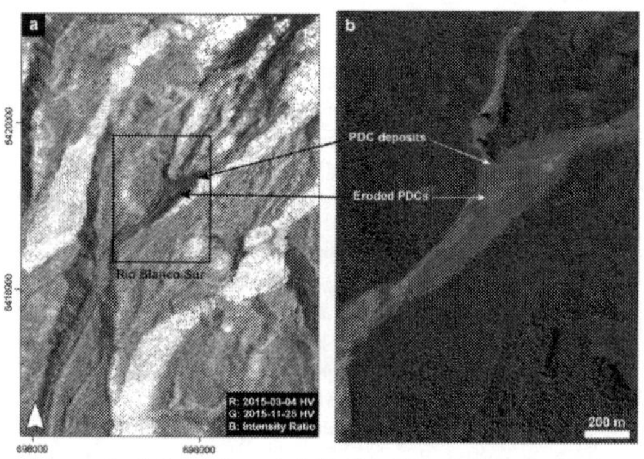

Figure 6. **a)** Georeferenced ALOS-2 false-color composite intensity ratio images of the Rio Blanco Sur valley at Volcán Calbuco (R: Earlier intensity image, G: Later intensity image, B: Result of the later intensity image divided by the earlier intensity image) obtained from a pre- and post-eruption image pair. **b)** Comparison with 0.5-m Pleiades image. The PDC deposits appear in red (decreased surface roughness) while the eroded PDCs from lahars appear in blue (increased surface roughness).

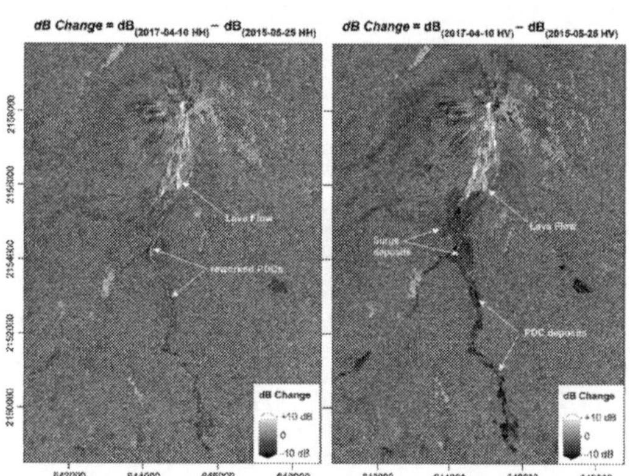

Figure 7. Georeferenced ALOS-2 ΔdB map of the south flank of Volcán de Colima obtained from differencing pre- and post-eruption dB images in **a)** HH-polarization and **b)** HV-polarization. Brown pixels indicate no-data values.

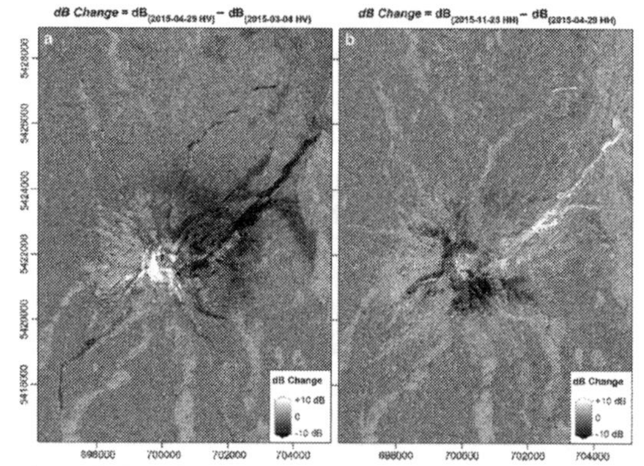

Figure 8. Georeferenced ALOS-2 ΔdB map of Volcán Calbuco obtained from differencing **a)** pre- and syn-eruption dB images in HV-polarization and **b)** syn- and post-eruption dB images in HH-polarization. Brown pixels indicate no-data values.

For different volcanic deposit types mapped from very high-resolution images and field data, we computed the average ΔdB signals (Table 1). HV polarization is more effective in displaying intensity evolution caused by the presence of volcanic deposits, while HH polarization is best suited for changes in the vegetation pattern and erosion. The ΔdB signal for volcanic flow deposits, either PDCs or lava flow, is about two times higher in HV than in HH polarization.

Volcano	Features	Polarization	
		HH	HV
Volcán de Colima	PDC deposits	↓ -4.2 ± 3.6 dB	↓ -8.0 ± 3.5 dB
	Ash-cloud surge deposits	↓ -1.2 ± 3.1 dB	↓ -3.9 ± 3.0 dB
	Lava flows	↑ +3.3 ± 4.6 dB	↑ +4.3 ± 4.8 dB
	Remobilized PDCs	↑ +2.5 ± 5.1 dB	↓ -0.2 ± 4.6 dB
Volcán Calbuco	PDC deposits	↓ -8.0 ± 3.3 dB	↓ -9.6 ± 3.5 dB
	Ash-cloud surge and/or tephra fall deposits	↓ -3.1± 3.3 dB	↓ -4.6 ± 3.6 dB
	Summit Changes	↑ +6.8 ± 6.4 dB	↑ +9.1 ± 6.2 dB
	Lahar erosion	↑ +7.7 ± 4.0 dB	↑ +8.0 ± 4.0 dB

Table 1. Average change in SAR intensity for individual features at Volcán de Colima and Volcán Calbuco, in co- and cross-polarization. The average intensity changes are measured in decibels (dB)

The results are comparable to other studies of PDCs mapped from SAR backscattering intensity changes at Montserrat volcano (Wadge et al., 2011), Merapi volcano in Indonesia (Solikhin et al., 2015) and Volcán de Fuego in Guatemala (Dualeh et al., 2021), which allowed us to develop "ΔdB signature fields" for volcanic flows (Figure 9).

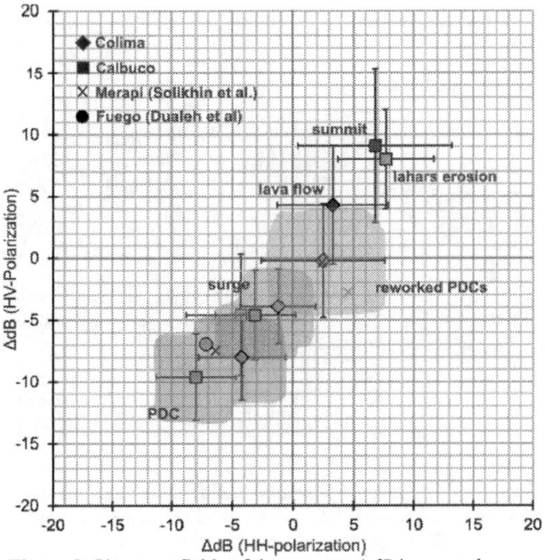

Figure 9. Signature fields of the average ΔdB in co- and cross-polarization for individual volcanic features at Volcán de Colima and Volcán Calbuco, and compared with the ΔdB measured at Merapi by Solikhin et al. (2015) and Fuego by Dualeh et al. (2021).

At Merapi volcano, Solikhin et al. (2015) also used ALOS L-band SAR backscattering intensity changes and found similar reduced surface roughness in the valleys from the deposition of PDCs, with comparable average values of ΔdB to those found at Volcán de Colima and Volcán Calbuco. At Montserrat and Volcán de Fuego, Wadge et al. (2011) and Dualeh et al. (2021) used X-band SAR basckattering intensity changes from TerraSAR-X and COSMO-SkyMed respectively. Both studies found similar results of changes in surface roughness from deposition of PDCs with an overall decrease in ΔdB.
Tracking lahars is more challenging because multiple lahars of different magnitude can occur over several days between SAR acquisitions. In the context of emergency mapping, the important information is knowing where fresh PDC deposits may become eroded and transformed into lahars during rainfall events. Finding erosional channels can provide this information. At Merapi, Solikhin et al. (2015) found increased surface roughness from reworked/eroded PDCs by lahars. They found that the higher water content resulted in $\Delta dB > 0$ in HH polarization but $\Delta dB < 0$ in HV polarization. This is consistent with findings for the reworked PDCs at Volcán de Colima, and the ΔdB values obtained by Solikhin et al. (2015) also fall within the range of the measured uncertainties. At Calbuco, however, the lahars resulted in ΔdB values closer to those measured for summit deformation. This may suggest that the changes in backscatter intensity observed cutting through the PDC deposits reflect the intense erosion, similar to what is observed at the summit, as opposed to watery reworked PDCs or lahar deposits. Dualeh et al. (2021) also found increased backscattering intensity where PDCs are being eroded by lahars. Their interpretation is that when deposits go from dry to wet, the radar penetration decreases and has more interactions with near-surface scatterers thus increasing the return signal towards the satellite. In the downstream areas of the volcanoes, the lahar deposits smooth the surface and show decreased ΔdB signals. Those results are all consistent with our findings at Volcán de Colima and Volcán Calbuco.

3.2 Probability Mapping

Probability maps (Figure 10) are generated from linear rescaling of the $|Q_{NDVI}|$ map and the SAR ΔdB map to probability values in the range [0, 1] using histogram thresholding estimated from mapping results such that:

- $P(X)_{SAR} = 0 < f(x) < 1,$

  $$\text{with } f(x) = \frac{(x - \Delta dB_{min})}{(\Delta dB_{max} - \Delta dB_{min})} \quad (3)$$

- $P(Y)_{Q_{NDVI}} = 0 < f(y) < 1,$

  $$\text{with } f(y) = \frac{(x - |Q_{NDVI\_min}|)}{(|Q_{NDVI\_max}| - |Q_{NDVI\_min}|)} \quad (4)$$

ΔdB_{min} and ΔdB_{max} are the lower and upper thresholds of intensity change for probabilities of PDC deposits between 0 and 1. For the lower threshold value for P(X) = 0, we use the average ΔdB measured for ash-cloud surge deposits in HV polarization, which is -3.5 dB. For the upper threshold value for P(X) = 1, we use the average ΔdB measured for PDC deposits in HV polarization, which is -8 dB.
$|Q_{NDVI\_min}|$ and $|Q_{NDVI\_max}|$ are the lower and upper thresholds of $|Q_{NDVI}|$ for PDC deposits set at 0.2 and 0.8 respectively.

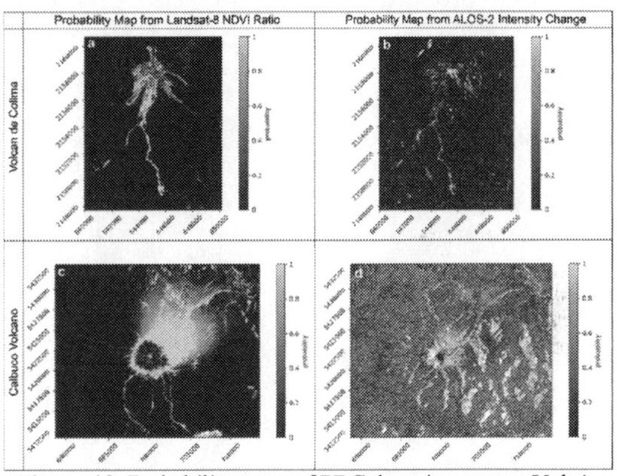

Figure 10. Probability maps of PDC deposit extent at Volcán de Colima (*top*) and Volcán Calbuco (*bottom*), from $|Q_{NDVI}|$ (**a** and **c**) and SAR ΔdB in HV polarization (**b** and **d**).

The NDVI probability map obtained with $|Q_{NDVI}|$ at Volcán de Colima (Figure 10a) shows *false positives* in the proximal area where rockfalls and lava flows were present. At Volcán Calbuco, the blanketing by tephra also covers a wider area than that of just the PDC deposits (Figure 10c). The SAR probability maps obtained with ΔdB, however, shows low probability values where lava flows were emplaced, due to the differences in ΔdB values between PDCs and lava flows (Figure 10b), but is also very noisy at Volcán Calbuco (Figure 10d). Therefore, we apply a method developed by Jo and Osmanoglu (2020) to map water flood extents by combining optical and SAR data into a weighted joint probability map. Here, the weighted joint probability algorithm is applied to combine the $|Q_{NDVI}|$ and ΔdB probability datasets to refine the mapping of PDC deposit extent (Figure 11) using Equation (5):

$$P(X|Y) = \frac{P(X) \cdot P(Y)}{P(X) \cdot P(Y) + [1 - P(X)] \cdot [1 - P(Y)]}, \quad (5)$$

where: P(X) = probability value on $|Q_{NDVI}|$ probability map
P(Y) = probability value on ΔdB probability map

https://doi.org/10.5194/isprs-archives-XLVIII-M-1-2023-175-2023 | © Author(s) 2023. CC BY 4.0 License.

The comparison of the histograms of the optical and SAR probability maps, with the joint probability histogram, are used here to check the performance of the technique. Background values in the $|Q_{NDVI}|$ probability map are significantly skewed towards 0, which is in agreement with the absence of any deposits. The histogram of ΔdB probability map, however, is spread over the whole range [0, 1], although with higher frequencies for lower probability values. This suggests higher background noise in the SAR probability map. Combining the results into the joint probability map is valuable to reduce the background noise and emphasize the extent of PDCs. Moreover, we find that it reduces the occurrence of *false positives* in the proximal area where lava flows and rockfalls showed as high probability pixels on the optical dataset, but with low probability values on SAR dataset.

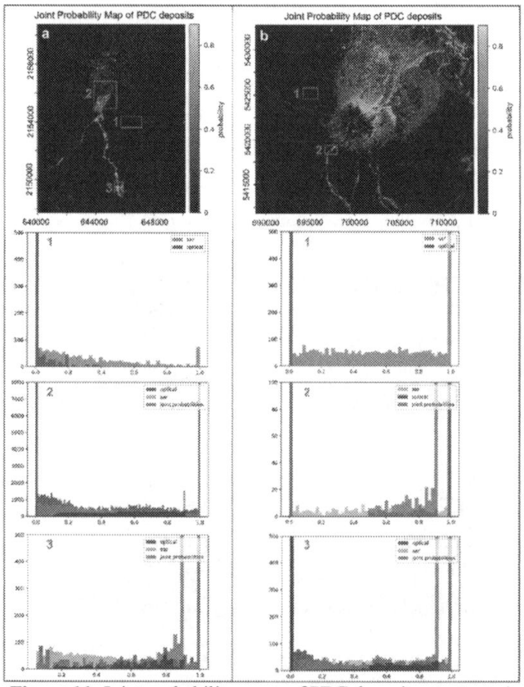

Figure 11. Joint probability maps of PDC deposit extents at **a)** Volcán de Colima and **b)** Volcán Calbuco. Histograms of the separate (optical and SAR) and joint probability maps for the areas of interests 1-3 are displayed below the respective map.

3.3 Application to the 2021 Eruption of Soufriere St Vincent

3.3.1 Soufrière St Vincent (1220 m a.s.l.), also known as "La Soufrière", is the northernmost, and youngest, stratovolcano of St Vincent and The Grenadines Island (Figure 12) located in the Lesser Antilles. Prior to 2021, its last known eruption was that of 1979 that destroyed a lava dome in the summit crater and generated several PDCs and lahars as well as tephra fall deposits and ended with the growth of a new lava dome (GVP, 2021a). After almost 42 years of quiescence, in December 2020, a new lava dome began growing SW of the 1979 dome, along with increased seismicity and gas-and-steam emissions (GVP, 2021a). Then, on April 9, 2021 La Soufrière began a series of explosive eruptions until April 22, 2021, which generated significant tephra fall deposits and numerous PDCs (GVP, 2021b). The heavy tephra fall resulted in widespread power and water outages on the island, as reported by the National Emergency Management Organization (NEMO), and some houses had collapsed due to significant ashfall. Rainfall in the following days and weeks remobilized a lot of fresh PDC and tephra deposits, generating numerous lahars (GVP, 2021b).

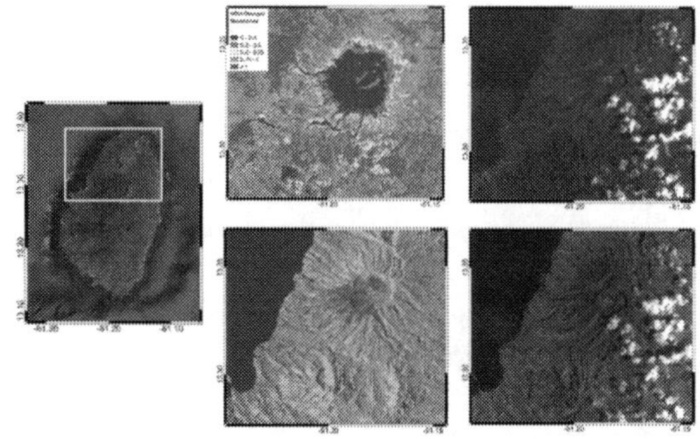

Figure 12. Soufrière St Vincent Volcano. **a)** Reclassified $|Q_{NDVI}|$ map using pre- and post-eruption Landsat-8 images showing volcanic deposits with $|Q_{NDVI}|<0.2$. **b)** False-color composite ALOS-2 ΔdB image showing the eruption products in purple from decreased SAR intensity. **c)** True color high-resolution PlanetScope image showing volcanic flows in grey going up to the sea, **d)** False-color composite PlanetScope image (R: NIR, G: Red, B: Green) showing the deposition of new volcanic products contrasting with vegetation in red.

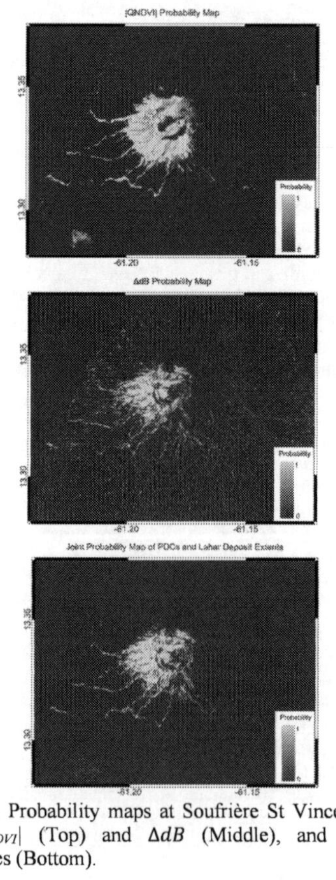

Figure 13. Probability maps at Soufrière St Vincent obtained from $|Q_{NDVI}|$ (Top) and ΔdB (Middle), and from joint probabilities (Bottom).

This contribution has been peer-reviewed.
https://doi.org/10.5194/isprs-archives-XLVIII-M-1-2023-175-2023 | © Author(s) 2023. CC BY 4.0 License.

3.3.1 Detecting Impacted Areas Post-Eruption

Landsat and ALOS-2 images are processed and analyzed similarly to the methodology described in section 2. Using similar threshold values for ΔdB and $|Q_{NDVI}|$ changes associated with PDC deposits (Figure 12), we generate a joint probability map (Figure 13) to estimate the extent of areas impacted during the April 2021 eruption of Soufrière St Vincent.
High-resolution images from PlanetScope are used for additional analysis of volcanic flow extent and validation of areas impacted by the flows (Figure 12).
Similarly to results obtained at Volcán de Colima and Volcán Calbuco, the joint probability map of SAR backscattering intensity and optical NDVI changes allows to refine the extent of areas impacted by volcanic flows (Figure 13).

4. CONCLUSIONS

A medium-resolution mapping technique using a combination of Optical and SAR data is developed using field-based and high-resolution validation data at Volcán de Colima and Volcán Calbuco volcanoes. The technique is then successfully applied during a volcanic crisis in 2021 at Soufrière St Vincent.
This method can be tuned to better fit each volcanic crisis, and although it remains at medium-resolution, it provides tremendous potential to supporting ground-based recovery and emergency efforts.
Continuation of this work currently involves developing algorithm with Machine Learning (ML) algorithm to automate the detection of surface changes from eruption, resolved on a set of multiple optical and SAR datasets. The ML process intends to remove the need for a priori information, focusing only on dense time-series to refine the detection of volcanic flows during a volcanic crisis.
As the list of available sensors grows, we hope to continue improving the use of multi-sensor analysis to reduce data processing latency and therefore increase disaster response efficacy. Testing this methodology at different spatial and temporal-resolution can also provide pointers to what will be relevant in future spaceborne and airborne missions.

REFERENCES

Arnold, D.W.D.; Biggs, J.; Wadge, G.; Mothes, P. 2018. Using satellite radar amplitude imaging for monitoring syn-eruptive changes in surface morphology at an ice-capped stratovolcano. *Remote Sens. Environ.* 209, 480–488.

Auker, M.R.; Sparks, R.S.J.; Siebert, L.; Crosweller, H.S.; Ewert, J. 2013. A statistical analysis of the global historical volcanic fatalities record. *J. Appl. Volcanol.* 2, 1–24.

Agencia Guatemalteca de Noticias. Available online: https://agn.com.gt/gobierno-de-guatemala-erogo-3609-millones-de-quetzales-para-atender-a-victimas-de-erupcion-del-volcan-de-fuego/ (accessed on 5 June 2019).

Bignami, C., Ruch, J., Chini, M., Neri, M., Buongiorno, M. F., Hidayati, S., et al. 2013. Pyroclastic density current volume estimation after the 2010 Merapi volcano eruption using X-band SAR. *J. Volcanol. Geoth. Res.* 261, 236–243.

Boccardo P., Giulio Tonolo, F. 2014. Remote Sensing Role in Emergency Mapping for Disaster Response. Lollino, G., Manconi, A., Guzzetti, F., Culshaw, M., Bobrowsky, P., Luino, F., *Engineering Geology for Society and Territory* - Volume 5. vol. Part I, p. 17-24, ISBN: 978-3-319-09047-4

Bunting, P., Clewley, D. 2018. Atmospheric and Radiometric Correction of Satellite Imagery (ARCSI). Available online: https://github.com/remotesensinginfo/arcsi (accessed on 28 November 2019).

Dualeh, E. W., Ebmeier, S. K., Wright, T. J., Albino, F., Naismith, A., Biggs, J., et al. 2021. Analyzing explosive volcanic deposits from satellite-based radar backscatter, Volcán de Fuego, 2018. *Journal of Geophysical Research: Solid Earth*, 126, e2021JB022250.

Global Volcanism Program, 2021a. Report on Soufriere St. Vincent (Saint Vincent and the Grenadines) (Bennis, K.L., and Venzke, E., eds.). *Bulletin of the Global Volcanism Network*, 46:3. Smithsonian Institution.

Global Volcanism Program, 2021b. Report on Soufriere St. Vincent (Saint Vincent and the Grenadines) (Bennis, K.L., and Venzke, E., eds.). *Bulletin of the Global Volcanism Network*, 46:5. Smithsonian Institution.

Harris, A. J. 2015. Basaltic lava flow hazard. In *Volcanic Hazards, Risks and Disasters*, Eds: J. F. Shroder, P. Papale,17–46. Amsterdam: Elsevier.

Head, E.N., Maclean, A.L., Carn, S.A. 2012. Mapping lava flows from Nyamuragira volcano (1967-2001) with satellite data and automated classification methods. *Geomatics, Natural Hazards and Risk*, 4(2), 119-144.

Jo, M., Osmanoglu, B. 2020. Generating flood probability map based on combined use of synthetic aperture radar and optical imagery. *IGARSS 2020*.

Krippner, J.B., Belousov, A.B., Belousova, M.G., Ramsey, M.S. 2018. Parametric analysis of lava dome-collapse events and pyroclastic deposits at Shiveluch volcano, Kamchatka, using visible and infrared satellite data. *Journal of Volcanology and Geothermal Research*, 354, 115-129.

Lu, D., Mausel, P., Brondizio, E., Moran, E. 2004. Change detection techniques. *International Journal of Remote Sensing*, 25, 2365-2407.

McAlpin, D., and Meyer, F.J. 2013. Multi-sensor data fusion for remote sensing of post-eruptive deformation and depositional features at Redoubt Volcano. *Journal of Volcanology and Geothermal Research*, 259, 414-423

Macorps, E., Charbonnier, S.J., Varley, N.R., Capra, L., Atlas, Z., Cabre, J. 2018. Stratigraphy, sedimentology and inferred flow dynamics from the July 2015 block-and-ash flow deposits at Volcán de Colima, Mexico. *Journal of Volcanology and Geothermal Research*, 349, 99-116.

Macorps, E. 2021. Field and Remote Sensing Analysis of the 2015 Pyroclastic Density Currents at Colima (Mexico) and Calbuco (Chile) Volcanoes: Implications for Hazard Assessment and Crisis Management (Doctoral dissertation, University of South Florida).

Motohka, T., Isoguchi, O., Sakashita, M., Shimada, M. 2018. July. Results of ALOS-2 PALSAR-2 calibration and validation after 3 years of operation. *In IGARSS 2018 IEEE International Geoscience and Remote Sensing Symposium*, pp. 4169-4170. IEEE.

This contribution has been peer-reviewed.
https://doi.org/10.5194/isprs-archives-XLVIII-M-1-2023-175-2023

Pallister, J., Wessels, R., Griswold, J., McCausland, W., Kartadinata, N., Gunawan, H., Budianto, A., Primulyana, S. 2019. Monitoring, forecasting collapse events, and mapping pyroclastic deposits at Sinabung volcano with satellite imagery. *Journal of Volcanology and Geothermal Research*, 382, 149-163.

Poland, M. 2015. Volcano monitoring from space. *In Global Volcanic Hazards and Risk*; Loughlin, S.C., Sparks, R.S.J., Brown, S.K., Jenkins, S.F., Vye-Brown, C., Eds.; Cambridge University Press: Cambridge, UK, pp. 311–316.

Pu, R., Gong, P., Tian, Y., Miao, X., Carruthers, R.I., Anderson, G.L. 2008. Using classification and NDVI differencing methods for monitoring sparse vegetation coverage: a case study of saltcedar in Nevada, USA. *International Journal of Remote Sensing*, 29(14), 3987 — 4011.

Solikhin, A., Thouret, J-C., Gupta, A., Harris, A.J., Liew, S.C. 2012. Geology, tectonics, and the 2002-2003 eruption of the Semeru volcano, Indonesia: Interpreted from high-spatial resolution satellite imagery. *Geomorphology*, 138(1), 364-379.

Solikhin, A., Pinel, V., Vandemeulebrouck, J., Thouret, J. C., & Hendrasto, M. 2015. Mapping the 2010 Merapi pyroclastic deposits using dual-polarization Synthetic Aperture Radar (SAR) data. *Remote Sensing of Environment*, 158, 180–192.

Terunuma, T., Nishida, K., Amada, T., Mizyama, T., Sato, I., Urai, M. 2005. Detection of traces of pyroclastic flows and lahars with satellite synthetic aperture radars. *Int. J. Remote Sens.*, 26, 1927-1942.

Thouret, J-C., Gupta, A., Lube, G., Liew, S.C., Cronin, S.J., Surono, 2010. The 2006 pyroclastic deposits of Merapi Volcano, Java, Indonesia: High-spatial resolution IKONOS images and complementary ground-based observations. *Remote Sensing of Environment*, 114, 1949-1967

Valade, S., Ley, A., Massimetti, F., D'Hondt, O., Laiolo, M., Coppola, D., et al. 2019. Towards global volcano monitoring using multisensor sentinel missions and artificial intelligence: the MOUNTS monitoring system. *Remote Sens.* 11:1528.

Wadge, G., Scheuchl, B., Stevens, N.F. 2002. Spaceborne radar measurements of the eruption of Soufriere Hills Volcano, Montserrat. In, Druitt, T.H. and Kokelaar, B.P. (Eds.), The Eruption of Soufriere Hills Volcano, Montserrat, from 1995 to 1999. *Geological Society, London, Memoir*, 21, pp. 583–594.

Wadge, G.; Cole, P.; Stinton, A.; Komorowski, J.C.; Stewart, R.; Toombs, A.C.; Legendre, Y. 2011. Rapid topographic change measured by high-resolution satellite radar at Soufriere Hills Volcano, Montserrat, 2008–2010. *J. Volcanol. Geotherm. Res.* 199, 142–152

Werner, C., Wegmuller, U., Strozzi, T., Weismann, A. 2000. Gamma SAR and interferometric processing software. In *Proceedings of the ERS-ENVISAT symposium,* Gothenburg, Sweden, 1620, 1620.

This contribution has been peer-reviewed.
https://doi.org/10.5194/isprs-archives-XLVIII-M-1-2023-175-2023 | © Author(s) 2023. CC BY 4.0 License.

CEREAL CROP IDENTIFICATION USING SENTINEL2 TIME SERIES, CASE OF SIDI BEL ABBES AREA

D. Mansour [1,2,*], D. Attaf [1], M. Ghabi[1]

[1]Algerian Space Agency, Centre des Techniques Spatiales, Algeria – (dalila.attaf, ghabi.inde)@gmail.com
[2]University of Oran, Geographical Space and Territorial Development Laboratory, Geography, BP 1015 EL M'naouer, Oran, Algeria, 31000 – mansour.djamel@univ-oran2.dz

KEY WORDS: Cereal profiles, Growing season, Sentinel2, NDVI, Time series, Weather conditions.

ABSTRACT:

The department of Sidi Bel Abbes is considered among the primary region producing cereal at the national level in Algeria. The study that we conducted is based on time series analysis using Sentinel2 images to monitor the cultivated cereal crops, combined with the field data including cereal plots for the growing season of 2020-2021.The methodology adopted in this study focused mainly on the processing of a Normalized Difference Vegetation Index (NDVI) time series. The analysis of the NDVI time series from October to July 2021, allowed us to determine the typical profiles of cereal crops, based on thresholding obtained to extract the areas cultivated by cereal crops. The geographical distribution of cereal in terms of area represents a high density of green biomass in the highlands and a low density of vegetation towards the plain and the highlands of the northwest area. The analysis of the profile's growth for the cereal crops, allowed us to understand the behavior of the cereal crop during its development and to catch the relationship of these behaviors, the meteorological conditions and the agricultural practices followed.

1. INTRODUCTION

In Algeria, the plain are the main cereal areas, which depends in most of these regions on the cultivation of cereal. This practice is spread in the major part of agricultural land including cereal crops. The objective of this study is to identified the cereal areas in the department of Sidi bel Abbes during the growing season 2020-2021 using sentinel2A time series images(Guechi et al., 2021), then to catch the relationship of crop duration from different plots at local scale according to growing season of NDVI values (field observations, May 2021) (Maselli & Rembold, 2001; Chehat, 2007).

To meet an estimation close to the expectations at local scale for the decision-makers, it is important to take into account some factors that influence the crop growth and the predicted yield, data such as climatic conditions, cultivation practices that can be related with recent data from remote sensing for crop growth monitoring.

The cereal sector is one of the main parts of agricultural production in Algeria, occupying an important place in the food system and the national economy. Cereal production, including fallow land, represents about 80% of the available agricultural area (Useful Agricultural Area: UAA).

Entire the country, the annual area sown to cereals is between 3 and 3.5 million hectares. The area harvested each year represents 63% of the area sown. It thus appears to be a dominant crop (Bessaoud et al., 2019)

Mainly, the average annual consumption of cereals per Algerian is207 kg of wheat in various forms (bread, couscous, pasta, etc.) (Chehat, 2007). The importance of consumption makes cereal a strategic product from the point of view of food security(Djermoun, 2009).

2. STUDY AREA AND USED DATA

2.1 Study area

The study area is located northwest of the country, at 470 m altitude, in the center of a vast plain between the Jebel Tessala in the north and the mountains of Daya in the south. The wilaya occupies a strategic central position and covers about 15% of the territory of the northwest region of the country or 9 150.63 km ^{2}. It is considered as a relay because of its privileged location as it is crossed by the main roads of this part of the country(Safia et al., 2015) .

The wilaya of Sidi Bel Abbes is bounded as follows (Fig.1):

- North by the wilaya of Oran.
- Northwest by the wilaya of Ain Témouchent.
- Northeast by the wilaya of Mascara.
- West by the wilaya of Tlemcen.
- East by the wilayas of Mascara and Saida.
- South by the wilayas of Nâama and El-Bayad.
- Southeast by the wilaya of Saida.

The wilaya of Sidi Bel Abbes has an agro-sylvo-pastoral vocation and is characterized by three distinct natural regions.

- The steppe, located in the south of the wilaya, is very important from the point of view of surface, it is dominated by esparto grass. It occupies an area of 323 944 (ha) or 35% of the area of the wilaya.

- The mountainous area (north and center) covered mainly by forests, has an area of 366 080 (ha) or 40% of the total area.

This contribution has been peer-reviewed.
https://doi.org/10.5194/isprs-archives-XLVIII-M-1-2023-183-2023

- The plains (plain of Sidi-Bel-Abbes and Telagh) with a dominance of cereal crops cover an area of 225 035 (ha) or 25% of the total area.

The majority of the land in the wilaya of Sidi Bel Abbes is destined to agriculture through a diversity of crops in particular: cereals, orchards, fruit plantations and market gardening.

The diachronic study of cereal crop cover in the Sidi Bel Abbes region is based on the use of a series of sentinel 2A images of the current agricultural year. The flowchart below traces different stages of analysis, use and processing performed in this study. The field visit at local scale in order to identify and collect GPS points of some cereal plots managed by the technical institute of field crops.

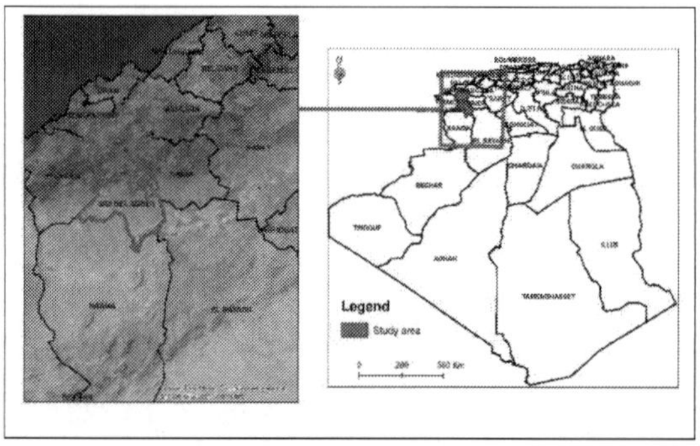

Figure 1. geographic extent of the study area.

2.2 Climate conditions

Agriculture is closely linked to climatic conditions, and cereal crops are no exception. In the case of Sidi Bel Abbes, Algeria, climatic conditions play an important role in the development of cereal crops.

It is important to note that Sidi Bel Abbes is located in a semi-arid region with a Mediterranean climate. Average temperatures range from 12°C in January to 26°C in August, with an average annual rainfall of about 350 mm (Fig.2).

The vegetative development of cereal crops in Sidi Bel Abbès depends on several factors, including temperature, humidity, soil quality, and the amount of sunlight. Although climatic conditions can be a challenge for growing cereals in this semi-arid region, farmers can adapt their farming practices to make the most of local conditions (Fig.2).

We notice average monthly variations in rainfall with a maximum in April and November (Fig.2). The minimum corresponds to the months of February, August and October. This confirms the characteristic of the semi-arid bioclimatic.

The comparison of the curve of variation of the average monthly precipitation, with the temperature, illustrates a significant decrease in the abundance of precipitation abundance of precipitation, which consequently has an influence on the distribution of the cereal crops.

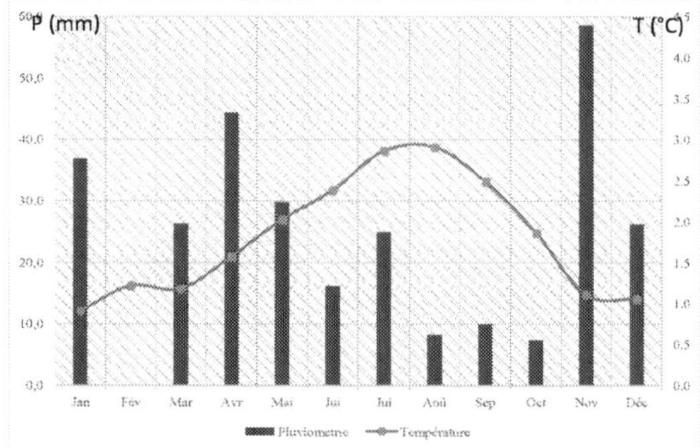

Figure 2. Rainfall across temperature variation in the season 2020-2021.

Moisture is also an important factor in the development of cereal crops. Cereals need an adequate amount of water to develop properly. In the case of Sidi Bel Abbes, the amount of annual precipitation is relatively low, which can have a negative impact on the growth of cereal crops. However, if cereals are grown in areas with good drainage, they can adapt to drier conditions.

Maximum of 70% of humidity is observed in October, winter season, and a and a minimum of 40% is observed in July, summer season (Fig.3).

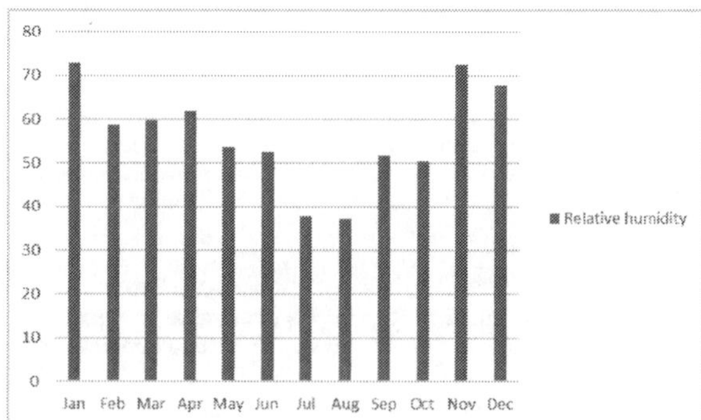

Figure 3. Relative humidity in the study area.

Assembling the vegetation indices (NDVI) of the current agricultural year. The determination of the NDVI values of the time series used relative to the collected GPS points, is done by superimposing these collected points in field on our satellite data to have the NDVI for each GPS point, and then we have the assemblies in a table, which represents the evolution of NDVI for each GPS point during the current agricultural year.

2.3 Used data

Sentinel-2 is a satellite imaging mission that is implemented by the European Commission (EC) and the European Space Agency (ESA) as a part of the Copernicus program (Fig.4). The two identical satellites (Sentinel-2A and Sentinel-2B) provide continually, multispectral, wide swath (290 km), high spatial resolution (four bands at 10 m, six bands at 20 m, and three bands at 60 m), and high revisit frequency (five days with combined satellites) image data.

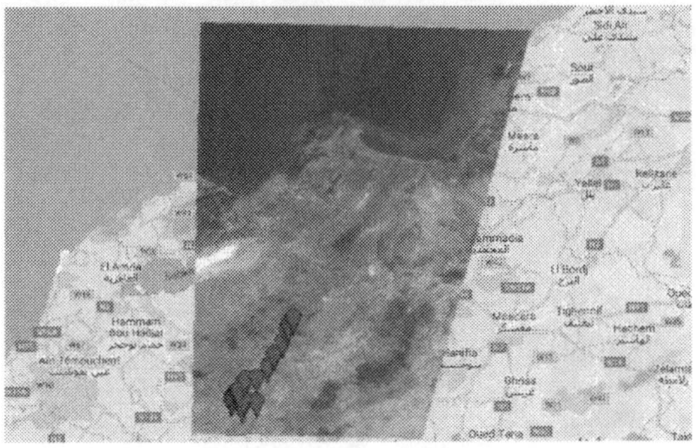

Figure 4. Sentinel 2 image of agricultural season 2020-2021 Data visualization of the field survey (GPS points collected from the field work in Sidi Bel Abbes 2021 (Red points).

The monitoring period of cereal crop cover in the Sidi Bel Abbes region is based on the use of 25 series of sentinel 2A images of the current agricultural year. The data were processed in Google Earth engine Platform (GEE)(Zhao et al., 2021).

The field visit at local scale in order to identify and collect GPS points of some cereal plots (Fig.4).

Assembling the vegetation indices (NDVI) of the current agricultural year. The determination of NDVI values of the time series used relative to the collected GPS points, is done by superimposing these collected points in field on our satellite data to have the NDVI for each GPS point, and then we have the assemblies in a table, which represents the evolution of NDVI for each GPS point during the current agricultural year.

3. METHODS

The methodology of this paper is detailed in two parts: monitoring crop phenology with Sentinel 2 time series and crop/ non crop mapping using thresholding algorithm.

The flowchart below traces the two parts of analysis, use and processing performed in this study (Fig.6).

First, Sentinel-2 time series are pre-processed on GEE by removing cloud and shadow to generate normalized difference vegetation index (NDVI) images. Then The monthly NDVI profiles are generated for the training GPS points for one-year 2021. On the other hand, The Otsu thresholding algorithm can obtain an optimal threshold automatically by maximizing the inter-class variance and minimizing the weighted within-class variance (Otsu, 1979).

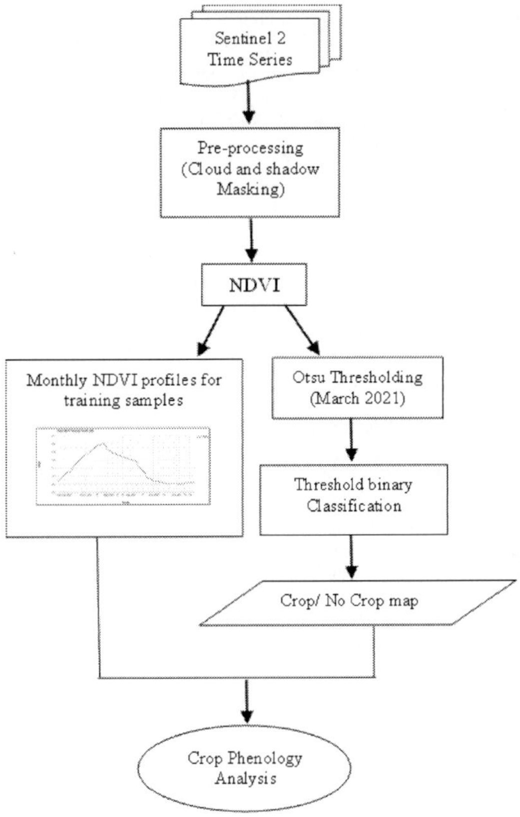

Figure 5. Flowchart of the proposed methodology.

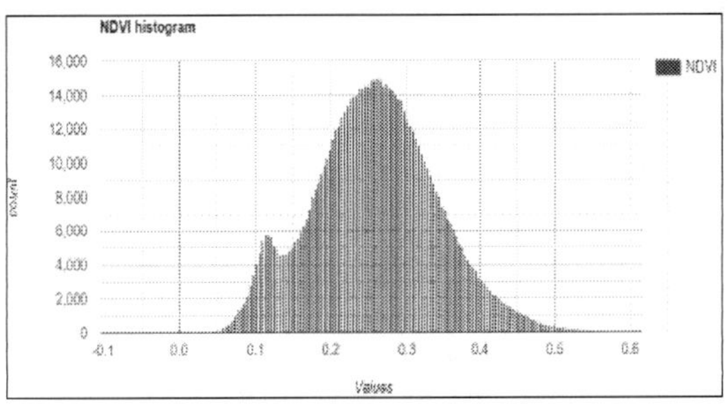

Figure 6. Thresholding NDVI values.

Binarization is the key step in extracting cereal crop land by using NDVI. During this process, it plays an important role. The (Otsu, 1979) as a common method of image segmentation is one of the most accurate and widely used methods (Sahoo et al., 1988). Otsu's method automatically obtains the optimal threshold through the histogram function of the grayscale image (Fig.6).

https://doi.org/10.5194/isprs-archives-XLVIII-M-1-2023-183-2023

The basic principle of this method is to calculate the threshold, which maximizes the variance between classes (Mizushima & Lu, 2013). It can be expressed as follows Equation (1):

$$\sigma_B^2(t) = \frac{[M_G P(t) - M(t)]^2}{P(t)[1 - P(t)]} \quad \text{..........(1)}$$

Where M_G is the average intensity of the entire image(t) is the cumulative average up to level t (t ∈ [a, b], P(t) is the cumulative sum of probabilities assigned to object (background). The value t is the optimal value that maximizes $\sigma^2_B(t)$ (equation (2)):

$$t^* = \underset{a \leq t \leq b}{\operatorname{argmax}} \, \delta_B^2(t) \quad \text{............................(2)}$$

In our study, the Otsu thresholding algorithm was applied to on the derived NDVI image of March 2021 to identify an optimal threshold for NDVI Sentinel 2 to separate crop from non-crop pixels for producing a crop maps.

Finally, crop phenology analysis is performed and detailed in the sections below.

Numerous studies have suggested that Otsu's method is an effective solution for extracting built-up or other land use types from various indices derived from remote sensing data (Estoque & Murayama, 2015; Shi et al., 2014). In this paper, we are based on Otsu's method, we implemented an algorithm under GEE to obtain the optimal thresholds.

4. EXPREMENTAL RESULTS

In this section, this study based on time series of images for the current growing season, allow us to follow the cereal growth by stacking the NDVI values during the agricultural year 2020-2021 using Google Earth Engine (Fig.6).

Figure 8. Cereal crop mask.

The obtained results show the binary classifier provides the lowest accuracy (k=0,56) for the experiments and the obtained classification maps have the less visual quality (high commission error), that can be explained by the difficulty in change of its parameters and to get over correct effect (difficulty thresholding according to different crops).

4.1 Evolution of NDVI values during the growing period

The temporal profiles were established for the pixels corresponding to the sampling points that were carried out on the plain of Sidi Bel Abbes.

The NDVI calculated in this study is generated for each month separately. Generally, vegetation covers take positive NDVI values globally above 0.1 or the densest vegetation have higher values. Bare soils have NDVI values below 0.1 and sometimes above 0.1 when the soil is rich in organic matter.

4.2 Interpretation of cereal crop profile during growing period

The analysis of changes in NDVI values of cereal plots collected from the field has allowed us to identify a relationship between the evolution of NDVI values of the current agricultural season and the cereal development cycle and climatic conditions.

4.2.1 Vegetative period

This is due to the fact that the vegetation cover at the vegetative stage is light and thin, which leads to a decrease in the amount of reflectance, therefore, this has affected negatively on the value of NDVI. This period is characterized by a decrease in temperature that is varied between 10 ° C to 15 ° C with a significant rainfall that reaches up to 58 mm for the month of December (Fig.7).

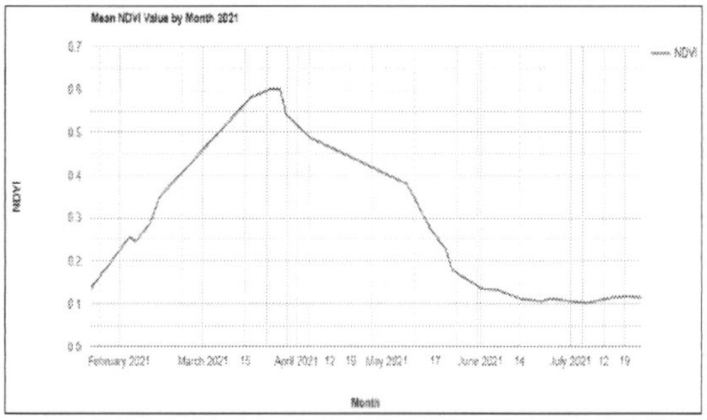

Figure 7. NDVI profile during the growing period (2020-2021).

4.2.2 Reproductive period

The high values of NDVI increase progressively from the end of December until the beginning of February, and from the end of February, we notice that there is a stop of increase and a stabilization of the values of NDVI until the middle of March where the NDVI restores their increase again, and arrives at its peak in the middle of April (Fig.7).

This contribution has been peer-reviewed.
https://doi.org/10.5194/isprs-archives-XLVIII-M-1-2023-183-2023 | © Author(s) 2023. CC BY 4.0 License.

The intensity occurs due to the important chlorophyll activity during the reproductive phase which consequently an intensity of green pigment of plant cover which positively influenced by the amount of reflectance. The breakdown of the increase and stabilization of NDVI values in the curves between mid-February and mid-March is influenced by the climate condition (frost during January in the case of durum wheat), which causes damage to the vegetation cover according to the climate observation during 2020-2021 (January is the coldest month known for frost).

4.2.3 Maturation period

From the end of April until June, we observe the beginning of a decrease in the value of NDVI. These decreases occur because the level of intensity of green pigment that corresponds to this period is low due to the change of color and the fall of leaves that according climatic conditions that affect damage that cause a sharp decrease in grain production (Fig.7).

The minimum values of the NDVI are located in this period corresponding to the driest period in the year.

4.3 Spatial distribution of Cereal crop

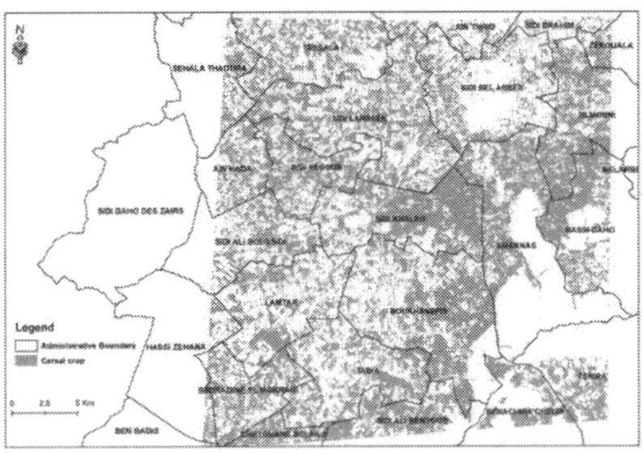

Figure 9. Geographical extent of cereal crop.

The first remark that can be extracted after reading the map of NDVI cover distribution in the department of Sidi bel Abbes (Fig.9). We can say that the vegetation cover is mainly and abundantly located in the plain area and they gradually decrease towards the north of the plain toward the tessala region. This difference in the distribution of vegetation cover is due to the nature of the soil and the climate condition in this region.

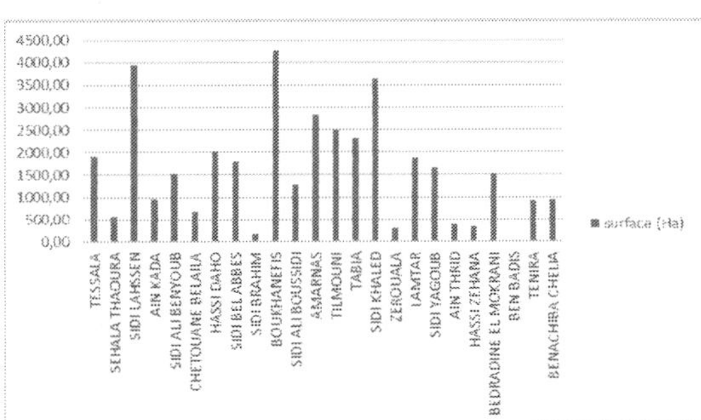

Figure 9. Graphical representation of the results obtained.

This graphical treatment brings out the geographical distribution of cereal growing in the 22-department covered by our results. The results obtained helped us to determine the geographical context of cereal growing and also to see the location of the potential in terms of surface. According to the graph, the potential communes in terms of surface area, we have recorded three communes in particular: Boukhanifis (4255 ha), Sidi lahssen (3943 ha) et Sidi Khaled (3634 ha) (Fig.10).

5. CONCLUSION

Analysis and determination of the periods of the cereal growth, made possible today using to the satellite imagery. The time series images during agricultural season (2020-2021), allows to determine the cereal zone of the region and to follow the growth for the different stage using thresholding NDVI values.

The results obtained from the processing of satellite images show that there is a disparity in the agricultural areas by municipality, and this is due to the geographical characteristics of the area, also the climate, which is one of the important factors.

Meteorological analysis, allowed us to determine the general characteristics of climate for the agricultural year (2020-2021), which is characterized by a continental climate in which the cold winter and hot summer and dry with some winds that can cause damage to agricultural production.

The NDVI values were successfully estimated to study the agricultural zone of the region. While the plain area has a high cereal cover.

The cereal phenology plays a necessary role in the analysis of vegetation cover, we must choose very carefully the date of extraction of the satellite image involved to assess this biological parameter in future research according to temporally data or satellite data rich in terms of spectral bands, spectral data or field measurements by spectrometer.

In terms of this work, this research requires field work in which the cereal-growing areas will be well identified to have a good characterization of this potential which is important for this country to preserve its food security.

ACKNOWLEDGEMENTS

I would like to thank the Algerian Space Agency (Algeria) for allowing me to participate in this scientific event. I would like to recognize my co-authors for their contribution to this work.

REFERENCES

Bessaoud, O., Pellissier, J.-P., Rolland, J.-P., & Khechimi, W. (2019). Rapport de synthèse sur l'agriculture en Algérie [PhD Thesis]. CIHEAM-IAMM.

Chehat, F. (2007). Analyse macroéconomique des filières, la filière blés en Algérie. Projet PAMLIM «Perspectives agricoles et agroalimentaires Maghrébines Libéralisation et Mondialisation» Alger, 7-9.

Djermoun, A. (2009). La production céréalière en Algérie : Les principales caractéristiques. Nature & Technology, 1, 45.

This contribution has been peer-reviewed.
https://doi.org/10.5194/isprs-archives-XLVIII-M-1-2023-183-2023 | © Author(s) 2023. CC BY 4.0 License.

Estoque, R. C., & Murayama, Y. (2015). Classification and change detection of built-up lands from Landsat-7 ETM+ and Landsat-8 OLI/TIRS imageries : A comparative assessment of various spectral indices. Ecological indicators, 56, 205-217.

Guechi, I., Gherraz, H., & Alkama, D. (2021). Correlation analysis between biophysical indices and Land Surface Temperature using remote sensing and GIS in Guelma city (Algeria). Bulletin de la Société Royale des Sciences de Liège.

Maselli, F., & Rembold, F. (2001). Analysis of GAC NDVI data for cropland identification and yield forecasting in Mediterranean African countries. Photogrammetric Engineering and Remote Sensing, 67(5), 593-602.

Mizushima, A., & Lu, R. (2013). An image segmentation method for apple sorting and grading using support vector machine and Otsu's method. Computers and electronics in agriculture, 94, 29-37.

Otsu, N. (1979). A threshold selection method from gray-level histograms. IEEE transactions on systems, man, and cybernetics, 9(1), 62-66.

Safia, A., Mohamed, B., Lotfi, B. T., Farouk, M., & Samir, M. M. (2015). Contribution to the Characterization of Peri-urban Natural Areas of Sidi Bel Abbes Town (Northwestern Algeria).
Sahoo, P. K., Soltani, S., & Wong, A. K. (1988). A survey of thresholding techniques. Computer vision, graphics, and image processing, 41(2), 233-260.

Shi, K., Huang, C., Yu, B., Yin, B., Huang, Y., & Wu, J. (2014). Evaluation of NPP-VIIRS night-time light composite data for extracting built-up urban areas. Remote Sensing Letters, 5(4), 358-366.

Zhao, Q., Yu, L., Li, X., Peng, D., Zhang, Y., & Gong, P. (2021). Progress and trends in the application of Google Earth and Google Earth Engine. Remote Sensing, 13(18), 3778.

This contribution has been peer-reviewed.
https://doi.org/10.5194/isprs-archives-XLVIII-M-1-2023-183-2023 | © Author(s) 2023. CC BY 4.0 License.

DETECTION OF PHYTOGEOGRAPHIC INDICATORS WITH REMOTE SENSING AND GEOARCHAEOLOGY FOR DEVELOPMENT OF A LONG-TERM CLIMATE MONITORING STATION IN ISRAEL

Neil Manspeizer[1*], Arnon Karnieli[1]

[1]Remote Sensing Laboratory, Blaustein Institutes for Desert Research, Ben Gurion University of the Negev, Sede Boker Campus 84990, Israel - (neilm@post.bgu.ac.il, akarnieli@bgu.ac.il).

KEY WORDS: geoarchaeology, remote sensing, vegetation dynamics, climate monitoring, Mediterranean, UN SDG 13

ABSTRACT:

The Mediterranean basin is a strong candidate for monitoring climate change successfully, a priority of UN SDG 13, because of its long settlement history. Human-environmental contact has typically hindered monitoring efforts because the natural indicators that enable dependable monitoring have been altered anthropogenically. This research describes a two-step process which turns that synanthropism into an advantage through geoarchaeology, remote sensing, GIS, meteorology, and floristic classification. First, archaeological survey data were employed in a semi-arid region of Israel to identify areas of more and less intensive land-use over 2300 years. Trend surfaces were derived in GIS from ancient periods of most intense monoculture agriculture (Hellenistic, Byzantine, and Ottoman). 2020 land-cover was classified using VENµS satellite wet / dry season NDVI imagery and an airborne LiDAR canopy height model. This was reclassed after fieldwork to three broad land-cover classes and cross-tabulated with the 2300-year land-use intensity model in GIS. The results indicate a land-use legacy whereby woody perennial species are more abundant where land-use was cumulatively less intense. Second, unmanned aerial vehicle surveys were conducted in these areas of less-intense land-use. There, on lightly grazed lands where woody encroachment occurs but chamaephytes regenerate due to the grazing, microplots were established. Phanerophyte and chamaephyte shrubs were identified based on Raunkiaer type and the species associated with chorotype (geographic origin). This was compared with aridity index values derived for the study area between 2010-2022. The study provides a prototype, using dependable phytogeographic indicators, that may be developed into a long-term climate monitoring station (LTCMS).

1. INTRODUCTION

One difficulty of challenging received wisdoms (Leach and Mearns, 1996) in this post-colonial period is establishing tools to develop new knowledge sets. That is true for climate science, especially regarding the decentralization of climate change data. Tools that enable nations to autonomously monitor climate change help the UN to achieve sustainability development goals (SDG) by promoting self-reliance and regional stability. The result is replicable science which utilizes remote sensing to develop knowledge sets from within the decentralized context (Ostrom, 2001). This paper addresses UN SDG goal 13 (climate change) through explanation of a prototype long-term climate monitoring station (LTCMS). The LTCMS, as a knowledge development tool, aims toward participatory development goals with a methodology that is transferable technologically. Climate monitoring would be ongoing under the SDG umbrella with data sharing through an "open access" data repository. The key lies in correct identification of the phytogeographic indicators of climate change within their anthropogenic-natural setting which is rigorous in scope. Typically, human disturbance is viewed as detrimental to use of phytogeographic indicators of climate change because the landscape has been altered. That is indicative of a well-known discussion in the literature concerning succession and dynamics of Mediterranean garrigue and maquis vegetation following disturbance (e.g., Giourga et al., 1998). The research presented here, as a methods paper, promotes the opposite idea, namely that anthropogenic activity associated with land-use (LU) can assist in identification of phytogeographic indicators for climate change monitoring.

The method incorporates utilization of synanthropic shrub-land vegetation, or flora developed in proximity to human activity, as indicators. That is accomplished by focusing geographically on two types of disturbance along the synanthropic axis: (1) long-term agricultural LU intensity that causes variable land-cover (LC) patterns due to differential primary and secondary succession of synanthropic vegetation following abandonment (Manspeizer and Karnieli, 2023); and (2) short term disturbance of woody perennial succession through a proximate cause, such as grazing, that promotes "arrested succession" of woody encroachment (Manspeizer, 2006) but also enables regeneration of sub-shrub species. That two-stage process of disturbance along the synanthropic axis is typical of many semi-arid rangelands in the Mediterranean because of the long human settlement history. Of the twenty-two countries around the Mediterranean basin, 15% of their total land area was considered permanent meadow or pasture in 2020 (FAO, 2023) which represents a potentially large-scale system to monitor climates. This synanthropic window between natural and human-disturbed landscapes in shrub-lands requires a multi-disciplinary approach for study: (1) archaeological survey data are developed into a geographic information systems (GIS) database and used to identify areas of more and less intensive LU over a long-term period; (2) comparison of current land-cover (LC), derived from remotely sensed imagery, with GIS derived LU intensity; and (3) areas of less intense long-term LU that demonstrate arrested succession from a proximate cause, such as light grazing, are focused upon for climate change study through species level shrub mapping.

At an even larger scale, semi-arid lands account for approximately 15% of the earth's land surface and are comprised of agricultural areas defined as rainfed, cropland and rangeland for almost 1/3 of the world's population (United Nations, 2011). Utilization of the Mediterranean basin as a prototype for how other semi-arid regions can monitor climate change represents an

* Corresponding author

https://doi.org/10.5194/isprs-archives-XLVIII-M-1-2023-189-2023

innovative use of rangelands and the potential to develop a global database. The hypothesis for the study is based on previous works in botany that describe phanerophyte shrub and chamaephyte sub-shrub presence along gradients related to climate such as elevation and temperature (Di Biase et al., 2021; Schmiedel et al., 2012). Climate related gradients can help to distinguish between plants based on physiological adaptations (Liphschitz and Lev-Yadun, 1986) and life forms (Danin and Orshan, 1990). For example, phanerophytes are associated with the moister part of the precipitation gradient and more semi-arid phytogeography. Chamaephytes, on the other hand, are associated with the drier part of the precipitation gradient and more arid phytogeography (Keshet et al., 1990; Margaris et al., 1992). The prototype LTCMS presented in this paper deals with climate gradients by correlating shrub life-form presence across geographic zones, for example from arid to semi-arid to sub-humid. In this manner, the chorotypes associated with species compositions in place may be associated with the convergence of climate systems reflected in local aridity. Therefore, shrubs as phytogeographic indicators can be monitored for significant changes to community composition to help define climate regimes and their changes.

1.1 Case Study: Maresha Hinterland

The best method to demonstrate this approach is a case study developed in the southern Judean foothills, Israel. Previous fieldwork describes that landscape as a Mediterranean phanerophyte maquis matrix with inter-space Mediterranean and Irano-Turanian chamaephytes associated with garrigue, hard-rock, and shrub-steppe habitats (Manspeizer and Karnieli, 2021). The area of study[1] lies in the hinterland south of *Tel* Maresha[2] (archaeological mound). The area is rich in settlement history with a wealth of material cultural remains in the landscape. Today, 96.7% of the hills in the study area lie within an overlapping military zone / nature reserve with light cattle grazing (for definitions see Kirk et al., 2019), which accounts for its character as "open space". The study area is also part of the foothills ecological corridor in Israel that connects the southern Negev Desert with the northern Galilee Mountains west of the Judean Mountains to promote wildlife migration and plant dispersal. Antiquities Authority (IAA) survey data were incorporated into a GIS database and cross-tabulated with a LC classification from 2020 to determine areas of more and less intense long-term LU. Second, unmanned aerial vehicle (UAV) surveys were conducted in areas of less-intense long-term LU, namely the Gad Hills nature reserve. There, on lightly grazed lands, where woody encroachment occurs and phytogeographic shrub indicators may regenerate, all woody perennials including phanerophyte and chamaephyte were mapped. These shrubs were classified according to percent presence of chorotypes and compared with the aridity index (AI) values between 2010-2022 derived from regional potential evapotranspiration (PET).

The three periods of most intensive ancient monoculture agriculture were chosen to represent major anthropogenic disturbance at the landscape scale. The location of Hellenistic oil presses, Byzantine winepresses, and Ottoman animal pens, were developed into a GIS database from IAA survey data (Dagan, 2006). Archaeological excavations at Tel Maresha and its hinterland to the south, revealed a large-scale olive oil industry from the Hellenistic period (333–63 BCE) in the study area then known as northern Idumea. Twenty-seven 'Maresha type' industrial-scale olive oil presses were found at Tel Maresha and another fifteen / sixteen at sites in the hinterland (Kloner and Sagiv, 1989). Based on the number and size of oil presses, estimates indicate that 450 mt of olive oil, or 500,000 l, could have been produced annually by northern Idumea. That would have required between 100,800-262,500 olive trees in the 42 km^2 study area based on sixteen to twenty-five olive trees per 0.4 ha (Vossen, 2007). During the Byzantine period (324–640 CE), a vast rural Christian population inhabited the study area. The agricultural change included a leap from thirty-seven winepresses found with Roman period pottery to 183 winepresses with Byzantine period pottery which conforms to descriptions of industrial scale viticulture in Byzanytine period Palaestina. Based on ancient methods of viticulture (Weber, 2009), the volume of wine capable of being produced in the study area during the Byzantine period is 2667 mt at 2 t of grapes per ha and 230 l of wine per ton of grapes. During the Ottoman period (1516–1917 CE) there is significant archaeological evidence of renewed LU after nearly a millenia of secondary succession during the Early Muslim and Medieval periods (640–1516 CE). Remains of several pastoralist movements were identifiable within the archaeological survey including twenty-six Ottoman period animal pens found in the study area (Dagan, 2006). Based on an average 9.5x7.5 m per pen and two-three sheep or goats per 1 m^2, average carrying capacity of the study area was between 3705 to 5558 animals/yr (USDA, 2006). The LU history is essential to understanding the current LC and while topoedaphic effects, associated with aspect, elevation, or soil type, are evident in the landscape they do not describe the most dominant landscape scale vegetation pattern. That is best described as a LU legacy (Foster et al., 2003) in which disturbances, such as long-term ancient agricultural LU, have affected the substrate causing a significant discernable modern LC.

Questions in the literature regarding Mediterranean vegetation dynamics concentrate on whether LC patterns are driven by anthropogenic or natural factors. For example, anthropogenic stress will initially in the short-term cause an increase in annual cover that is replaced by shrub species (Giourga et al., 1998: 589-596). The case study presented here was established to help answer these questions through juxtaposition of the two drivers. Results of the initial research (Manspeizer and Karnieli, 2021) demonstrate the LU legacy which contributes to the debate on Mediterranean vegetation. That study stated that long-term anthropogenic stresses, as demonstrated through the archaeological survey work, drives a more dominant long-term LC state than described in previous literature. However, this left questions regarding the role of climate and other environmental factors that were described in more detail in a second study (Manspeizer and Karnieli, 2023). There, it was clarified, following UAV vegetation mapping, that anthropogenic stresses do produce different LC patterns of the same floristic association (phytogeographic state) given continuity in climate regime. Thus, Tzanopoulos et al. (2005: 27-38) write that maquis species are replaced by sub-shrub species if certain thresholds of anthropogenic use are exceeded. However, in addition to synanthropic relations, changes in climate regime were hypothesized to also drive changes in floristic composition. Monitoring climate change could therefore be accomplished through the LTCMS, especially with use of remote sensing methods to detect appropriate phytogeographic indicators. This paper offers an empirical approach to deconstruct the shrub composition and the temporal aspects of climate change (variability / flucturation) as they relate to the phanerophyte maquis matrix compared with the chamaephyte garrigue.

[1] 42 km^2 rectangle centered at Lat. 31.54135, Long. 34.88722

[2] Tel Maresha was inscribed a UNESCO world heritage site in 2014.

This contribution has been peer-reviewed.
https://doi.org/10.5194/isprs-archives-XLVIII-M-1-2023-189-2023 | © Author(s) 2023. CC BY 4.0 License.

2. ANALYSIS

2.1 Data and Methodology

The methodology comprised two stages: (1) LU/LC analysis with geoarchaeology, remote sensing and GIS to isolate less-anthropogenically influenced vegetation within the synanthropic gradient; and (2) climate analysis based on vegetation mapping in less-intensely used areas and species correlation with chorotype and climate variables. Four materials were assembled within a database used for the fieldwork and analysis: (1) Israel Antiquities Authority (IAA) archaeological survey data for ancient land-use; (2) Remotely sensed data (VENμS spaceborne, LiDAR airborne, and UAV for land-cover; (3) climate related data from the Israel Meteorological Service (IMS) and US Department of Energy (DOE); and (4) field survey for floristic classification and literature review of the floristic association

2.1.2 LU/LC Analysis was comprised of land-use and land-cover components with IAA archaelogical data and 2020 remotely sensed land-cover classification. The IAA survey of the study area, conducted between 1982-1986 by Dagan (2006), included the Amazya and Lakhish sheets (1:20000 scale). The survey maps contain information organized as 'sites' where archaeological remains were found based on the maximalist approach. The IAA surveyors described the material remains (e.g., pottery, agricultural installations, building stones) that were identified and recorded according to location and historical periodization. In the study area were 124 relevant archaeological survey sites on the Amazya sheet and 72 on the Lakhish sheet. Sites were categorized based upon: (1) type of remains, such as settlement, farmstead, building(s), and agricultural installations (presses, terraces and kilns); and (2) pottery scatter per period associated with each find. 196 IAA survey sites in that study area were digitized and stratified in the GIS database according to the relevant findings between the Hellenistic and Ottoman periods.

The three ancient periods of most intensive monoculture agriculture were chosen to represent major historical anthropogenic disturbance. This decision was based on the agricultural extensification visible in the material culture remains. Associated installations, namely subterranean Hellenistic oil presses, rock-hewn surface Byzantine winepresses, and stone construction Ottoman animal pens, were digitized in GIS as point files from the IAA survey data. Distance images from the agricultural installations for each of the three periods were derived. A reciprocal of the distance image was derived to represent land-production pressure in proximity to installation. These images were stretched between 0-255 such that a buffer around the installation reflected an impact that decreased along a gradient as distance increased. This notion has been described in range ecology literature for grazing impacts around pens or watering holes (Sasaki et al., 2008). A ninth-order polynomial surface trend image was derived from the distance images and stretched again (0-255) for each period. The three stretched trend images were then averaged into one image that describes the cumulative land-use intensity over 2300 years.

2020 land-cover data were derived from two sources: (1) spaceborne Vegetation and Environment monitoring New Micro Satellite (VENμS) imagery; (2) airborne Light Detection and Ranging (LiDAR) imagery. Two 2020 VENμS satellite images of the study area were acquired with 10 m spatial resolution that represent vegetation peak (March) following the wet winter season in the Mediterranean region and vegetation trough (September) following the dry summer season. Red (666.2 nm center) and Near Infrared (861.1 nm center) bands were chosen, and the two dates analyzed using the normalized difference vegetation index (NDVI). NDVI, a sum-difference ratio, has been used to discern vegetation from bare soil areas studies and assist study of seasonal differences in vegetation cover (Karnieli, 2003; Karnieli et al., 2002).

$$NDVI = (NIR-Red) / (NIR+Red) \quad (1)$$

LiDAR is an active laser system that detects canopy height and topographical differences with high spatial resolution. The two common LiDAR data products are the Digital Terrain Model (DTM) that represents elevation and the Digital Surface Model (DSM) that provides the height of the vegetation at each point. The LiDAR data was acquired between February 2015 and July 2016 at 1 m spatial resolution. The DSM was smoothed (DSMs) and the DTM subtracted to create a normalized Digital Surface Model (nDSM), i.e., the vegetation height.

$$nDSM = DSMs - DTM \quad (2)$$

The nDSM image and two NDVI images from March and September 2020 were used as three bands in an unsupervised LC classification with TerrSet (2017). The resulting classification, with fifteen classes based on spectral reflectance and canopy height, was used to ground-truth the LC in situ at the end of the rainy season (March 2021). Twenty-two points were chosen to ground-truth the classification and the fifteen classes were reclassified into three vegetation classes: bare, shrub, and tree.

The intermediate output of the GIS work was one model of ancient LU intensity and one model of 2020 LC. GIS analysis was standardized spatially between LU / LC at 1 m resolution, All alluvial areas, territory in the Palestinian Authority, and a high-power line were masked which left the hills of the study area. The trend surface image, 2300 years average cumulative LU intensity, was stretched as an integer binary image to 255 classes. The lowest (0-92) and highest (207-255) levels of LU intensity contained no assignments as they occurred in the masked areas. Crosstabulation, a GIS method of a database query, was conducted between the cumulative trend image and 2020 LC. Spatial crosstabulation, a statistical tool to compare categorical data sets, allows each new category to represent the original data sets. All old pixel categories are preserved in the analysis and re-sorted into new combined categories such that each new pixel has two old values (LU intensity and vegetation class). The results of the crosstabulation reflect the range of values between 92-207 on 55% of the image or 19.8 sq km within the hills of the study area.

2.1.3 Climate Analysis was conducted in Gad Hills Nature Reserve, after the LU/LC analysis determined that the reserve corresponded with less intensely used land based on archaeological data[3]. Four plots in the hills were selected based on stratified-random sampling that contained garrigue / maquis communities only rather than species related to the tributaries. Aerial surveys were flown at the end of the winter wet season during NDVI peak between April 10 – May 7, 2022 (Karnieli, 2003). Four microplots in each plot were photographed at nadir from a height of 50 m using the DJI mini2 UAV and autonomous flight plans in Litchi App. Ground sample distance (GSD) was calculated at 1.82 cm based on flight height, focal length, and fields of view (FOV). All aerial photos were georeferenced in QGIS (2022) with 0.125m georeferenced orthophotos from the Survey of Israel. Twenty ground control points were established for each multiplot with average residual (RMS) error of 15.56

[3] Israel National Nature and Parks Authority Permit #043-2022

which translates to an average 28.31 cm on the ground (GSD * RMS error) for the four plots.

The goal was to map percent presence of shrub species and the UAV photos assisted only in the detailed digitization of the vegetation data. Two microplots were chosen from each of four plots (eight total) and all woody vegetation within were mapped over the wet season 2022-2023. The final percent presence included only phanerophyte shrubs and chamaephyte sub-shrubs that were not also climbers or parasites. In all 13411m^2 was surveyed in the eight microplots and the coverage checked through GPS tracking. Chamaephyte regeneration due to wet season changes was accounted for by fieldwork conducted three different times between December 2022 – February 2023. The published Ceratonio-Pistacion lentisci typicum association that describes the regional area was first developed by Zohary and Orshan (1959: 288-290). It was cross-referenced with the online database by Danin and Fragman-Sapir (2016) to build a "published cross-referenced association" of potential shrub species along with their chorotypes (region of origin) and habitat.

Specific temperature and precipitation data do not exist for the study area so seven nearby meterological stations (Kiryat Gat, Kibbutz Halamed'hay, Lehavim, Dorot, Negba, Shani and Rosh Tzurim) were used to provide a proxy. The climate data for the seven stations was established from two sources, the IMS climate archive data (I.M.S., 2023) and the DOE solar position calculator (SOLPOS, 2023). Potential evapotranspiration (PET_{HS}) based on Hargreaves-Samani (Almorox and Grieser, 2016) was derived between 2010-2022 as the climate variables were existent for all seven stations during that period (3). AI values were then derived from precipitation and PET_{HS} (4) for the seven stations.

$$PET_{HS} = 0.135*KT*(T_{av}+17.8)*(T_{max} - T_{min})0.5 *R_a*d \quad (3)$$

where $KT = 0.17$ (coefficient default)
T_{av} = Average Temp / day (C°),
$T_{max/min}$ = Temp / day (C°)
R_a = Extraterrestrial solar radiation in mm/d
d = number days in each iteration (1 = daily)
Avg PET / day (mm day^-1)

AI is calculated as:

$$AI = P / PET_{HS} \quad (4)$$

where P = Precipitation / yr (mm)

GIS analysis was used to develop AI surfaces between 2010-2022 at the meteorological stations. The surfaces were interpolated at 10 m spatial resolution for the 7 points using TerrSet (2017) with no six-point radius to smooth the results. AI for the study area specifically was derived from the interpolated surfaces (2010-2022). Aridity was compared with the chorotypes of shrub species identified and the chorotype zones as mapped previously by Danin and Orshan (1999: 16). A more precise representation of the relationship between aridity and chorotype would reflect shifting chorotype contours as vegetation compositions changed due to climate variability and fluctuation. That is beyond the scope of this paper, therefore a functional representation was derived by extracting statistical values from the AI surfaces in GIS between 2010-2022 for the two chorotypes found there. Chorotype shift, a reflection of the climate changes evident in the vegetation composition, may be expressed statistically as a range between: (1) max study area AI value minus min chorotype one AI value; and (2) min study area AI value minus max chorotype two AI value.

$$R : \{(AI_{S"A\_MIN} - AI_{CH1\_MAX}), (AI_{S"A\_MAX} - AI_{CH2\_MIN})\} \quad (5)$$

where R = Range of chorotype shift in study area
$AI_{S"A\_MAX/MIN}$ = max/min AI value (Study Area)
AI_{CH1_MAX} = chorotype one min AI value (M-IT)
AI_{CH2_MIN} = chorotype two max AI value (M)

This representation enables research along a gradient as conditions move between favorable and unfavorable over time relative to chorotype and species (sect. 3.5).

3. RESULTS AND DISCUSSION

Results of the study are limited to: (1) LC classification in 2020; (2) LULCC crosstabulation; (3) Climate analysis; (4) UAV vegetation mapping; and (4) Phytogeographic-Climate correlation.

3.1 Land-cover Classification

The 2020 classification used VENµS NDVI seasonal imagery and vegetation height from the LiDAR nDSM. That classification described fifteen distinct spectral classes and three vegetation covers: (1) five classes of low annual cover between 0.09 to 0.13 m; (2) five classes of shrub between 0.10 to 2.42 m; and (3) five classes of tree species from 5.5 to 7.2 m. The masked 2020 classification included 56% annuals (grasses and flowers) and bare areas (rocky), 43% shrubs, and 1% trees.

3.2 Land-use Legacy

Cumulative LU intensity was crosstabulated with the 2020 LC images using a hard GIS crosstabulation. After normalization, the crosstabulation indicates that areas of least intense LU contained proportionally higher ratios of shrub and tree cover in 2020 than areas of most intense LU that contained higher ratios of bare regions in 2020. The results demonstrate that anthropogenic disturbance produces long-term LC effects identifiable in garrigue and maquis communities that are not due to topo-edaphic effects. That was reported previously in detail (Manspeizer and Karnieli, 2021, 2023), concurs with dominant theories in disturbance ecology (White and Jentsch, 2001), and demonstrates a LU legacy (Foster et al., 2003).

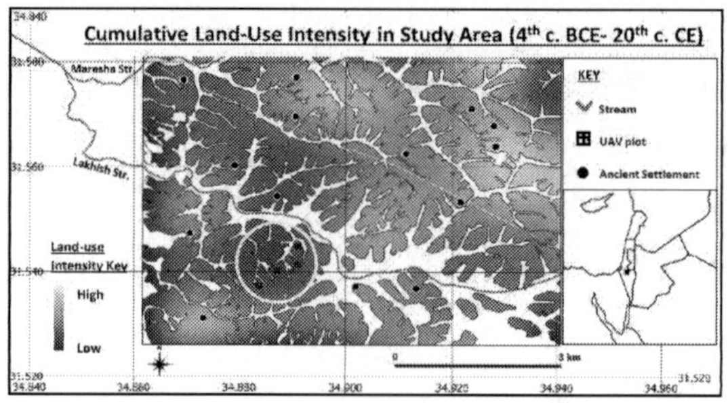

Figure 1. 2300 years cumulative land-use intensity in the study area from trend analysis with major ancient settlements and location of UAV vegetation plots in Gad Hills reserve circled.

3.3 Climate Analysis

The climate classification model for this paper is constructed around the aridity index derived from P and PET_{HS} calculations. Like Koppen-Trewartha (KT), it is based on: (1) climate analysis of meteorological variables (precipitation, solar irradiance and temperature); and (2) vegetation compositions with associated species chorotypes. This classification process requires several steps to utilize phytogeographic indicators of climate change, namely: (a) development of aridity indices over time; (b) shrub

This contribution has been peer-reviewed.
https://doi.org/10.5194/isprs-archives-XLVIII-M-1-2023-189-2023 |

species and chorotype mapping; and (c) ultimately development of a chorotype based atmospheric circulation model (future goal). In step one, regional AI were developed from PET_{HS} data. Figure 2 shows AI of the seven meteorological stations and study area according to the 12-year period of recorded data (2010-2022). The non-normal distribution of data demonstrates the seasonal conditions described by Raunkiaer as favorable or unfavorable (Danin and Orshan, 1990). For example, the study area is defined as semi-arid (0.2-0.5) based on the results, although in 2010-2011 and 2016-2017 the AI dipped into arid conditions (<0.2). This alone is insufficient to declare whether the conditions are favorable or unfavorable phytogeographically without inclusion of chorotype regions which is a spatial exercise.

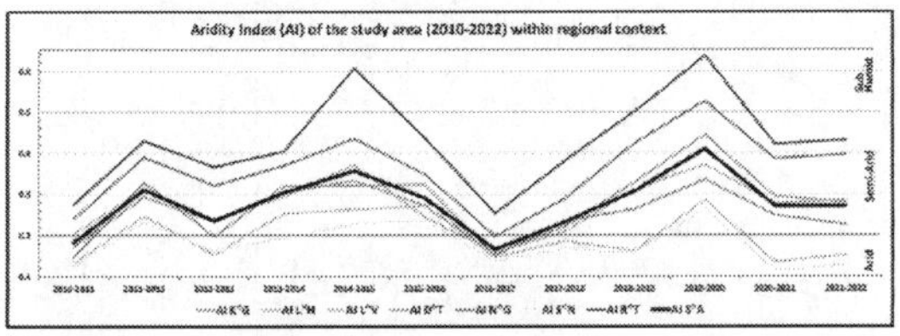

Figure 2. AI of the study region between 2010-2022 (Sources: I.M.S., 2023; SOLPOS, 2023). The seven stations are Kiryat Gat (K"G), Kibbutz Halamed'Hay (L"H), Lehavim (L"V), Dorot (D"T), Negba (N"G), Shani (S"N) and Rosh Tzurim (R"T). The Study Area (S"A) is depicted in thick black line.

Therefore, GIS interpolation was used to develop AI surfaces between 2010-2022 based on variables at the seven meteorological stations so that could be compared spatially with chorotype region. The shifting AI between 2010-2022 of the study region, evident in figure 2 as non-normal data trends, are viewed as interpolated surfaces in figure 3. Chorotype contours were digitized from phytogeographic maps published by Danin and Orshan (1999: 16) so that they may be associated with the aridity index surfaces. The chorotypes include Mediterranean (M), Irano-Turanian (IT), and Saharo-Arabian (SA) but may often described as a combination between them. The southern Levant is a convergence zone between these three large chorotype regions and locally that convergence demonstrates itself in mixed spatial zones. One goal of the larger project is to understand the percent presence of shrub indicators as related to chorotype in order to redefine these mixed zones and regions as they change over time due to climate change.

The chorotype contours in figure 3 do not reflect any change over the 12 year period and were initially derived in 1987 following a significant mapping effort (Danin and Orshan, 1999). A more ideal representation would reflect shifting chorotype contours because they would represent changes in species composition due to changing climate conditions. That necessitates a more elaborate project, such as the proposed LTCMS, however a provisional method is described in section 3.5. Statistics derived from chorotype regions based on the AI are correlated so the current conditions may be understood relative to favorable / unfavorable. Figure 3 is significant because it is a visual representation of the climate classification system demonstrating both meteorological variables and chorotypes. On its own, that method would suffice as a beginning climate classification systems akin to KT. However the vegetation mapping is necessary to relate the shifting chorotypes, phytogeographic indicators and aridity with climate changes. In that way, chorotype regions, defined by the vegetation that comprises them, may be linked with the climate conditions. To that end, sections 3.4 and 3.5 describe how UAV vegetation mapping may be linked to chorotypes and associated meteorological variables.

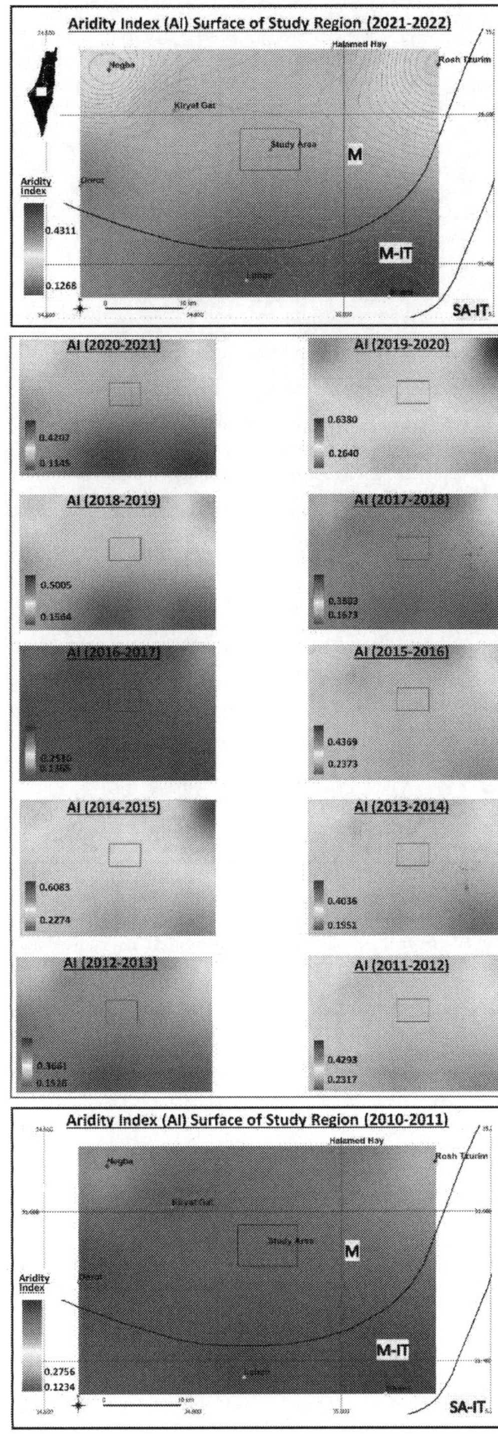

Figure 3. The study area (rectangle) within larger regional chorotype-aridity context derived from interpolation analysis in GIS. Chorotypes M, IT and SA and their combinations after Danin and Orshan (1999: 16) phytogeographical map.

3.4 UAV Vegetation Mapping

Results of the UAV shrub mapping are represented in table 1 according to phanerophyte / chamaephyte species identified in eight microplots during the wet season 2022-2023. 4006 woody plant units were mapped including 3505 phanerophytes and chamaephytes. Percent presence is 64.3% M, 14.3% IT, 14.3% M-IT and 7.1% M-SA with n species = 14 (fig. 4).

Genus species	Name	Raunkiaer	Chorotype	Commonality	Habitat
Pistacia lentiscus	Mastic Bush	Phanerophyte	Mediterranean	V. common	Maquis
Rhamnus lycioides	Palestinian Buckthorn	Phanerophyte	Mediterranean	V. common	Maquis
Quercus coccifera	Kermes Oak	Phanerophyte	Mediterranean	Common	Maquis
Sarcopoterium spinosum	Thorny Burnet	Chamaephyte	Mediterranean	V. common	Garrigue
Ballota undulata	Undulate Horehound	Chamaephyte	Mediterranean	V. common	Garrigue
Chiliadenus iphionoides	Sharp Varthemia	Chamaephyte	Mediterranean	V. common	Hard-rock
Micromeria nervosa	Veined Savory	Chamaephyte	Mediterranean	V. common	Hard-rock
Teucrium divaricatum	Aegean Sage	Chamaephyte	Mediterranean	Common	Garrigue
Fumana arabica	Rock Rose	Chamaephyte	Mediterranean	Common	Garrigue
Teucrium capitatum	Cat-thyme Germander	Chamaephyte	Med-Irano-Turanian	V. common	Garrigue
Phagnalon rupestre	African Fleabane	Chamaephyte	Med-Irano-Turanian	V. common	Hard-rock
Phlomis brachyodon	Short-tooth Phlomis	Chamaephyte	Irano-Turanian	V. common	Garrigue
Noaea mucronata	Thorny Saltwort	Chamaephyte	Irano-Turanian	Common	Shrub-steppe
Kickxia aegyptiaca	Egyptian toadflax	Chamaephyte	Med-Saharo-Arabian	V. common	Garrigue

Table 1. Results of UAV mapping study in the Gad Hills Nature Reserve. Phanerophytes / chamaephytes reflect presence in the winter 2022/23 based on 5 plant leaf-on minimum.

Very common / common shrub species of the published cross-referenced association (sect. 2.1.3) are defined as 75% M, 15% M-IT, 5% IT, and 5% M-SA with n species = 20. A simple linear regression between the percent presence of the published cross-referenced association and 2022-2023 field survey was conducted and reflects the correlation (p-value = 0.009 and R Square = 0.982). However, any climate change that this reflects must be correlated with aridity, which is discussed in section 3.5.

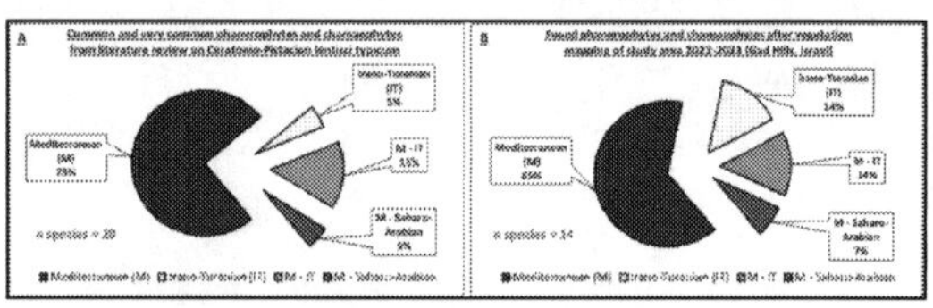

Figure 4. Percent presence of common and very common phanerophytes and chamaephytes by chorotype in the study area: (A) from the published cross-referenced association; and (B) UAV assisted mapping 2022-2023 in Gad Hills, Israel.

The vegetation mapping also provided significant findings regarding the landscape mosaic (Forman, 1995). The microplots represent a mix of garrigue / maquis habitats with a stable patch matrix of phanerophytes and inter-patch chamaephyte species (see fig 5). Based on Raunkiaer type, the phanerophytes are homogenously 100% M chorotype and maquis, while chamaephytes are heterogenoously 55% M, 18% M-IT, 18% IT, 9% M-SA and represent garrigue, hard rock, and shrub-steppe habitats. The results demonstrate that light grazing as a proximate cause of arrested succession helps to slow woody encroachment and encourage chamaephyte regeneration. This is important because distinction between phanerophyte and chamaephyte dynamics can help to focus climate change analysis more efficiently by temporal responses. This is also discussed further in section 3.5, although for the purposes of mapping the mosaic, it is crucial to isolate the background phanerophyte matrix from the regenerating chamaephytes. In that way, once the phanerophyte matrix mapping is complete, monitoring the composition change in the LTCMS is dependent on following the inter-patch chamaephyte regeneration over time. The differentiation may distinguish between evergreen sclerophyll phanerophyte and seasonal dimorphic chamaephyte response to climate changes with regard to variability and fluctuation.

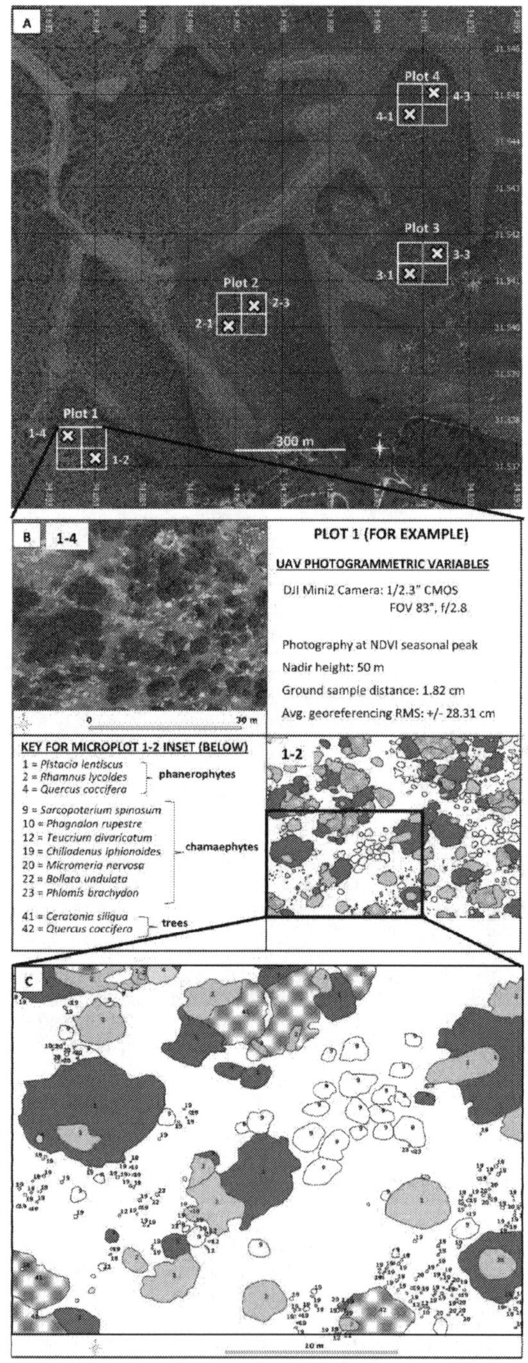

Figure 5. The UAV assisted mapping at scale from (A) the four plots shown on aerial photography; (B) plot 1, for example, with two microplots: microplot 1-4 shown as the UAV nadir photo and microplot 1-2 shown as the digitized product after mapping the vegetation; (C) inset from microplot 1-2, for example, with a legend to demonstrate the phanerophyte matrix with inter-patch chamaephyte species.

https://doi.org/10.5194/isprs-archives-XLVIII-M-1-2023-189-2023

3.5 Phytogeographic-Climate correlation

This section summarizes the results of the study and describes the phytogeographic-climate correlation. That is assisted by theorizing biogeographical regions at scale while mapping the floristic association as a geographic field exercise. The study area, within the Tethyan and African floristic subkingdoms, is comprised of three main bioregions, namely M, IT and SA chorotypes (Loidi, 2021; Takthajan, 1986). This is expressed, in-situ, through vegetation compositions in which the homogenous and mixed chorotypes change over time according to climate conditions. Vegetation is classified according to content of the compositions as percent presence based on plant taxa and units. The unit of the study area has been described as the Ceratonio-Pistacion lentisci typicum association or part of the larger class of carob maquis in Israel (Zohary and Orshan, 1959). Following fieldwork in 2022-2023, the study area is currently defined by percent chorotype presence of phanerophytes and chamaephytes as 64.3% M, 14.3% IT, 14.3% M-IT and 7.1% M-SA (fig. 5).

Due to prolonged synanthropic relationship of the area, there are two main interactions between humans and the landscape. The first is the long-term historical LU impact from monoculture agricultural practices that produced a LU legacy in the garrigue / maquis LC. The second is the modern LU of light grazing that holds the current carob maquis from woody encroachment in a state of arrested succession. The grazing enables chamaephyte sub-shrub regeneration between the patches of phanerophyte shrubs that grow there based on the LU legacy. Shrub species may be linked with chorotype and the evolutionary adaptations associated with climate regime at the chorotype geo-origin. This provides indication of the favorable / unfavorable conditions that limit and enable regeneration for each species over time. This notion has been discussed regarding lag times of vegetation in response to climate change (Wu et al., 2015) in which vegetation demonstrates temporal response, positive and negative, to climate conditions. Figure 6 provides a view of this approach as conditions in the study area move between favorable and unfavorable over time relative to chorotype.

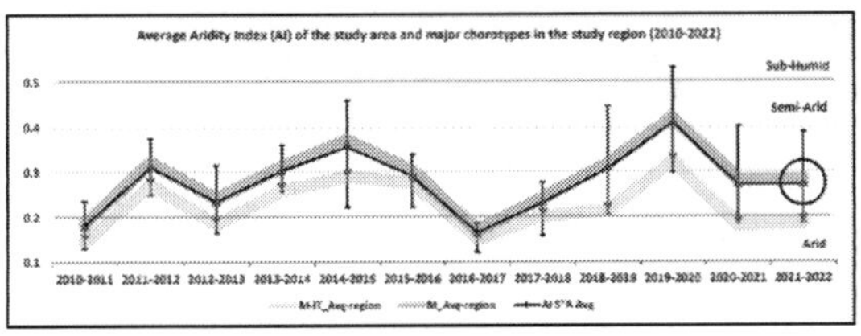

Figure 6. Range of AI in the study area (black line with high-low bars) between M-IT chorotype (yellow line) and M chorotype (green line). Current vegetation mapping circled.

The conditions described are AI values associated with chorotype in the study region. For example, in figure 6, the average M and M-IT AI values were extracted from the spatial range of the chorotypes between 2010-2022. Within that larger region, the conditions in the study area are represented by a range of values (high-low bars) that describe the range of favorable / unfavorable conditions that limit and enable the vegetation composition found there. The results are divided into three: (1) avg AI for M chorotype (2) avg AI for M-IT chorotype; and (3) the range of values between the min study area AI value and max M-IT AI value and the max study area AI value and min M AI value. Favorable / unfavorable conditions in the study area may thus be modeled conducive relatively to the shifting aridity. Where the high bars extend into the wetter M chorotype region there are more favorable conditions for M chorotype species which will flourish. But, when the low bars extend to the drier M-IT chorotype region, the unfavourable conditions to M species restrict their regeneration but enable dispersal of IT chorotype related species.

Vegetation presence changes relative to the climate conditions which indicates a strong relationship between aridity and floristic composition. UAV assisted mapping of woody perennials at the species level, as presented in this paper, enabled further differentiation in temporal response to climate conditions between phanerophytes and chamaephytes. Based on research in climate gradients, discussed in section 1, phanerophytes and chamaephytes respond differently to climatic pulses. This is evident perhaps in the homogeneity of the phanerophyte matrix in comparison with the heterogeneous inter-patch chamaephytes. To demonstrate how the change in species composition can indicate changes in climate based on aridity, chamaephyte and phanerophyte presence (M and IT chorotypes) was extracted from the two data sets. The published cross-referenced association recorded 3 M related phanerophytes, and 15 M / 4 IT related chamaephytes. The recent mapping recorded 3 M related phanerophytes, and 8 M / 4 IT related chamaephytes. Phanerophyte and IT related chamaephyte presence remained constant while M related chamaephyte presence changed by nearly half since the initial mapping of the Ceratonio-Pistacion lentisci typicum association (Zohary and Orshan, 1959).

One explanation may be viewed in figure 6 as the chorotypes shift independently relative to the aridity over time and the vegetation composition in the study area is affected by these conditions. 2010-2011 and 2016-2017 were both years in which the M chorotype aridity dropped to arid conditions that are more conducive to the IT species. These are the unfavourable seasons defined by Raunkiaer and others to describe adaptive strategies in plant life-forms (Danin and Orshan, 1990; Keshet et al., 1990) within a larger evolutionary process. Further knowledge of the range of favorable / unfavorable conditions that limit and enable each species, as well as their geographic ranges, is necessary to clarify how evolutionary adaptations may be associated with climate regime at the chorotype geo-origin. That approach can offer assistance to human related climate change issues by identifying conditions that may also prove limiting over time anthropogenically. The method presented in this paper can lead to scaling up the project successfully, spatially and temporally, through remote sensing platforms and statistical analysis that leads to more conclusive results.

4. CONCLUSION

This methods paper demonstrates that utilization of phytogeographic shrub species in synanthropic conditions as indicators of climate change is meritorious. Thus, floristic association dynamics can continuously be monitored through remote sensing efforts and fieldwork. Those dynamics can indicate the unfavorable or favorable conditions of critical climate changes and inform regarding the human response. We recommend development of a LTCMS in the Gad Hills reserve, Israel, as a prototype for the UN SDG 13 agenda and example for the Mediterranean region. The rigorous scope presented in this paper traverses two complex geographic effects: (1) long and short-term anthropogenic disturbances that have differential effects on LC and the substrate; and (2) notions that vegetation presence is a natural indicator of climate conditions interpretable based on chorotype. Further work intends to examine this disjointed trajectory in more depth because it accurately depicts the nexus in human-environmental relations. Ultimately, the process will be linked to atmospheric circulation models for a greater understanding of climates.

https://doi.org/10.5194/isprs-archives-XLVIII-M-1-2023-189-2023

REFERENCES

Almorox, J., Grieser, J., 2016. Calibration of the Hargreaves–Samani method for the calculation of reference evapotranspiration in different Köppen climate classes. *Hydrology Research*, 47(2), 521-531.

Dagan, Y., 2006. Archaeological Survey of Israel, Map of Amazya (109) and Lakhish (98). Israel Antiquities Authority. survey.antiquities.org.il/index_Eng.html#/MapSurvey/18 (30 January 2023).

Danin, A., Fragman-Sapir, O., 2016. Flora of Israel. flora.org.il/en/en/ (30 January 2023).

Danin, A., Orshan, G., 1990. The distribution of Raunkiaer life forms in Israel in relation to the environment. *Journal of vegetation science*, 1(1), 41-48.

Danin, A., Orshan, G. 1999: *Vegetation of Israel* (Vol. 1). Backhuys Publishers, Leiden.

Di Biase, L., Pace, L., Mantoni, C., Fattorini, S., 2021. Variations in Plant Richness, Biogeographical Composition, and Life Forms along an Elevational Gradient in a Mediterranean Mountain. *Plants*, 10(2090), 1-20.

FAO, 2023. FAOSTAT, Land-Use. Food and Agriculture Organization of the United Nations (FAO). fao.org/faostat/en/#data/RL (1 February 2023).

Forman, R. T., 1995. Some general principles of landscape and regional ecology. *Landscape ecology*, 10, 133-142.

Foster, D., Swanson, F., Aber, J., Burke, I., Brokaw, N., Tilman, D., Knapp, A., 2003. The Importance of Land-Use Legacies to Ecology. *BioScience*, 53(1), 77-88.

Giourga, H., Margaris, N., Vokou, D., 1998. Effects of grazing pressure on succession process and productivity of old fields on Mediterranean islands. *Environm. Management*, 22(4), 589-596.

I.M.S., 2023. Israel Meteorological Service Database Israel Meteorological Service. ims.gov.il/en/data_gov (30 January 2023).

Karnieli, A., 2003. Natural vegetation phenology assessment by ground spectral measurements in two semi-arid environments. *International Journal of Biometeorology*, 47(4), 179-187.

Karnieli, A., Gabai, A., Ichoku, C., Zaady, E., Shachak, M., 2002. Temporal dynamics of soil and vegetation spectral responses in a semi-arid environment. *International Journal of Remote Sensing*, 23(19), 4073-4087.

Keshet, M., Danin, A., Orshan, G., 1990. Distribution of ecomorphological types along environmental gradients in Israel: 1. Renewal bud location and leaf attributes. *Ecologia mediterranea*, 16(1), 151-161.

Kirk, D. A., Hébert, K., Goldsmith, F. B., 2019. Grazing effects on woody and herbaceous plant biodiversity in northern Tunisia. *PeerJ*, 7, e7296.

Kloner, A., Sagiv, N., 1989. בתי בד מתקופה ההליניסטית במערות מרישה [Maresha: Olive Oil Production in the Hellenistic Period]. *Niqrot Zurim*, 15, 17-65.

Leach, M., Mearns, R. 1996: *The lie of the land: challenging received wisdom on the African environment*. James Currey Ltd, Oxford.

Liphschitz, N., Lev-Yadun, S., 1986. Cambial activity of evergreen and seasonal dimorphics around the Mediterranean. *IAWA journal*, 7(2), 145-153.

Loidi, J., 2021: Mapping of Biogeographical Territories: Flora, Vegetation and Landscape Criteria. In F. Pedrotti & E. Box (Eds.), *Tools for Landscape-Scale Geobotany and Conservation* (pp. 21-36). Springer, Cham.

Manspeizer, N., 2006. The Mediation of a Functional-Aesthetic Landscape in New England: New Braintree, Massachusetts and the Identity of Place. Middle States Geographer. *Middle States Geographer*, 39, 84-92.

Manspeizer, N., Karnieli, A. 2021. Identification of a long-term land-use legacy in the southern Judean foothills, Israel [Master's Thesis]. Ben Gurion University.

Manspeizer, N., Karnieli, A., 2023. Isolating phytogeographic shrub indicators through geoarchaeology and exploration of a long-term land-use legacy in Israel for aridity-chorotype climate classification. *Geomorphologie* (pending publication).

Margaris, N., Giourga, C., Koutsidou, E., 1992. Adaptive strategies of plants dominating phryganic ecosystems. *Ekistics*, 59(356/357), 301-305.

Ostrom, E., 2001: Decentralization and development: The new panacea. In K. Dowding, Hughes, J., Margetts, H. (Ed.), *Challenges to Democracy. Political Studies Association Yearbook* (pp. 237-256). Palgrave Macmillan, London.

Sasaki, T., Okayasu, T., Jamsran, U., Takeuchi, K., 2008. Threshold changes in vegetation along a grazing gradient in Mongolian rangelands. *Journal of Ecology*, 96(1), 145-154.

Schmiedel, U., Dengler, J., Etzold, S., 2012. Vegetation dynamics of endemic-rich quartz fields in the Karoo, South Africa, in response to recent climatic trends. *Journal of Vegetation Science*, 23(2), 292-303.

SOLPOS, 2023. Solar Position and Intensity, Version 2.0. US-DOE. National Energy Lab. midcdmz.nrel.gov/solpos/solpos.html (30 January 2023).

Takthajan, A. 1986: *Floristic regions of the world*. The University of California Press, Berkeley.

TerrSet, 2017. TerrSet Geospatial Monitoring and Modeling System, Version 18.31. Clark University. clarklabs.org/terrset/ (30 January 2023).

Tzanopoulos, J., Mitchley, J., Pantis, J., 2005. Modelling the effects of human activity on the vegetation of a northeast Mediterranean island. *Applied Vegetation Science*, 8(1), 27-38.

United Nations, 2011. Decade for Deserts and the fight against Desertification. United Nations. un.org/en/events/desertification_decade/whynow.shtml (30 January 2023).

USDA, 2006. Sheep and Goat Handling and Facility Options. US Department of Agriculture, 5. URL available upon request (30 January 2023).

Vossen, P., 2007. Olive oil: history, production, and characteristics of the world's classic oils. *HortScience*, 42(5), 1093-1100.

Weber, E., 2009. How Much Wine Can a Small Vineyard Produce? UC Davis Small Vineyard Series 1. URL available upon request (30 January 2023).

White, P. S., Jentsch, A., 2001. The search for generality in studies of disturbance and ecosystem dynamics. *Progress in botany*, 62, 399-450.

Wu, D., Zhao, X., Liang, S., Zhou, T., Huang, K., Tang, B., Zhao, W., 2015. Time-lag effects of global vegetation responses to climate change. *Global change biology*, 21(9), 3520-3531.

Zohary, M., Orshan, G., 1959. The maquis of Ceratonia siliqua in Israel. *Vegetatio*, 8(5/6), 285-297.

This contribution has been peer-reviewed.
https://doi.org/10.5194/isprs-archives-XLVIII-M-1-2023-189-2023

SOYBEAN CROP YIELD PREDICTION BY INTEGRATION OF REMOTE SENSING AND WEATHER OBSERVATIONS

J.D.Mohite[1]*, S.A.Sawant[2], A.Pandit[3], R.Agrawal[3], S. Pappula[4]

[1] TCS Research and Innovation, Tata Consultancy Services, Mumbai, India - jayant.mohite@tcs.com
[2] TCS Research and Innovation, Tata Consultancy Services, Pune, India - suryakant.sawant@tcs.com
[3] TCS Research and Innovation, Tata Consultancy Services, Indore, India - (ankur.pandit, rishabh.agrawal3)@tcs.com
[4] TCS Research and Innovation, Tata Consultancy Services, Hyderabad, India - srinivasu.p@tcs.com

KEY WORDS: Yield Forecasting, Soybean Crop, Remote Sensing, Weather Data, Random Forest Regression

ABSTRACT:

The main objective of this study is the in-season forecasting of soybean crop yield using the integration of satellite remote sensing and weather observations. The study was carried out in the Paraná state of Brazil. The soybean crop in the study region is sown during Oct.–Nov. month and harvested between Feb.–Mar. of the next year. Municipality-level soybean yield data for 15 municipalities was obtained from the AGROLINK portal of Brazil, from the 2005–06 season to the 2020–21 season. The crop yield data constituted yearly municipality-wise yield in kg/ha. Remote sensing-based indicators such as the Normalized Difference Vegetation Index (NDVI), Enhanced Vegetation Index (EVI) and Land Surface Temperature (LST), and Rainfall data from CHIRPS was considered in the study. Regression modelling was carried out between municipality-level yield as the dependent variable and features generated from remote sensing and weather observations as independent variables. Performance evaluation of tuned random forest regression (RFR) and tuned support vector regression (SVR) were performed against multiple linear regression (MLR). A comparison of results in terms of algorithms shows that RFR performed better than SVR and MLR. Further, a root-mean-square-error (RMSE) of 414 kg/ha and an R^2 value of 0.748 were achieved by the best RFR model. Validation of developed RFR model was performed on the data from the new soybean season, i.e., 2020–21. We have achieved an R^2 value of 0.693 with a RMSE of 585 kg/ha. Although the model performance on the data of 2020-21 season is slightly reduced, R^2 and RMSE are in good agreement with test results. This study showed that, integration of remote sensing and weather observations would be useful for in-season yield forecasting of soybean at municipality level.

1. INTRODUCTION

Soybean (Glycine max (L.) Merrill) is one of the most important sources of protein and oil. United States, Brazil, Argentina, and China account for around 90% of the global soybean production (Embrapa, 2018, USDA, 2019). Brazil is the second largest producer of soybean in the world, after the United States, and the crop is grown in almost all of the country's states. The production of soybean in Brazil has shown a steady increase over the years, driven by factors such as technological advancements, favourable weather conditions, and government policies that promote agricultural development. According to data from the Brazilian National Company for Food Supply Companhia Nacional de Abastecimento (CONAB), Brazil's soybean cultivation area has increased from around 15 million hectares in year 2000 to more than 38 million hectares in the 2020-2021 harvest season (CONAB, 2021). The majority of soybean cultivation in Brazil is concentrated in the southern and central regions of the country, which have favourable soil and weather conditions for the crop. Parana is the second largest producer of soybeans in Brazil, accounting for approximately 19% of the country's total soybean production in 2020-21 harvest season. The largest producer is Mato Grosso, which accounts for approximately 28% of the country's total soybean production (CONAB, 2021). The Parana state has favourable soil and weather conditions for soybean production, and has invested heavily in research and development of the crop, including the development of high-yielding soybean varieties and advanced production techniques. The area under soybean cultivation in Parana has also increased significantly over the years, from around 2 million hectares in year 2000 to more than 5 million hectares in year 2021 (CONAB, 2021).

Accurate and timely statistics on soybean crop area and yield in Brazil and Parana are crucial for crop management, food security, and economic planning. Many approaches have been developed in the past for in-season yield assessment ranging from field based survey, physical process based models (Jones et al., 2003), use of non-invasive technologies, etc. The Brazilian government collects and publishes regular data on soybean production, including crop area, yield, and production volume, which are widely used by farmers, policymakers, and analysts. However, manually collecting the data in yield is costly and time consuming process. Alternatively, crop cutting experiments are being conducted by various state and private agencies, which have limited scalability and involves huge cost and time (Mohite et al., 2019).

Over the years, there has been significant progress in the use and development of physical process-based crop growth simulation models. Few examples of crop growth simulation models being used for soybean yield prediction include Agricultural Production System Simulator (APSIM) (Keating et al., 2003), Decision Support System for Agro-Technology Transfer (DSSAT) (Jones et al., 2003), SoySim (Wilhelm et al., 2004), STICS (Brisson et al., 1998). Crop growth simulation models basically uses mathematical equations to simulate the growth and development of crops and used to assess the response of various man-

* Corresponding author

https://doi.org/10.5194/isprs-archives-XLVIII-M-1-2023-197-2023 | © Author(s) 2023. CC BY 4.0 License.

agement practices on yield. Although, physical process based models are quite accurate, and a large amount of ground data is needed to run the simulations. Model performance is largely affected by the quality and quantity of ground data available.

On the other hand, non-invasive technologies mainly involve the use of remote sensing observations in machine learning based models for in-season yield forecast (Peralta et al., 2016, Schwalbert et al., 2018, Xia et al., 2020, Li et al., 2020). With the advent of global coverage and openly available remote sensing satellite data in Optical and Synthetic Aperture Radar electromagnetic spectrum at regular intervals. Thereby, resulted into ease in accessing the time series of satellite data and couple machine learning algorithms to provide value added insights into the agriculture sector. Additionally the compute platform such as Google Earth Engine (GEE) (Gorelick et al., 2017) facilitated the spatio-temporal scalability. Remote sensing data can be used to estimate various crop-related parameters, such as vegetation indices, surface temperature, and soil moisture, which are closely related to crop growth and yield. These parameters can be used to develop models for predicting crop yield. In recent years, several studies have been conducted on soybean crop yield prediction using remote sensing data in Brazil. These studies have focused on different aspects of the problem, such as the choice of remote sensing data sources, the development of appropriate models, and the validation of the results. Moreover, integration of weather data along with remote sensing observations has proven effective in improving the model's performance (Peng et al., 2018, Cai et al., 2019). Considering the advantages and limitations of the prior art, we have attempted to use the integration of satellite and weather data for in-season yield forecasting.

The main objective of this study is the in-season forecasting of soybean crop yield using the integration of satellite remote sensing and weather observations. The study wasconducted in selected municipalities of the Paraná state of Brazil.

2. MATERIALS AND METHODS

This section covers information about the study area, insights into the satellite and ground reference data, and a detailed approach for yield prediction using various machine learning algorithms.

2.1 Study Area

We have conducted this study in selected municipalities in Parana State, Brazil 1. Parana is one of the southern states of Brazil, bordered on the north by Sao Paulo state, on the east by the Atlantic Ocean, on the south by Santa Catarina state and the province of Misiones, Argentina, and on the west by Mato Grosso do Sul and Paraguay, with the Parana River as its western boundary line (Chisholm, 1911). The annual mean air temperature ranges between 15 and 24 °C, with the highest temperatures found in the northwest and the lowest around Palmas (Aparecido et al., 2016). Precipitation is less than 1,200 mm (47 in) a year in the north of the state, rising to above 1,800 mm (71 in) in the southwest and southeast of the state (Aparecido et al., 2016). Agriculture is one of the main economic drivers of the state, and about 15% of Parana's GDP comes from agriculture. Parana is a leading producer of crops such as soybeans, corn, coffee, and wheat.

2.2 Satellite Datasets Used

We have mainly used satellite data products from the Moderate Resolution Imaging Spectroradiometer (MODIS) and rainfall data from the Climate Hazards Group InfraRed Precipitation with Station data (CHIRPS).

2.2.1 MODIS Based Vegetation Indices MOD09GA and MYD09GA are two products of the Moderate Resolution Imaging Spectroradiometer (MODIS) sensors onboard the Terra and Aqua satellites, respectively (Vermote and Wolfe, 2015). These products provide daily surface reflectance data at a 1-kilometer spatial resolution. The data are corrected for atmospheric effects and provide a consistent surface reflectance time series over the lifetime of the MODIS sensors. Version 6 of the products include improvements such as better cloud and cloud shadow screening, and updated aerosol retrieval algorithms. The surface reflectance bands, such as Red, NIR were used to estimate the Normalized Difference Vegetation Index (NDVI), However, surface reflectance in blue, red and NIR bands was used to estimate Enhanced Vegetation Index (EVI). The data was accessed from the GEE during the Soybean crop season i.e. for the months November, December and January between 2005-2021. Monthly mean composites at 1 km spatial resolution were generated and used in the analysis.

2.2.2 MODIS Based Land Surface Temperature We have utilized land surface temperature data from MODIS (MOD11A1 and MYD11A1 version 6 products) (Wan, 2015). The MODIS Land Surface Temperature (LST) product is a remotely sensed dataset that provides information about the temperature of the Earth's surface (Wan et al., 2002). These products are generated using a split-window algorithm that utilizes the 3.7 μm and 11 μm thermal bands to retrieve LST. The data is available at a daily temporal scale and 1 km spatial resolution during daytime and nighttime modes. We have utilized the data available in the daytime mode. LST data was accessed from GEE for the months of November, December, and January between 2005 and 2021 and used in the analysis.

2.2.3 Rainfall Data from CHIRPS The Climate Hazards Group InfraRed Precipitation with Station data (CHIRPS) daily rainfall product is a remotely sensed dataset that provides information about daily rainfall amounts across the globe. The product is derived from a combination of satellite data and ground station observations (Funk et al., 2015) and has a 30+ year quasi-global coverage. The product provides daily global coverage of rainfall amounts at a spatial resolution of 0.05 degrees, or approximately 5 km. We created monthly accumulated composites from the daily data and resampled it to 1 km spatial resolution to match it with MODIS-based vegetation indices and LST. We accessed the data for the months of November, December, and January between 2005 and 2021 from the GEE.

2.3 Ground Reference Data on Yield

AGROLINK is a web portal focused on the agricultural sector in Brazil. The portal was created in year 1996 and provides a wide range of information and services related to agriculture, livestock, agribusiness, and rural development (AGROLINK, 2019). AGROLINK offers a variety of content, including news, articles, interviews, market analyses, and technical information. The portal covers a wide range of topics, such as crop production, livestock management, pest and disease control, irrigation, precision agriculture, biotechnology, sustainability, and rural

This contribution has been peer-reviewed.
https://doi.org/10.5194/isprs-archives-XLVIII-M-1-2023-197-2023 | © Author(s) 2023. CC BY 4.0 License.

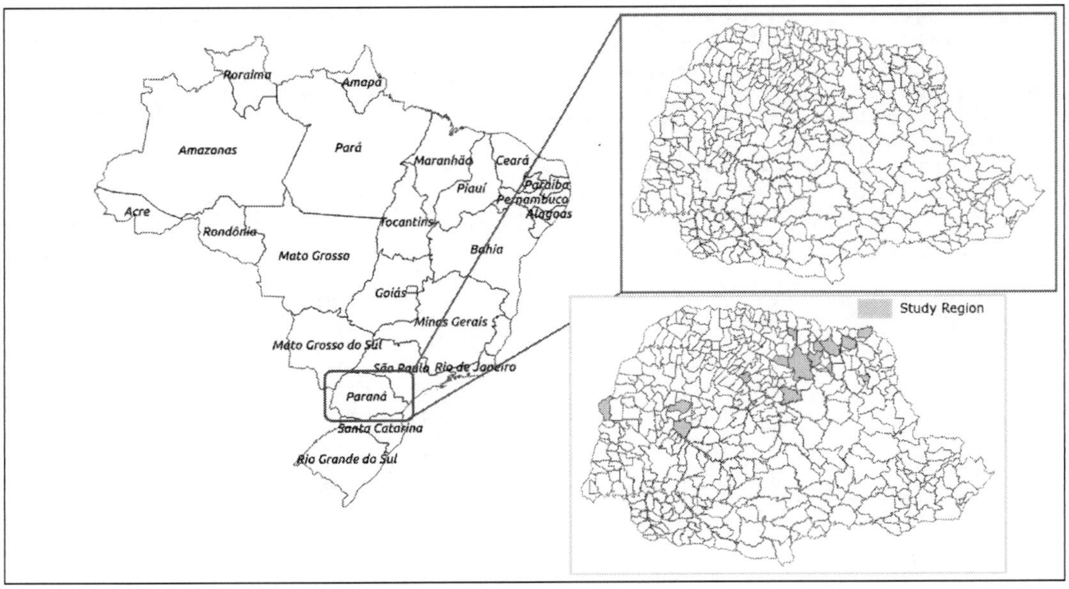

Figure 1. Study area

policies. AGROLINK provides crop yield data at the municipality level for various agricultural crops for various historical years. We accessed the yield data for the soybean crop for 15 municipalities in the state of Parana, Brazil, for the soybean crop seasons from 2005-06 to 2020-21. The data was represented as the yield value per hectare per municipality for specific year.

2.4 Overall Framework

We obtained municipality-level soybean yield data for 15 municipalities from AGROLINK (AGROLINK, 2019) for the soybean crop seasons from 2005-06 to 2020-21. The data was represented as yield value per hectare per municipality for specific year and used as ground truth for modeling purposes. For the municipality-level analysis, we considered observations from MODIS satellite and remote sensing-based indicators, such as Normalized Difference Vegetation Index (NDVI), Enhanced Vegetation Index (EVI), and Land Surface Temperature (LST), as well as rainfall data from CHIRPS. For NDVI and EVI, we considered the municipality-level maximum, minimum, and average values for the period from 1 Nov. to 31 Jan.. For LST, we estimated the municipality-level maximum, minimum, and average values for the months of November, December, and January, and accumulated rainfall for Nov., Dec., and Jan. was considered features in the study. We utilized a total of 18 features, including 3 for NDVI, 3 for EVI, 9 for LST, and 3 for rainfall. The dataset contained 248 samples from various municipalities and seasons. Regression modeling was performed between municipality-level yield and features derived from remote sensing and weather observations. Municipality-level yield data was considered as the dependent variable, while features derived from MODIS and CHIRPS data were used as independent variables in the regression analysis. The modeling approach involved dividing the data into training (80%) and validation (20%) sets. The performance of tuned Random Forest Regression (RFR) and tuned Support Vector Regression (SVR) was evaluated against the Multiple Linear Regression (MLR). Finally, the best model was used to generate yield maps for the study region.

2.5 Machine Learning Algorithms used in the study

We mainly used three different Machine Learning algorithms, namely Multiple Linear Regression (MLR), Random Forest Regression (RFR), and Support Vector Regression (SVR).

2.5.1 Multiple Linear Regression MLR also known as Multiple regression is a statistical method that uses multiple explanatory variables to predict the outcome (Stepanov et al., 2020). MLR models the linear relationship between explanatory variables and target variables. We used MLR as the baseline regression model in our study, where NDVI, EVI, LST, and rainfall were used as explanatory variables with yield as the target variable.

2.5.2 Random Forest Regression Random Forest is a well-known machine learning algorithm used for ensemble learning, which combines multiple decision trees to make more accurate predictions. This algorithm generates a large number of decision trees from a subset of random training data, and the output from individual trees is used to make the final decision (Breiman, 1996). The method is widely used for various remote sensing applications, including crop yield prediction, due to its capability to resist over-fitting and require less computational time (Inglada et al., 2015, Belgiu and Csillik, 2018, Were et al., 2015). However, like many other algorithms, RFs are sensitive to the choice of hyperparameters and the training data used. In this study, we followed a K-fold cross-validated grid search approach. In this approach, the dataset was first randomly split into a training set (80% of the total dataset) and a test set (20% of the total dataset). The test set was kept for final evaluation, while the training set was further split into three subfolds. The model was iteratively trained on K-1 (2) of the folds and validated on the remaining one to optimize the performance by tuning RF hyperparameters, such as the number of trees and minimum number of variables available at each split.

https://doi.org/10.5194/isprs-archives-XLVIII-M-1-2023-197-2023

2.5.3 Support Vector Regression SVR is a type of supervised learning that can handle non-linear relationships between variables and high-dimensional datasets, making it a powerful tool for yield prediction. SVR has been extensively used for various agricultural applications (Khosla et al., 2020, Anandhi and Chezian, 2013, Brdar et al., 2011). Similar to other ML algorithms, the performance of SVR could be improved by having the optimum choice of hyperparameters. In this study, we have tuned the SVR for two hyperparameters, namely C and Sigma. The train-test splitting and modeling strategy were kept similar to RFR.

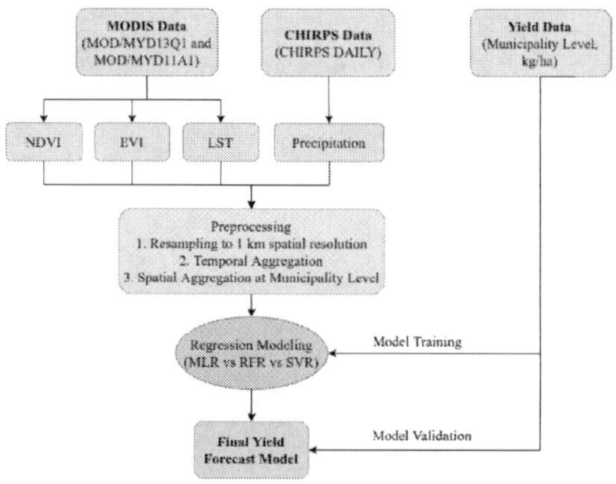

Figure 2. Overall analysis approach

3. RESULTS AND DISCUSSION

This section describes the performance of various ML models, such as MLR, RFR, and SVR. Additionally, we have also discussed the predicted versus actual yields at the municipality level for a new season.

3.1 Implementation of Various ML Models

We mainly used three algorithms in our study: MLR, RFR, and SVR. There are no hyperparameters for MLR, but algorithms such as RFR and SVR are sensitive to the choice of hyperparameters and training data. The right selection of these hyperparameters is crucial for optimal results. For hyperparameter selection, we followed a 3-fold cross-validated grid search strategy for RFR and SVR. In this approach, the dataset was first randomly split into a training set (80% of the total dataset) and a test set (20% of the total dataset). The test set was kept for final evaluation, while the training set was further split into 3 sub-folds. The model was iteratively trained on K-1 (2) of the folds and validated on the remaining one to optimize the performance by tuning hyperparameters. RFs were tuned for two parameters: the number of trees (ntree) and the minimum number of variables available at each split (mtry). We tried ntree values between 50-400 with an interval of 50 (8 ntree values) and mtry between 2-8 (7 mtry values) with an interval of 1. A total of 56 models were trained, considering each combination of ntree x mtry, and the best model was considered for testing on the remaining 20% data. SVR was tuned for two hyperparameters: C and Sigma. We tried C values of 0.1, 1, 10, 100 and Sigma values of 0.001, 0.01, 0.1, and 1. Therefore, 16 models were trained, considering each combination of C x Sigma, and the best model was considered for testing on the remaining 20% data. In the case of MLR, we used all 80% of the data for model training and the remaining data for testing of the model. Finally, we chose the best model from MLR, RFR, and SVR, predicted the yield for the season 2020-21, which was not considered in the modeling process, and validated our results.

Table 1. Performance of MLR

SN	TrainSet	R^2	RMSE
1	TrainSet 1	0.578	822
2	TrainSet 2	0.545	854
3	TrainSet 3	0.549	853

3.2 Performance of Various ML Models

Table 1 shows the performance of MLR for three different simulations (training sets). Since the training data was randomly selected, we ran the models three times to assess the performance of MLR with changing training data. The average R^2 value was found to be 0.557 and RMSE of 843 kg/ha with a standard deviation of 0.018, indicating that the model is not sensitive to training data and performed equally well when the training data was changed. However, the overall ^{2} value across all the models is relatively lower.

Tables 2 and 3 show the performance of the RFR model in terms of R^2, RMSE for various combinations of ntree and mtry. The results show that the best RFR model achieved a Root Mean Square Error of 414 kg/ha and an R^2 value of 0.748, which was obtained for a model trained using ntree=200 and mtry=5. It can be observed that as the number of trees increased, there was a steady increase in R^2 from 50 to 350 for mtry values of 2-4. However, for mtry values between 5-8, there was an increase in R^2 only for ntree values from 50-200, and R^2 either reduced or remained constant after an ntree of 200. Comparing MLR and RFR, it can be seen that there was a significant improvement in R^2 for all the combinations of ntree x mtry.

Table 4 and Table 4 show the performance of the SVR model for various combinations of C and Sigma. The best R^2 value of 0.722 was achieved for a C value of 10 and a Sigma value of 0.01. For the same combination of C and Sigma, the RMSE reached up to 502 kg/ha. Furthermore, there was a significant improvement in performance compared to MLR; however, the results were not as good as RFR.

Finally, we chose the best RFR model (trained with ntree=200 and mtry=5) and applied it to the data from the 2020-21 season for yield prediction. We estimated the R^2 and RMSE between the predicted yield and actual yields for the season provided by AGROLINK. We achieved an R^2 value of 0.693 with an RMSE of 585 kg/ha. Although the model's performance on the data from the 2020-21 season is slightly reduced, the R^2 and RMSE are in good agreement with the test results. The figure 3 shows the percent change between the ground reference yield and predicted yield at the municipality level. We can observe that half of the municipalities are within a 10% deviation. However, remaining half of the municipalities are within a 15% deviation, and for the one municipality, the deviation goes beyond 15%.

4. SUMMARY AND CONCLUSIONS

We have developed in-season yield forecasting models for soybean crop in the Parana state of Brazil. The models were developed using historical municipality level yield data from the

Table 2. Performance of RFR in terms of R^2

ntree ↓ \ mtry →	2	3	4	5	6	7	8
50	0.63	0.644	0.667	0.68	0.678	0.634	0.661
100	0.63	0.661	0.678	0.684	0.689	0.699	0.699
150	0.645	0.673	0.671	0.691	0.711	0.727	0.7
200	0.68	0.701	0.712	**0.748**	0.733	0.73	0.732
250	0.69	0.691	0.718	0.744	0.733	0.738	0.74
300	0.716	0.72	0.733	0.739	0.727	0.744	0.73
350	0.722	0.733	0.74	0.741	0.734	0.742	0.74
400	0.689	0.69	0.692	0.721	0.704	0.712	0.698

Table 3. Performance of RFR in terms of RMSE

ntree ↓ \ mtry →	2	3	4	5	6	7	8
50	731	722	700	678	623	710	723
100	701	689	659	601	600	621	700
150	700	677	622	507	588	521	631
200	685	612	572	**414**	522	471	552
250	633	606	512	434	462	464	498
300	544	576	473	455	445	430	477
350	498	502	444	471	438	444	456
400	512	555	502	498	476	465	578

Table 4. Performance of SVR in terms of R^2

C ↓ \ Sigma →	0.001	0.01	0.1	1
0.1	0.645	0.666	0.665	0.678
1	0.689	0.7	0.7	0.7
10	0.712	0.722	0.71	0.71
100	0.682 3	0.655	0.642	0.642

Table 5. Performance of SVR in terms of RMSE

C ↓ \ Sigma →	0.001	0.01	0.1	1
0.1	722	687	555	598
1	643	600	515	543
10	599	502	509	522
100	576	544	536	519

AGROLINK portal of Brazil, as well as remote sensing and weather-based indices. We used regression modeling with municipality level yield data as the dependent variable and remote sensing and weather indices as independent variables. The performance of tuned Random Forest Regression (RFR) and tuned Support Vector Regression (SVR) was evaluated against the Multiple Linear Regression (MLR). The results showed that RFR outperformed SVR and MLR, achieving a Root Mean Square Error of 414 kg/ha and an R^2 value of 0.748 for the best RFR model trained with ntree=200 and mtry=5. We validated the developed RFR model with data from the 2020-21 soybean season, achieving an R^2 value of 0.693 with an RMSE of 585 kg/ha. Although the model performance on the data from the 2020-21 season was slightly reduced, the R^2 and RMSE were in good agreement with the test results. This study showed that the integration of remote sensing and weather observations would be useful for in-season yield forecasting of soybean at the municipality level.

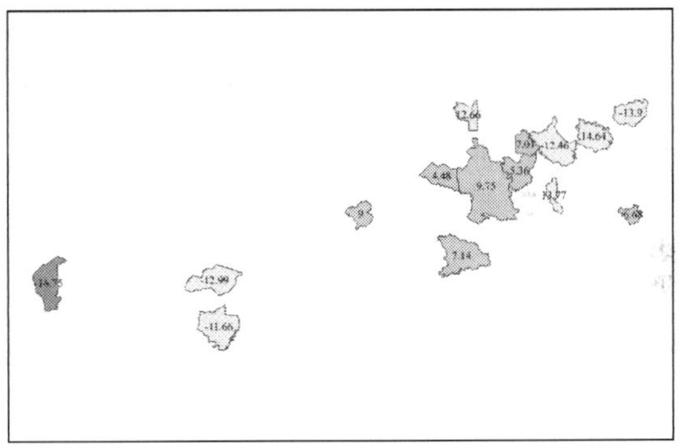

Figure 3. Percent deviation between actual and predicted yield at municipality level

5. FUTURE WORK

In the existing work, we have considered municipality-level data from 15 municipalities of Parana state in Brazil and used well-known machine learning algorithms such as MLR, RFR, SVR. In the future, we would like to assess the performance of deep learning algorithms by adding data from other municipalities in Parana and other states of Brazil. Additionally, we would like to validate the developed approach for upcoming seasons.

REFERENCES

AGROLINK, 2019. AGROLINK - O Portal do conteúdo Agropecuário. https://www.agrolink.com.br/.

Anandhi, V., Chezian, R. M., 2013. Support vector regression in forecasting. *International Journal of Advanced Research in Computer and Communication Engineering*, 2(10), 4148-4151.

Aparecido, L. E. d. O., Rolim, G. d. S., Richetti, J., Souza, P. S. d., Johann, J. A., 2016. Koppen, Thornthwaite and Camargo climate classifications for climatic zoning in the State of Parana, Brazil. *Ciencia e Agrotecnologia*, 40(4), 405-417.

Belgiu, M., Csillik, O., 2018. Sentinel-2 cropland mapping using pixel-based and object-based time-weighted dynamic time warping analysis. *Remote sensing of environment*, 204, 509–523.

Brdar, S., Culibrk, D., Marinkovic, B., Crnobarac, J., Crnojevic, V., 2011. Support vector machines with features contribution analysis for agricultural yield prediction. *Second International Workshop on Sensing Technologies in Agriculture, Forestry and Environment (EcoSense 2011), Belgrade, Serbia*, 43–47.

Breiman, L., 1996. Bagging predictors. *Machine learning*, 24, 123–140.

Brisson, N., Mary, B., Ripoche, D., Jeuffroy, M.-H., Ruget, F., Nicoullaud, B., Gate, P., Devienne-Barret, F., Antonioletti, R., Durr, C. et al., 1998. STICS: A generic model for the simulation of crops and their water and nitrogen balances. I. Theory and parameterization applied to wheat and corn. *Agronomie*, 18(5-6), 311–346.

Cai, Y., Guan, K., Lobell, D., Potgieter, A. B., Wang, S., Peng, J., Xu, T., Asseng, S., Zhang, Y., You, L. et al., 2019. Integrating satellite and climate data to predict wheat yield in Australia using machine learning approaches. *Agricultural and forest meteorology*, 274, 144–159.

Chisholm, H., 1911. Paraná. *Encyclopædia Britannica*, 20, Cambridge University Press, 766.

CONAB, 2021. Acompanhamento da safra brasileira de grãos: Safra 2020/2021 - décimo levantamento. Retrieved from http://www.conab.gov.br/info-agro/safras/graos/boletim-da-safra-de-graos.

Embrapa, 2018. Soja em números. https://www.embrapa.br/soja/cultivos/soja1/ dados-economicos.

Funk, C., Peterson, P., Landsfeld, M., Pedreros, D., Verdin, J., Shukla, S., Husak, G., Rowland, J., Harrison, L., Hoell, A., Michaelsen, J., 2015. The climate hazards infrared precipitation with stations-a new environmental record for monitoring extremes. *Scientific data*, 2, 150066.

Gorelick, N., Hancher, M., Dixon, M., Ilyushchenko, S., Thau, D., Moore, R., 2017. Google Earth Engine: Planetary-scale geospatial analysis for everyone. *Remote sensing of Environment*, 202, 18–27.

Inglada, J., Arias, M., Tardy, B., Hagolle, O., Valero, S., Morin, D., Dedieu, G., Sepulcre, G., Bontemps, S., Defourny, P. et al., 2015. Assessment of an operational system for crop type map production using high temporal and spatial resolution satellite optical imagery. *Remote Sensing*, 7(9), 12356–12379.

Jones, J. W., Hoogenboom, G., Porter, C. H., Boote, K. J., Batchelor, W. D., Hunt, L., Wilkens, P. W., Singh, U., Gijsman, A. J., Ritchie, J. T., 2003. The DSSAT cropping system model. *European journal of agronomy*, 18(3-4), 235–265.

Keating, B. A., Carberry, P. S., Hammer, G. L., Probert, M. E., Robertson, M. J., Holzworth, D., Huth, N. I., Hargreaves, J. N., Meinke, H., Hochman, Z. et al., 2003. An overview of APSIM, a model designed for farming systems simulation. *European Journal of Agronomy*, 18(3-4), 267–288.

Khosla, E., Dharavath, R., Priya, R., 2020. Crop yield prediction using aggregated rainfall-based modular artificial neural networks and support vector regression. *Environment, Development and Sustainability*, 22, 5687–5708.

Li, Y., Wang, X., Li, J., Wu, B., Zhang, C., 2020. Soybean Yield Prediction Using Remote Sensing Data and Machine Learning Algorithms in Northeast China. *Sensors*, 20(12), 3522.

Mohite, J., Sawant, S., Sakkan, M., Shivalli, P., Kodimela, K., Pappula, S., 2019. Spatialization of rice crop yield using sentinel-1 sar and oryza crop growth simulation model. *2019 8th International Conference on Agro-Geoinformatics (Agro-Geoinformatics)*, IEEE, 1–6.

Peng, B., Guan, K., Pan, M., Li, Y., 2018. Benefits of seasonal climate prediction and satellite data for forecasting US maize yield. *Geophysical Research Letters*, 45(18), 9662–9671.

Peralta, N. R., Assefa, Y., Du, J., Barden, C. J., Ciampitti, I. A., 2016. Mid-season high-resolution satellite imagery for forecasting site-specific corn yield. *Remote Sensing*, 8(10), 848.

Schwalbert, R. A., Amado, T. J., Nieto, L., Varela, S., Corassa, G. M., Horbe, T. A., Rice, C. W., Peralta, N. R., Ciampitti, I. A., 2018. Forecasting maize yield at field scale based on high-resolution satellite imagery. *Biosystems engineering*, 171, 179–192.

Stepanov, A., Dubrovin, K., Sorokin, A., Aseeva, T., 2020. Predicting soybean yield at the regional scale using remote sensing and climatic data. *Remote Sensing*, 12(12), 1936.

USDA, 2019. USDA foreign agricultural service. https://www.fas.usda.gov/regions/ brazil.

Vermote, E., Wolfe, R., 2015. MOD09GA MODIS/Terra Surface Reflectance Daily L2G Global 1km and 500m SIN Grid V006 [data set].

Wan, Z., 2015. MOD11A1 MODIS/Terra land surface temperature and the emissivity daily l3 global 1km sin grid. NASA LP DAAC.

Wan, Z., Zhang, Y., Zhang, Q., Li, Z., 2002. Quality assessment and validation of the MODIS global land surface temperature. *International Journal of Remote Sensing*, 23(14), 2661–2674.

Were, K., Bui, D. T., Dick, Ø. B., Singh, B. R., 2015. A comparative assessment of support vector regression, artificial neural networks, and random forests for predicting and mapping soil organic carbon stocks across an Afromontane landscape. *Ecological Indicators*, 52, 394–403.

Wilhelm, W. W., Johnson, J. M., Hatfield, J. L., Voorhees, W. B., Linden, D. R., Berry, L. D., 2004. SoySim: simulation model of soybean growth and management. *Agronomy Journal*, 96(1), 139–149.

Xia, H., Wang, J., He, M., Chen, J., 2020. Soybean yield prediction using remote sensing and machine learning techniques: A case study in the Heilongjiang Province, China. *Journal of Applied Remote Sensing*, 14(2), 026506.

The International Archives of the Photogrammetry, Remote Sensing and Spatial Information Sciences, Volume XLVIII-M-1-2023
39th International Symposium on Remote Sensing of Environment (ISRSE-39) "From Human Needs to SDGs", 24–28 April 2023, Antalya, Türkiye

EXPERIMENTAL ECOSYSTEM NATURAL CAPITAL ACCOUNTS IN THE REPUBLIC OF GUINEA

J.-A. Morand [1], G. Jaffrain [1], C. Sannier [1], J.-L. Weber [2]

[1] IGN FI, Paris - Bastille, France - (jamorand, gjaffrain, csannier)@ignfi.fr
[2] Independent consultant of ENCA - jlweber45@gmx.fr

KEY WORDS: remote sensing, ecosystem accounting, environmental indicator, landcover, landcover change, carbon, water, watershed, Guinea

ABSTRACT :

Over the past thirty years (Rio Conference, 1992), the climate has become an important issue in world politics. With the Kyoto Protocol (1997), atmospheric carbon accounting has gradually been introduced with the aim of raising awareness of national and international decision-making systems for the implementation of an energy transition. However, this carbon approach does not take into account ecosystems despite their fundamental role in climate regulation. A global assessment of all natural resources and ecosystem services, known as natural capital, is necessary from a sustainable development perspective. This assessment must then be taken into account in national accounting systems. The aim of this article is to test, on the national territory of the Republic of Guinea, the ecosystem-based natural capital accounting method developed by Jean-Louis Weber (Weber, 2014). Based on three accounts (ecosystem infrastructure, ecosystem carbon and water resources), this method aims to measure the sustainable capacity or 'sustainability' of ecosystems to provide services. Based on land use & land cover layers produced in the framework of the agroecological zoning project (Jaffrain et al., 2021), we have operationalised this ecosystem accounting methodology in the Republic of Guinea to calculate the total sustainability of the ecosystem. The land cover layers are the basic structural data for monitoring and describing the evolution of the territory at different temporal intervals. Thus, several environmental indicators were defined from these combined geospatial data and eventually allowed to define the evolution of the total sustainability of the territory's ecosystem between 2005 and 2015. A clear degradation of this sustainability value was identified, which reflects the numerous land use changes affecting the country in the recent period (2005-2015).

1. INTRODUCTION

1.1 From national land cover to ecosystem accounting

IGN FI is participating in the implementation of an ecosystem-based accounting of natural capital (ENCA) based on its land-use maps produced as part of the Guinea Agroecological Zoning Project (ZAEG) (Jaffrain et al., 2021). The project, carried out in collaboration with the Guinean Ministry of Agriculture, aims to provide a decision-making tool for the government by mapping land use and land use dynamics over a 10-year period and identifying and assessing agricultural potential. Within the framework of this project, two land use maps were produced (2005 and 2015) thanks to photointerpretation and remote sensing work carried out on site by a national team of Guinean technicians trained and supervised by IGN FI experts. The land use data produced from recent and old satellite images are the basis of the ecosystem accounting of natural capital developed by Jean-Louis Weber (Babin & Weber, 2019). It is worth recalling here that the Convention on Biological Diversity (CBD) published a manual in 2014 to support countries in the implementation of ENCA .

The Republic of Guinea (Fig.1), a West African country whose capital Conakry is located on the Atlantic coast, is therefore the territory chosen for our experimentation with Ecosystemic Accounting of Natural Capital.

With a surface area of 245,857 km², the country is subdivided into four large geographical zones, otherwise known as natural regions: Lower Guinea (coastal zone), Middle Guinea (mountainous zone), Upper Guinea (savannah zone located in the north-east) and Forest Guinea (dense rainforest zone).

With a rapidly growing population and the largest bauxite and iron reserves in the world, its territory has been marked in recent years by numerous changes in land use (artificialization, agricultural expansion, major development projects, etc.).

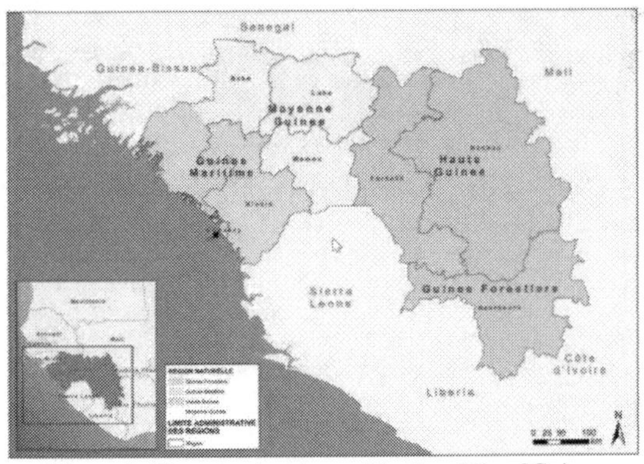

Figure 1. The natural regions of the Republic of Guinea

1.2 Accounting in line with the Sustainable Development Goals (SDGs)

Since the end of the 20th century, climate has become an important policy issue worldwide. Numerous scientific studies point to climate change and a rapid decline in biodiversity. In 1997, at the third Conference of the Parties (COP 3) of the United Nations Framework Convention on Climate Change (UNFCCC), the Kyoto Protocol committed to a global policy to reduce greenhouse gas emissions, as the increase of carbon dioxide (CO_2) in the atmosphere is one of the main causes of climate

This contribution has been peer-reviewed.
https://doi.org/10.5194/isprs-archives-XLVIII-M-1-2023-203-2023

change. Since then, a climate accounting system based on the guidelines of the Intergovernmental Panel on Climate Change (IPCC) has been set up around a unit of measurement: the CO2 equivalent. However, this accounting does not take into account the functions of ecosystems or their role in regulating the climate. Ecosystems are therefore only considered as carbon stocks, even though they are important assets in adapting to climate change, since they are at the origin of biological, biophysical and biogeochemical processes. Their state of health is not taken into account in the wealth of States (J.-L. Weber, 2022). However, the degradation of ecosystems (reduction in biodiversity, reduction in sequestration functions, etc.), as well as the emission of greenhouse gases, is also one of the causes of global warming (Delangue & Teillac-Deschamp, 2019).

The need to integrate natural ecosystems into international economic exchanges and policies has been apparent for several years. As early as 1992, the Convention on Biological Diversity (CBD), signed in Rio de Janeiro, expressed concern about the rapid degradation of biodiversity and set a number of main objectives: "the conservation of the various forms of life, the sustainable use of its components so as not to jeopardise the renewal capacity of natural environments and access to genetic resources, as well as the fair sharing of the benefits arising from their use" (Lévêque & Duhautois, 2006). In 2010, within the framework of the United Nations Convention on Biological Diversity in Nagoya, Japan, a Strategic Plan for Biological Diversity 2011-2020 was adopted. In it, 20 biodiversity targets called Aichi targets are established and organised into 5 strategic goals.
The integration of "biological diversity throughout government and society" (Convention on Biological Diversity & UNEP, 2010) is at the heart of the first strategic goal. The second Aichi Goal proposes to integrate biodiversity "into national and local development strategies and planning processes" and to incorporate it "into national accounts" (Convention on Biological Diversity & UNEP, 2010). In order to keep up with the evolution of society and respond to the challenges posed by climate change, the Sustainable Development Goals (SDGs) were adopted by the United Nations General Assembly in September 2015. 17 SDGs are set out in the 2030 Agenda for Sustainable Development, covering all development issues (climate, biodiversity, energy, agriculture, etc.).
Thus, SDG 15.9 (target 9), inspired by the second Aïchi Goal, deals with the integration of "the protection of ecosystems and biodiversity into national planning, development mechanisms and accounting"(United Nations, 2015). SDG 17.19 (target 19) leads to a reflection on the construction of indicators of progress in sustainable development, enriching the Gross Domestic Product (GDP) and strengthening the statistical capacities of developing countries (Babin & Weber, 2019). In this context, ENCA, developed by Jean-Louis Weber, is an approach that fits perfectly into the global political framework by responding to several Sustainable Development Goals such as the Convention on Biodiversity (SDG 15.9), the Convention on Combating Desertification and Land Degradation Neutrality (SDG 15.3) and the 2015 Paris Agreement on Climate Change... Indeed, ENCA appears to be a decision-making tool for countries in terms of environmental protection and management.
In the course of this work, it was therefore necessary to carry out a first experimentation of ENCA on the territory of the Republic of Guinea. The main accounting methods are reviewed, and the methodology of ecosystem accounting is then detailed. Finally, the first results of the application of this method in this West African territory are presented, accompanied by maps.

2. DATA AND METHODS

2.1 Accounting tool

The economic performance of societies is one of the main priorities of governments worldwide. The economic activities of human societies are highly dependent on the services provided by ecosystems. Yet the "natural capital" that refers to ecosystems is not taken into account in national accounting. This "oversight" on the part of national decision-makers is one of the main factors behind the decline in biodiversity. To counteract this trend, initiatives have been launched since the end of the 20th century to develop accounting systems that combine environmental and economic dimensions. The term natural capital, at the crossroads of ecology and economics, is defined as "the natural wealth that provides society with renewable and non-renewable resources and ecosystem services" (Ten Brink, 2016).

In order to take ecosystems into account within natural capital, multiple measurement and quantification approaches are being developed. Among these tools, the Artificial Intelligence For Ecosystem Services (ARIES) developed by the University of Vermont makes it possible to map and quantify ecosystem services and their beneficiaries at the scale of a territory (IONESCU et al., 2019). Another example is the Integrated Valuation of Ecosystem Services and Tradeoffs (InVEST), a suite of software models based on ecosystem services. These models are designed for different types of ecosystems (terrestrial, aquatic, coastal ...).

Governmental" natural capital accounting tools account for the variation and evolution of ecosystems at the scale of an entire territory. These integrated accounts seek to aggregate biophysical and socio-economic information in order to guide and provide a decision-making tool for public policies on ecosystems and also to inform the population on the ecosystem services provided by a territory (Feger & Mermet, 2021).

Initiated in 2003 by the feasibility study on land and ecosystem accounts conducted by the European Environment Agency (J.-L. Weber & European Environment Agency, 2006), the ENCA approach developed by Jean-Louis Weber is in line with a perspective that favours the safeguarding of natural capital. The ENCA quick Start Package published by the Convention on Biological Diversity in 2014 (Weber, 2014) to assist countries wishing to carry out ecosystem accounting.
In addition, the approach is recognised by the United Nations Statistical Commission as fitting within the broad ecosystem accounting framework of the System of Environmental and Economic Accounts (SEEA). ENCA is a comprehensive approach to integrating and synthesising biophysical and socio-economic data on sustainability and potential across all ecosystems (continental, coastal, natural, man-made...) in a territory or country. It aims to measure the capacity of ecosystems to provide services (ecosystem potential) both in the short term and in the future.

2.2 General methodology ENCA

This ecosystem capacity of ecosystem capital is calculated, at a given date for a defined area, on the basis of three accounts: ecosystem carbon, water resources and ecosystem infrastructure (biodiversity and river, land infrastructure, etc.).
The stocks and biophysical flows of natural capital are compared between two dates to better understand their evolution over time 'resulting from natural renewal and resource use flows' (Babin & Weber, 2019) (see Fig. 2 below).

https://doi.org/10.5194/isprs-archives-XLVIII-M-1-2023-203-2023 | © Author(s) 2023. CC BY 4.0 License.

ENCA therefore consists of an inventory and then a diagnosis of the degradation or improvement of ecosystems by establishing biophysical and ecological balances on geo-referenced land use bases.

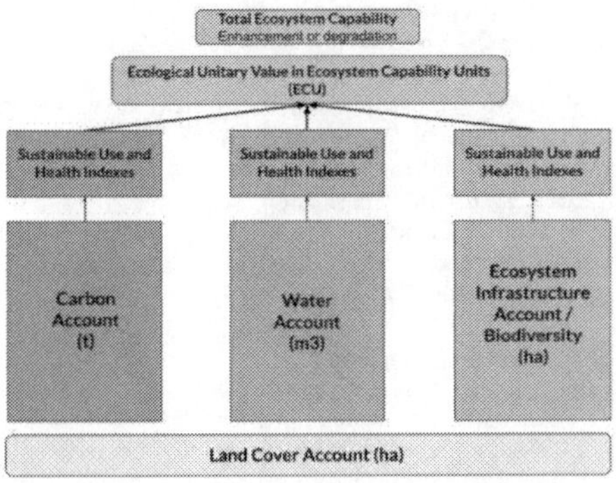

Figure 2. Overview of the Ecosystem Natural Capital Accounting (ENCA) Framework

As with carbon accounting and the CO2 equivalent unit, the aim is to establish an Ecosystem Capability Unit (ECU) to assess the sustainability of ecosystems. This ECU measure should, among other things, make it possible to measure the resource that is accessible, i.e. that can be used without causing ecosystem degradation. Subsequently, ENCA plans to transform this ECU into a monetary value based on the costs of protection and restoration, among other things (Weber, 2022). The resilience of ecosystems is therefore at the heart of this approach, which is in line with a strong sustainability trend.

2.3 A robust infrastructure

The basic functional unit of ENCA is the Socio-Ecological Landscape Unit (SELU). It is a spatialized unit (Fig. 3) built around the combination of two dimensions: the dominant landscape type and the belonging to a watershed. It thus integrates an essential geographical element (the catchment area) and makes it possible to describe various variables such as the water resource and its accessibility on a territory. In order to produce it, the start-up manual of the Convention on Biological Diversity (Weber, 2014) proposes several methods. The one chosen for the Republic of Guinea consists in using the small river basins as the basic boundary of the units and assigning to them the dominant landscape type of land cover.

Figure 3 shows the socio-ecological landscape units selected for Guinea. These are based on the Hydroshed 10 (HYBAS 10) level retrieved from the HydroBASINS database which is composed of a series of vectorised polygon layers that follow the boundaries of sub-catchments of different sizes (Lehner & Grill, 2013).

ENCA requires a large amount of collected data (freely available online) with different resolutions (from 10-30 m to 250 m). A spatial resolution of 100 m by 100 m was chosen over the territory of the Republic of Guinea.

As a result, all geographic data are resampled to a 1 ha grid corresponding to the project grid. A common projection system is also chosen for the project, which may lead to a reprojection of some geographical data. SAGA software (Conrad et al., SAGA, 2015) is used to carry out most of the ENCA process.

ENCA is produced from various types of national and international data. Geographic reference layers (administrative boundaries, road networks, etc.) are easily accessible online via national mapping agencies or ministries, such as those in charge of spatial planning (Weber, 2014).

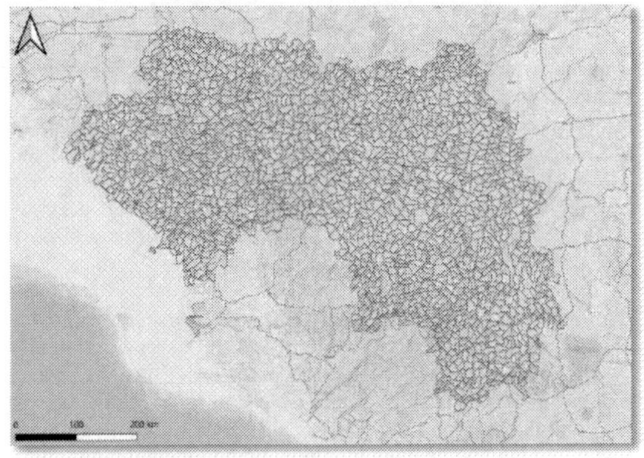

Figure 3. SELU cover the République of Guinea (HydroBASINS, Lehner & Grill, 2013)

Examples include watershed data available in HydroBASINS or river maps obtained from WWF-HydroSHEDS/FAO-AQUASTAT sources. Socio-economic statistical data are used in the calculation of ENCA, notably national agricultural statistics (FAOSTAT). Thus, apart from the land use maps produced by IGN FI, all the data used come from global and national open access databases.

Depending on the availability of the data, the scale of the information is adapted. Thus, for local applications, more precise monitoring data or even field observation data can be integrated to enrich the calculation of the accounts. If the data is not available on a local/regional scale, the use of national and international databases is preferred. The latter are constantly being improved and updates are to be expected over time, particularly with the availability of regularly updated high-resolution satellite images (Sentinel, Landsat 8, SPOT, etc.) and the development of more efficient monitoring systems.

2.4 Land accounts : the foundation for the account

In the ENCA system, land cover data are central, as they reflect biophysical characteristics (vegetation cover, density, height...) and land use. According to the FAO, land cover indicates the physical cover of the land such as forests and wetlands. Land use indicates the use of the land cover by humans activities (such as crops or fallow land).

Within the framework of the Guinea agro-ecological zoning project, land use maps are produced based on the CORINE Land Cover nomenclature (Jaffrain in Feranec, European landscapes dynamics, 2016) and integrating modifications linked to regional specificities, notably by relying on the Yangambi classification (Aubréville, 1957). Two land coverages were carried out by remote sensing and photointerpretation for the years 2005 and 2015. SPOT 4/5 and SPOT 6/7 satellite data were used for 2005 and 2015, respectively, over the entire territory of Guinea (Jaffrain et al., 2021).

The land account based on information from the land cover databases provides a first assessment of land cover changes over a period of time. It is the basic account required for the construction of the ENCA. Based on the gains and losses of areas (in ha) between two dates, it first allows the establishment of a matrix of changes in terms of land cover classes. This matrix is produced from the SAGA GIS software.

This contribution has been peer-reviewed.
https://doi.org/10.5194/isprs-archives-XLVIII-M-1-2023-203-2023 |

The interpretation of land cover classification changes allows the identification of the processes at their origin and thus to group them into fluxes. A land cover flux is thus 'a grouping of changes of the same nature' (Jaffrain et al., 2021) which allows a better understanding of the changes that have occurred by integrating information on the land cover prior to the change. These main flows can then be detailed at finer sub-levels according to the specificities of the regional context. This requires a land cover classification also detailed at different levels. The realization of the land account already allows the characterization of some indicators such as deforestation, artificialisation, etc.

The ENCA method combines qualitative (e.g. ecosystem health indicators) and quantitative (e.g. soil carbon content in tonnes) measures on three basic accounting balances: biocarbon, water and ecosystem infrastructure. These accounts are built around land cover as a foundation. The combination of these accounts makes it possible to measure the total ecosystem capacity of natural capital (unit ecological value) in order to gauge its evolution (improvement or degradation) over a given territory and time interval. The integrated nature of the accounting framework makes it possible to calculate, for each of the thematic accounts, an index of sustainable use and an index of resilience or health.

2.5 Ecosystem infrastructure account

According to the Millennium Ecosystem Assessment (Millennium Ecosystem Assessment (Program), 2005), ecosystems are multifunctional, providing a range of both tangible and directly measurable services (e.g. provisioning services) and intangible services (e.g. cultural and regulatory services). In ENCA, the latter are included in the ecosystem account of functional landscapes (also called "ecosystem infrastructure") which measures the capacity of ecosystems to provide these intangible services. It includes the services of landscapes, rivers and coastal marine areas to provide services that cannot be directly measured as physical quantities. The ecosystem infrastructure account describes changes in stocks between two dates. Its balance sheet is structured in four tables. It focuses first on a quantitative description of the extent of ecosystems with the calculation of stock, land cover in hectares, and a balance of rivers in terms of linear lengths.
The potential of the ecosystem infrastructure to provide services is measured by a combination of several types of environmental indicators. On the one hand, indicators on the biophysical characteristics of the ecosystem infrastructure and on the other hand indicators on the health of the ecosystem.

The first type of indicator consists of the Net Landscapes Ecosystem Potential (NLEP) and Rivers (NREP). These two composite indicators are classified and take into account an index of sustainability of terrestrial and river environments, based on land use and forest density in the case of the PENP. Combined with this, an index specifying the natural value of the territory based on the status of natural protection (wildlife reserves, national parks, etc.) extracted from national and international maps. Also taken into consideration is the degree of artificial fragmentation of the territory linked to road and rail infrastructures and dams (on rivers), which at a certain level limits exchanges between ecosystems (reduces the movement of species and the connectivity of habitats). The NLEP and NREP are finally aggregated into the Total Ecosystem Infrastructure Potential (TEIP).

In order to complete this account of the biophysical integrity of ecosystems, a table dealing with their state of health is included. This includes, among others, an index of local biodiversity integrity, the Local Biodiversity Intactness Index (Local Biodiversity Intactness Index – GEO BON, s. d.), which estimates the proportion of biodiversity of a terrestrial environment in relation to human use.

The overall access to these functional ecosystem services is based on the proximity between the ecosystem infrastructure and people. A series of indicators (local access of the population to the TEIP, access to water regulation services, etc.) is grouped together in another accounting table. These indicators combine the different potentials of the ecosystem infrastructure with demographic data (Babin & Weber, 2019).

2.6 Ecosystem carbon account

The ecosystem carbon account records carbon stocks and flows. It aims to assess the sustainable capacity of ecosystems to produce biomass (measured in biocarbon). The evolution of this biomass value is also taken into account through various biophysical processes: agriculture and crops, soil erosion, forest fires, etc. This account therefore deals with the entire carbon cycle, with the exception of fossil carbon resources. As with the ecosystem infrastructure account, each element of the carbon cycle is considered in structured tables. For the time being, only carbon stored in the biosphere has been considered. Atmospheric carbon will be considered in an additional table at a later stage.

First, carbon stocks are recorded for vegetation (living above-ground biomass), litter and roots and soil organic carbon. Forestry data from ESACCI (ESA Climate Change Initiative) and Hansen (University of Maryland) are used to estimate forest biomass at both dates.
The International Soil Reference and Information Centre (ISRIC) provides soil biocarbon data per hectare at different depths. Terrestrial animal biocarbon is also taken into account, starting with livestock (FAO data). All data produced is converted to tonnes of carbon using the equivalence of 50% biocarbon in biomass.
The account then describes biocarbon flows within ecosystems. Inputs are mainly based on net primary production (NPP) data. Biocarbon outputs include agricultural and forestry harvests (tree felling), soil erosion, losses due to land use changes (such as artificialisation), wood burning, forest fires. FAO data and national statistics are used to determine agricultural crops.
Forest biomass stocks, previously calculated for carbon stocks, are used to estimate the extraction of roundwood from the forest and the residues of forestry (dead leaves, branches, bark...).
In ENCA, carbon inputs are also taken into account in the "input" flows. Indeed, part of the biocarbon used by human activities returns to the ecosystem (production returns): crop residues, forestry residues, livestock manure. The carbon content of manure is estimated on the basis of livestock units (LU).

Thus, the ENCA ecosystem carbon account starts with a baseline ecosystem carbon balance (total inputs) and then records the accessible resource, i.e. the measure of ecosystem carbon that can be used in a sustainable manner.

Thereafter, the ENCA method deals with total ecosystem biocarbon use presented as the sum of total removals and indirect anthropogenic net losses of biocarbon due to land use (such as artificialisation). The accounting concludes with a table of indices of use intensities and ecosystem health. In particular, an index of the stability of carbon pools in forest ecosystems and an index of soil resistance to erosion are calculated in the implementation done on the Republic of Guinea.

2.7 Water accounts

Water accounts (especially in the form of water balance) are commonly used in hydrology and agronomy around the world. As early as the 1980s, accounts of water quantities per catchment area were produced in France and Spain, taking into account both the volume and the quality of the water. These experiences allow the SEEA to integrate, in 2007, a sub-system of economic and environmental water accounting focusing on the use of water by the economy. In ENCA, the objective is to diagnose the components of the ecosystem and thus to assess the impact of a degradation of the water resource on the whole ecosystem. The ecosystem water account is therefore an extension to the SEEA water accounts (Weber, 2014).

The ENCA water account addresses all the interactions and exchanges of the water cycle in a territory by recording the different stocks and flows. The hydrological system is treated as a whole as well as the flows of water use by human activities. To begin with, the basic ecosystem water balance records the water stocks of lakes, rivers, soil and vegetation. In the infrastructure account, rivers are treated only in terms of their accessibility through the River Accessibility Weighted Index (RAWI). However, in both the infrastructure and water accounts, rivers are calculated in terms of their potential, not just their volume of water. The Standard River Measurement Unit (SRMU) allows for the length as well as the flow of rivers. A single unit of flow can only provide information at a single point on the river, whereas a river is a continuum of points. The concept of a "standardized river-kilometer" was suggested (Heldal & Østdahl, 1984).
It is defined as a 1 km long stretch of watercourse (river, stream) that has a flow of 1 m^3/s at any point along its course. The basic balance then describes the inflow and outflow of rivers and groundwater following the series "precipitation, evapotranspiration, infiltration and runoff" (Weber, 2021).
Data are mainly taken from global meteorological databases (such as WorldClim, n.d.) but local monitoring stations can also be used. Natural inflows and outflows from HYBAS 10 catchments (on which the SELU is based) are measured, as well as water flows related to irrigation and other uses (hydropower dams, cooling water, etc.), based on national statistical data where available. The water use account summarises all recorded water uses. In order to assess the sustainability of the water resource, it is important to know the accessible water resource, i.e. the water whose use by humans does not endanger the provision of the service by the ecosystem.
One of the final calculations in the water accounts is the net accessible water surplus balance, which describes the exploitable water resource, taking into account potential limitations on use (Weber, 2014). Finally, as with the other accounts, the water accounts end with a table on the intensity of use index, which specifies the degradation (below 1) or not of ecosystems. It is defined by the ratio of the exploitable water resource of the ecosystem to the total water use. In addition to this, a composite indicator on the change in ecosystem water health (including biochemical water quality) is calculated (Babin & Weber, 2019). At the end of each account (water, carbon, infrastructure) the internal ecological unit value per SELU is calculated by combining these two indicators, a health status indicator, and a change indicator. This value calculated for each ecosystem component reflects its integrity, health and resilience.

3. RESULTATS

As explained above, the years 2005 and 2015, for which land cover data are available, were selected for the implementation of the ENCA for the Republic of Guinea. All indicators were therefore calculated for both dates. For the sake of simplicity, only a few relevant and representative indicators for the territory dealing with 2015 data appear in this section.

3.1 Some indicators for the infrastructure account

The Net Landscapes Ecosystem Potential (NLEP) is composed of the Green Background Landscape Index (GBLI). This estimates the naturally sustainable biomass of various land cover types. For example, natural forests with little anthropogenic activity and wetlands have a high index, whereas a large-scale monoculture, because of its dependence on human inputs (seeds, fertilisers, etc.), has a low GBLI. In order to calculate this index, a score from 0 to 100 is assigned to the different land cover classes. To improve the GBLI, we combined it with a forest density index produced by the University of Maryland (Hansen data). By averaging these two indices, a synthetic index (see Fig. 4 below) is obtained that is more representative of changes in the natural landscape and considers density variations within classes (especially in mixed landscapes).

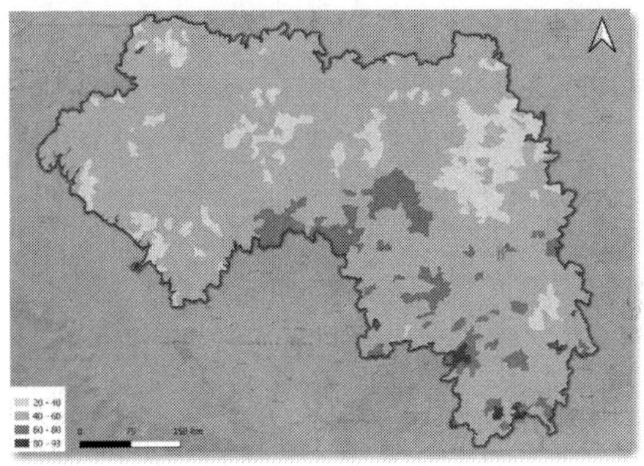

Figure 4. Map of Green background landscape Indicator (GBLI) per SELU in 2005

The Landscape High Nature Value Index (LHNVI) is based on maps of protected areas. The aim is to get an indication of the natural conservation value of the area concerned. Indeed, the index should reflect the level of protection of the area.
Similarly, where an area has more than one type of protection, it is assumed that it has a higher value and therefore the protections add up. An area that is not protected or not designated for its high nature value has a value of 1.
In this way, one level of protection provides a score of 2, and so on. In the case of Guinea, there are 3 levels of protection as set out below:

- Level 1: Unprotected areas
- Level 2: Sacred forests; Wildlife reserves
- Level 3: Classified forests; Strict reserves, National parks

This contribution has been peer-reviewed.
https://doi.org/10.5194/isprs-archives-XLVIII-M-1-2023-203-2023

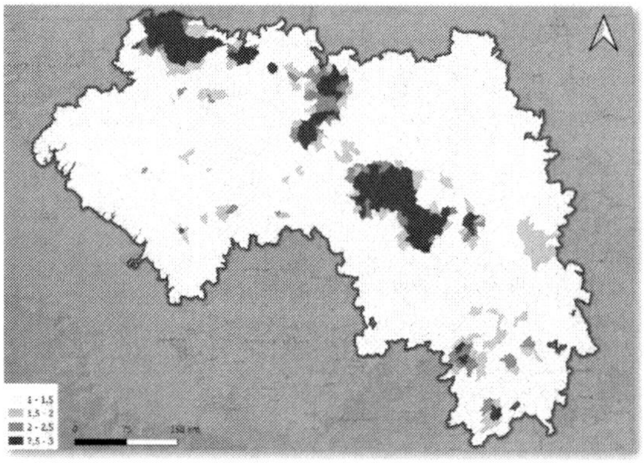

Figure 5. Map of the Landscape High Nature Value Index (LHNVI) by UPSE

According to the maps produced (Fig. 5), Forest Guinea, a natural region in the south-east of the country, has a high green landscape background index. This could be correlated with the existence of several protection zones. However, this correlation between the GBLI and LHNVI indices does not necessarily hold true for the whole country.

In ENCA, the landscape fragmentation index is used to adjust the green and high nature value landscape fund indices by taking into account the barrier effects that can limit interactions between ecosystems. Produced from maps of the main road and rail networks, it represents the "strong" fragmentation of the landscape, i.e. that which induces significant negative effects (isolation of populations, disruption of trophic chains, etc.) on the ecosystems. Thus, the landscape fragmentation index reflects the degree of fragmentation of natural (non-artificial) land cover areas. An unfragmented landscape unit (SELU) has a value of 1. The lower the mesh value, the more fragmented the natural areas within the unit are. On the map below (Fig. 6), a low mesh size value and therefore a high fragmentation index (red on the map) can be observed in SELUs with dense transport infrastructures, such as the capital, Conakry.

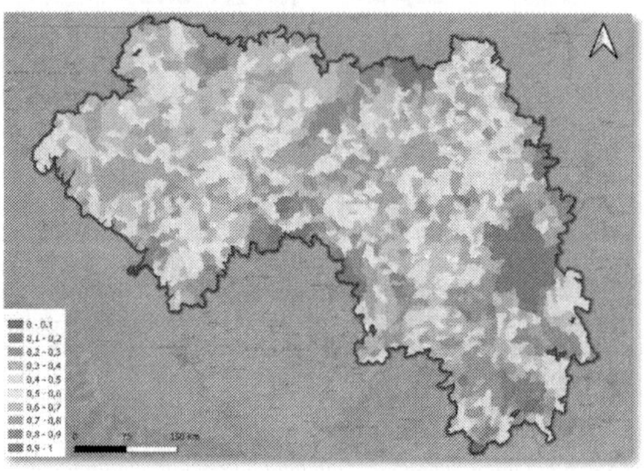

Figure 6. Map of the Landscape Fragmentation Indicator, by SELU –data in 2005

3.2 Some indicators of carbon accounts

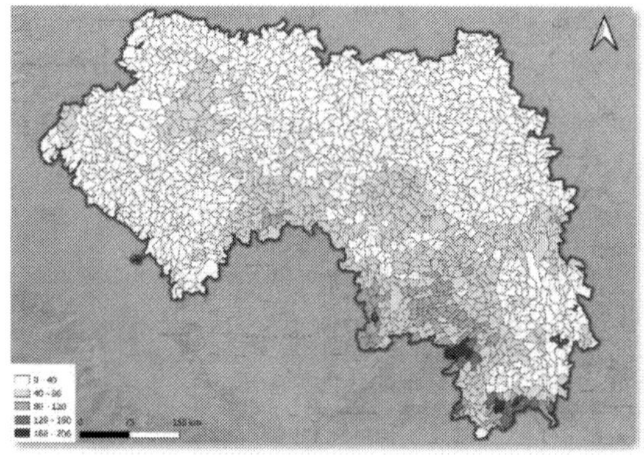

Figure 7. Living above-ground biomass in 2015 (t/ha)

The main causes of variation in ecosystem carbon stocks are variations in above-ground biomass stocks (living biomass and litter, dead wood) and soil organic carbon (also includes living roots, litter and dead wood).
In the Republic of Guinea, the levels of tree biomass stocks are particularly high in the natural region of Forest Guinea.
Up to 200 t/ha of tree biomass (Fig. 7) are observed in the classified forest of the Ziama massif, designated as a biosphere reserve in 1980 by UNESCO (The Ziama Massif: a vestige of the decline of forest ecosystems in High Guinea, West Africa, s. d.)

From this biomass value in tons is estimated the biocarbon stock (Figure 8) using the ratio of ½ of biocarbon in the biomass.

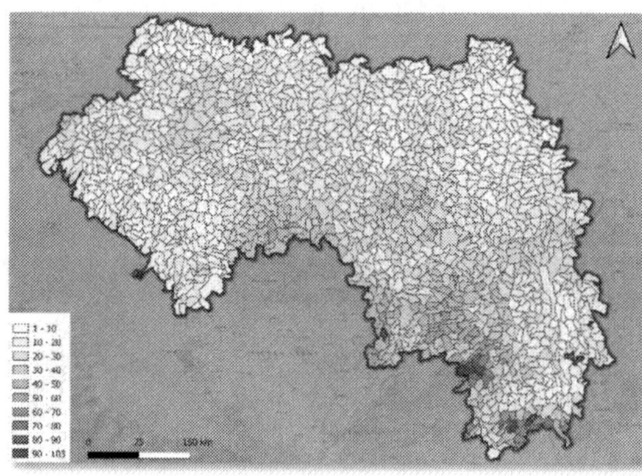

Figure 8. Carbon in living above-ground biomass in 2015 (t/ha)

Based on data from the International Soil Reference and Information Centre (ISRIC), the soil organic carbon stock is essentially the result of residues of plant and animal matter, decomposed under the influence of temperature, humidity, environmental conditions, and soil micro-organisms.
Soil organic carbon is heterogeneously distributed over the Guinean territory.
By cross-referencing this map with the land use layer, it can be seen that the coastal mangrove areas have the highest carbon rates per hectare (300 to 400t/ha).
Land use changes, particularly agricultural expansion dynamics in these areas, should therefore have an impact on soil organic carbon stocks.

This contribution has been peer-reviewed.
https://doi.org/10.5194/isprs-archives-XLVIII-M-1-2023-203-2023 | © Author(s) 2023. CC BY 4.0 License.

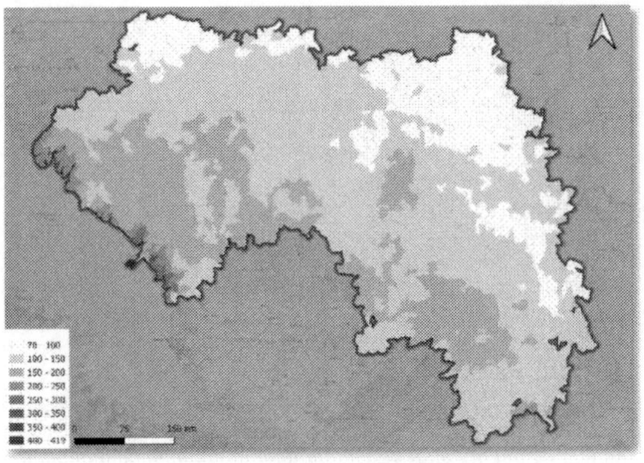

Figure 9. Soil Carbon at 1m depth (T/ha)

One of the main biocarbon input streams into the system is the net primary production of ecosystems, which is based on the overall dry matter productivity directly related to the growth rates of vegetation biomass. Net primary productivity is higher in 2015 compared to 2005.

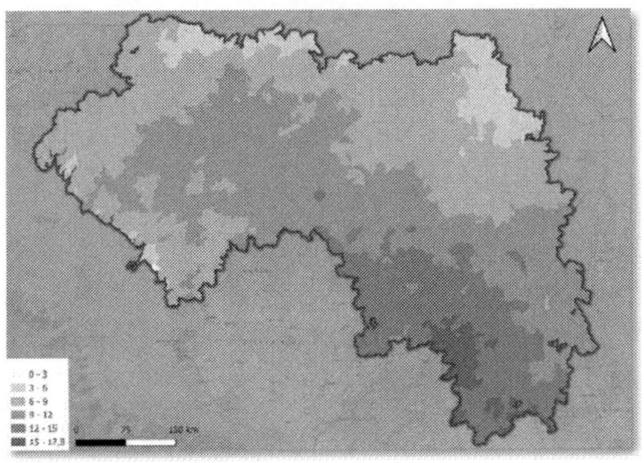

Figure 10. Net Primary Productivity (NPP), in tonnes of carbon per ha, by UPSE – data in 2015

3.3 Some indicators of water accounts

The ENCA water account is calculated from available input data on stocks (lakes and reservoirs, rivers) and input and output flows (rainfall, evapotranspiration, irrigation abstraction).

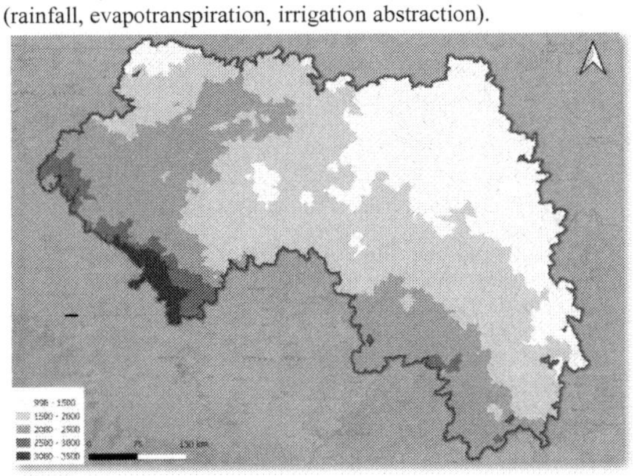

Figure 11. Average rainfall (mm) for 2005 (top) and 2015 (bottom)

The main input to the water account is rainfall. An increase in average annual rainfall in Guinea between 2005 and 2015 was observed, which partly explains the increase in net primary productivity (see Fig. 10) observed in the area. The distribution of rainfall is heterogeneous and varies across the country (Fig. 11). The coast is wetter, while the north-east of the country is drier. In order to smooth out interannual variations, 3-year averages were used.

3.4 Final summary of ecosystem potentials

From the table below (Fig.12), we see a decrease in the total ecosystem potential at the scale of Guinea. The value went from $1.03.10^9$ to $1.01.10^9$. This variation can be explained by the numerous land use and land cover changes that took place in the territory between 2005 and 2015.

Potentials		Years	
		2005	2015
Carbon Ecosystem Capability	**C_EC_capab**	1,56E+08	2,03E+08
Water Ecosystem Capability	**W_EC_capab**	1,23E+08	1,39E+08
Ecosystem Infrastructure Capability	**EI_EC_capab**	7,49E+08	6,67E+08
Total Ecosystem Capability	***TEC_capability***	***1,03E+09***	***1,01E+09***

Figure 12 : Total ecosystem capability (TEC) of Guinea for the years 2005 and 2015

4. DISCUSSION

Ecosystem-based natural capital accounting allows the evolution of the capacity of ecosystems to provide sustainable services to be assessed. This inherent capacity is an integral part of measuring the ecological value of each of the ecosystem components. The land cover produced by the IGN FI teams provided a perfect testing ground for the CECN at the scale of a national territory.

The results of this work in Guinea already demonstrate the importance of a policy to maintain the integrity of ecosystems with regard to the degradation of the total ecosystem potential linked to that of the internal unit value of the ecosystem infrastructure.

ENCA process thus appears to be an interesting decision-making tool for countries on environmental issues.

From an economic point of view, when this total capacity reflects a degradation of the ecosystem, it leads to unpaid ecological costs produced by the perpetrator(s) of the degradation. These ecological debts could be measured to estimate the restoration value of the degraded ecosystem (Babin & Weber, 2019). In addition, this experimentation makes it possible to advance on

https://doi.org/10.5194/isprs-archives-XLVIII-M-1-2023-203-2023 | © Author(s) 2023. CC BY 4.0 License.

the problems of implementing such a method in the Republic of Guinea.

As the ENCA method depends on the data used, the latter play an essential role in the smooth running of the method. The quality and availability of the data for the periods studied (2005 and 2015) sometimes raise difficulties. Indeed, some data, such as the OpenStreetMap road network map, are too heterogeneous and/or undated, making the resulting indicator (of landscape fragmentation) less representative of the reality on the ground and more static (available only for one date).

The ENCA protocol is currently being improved, in particular to better integrate the morphology of the territory and the relief in the definition of socio-ecological landscape units (SELU), or to better take into account marine coastal ecosystems in the accounts.

5. REFERENCES

Aubréville, A., 1957. Accord à Yangambi sur la nomenclature des types africains de végétation. Bois et Forêts des Tropiques. https://agritrop.cirad.fr/443212/

Babin, D., & Weber, J.-L., 2019. Économie et gestion de l'environnement et des ressources naturelles. 36.

Conrad, O., Bechtel, B., Bock, M., Dietrich, H., Fischer, E., Gerlitz, L., Wehberg, J., Wichmann, V., & Böhner, J., 2015. System for Automated Geoscientific Analyses (SAGA) v. 2.1.4. Geoscientific Model Development, 8(7), 1991-2007. https://doi.org/10.5194/gmd-8-1991-2015

Convention sur la diversité biologique, & PNUE., 2010. Plan stratégique pour la diversité biologique 2011-2020 et les Objectifs d'Aichi.

Delangue, J., & Teillac-Deschamp, P., 2019. IUCN - Nature-based Solutions for Climate Change Adaptation & Disaster Risk Reduction. https://ideas4development.org/preserver-ecosystemes-contre-changement-climatique/

Feger, C., & Mermet, L., 2021. Innovations comptables pour la biodiversité et les écosystèmes : Une typologie axée sur l'exigence de résultat environnemental. Comptabilité Contrôle Audit, 27(1), 13-50. https://doi.org/10.3917/cca.271.0013

Feranec, J., & al., 2016. European landscape dynamics: CORINE land cover data, 337 P. Taylor & Francis Group. https://doi.org/10.1201/9781315372860

Heldal, J., & Østdahl, T., 1984. Synoptic monitoring of water quality and water resources. A suggestion on population and sampling approaches. Statistical Journal of the United Nations Economic Commission for Europe, 2(4), 393-406. https://doi.org/10.3233/SJU-1984-2406

IONESCU et al., C., 2019. Capital naturel et stratégie des organisations : Une visite guidée des outils. Paris: WWF France.

Jaffrain, G., Leroux, A., Quang, A. V., Pinet, C., Dessagboli, B., Gauthier, Y., & Moiret, A., 2021. Suivi de la dynamique de l'occupation du sol en République de Guinée par imagerie satellitaire Spot : Transfert technologique pour le développement d'outils performants d'aide à la décision. Revue Française de Photogrammétrie et de Télédétection, 223, 59-80. https://doi.org/10.52638/rfpt.2021.563

Le massif de Ziama : Vestige de la diminution de l'écosystème forestier de Haute-Guinée | Afrique occidentale. (s. d.). Consulté 28 juillet 2022, à l'adresse https://eros.usgs.gov/westafrica/case-study/ziama-massif-relic-diminishing-upper-guinean-forest-ecosystem

Lehner, B., & Grill, G., 2013. Global river hydrography and network routing : Baseline data and new approaches to study the world's large river systems. Hydrological Processes, 27(15) : 2171–2186. www.hydrosheds.org

Lévêque, A., & Duhautois, L., 2006. L'environnement en France (édition 2006). Chapitre « La biodiversité ». (p. 273-296).
Local Biodiversity Intactness Index – GEO BON. (s. d.). Accessed on July 19, 2022, à l'adresse https://geobon.org/ebvs/indicators/local-biodiversity-intactness-index/

Millennium Ecosystem Assessment (Program) (Éd.), 2005. General synthesis. Island Press. http://www.millenniumassessment.org/documents/document.356.aspx.pdf

Nations Unies., 2015. Programme de développement durable à l'horizon 2030 : Résolution A/RES/70/1 de l'Assemblée générale des Nations Unies Transformer notre monde : Le Programme de développement durable à l'horizon 2030. https://www.un.org/ga/search/view_doc.asp?symbol=A/70/L.1&Lang=F

Ten Brink, P., 2016. Qu'est-ce que le capital naturel ? La revue du CGDD - Nature et richesses des nations | Mission Économie de la Biodiversité, 43-52.

Weber, J.-L., 2014. Ecosystem natural capital accounts : A quick start package - for implementing Aichi Biodiversity Target 2 on Integration of Biodiversity Values in National Accounting Systems in the context of the SEEA Experimental Ecosystem Accounts.

Weber, J.-L., 2021. CECN-TDR: KANGARÉ Tutorial version3

Weber, J.-L., 2022. L'émergence d'une comptabilité écosystémique. L'Économie politique, 93(1), 50-62.

Weber, J.-L., & European Environment Agency., 2006. Land accounts for Europe 1990-2000 / Towards integrated land and ecosystem accounting | Land and Ecosystems Accounts. https://projects.eionet.europa.eu/leac/library/land_account_wwwpdf

CHARACTERIZING MOVEMENT PATTERNS OF NOMADIC PASTORALISTS AND THEIR EXPOSURE TO RIFT VALLEY FEVER IN KENYA

G. Mosomtai[1,2*], J.L. Kasiiti[3], R.M. Murithi[3], P. Sandström[4], T. Landmann[2], O.W. Lwande[5], O.A. Hassan[5], C. Ahlm[5], R. Sang[2], M. Evander[5], Z. Szantoi[1], G. Ottavianelli[1]

[1] ESA, Directorate of EO Programmes, Science Applications & Climate Department, Frascati, Italy – Gladys.Mosomtai@ext.esa.int, zoltan.szantoi@esa.int & giuseppe.ottavianelli@esa.int
[2]International Centre of Insect Physiology and Ecology, Nairobi, Kenya - rsang@icipe.org, tlandmann@icipe.org
[3]Directorate of Veterinary Services, Ministry of Agriculture, Livestock and Fisheries, Kabete, Kenya - kasiiti.orengo@gmail.com,
[4]Department of Forest Resource Management, Faculty of Forest Sciences, Swedish University of Agricultural Sciences, Umea°, Sweden - per.sandstrom@slu.se
[5]Department of Clinical Microbiology, Umeå University, Sweden - magnus.evander@umu.se, clas.ahlm@umu.se, osama.a.hassan@umu.se, olivia.lwande@umu.se

KEY WORDS: Nomadic pastoralism, Rift Valley fever, Movement ecology, Seroprevalence, Mosquito vectors.

ABSTRACT:

The role of animal movement in spreading infectious diseases is highly recognized by various legislations and institutions such as the World Organisation for Animal Health and the International Animal Health Code. The increased interactions at the nexus of human-animal-ecosystem interface have seen an unprecedented introduction and reintroduction of new zoonotic diseases with high socio-economic impacts such as the COVID-19 pandemic. Rift Valley fever (RVF) is a zoonotic disease that affects both humans and animals and is transmitted by Aedes mosquitoes or through contact with the body fluids of infected animals. This study seeks to characterize movement patterns of pastoralist and how this movement behaviour increases their susceptibility to RVF virus exposure. We levarage on a rapidly growing field of movement ecology to monitor five herds collared from 2013 – 2015 in an RVF endemic semi-arid region in Kenya. The herds were also sampled for RVF antibodies to assess their exposure to RVF virus during the rainy seasons. *adehabitatLT* package in R was used to analyze the trajectory data whereas the first passage time (FPT) analysis was used to measure the area utilized in grazing. Sedentary herds grazed within 15km radius while migrating herds presented restricted space use patterns during the dry seasons and transient movement during the start and end of the rainy season. Furthermore, RVF virus antibodies were generally low for sedentary herds whereas the migrating herds recorded high levels during their transition periods. This study can be used to identify RVF risk zones for timely and targeted management strategies.

1. INTRODUCTION

Rift Valley fever (RVF) is a zoonotic disease that affects both humans and animals alike with high socio-economic impact. The infection can cause a severe hemorrhagic fever in humans and in ruminants such as cattle, sheep and goats with young animals being more vulnerable. It is also known to cause abortions in livestock resulting in huge losses to farmers who depend on livestock production for their livelihoods (Baudin et al., 2016).

The disease was first confirmed in 1931 in Kenya after a successful virus isolation (Daubney et al., 1931). RVF virus (RVFV) belongs to genus *Phlebovirus* of family *Phenuiviridae* viruses and RVFV is transmitted through bites by infected mosquitoes or by coming in contact with body fluids of infected animals (Sang et al., 2010). *Aedes* species are the primary vectors that maintain the virus through transovarial transmission during interepizootic periods while secondary vectors of *Culex, Anopheles* and *Mansonia* species amplify RVFV spread resulting in outbreaks during heavy rainfall seasons (Fafetine et al., 2013; Sang et al., 2010).

The World Organization for Animal Health classifies RVFV as high impact transboundary pathogen (Nanyingi et al., 2015). Since its discovery in Kenya, several epidemics have been reported in 15 countries while virus isolaton and serologial evidence have been detected in several countries in Sub Saharan Africa (Balenghien et al., 2013). The transboundary nature of the disease became more evident when it crossed over from continental Africa to Madagascar, Mayotte, Islands of Comoros and most recently to Arabian Peninsula (Nanyingi et al., 2015). Furthermore, phylogenetic studies on RVFV strains reveal this transboundary nature as multiple re-introductions of RVFV strains, e.g. in Senegal and Mauritania the strains were genetically related to South African and Zimbabwean strains (Soumaré et al., 2012). Similarly, an RVFV strain in Madagascar has genetic linkages with Egyptian, Zimbabwean and Kenyan isolate while the Arabian Penisulan isolate was closely related with the Kenyan isolate (Balenghien et al., 2013).

The long range dispersal of RVFV is linked to movement of viremic livestock through local and international trade (legal and illegal) or through exportation of infected mosquitoe vectors through airlines and cargo (Soumaré et al., 2012). This poses a threat to RVFV introduction to RVF-free continents where similar mosquito vectors found in endemic countries and susceptible ruminants are present (Golnar et al., 2014; Chevalier, 2013). Furthermore, the risk is hightened by climate change and the ability of the vector to occupy diverse ecological zones ranging from arid to semi arid regions in West Africa, Horn of Africa and Arabian Penisula to mountainous regions in Madagascar and highlands of Kenya (Chevalier, 2013).

Cofounding factors that maintains virus circulation during epizootic and enzootic periods have been an area of research interest in many studies where both biotic and abiotic factors

This contribution has been peer-reviewed.
https://doi.org/10.5194/isprs-archives-XLVIII-M-1-2023-211-2023 | © Author(s) 2023. CC BY 4.0 License.

have been identified (Anyamba et al., 2002; Anyangu et al., 2010; Olive et al., 2016; Pepin et al., 2010). However, studies that characterize and quantify the influence of movement of viremic animals especially, nomadic pastoralism in spreading RVFV is still limited and often proxy methods such as social network analysis and network modelling are used to quantify movement (Craft, 2015; Kim et al., 2018; Nicolas et al., 2014). In this study, we seek to deepen insights on the role of nomadic pastoralism in maintaining RVFV circulation.

We levarage on a rapidly growing field of movement ecology to characterize livestock movement patterns of pastoralist in an RVF endemic region in Kenya to understand how they move in space and time and how this movement behaviour increase their susceptibility to RVFV exposure (Nathan, 2008). Movement ecology has grown in the last two decades due to the advancement in tracking technology such as GPS and radio frequency at high spatial-temporal resolution and in data computation with applications in human, animal, and plant mobility (Miller et al., 2019).

2. MATERIALS AND METHODS

2.1 Study area and livestock collaring

The study was conducted in an RVF endemic region in Northeast Kenya where five livestock herds were GPS-collared from 2013 to 2015. Two herds were in Tana River County herein, herd4168 and herd4169 and three in Isiolo County herein, herd4006, herd4167 and herd4170 (Figure 1). The Tana River is the eastern border for the Tana River County, which stretches from Isiolo County in the north to the coast of Indian Ocean in the south. The two counties are in the semi-arid to arid regions characterized by average daily temperatures ranging from 30 – 35°C with a bimodal rainfall pattern where the long rains occur from March to May and the short rains in October to December with occasional variation. The rainfall amount ranges from 300 – 800 mm annually with extreme cases occurring during El Niño that results in excessive flooding.

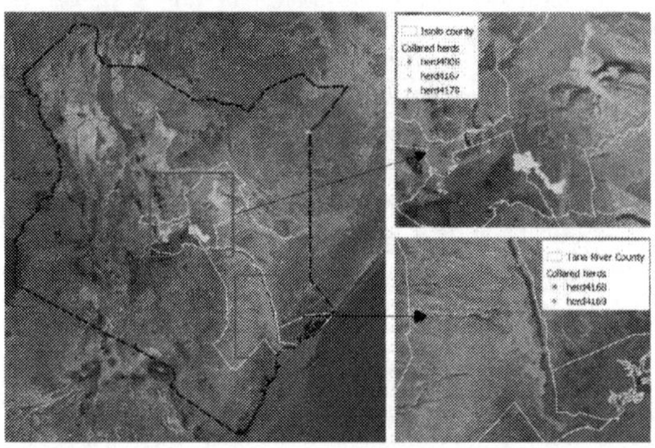

Figure 1: Study location with telemetry data points of five collared herds in Isiolo and Tana River Counties, Kenya.

The vegetation is mainly open shrubland with woody acacias and the community are mainly pastoralist who settle around water points and *dambos* (shallow depression on land that provide breeding for mosquitoes during rainy season) and live closely with their animals, increasing chances for zoonotic disease contraction (Owange et al., 2014). During the tracking period some of the GPS collars failed to capture data due to battery life while in some instances, the animals were out of reach with the GPS signal, resulting in some missing data points. Despite recording over 20 RVF outbreaks since 1961, Tana River and Isiolo counties have varying RVF risk levels (Murithi et al., 2011; Sang et al., 2017). Tana river county is classified as a high-risk county while Isiolo is a medium risk area.

2.2 Sero surveys along the migratory routes

The GPS-monitored herds were surveyed for anti-RVF immunoglobulin M (IgM) antibodies in their blood to assess their exposure to RVFV along their migration routes. Cattle aged not more than three years old were randomly sampled from each herd every time sampling was done. The strategy of sampling animals (< 3 years of age) was meant to leave out those with possibility of having RVFV antibodies due to previous outbreaks, the last one being experienced in 2007. In this study, we used results obtained from laboratory analysis of the serum survey conducted at the onset of the long and short rainy seasons and details regarding sampling will be presented in another report (manuscript in preparation).

2.3 Characterizing livestock movement pattern

We used *adehabitatLT* package in R (version 4.0.2) software to characterize the movement patterns of the herds and to understand how pastoralist utilize their landscape in time and space across landscape scales. *adehabitatLT* first generates several descriptive statistics of the trajectory for each successive relocation points (i.e., j, $j+1$) which forms the basis for further analysis (Calenge et al., 2008). Of interest to this study was characterizing the trajectory of the herds to identify change in the movement behaviour and the scale at which this behaviour was evident. Firstly, we summarized the speed the herds moved daily and monthly by dividing the step length with time taken between points. This was particularly important for understanding how pastoralist move on their landscape across time and seasons. Secondly, we used first passage time (FPT) analysis to identify the scale at which the behaviour occurred and to measure the area used for grazing based on the time the herds took to enter and leave a specified radius. If the circle is greater than the defined radius, intensely used spaces results in high FPT plateau while transient movements have smaller and often varying FPT (Fauchald & Tveraa, 2003).

3. RESULTS

3.1 Movement characterization

Figure 2 shows monthly (A and B) and hourly (C and D) mean speed at which the herds moved on their landscape during the collaring period. The herds presented similar daily movement patterns, especially the Tana Rivers herds (Figure 2D), with the highest speed being recorded at around 11:00hrs and 17:00hrs followed by a slower speed in between from 13:00hrs – 16:00hrs. Except for herd4006 in Isiolo, all the herds had a gradual increment of speed from 07:00hrs to 10:00hrs and a gradual decrease from 17:00hrs – 19:00hrs. In contrast, herd4006 had the highest speed in the morning and in the evening (Figure 2C). At monthly scale (Figure 2A and B), all the herds had varying movement speed with a notable contrast from herd4006 in Isiolo. The lowest speed of herd4006 (less than 0.1m/s) was recorded in January to February and in May to July with the highest speed occurring in April and November. Despite little variation in the daily speed, herd4006 showed a cyclic pattern with peak speed occurring in mid-February and August and lower speed from April to June and in October. On the other hand, herd4167 which had similar daily speed pattern with herd4170, showed little variation across the year. The herd maintained a speed greater than 0.2m/s across the year with peak speed occurring from June to September and lower speed from February to April. For Tana River herds, their monthly patterns also contrasted despite having identical daily movement pattern (Figure 2B). Herd4169 presented a cyclic pattern with peak speed occurring in February, April, July, and October while the lowest speed occurred in March, June, and August. For herd4168, peak speed occurred in March and October and a gradual decline in sped from May with the lowest speed occurring from July to September.

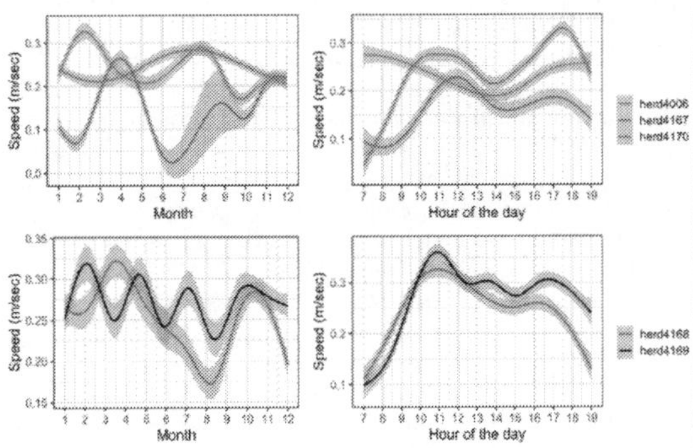

Figure 2: Monthly (A and B) and hourly (C and D) mean speed at which the herds moved on their landscape during the collaring period.

At monthly scale (Figure 2A and B), all the herds had varying movement speed with a notable contrast from herd4006 in Isiolo. The lowest speed of herd4006 (less than 0.1m/s) was recorded in January to February and in May to July with the highest speed occurring in April and November. Despite little variation in the daily speed, herd4006 showed a cyclic pattern with peak speed occurring in mid-February and August and lower speed from April to June and in October. On the other hand, herd4167 which had similar daily speed pattern with herd4170, showed little variation across the year. The herd maintained a speed greater than 0.2m/s across the year with peak speed occurring from June to September and lower speed from February to April. For Tana River herds, their monthly patterns also contrasted despite having identical daily movement pattern (Figure 2B). Herd4169 presented a cyclic pattern with peak speed occurring in February, April, July, and October while the lowest speed occurred in March, June, and August. For herd4168, peak speed occurred in March and October and a gradual decline in sped from May with the lowest speed occurring from July to September.

3.2 Scale and space use intensity

The movement behaviour (i.e., foraging and transient movement) of the herds in space and time are summarized in Figure 4. At large scale (15 km radius), the movement behaviour of the herds was well classified by FPT analysis than at local scale (5 km radius). herd4170, herd4168 and herd4169 maintained similar movement patterns across scales while herd4006 and herd4167 showed contrasting patterns. At 15 km radius, herd4006 and herd4167 presented restricted movement patterns whereas at 5 km radius, the space use pattern varied with more transient movement occurring during the dry season (June to September 2014) and restricted space use occurring during the long and short rains. For the migrating herds, the movement patterns of herd4168 and herd4169 were similar with the highest FPT plateau recorded from July to November 2014 while herd4170 was from December 2013 to early March 2014, which all coincided with the dry seasons in Kenya whereas the transient movements occurred at the start and end of each rainy season.

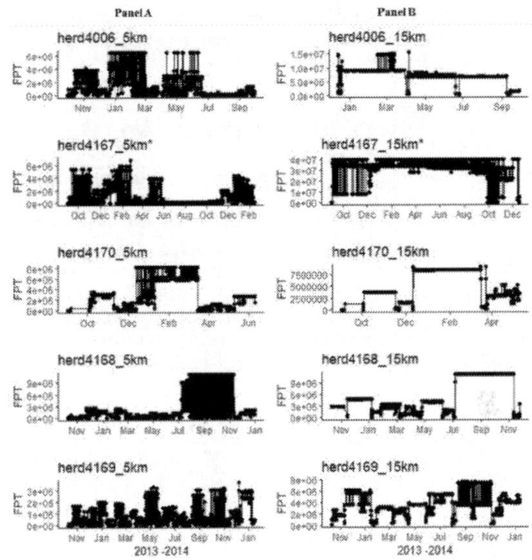

Figure 3: Space use intensity at 5 km and 15 km radius of Isiolo (herd4006, herd4167 and herd4070) and Tana River (herd4168 and herd4169) herds collared from September 2013 to 2014 except for herd4167* which was collared up to February 2015

3.3 Livestock movement and exposure to RVFV

The FPT value for all the herds at local scale (FPT_5km) were low in in all the time of sampling perid indicating that animals were in transient movement while at the large scale, only herd4168 in Tana River and herd4170 in Isiolo were sampled in locations were the FPT values were high indicating that they had spent a long time grazing there.

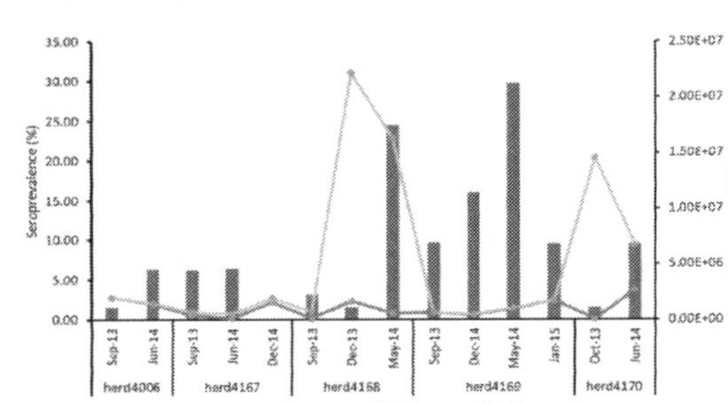

Figure 4: Relationship between RVFV serology survey and intensity of space use at the time of sampling. The movements of herds and seroprevalence were measured in Isiolo County (herd4006, herd4167 and herd4170) and Tana River County (herd4168 and herd4169), Kenya. High first passage time (FPT) alludes to intensely used space while low values indicate transient movements.

Overall, the herds in Isiolo generally recorded lower levels of RVF virus antibodies while the migrating herds in Tana River recorded the highest levels of RVF antibodies. herd4169 in Tana River county recorded the highest seroprevalence levels in all the sampling period, especially in May 2014 (30%) and this coincided with low FPT both at local and large scales whereas the counterpart, herd4168 recorded the highest seroprevalence but only in May 2014 (25%) (Figure 4). The Isiolo county herds recorded seroprevalence levels less than 10% despite the low FPT value both at local and large scale.

4. DISCUSSION AND CONCLUSION

Movement of viremic animals during active virus transmission play a critical role in the spread of RVF virus over short and long distances however quantifying its impact or understanding the dynamics of how it contributes to the spread of the disease is unclear (Balenghien et al., 2013). Role of livestock movement especially because of trade in spreading diseases has been exemplified by a review done by (Fèvre et al., 2006) and its effect quantified by (Adkin et al., 2016). Developed countries are in a better position in quantifying these dynamics due to established infrastructure unlike in many developing countries (Bajardi et al., 2011). We innovatively examined the dynamics of livestock movement in the context of nomadic pastrolism in an RVF endemic region in Kenya. Using GPS collars and conducting serum surveys to check on RVFV antibodies in herds collared in Tana River and Isiolo counties in Kenya, we established the relationships between movement and RVFV exposure.

Despite the two counties being endemic to RVF since 1961, severity levels have varied in the outbreaks with Tana River being classified as a high risk whereas Isiolo as medium risk. Our serum surveys confirmed this variation since we recorded the highest seroprevalance in Tana River herds than in Isiolo herds. The factors causing this variation is not well understood and a study by Sang et al., (2017) began by examing the diversity and abundance of primary RVF vectors in these two counties. They noted that *Aedes ochraceus,* a key primary RVF vector was missing in Isiolo but *Aedes mcintoshi* was present whereas in Tana River both the vectors were present in high abundance. Our study examined another factor that could be contributing to the variation by looking at the movement patterns of pastoralists and we first noted differences in the general movement patterns of pastoralist in the two counties. Tana River herds exhibited a nomadic pattern utilizing their landscape at a large scale (greater than 15km radius) whereas Isiolo herds exhibited two type of pastoralism, one that migrated and another that was more sedentary. Two of the Isiolo herds presented a restricted space use pattern at less than 15 km radius but at local scale (5 km radius) there was high variability in their movemnt patterns.

We analyzed how the movement speed of the herds varied across time scales. Daily movement patterns were characterized by increasing speed in the mid-mornings when the herds leave where they rested over night ('*boma*') to look for pastures and decreased in the evening when going back to rest in their '*boma*'. In the afternoon, the herds slowed their speed and we associate this slowed speed to an increase in foraging due to pasture availability and possibly herds being around water points. We hypothesize that the herds' risk to RVFV infection increased during this period and this is supported by an interview conducted by Owange et al., (2014) on what the pastoralist perceive as a key risk pathway for RVFV exposure. This is because RVFV vectors have been shown to prefer habitats with green vegetation and shallow water for breeding (Mosomtai et al., 2016). Furthermore, convergence of different herds at the watering point presents an opportunity for virus transmission among different herds from different locations to occur.

Looking at monthly movement speed of the herds, seasonal influence was evident however, every herd presented unique movement patterns including Tana River herds which had identical daily movement patterns. This unique annual movement patterns points to the various decisions pastoralist make regarding grazing strategies for their livestock and the seemigly similar movement pattern may actually vary across the year and this can have an influence on RVFV exposure. For instance, herd4169 was very mobile than the rest of the herds and from the serum surveys, it recorded the highest seroprevalence in all the sampling periods whereas herd4168 only recorded the highest seroprevalence yet both herds grazed in closed proximity but using different grazing strategies, with herd4168 spending much of its time in one area. Further studies however, need to be conducted to examine why in one specific sampling period (May 2014) the seroprevalence was generaly high for all the Tana Rivers herds.

In the study area, the livestock movement pattern involved pastoralists returning to their homes during the long and short rains, whereas during the dry season, they migrated to regions where pasture was available (Gomes, 2006). Availability of pasture and water is largely influenced by both human and environmental factors. The erratic climatic conditions due to climate change threatens the livelihood of pastoralist and increase their susceptability to zoonotic diseases. It is now established that RVF outbreaks are linked to El Nino rains that leads to flooding, providing unlimited habitat for mosquitoes to breed and amplify the virus (Anyamba et al., 2009). On the other hand, landscape modification due to human activities such as irrigation schemes in the arid regions, increase the risk for RVFV infections (Sang et al., 2018). The study area has also faced frequent drought events and the ongoing 2020 – 2022 drought has resulted in massive death of livestocks and the community is facing extreme mulnutrition, which has forced pastoralist to migrate to the neighbouring countries (WHO, 2022).

This study broaden our understanding on how pastoralist move and how it influence their exposure to RVFV. We characterized livestock movement patterns across spatial and temporal scale in two RVF endemic counties with varying RVF severity in Northeastern Kenya. Tana River county recorded high seroprevalence than Isiolo county and this was associated to the increased livestock movement pattern. This study shows the potential of levarging movement ecology to understand RVF disease ecology better by incorporating the aspect of movement of viremic animal that is often understudied.

ACKNOWLEDGEMENTS

The authors gratefully acknowledge the financial support for this research by the European Space Agency (ESA) under the African Research Fellowship, Swedish Research Council (2013-06257) and the Swedish International Development Cooperation Agency (SIDA) (SWE-2011-016). We also acknowledge the International Centre for Insect Physiology and Ecology (*icipe*) and ESA, ESRIN for supporting the implementation of this work.

This contribution has been peer-reviewed.
https://doi.org/10.5194/isprs-archives-XLVIII-M-1-2023-211-2023 | © Author(s) 2023. CC BY 4.0 License.

REFERENCES

Adkin, A., Brouwer, A., Downs, S. H., & Kelly, L. (2016). Assessing the impact of a cattle risk-based trading scheme on the movement of bovine tuberculosis infected animals in England and Wales. *Preventive Veterinary Medicine*, *123*, 23–31.

Bajardi, P., Barrat, A., Natale, F., Savini, L., & Colizza, V. (2011). Dynamical patterns of cattle trade movements. *PLoS ONE*, *6*(5), 1–51. https://doi.org/10.1371/journal.pone.0019869

Balenghien, T., Cardinale, E., Chevalier, V., Elissa, N., Failloux, A., Jean Jose Nipomichene, T., Nicolas, G., Rakotoharinome, V., Roger, M., & Zumbo, B. (2013). Towards a better understanding of Rift Valley fever epidemiology in the south-west of the Indian Ocean. *Veterinary Research*, *44*(1), 78. https://doi.org/10.1186/1297-9716-44-78

Baudin, M., Jumaa, A. M., Jomma, H. J. E., Karsany, M. S., Bucht, G., Näslund, J., Ahlm, C., Evander, M., & Mohamed, N. (2016). Association of Rift Valley fever virus infection with miscarriage in Sudanese women: a cross-sectional study. *The Lancet Global Health*. https://doi.org/10.1016/S2214-109X(16)30176-0

Calenge, C., Dray, S., & Royer-carenzi, M. (2008). The concept of animals' trajectories from a data analysis perspective. *Ecological Informatics*, *4*(1), 34–41. https://doi.org/10.1016/j.ecoinf.2008.10.002

Chevalier, V. (2013). Relevance of Rift Valley fever to public health in the European Union. *Clinical Microbiology and Infection*, *19*(8), 705–708. https://doi.org/10.1111/1469-0691.12163

Craft, M. E. (2015). Infectious disease transmission and contact networks in wildlife and livestock. *Philosophical Transactions of the Royal Society B: Biological Sciences*, *370*(1669). https://doi.org/10.1098/RSTB.2014.0107

Edelhoff, H., Signer, J., & Balkenhol, N. (2016). Path segmentation for beginners: an overview of current methods for detecting changes in animal movement patterns. *Movement Ecology 2016 4:1*, *4*(1), 1–21. https://doi.org/10.1186/S40462-016-0086-5

Fafetine, J., Neves, L., Thompson, P. N., Paweska, J. T., Rutten, V. P. M. G., & Coetzer, J. A. W. (2013). Serological Evidence of Rift Valley Fever Virus Circulation in Sheep and Goats in Zambézia Province, Mozambique. *PLoS Neglected Tropical Diseases*, *7*(2), 2065. https://doi.org/10.1371/journal.pntd.0002065

Fèvre, E. M., Bronsvoort, B. M. D. C., Hamilton, K. A., & Cleaveland, S. (2006). Animal movements and the spread of infectious diseases. *Trends in Microbiology*, *14*(3), 125–131. https://doi.org/10.1016/j.tim.2006.01.004

Golnar, A. J., Turell, M. J., LaBeaud, A. D., Kading, R. C., & Hamer, G. L. (2014). Predicting the Mosquito Species and Vertebrate Species Involved in the Theoretical Transmission of Rift Valley Fever Virus in the United States. *PLoS Neglected Tropical Diseases*, *8*(9). https://doi.org/10.1371/journal.pntd.0003163

Gomes, N. (2006). *Access to water, pastoral resource **management and pastoralists' livelihoods**.*

Hassan, O. A., Ahlm, C., Sang, R., & Evander, M. (2011). The 2007 Rift valley fever outbreak in Sudan. *PLoS Neglected Tropical Diseases*, *5*(9), 1–7. https://doi.org/10.1371/journal.pntd.0001229

Kim, Y., Dommergues, L., M'sa, A. ben, Mérot, P., Cardinale, E., Edmunds, J., Pfeiffer, D., Fournié, G., & Métras, R. (2018). Livestock trade network: potential for disease transmission and implications for risk-based surveillance on the island of Mayotte. *Scientific Reports*, *8*(1). https://doi.org/10.1038/S41598-018-29999-Y

Miller, H. J., Dodge, S., Miller, J., & Bohrer, G. (2019). Towards an integrated science of movement: converging research on animal movement ecology and human mobility science. *International Journal of Geographical Information Science*, *33*(5), 855–876. https://doi.org/10.1080/13658816.2018.1564317

Mosomtai, G., Evander, M., Sandström, P., Ahlm, C., Sang, R., Hassan, O. A., Affognon, H., & Landmann, T. (2016). Association of ecological factors with Rift Valley fever occurrence and mapping of risk zones in Kenya. *International Journal of Infectious Diseases*, *46*, 49–55. https://doi.org/10.1016/j.ijid.2016.03.013

Murithi, R. M., Munyua, P., Ithondeka, P. M., Macharia, J. M., Hightower, a, Luman, E. T., Breiman, R. F., & Njenga, M. K. (2011). Rift Valley fever in Kenya: history of epizootics and identification of vulnerable districts. *Epidemiology and Infection*, *139*(3), 372–380. https://doi.org/10.1017/S0950268810001020

Nanyingi, M. O., Munyua, P., Kiama, S. G., Muchemi, G. M., Thumbi, S. M., Bitek, A. O., Bett, B., Muriithi, R. M., & Njenga, M. K. (2015). A systematic review of Rift Valley Fever epidemiology 1931–2014. *Infection Ecology & Epidemiology*, *5*, 1–12. https://doi.org/10.3402/iee.v5.28024

Nathan, R. (2008). An emerging movement ecology paradigm. *Proceedings of the National Academy of Sciences of the United States of America*, *105*(49), 19050–19051. https://doi.org/10.1073/PNAS.0808918105

Nicolas, G., Chevalier, V., Tantely, L. M., Fontenille, D., & Durand, B. (2014). A Spatially Explicit Metapopulation Model and Cattle Trade Analysis Suggests Key Determinants for the Recurrent Circulation of Rift Valley Fever Virus in a Pilot Area of Madagascar Highlands. *PLoS Neglected Tropical Diseases*, *8*(12). https://doi.org/10.1371/JOURNAL.PNTD.0003346

Olive, M.-M., Chevalier, V., Grosbois, V., Tran, A., Andriamandimby, S.-F., Durand, B., Ravalohery, J.-P., Andriamamonjy, S., Rakotomanana, F., Rogier, C., Heraud, J.-M., Chevalier, V., Pepin, M., Bouloy, M., Bird, B., Kemp, A., Paweska, J., Nicolas, G., Durand, B., … Nichol, S. (2016). Integrated Analysis of Environment, Cattle and Human Serological Data: Risks and Mechanisms of Transmission of Rift Valley Fever in Madagascar. *PLOS Neglected Tropical Diseases*, *10*(7), e0004827. https://doi.org/10.1371/journal.pntd.0004827

Owange, N. O., Ogara, W. O., Kasiiti, J., Gathura, P. B., Okuthe, S., Sang, R., Affognon, H., Onyango-Ouma, W., Landmann, T. T. O., & Mbabu, M. (2014). Perceived risk factors and risk pathways of Rift Valley fever in cattle in Ijara district, Kenya. *Onderstepoort J Vet Res*, *81*(1), 1–7. https://doi.org/10.4102/ojvr.v81i1.780

Pepin, M., Bouloy, M., Bird, B. H., Kemp, A., & Paweska, J. (2010). Rift Valley fever virus(Bunyaviridae: Phlebovirus): an update on pathogenesis, molecular epidemiology, vectors, diagnostics and prevention. *Veterinary Research*, *41*(6), 61. https://doi.org/10.1051/VETRES/2010033

Sang, R., Arum, S., Chepkorir, E., Mosomtai, G., Tigoi, C., Sigei, F., Lwande, O. W., Landmann, T., Affognon, H., Ahlm, C., & Evander, M. (2017). Distribution and abundance of key vectors of Rift Valley fever and other arboviruses in two ecologically distinct counties in Kenya. *PLoS Neglected Tropical Diseases*, *11*(2), e0005341. https://doi.org/10.1371/journal.pntd.0005341

Sang, R., Kioko, E., Lutomiah, J., Warigia, M., Ochieng, C., O'Guinn, M., Lee, J. S., Koka, H., Godsey, M., Hoel, D., Hanafi, H., Miller, B., Schnabel, D., Breiman, R. F., & Richardson, J. (2010). Rift Valley fever virus epidemic in Kenya, 2006/2007: The entomologic investigations. *American Journal of Tropical Medicine and Hygiene*, *83*(2 SUPPL.), 28–37. https://doi.org/10.4269/ajtmh.2010.09-0319

Soumaré, P. O. L., Freire, C. C. M., Faye, O., Diallo, M., de Oliveira, J. V. C., Zanotto, P. M. A., & Sall, A. A. (2012). Phylogeography of Rift Valley Fever virus in Africa reveals multiple introductions in Senegal and Mauritania. *PLoS ONE*, *7*(4), 23–26. https://doi.org/10.1371/journal.pone.0035216

WHO. (2022). *Greater Horn of Africa (Food Insecurity and Drought)*.

This contribution has been peer-reviewed.
https://doi.org/10.5194/isprs-archives-XLVIII-M-1-2023-211-2023

A DEBIASING VARIATIONAL AUTOENCODER FOR DEFORESTATION MAPPING

M. X. Ortega Adarme[1,4]*, P. J. Soto Vega[2], G. A. O. P. da Costa[3], R. Q. Feitosa[4], C. Heipke[1]

[1] Institute of Photogrammetry and GeoInformation, Leibniz Universität Hannover, Germany - (ortega, heipke)@ipi.uni-hannover.de
[2] LaTIM, INSERM, UMR 1101, University Brest, Brest, France - sotovega@univ-brest.fr
[3] State University of Rio de Janeiro (UERJ), Rio de Janeiro, Brazil - gilson.costa@ime.uerj.br
[4] Pontifical Catholic University of Rio de Janeiro (PUC-Rio), Brazil - raul@ele.puc-rio.br

Commission III, WG III/4

KEY WORDS: Debiasing Variational Autoencoder, Deforestation Detection, Deep Learning, Semantic Segmentation.

ABSTRACT:

Deep Learning (DL) algorithms provide numerous benefits in different applications, and they usually yield successful results in scenarios with enough labeled training data and similar class proportions. However, the labeling procedure is a cost and time-consuming task. Furthermore, numerous real-world classification problems present a high level of class imbalance, as the number of samples from the classes of interest differ significantly. In various cases, such conditions tend to promote the creation of biased systems, which negatively impact their performance. Designing unbiased systems has been an active research topic, and recently some DL-based techniques have demonstrated encouraging results in that regard. In this work, we introduce an extension of the Debiasing Variational Autoencoder (DB-VAE) for semantic segmentation. The approach is based on an end-to-end DL scheme and employs the learned latent variables to adjust the individual sampling probabilities of data points during the training process. For that purpose, we adapted the original DB-VAE architecture for dense labeling in the context of deforestation mapping. Experiments were carried out on a region of the Brazilian Amazon, using Sentinel-2 data and the deforestation map from the PRODES project. The reported results show that the proposed DB-VAE approach is able to learn and identify under-represented samples, and select them more frequently in the training batches, consequently delivering superior classification metrics.

1. INTRODUCTION

The popularity of Deep Learning (DL) techniques has increased enormously (Wang et al., 2020) in recent years. Indeed, DL-based approaches are associated with exceptional advances across a wide range of applications, in a diverse variety of fields, such as medical image analysis and diagnosis, biometry, environmental monitoring, autonomous driving, and natural language processing among others. These systems need to be carefully conceived and properly developed, ideally taking into consideration the concept of fairness (Osoba, Welser IV, 2017), as the solutions have a direct impact on society. In that context, providing safe and fair systems has become a relevant topic in a research community concerned with guaranteeing long-term, successful implementation of such systems, while minimizing potential negative side effects of their continuous use (Mehrabi et al., 2021, Amini et al., 2019).

Conventionally, DL approaches achieve good performances in scenarios with enough labeled training data (Vardi, 2022). In that case, they are able to generalize successfully on the unobserved test data. Nevertheless, the lack of large sets of labeled data is inherent to many real-world problems, and such a condition can lead to the creation of biased systems.

In addition, the class imbalance problem is quite common in different DL applications (Thabtah et al., 2020). In such cases, the trained classifier may be biased towards the majority classes, and it is possible to reach a high overall accuracy by just assigning those classes to the test samples (Ali et al., 2019). For the minority classes, however, the classification performance may be very poor. We observe that such issue is particularly significant in change detection, as usually the instances of the changed class are much scarcer than those of the unchanged samples. The capacity to properly learn from imbalanced data is, therefore, of paramount importance for change detection applications.

To address the fairness and the bias issues, several methods have been proposed thus far. Typically, those methods are categorized based on the stage where the bias mitigation is performed. Three different categories have been considered (Friedler et al., 2019): pre-process, in-process; and post-process.

Pre-process algorithms involve the transformation of the training data before feeding them into the model. For instance, the authors of (Naseriparsa et al., 2020, More, 2016) proposed resampling techniques to generate synthetic samples of the minority class. Those methods, however, do not properly consider within-class variability, and can also oversample uninformative or noisy samples (Jiang et al., 2021). The authors of (Kamiran, Calders, 2012, Luong et al., 2011) propose reweighing or changing the labels of some samples to train a classifier with non-discrimination constraints. The target samples in those works are those closest to the decision boundary, which are usually critical in the classification prediction. Similarly, works for learning the latent structure of data have also been studied, e.g., (Feldman et al., 2015, Calmon et al., 2017, Louizos et al., 2015). Those include latent SVM (Felzenszwalb et al., 2008) and fair dimensionality reduction with Principal Component Analysis (PCA) (Samadi et al., 2018). More recently, DL-based approaches for generating fair representations have been proposed. In that group, Variational Autoencoders (VAE) (Kingma, Welling, 2013, Louizos et al., 2015, Liu et al., 2022) and adversarial

* Corresponding author

https://doi.org/10.5194/isprs-archives-XLVIII-M-1-2023-217-2023

training-based methods (Edwards, Storkey, 2015, Xie et al., 2017, Madras et al., 2018, Feng et al., 2019, Ruoss et al., 2020) are commonly used.

For in-process mechanisms, fairness is addressed during the training stage. The authors of (Kamishima et al., 2012) included a regularization term representing a trade-off between classification accuracy and fairness in the loss function. Likewise, (Bechavod, Ligett, 2017) included terms into the loss function that penalize unfairness by minimizing the differences between the false positive and false negative rates.

Finally, the post-process mechanisms make decision fairer by further processing the output scores of the classifiers. As an example, (Hardt et al., 2016) ensures a non-discriminatory prediction by flipping some decisions of a classifier to equalize predictor odds. Additionally, a decoupling system is proposed in (Corbett-Davies et al., 2017) to train independent classifiers for each class, and combine transfer learning approaches to learn from samples out of all classes.

In this work, we adapted the Debiasing Variational Autoencoder (DB-VAE), proposed by (Amini et al., 2019), for the task of semantic segmentation. Our approach is based on an end-to-end training scheme and employs the learned latent variables to adjust the individual sampling probabilities of data points during the training process. Thus, the method is able to identify types of samples that are under-represented in the training set, and to increase the likelihood that such instances are sampled during training. For semantic segmentation, we added an extra enconder-decoder module to the original architecture, composed of a series of convolutional operations followed by a softmax layer, which infers the probability of each class in a dense labeling fashion. We evaluated the proposed method on a deforestation detection application, using Sentinel-2 data from a region in the Amazon rain forest.

This paper is organized as follows: Section 2 describes the fully convolutional DB-VAE method used for deforestation detecting. Next, Section 3 presents the study area, the detailed experimental protocol, and the analysis of the obtained results. Finally, conclusions and final remarks are presented in Section 4.

2. FULLY DB-VAE

This section presents a brief description of the DB-VAE, originally proposed in (Amini et al., 2019), as well as the adaptation of the method to the semantic segmentation task.

Formally, let's consider a classification problem with d classes, and n training samples $\mathcal{D} = \{(\mathbf{x}_i, \mathbf{y}_i)\}_{i=1}^{n}$, where $\mathbf{x}_i$ indicates the features ($\mathbf{x} \in \mathbb{R}^m$), and $\mathbf{y}_i$ denotes the class label ($\mathbf{y} \in \mathbb{R}^d$) of the i-th sample. The goal is to find a nonlinear mapping $f_\theta : \mathbf{X} \rightarrow \mathbf{Y}$ parameterized by θ which minimizes a certain loss $\mathcal{L}$ over the entire training dataset. Then, given a new unobserved sample $(\mathbf{x}_i, \mathbf{y}_i)$, the classifier should ideally output $\hat{\mathbf{y}}_i = f_\theta(\mathbf{x}_i)$, where $\hat{\mathbf{y}}_i$ is "proximate" to $\mathbf{y}_i$, where the proximity is defined by the loss function. In addition, let's assume that each sample has an associated latent-space representation of dimension k ($\mathbf{z}_i \in \mathbb{R}^k$ and $k < m$), which compresses the feature information of the original input.

Inspired by (Amini et al., 2019), we propose a fully convolutional architecture for semantic segmentation. In particular, we focus on a change detection application, i.e., deforestation detection. Accordingly, the input of the neural network comprises a co-registered pair of images acquired at different dates: I_{t_0} and I_{t_1}. The images were concatenated along their spectral dimension, resulting in a tensor $\mathbf{I} \in \mathbb{R}^{H \times W \times C}$, where H and W refer to the spatial dimensions, and C to the number of image channels (two times those of each individual image).

Figure 1 shows the general design of the Fully DB-VAE. The encoder module of the VAE compresses the input samples and learns an approximation $q_\phi,(\mathbf{z}|\mathbf{x})$, parametrized by ϕ of the true distribution of the latent variables given a data point $\mathbf{x}$. The decoder reconstructs the input using the latent space representation by approximating $p_\psi(\hat{\mathbf{x}}|\mathbf{z})$, parametrized by ψ. VAEs parameterize the outputs as a normal distribution $\epsilon \sim \mathcal{N}(0,(I))$ and compute $\mathbf{z} = \mu(\mathbf{x}) + \sum^{1/2}(\mathbf{x}) \circ \epsilon$, where μ and σ are the mean and standard deviation of the latent variable distribution.

As we address a semantic segmentation approach, we added an extra encoder-decoder network to the architecture. That fully convolutional network is responsible for the pixel-wise classification, i.e., for assigning a class label to each input pixel location.

The whole model is trained in an end-to-end fashion; the loss function contains three components: a supervised latent loss (i.e., cross-entropy loss), a reconstruction loss (i.e., L2 norm), and a latent loss for the unsupervised variables (i.e., KL-divergence). The total loss $\mathcal{L}$ is a weighted sum of the three terms, and it is defined as follows:

$$\begin{aligned}\mathcal{L} = \lambda_s \left[\sum_{i=0}^{d} y_i log\left(\hat{y}_i\right)\right] + \lambda_r \left[\left\|\mathbf{x} - \hat{\mathbf{x}}\right\|_2\right] + \\ \lambda_u \left[\frac{1}{2}\sum_{i=0}^{k}\left(\sigma_i + \mu_i^2 - 1 - log\left(\sigma_i\right)\right)\right]\end{aligned} \tag{1}$$

where λ_s, λ_r, and λ_u, represent the weights of the supervised latent loss, the reconstruction loss, and the unsupervised loss, respectively.

At each training epoch, we compute the histogram $\hat{\mathcal{Q}}_i$ of each latent variable $z_i(\mathbf{x})$ delivered by the VAE encoder. We further assume that the latent variables are statistically independent so that the probability of the latent representation $\mathbf{z}(\mathbf{x}) = [z_1(\mathbf{x}), ..., z_k(\mathbf{x})]$ of sample $\mathbf{x}$ is given by $\prod_i \hat{\mathcal{Q}}_i$.

The goal is that each subcategory of the training samples is equally represented in each batch. Therefore, the DB-VAE adopts the following probability of selecting a sample $\mathbf{x}$ for the next batch:

$$\mathcal{W}(\mathbf{z}(\mathbf{x}) \mid X) \propto \prod_i \frac{1}{\hat{\mathcal{Q}}_i\left(z_i(\mathbf{x}) \mid X\right) + \alpha} \tag{2}$$

where α is the debiasing parameter, which tunes the degree of debiasing during training. When $\alpha \rightarrow 0$, the samples of the training batch tend to follow a uniform distribution over the latent variables z. On the other hand, when $\alpha \rightarrow \infty$, the training batch follows a random uniform sampling of the original training dataset.

https://doi.org/10.5194/isprs-archives-XLVIII-M-1-2023-217-2023 | © Author(s) 2023. CC BY 4.0 License.

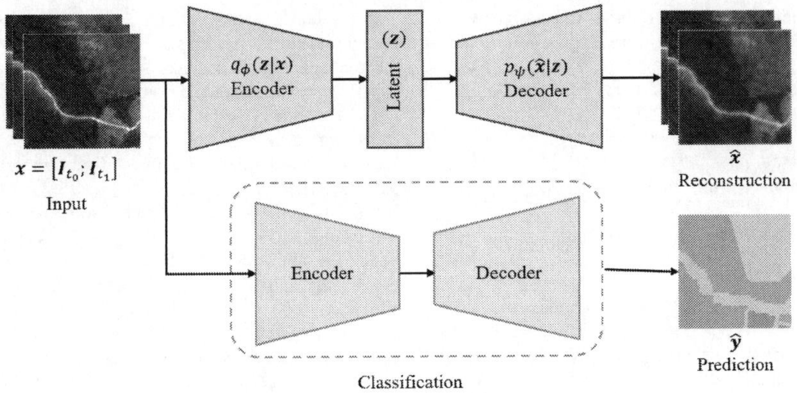

Figure 1. Fully Debiasing Variational Autoencoder (Fully DB-VAE) for semantic segmentation.

3. EXPERIMENTS

In this section, we report the results obtained with the Fully DB-VAE method in the deforestation detection application. We start by providing information about the dataset used in the experiments. Next, we describe the experimental setup, and finally we analyze the results in terms of classification metrics and visual interpretation. Furthermore, we compare the Fully DB-VAE with the conventional U-Net (Ronneberger et al., 2015), which was chosen as a baseline model.

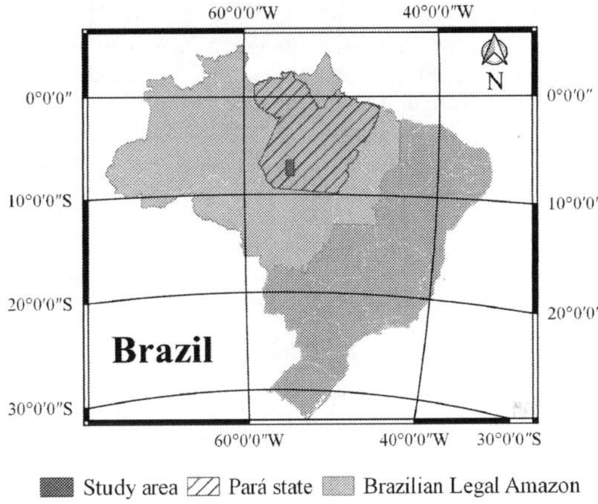

Figure 2. Geographical localization of the study area.

3.1 Study Area

The study area comprises a Sentinel-2 scene, with a size of 17730 × 9200 pixels The region is located in Pará State, Brazil. That State recorded one of the highest deforestation rates in 2021, according to the PRODES monitoring system (Assis et al., 2019) from the Brazilian National Institute for Space Research (INPE). We used all Sentinel-2 bands with spatial resolutions of 10m and 20m. The 20m bands were resampled to 10m using nearest neighbor interpolation. The images were downloaded from the Google Earth Engine (GEE) platform (Gorelick et al., 2017). we used Level-1C images, which means they are orthorectified and contain top-of-atmosphere reflectance data. The change map was downloaded from the INPE web site, which is freely available at the PRODES database [1].

3.2 Experimental Setup

In all experiments, the input was a tensor resulting from the concatenation of a bitemporal image pair, acquired at dates t_0 and t_1, along the spectral dimension. Next, each band was normalized in the range of $[-1, 1]$. The study area was divided into 20 large tiles, following a distribution of $40\% - 10\% - 50\%$ for training-validation-test. The input to the encoder networks were patches of size 128 × 128 pixels, extracted using a sliding window procedure with a stride equal to 32.

Table 1 shows the acquisition dates of the Sentinel-2 data and the class distribution of the study area. We are interested in mapping the deforestation that occurred between 2020 and 2021. The dates of selected images are based on the PRODES project, which uses images during the dry season, with the minimum cloud cover. For date t_1, the image data is a mosaic composed of two Sentinel-2 scenes. The class deforestation represents samples mapped as forest at epoch t_0, and no forest at epoch t_1, the class no-deforestation corresponds to samples mapped as forest at epoch t_0 and t_1, and the class past-deforestation corresponds to all regions deforested before epoch t_0. It's worth mentioning that this application presents a

[1] Available at: http://terrabrasilis.dpi.inpe.br/map/deforestation

high level of class imbalance. Table 1 shows that less than 2% of the study area corresponds to the deforestation class.

Date t_0	July 15, 2020
Date t_1	July 25, 2021 August 4, 2021
Class deforestation (%)	1.86
Class no-deforestation (%)	56.40
Class past-deforestation (%)	41.74

Table 1. Image acquisition dates and class distribution of the study area.

Table 2 presents the Fully DB-VAE architectures, with information in the corresponding layers of the encoders, bottlenecks, and decoders. Furthermore, the following training parameters were defined: batch size equal to 16; Adam optimizer with learning rate equal to $2e^{-4}$, and β equal to 0.5. Also, the early stopping strategy was used to avoid over-fitting. The values of the loss function weights λ_s, λ_r, and λ_u were set to 1.0, 1.0, and $5e^{-4}$, respectively.

Following the PRODES methodology, during training and testing, we disregarded predictions within a two-pixel wide buffer in the internal and external edges of all polygons classified as deforestation in the reference change map. Similarly, pixels corresponding to the past-deforestation (before 2020) and polygons smaller than 625 pixels (6,25 ha) were ignored. The inference was carried out tile-wise, and each experiment was run five times.

3.3 Results and Discussion

In this section, we present the obtained results using the Fully DB-VAE. The classification performance was measured in terms of *Recall*, *Precision*, and *F1-score* for the deforestation class. Furthermore, for a visual interpretation, we present the deforestation probability maps.

Figure 3 shows the accuracy values obtained in each experiment. The first bar group represents the results for the baseline, which was a conventional U-Net, using a random sampling strategy for selecting the training samples. The other bar groups present the metrics obtained for the Fully DB-VAE with different values of α, starting from $\alpha = 1e{-}1$ to $\alpha = 1e{-}10$. According to the figure, one can notice that the Fully DB-VAE obtained the highest *Recall* scores. Also, the false deforestation predictions were lower than the baseline, but *Precision* was also lower. However, in all the cases, it exceeded 80%. In terms of *F1-score*, the best result was achieved for $\alpha = 1e{-}7$, yielding a score of 79.5, about 5% higher than the baseline.

As an additional experiment, we compared the performance of the Fully DB-VAE with the U-Net architecture using an arbitrary selection criterion in the training procedure: similar to (Ortega Adarme et al., 2020, Ortega et al., 2021), we only selected patches with at least 2% of pixels belonging to the class deforestation. Figure 4 summarizes the results of the U-net and the Fully DB-VAE with the two strategies for training sample selection. For the experiments with the Fully DB-VAE we defined $\alpha = 1e{-}7$, which presented the best performance in terms of the *F1-score* of previous results. We can observe the Fully DB-VAE still yields the best performance, although the results using the 2% selection criterion were better than with the random selection.

For a visual analysis, Figure 5 illustrates the deforestation probability maps of a snip in the test set. The maps represent the average prediction map. They were generated for all experiments using the U-Net as a baseline and the Fully DB-VAE with different values of α. In the first line, the RGB compositions of t_0 and t_1 are shown. Next, the class label map with classes deforestation, no-deforestation and past-deforestation is presented, as well as the outcome of the U-Net using the training samples from the random selection and from the 2% selection criteria. The second and third lines depict the outcomes of the Fully DB-VAE modifying the value of α. Red and blue colors represent the highest and lowest probability of belonging to the deforestation class. The black color symbolizes the past- deforestation class. Similar to the classification metrics, we note that the Fully DB-VAE with $\alpha = 7$ delivered the most accurate and reliable outputs, i.e., with probabilities close to one. In addition, we observed that the outcomes from the U-Net with random selection samples and the Fully DB-VAE with $\alpha = 1e^{-1}$ and $\alpha = 1e^{-2}$ are farther from the reference, delivering less reliable and less accurate definition of the polygons of the class deforestation.

4. CONCLUSIONS

This work presented an extension of the debiasing variational autoencoder (DB-VAE) architecture based on Fully Convolutional Neural Networks (FCN), so called Fully DB-VAE. The method was adapted to perform semantic segmentation in the context of detection of deforestation spots in a region of the Amazon Brazilian biome. In order to find the optimal value of the debiasing parameter α, an ablation study was carried out. The obtained results were compared with a conventional U-Net model using two strategies for training sample selection. The first one is a random approach, and the second one only patches with at least 2% of pixels belonging to the class deforestation were used. According to the experiments, the best performance in terms of $F1 - score$ was achieved with $\alpha = 1e^{-7}$. In that case, the Fully DB-VAE outperformed both baselines and delivered the most reliable values in the deforestation probability maps.

5. ACKNOWLEDGEMENTS

The authors would like to thank the German Academic Exchange Service (DAAD), the Conselho Nacional de Desenvolvimento Científico e Tecnológico (CNPq), the Coordenação de Aperfeiçoamento de Pessoal de Nível Superior (CAPES), and the Fundação de Amparo à Pesquisa do Estado do Rio de Janeiro (FAPERJ) for the financial support.

REFERENCES

Ali, H., Salleh, M. N. M., Saedudin, R., Hussain, K., Mushtaq, M. F., 2019. Imbalance class problems in data mining: A review. *Indonesian Journal of Electrical Engineering and Computer Science*, 14(3), 1560–1571.

Amini, A., Soleimany, A. P., Schwarting, W., Bhatia, S. N., Rus, D., 2019. Uncovering and mitigating algorithmic bias through learned latent structure. *Proceedings of the 2019 AAAI/ACM Conference on AI, Ethics, and Society*, 289–295.

Assis, F., Fernando, L., Ferreira, K. R., Vinhas, L., Maurano, L., Almeida, C., Carvalho, A., Rodrigues, J., Maciel, A.,

Architecture	Encoder	Bottleneck	Decoder	Output
Autoencoder	MP(C(3 × 3, 16)) MP(C(3 × 3, 32)) MP(C(3 × 3, 64)) MP(C(3 × 3, 128)) MP(C(3 × 3, 128))	Flatten() Dense(latentDim) Dense(latentDim*2)	Reshape(Dense(.)) US(C(3 × 3, 128)) US(C(3 × 3, 128)) US(C(3 × 3, 64)) US(C(3 × 3, 32)) US(C(3 × 3, 16))	Tanh (C(1 × 1, #IC))
Classifier	MP(C(3 × 3, 16)) MP(C(3 × 3, 32)) MP(C(3 × 3, 64)) MP(C(3 × 3, 128)) MP(C(3 × 3, 128))	-	US(C(3 × 3, 128) US(C(3 × 3, 128) US(C(3 × 3, 64) US(C(3 × 3, 32) US(C(3 × 3, 16)	Softmax (C(1 × 1, #CL))

Table 2. Network Architectures. Symbols: C (strided convolution), MP (max-pooling), US (bilinear up-sampling), IC (Input Channels), CL (Output classes). The values in parentheses refer to (kernel width x kernel height, number of filters).

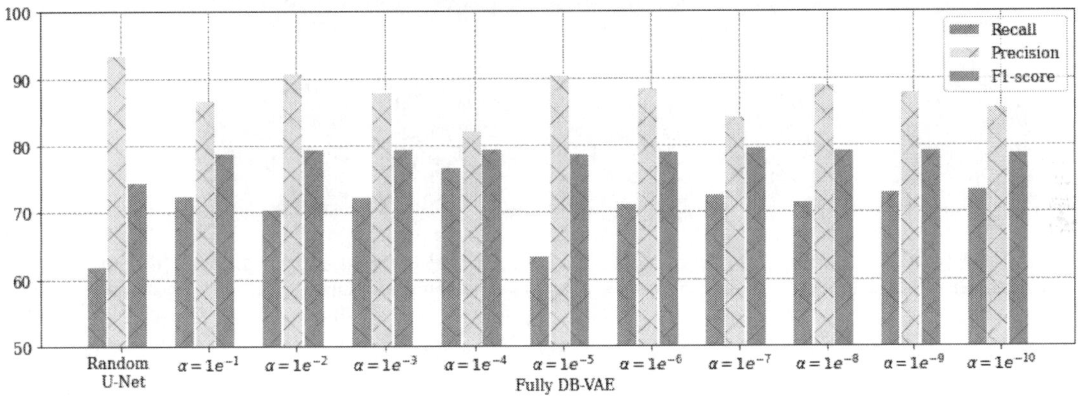

Figure 3. Classification metrics in [%], for the U-Net with random selection sampling and Fully DB-VAE with different values of α.

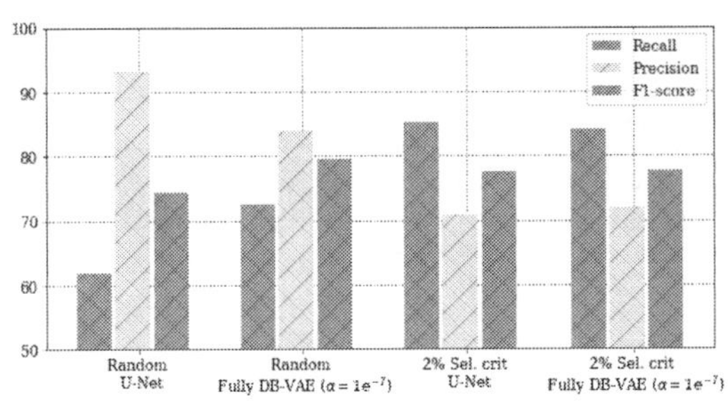

Figure 4. Classification metrics in [%], for the U-Net and Fully DB-VAE with $\alpha = 1e^{-7}$, using random samples and the 2% selection criteria.

Camargo, C., 2019. TerraBrasilis: A Spatial Data Analytics Infrastructure for Large-Scale Thematic Mapping. *ISPRS International Journal of Geo-Information*, 8(11), 513.

Bechavod, Y., Ligett, K., 2017. Penalizing unfairness in binary classification. *arXiv preprint arXiv:1707.00044*.

Calmon, F., Wei, D., Vinzamuri, B., Natesan Ramamurthy, K., Varshney, K. R., 2017. Optimized pre-processing for discrimination prevention. *Advances in neural information processing systems*, 30.

Corbett-Davies, S., Pierson, E., Feller, A., Goel, S., Huq, A., 2017. Algorithmic decision making and the cost of fairness. *Proceedings of the 23rd acm sigkdd international conference on knowledge discovery and data mining*, 797–806.

Edwards, H., Storkey, A., 2015. Censoring representations with an adversary. *arXiv preprint arXiv:1511.05897*.

Feldman, M., Friedler, S. A., Moeller, J., Scheidegger, C., Venkatasubramanian, S., 2015. Certifying and removing disparate impact. *proceedings of the 21th ACM SIGKDD international conference on knowledge discovery and data mining*, 259–268.

Felzenszwalb, P., McAllester, D., Ramanan, D., 2008. A discriminatively trained, multiscale, deformable part model. *2008 IEEE conference on computer vision and pattern recognition*, Ieee, 1–8.

Feng, R., Yang, Y., Lyu, Y., Tan, C., Sun, Y., Wang, C., 2019. Learning fair representations via an adversarial framework. *arXiv preprint arXiv:1904.13341*.

Friedler, S. A., Scheidegger, C., Venkatasubramanian, S., Choudhary, S., Hamilton, E. P., Roth, D., 2019. A comparative study of fairness-enhancing interventions in machine learning. *Proceedings of the conference on fairness, accountability, and transparency*, 329–338.

Gorelick, N., Hancher, M., Dixon, M., Ilyushchenko, S., Thau, D., Moore, R., 2017. Google Earth Engine: Planetary-scale geospatial analysis for everyone. *Remote sensing of Environment*, 202, 18–27.

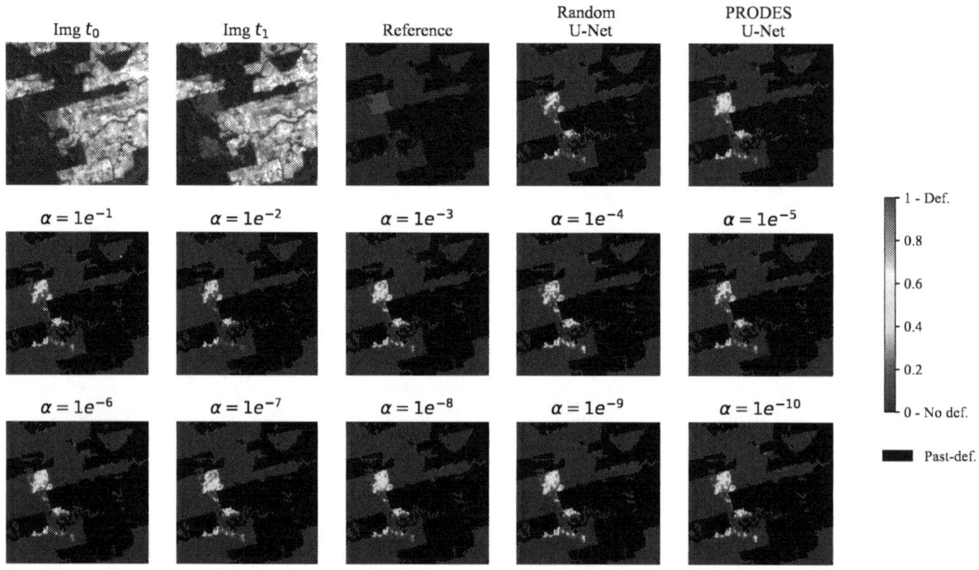

Figure 5. Visual example of a snip from the test set. The first line illustrates the RGB compositions of the snip in t_0 and t_1, the class label map, and the baselines, using the U-Net architecture with two strategies of training sample selection. The second and third lines presents the probability maps varying the values of α in the Fully DB-VAE. Red and Blue color represent the highest and lowest probability of belonging to the deforestation class. Black color symbolize the past deforestation.

Hardt, M., Price, E., Srebro, N., 2016. Equality of opportunity in supervised learning. *Advances in neural information processing systems*, 29.

Jiang, Z., Pan, T., Zhang, C., Yang, J., 2021. A new oversampling method based on the classification contribution degree. *Symmetry*, 13(2), 194.

Kamiran, F., Calders, T., 2012. Data preprocessing techniques for classification without discrimination. *Knowledge and information systems*, 33(1), 1–33.

Kamishima, T., Akaho, S., Asoh, H., Sakuma, J., 2012. Fairness-aware classifier with prejudice remover regularizer. *Machine Learning and Knowledge Discovery in Databases: European Conference, ECML PKDD 2012, Bristol, UK, September 24-28, 2012. Proceedings, Part II 23*, Springer, 35–50.

Kingma, D. P., Welling, M., 2013. Auto-encoding variational bayes. *arXiv preprint arXiv:1312.6114*.

Liu, S., Sun, S., Zhao, J., 2022. Fair Transfer Learning with Factor Variational Auto-Encoder. *Neural Processing Letters*, 1–13.

Louizos, C., Swersky, K., Li, Y., Welling, M., Zemel, R., 2015. The variational fair autoencoder. *arXiv preprint arXiv:1511.00830*.

Luong, B. T., Ruggieri, S., Turini, F., 2011. k-nn as an implementation of situation testing for discrimination discovery and prevention. *Proceedings of the 17th ACM SIGKDD international conference on Knowledge discovery and data mining*, 502–510.

Madras, D., Creager, E., Pitassi, T., Zemel, R., 2018. Learning adversarially fair and transferable representations. *International Conference on Machine Learning*, PMLR, 3384–3393.

Mehrabi, N., Morstatter, F., Saxena, N., Lerman, K., Galstyan, A., 2021. A survey on bias and fairness in machine learning. *ACM Computing Surveys (CSUR)*, 54(6), 1–35.

More, A., 2016. Survey of resampling techniques for improving classification performance in unbalanced datasets. *arXiv preprint arXiv:1608.06048*.

Naseriparsa, M., Al-Shammari, A., Sheng, M., Zhang, Y., Zhou, R., 2020. RSMOTE: improving classification performance over imbalanced medical datasets. *Health information science and systems*, 8, 1–13.

Ortega Adarme, M., Queiroz Feitosa, R., Nigri Happ, P., Aparecido De Almeida, C., Rodrigues Gomes, A., 2020. Evaluation of deep learning techniques for deforestation detection in the Brazilian Amazon and cerrado biomes from remote sensing imagery. *Remote Sensing*, 12(6), 910.

Ortega, M. X., Feitosa, R. Q., Bermudez, J. D., Happ, P. N., De Almeida, C. A., 2021. Comparison of optical and sar data for deforestation mapping in the amazon rainforest with fully convolutional networks. *2021 IEEE International Geoscience and Remote Sensing Symposium IGARSS*, IEEE, 3769–3772.

Osoba, O. A., Welser IV, W., 2017. *An intelligence in our image: The risks of bias and errors in artificial intelligence*. Rand Corporation.

Ronneberger, O., Fischer, P., Brox, T., 2015. U-net: Convolutional networks for biomedical image segmentation.

Medical Image Computing and Computer-Assisted Intervention–MICCAI 2015: 18th International Conference, Munich, Germany, October 5-9, 2015, Proceedings, Part III 18, Springer, 234–241.

Ruoss, A., Balunovic, M., Fischer, M., Vechev, M., 2020. Learning certified individually fair representations. *Advances in neural information processing systems*, 33, 7584–7596.

Samadi, S., Tantipongpipat, U., Morgenstern, J. H., Singh, M., Vempala, S., 2018. The price of fair pca: One extra dimension. *Advances in neural information processing systems*, 31.

Thabtah, F., Hammoud, S., Kamalov, F., Gonsalves, A., 2020. Data imbalance in classification: Experimental evaluation. *Information Sciences*, 513, 429–441.

Vardi, G., 2022. On the implicit bias in deep-learning algorithms. *arXiv preprint arXiv:2208.12591*.

Wang, X., Zhao, Y., Pourpanah, F., 2020. Recent advances in deep learning. *International Journal of Machine Learning and Cybernetics*, 11, 747–750.

Xie, Q., Dai, Z., Du, Y., Hovy, E., Neubig, G., 2017. Controllable invariance through adversarial feature learning. *Advances in neural information processing systems*, 30.

This contribution has been peer-reviewed.
https://doi.org/10.5194/isprs-archives-XLVIII-M-1-2023-217-2023 | © Author(s) 2023. CC BY 4.0 License.

The International Archives of the Photogrammetry, Remote Sensing and Spatial Information Sciences, Volume XLVIII-M-1-2023
39th International Symposium on Remote Sensing of Environment (ISRSE-39) "From Human Needs to SDGs", 24–28 April 2023, Antalya, Türkiye

Benefit Assessment of Commercial Synthetic Aperture Radar Observations for NASA's Surface Deformation and Change Mission Study

B. Osmanoglu[1*], S. A. Huang[1], C. A. Jones[2], B. Scheuchl[3], A. Khazendar[4], J. Sauber[1], K. Tymofyeyeva[4], M.J. Jo[1]

[1] NASA Goddard Space Flight Center, Greenbelt, MD, U.S.A. - (batuhan.osmanoglu, stacey.a.huang, jeanne.m.sauber-rosenberg, minjeong.jo)@nasa.gov
[2] Langley Research Center, Hampton, VA, U.S.A. - christopher.a.jones-2@nasa.gov
[3] University of California, Irvine, CA, U.S.A. - bscheuch@uci.edu
[4] Jet Propulsion Laboratory, California Institute of Technology, Pasadena, CA, U.S.A. – (ala.khazendar, katia.tymofyeyeva)@jpl.nasa.gov

KEY WORDS: Synthetic Aperture Radar, Commercial SAR, Surface Deformation and Change.

ABSTRACT:

The Surface Deformation and Change (SDC) mission study is investigating a synthetic aperture radar (SAR) mission that is expected to launch in the next decade, building on the foundation established by the NASA ISRO Synthetic Aperture Radar (NISAR) mission. Since 2019, the SDC study team has updated the observation needs identified by the 2017 Earth Science Decadal Survey, developing a Science and Applications Traceability Matrix (SATM) that includes an expanded set of geophysical observables (GOs). These needs were further refined by a team of discipline experts, resulting in 48 GOs. For each GO, imaging characteristics such as revisit, accuracy, resolution, polarisation, data latency, are defined in the SATM. This paper describes the benefit assessment methodology, provides an example to generate current commercial feasibility scores for each GO in the SATM, even though the SDC mission will not be launched until the next decade. This methodology generates a quantitative assessment of commercial SAR data in meeting the measurement needs of a GO defined in SDC's SATM. Our assessment suggests that current commercial SAR data are particularly useful for constraining geophysical processes that benefit from short-repeat acquisition times and high spatial resolution.

1. INTRODUCTION

Guided by the SATM (Horst et al. 2021, Khazendar et al. 2021), the SDC mission study is evaluating several architectures having different numbers of spacecraft, orbital configurations, sensor characteristics, and geometries. These configuration differences result in distinct core capabilities, which are evaluated in a consistent value framework.

Some of these core capabilities can benefit from commercial SAR constellations filling in data gaps, especially in terms of high temporal and spatial resolutions. In recent years, commercial sector capabilities in SAR imaging have been expanding rapidly, however, the interferometric capability is currently still limited. The feasibility of commercial capabilities to support the GOs defined in SDC's SATM are being analysed to identify the strengths and weaknesses of commercial SAR data when applied to SDC observables. In addition to the parameters defined in the SATM, our analysis also looks at the interferometric capability of commercial constellations, along with their swath width and their global sampling coverage rate. The GOs are described in detail in the SATM, and a summary of the relevant parameters used in the scoring is provided in the appendix. To achieve a result that is resilient in the face of market fluctuations, the commercial assessment relies on studying industry capabilities as a whole instead of individually focusing on each constellation.

2. METHODOLOGY

The methodology is designed to assess possible strengths and weaknesses of commercial SAR data in meeting the GOs defined in SDC's SATM. The scoring equation uses eight key factors to parameterize the benefit of commercial data for each GO.

2.1 Parameters

All parameters defined in this assessment are defined as a percentage, represented as a number between 0 and 1 for low and high scores, respectively. Table 1 lists these parameters, how they affect the score in (1), their description, and the equation to calculate each factor.

Parameter	Type	Description	Eq.
Polarisation	Additive	Percent channels available vs. desired in SATM	$min(1, \frac{cc}{satm})$
Resolution	Additive	Inverse percentage of available resolution vs. SATM.	$min(1, \frac{satm}{cc})$
Revisit	Additive	Inverse percentage of available revisit vs. SATM.	$min(1, \frac{satm}{cc})$
Latency	Additive	Inverse percentage of available revisit vs. SATM.	$min(1, \frac{satm}{cc})$
Coverage	Multipli-cative	Percentage of available vs. requirement	$min(1, \frac{cc}{req.})$
Large Swath	Multipli-cative	Percentage of large swath capable constellations	No = 1.0 Yes & No = $\frac{1+perc}{2}$ Yes = $perc$
InSAR	Multipli-cative	Percentage of SAR capable constellations.	No = 1.0 Yes & No = $\frac{1+perc}{2}$ Yes = $perc$
*Accuracy	Minimum	Accuracy percentage for the given geophysical observable either through heuristic or literature survey.	$min(1, \frac{satm}{lit.svy})$

Table 1. Parameters used in the scores.
*The equation is only used for literature survey results when heuristic scores are not available.

Benefit scores are calculated using the following equations:

$$score_{int} = \frac{1}{N}\sum_{i}^{N} S_i \times \prod_{j}^{M} C_j \quad (1)$$

$$score_{fin} = min(acc, score_{int}) \quad (2)$$

where $score_{int}$ is the interim score, N is the number of parameters that are arithmetically averaged, and M is the number of multiplicative parameters (defined in Table 1). S and C represent the partial scores, $score_{fin}$ is the final score, and acc is expected measurement accuracy (relative to the desired capability in the SATM, see example in Table 4).

In (1), arithmetic averaging is used in the partial scores S and multiplication is used in the partial scores C due to the perceived difference in the impact of these parameters. For example, if data can be acquired every two weeks, but the desire is to collect data every week, the revisit score would be 50%. We equally weighted revisit, polarisation, resolution, and latency, therefore arithmetic averaging is used for a combined S score. On the other hand, coverage, large swath imaging, and InSAR capabilities are more limiting compared to parameters contributing to the S score. For example, if global coverage is needed every two weeks, and there is only enough commercial capacity to image 50% of the globe every two weeks, this should not result in a C score that is higher than the lowest factor, even if InSAR and large-swath capacity are fully met. Therefore the C scores are multiplicative as shown in (2).

As part of the commercial SAR feasibility assessments, the team identified the main observational parameters and associated representative capabilities that commercial SAR can currently provide, as shown in Table 2.

Parameter	Value	Justification
Polarisation	Single	Commercial market capability currently focuses on co-pol observations
Resolution	5m	Commercial imagery is capable of delivering higher resolution at the expense of measurement noise level.
Revisit	1 day	Commercial constellations are aiming to deliver data at a daily frequency anywhere in the world. At the moment this is aspirational for interferometry and demonstrated for radiometry.
Latency	4 hours	Most commercial vendors provide low latency downlink
Coverage	2.21M km^2/ day	At the time of the assessment, the commercial market had a combined capacity of over 2.21M km^2 per day.
Large Swath	0.43	Four out of the seven active constellations support collecting >100 km wide imagery for large area coverage
InSAR	0.43	Four out of the seven active constellations support collecting interferometric data at least in part of their constellation.
Accuracy	Variable	The value ranges from 0% to 100% based on the expected performance.

Table 2. An overview of currently available representative commercial capabilities used in the commercial SAR feasibility estimation

These capabilities are derived from publicly available information from the commercial constellations operated by Capella Space, ICEYE, iQPS, NEC, Surrey Sat. Tech. Ltd, Synspective, and Umbra. One important parameter that is not considered in the above set of parameters is the satellite frequency band. This is due to the fact that the bulk of the available commercial data is in X-band. Rather than excluding GOs that desire a specific band, we estimate expected performance at X-band for all GOs with the assumption that any GO can be satisfied under specific observation scenarios or locations, even if they cannot be satisfied globally at X-band. However, depending on the target of interest, X-band sensors can fall short for Ecosystems and Solid Earth observables due to shallow penetration depth and rapid decorrelation (Zebker and Villasenor, 1992; Wei and Sandwell, 2010; Hajj et al., 2019). In some cases, imaging might still be possible, albeit with significantly less coverage and accuracy compared to a longer wavelength sensor; in other cases, however, the sensors may completely fail to retrieve any meaningful measurements. In particular, long-wavelength radar can be critical for observing change mapping in wetlands, forests, and coastal zones (Ottinger and Kuenzer, 2020), and for imaging subtle movements of faults in heavily vegetated areas (Nikolaeva et al., 2014; Tong et al., 2018). Here, we have not explicitly accounted for the cases in which X-band sensors would completely fail to return meaningful data, but these limitations are important to consider in the overall evaluation of commercial SAR capabilities. For the GOs that are not assessed by the performance tool, a literature review was conducted to approximate the measurement capability at X-band, and for heuristic scores the expert panel provided accuracy estimates considering X-band. Ultimately, a percentage of desired capability is calculated based on this value and the SATM is used in the assessment (see Table 4).

Coverage may require special attention in this calculation as this capability is defined by km^2/day for the average commercial market, while the SATM did not define such a metric, assuming a global need. Unlike the value framework assessment for SDC architectures, a coverage area is defined in the commercial assessment for each geophysical observable, as shown in Table 3.

Global	129 M km^2	Total land area
Continental	25 M km^2	
Regional	9 M km^2	About the size of the US or EU
Local	1 M km^2	

Table 3. Coverage bins used in this study for each GO.

Table 3 is used to check the capability of the commercial sector to see if such an area can be covered during the required revisit time defined for the GO.

Even though the same SATM is used for both the value framework analysis and the commercial SAR assessment, the assessments are different to compensate for the differences between commercial capabilities and SDC architectures. For example, all SDC architectures provide coverage of all land and ice areas every 12 days or faster, and it is not feasible to acquire global coverage using commercial constellations due to coverage limitations and cost. Similarly, all SDC architectures are capable of interferometry, while only three out of seven commercial constellations provide that capability. Such differences necessitate the use of additional parameters for the commercial SAR assessment.

2.2 Scoring

The assessment scores each GO against the needs defined in the SATM. For example, the ninth ecosystem (E9) GO calls for "measuring inland coastal wetland areas" and mentions the need for dual polarisation imagery at 20 m resolution with a revisit and latency of 14 days (Table 4). Similarly, the need for coverage is classified as regional (9 million km^2) every 14 days, and neither large swath imagery nor interferometry are necessary (Table 5). Our panel of experts assessed the X-band performance of wetland mapping as 35%.

Parameter	Polar. (#)	Resol. (m)	Revisit (day)	Latency (day)	Arithmetic Average
Need	2	20	14	14	
Capability	1	5	1	0.16	
Part. Score	0.5	1	1	1	0.875

Table 4. Example estimation of percent capabilities addressing the desired needs for parameters defined in the SATM (GO E9).

Parameter	Coverage (km^2/day)	Large Area Imaging	InSAR	Multiplication Result
Need	Regional @14 days (revisit)	No	No	
Capability	2.2e6	0.43	0.43	
Part. Score	1	1	1	1.0

Table 5. Example estimation of percent capabilities addressing the desired needs for Commercial SAR assessment specific parameters (GO E9).

$$score_{int} = 0.875 \times 1.0 = 0.875 \quad (3)$$

$$score_{fin} = min(0.35, 0.875) = 0.35 \quad (4)$$

Where $score_{int}$ is calculated based on the multiplication of partial scores in Table 4 and Table 5. By design of (2) the final score is limited with the accuracy parameter.

3. RESULTS

This methodology was applied to all of the GOs in the SATM to produce a self-consistent set of commercial benefit scores. These results allow for identification of GOs that would benefit more than others in a relative sense.

3.1 Comparative Assessment

In order to better understand the benefit of commercial data, an architecture based on a simplified NISAR design (Kellog et al. 2020, Rosen et al. 2021), which we call as L1A, was assessed. The relevant parameters of this design are summarised in Table 6.

Parameter	Pol. (#)	Res. (m)	Rev. (day)	Lat. (day)	Cov. (km^2/d)	Large Area	InSAR
Capability	4	5	12	1	10.8e6	1.0	1.0

Table 6. Conservative capability estimate of L1A architecture for this analysis.

The accuracy scores for L1A came from a Value Framework in development for assessing the SDC architectures with respect to Science Benefit and other components of Value. In the Value Framework, accuracy metrics for each GO can depend on either or both vertical and horizontal accuracy needs. In cases where more than a single accuracy score was provided, these scores were averaged to obtain GO specific L1A accuracy score. These scores then compare the capability of an architecture to the desired capability in the SATM for each GO, and quantify the degree to which that architecture achieves the SATM's needs.

Figures 1 and 2 show the relationship between the commercial constellations vs. L1A scores, separating the GOs based on the source of the accuracy metric, heuristic, and performance tool, respectively. There are two reasons for this: 1) the performance tool accuracy calculation is holistic and accounts for parameters such as revisit rates as well; and 2) due to the very different nature of accuracy estimates and the strong control of accuracy on the final outcome, the behaviour of GOs for both cases needs to be investigated.

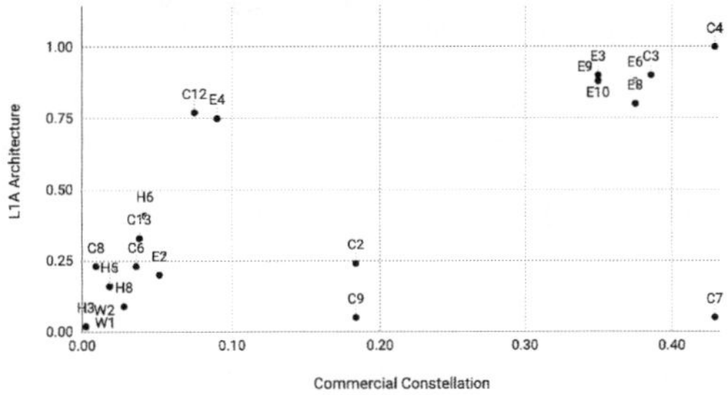

Figure 1. Commercial assessment scores against L1A architecture for GOs with heuristic accuracy.

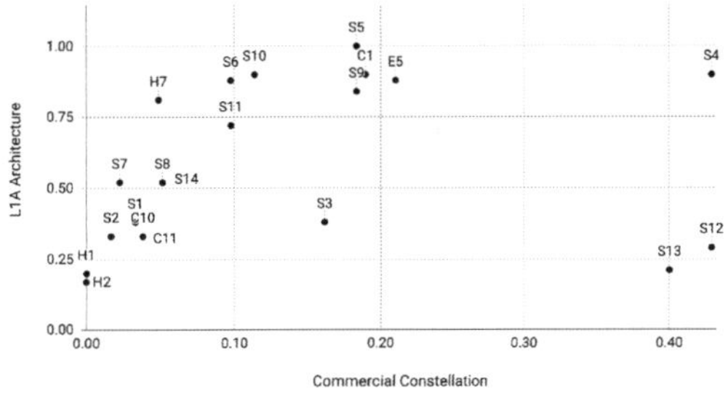

Figure 2. Commercial assessment scores against L1A architecture for GOs with performance tool based accuracy.

Figures 1 and 2 demonstrate a logarithmic nature, with a sharp rise close to the origin and relative flattening at the far end of the x-axis. This behaviour is perhaps indicative of the fact that both SAR systems are capable of addressing the SATM albeit less so for commercial constellations. Furthermore, the different range of scores are evident, as commercial assessment scores achieve a high of only 0.43 while the L1A architecture does achieve a full score (1.0) for some GOs. On average, L1A scores an average of 0.49, while commercial constellations achieve a score of 0.15 in realising the SDC SATM. Despite this, there are some GOs that achieve higher assessment scores for the commercial constellation compared to L1A, such as cryospheric GOs C7 and C9, as well as solid Earth GOs such as S12, S13.

https://doi.org/10.5194/isprs-archives-XLVIII-M-1-2023-225-2023

4. DISCUSSION

The effort documented here covers only an augmentation of a core SDC capability. The SDC team has also reached out to the industry through a Request For Information (RFI) and is planning to release a Request for Proposal (RFP) in the first quarter of 2023. The RFI/RFP activities include public-private partnership solutions, but these are outside the scope of this whitepaper.

4.1 Assessment Scores

The assessment scores obtained here are based on the commercial capabilities at the end of 2022. It is a fair assumption that commercial capabilities will increase significantly by the time of the SDC mission. However, it is difficult to project commercial capabilities into the 2030s, as the market is currently going through a rapid growth. It is unclear how the growth rate will change in the next ten or so years. Therefore, a conservative approach is taken by assessing the current capabilities and repeating the assessment over time before key decision points for SDC.

As it can be seen from Tables 4 and 6, the horizontal resolution capability of both commercial constellations and the L1A architecture is defined as 5 m. This is due to the fact that the SATM does not include any resolution needs beyond 5 m, and any architecture providing a resolution of 5 m or better automatically obtains the highest partial score of 1.

The Earth Science Decadal Survey (National Academies, 2019) emphasised the need for interferometric repeat pass at weekly to daily rates, resolution between 5-15 m, sensitivity to height changes between 1 and 10 mm, with a time series measurement accuracy of between 1 mm/week and 1 mm/year (depending on the duration of the time series) all the while providing continuous global monitoring of all land and coastal areas. Architectures that obtain daily global monitoring of all land and coastal areas easily exceed the cost cap due to increased in orbit duty cycles (or number of satellites) necessary. Therefore, if an SDC architecture is selected such that it provides a generally accepted solution for global monitoring, albeit at longer end of the decadal survey revisit requirements, augmentation of SDC with commercial data can reduce the revisit for areas of interest whenever necessary (e.g. geohazards).

The GOs that score higher than SDC architectures indicate potential benefits to the science community beyond routine monitoring, all of which have a higher X-band accuracy score and are regional in nature (Table 7). In other words, if areas of interest can be narrowed down using non-commercial data, these GOs may benefit from routine commercial data collections for the specific applications.

4.2 Ice-Sheet Speckle Tracking and InSAR Analyses

Ice sheets are acknowledged by the World Meteorological Organization (WMO) and the United Nations entity tasked with supporting the global response to the threat of climate change (UNFCCC) as an Essential Climate Variable (ECV) needed to make significant progress in the generation of global climate models. Several national and international programs (NASA MEaSUREs, ESA Climate Change Initiative) fund efforts to generate high quality geoinformation products for Antarctica and Greenland based on satellite remote sensing data. The SDC SATM builds on the ECV requirements.

		Accuracy		Scores	
GO	**DS Science Variable**	**L1A**	**C.C.**	**L1A**	**C.C.**
C7	Tidal flexure in the grounding zones (50 km upstream and all floating)	0.05	0.64	0.05	0.43
C9	Elevation signal of interior sub-glacial lake drainage and filling	0.05	0.64	0.05	0.18
S12	Deformation from fluid fluxes in shallow aquifers	0.29	1.0	0.29	0.43
S13	Vertical surface deformation – impact of human activities and water flow on earthquakes	0.21	0.4	0.21	0.40

Table 7. Comparative assessment of scores for the L1A architecture and commercial constellations (C.C.) for the GOs that do not follow the logarithmic trend.

Interferometric SAR data are particularly useful for ice sheet science with respect to surface velocity and grounding line measurements. Ice velocity (C1, C2, C3), and strain fields (C4, C5) are typically measured using speckle tracking (Rignot et al., 2011), where surface parallel 2D motion can be measured using a pair of acquisitions. A more accurate method is to exploit the InSAR phase; this approach requires interferograms acquired in different viewing geometries to resolve 2D motion (Mouginot et al., 2019). Figure 3 shows an example of a feature tracking result of high resolution X-band data on an Antarctic glacier. In this case data were acquired in a near-repeat track orbit, but without controlling the interferometric baseline. The high resolution data provide sufficient surface features on the glacier itself for offset tracking to work. The ultimate goal for a comprehensive data assessment is to obtain InSAR data, however, to assess the full potential of a mission, as demonstrated by the following example.

The ice-ocean interface of a glacier is a critical boundary. The grounding line delineates where ice detaches from the bed and becomes afloat and frictionless at its base. Using double difference interferometry, the flexing of the ice shelf due to differential tide levels at four acquisition times results in a dense band of fringes in the interferogram due to the vertical displacement. The upstream boundary of this fringe band is interpreted as the InSAR grounding line (related to C7 in Table 7). The approach requires the availability of 2 interferograms with the same geometry. Fast glaciers can pose a problem for missions with a longer repeat orbit due to phase decorrelation. Commercial high resolution InSAR missions with short repeat orbits do not generally satisfy the spatial coverage requirement, however, they can be used to augment agency missions like SDC with a more global coverage approach. Early results highlighting the value of a commercial InSAR data for grounding line studies are presented in (Figure 3, Ciraci et al., 2022).

4.3 Analysis of Potential Persistent Scatterers

Preliminary analysis on the interferometric potential of commercial X-band data for deformation monitoring over semi-arid land and exposed rock was also conducted through the analysis of a set of available scenes from Capella Space over the West Angelas Mine in western Australia, shown in Figure 4. The study area is a desert devoid of vegetation, which presents a good opportunity for image analysis without confounding issues of decorrelation and terrain type. Interferometric techniques can also supplement traditional change detection methods for

This contribution has been peer-reviewed.
https://doi.org/10.5194/isprs-archives-XLVIII-M-1-2023-225-2023

monitoring environmental health in and around active mines; interferometry is more sensitive to small changes on the ground and is capable of monitoring hydrological activity such as groundwater extraction through measurement of deformation as a proxy.

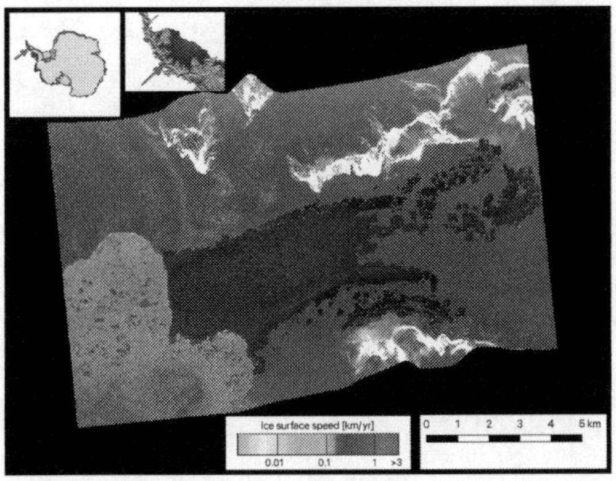

Figure 3. Feature tracking result of imagery from Capella Space over Murphy Glacier, Antarctic Peninsula.

Figure 4. Coverage of imagery from Capella Space over the West Angelas Mine; area in western Australia shown in inset.

The potential performance of the scenes for use in persistent scatterer InSAR (PS-InSAR) was evaluated through the estimation of the density of potential persistent scatterer (PS) candidates. Because the scenes were described to be interferometry-feasible but were not acquired to be interferable, this preliminary analysis evaluated PS candidates not using phase-based methods but instead by using two common amplitude-based methods that are used in existing PS identification algorithms as initial filtering steps before detailed phase analysis. Results were compared with Sentinel-1B imagery that was acquired over the same area and over the same time period. Dataset parameters are shown in Table 8.

Capella imagery was corrected based on the WGS84 ellipsoid and geocoded to a 20 x 20 m grid using the GAMMA software (Wegnüller et al., 2016). Sentinel-1 imagery was processed using the 20 x 20 m Copernicus DEM for topographic correction and geocoding using a backprojection processor (Zebker, 2022). The Sentinel-1 imagery was then masked to the approximate size and area of the Capella acquisitions. Preliminary analysis of PS candidates was conducted by thresholding of calculated amplitude dispersion and signal-to-clutter ratio (SCR) of each stack, defined respectively as:

$$D_a = \frac{\sigma_a}{\mu_a} \quad (5)$$

$$\gamma = \frac{P_d}{P_c} \quad (6)$$

where σ_a is the standard deviation and μ_a the mean of the amplitude of a pixel across the image stack, and D_a is the amplitude dispersion; in (6), P_d is the power of the dominant scatterer, taken as the power of a single pixel, P_{cc} is the power of the clutter, estimated by taking the average of the immediately surrounding pixels, and γ is the SCR.

Parameter	Capella Space	Sentinel-1B
Number of scenes	6	7
Imaging mode	Spotlight	Interferometric Wide swath (IW)
Orbit pass direction	Descending	Descending
Dates	2021-07-01 to 2021-09-10	2021-07-05 to 2021-09-1
Native resolution	0.4 m x 0.5 m	5 x 20 m

Table 8. Imaging parameters for Capella Space and Sentinel-1B datasets used in the mining case study.

Figure 5 shows the comparison between the Capella and Sentinel stacks with an amplitude dispersion of 0.4 and 0.25, which are the thresholds used for initial pixel elimination during the first step of StaMPS and the original PSI algorithms, respectively (Hooper et al. 2004, Ferretti et al. 2001). Figure 6 shows the comparison between the two stacks with an SCR of 2, which is the threshold used for initial candidate selection in the STUN algorithm (Kampes, 2005), and an SCR of 10, a much more stringent threshold. The results are mixed, showing a higher estimated PS candidate density in Capella data when comparing SCR, but a higher estimated PS candidate density in Sentinel-1 data when comparing amplitude dispersion.

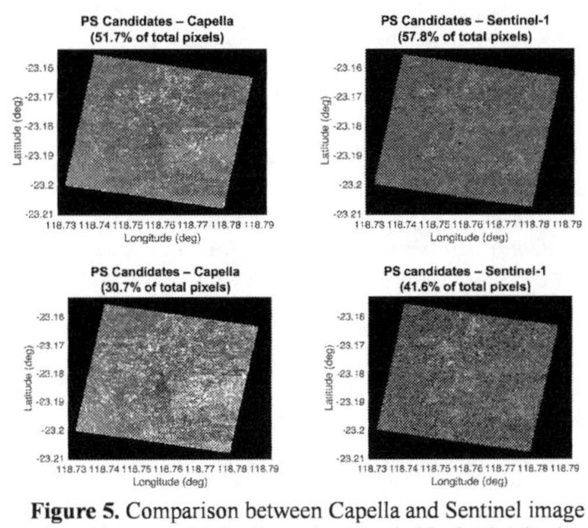

Figure 5. Comparison between Capella and Sentinel image stacks with an amplitude dispersion threshold of < 0.4 (top) and < 0.25 (bottom). The number in parentheses indicates the percentage of total pixels identified as PS candidates.

https://doi.org/10.5194/isprs-archives-XLVIII-M-1-2023-225-2023 | © Author(s) 2023. CC BY 4.0 License.

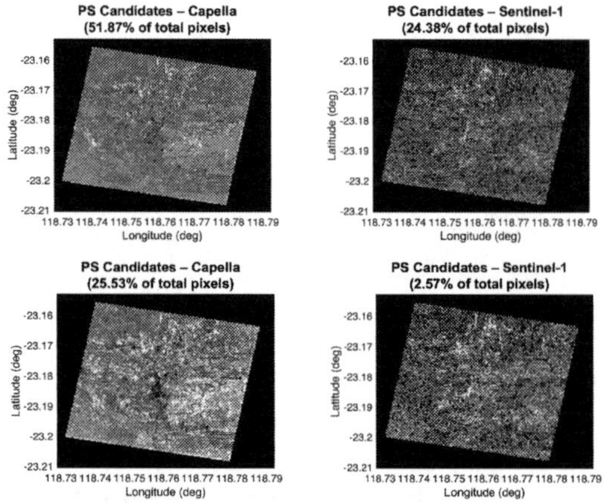

Figure 6. Comparison between Capella and Sentinel image stacks with an SCR threshold of > 2 (top) and > 10 (bottom). The number in parentheses indicates the percentage of total pixels identified as PS candidates.

The discrepancy between the two tested measures highlights the limitations of purely amplitude-based measures as effective phase proxies, given that PS candidate density is expected to scale with resolution regardless of wavelength (Huang and Zebker, 2018). The lower PS candidate density for Capella data using amplitude dispersion, in particular, is also likely due to the fact that the commercial imagery was not acquired for interferometry, resulting in an uncontrolled baseline that lowers phase stability. SCR, by contrast, is a more direct measure of relative scatterer brightness, which is higher in the higher-resolution Capella data compared to Sentinel-1 data. Capella data should show an improvement in PS candidate density detected using amplitude dispersion if acquired in orbit-controlled passes designed for interferometry. Ultimately, while this preliminary analysis indicates promise for the use of commercially acquired high-resolution SAR data in interferometric applications, future analysis with high-resolution DEMs as well as with an interferometric dataset, when available, will be required to make a further determination on the capabilities and limitations of commercially available data to augment observation of SDC GOs.

5. CONCLUSION

In conclusion, this quantitative feasibility study of the commercial sector indicates that, as is, the X-band commercial systems can have some role in meeting the observational capabilities defined in SDC's SATM. Some GOs in the SATM can benefit from commercial data either to supplement the observations of an SDC architecture or to fill a gap that is left by such an architecture. Assessing all of the SDC architectures through this framework would identify the relative benefits of commercial data for each architecture. It is important to note that, as the commercial constellation capability increases, many more GOs will benefit from this data-rich environment.

REFERENCES

Ciracì, E., Rignot, E., Milillo, P. and Dini, L., 2022, May. Evaluating Petermann Gletscher ice-shelf basal melt and ice-stream dynamics from high-resolution TanDEM-X elevation data. In *EGU General Assembly Conference Abstracts* (pp. EGU22-10979).

Ferretti, A., Prati, C., & Rocca, F., 2001: Permanent scatterers in SAR interferometry. *IEEE Transactions on geoscience and remote sensing*, 39(1), 8-20. doi.org/10.1109/36.898661

Hooper, A., Zebker, H., Segall, P., & Kampes, B., 2004: A new method for measuring deformation on volcanoes and other natural terrains using InSAR persistent scatterers. *Geophysical research letters*, 31(23). doi.org/10.1029/2004GL021737

Horst, S., Chrone, J., Deacon, S., Le, C., Maillard, A., Molthan, A., Nguyen, A., Osmanoglu, B., Oveisgharan, S., Perrine, M., Others, 2021. NASA's Surface Deformation and Change Mission Study, in: 2021 IEEE Aerospace Conference (50100).

Huang, S., & Zebker, H., 2019: Persistent scatterer density by image resolution and terrain type. *IEEE Journal of Selected Topics in Applied Earth Observations and Remote Sensing*, 12(7), 2069-2079. doi.org/10.1109/JSTARS.2019.2896038

Kampes, B., & Adam, N., 2005: The STUN algorithm for persistent scatterer interferometry. *Proceedings of FRINGE 2005*, 1-14.

Kellogg, K., Hoffman, P., Standley, S., Shaffer, S., Rosen, P., Edelstein, W., Dunn, C., Baker, C., Barela, P., Shen, Y., Guerrero, A.M., Xaypraseuth, P., Sagi, V.R., Sreekantha, C.V., Harinath, N., Kumar, R., Bhan, R., Sarma, C.V.H.S., 2020. NASA-ISRO Synthetic Aperture Radar (NISAR) Mission, in: 2020 IEEE Aerospace Conference. ieeexplore.ieee.org, pp. 1–21.

Khazendar, A., Molthan, A., Sauber, J., Bell, J., Carabajal, C., Glasscoe, M., Le Roy, A., Lucey, R., Rogers, L., Sylak-Glassman, E., Gardner, A., Jones, C., Owen, S., Reager, Rieger J.T., Siqueira, P. et al., 2021. Science and Applications Traceability Matrix [WWW Document]. Surface Deformation and Change (SDC). URL https://science.nasa.gov/science-pink/s3fs-public/atoms/files/SDC_SATM_Phase1_Final_0.pdf (accessed 1.24.23).

National Academies of Sciences, Engineering, and Medicine, Division on Engineering and Physical Sciences, Space Studies Board, Committee on the Decadal Survey for Earth Science and Applications from Space, 2019. Thriving on Our Changing Planet: A Decadal Strategy for Earth Observation from Space.

Rignot, E., Mouginot, J. and Scheuchl, B., 2011. Ice flow of the Antarctic ice sheet. *Science*, *333*(6048), pp.1427-1430.
Mouginot, J., Rignot, E. and Scheuchl, B., 2019. Continent-wide, interferometric SAR phase, mapping of Antarctic ice velocity. *Geophysical Research Letters*, *46*(16), pp.9710-9718.

Rosen, P.A., Kumar, R., 2021. NASA-ISRO SAR (NISAR) Mission Status, in: 2021 IEEE Radar Conference (RadarConf21). pp. 1–6.

Wegnüller, U., Werner, C., Strozzi, T., Wiesmann, A., Frey, O. and Santoro, M., 2016: Sentinel-1 support in the GAMMA software. *Procedia Computer Science*, 100, 1305-1312. doi.org/10.1016/j.procs.2016.09.246

Zebker, H., 2022: Sentinel-1 Analysis Ready Data – A Convenient and Easy to Use System Producing Common-coordinate Timeseries. In *AGU Fall Meeting Abstracts 2022*, G42D-0257.

This contribution has been peer-reviewed.
https://doi.org/10.5194/isprs-archives-XLVIII-M-1-2023-225-2023 | © Author(s) 2023. CC BY 4.0 License.

APPENDIX

In the table below, the needs are grouped for display purposes. Needs 1-8 correspond to:

1. Polarisation (Number of polarimetric channels)
2. Resolution (metres)
3. Revisit (Number of days between revisits)
4. Latency (Number of days between acquisition and data delivery)
5. Accuracy (Percent accuracy based on performance tool, literature survey or heuristic evaluation)
6. Coverage:
    - G = Global
    - C = Continental
    - R = Regional
    - L = Local
7. Large area imaging capability (Yes if needed for the GO or No)
8. InSAR capability (Yes if needed for the GO or No)

"&" is entered for GOs that have multiple methods that and not all of them require InSAR. Same thing is true for Large area imaging.

Modifiers "k" indicate "times 1000".

GO	DS Science Variable	Parameters							
		1	2	3	4	5	6	7	8
C1	Mean flow of fast flowing outlet glaciers, grounded and floating (>50 m/yr)	1	5	7	90	1.00	C	&	Y
C2	Mean flow of slow flowing ice-sheet interiors (<50 m/yr)	1	500	365	90	1.00	C	Y	Y
C3	Mean flow of large mountain glaciers	1	3	60	90	1.00	R	N	Y
C4	Strain rates at shear margins (grounded ice)	1	5	30	90	1.00	R	N	Y
C5	Strain rates associated with fracture and calving and ice mélange	1	5	0.25	90	1.00	R	&	N
C6	Geometry of ice mélange	1	10	0.25	90	0.5	R	N	&
C7	Tidal flexure in the grounding zones (50 km upstream and all floating)	1	50	7	90	0.64	R	N	Y
C8	Basal friction inverted from temporal changes in velocity	1	5	0.25	90	1.00	R	Y	Y
C9	Elevation signal of interior sub-glacial lake drainage and filling	1	50	7	90	0.64	R	Y	Y
C10	Circumpolar fine scale sea ice motion & deformation	1	100	1	90	0.65	C	Y	N
C11	Sea ice ridge and lead evolution, polynya formation, brine formation, and heat flux	1	100	1	90	0.65	C	Y	N
C12	Marginal/coastal sea ice zone mechanics and fluxes	1	100	1	90	0.65	R	Y	&
C13	Sea ice formation and distribution	1	100	1	90	0.65	C	Y	N
E1	Absolute Soil Moisture	1	20	3	1	0.35	G	N	N
E2	Relative Soil Moisture	1	20	3	1	0.35	G	N	N
E3	Freeze/Thaw Boundary Dynamics (Frozen/Not Frozen)	1	20	7	7	0.35	R	N	N
E4	Vegetation Structure (canopy height & vertical distribution)	2	20	14	14	0.35	G	N	Y
E5	Above ground vegetation biomass & biomass change	2	20	14	14	0.6	G	N	N
E6	Change in land cover due to flooding, wildfire, wind, insects, anthropogenic	2	20	14	14	0.5	L	Y	N
E7	Agricultural Management (Active crop area, crop class, crop yield, crop practices)	2	20	14	14	0.5	R	Y	N
E8	Classification of rice and aquaculture	2	20	14	14	0.5	R	Y	N
E9	Measure inland and coastal wetlands areas	2	20	14	14	0.35	R	N	N
E10	Measure relative water level changes in wetlands	2	20	14	14	0.35	R	N	Y
H1	Snow Water Equivalent (SWE)	1	4k	3	7	0	G	N	Y
H2	SWE	2	4k	3	7	0	G	N	N
H3	Latent heat flux (soil moisture)	4	200 k	0.5	7	0.35	G	Y	N
H4	Recharge rates (soil moisture)	4	200 k	3	7	0.35	G	Y	N
H5	Fire prediction (fuel load)	4	200 k	3	7	0.35	G	Y	N
H6	Drought monitoring (soil moisture)	4	1k	7	7	0.35	G	Y	N
H7	Recharge rates (Groundwater subsidence)	1	100	3	1	1	C	Y	Y
H8	Inundated area	2	10	0.5	0.1	0.35	C	N	N
S1	Volcanic Systems and Hazards, Land surface deformation	1	10	1	365	0.03	L	N	Y
S2	Earthquake Cycle and Hazards, Land surface deformation	1	50	1	365	1.00	C	Y	Y
S3	Landslides Hazards, Land surface deformation	1	10	30	365	0.16	R	N	Y
S4	Rapid Deformation Map Acquisitions	1	10	7	2	0.52	L	N	Y
S5	Sea Level Rise, 3D Surface deformation vectors on ice sheets, Ice Velocity	1	100	30	365	1.00	C	Y	Y
S6	Sea Level Rise, Vertical motion of land along coastlines	1	50	6	365	0.31	C	Y	Y
S7	Landscape Change – Quantify G decadal landscape change, Land Surface, Soil Moisture Deformation	1	10	7	365	0.5	G	Y	Y

This contribution has been peer-reviewed.
https://doi.org/10.5194/isprs-archives-XLVIII-M-1-2023-225-2023

GO	DS Science Variable	Parameters							
		1	2	3	4	5	6	7	8
S8	Landscape Change – Content of Near Surface Materials, Reflectance for freeze/thaw, Reflectance for snow depth/snow water equivalent	1	100	7	365	1	G	Y	N
S9	Energy Change – Effect of Convection, Plate motion and deformation	1	100	30	365	0.41	C	Y	Y
S10	Groundwater Flow and its impact on geological processes and water supply, Land surface deformation	1	10	7	365	1	C	Y	Y
S11	Measure Fluxes in and out of groundwater system	1	10	6	365	0.60	C	Y	Y
S12	Deformation from fluid fluxes in shallow aquifers	1	5	7	365	1	R	N	Y
S13	Vertical surface deformation – impact of human activities and water flow on earthquakes	1	5	7	365	0.4	R	N	Y
S14	Discovery & Management – Map energy, mineral, agricultural and natural resources for improved management	1	30	7	365	0.24	G	Y	N
W1	Energy balance (soil moisture)	4	1k	0.5	7	0.35	G	Y	N
W2	Weather prediction (soil moisture)	4	1k	0.5	1	0.35	G	Y	N
G1	Geohazards: Volcanic Systems and Hazards, Land surface deformation	1	10	7	0.1	0.03	L	N	Y
G2	Geohazards: Earthquake Cycle and Hazards, Land surface deformation	1	10	1	0.1	1.00	L	N	Y
G3	Geohazards: Landslides Hazards, Land surface deformation	1	10	30	0.5	0.16	L	N	Y
G4	Geohazards: Rapid Deformation Map Acquisitions	1	10	1	0.1	1.00	L	N	Y

This contribution has been peer-reviewed.
https://doi.org/10.5194/isprs-archives-XLVIII-M-1-2023-225-2023 | © Author(s) 2023. CC BY 4.0 License.

DESIGN OF A SPACE-BORNE HYPERSPECTRAL IMAGER FOR MICRO-SATELLITES

Zeynep Nilüfer Öztürk[1,2], Mustafa Ekinci[1], Özgür Karcı[3]

[1] Tübitak Space Technologies Research Institute, Ankara, Türkiye (nilufer.ozturk, mustafa.ekinci@tubitak.gov.tr)
[2] Middle East Technical University, Physics Department, Ankara, Türkiye (e160432@metu.edu.tr)
[3] Turkish Aerospace Industry, Ankara, Türkiye (ozgur.karci@tai.com.tr)

KEYWORDS: Space-borne hyperspectral imagers, earth observation, three mirror anastigmat, imaging spectroscopy, remote sensing

ABSTRACT:

We present an optical design of a space-borne hyperspectral imager (HSI) for Earth observation micro-satellite platforms. Space-borne hyperspectral imaging has many crucial applications in fields such as agriculture, water management, environmental monitoring, mineralogy, and remote sensing. An HSI system capable of ground sampling distance (GSD) less than 15 m with a swath width greater than 15 km, spectral resolution less than 10 nm and operating in Low Earth Orbit (LEO) is designed. The system dimensions are restricted to fit in a volume of less than 0.125 m^3. A commercial, cooled HgCdTe type imaging sensor is selected to operate 400 – 2500 nm of the spectrum for the designed imager. The HSI optical design comprises of an off-axis three mirror anastigmat (TMA) type telescope and a modified-Offner type spectrometer. Modified-Offner type spectrometer design with two Féry prisms as a diffractive element is used. The overall HSI system design meets the performance goals described in this paper work.

1. INTRODUCTION

Hyperspectral space-borne imaging was introduced in the late 1980s for airborne mineral mapping by Goetz (Eismann, 2012). Hyperspectral imaging is a fast-growing and promising area of remote sensing, combining two main remote sensing techniques which are imaging and spectroscopy. Hyperspectral imagers allow for capturing of images with much higher spectral resolution than traditional multispectral cameras that capture images in the red, green, blue (RGB) and NIR spectrum. They collect images across a broad spectrum of wavelength, typically including the visible and near-infrared (NIR) or shortwave infrared (SWIR) spectrum. One of the main advantages of hyperspectral imaging is its ability to capture detailed spectral information from a scene. This allows researchers to identify and distinguish between different materials based on their unique spectral signatures. For example, in agriculture, HSIs can be used to identify the health and growth of crops by detecting subtle changes in chlorophyll content and soil moisture levels. In geology, these imagers can help identify minerals or other materials based on their reflectance spectra. However, the technology is still relatively expensive and requires specialized training to operate and analyse data effectively. Nonetheless, advancements in sensor design and data analysis algorithms are making hyperspectral imaging more accessible and practical for a growing number of research and industrial applications. The demands of hyperspectral imaging for Earth's surface, atmosphere observation, and environmental monitoring applications from space push the limits of the telescopes and imaging sensors forward.

Numerous hyperspectral designs have been studied in literature. Optical system parameters of published HSI designs, which have similar requirements with this study are shown and compared in Table 1. Some of the published HSI designs are investigated at this part of the paper. Hyperion is a hyperspectral imager designed and built by NASA (Calin et al. , 2021). The HSI instrument is mounted on the Earth Observing-1 (EO-1) satellite to operate in LEO orbit, which was launched in 2000. Hyperion captures images with a swath width of 7.7 km. The Compact High-Resolution Imaging Spectrometer (CHRIS) is a hyperspectral imager developed by the European Space Agency (ESA) (Escalante-Ramírez et al., 2012). It was launched in 2001 on board the PROBA-1 satellite and is still in operation in LEO orbit today. One of the key advantages of the CHRIS hyperspectral imager is its high spatial resolution. The Advanced Responsive Tactically Effective Military Imaging Spectrometer (ARTEMIS) is a hyperspectral imager designed for remote sensing applications (Qian, 2021). The ARTEMIS, an advanced hyperspectral imager, was included in the TacSat-3 satellite, which was launched in 2009. The ARTEMIS hyperspectral imager used a single 2-D HgCdTe detector array that covered both VNIR and SWIR spectral regions, providing a 4 km swath width. The Hyperspectral Environment and Resource Observer (HERO) is an HSI designed by Canadian Space Agency (CSA) (Bergeron et al., 2008). The HERO is planned as an operational hyperspectral mission capable of covering an area of more than 600,000 km^2 daily. The resulting image consists of radiance spectra that can be transformed into reflectance spectra by correction of the influence of the atmosphere. The total instrument MTF is at least 0.3 at Nyquist frequency at all wavelengths and fields. The smile and keystone distortions are less than 10%. The Copernicus Hyperspectral Imaging Mission for the Environment (CHIME) (Celesti et al., 2022) is a satellite mission designed to provide detailed imaging of the Earth's surface using hyperspectral technology. The mission is being developed by ESA and is part of the European Union's Copernicus program. The HSI instrument has a spectral resolution of 10 nanometers and is able to image in over 200 bands over a wavelength range from 400 nm - 2500 nm in the visible (VIS), near infrared (NIR), and short-wave infrared (SWIR) spectrum. CHIME HSI will be able to measure at a ground resolution of 30 m for a swath width of 130 km and will operate in LEO orbit. CHIME is planned to be launched in 2028.

This contribution has been peer-reviewed.
https://doi.org/10.5194/isprs-archives-XLVIII-M-1-2023-233-2023

Parameter	Hyperion	Chris	Hero	HSI
Telescope Design	TMA	Two-mirror	TMA	TMA
Spectrometer Design	Offner	Modified Offner	Offner Dyson	Modified Offner
Effective Focal Length	1320 mm	746 mm	704 mm	712 mm
Aperture Diameter	120 mm	120 mm	320 mm	200 mm
GSD	30 m	17 m	30 m	14.7 m
Altitude	705 km	700 km	700 km	700 km
Spectral Resolution	10 nm	1.25 nm	7.5 nm	8.3 nm
Bandwidth	400-2500 nm	400-1050 nm	400-2500 nm	400-2500 nm

Table 1. Comparison of HSI design parameters

The HSI design presented in this study is aimed for environmental monitoring applications. The resulting design is restricted to fit in a volume less than 0.2 m^3 to be used in a micro-satellite platform. Our study showed that, it is not possible to optimize the system with standard mirror surface forms according to the performance goals of the design in a compact volume for micro-satellite platforms. Therefore, even aspheric surface forms need to be defined in the design to fit in compact volume. In this paper, all the system engineering calculations and each parameter of HSI design are investigated in detail.

2. SYSTEM OVERVIEW

Performance goals of the project are shown in Table 2. Imaging sensors should also be decided according to the performance goals before the optical design process. The imaging sensor is chosen due to three main parameters: cost, availability, and performance. In Table 3 chosen sensor parameters are shown.

Parameter	Value
Orbital Altitude *(H)*	700 km [LEO orbit]
Image Scanning Technique	Push-broom
Ground Sampling Distance *(GSD)*	≤ 15m
Spectral Resolution	≤ 10 nm
Swath Width	>15 km
Wavelength Range	400-2500 nm
Spatial Distortion (Smile)	≤ 15 µm
On-ground pixel smile	1.5 m
Spectral Distortion (Keystone)	≤ 15 µm
Modulation Transfer Function (MTF)	≥36

Table 2. Performance goals of the hyperspectral imager

Parameter	Value
Pixel Size	15 x 15 µm
Number of pixels	1024 x 256
Number of spectral pixels *(#sp)*	256
Number of spatial pixels *(#stp)*	1024
Bandwidth	400-2500 nm
Detector	HgCdTe
Scanning sensor type	Push-broom

Table 3. Specifications of a customized imaging sensor

3. THEORETICAL BACKGROUND OF THE SYSTEM DESIGN PARAMETER CALCULATIONS

Optical system design parameter calculation of the project is the second step of the HSI imager design process. Field of View (*FOV*) value across the flight direction and along the flight direction, swath width (*sw*), ground sampling distance (*GSD*), and focal length (*f*) values are calculated by the equations (1-6) before the optical design process (Shaoyuan et al.,2022, Öztürk et al. , 2023). In orbital push-broom imaging sensors, a line of on-ground pixel images arranged perpendicular to the flight direction of the spacecraft as can be seen in Figure 1. In the across flight direction, on ground smile is restricted to be the 10% of the pixel projection. The system design parameters are calculated as the following.

$$GSD = \frac{H.p}{f} \tag{1}$$

$$sw = GSD \ x \ \#sp \tag{2}$$

$$\theta_{along} = \pm tan^{-1}\left(\frac{sw}{2.H}\right) \tag{3}$$

$$\theta_{across} = tan^{-1}\left(\frac{1.5\ m}{H}\right) \tag{4}$$

$$d_{spec} = p * \#sp \tag{5}$$

$$d_{spat} = p * \#stp \tag{6}$$

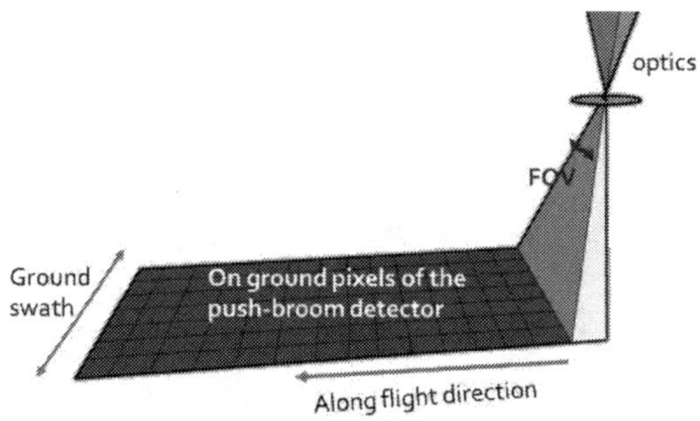

Figure 1. Scheme of the field of view (FOV) geometry and the projection of the detector array on the ground

Parameter	Symbol	Value
Ground Sampling Distance	*GSD*	≤ 15 m
Focal Length	*f*	712 mm
FOV along flight direction	θ_{along}	± 0.61 °
FOV across flight direction (Spectral field)	θ_{across}	1.22 10^{-4} °
Swath width	*sw*	15.3 km
Spectral detector width	d_{spec}	3.84 mm
Spatial detector width	d_{spat}	15.36 mm

Table 4. Calculated system parameters

4. OPTICAL SYSTEM DESIGN OF THE HYPERSPECTRAL IMAGER

Hyperspectral spaceborne imager designs comprise of two parts: telescope and spectrometer. Off-axis TMA type telescope is selected for the telescope design and modified-Offner type spectrometer is selected for the spectrometer design. In the spectrometer design two Féry prisms are used as a diffractive element. A real ray-tracing software (i.e., Zemax OpticStudio) is used to optimize the HSI.

4.1 Telescope Design

The design of a TMA telescope is more complex than other types of telescopes, but it offers several advantages, including a larger field of view, a flatter focal plane, and better image quality across

https://doi.org/10.5194/isprs-archives-XLVIII-M-1-2023-233-2023 | © Author(s) 2023. CC BY 4.0 License.

the entire field of view (FOV) (Chang et al., 2016, (Wang et al., 2022). TMA is corrected for the third-order aberrations of spherical aberration, coma, and astigmatism. Moreover, off-axis TMA design is freeform without obscuration which allows for more signal transferred towards sensor.

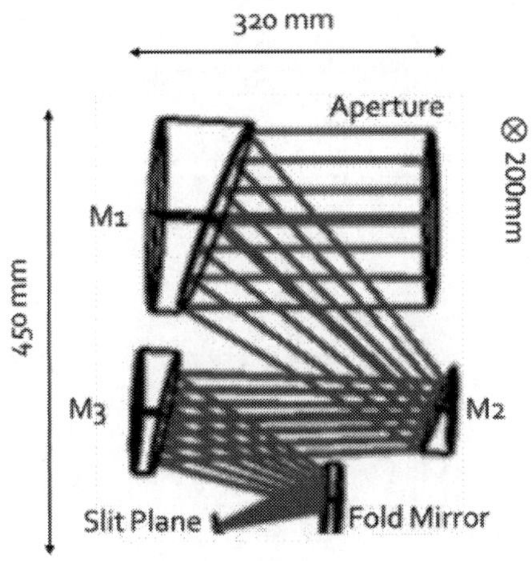

Figure 2. TMA telescope design layout of the HSI

Optimization process started with the standard aspheric surface. Maximum total length of the telescope is constrained to fit in 320 mm. However, optimization is not possible with the standard aspheric surface form with the restriction of the TMA telescope dimensions. Even aspheric surfaces are often used in the design of lenses, mirrors, and other optical components to improve their performance (Amani et al., 2020). They can help to reduce aberrations, improve image quality, and increase the efficiency of light transmission through the optical system. Even aspheric surface type is chosen to optimize the system in limited dimensions. Even aspheric surface sag equation is given by (7):

$$z(r) = \frac{cr^2}{1+\sqrt{1-(1+k)c^2r^2}} + A_2r^2 + A_4r^4 + A_6r^6 + A_8r^8 + A_{10}r^{10} \quad (7)$$

where, z is the sag equation, c (1/R) is the curvature of the mirror, k is the conic constant (in our case k = 0), r is the radial mirror coordinate, A_n are coefficients of the higher order terms. Three even aspheric surface mirrors and a folding mirror is used for the minimal design dimensions. TMA Telescope design optical layout and system parameters are shown in Figure 2 and Table 5 respectively. Optical design prescription of the TMA telescope design is given in Table 6. The image plane of the TMA telescope is the slit plane in the design shown in Figure 3. Slit length of the spatial direction (x-direction) is calculated from the multiplication of number of spatial pixels and pixel size. Spatial slit dimension equals to the spatial detector dimension. After the slit plane spectral sampling will be operated for each wavelength.

Parameter	Value
Aperture Diameter	200 mm
Effective Focal Length	712 mm
Dimensions	450 mm x 320 mm x200 mm
Mirror Surface Type	Even Asphere
Bandwidth	400-2500 nm
Slit Dimensions	15.3 mm x 15 µm
MTF @Nyquist	0.78
Nyquist Frequency	33 cycles per mm

Table 5. Off-axis TMA optical design parameters

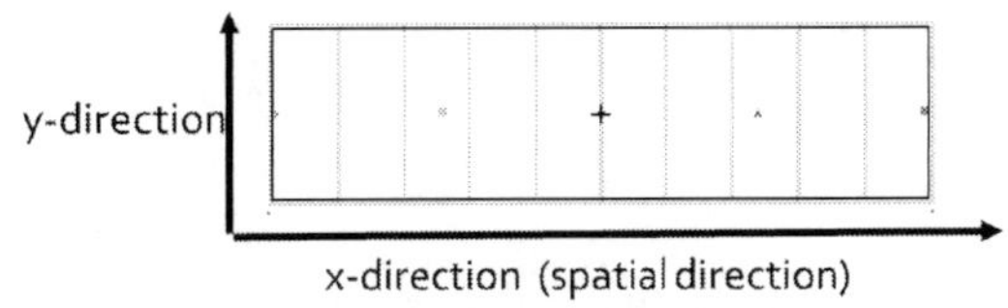

Figure 3. Footprint Diagram of the slit plane

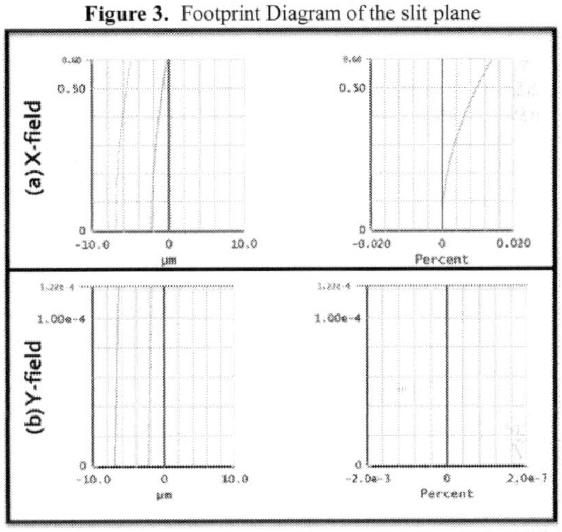

Figure 4. Field curvature and distortion results of the TMA design (a) X-field and (b) Y- field

Field curvature and distortion percentages of the TMA design for X-field and Y-field are shown in Figure 4. Field curvature values of X- and Y-field are less than 15 µm and distortion percentages are less than 0.02%. MTF of the final optimized TMA telescope is shown in Figure 5. Nyquist frequency is calculated from the eq. (8). Resulting MTF value of the designed TMA telescope at Nyquist is 0.79. Resulting Spot diagram of the TMA telescope is shown in Figure 6.

$$f_{Nyquist} = \frac{1}{2 * pixel\ size\ (mm)} \quad (8)$$

https://doi.org/10.5194/isprs-archives-XLVIII-M-1-2023-233-2023 | © Author(s) 2023. CC BY 4.0 License.

Surface	Surface Type	Radius (mm)	Thickness (mm)	Semi Diameter (mm)	Conic Constant	A_2	A_4	A_6	A_8	A_{10}
Object	Plane	-	320	100	-	-	-	-	-	-
M1	Even Aspheric	-1100.2	-300	110	0	$-6.4\ 10^{-5}$	$1.3\ 10^{-10}$	$5.2\ 10^{-17}$	$-7.7\ 10^{-23}$	$1.5\ 10^{-28}$
M2	Even Aspheric	301.9	-200	50	0	$1.2\ 10^{-4}$	$4.4\ 10^{-9}$	$1.4\ 10^{-14}$	$-2.7\ 10^{-19}$	$5.1\ 10^{-24}$
M3	Even Aspheric	-485.7	120	70	0	$1.2\ 10^{-4}$	$3.3.\ 10^{-10}$	$4.1\ 10^{-21}$	$4.4\ 10^{-27}$	$4.4\ 10^{-27}$
Image Plane	Plane	-	-	7.1	-	-	-	-	-	-

Table 6. Optical prescription of the TMA telescope design

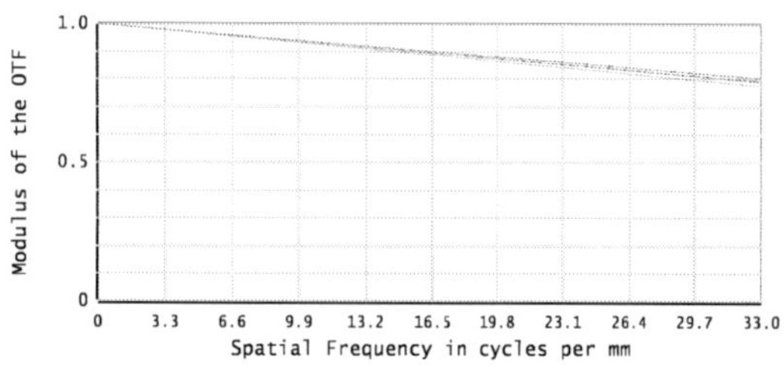

Figure 5. Resulting MTF graph of the TMA Telescope Design [400-2500 nm]

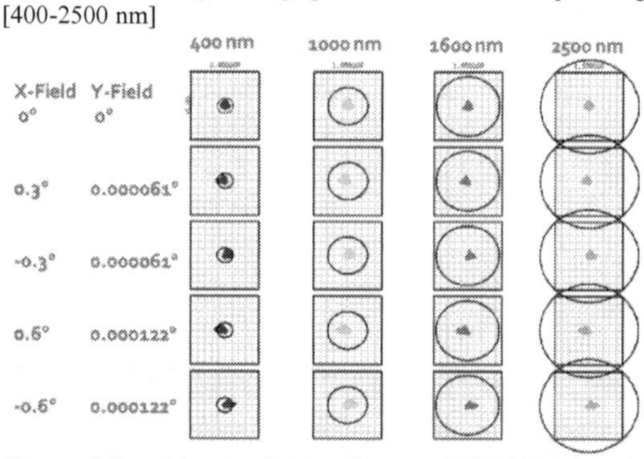

Figure 6. Resulting matrix spot diagram of the TMA telescope design (black circles represent Airy disk, black squares represent 15 μm square pixel)

Full Field Displays (FFDs) of HSI TMA design for the third-order spherical aberration, astigmatism and coma are shown (Thompson, 1985). FFDs show the field dependencies of the third-order aberrations for the full FOV.

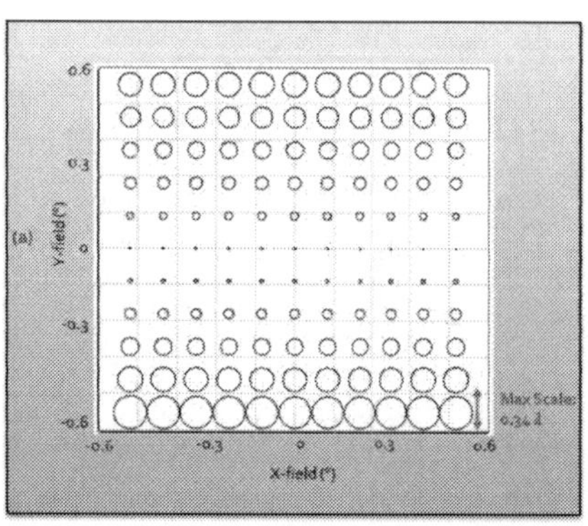

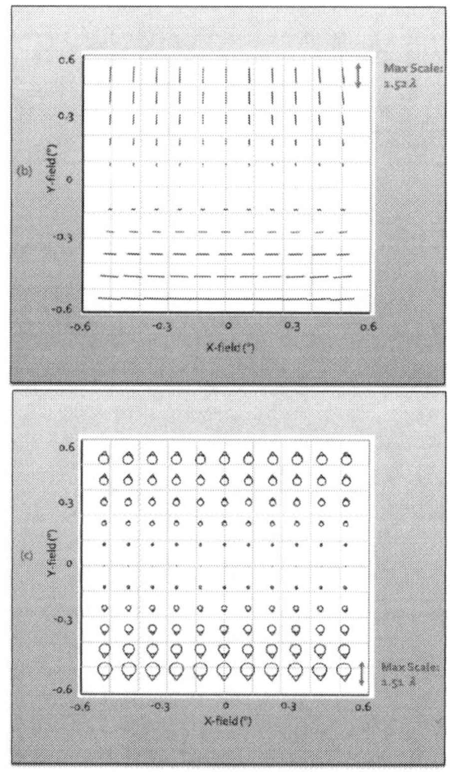

Figure 7. FFDs of the off-axis TMA for the third-order: **(a)** spherical aberration, **(b)** astigmatism, **(c)** coma

4.2 Spectrometer Design

Spectrometer design process started with the Offner type spectrometer. The Offner spectrometer was invented by Horst Offner in 1960 (Prieto-Blanco et al., 2006). It is a type of diffraction grating spectrometer, which uses a diffraction grating as a diffractive element. The grating consists of a series of closely spaced parallel lines or grooves, which diffract the light at different angles depending on the wavelength. However, Offner design with diffraction grating has low light efficiency due to the diffraction of light into its orders and does not able to optimize large FOV in a compact form. Then, Modified-Offner spectrometer with Fery prism is used, which was invented by C. Féry in 1911 (Nie et al., 2018). The difference between traditional prism and Féry prism is the change in the surfaces of prism. Féry prism has spherical type of surfaces instead of plane surface that can focus the spectral image on the detector plane effectively.

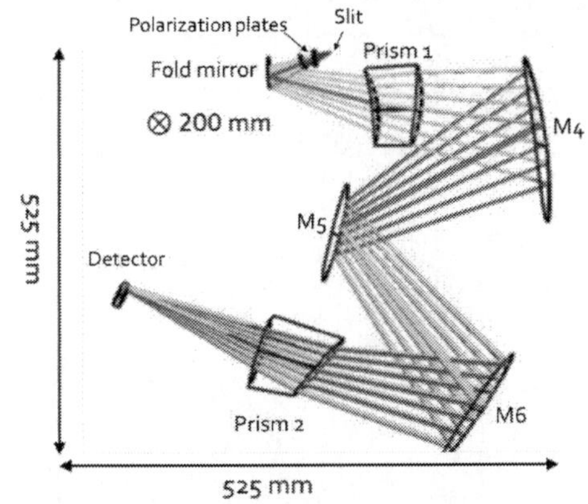

Figure 8. Design layout of the HSI spectrometer

Parameter	Value
Dimensions	525 mm x 525 mm x 200 mm
Mirror Surface Type	Even Aspheric
Bandwidth	400-2500 nm
Prism type	Fery
Prism Material	CaF2
Polarization Plate Material	Silica
Average Spectral Sampling	8 nm
Detector vacuum window material	Silica

Table 7. Spectrometer specifications

Parameter	M4	M5	M6
Radius (mm)	488.4	-323.3	488.4
A_2	-6.9 10^{-5}	1.5 10^{-4}	-6.9 10^{-5}
A_4	-8.9 10^{-11}	4.2 10^{-9}	-8.9 10^{-11}
A_6	-3.6 10^{-16}	3.9 10^{-14}	-3.6 10^{-16}
A_8	-3.7 10^{-22}	2.8 10^{-18}	-3.7 10^{-22}

Table 8. Spectrometer mirror specifications

Parameter	Prism 1	Prism 2
Radius front (mm)	106.4	386.4
Radius back (mm)	136.1	286.1
Thickness (mm)	32	25
Apex angle (°)	13	24

Table 7. Spectrometer prism specifications

Spectrometer part starts with the slit. After the slit, there are two flat polarization plates aiming to make hyperspectral design equally sensitive for both s-polarized and p-polarized light. These plates are placed under opposite angles with the optical axis and positioned between the slit and the folding mirror. In the HSI spectrometer design three concentric even aspheric surface type mirrors are used. Likewise, the Offner kind of mirror configuration three mirrors share the same optical center M4 and M6 mirrors are the same mirrors as can be seen from Table 8 (Nie et al., 2018). To reduce the weight of the imager modified-Offner spectrometer mirror is divided into two mirrors in the design and the unused region of the mirror is subtracted from the design. Moreover, two Féry prisms are preferred as a diffractive element in the design. Specifications of the prisms are shown in Table 8. A folding mirror is used to make the spectrometer fit in the limited dimensions. Finally, detector is designed to be in a vacuum box with a window defined in the design. This vacuum window makes the detector operate in low temperature conditions. Spectrometer part of the HSI design optical layout is shown in Figure 8. Spectrometer specifications are given in Table 7.

5. ANALYSIS AND RESULTS

Overall HSI design is shown in Figure 9. MTF requirement of the study is 0.36. Minimum overall resulting system MTF at Nyquist frequency is 0.47 as shown in Figure 10 for 2500 nm. Best overall resulting system MTF at Nyquist frequency is 0.72 for 400 nm, shown in Figure 11. Nyquist frequency value for 15 μm detector pixel is calculated to be 33.

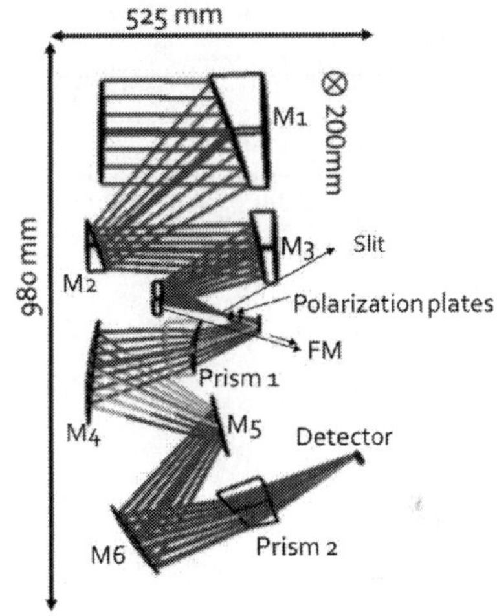

Figure 9. Overall HSI optical design layout

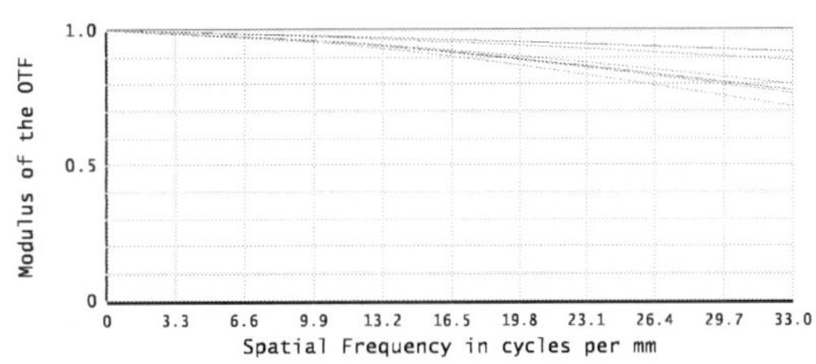

Figure 10. Resulting MTF graph of the overall HSI for 400 nm

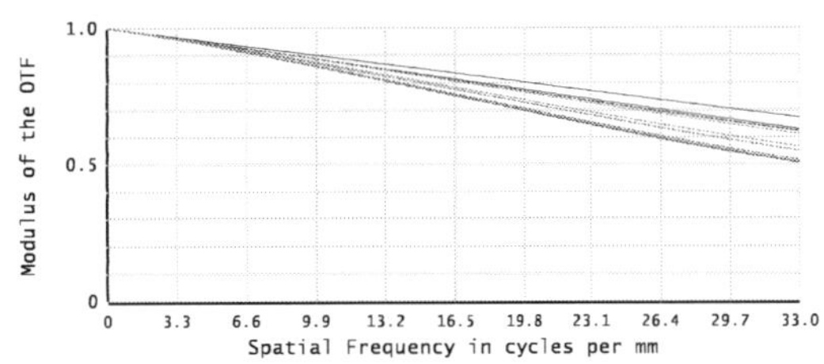

Figure 11. Resulting MTF graph of the overall HSI for 2500 nm

Field curvature and distortion values are shown in Figure 12 and Table 10, which meet the project performance goal requirements.

https://doi.org/10.5194/isprs-archives-XLVIII-M-1-2023-233-2023

Smile and keystone distortion values are calculated from the spot diagram. Change in the spatial direction image width results in keystone distortion, while change in spectral direction image width results in smile distortion as can be seen in Figure 13.

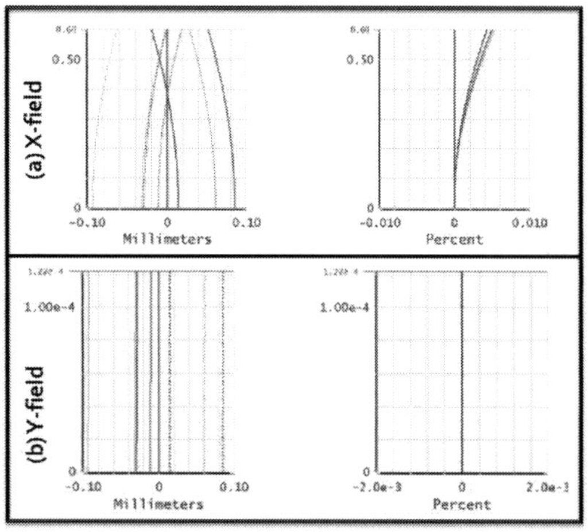

Figure 12. Field curvature and distortion results of the overall HSI design

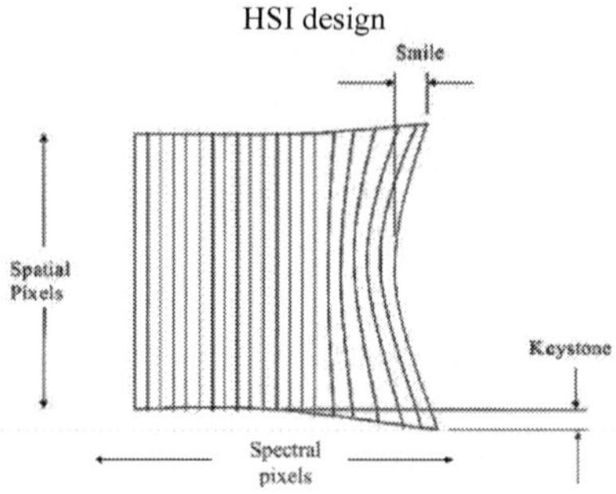

Figure 13. Keystone and smile distortion

Field Angle (°)	Smile Distortion		Keystone Distortion
	400 nm	2500 nm	
±0.3	1.98 mm	2.01 mm	2.6 mm
±0.6	8.17 mm	8.98 mm	5.3 mm

Table 8. Keystone and smile distortion results

Average spectral sampling is calculated from the total bandwidth divided by total number of spectral pixels. Average spectral sampling is calculated as 8 nm. Dimensions of the detector is calculated from total spectral and spatial number of pixels with pixel size. Spectrometer spatial magnification equals to one to avoid distortions. Moreover, spectrometer spectral sampling is tuned by the angles of the prisms and the HSI imager is optimized to fit in the spectral dimension of the detector as can be seen in Figure 14.

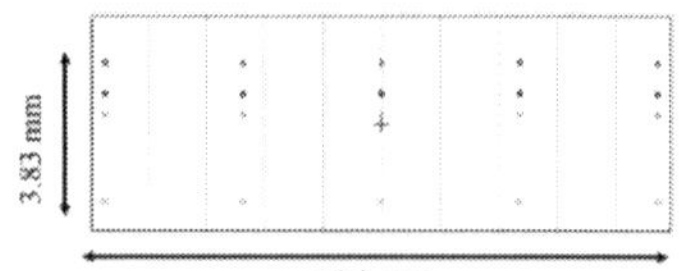

Figure 14. Footprint Diagram of the image plane

Spot diagram of the overall HSI design is shown in Figure 15. The overall HSI design is concluded to be in the diffraction limited performance.

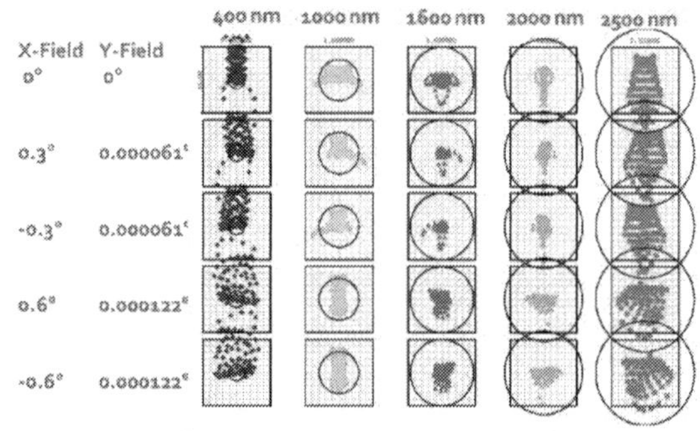

Figure 15. Spot diagram of the overall HSI design (Black circles represent Airy disk, black squares represent 15 µm square pixel)

6. CONCLUSION

In conclusion, an overall HSI design is developed and presented in this work. Off-axis TMA telescope with even aspheric mirror surface forms has been designed. This type of design avoids obscuration and satisfies the required full FOV value of 1.2 degrees with 14.7 m resolution and 15.3 km swath width. Overall optical design is corrected for the third-order spherical aberration, astigmatism, and coma. The minimum MTF value of the design is calculated to be 0.49 at Nyquist frequency. Spot diagrams show that resulting spots on the detector plane satisfies the Rayleigh criteria on each pixel of the detector. Keystone and smile distortion values are less than 15 µm requirement. Modified-Offner type spectrometer design with Féry prisms and three even aspheric form mirrors has been designed such a way that they share the same optical centre works between 400 to 2500 nm. The resulting average spectral resolution of 8.3 nm. Compactness of the design is also considered; the overall optical system can fit in a box with dimensions 980 mm x 525 mm x 200 mm (Volume of the design= 0.1 m^3). The HSI design is concluded to meet the performance goals of the project which aims to be used in monitoring of Earth's surface for environmental purposes operating in LEO. Hyperspectral design is planned to be realized micro-satellite platforms in near future.

7. REFERENCES

Amani A., Bai J., Huang X., 2020. Dual-view catadioptric panoramic system based on even aspheric elements. *Applied Optics*, 59(25), pp. 7630-7637.

Bergeron M., Hollinger A., Staenz K., Maszkiewicz M., Neville R.A., Qian S.-E., Goodenough D. G., 2008. Hyperspectral Environment and Resource Observer (HERO) mission. *Can. J. Remote Sensing*, p. S1–S11.

Calin M. A., Calin C. A. , Nicolae D. N. , 2021. Application of airborne and spaceborne hyperspectral imaging techniques for atmospheric research: past, present, and future. *Applied Spectroscopy Reviews,* Volume 56, pp. 289-323.

Chang S. T. ,Lin Y. C. , Wu K.H. ,Lien C.C. , Huang T. M., Tsay H. L., Chan C.Y., 2016. *The design and assembly of aluminum mirrors of a three-mirror-anastigmat telescope.* Biarritz, Proceedings of the SPIE, pp. 1-6.

Eismann, M. T., 2012. *Hyperspectral Remote Sensing.* Washington: SPIE Press.

Escalante-Ramírez, B., 2012. *Remote Sensing Advanced Techniques and Platfroms.* Croatia : InTech .

K. P. Thompson, 1985. *Beyond optical design Interaction between the lens designer and the real world.* s.l., SPIE Proceedings, pp. 426-438.

M. Celesti, M. Celesti, J. Adams, V. Boccia, F. Gascon, C. Isola, J. Nieke, 2022. The Copernicus Hyperspectral Imaging Mission for the Environment (CHIME): Status and Planning. *IEEE International Geoscience and Remote Sensing Symposium,* pp. 5011 -5014.

Nie Y., Zhang J., Wang J., Xiang L., 2018. Optical design and performance comparison of various hyperspectral imagers based on Fery prisms. *Journal of Advanced Optics and Photonics,* 1(4), pp. 279-290.

Öztürk Z. N., Başlar İ., Selimoğlu Ö., Karcı Ö., 2023. Signal-to-noise ratio model in Python for high-resolution space-borne electro-optic imagers. *SPIE JARS,* 17(1), pp. 1-18.

Qian, S.-E., 2021. Hyperspectral Satellites Evolution and Development History. *IEEE JARS,* Volume 14, pp. 7032-7056.

Shaoyuan, C., 2022. *Geometric parameters analysis of two large FOV space-borne optical.* Beijing, Proc. SPIE , pp. 1-6.

Wang J., He X., Zhang X., Ma M., Cao Z., 2022. Wave Aberration Correction for an Unobscured Off-Axis Three-Mirror Astronomical Telescope Using an Aberration Field Compensation Mechanism. *MPDI Appl. Sci.* , 12(21), p. 10716.

X. Prieto-Blanco, C. Montero-Orille, B. Couce, R. de la Fuente, 2006. Analytical Design of an Offner Imaging Spectrometer. *Optics Express,* 14(20), pp. 9156-9168.

This contribution has been peer-reviewed.
https://doi.org/10.5194/isprs-archives-XLVIII-M-1-2023-233-2023

GROUND TRUTHING: STUDYING PERI-URBANIZATION AND LAND-BASED LIVELIHOODS IN MANKWENG AND ITS ENVIRONS (LIMPOPO, SOUTH AFRICA)

D. Patric*, G. Schaab

Faculty of Information Management and Media / Institute of Applied Research, Karlsruhe University of Applied Sciences, Moltkestrasse 30, 76133 Karlsruhe, Germany - (Deepthi_Jeslet.Patric, Gertrud.Schaab) @h-ka.de

KEY WORDS: Former homeland, Ground truth, WorldView-3, Peri-urban, Land use/cover, Informants.

ABSTRACT:

Peri-urbanization is often referred to as a transition zone following either an urban-rural gradient or an inner and outer peri-urban zone. Contrary to these notions and linked to its history in a former homeland, peri-urbanization in Mankweng appears as a complex and erratic process in reaction to shifting livelihoods, in-migration, and climate change. The process of ground truth for the study was designed by accommodating a combination of remote sensing images of WorldView-3 for the densely, vastly changing populated area (200 km^2) and of Sentinel-2 for the overall study area of 400 km^2. For the purpose of intensive data gathering, 16 plus three additional sites were selected in the area covered with the very high-resolution imagery as well as transect drives in the rural part. The study sites were selected and well distributed based on specific parameters such as building density, home garden size, road type with the goal of better interpreting what can be distinguished on the imagery and what influences land-based livelihoods. Additional information was gathered from informants during the ground truth through open questions and informal discussions. A solid knowledge of the land uses and land cover, including housing structures, agricultural practices, and challenges of land-based subsistence, was attained. The results revealed influential change drivers, settlement structure types, distinct development patterns, changes in traditional land-based livelihood specifically in terms of agriculture in the region and indicators for a functional zoning of the area. The ground truth provides a thorough basis for the object-based image analysis of the 200 km^2 peri-urban area being next.

1. INTRODUCTION

Peri-urban regions of expanding cities present a highly dynamic landscape with multifarious socio-economic systems, land use patterns, and functions, thereby posing a significant spatial planning challenge in the present times (Karg et al., 2019; Nilsson et al., 2013). In the case of South Africa, the peri-urbanization landscape transformation is intrinsically linked to a complex interplay of post-apartheid political and economic factors (Smith, 2015). To tackle the complex, ever-changing nature of peri-urban areas, there is a pressing need for advanced technologies and techniques to ensure efficient and effective spatial planning. Remote sensing has demonstrated remarkable potential as a tool for the study of peri-urban landscapes, with multiple use cases showing its efficacy in identifying settlement structures, land use covers, and spatial drivers that contribute to the rapid urbanization and transformation of such areas (Addo, 2010; Kemeling, 2001). However, to augment the accuracy of remote sensing techniques, it is crucial to complement them with ground truth data. Ground truth data provides a critical reference point for interpreting satellite imagery, and it helps in the understanding of the context of land-based livelihoods, farming practices, and settlement structures (Dinis et al., 2010; Kedar et al., 2020). In the present study, our objective is to develop a typology of settlements for sustainable livelihood scenarios in peri-urban areas using very high-resolution (VHR) imagery as a starting point. Developing a ground truthing methodology was essential for fully visually interpreting the satellite imagery and for gaining insight into the contexts of land use and settlement structures. The study approach adopted is to offer rich knowledge and understanding of the peri-urban landscape under study.

2. STUDY AREA

The study area selected for the WoPedyP research project (https://www.imm.hs-karlsruhe.de/WoPedyP/), targeting four designated study sites, is situated in Limpopo province, approximately 40 km east of Polokwane. The ground truthing activities were carried out in the peri-urban and rural outskirts of Mankweng, covering an area of 400 km^2, characterized by hills (rocky outcrops, known as 'koppies') amidst mostly flat terrain. Mankweng is a region located in one of South Africa's former homelands, which developed around the University of Limpopo (McCusker & Ramudzuli, 2007) from 1959 onwards. Settlement development accelerated with the end of the apartheid with peri-urban settlements stretching south and north of the two roads which travers the area west to east (Figure 1). Agriculture in the area is predominantly monocropping, with maize being the dominant crop (Makgoba et al., 2021), and some occasional cultivation of other crops during the wet season, such as leafy greens (known as 'morogo'), carrots, sweet potatoes, cowpeas, and lemons (Sekgobela, 2019). The region is home to communal agricultural land use, locally known as 'mashemo', and encompasses also grasslands, shrubs, trees, as well as the Turfloop Nature Reserve. Mankweng and its surroundings have a semi-arid climate (BSk in the Köppen climatic classification) with the highest precipitation occurring during the summer months from October to March (Rwanga, 2017).

* Corresponding author

This contribution has been peer-reviewed.
https://doi.org/10.5194/isprs-archives-XLVIII-M-1-2023-241-2023 | © Author(s) 2023. CC BY 4.0 License.

Figure 1: Study area (-23.90 Lat, 29.72 Long) with the four WoPedyP study sites (orange), ground truth sites delineated by square or circular polygons (white), ground truth points (green from 1st visit, yellow from 2nd visit) and tracks (yellow) overlaid on WorldView-3 (20 March 2022; bands 7,5,3) and Sentinel-2 (11 April 2021; bands 10,7,2) imagery.

3. MATERIALS AND METHODS

Sentinel-2 imagery (11 Apr 2021, Level-2a) is used for visualizing the overall study area of 400 km². Out of the 13 stacked bands, bands 10 (SWIR), 7 (Red Edge) and 2 (Blue) are used in the false colour composite for during ground truth. WorldView-3 (20 Mar 2022, Level-2) is utilized for the coverage of the peri-urban area of Mankweng and surroundings. Out of the eight bands, bands 7 (Near-IR), 5 (Red) and 3 (Green) are used as colour composite. The image composites were created to distinguish between settlement structure types in the study area (Figure 1). For image processing, *ERDAS IMAGINE 2022* and *ArcGIS Pro 2.9* were used. *ODKCollect*, in combination with *KoboToolbox*, served as data collection tool in the field, which allows for form-based data collection (Cai et al., 2017). *OSMAnd* was used for navigation to the predetermined ground truth sites and points, while *OSMTracker* for daily route logging.

The project's ground truth procedure can be broadly divided into four phases: prerequisite research, lab preparations, fieldwork, and post-analysis of ground truth (Lübker & Schaab, 2008a). Prerequisite research included a literature review on the region's geography. Relevant geodata sets and demographical datasets were pooled for adding to the spatial understanding of the area (Openshaw, 1997). WorldView-3 satellite imagery had to be acquired. Based on previous comparison studies and a review of the literature on PAN sharpening approaches, Tu et al. (2012) and Belfiore et al. (2016) determined IHS as the method most suited for WorldView-3 imagery. For our study the PAN sharpening approaches of NNDiffuse Resolution Merge, HPF Resolution Merge, Subtractive Resolution Merge, and Modified IHS Resolution Merge were run on a subset of the WorldView-3 image data (cp. Lübker & Schaab, 2008b). Finally, the Modified IHS Resolution Merge algorithm was chosen as the best for the WorldView-3 scene. The result was visually acceptable and minimized unwanted artefacts in the image that were causing pixel dissolution due to high reflectance values (Wang et al., 2021).

Site selection prior to ground truthing serves as an optimization step for efficient ground truthing (Hegarty-Craver et al., 2020). It helps with the goal of better interpreting what can be distinguished on the imagery, for identifying the region's land cover, and for understanding the land-based livelihoods in this former homeland area. The actual ground truthing was done in two parts, at first intensively in areas with WorldView-3 image coverage and later in the more rural areas with Sentinel-2 coverage. To conduct the ground truth in the area of WorldView-3 coverage, a total of 16 ground truth sites were selected with a square area of 1 km² each (Figure 1), except for site 1 which is rectangular in shape due to the need to fit within the demarcated WoPedyP study site, and further three circular sites for covering additional aspects of ground truth. The 16 sites were selected with the main intention of identifying settlement structure types. Here, 15 variables specific to the peri-urban landscape guided the selection (cp. Lübker & Schaab, 2008a) The variables considered were: arrangement of parcels, parcel size, size of main house, location of main house, building density, type of dwelling and roof, parcel boundary demar-

https://doi.org/10.5194/isprs-archives-XLVIII-M-1-2023-241-2023 | © Author(s) 2023. CC BY 4.0 License.

cation, road type, size and type of agricultural plot, type of main crop, plot layout of mashemo, abundance of trees, other natural vegetation, and other land use/cover (LUC). A preliminary zoning was conducted to identify the unique settlement structures in the study area and to ensure that all were covered by the placement of the 16 sites. The sites are therefore representative and reflect the strong heterogeneity of the area covered by the WorldView-3 image data.

In addition to selecting the ground truth sites, prior selection of points aids in improved data gathering and correlation later (Joyce, 1977). A total of 261 points of 18 distinct LUC classes, reflecting mainly vegetation and land cover types, crops, and settlement structure components, were marked inside the ground truth sites on the WorldView-3 data to serve as a basis for the field survey (Bhatt et al., 2018). In light of the subsequently placed Object-based Image Analysis (OBIA) categorization, a complete LUC classification scheme was designed and implemented, annotations added of what additional details to capture, and maps of the ground truth sites and points created (Maktav et al., 2020). Finding a suitable smartphone-based survey application for the ground truth is important. The application was chosen with anticipated challenges in mind, such as the ability to work in offline conditions, to help with navigation, a decent bandwidth requirement for uploading the collected points to a server, the customizing of forms for point collection, and the function of adding the WorldView-3 imagery as background layer (Manandhar et al., 2010). For planning of time in the field, a routing to the ground truth sites was calculated based on the OpenStreetMap road network (Hegarty-Craver et al., 2020).

Approximately 40 questions were formulated for addressing potential informants to better comprehend the livelihood of the region, including the dependency on natural resources, the accessibility of services, the age of settlements, agricultural practices, changes in crop yield, variations in rainfall throughout the years, crop diseases, and particularly land-based livelihoods in the area. It included also the roles of the traditional and municipal authorities. The combination of gathering local knowledge while reading the remote sensing data was employed to obtain contextually relevant data and to gain a comprehensive understanding of the factors driving land use change, as recommended by Hoover et al. (2017). In the first stage of ground truthing, 16 plus three predetermined locations in the peri-urban area and covered by WorldView-3 imagery were targeted and data was collected over a period of 17 days (23 Jun – 16 Jul 2022). In the second stage (31 Aug – 02 Sep & 09–16 Sep 2022), transect drives were conducted to cover the remaining 200 km^2 of more rural area over a period of eleven days. During both stages (Figure 1), field assistants assisted with tasks such as capturing photographs and their compass directions, to help with navigation, translation, administering the questions, and recording all information collected in the *ODKcollect* mobile application, a booklet or on the printed maps.

Table 1: Ground truth efforts: number of points captured and number of informants spoken to per site. For location of sites see Figure 1.

Name of ground truth site	Planned no of points	Points captured 1st + 2nd visit	Informants 1st + 2nd visit
Nchechane (1)	28 (19)	38 + 14	14 + 6
Nobody (2)	31 (28)	45 + 5	17 + 8
Ramogale (3)	16 (22)	42 + 2	19 + -
Ga-Mothabo mashemo (4)	8 (0)	4 + 2	5 + -
Mankweng Unit F (5)	28 (24)	32 + 3	12 + 10
Total Nobody (6)	23 (18)	20 + -	6 + -
Morongoa (7)	25 (20)	25 + 5	10 + 7
Seckline (8)	16 (8)	18 + 2	15 + -
Ga-Kama (9)	23 (12)	15 + -	7 + -
Iraq & Thabang (10)	17 (14)	23 + 3	8 + -
Ga-Makanye (11)	19 (16)	27 + 2	ca 7 + -
Makanye Extension & Magowa (12)	19 (12)	19 + 1	13 + -
Makanye mashemo (13)	18 (0)	1 + -	2 + -
Mentz Moshongo (14)	27 (14)	23 + 4	5 + 7
Matshelapata (15)	28 (31)	40 + -	9 + -
Mamabolo Village (16)	25 (17)	26 + -	12 + -
Tshware (Misc1)	9 (6)	7 + -	7 + -
Mothibaskraal (Misc2)	5 (0)	0 + -	3 + -
University Farm (Misc3)	3 (0)	2 + -	1 + -
Larger study area	-	- + 28	- + 14
Σ	**368**	**478**	**ca 225**

() no of points captured of those initially planned

The post-analysis of the ground truth data is a crucial step in the research process (Zhang et al., 2016). It was conducted to ensure accuracy and relevance of the collected data. It involved daily data cleaning and sorting to verify the correct locating of the data points. This step proved important for enhancing the visual interpretation of the WorldView-3 imagery, i.e. the learning process of reading the VHR satellite imagery. The photographs taken in the field were also organized and archived appropriately to facilitate future analysis.

4. RESULTS OF THE GROUND TRUTHING

4.1 Ground truth efforts

The study involved two separate field visits, which were 17 and 11 days long, respectively, for a total of 28 days of fieldwork. An overall distance of approximately 700 km were covered during these visits, by car and on foot, in order to collect data in the field. The primary goal of the ground truthing efforts was to check features in the field and to gather additional information in order to add meaning to the findings from remote sensing data. In total, 478 points were captured and ca 225 informants were spoken to. The local names for the various ground truth sites (mostly villages) were collected to provide further context. Table 1 gives a summary on the ground truthing efforts in comparison to the number of points originally planned to visit.

4.2 Hierarchical LUC classification scheme

The LUC classification scheme initially developed was based on research literature and the preliminary visual interpretation of the WorldView-3 satellite data. However, the ground truthing process revealed some shortcomings in the initial interpretation, such as e.g. the omission of wetlands. As a result, the LUC classification scheme was revised to better reflect area-specific classes. The revised scheme (Table 2) includes additional agricultural crops, fruit trees, and other classes relevant to land-based livelihoods. Overall, seven classes were removed, and ten added. Some of these 56 classes (belonging to 9 hyper-classes) may not be detectable from the WorldView-3 satellite imagery due to its acquisition towards the end of the raining season.

4.3 Zoning of settlements

Prior to placing the ground truth sites, a zoning of settlement structure types was created. This was based on knowledge from the literature as well as the preliminary visual image interpre-

This contribution has been peer-reviewed.
https://doi.org/10.5194/isprs-archives-XLVIII-M-1-2023-241-2023 | © Author(s) 2023. CC BY 4.0 License.

tation of the WorldView-3 imagery. The settlement structure types were visually grouped distinguishing differences in building densities, building alignment, and other typical feature arrangements and patterns. Eleven zones were highlighted on the VHR imagery. With a thorough understanding of the settlement types based on details as gathered from informants and with the ground perspective in mind, this number reduced to eight. Both zoning approaches can be seen in Figure 2.

Table 2: Hierarchical LUC classification scheme, updated post ground truthing.

Homestead (Roof of building)	**Agriculture**
Metal roof	Maize
Concrete roof ^{x}	Sorghum ^{x}
Sheet roof ^{x}	Wheat ^{x}
Slate/tile roof $^{+}$	Beans
Thatched roof $^{+}$	Beetroots
Unfinished roof $^{+}$	Cabbage
Homestead (Purpose of building)	Carrots
Occupied dwelling	Chillies $^{+}$
Vacant dwelling	Cowpeas
Cattle shed	Groundnuts
Garage shed	Lettuce
Plot boundaries	Onions
Fence	Spinach
Wall	Tomatoes
Hedge ^{x}	Fruit tree(s) ^{x}
Miscellaneous LC in settlement	Avocado $^{+}$
Pavement	Lemon $^{+}$
Grass meadow	Mango $^{+}$
Unused land	Orange $^{+}$
Other (urban) LU	Fallow land
Car park	Burnt area
Cemetery	Other
Dumpsite	**Roads**
Mining $^{+}$	Tarmac road
Erosion $^{+}$	Paved road ^{x}
Golf club	Dirt road
Industry	**Natural vegetation**
Bare ground	Forest
Kraal $^{+}$	Bushland
Water bodies	Shrubland
Stream	Rangeland
Dry riverbed	Grassland
Natural pond	Single tree(s)
Water dam	Outcrop
Water tank (closed)	
Wetland $^{+}$	

$^{+}$ indicates classes that were added, ^{x} those eliminated

Figure 2: Zoning pre and post ground truth: Initial zoning pre ground truth considered eleven narrow zones indicated by thicker lines, while the zoning post ground truth contains eight broader zones shown by thinner lines.

4.4 Conceptual timeline of settlement development

Based on the field visits and WorldView-3 image interpretation, a conceptual timeline of settlement development was developed. Figure 3 demonstrates this progression of settlements from the first transformation of mashemo land to the stage of structured settlements as observed in the WorldView-3 imagery. The timeline's objective is to show how the area evolves through time, from the initial makeshift tin shacks to the creation of larger, permanent, and orderly houses. The following are the stages of peri-urbanization: In the mashemos (a) the Traditional Authorities (TA) allocate land as stands for residential dwellings. The stands may or may not be cultivated at the time of allocation. In the first stage of construction (b), temporary shacks with tin walls and roofing are built on the designated stands. Demarcated by fences, the TA gives the residents Permission To Occupy (PTO) which is considered to be an informal right to land (Polokwane Local Municipality, 2022). These stands must remain unoccupied for more than a year before residents can move in. Services are absent at this stage of settlement growth. Plenty of bare ground is typically observed around the metal shacks and some agriculture may be present. Occupants of these shacks are seen to take up piece jobs, with a majority being unemployed. It was observed that some shacks at this stage are rented out. In the next stage of settlement development, there is a densification of the shacks (c), but amenities such as water, electricity, and sanitation remain scarce. The bare ground around the home structures is decreasing and only a little amount of original agriculture is remaining. A massive boom in construction can be seen, including many unfinished roofs of larger structures. Occupants at this stage are either seen taking up seasonal piece jobs or traveling to Polokwane for jobs. In the third stage of settlement growth, a significant level of diversification (d) can be recognized. Communities consist of mixed constructions, including still some shacks and many more permanent, larger residences. At this stage, both bare ground and home gardens around the house are relatively limited, but in some cases more fruit trees can be seen and partial provision of services can be noted. At this point, the settlements are identified as 'less-formal' by the municipal authority (Polokwane Local Municipality, 2022).

Regularization of the settlements (dito) occurs later on (e). At this stage, housing structures exhibit a distinct pattern, as evidenced by the alignment of the settlements to the roads. Housing looks more orderly, with tiled roofs and concrete walls, and a good number of houses have municipal approval. The bare area around the home may be cemented or have grass meadows surrounding it. Home gardens are not apparent, although more fruit trees can be observed across the land. The services of electricity, water, and sanitation are more easily accessible in the settlement and the residents tend to have regular occupations.

The Mankweng region features two additional distinct types of settlement shown as special cases in Figure 3: (f) Mankweng itself experienced a rapid and unpredictable expansion and development, influenced by its proximity to the university and administration offices (McCusker & Ramudzuli, 2007). It looks rather planned and has the most stable service provision. Due to the housing density, there are no home gardens and this suburban area has a dynamic student and migrant population. Villagization settlements (g), the result of the betterment scheme during apartheid (De Wet, 1991), exhibits a higher dependence on mashemo. Ground truthing indicates that the population in these villages consists primarily of older individuals. They have their home gardens and access to services, although limited.

https://doi.org/10.5194/isprs-archives-XLVIII-M-1-2023-241-2023 | © Author(s) 2023. CC BY 4.0 License.

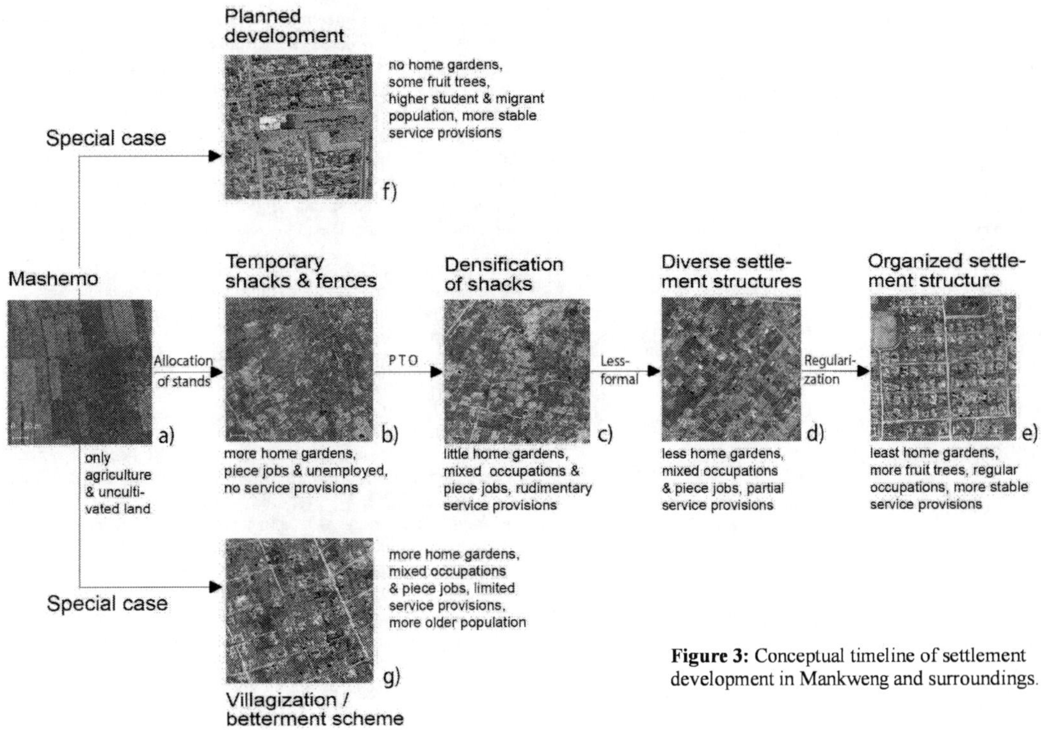

Figure 3: Conceptual timeline of settlement development in Mankweng and surroundings.

4.5 Conceptual framework of the anticipated study

The anticipated study's conceptual framework, as illustrated in Figure 4, serves to elucidate the study and is rooted in the understanding of livelihoods as obtained through ground truthing. Given the complexity of the region, the framework was designed to provide insight into the spatial and other parameters that directly or indirectly contribute to the peri-urbanization in such an unplanned and erratic manner. It neither follows the notions of an urban-rural gradient (Schlesinger, 2013) nor of an inner and outer peri-urban zone (Karg et al., 2019), but appears as a complex and erratic process in reaction to shifting livelihoods, in-migration, and climate change. Therefore, the ultimate goal of the remote sensing analysis is to comprehend the current settlement structure types. For that, the framework establishes foci and bounds for the subsequent methodical research. Besides gaining details on the current pattern of land-based activities (covering agriculture, livestock and wild natural resource use), additional understanding to add

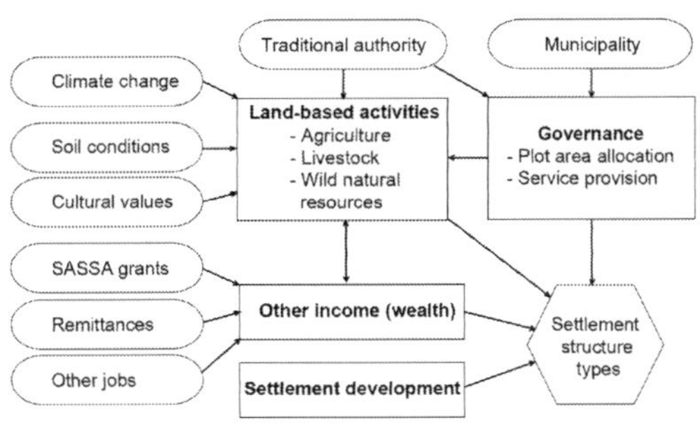

Figure 4: Conceptual framework of the research.

is related to governance (plot allocation, service provision) as well as other income sources. SASSA grants, which is social grant provision in South Africa (Lund, 2006), and remittances from relatives in cities elsewhere impact the livelihoods. Here, particular wealth can be inferred indirectly via indicators. Climate change has an overall impact but will not be explicitly studied. But for concluding on settlement structure types and associated typical livelihoods, the inclusion of a remote sensing-based time-series analysis (extending and supplementing the LUC time series of McCusker & Ramudzuli, 2007) is crucial.

4.6 LUC classes for identification in OBIA

With a framework in hand and the knowledge that the detailed, agriculture-related LUC classes (Table 2) cannot be distinguished in the WorldView-3 imagery, Table 3 provides the LUC classes, which OBIA should focus on. By defining rules, classes will go beyond broadly defined area-specific land uses and covers in order to help in quantifying information related to livelihoods. For example, building occupancy can be inferred from the presence of bare soil surrounding a home structure. Indicators for wealth of households are the roof type in combination with the presence of fruit trees, green lawns and water tanks on the compound. Areas used to harvest wild natural resources can be derived from the distribution of footpaths leading to and along dry riverbeds and bushland. And the distinct pattern of spots seen on grassland in the imagery was identified as termite mounds, which point to former woodlands.

5. DISCUSSION

Mankweng and surroundings is characterized by the disappearance of grassland and agricultural area, and a high level of urbanization, resulting from a dramatic pattern of increasing spatial

https://doi.org/10.5194/isprs-archives-XLVIII-M-1-2023-241-2023 |

Table 3: LUC classes aimed at in OBIA.

Vegetation	Land uses inferred via rules
Maize	*Farming methods:*
Tree(s)	- mashemo
Bushland	- commercial farm
Riverine vegetation	- home garden
Grassland	
Fallow	*Uses (built-up):*
Burnt area	- dwelling
Land uses	- industry
Dry riverbed	- school
Cemetery	- shed
Dumpsite	- water tank
Mining	
Roads & infrastructure	*Building occupancy:*
	- bare ground around the structure vs. overgrown with grass
Tarmac road	
Dirt road	
Footpath	*RDP houses:*
Car park	- based on roof size and shape
Other land cover	
Bare ground	*Wealth of household (indicators):*
Outcrop	- slate/tile roof
Waterbody (dam)	- fruit trees
Wetland	- greenness of lawn
Forest	- water tank(s)
Built-up	
Building:	*Kraal:*
- with metal roof	- black soil colour and closeness to house
- with slate/tile roof	
- with thatched roof	
- with unfinished roof	*Wild natural resource use:*
Compound demarcation:	- based on footpaths
- wall	*Former woodlands:*
- fence	- based on termite mounts

concentration (McCusker & Ramudzuli, 2007). The WorldView-3 imagery being used for the peri-urban part of the study area was captured on 20 March 2022 (Figure 1) with the hope to in particular also learn about the peri-urban land-based activities. One could argue that the timing of image acquisition towards the end of the rainy season was not the best, with rotation and mixed cropping during the wet season not being captured. However, cloud-free imagery was given priority. And as most of the communal agricultural land and the home gardens is used for maize planting (acc. to Dovie et al., 2007 e.g. 96% of the households), the additional information would have been only minor. Instead, the ground truthing was a combination of visually studying the area in comparison to the satellite imagery and obtaining relevant information from residents and locals of the ground truth sites (Table 1). Thus, very valuable information for complementing the image-based information and for understanding the complexity and diversity of land-based livelihoods in the area could be gained. Here, it proved sufficient to limit the VHR satellite imagery to the more diverse peri-urban area of the study site (Figure 1).

From ground truth investigation it was noted that the farming patterns in Mankweng is seen as home gardens attached to the residential property, mashemo (or community farms) and some commercial agriculture plots. Maize is the most important crop grown throughout the country and grown under diverse conditions (Baloyi, 2011). During the ground truthing, rarely diversified agriculture practices were observed in the region. As challenges, the lack of inputs such as irrigation facilities, financial aids and capacity building (cp. Oluwatayo & Rachoene, 2017) in addition to poor quality of soil, the size of land holding, crop diseases and the depleting quality of maize seeds were identified. However, these factors cannot be identified on the VHR imagery. Besides, in particular the younger population are not keen in pursuing a farming life. Cattle and poultry are prevalent in the areas ground truthed, but it is for self-subsistence only and seen as a status symbol within the community. Large cattle herds were observed more often in the rural part of the study area, which is in agreement with Puttick et al. (2014) who point to the diminishment of ranging lands. It is hoped that the census 2022 data will become available in time to complement the kraals to be detected in the WorldView-3 image (Table 2 and 3) and thus in half of the study area only.

Settlement structure types are vastly studied using the aid of VHR satellite imagery (e.g. Niebergall et al., 2008). Sentinel-2 imagery is widely used for studies in landcover as well as peri-urban and human settlement layer contexts (Helber et al., 2019; Phiri et al., 2020). For this present study the VHR imagery helped in identifying several housings structure types, particularly via four roof types such as slated roof, metal roof, thatched and unfinished roofs as listed in Table 2 and 3. This can be cautiously used as an identifier of household wealth in combination with other features like fruit trees, irrigated lawns and water tanks (see Table 3). Such an indirect assessment of socio-economic conditions will help in testing and strengthening the conceptual timeline of settlement development as visualized in Figure 3. However, the challenge remains that the areas are mixed in nature. Even the core peri-urban areas of Mankweng comprise of large permanent houses besides shacks of temporary materials. During the ground truthing subsidised housing, known as RDP houses, were seen clustered or scattered in the peri-urban and rural areas. These houses are hoped to be distinguished by their smaller size and simplicity in shape (Table 3). Settled areas are formally categorized as A, B, C, D and only unplanned settlements (D) without prior approval from the authorities are categorized as 'illegal settlements', while the term 'less-formal' is used for all settlements initiated by the traditional authorities (Polokwane Local Municipality, 2022). Distinguishing them based on their structure types is not easy due to their heterogeneity, especially when being built non-compliant to the formal regulations. Here the conceptual timeline (Figure 3) in combination with a time-series analysis (see Figure 4) will aid to understanding and distinguishing between settlement structure types.

The use of wild natural resources such as fuel wood, edible insects, wild herbs and wild fruits is common in many of the developing regions (Twine et al., 2003). Ward & Shackleton (2016) presented a detailed study on wild resource use along an urban-rural gradient in a place close by. In agreement to their paper, during ground truth it was seen that regardless of the economic status, the use of fuel wood collected or bought was prevalent. Despite improvements in electrification and government efforts, most South African households still rely on firewood for energy (Masekela & Semenya, 2021). Cow dung was also seen to be used in some sites. Traditional healers were found to be collecting wild herbs and plants. Seasonally, some wild edible fruits such as prickly pears are collected for sale or self-consumption. Identifying collection sites can be achieved with rule-based classification as indicated in Table 3. Sand mining has grown popular due to the increasing demand for construction in and around peri-urban regions (Malebana, 2021). Around Mankweng, illegal sand mining was seen to form an important livelihood from natural resources linked to unused mashemos and dry riverbeds, although studies state this can lead to loss of land and livelihood (Shackleton, 2020). Land degradation as a consequence, but also due to erosion, poor sanitation and illegal waste dumping, was extensively seen throughout the area. Some large garbage cans were observed in the ground truth sites, however, proper sanitation facilities are missing (Maphosa, 2017).

This contribution has been peer-reviewed.
https://doi.org/10.5194/isprs-archives-XLVIII-M-1-2023-241-2023

Services provided by the municipality in the region is notably running water and waste removal, while Eskom provides electricity. Despite the willingness of residents to pay for the above-mentioned services, there is a lack in infrastructure for providing the services (Nkoana et al., 2019). Similar challenges were found during ground truthing in multiple sites majorly categorized as 'less-formal' settlements. In certain sites, which have undergone regularization, water distribution happens bi-weekly in trucks provided by the municipality. Also private vendors were seen selling water in large cans. The influence of service provision forms an important part of settlement development (Figure 3) and was also reflected in the rezoning of the settlements (Figure 2). Only after OBIA of the World-View-3 imagery, a final zoning of the settlements of Mankweng and surroundings will be available, then thoroughly combining the remote-sensing based details with other geodatasets and the knowledge coproduced by involving informants.

6. CONCLUSION AND OUTLOOK

The findings of the ground truthing highlight the significant role that social grants, traditional farming practices, and the availability of services play in the spatial pattern which has evolved in the area and is constantly and rapidly changing. The peri-urban landscape has witnessed a reduction in dependence on land-based activities, with livelihood diversification moving away from indigenous practices, this particularly in Mankweng itself. The decline of land-based livelihoods is evident, with an increasing demand for residential areas paralleled by a decrease in farming land and in the interest to still rely on farming. Here, provision of services plays a vital role in area development and thus when identifying settlement structure types. The ground truth efforts provide a thorough basis for the subsequent OBIA of the WorldView-3 imagery and the scaling to the entire study area based on a Landsat/Sentinel-2 time series analysis. However, it has to be noted, that the analysis will only allow to provide a snapshot in time which is already out-dated, a fact which became evident during ground truth.

ACKNOWLEDGEMENTS

This work is based on the research supported wholly by the National Research Foundation of South Africa (Grant Number 136090) and the German Federal Ministry of Education and Research (Grant Number 01DG21049). The authors acknowledge the help of the student assistants Tshilidzi, Melon, Willem and the senior lab assistant Prudence who helped with translation and data collection. Special thanks to Izelque Botha for the arrangements and logistics on the ground.

REFERENCES

Addo, K (2010). Urban and peri-urban agriculture in developing countries studied using remote sensing and in situ methods. *Remote Sensing*, 2(2), 497–513, doi: 10.3390/rs2020497.

Baloyi, RT (2011). *Technical Efficiency in Maize Production by Small-scale Farmers in Ga-Mothiba, Limpopo Province, South Africa*. Master thesis, University of Limpopo http://ulspace.ul.ac.za/handle/10386/1253.

Belfiore, O, Meneghini, C, Parente, C & Santamaria, R (2016). Application of different pan-sharpening methods on WorldView-3 images. *ARPN Journal of Engineering and Applied Sciences*, 11(1), 490-496, http://www.arpnjournals.org/jeas/research_papers/rp_2016/jeas_0116_3363.pdf (01.03.2023).

Bhatt, MP, Kumar, P & Bisht, G (2018). Assessment of LULC changes using remote sensing and GIS techniques: A case study of Mussoorie town, India. *Journal of Indian Society of Remote Sensing*, 46(8), 1201-1211.

Cai, X, Magidi, J, Nhamo, L, & van Koppen, B (2017). Mapping irrigated areas in the Limpopo Province, South Africa. *IWMI Working Paper, 172, Colombo (Sri Lanka)*, doi: 10.5337/2017.205.

De Wet, C (1991). Some socio-economic consequences of villagezation schemes in Africa, and the future of "betterment villages" in the "new South Africa". *Development Southern Africa*, 8(1), 3–17, doi: 10.1080/03768359108439566.

Dinis, J, Navarro, A, Soares, F, Santos, T, Freire, S, Fonseca, A, Afonso, N & Tenedório, J (2010). Hierarchical object-based classification of dense urban areas by integrating high spatial resolution satellite images and lidar elevation data. *The International Archives of the Photogrammetry, Remote Sensing and Spatial Information Sciences*, Vol. XXXVIII-4/C7, https://www.isprs.org/proceedings/xxxviii/4-c7/pdf/Dinis_66.pdf (01.03.2021).

Dovie, DBK, Witkowski, ETF & Shackleton, CM (2003). Direct-use value of smallholder crop production in a semi-arid rural South African village. *Agricultural Systems*, 76, 337–357.

Hegarty-Craver, M, Polly, J, O'Neil, M, Ujeneza, N, Rineer, J, Beach, RH, Lapidus, D & Temple, DS (2020). Remote crop mapping at scale: Using satellite imagery and UAV-acquired data as ground truth. *Remote Sensing*, 12(12), 1–15, doi: 10.3390/rs12121984.

Helber, P, Bischke, B, Hees, J & Dengel, A (2019). Towards a Sentinel-2 based human settlement layer. *Proceedings of International Geoscience and Remote Sensing Symposium (IGARSS)*, Yokohama (Japan), 28 July – 02 August 2019, 5936–5939, doi: 10.1109/IGARSS.2019.8898172.

Hoover, JD, Leisz, SJ & Laituri, ME (2017). Comparing and combining Landsat satellite imagery and participatory data to assess land-use and land-cover changes in a coastal village in Papua New Guinea. *Human Ecology*, 45(2), 251–264. doi: 10.1007/S10745-016-9878-X/FIGURES/6.

Joyce, A (1977). *Earth Resources Laboratory Procedures for Gathering Ground Truth Data*. NASA Technical Memorandum, 163, Houston (TX, USA).

Karg, H, Hologa, R, Schlesinger, J, Drescher, A, Kranjac-Berisavljevic, G & Glaser, R (2019). Classifying and mapping periurban areas of rapidly growing medium-sized Sub-saharan African cities: A multi-method approach applied to Tamale, Ghana. *Land*, *8*(3), 1-22, doi: 10.3390/land8030040.

Kedar, P, Gaikwad, S & Kulkarni, M (2020). *Importance of Ground Truthing in Satellite Remote Sensing Technique for Crop Assessment*. Technical report, Resources Engineering Center, Maharashtra Engineering Research Institute, Nashik (India), http://www.mobileappscircle.com/upload_files/ research/rec-paper-12.pdf (01.03.2021).

Kemeling, I (2001). *Mapping Urban and Peri-Urban Agricultural Areas in Ouagadougou, Burkina Faso*. Master thesis, University of Wageningen (The Netherlands).

Lübker, T & Schaab, G (2008a). „Ground truthing" – Erfassung von Referenzdaten für VHR-Satellitenbild-daten in einem ländlichen Gebiet Westkenias. *Angewandte Geoinformatik 2008, Bei-*

https://doi.org/10.5194/isprs-archives-XLVIII-M-1-2023-241-2023

träge zum 20. AGIT-Symposium in Salzburg (J Strobl, T Blaschke & G Griesebner, eds.), Wichmann Verlag / VDE, Heidelberg, 28-33.

Lübker, T & Schaab, G (2008b). Identifying benefits of pre-processing large area QuickBird imagery for object-based image analysis. *Object-based Image Analysis. Spatial Concepts for Knowledge-Driven Remote Sensing Applications* (T Blaschke, S Lang & GJ Hay, eds.), Springer, Berlin, Heidelberg, 203-219.

Lund, F (2006). The impact of social assistance on households in South Africa. *Development Southern Africa*, 23(5), 647-664.

Makgoba, MC, Tshikhudo, PP, Nnzeru, LR & Makhado, RA (2021). Impact of fall armyworm (Spodoptera frugiperda) (J.E. Smith) on small-scale maize farmers and its control strategies in the Limpopo province, South Africa. *Jàmbá: Journal of Disaster Risk Studies*, 13(1), 1–9, doi: 10.4102/JAMBA. V13I1.1016

Maktav, D, Küçük, Ö & Kavzoglu, T (2020). A comparison of object-based and pixel-based classification methods for land use/ land cover mapping. *GIScience & Remote Sensing*, 57(5), 622-640.

Malebana, DS (2021). *Sand Mining, Land Degradation and Rehabilitation in Rural Areas of South Africa: A Case of Mentz Village, Limpopo Province.* Master thesis, University of Limpopo, http://ulspace.ul.ac.za/handle/10386/3772.

Manandhar, R, Odeh, I & McAlpine, C (2010). An assessment of the effectiveness of a smartphone survey application for ground truth data collection. *The Journal of Open Source Software*, 5(49), 1813-1823.

Maphosa, E (2017). *Public Perception on the Environmental Effect of Sanitation: A Case Study of the Polokwane Local Municipality in the Limpopo Province.* Master thesis, University of Limpopo, http://hdl.handle.net/10386/1975.

Masekela, ME & Semenya, K (2021). Factors influencing the use of firewood post-electrification in rural South Africa: The case of Ga-Malahlela village. *Journal of Energy in Southern Africa*, 32(3),24–40, doi: 10.17159/2413-3051/2021/ V32I3A7781.

McCusker, B & Ramudzuli, M (2007). Apartheid spatial engineering and land use change in Mankweng, South Africa: 1963-2001. *Geographical Journal*, 173(1), 56–74, doi: 10.1111/ J.1475-4959.2007.00222.X.

Niebergall, S, Loew, A & Mauser, W (2008). Integrative assessment of informal settlements using VHR remote sensing data – The Delhi case study. *IEEE Journal of Selected Topics in Applied Earth Observations and Remote Sensing*, 1(3), 193–205, doi: 10.1109/ JSTARS.2008.2007513.

Nilsson, K, Pauleit, S, Bell, S, Aalbers, C, Nielsen, TS (2013). Introduction. *Peri-urban Futures: Scenarios and Models for Land Use Change in Europe* (K Nilsson, S Pauleit, S Bell, C Albers, T Sick Nielsen, eds.), Springer, Berlin, Heidelberg, doi: 10.1007/978-3-642-30529-0_1.

Nkoana, MA, Cholo, MS, Hlongwane, JJ & Belete, A (2019). Determinants of households' willingness to pay for water and electricity in Moletjie under Aganang municipality, Limpopo Province, South Africa. *Journal of Agribusiness and Rural Development*, 51(1), 43–50, doi: 10.17306/J.JARD.2019.01143.

Oluwatayo, IB & Rachoene, MA (2017). Effect of agricultural commercialization on food security among smallholder farmers in Polokwane municipality, Capricorn district of Limpopo province, South Africa. *Journal of Agribusiness and Rural Development*, 16(1), 143–156, doi: 10.17306/j.jard.2017.00277.

Openshaw, S (1997). The truth about Ground truth. *Transactions in GIS*, 2(1), 7–24, doi: 10.1111/J.1467-9671. 1997.TB00002.X.

Phiri, D, Simwanda, M, Salekin, S, Nyirenda, VR, Murayama, Y & Ranagalage, M (2020). Sentinel-2 data for land cover/use mapping: A review. *Remote Sensing*, 12(14), 2291, doi: 10.3390/ RS12142291.

Polokwane Local Municipality (2022). *Development of an Integrated Land Use Scheme. Draft Status Quo Analysis Report.* Polokwane (South Africa).

Puttick, JR, Hoffman, MT & Gambiza, J (2014). The influence of South Africa's post-apartheid land reform policies on bush encroachment and range condition: A case study of Fort Beaufort's municipal commonage. *African Journal of Range & Forage Science*, 31(2), 135–145. doi: 10.2989/10220119. 2014.880943.

Rwanga, N (2017). Approach to quantify groundwater recharge using GIS Based water balance model: A review. *International Journal of Research in Chemical, Metallurgical and Civil Engineering*, 4(1), 166-172, doi: 10.15242/IJRCMCE.

Schlesinger, J (2013). *Agriculture Along the Urban-rural Continuum: A GIS-based Analysis of Spatio-temporal Dynamics in Two Medium-sized African Cities.* PhD dissertation, Albert-Ludwigs-Universität Freiburg, doi: 10.6094/UNIFR/9116.

Sekgobela, MM (2019). *Performance of Elite Cowpea (Vigna Unguiculata) Genotypes at Mankweng and Bela-Bela, Limpopo Province.* Master thesis, University of Limpopo, http://ulspace.ul.ac.za/ handle/10386/2907.

Shackleton, RT (2020). Loss of land and livelihoods from mining operations: A case in the Limpopo Province, South Africa. *Land Use Policy*, 99, 104825. doi ,10.1016/j.landusepol.2020.104825.

Smith, JL (2015). *Uneven Development and Peri-urbanization in South Africa: Exploring Landscape Change in Polokwane, South Africa.* PhD Dissertation, University of West Virginia, doi: 10.33915/ etd.6671.

Tu, TM, Hsu, CL, Tu, PY & Lee, CH (2012). An adjustable pan-sharpening approach for IKONOS/QuickBird/GeoEye-1/ World-View-2 imagery. *IEEE Journal of Selected Topics in Applied Earth Observations and Remote Sensing*, 5(1), 125-134, doi: 10.1109/ JSTARS.2011.2181827.

Twine, W, Moshe, D, Netshiluvhi, T & Siphugu, V (2003). Consumption and direct-use values of savanna bio-resources used by rural households in Mametja, a semi-arid area of Limpopo province, South Africa. *South African Journal of Science*, 99, 467-473.

Wang, W, Liu, H & Xie, G (2021). Pansharpening of WorldView-2 data via graph regularized sparse coding and adaptive coupled dictionary. *Sensors*, 21(11), 3586. doi: 10.3390/s21113586.

Ward, CD & Shackleton, CM (2016). Natural resource use, incomes, and poverty along the rural–urban continuum of two medium-sized, South African towns. *World Development*, 78, 80–93, doi: 10.1016/j.worlddev.2015.10.025.

Zhang, J, Zhao, Y, Zhou, J, Yang, J & Zheng, W (2016). Utilizing ground truth data for high-resolution urban land use classification: A case study in Shanghai. *ISPRS International Journal of Geo-Information*, 5(2), 11.

MULTICRITERIA EVALUATION OF THE POTENTIAL OF P-BAND AIRBORNE RADAR DEMs FOR 3D CHARACTERISATION OF HYDROGRAPHY IN TROPICAL FORESTS

L. Polidori[1,2], C.R.T. Caldeira[1], L. dos Reis[1], N. Eco[3], M. El Hage[4]
[1] PPGG, Instituto de Geociências, UFPA, Belém, Brazil – laurent.polidori@ufpa.br, caldeiracrt@gmail.com leo2reisgeo@gmail.com
[2] CESBIO, 18 av. Edouard Belin, bpi 2801, 31401 Toulouse cedex 9, France – laurent.polidori@ird.fr
[3] LASTIG, Institut Géographique National, Saint-Mandé, France – econejla@gmail.com
[4] Geospatial Studies Laboratory, Lebanese University, Tripoli, Lebanon – mhamad.elhage@ul.edu.lb

KEYWORDS: Amazon, hydrography, scale, P-band radar, digital elevation model, quality

ABSTRACT:

Relief mapping through dense tropical forest is a challenge, which can be met by processing P-band radar images. Digital terrain models (DTMs) obtained over three sites in the Amazon region (French Guiana and North Brazil) are evaluated according to two types of quality criteria: on the one hand, the accuracy of elevations and slopes, calculated by comparison with lidar surveys used as reference data; on the other hand, the hydrographic coherence of the DTM, revealed by its compliance to some universal rules like "all rivers flow downhill", or the fact that landforms shaped by water have a fractal behaviour. The results depend on the scale, the effect of which is addressed. Overall, the results confirm the potential of P-band airborne radar for 3D characterisation of hydrography in tropical forested areas.

1. INTRODUCTION

The suitability of a digital elevation model (DEM) for hydrological modelling, and in particular for watershed delineation and drainage network extraction, must be evaluated according to multiple quality criteria, because the DEM must be close to the topographic surface (i.e., minimize elevation error) and at the same time respect the shapes (i.e., minimize slope error due to random variations of elevation error), among other requirements (Polidori and El Hage, 2020). This evaluation can be based on the comparison of the DEM with a more accurate local DEM or a set of ground control points (GCPs). However, even when no GCPs are available, a detection of geomorphological inconsistencies can contribute to DEM quality assessment (Polidori et al., 2014).

In the case of tropical forest regions, such as the Amazon, these requirements are more difficult to meet. Indeed, in addition to difficult field access and to the cloud cover, the presence of forests, with an average height of about 30 m and a very irregular texture, makes it impossible to observe the ground with optical techniques (such as photogrammetry) and even with short wavelength radar interferometry (Polidori *et al.* 2022). Therefore, those methods provide a model of the canopy elevation, which turns very uncertain the description of hydrography due to important elevation and slope errors, especially in areas with moderate relief. This is the case, for example, of the widely used SRTM model (Bamler, 1999).

To overcome this limitation, airborne lidar is commonly considered as the most accurate method to measure topographic points through dense forest foliage (Kraus and Pfeifer, 1998), which can be interpolated to generate a digital terrain model (DTM). However, this method is only accurate at the cost of a low and slow flight, which makes it very expensive and time consuming for large areas.

Another solution is P-band airborne radar, which has a wavelength close to 70 cm that can penetrate the forest cover and reach the ground. The backscattered signal can be processed by interferometry (InSAR) or tomography (TomoSAR) depending on the acquisition configuration, offering the possibility to model the relief through the forest. Acquisitions have taken place in several states of the Brazilian Amazon as part of the *Radiografia da Amazônia* project to update deficient mapping (Correia 2011), and more recently in the state of Amapá, bordering French Guiana (Guimarães Filho and Borba, 2020). In addition, P-band airborne radar data were acquired in a similar ecosystem in Paracou (French Guiana) as part of the TropiSAR project (Dubois-Fernandez et al., 2010), for the calibration-validation activities of the European Space Agency's Biomass space mission, which will carry onboard a P-band imaging radar for the first time (Le Toan et al., 2011). As one of the foreseen products of the Biomass mission is a quasi-global digital terrain model, data from the TropiSAR project have been used to evaluate the potential of P-band radar tomography for this application (Mariotti D'Alessandro and Tebaldini, 2019, El Hage et al. 2022).

However, the potential of these data for scientific exploitation remains poorly understood. DTMs obtained from P-band radar in three areas located in French Guiana and Amapá have been analysed in recent experimental studies, most of which are reported in El Hage et al. (2022), Eċo (2022), Reis (2023). They are compared to aerial lidar data, which are considered as more accurate in forested areas and therefore taken as reference data, and to other more widely used models, leading to preliminary results about DTM quality. This paper brings together some of these results to evaluate the suitability of P-band radar DTMs for 3D characterisation of hydrography in tropical forest environments like the Amazon region.

2. MATERIALS AND METHODS

2.1. Study areas and data

P-band airborne radar data were acquired in 3 areas of the North coast of South America (figure 1): Paracou in French Guiana (A), a steep relief area along the Oiapoque border river (B) and a floodplain in the lower Amazon in the Amapá state, Brazil (C). Areas A and B are covered by dense forest, while area C is a partially flooded plain with low vegetation and some woodland. In area (A), the DTM was obtained by tomographic processing of SETHI radar data from ONERA (Dubois-Fernandez et al., 2010) as part of a calibration-validation experiment for the ESA Biomass space mission. The data from areas B and C are part of the *Base Cartográfica Digital e Contínua do Amapá* (BCDCA) of the Amapá state. The images

This contribution has been peer-reviewed.
https://doi.org/10.5194/isprs-archives-XLVIII-M-1-2023-249-2023 | © Author(s) 2023. CC BY 4.0 License.

acquired in P and X bands were processed by interferometry to produce a DTM (digital terrain model) and a DSM (digital surface model, which describes the upper envelope of the canopy) respectively, at 5 meter ground sampling (Guimarães Filho and Borba, 2020). In areas A and B, high resolution DTMs obtained from airborne lidar are used as reference data. The SRTM product, derived from interferometric processing of C-band radar data at 30 m resolution (Bamler, 1999), and the Topodata model obtained by densifying the initial version of SRTM from 90 to 30 m over the entire Brazilian territory (Valeriano and Rossetti, 2012), are also used in areas B and C. In area C, several other open access DEMs are used, including ASTER GDEM v3, obtained by spatial photogrammetry at 30 m (Abrams et al. 2020).

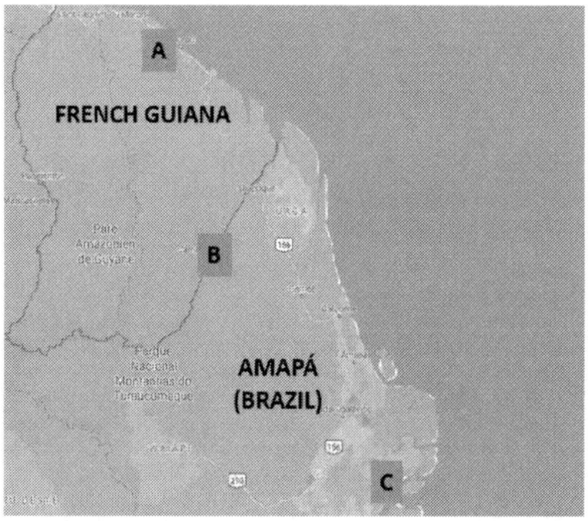

Figure 1. Location of study areas in French Guiana and Amapá

2.2. Absolute accuracy assessment

A visual comparison of the DTM obtained from P-band radar with the lidar reference DTM and other DEMs along a river in area B provides a preliminary evaluation of the ability of this DTM to describe hydrographic shapes.

In more quantitative terms, the ability of a DTM to describe a hydrographic system is conditioned by the accuracy of elevations and slopes, which may be degraded in forested areas. The aerial lidar survey is used as a reference in areas A and B to assess the DTM accuracy based on a statistical analysis of the discrepancies between the two DTMs. The bias and the random error of the DTM produced by P-band radar interferometry are calculated in terms of both elevation and slope. The effects of the local terrain shape (i.e., slope and orientation) and of the incidence angle (i.e., the angle between the radar line of sight and the perpendicular to the ground surface) are studied for both elevation and slope errors. Finally, the statistical distribution of slope, which is generally Gaussian over most continental surfaces, is analysed for the P- and X-band radar DEMs at the same scale (5 m pixel).

Area C is used to assess the ability of the P-band radar DTM to model the relief of a floodplain, assumed horizontal, where vegetation potentially forms shapes that can be misinterpreted as relief. In this case, the reference is implicit, since the surface is assumed to be horizontal. The DTM is also compared to other available models.

2.3. Requirements related to hydrography

The quality of a DTM has an impact on the extracted hydrographic information, such as the 2D drainage network, so that the network extracted automatically from the DTM may be used as an indicator of DTM quality, as illustrated in figure 2. Indeed, the terrestrial reliefs shaped by water do not have a random behaviour. They respect some universal rules. The non-compliance of these rules may reveal errors in the DTM (Polidori *et al.* 2014). It may be noted in figure 2 that the detection or unrealistic shapes or impossible topology does not require reference data, while an absolute location error or a wrong topology can only be revealed by comparison with a reference drainage network.

One of the rules is the fact that all rivers flow downhill. The respect of this rule is verified by assessing the number and depth of the sinks detected along the drainages. A reasonable requirement is that the mean depth is statistically lower that the overall accuracy of the DTM, ensuring that the longitudinal profile of the rivers is well preserved despite the limitation of the DTM accuracy.

Another rule most often verified in terrestrial relief is that landforms shaped by water have a fractal behaviour (Rodriguez-Iturbe and Rinaldo, 2001). In this study, the DTMs are analysed to verify that they respect Horton's law, according to which the logarithm of the number of drainage sections with Strahler order N decreases linearly when N increases.

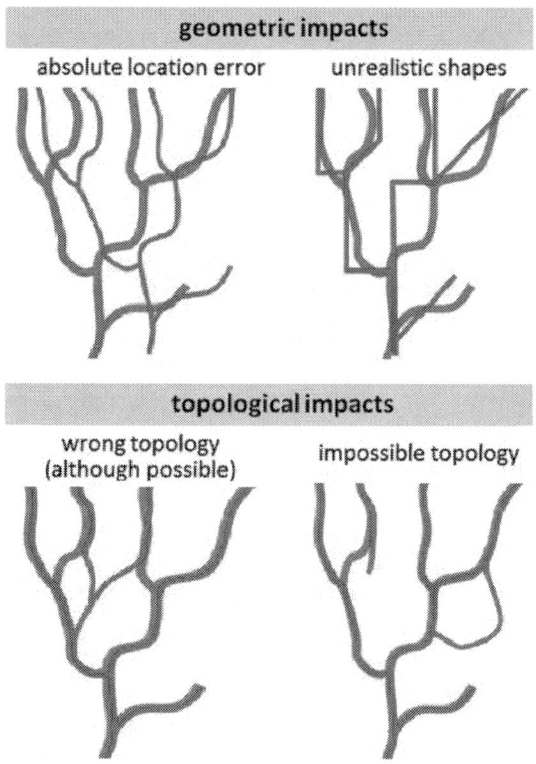

Figure 2. Effects of main DTM error behaviours on extracted drainage errors, with ground truth in blue and extracted network in red (adapted from Polidori and El Hage, 2020)

2.4. Influence of scale

The resampling period, i.e., the scale at which the relief is modelled, has an impact on slope accuracy (Polidori and Simonetto, 2014) and therefore on the respect of hydrographic shapes. Therefore, the hydrographic indicators based on slope are unstable when the scale is changed (Santos et al., 2017).

For example, the Gravelius index, which characterises the shape of a watershed in terms of compactness, depends on both the area (which remains almost stable when changing scale) and the perimeter (which increases and may diverge when the scale becomes more detailed) :

$$K_G = \frac{P}{2\sqrt{\pi A}} = 0.28\frac{P}{\sqrt{A}} \quad (1)$$

where P is the perimeter and A is the area. Consequently, the Gravelius index is expected to be unstable with regards to scale. The distribution of Gravelius indices obtained at two different scales (5 and 30 m) are compared in order to analyse the effect of scale, for both X- and P-bands.

3. RESULTS

A preliminary visual analysis shows the similarity of the DTM obtained from P-band radar with regards to the reference lidar DTM, though with less hydrographic details due a coarser resolution. It is clearly a DTM, as compared with the X-band radar DSM extracted from the same data base, and with SRTM, which both describe the upper surface of the canopy as can be observed in figure 3.

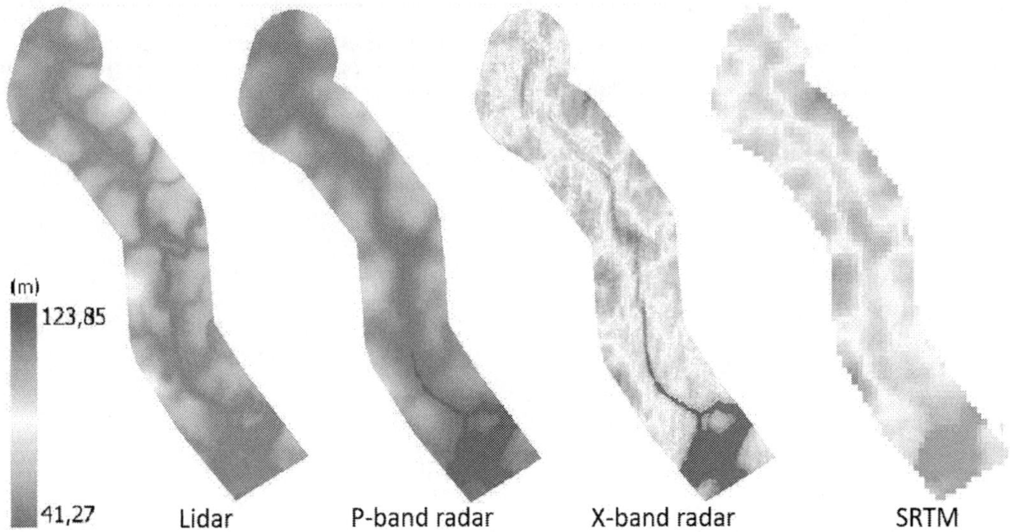

Figure 3. Comparison of the P-band radar DTM with the reference lidar DTM and other DEMs in color hypsometry

In terms of absolute accuracy, ongoing studies carried out in areas A and B show that the RMS error ranges from 2 to 3 m for elevations, with no significant bias, and from 7 to 15 degrees for slopes. However, this error is not homogeneous, and results obtained in area A show that it is a function of the local slope, the surface orientation with regards to the radar viewing direction, and the incidence angle. Indeed, both elevation and slope RMS errors increase as the local slope increases, they are more important on slopes facing away from the radar, and they increase when the incidence angle decreases (El Hage et al., 2022).

In area B, it can be observed that in the DSM obtained from X-band radar, besides the elevation bias due to the fact that short wavelengths do not penetrate the vegetation, the slopes have a very different statistical distribution as compared with the DTM obtained from P-band radar (figure 4). They have a much higher mean and standard deviation, because of the irregular geometry of the trees at this resolution (5 m), as confirmed by the profile shown in figure 6. This difference disappears when the resolution is significantly degraded, as shown in a previous study (Polidori and Simonetto 2014), suggesting that it is at high resolution that the superiority of the P-band radar DTM is more obvious.

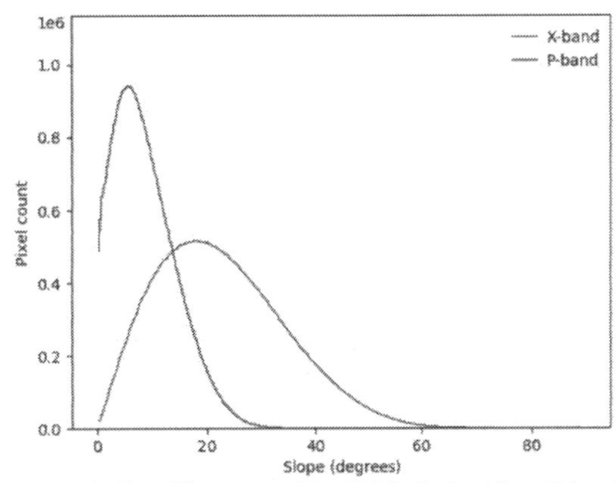

Figure 4. Slope histograms of the DEMs obtained from P-band and X-band radar in area B (adapted from Ećo 2022)

https://doi.org/10.5194/isprs-archives-XLVIII-M-1-2023-249-2023 | © Author(s) 2023. CC BY 4.0 License.

Floodplain areas present an additional difficulty, since non-topographic objects (like vegetation) and various error sources have a much larger impact on elevation measurements than the natural variations in ground and water elevation. In area C, figure 5 shows that the P-band radar DTM – profile A – clearly describes the flat horizontal shape of the floodplain, as well as a small hill on the right side. It is very close to the ground and water surface since it is almost not affected by the presence of vegetation. Indeed, the vegetation identified in optical imagery has no impact on the elevation in the P-band radar DTM. On the contrary, all other models create ghost shapes due to vegetation, with errors around 5-10 m in the case of shrub vegetation and 20-25 m in the case of dense forest (with the exceptions of the X-band radar DEM – profile B – which is explicitly designed to provide a DSM), while ASTER GDEM – profile D – exhibits gross altimetric errors whatever the land cover.

Figure 6 compares several DEMs with the reference lidar DTM along a tributary of the Oiapoque River (area B), showing that the P-band radar DTM meets two essential quality requirements for hydrological modelling. On the one hand, the similarity of this model with the reference. On the other hand, the fact that the river flows downhill, the presence of depressions being indicative of local elevation errors. Similarly, a quantitative analysis performed in area A showed that depressions affect only 2% of the DTM with a mean of 0.38 m and a standard deviation of 0.60 m, much below the DTM elevation error.

Figure 6 also shows that SRTM and Topodata, with a 30 m resolution and based on a short radar wavelength which is unable to penetrate through the vegetation layer, only describe the upper surface of the trees. The X-band radar DSM has the same behaviour in the case of dense forest due to its short wavelength, but the higher resolution allows the radar signal to reach the water surface when it is not obscured by trees, resulting in the same elevation as in the P-band radar DTM.

Horton's law is verified in the data of area A with an almost perfect linearity (R^2 = 0,99832), which means that the hydrographic network extracted from this DTM has a fractal behaviour. Similar results are found in other areas. This confirms the hydrographic coherence of the DTM obtained by P-band radar.

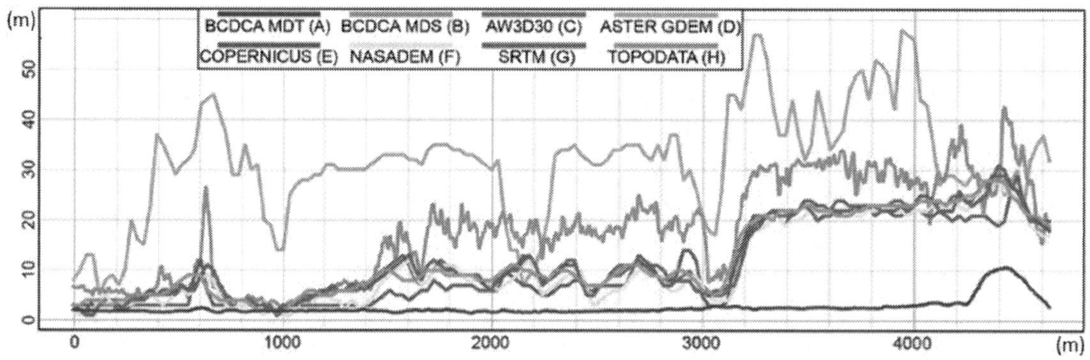

Figure 5. Elevation profiles of different DEMs in a floodplain in area C. The black profile (A) representing the P-band radar DTM analysed in this study is compared with the X-band radar DSM (B), the ALOS WORLD 3D DEM (C), ASTER GDEM v3 (D), the Copernicus DEM (E), The NASADEM (F), SRTM (G) and Topodata (H)

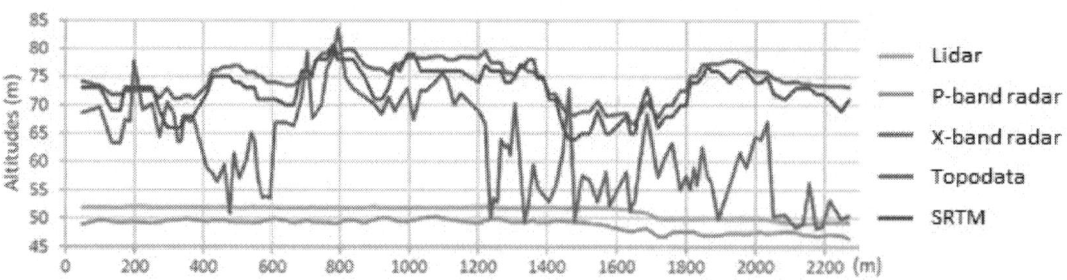

Figure 6. Elevation profiles of DEMs along a river in area B.

The comparison of Gravelius compactness indices computed in the P-band radar DTM and in the X-band radar DSM shows that watersheds are statistically more compact in the DTM, as shown in figure 7. This difference is due to the fact that for a given area, the perimeter is longer in the DSM, as a consequence of the small ghost shapes created by the trees at 5 meters. This leads to a smaller value of K_G in the P-band radar DTM according to equation (1).

This difference between X and P-band is reduced when the two models are subsampled fron 5 to 30 meters, as the smallest shapes tend to disappear. Consequently, the calculated perimeter becomes shorter in both DEMs, leading to more compact basins. This confirms that the scale of the model influences the description of hydrography and therefore should be specified carefully.

https://doi.org/10.5194/isprs-archives-XLVIII-M-1-2023-249-2023 | © Author(s) 2023. CC BY 4.0 License.

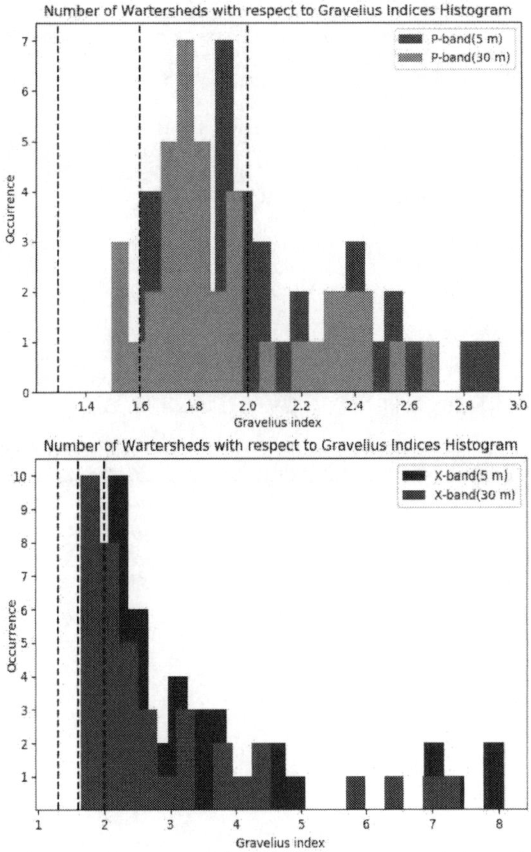

Figure 7. Histograms of the Gravelius compactness indices of drainage basins extracted from P-band radar DTM (top) and X-band radar DSM (bottom) in area B at 5 and 30 m.

CONCLUSION

The results of this multi-criteria analysis confirm the potential of DTMs obtained by interferometric or tomographic processing of P-band airborne radar data for the 3D characterisation of hydrography in tropical forests. On the one hand, their accuracy is around 2-3 meters with no significant biais, which is much better than that of most available DEMs, produced by photogrammetry or by short wavelength radar interferometry, which in fact describe the canopy. On the other hand, they account for the geometric characteristics of continental hydrography, such as the longitudinal profile of rivers and the fractal behaviour of the drainage network. This assessment is made possible by the fact that the landforms shaped by water obey certain rules that can be easily checked, thus allowing the hydrographic consistency of DTMs to be controlled.

Beyond this overall assessment of the quality of P-band radar DTMs, ongoing studies ae revealing that the error is not homogeneous, with local variations depending on the slope, azituth and local incidence angle. Further work should also consider the possibility to improve a raw DTM by taking into account the known properties of the terrestrial hydrography.

These satisfactory results are due to both the physical properties of P-band radar waves, which penetrate through dense forest cover and reach the ground, and the availability of rigorous algorithms for interferometric and tomographic processing of radar images. This makes DTMs obtained from P-band airborne radar suitable for hydrology.

The studies that contributed to this paper are ongoing, and are expected to contribute both to improving the data used in hydrology, and to using the known behaviour of water on continental surfaces to improve the production and validation of digital terrain models.

ACKNOWLEDGEMENTS

The work carried out on area A was funded by CNES (France) as part of the TOSCA programme within the framework of the CALVAL experiments of the ESA Biomass mission. The work carried out on zones B and C benefited from study grants awarded by IGN (France) and CAPES (Brazil), respectively, and benefited from radar data made available by SEMA-AP (Macapá, Brazil) and lidar data made available by CTG (Cayenne, French Guiana), respectively.

REFERENCES

Abrams, M., Crippen, R., Fujisada, H., 2020. ASTER Global Digital Elevation Model (GDEM) and ASTER Global Water Body Dataset (ASTWBD). *Remote Sensing*, 12, 1156.

Bamler, R., 1999. The SRTM mission: a world-wide 30 m resolution DEM from SAR Interferometry in 11 days. In: Fritsch, D.; Spiller, R. (Ed.). *Photogrammetric Week 99*. Wichmann Verlag, Heidelberg, 145-154.

Correia, A.H., 2011. Metodologias e resultados preliminares do projeto Radiografia da Amazônia. *Anais XV Simpósio Brasileiro de Sensoriamento Remoto*, INPE, Curitiba, p.8083-8090 (http://marte.sid.inpe.br/col/dpi.inpe.br/marte/2011/06.27.19.46/doc/p1032.pdf). Accessed on 15 Jan 2023.

Dubois-Fernandez, P., Oriot, H., Coulombeix, C., Cantalloube, H., Plessis, O.R., Le Toan, T., Daniel, S., Chave, J., Blanc, L., Davidson, M., 2010. TropiSAR, a SAR Data Acquisition Campaign in French Guiana. *Proc. 8th European Conference on Synthetic Aperture Radar*, 1-4.

Eĉo, N., 2022. *The Study of the Potential of P-band Radar for Tropical Forest Relief Mapping–French Guyana*. Master Internship Report, Université de Paris (Paris, France).

El Hage, M., Villard, L., Huang, Y., Ferro-Famil, L., Koleck, T., Le Toan, T., Polidori, L., 2022. Multicriteria Accuracy Assessment of Digital Elevation Models (DEMs) Produced by Airborne P-Band Polarimetric SAR Tomography in Tropical Rainforests. *Remote Sensing*, 14, 4173, https://doi.org/10.3390/rs14174173.

Guimarães Filho, A.G., Borba, P., 2020. Methodology for land mapping of Amapá state – a special case of the Amazon Radiography project. *Proc. IGARSS'2020*, 1540-1543.

Kraus, K.; Pfeifer, N. 1998. Determination of terrain models in wooded areas with airborne laser scanner data. *ISPRS Journal of Photogrammetry and Remote Sensing*, 53, 193-203.

Le Toan, T., Quegan, S., Davidson, M.W.J., Balzter, H., Paillou, P., Plummer, S., Rocca, F., Saatchi, S., Shugart, H., L., Ulander,

L., 2011. The BIOMASS mission: mapping global forest biomass to better understand the terrestrial carbon cycle. *Remote Sensing of Environment,* 11, 2850-2860.

Mariotti D'Alessandro, M., Tebaldini, S. 2019. Digital terrain model retrieval in tropical forests through P-Band SAR tomography. *IEEE Transactions on Geoscience and Remote Sensing*, 57, 6774-6781.

Polidori, L., Simonetto, E., 2014. Effect of scale on the correlation between topography and canopy elevations in an airborne InSAR product over Amazonia. *Procedia Technology*, 16, 180-185.

Polidori, L., El Hage, M., Valeriano, M.M., 2014. Digital elevation model validation with no ground control; application to the Topodata DEM in Brazil. *Boletim de Ciências Geodésicas*, 20, 467-479.

Polidori, L., El Hage, M. 2020. Digital elevation model quality assessment methods: A critical review. *Remote Sensing*. 12(21), 3522.

Polidori, L., Caldeira, C.R.T., Smessaert, M., El Hage, M., 2022. Digital Elevation Modeling through Forests: The Challenge of the Amazon. *Acta Amazonica*, 52, 69-80, doi:10.1590/1809-4392202103091.

Reis, L. N., 2023. *Caracterização morfológica de planícies inundáveis na costa do Amapá: uma análise a partir de modelos digitais de elevação.* Master Internship Report, Universidade Federal do Pará (Belém, Brazil).

Rodriguez-Iturbe, I., Rinaldo, A., 2001. *Fractal River Basins, Chance and Self-Organization*. Cambridge University Press, Cambridge, 570 p.

Santos, V.C. El Hage, M., Polidori, L., Stevaux, J.C., 2017. Effect of digital elevation model mesh size on geomorphic indices: a case study of the Ivaí river watershed – state of Paraná, Brazil. *Boletim de Ciências Geodésicas*, 23, 684-699.

Valeriano, M.M., Rossetti, D.F., 2012. Topodata: Brazilian full coverage refinement of SRTM data. *Applied Geography,* 32, 300.

ASSESSMENT OF CROPLAND TRANSITION TO SOLAR FARMS AND OTHER LAND USE/COVER USING RS, GIS AND ANN-CA

J. A. Principe [1]*, M. E. Sotto [2], I. B. Benitez [2]

[1] Dept. of Geodetic Engineering, College of Engineering, University of the Philippines Diliman, Philippines - japrincipe@up.edu.ph
[2] National Engineering Center, University of the Philippines Diliman, Philippines - (mesotto, ibbenitez)@up.edu.ph

KEY WORDS: cropland, solar farm, ANN-CA, MOLUSCE, MCD12Q1, nighttime lights, LandScan

ABSTRACT:

The balance between food security and energy security is a national issue of extreme importance. A more stable supply of electricity could be achieved as solar farms expand but at the expense of losing some of the prime agricultural lands which endangers availability of sufficient agricultural produce. This study aims to use ANN-Cellular Automata (CA) via the Geographic Information System (GIS) platform and remote sensing (RS) data to assess the impact of cropland transition to solar farms and other land use/land cover (LULC). Several remotely sensed data were processed including MODIS land cover data (MCD12Q1), VIIRS nighttime lights (VNL v2.1), Advanced Himawari Imager Shortwave Radiation (AHI-SWR) product, and population density (LandScan) as inputs to the Cellular Automata-Artificial Neural Network (CA-ANN) model to simulate LULC changes in Tarlac Province, Philippines via the Modules for Land Use Change Evaluation (MOLUSCE) plugin in QGIS. For years 2019, 2023 and 2027 with 2015 as the base year, results showed an increasing trend for savannas and grassland with ΔLULC values of +11.4% to +15.1% and +0.2% to 3.5%, respectively. Meanwhile, a decreasing trend is observed for built-up/water, forest, and cropland with ΔLULC values of -3.0% to -6.3%, -8.5% to -21.1%, and -3.9% to -4.2%, respectively. Results also showed a conversion of a 100-ha area of croplands to solar farm from year 2019 to 2023 which translates to an estimated monetary loss from agricultural produce due to solar farm conversion amounting to Php 7,584,720.00 (~USD 138,000) which is equivalent to the total average annual income of about 67 families in Tarlac. Lastly, the simulated 2027 LULC map showed pixels with unrealistic conversions from solar farm (year 2023) to cropland (year 2027). To improve the model, it is recommended to add more spatial data to effectively capture factors that may contribute to the expansion of solar farms in the future. Moreover, high resolution LULC maps (vector maps if available) can be used instead of a course resolution satellite-derived raster data. Nonetheless, this study has demonstrated the use of RS, GIS and machine learning techniques to model cropland conversion to solar farms and other LULC classes. Results from this study can provide scientific data to policy makers, solar industry players and other relevant stakeholders in doing technoeconomic assessment of solar farm development and expansion considering its effect on energy security and food security towards national sustainable development.

1. INTRODUCTION

Food security and energy security are two of the most pressing global issues of today. The United Nations Sustainable Development Goals (SDGs) seek to address these concerns by providing goals targeting to end all forms of poverty everywhere, achieve food security, and guarantee everyone's access to affordable, reliable, sustainable, and modern energy (United Nations, 2015). To meet the growing demand for energy, countries around the world have been utilizing renewable sources of energy. However, conversion of agricultural lands to solar power plants poses serious concerns with regards to food security. It is for this reason that the Philippine Department of Agrarian Reform (DAR) proposed a two-year moratorium on land conversions (Gomez, E. J., 2019). Therefore, there is a need to assess how much agricultural areas have been converted to different uses (e.g., energy use) and how it can negatively affect food security while trying to meet legitimate needs of the nation (e.g., energy supply).

The integration of remotely-sensed data and geographic information system (GIS) has become a powerful tool in land use/land cover (LULC) change analysis for different applications including, but not limited to, its impact on vegetation cover, urban expansion, deforestation, food security and energy security (Hussain, S. et al., 2020; Guo, L. et al., 2021; Hassan, Z. et al., 2016; de Luna, A. D. et al., 2021). Meanwhile, machine learning algorithms such as artificial neural networks (ANN) have been extensively used for LULC change studies with satellite data as important inputs (Muhammad, R. et al., 2022; Kamaraj, M. and Rangarajan, S., 2022; Megahed, Y. et al., 2015).

The Advanced Himawari Imager (AHI-8) is a multipurpose moderate resolution geostationary imager which operators across the 16 channels from the visible to the infrared regions with wavelengths of 470 nm to 13.30 μm and bandwidth of 20 nm to 0.37 μm (JAXA-EORC, 2015; Bessho et al., 2016). Shortwave Radiation / Photosynthetically Available Radiation (SWR/PAR) product of AHI-8 is derived from a model proposed by Frouin and Murakami (2007) which uses plane-parallel radiative transfer theory (Frouin and Murakami, 2007; Takenaka et al., 2011). Studies of Yu et al (2018) and Zhang et al (2022) showed that AHI-8 product exhibits the highest level of accuracy when validated with ground data and compared to other satellite data such as MERRA-2, CERES-SYN and ERA-Interim.

Machine learning (ML) techniques such as random forest (RF), neural networks (NN), support vector machine (SVM), maximum likelihood classifier (MLC), decision trees, and K-Nearest Neighbors (KNN) (Gislason et al., 2006; Svoboda et al, 2022; Kulkarni and Lowe, 2016; Storie and Henry, 2018; Alshari et al., 2023; Taati, A. et al. 2015; Norovsuren, B. et al., 2019; Friedl and Brodley, 1997; Srimani and Prasad, 2012; Upadhyay et al., 2016) have been used for land use/land cover

This contribution has been peer-reviewed.
https://doi.org/10.5194/isprs-archives-XLVIII-M-1-2023-255-2023 | © Author(s) 2023. CC BY 4.0 License.

mapping due to its high accuracy and efficiency. Moreover, NN and Artificial Neural Networks (ANN) are often used interchangeably on machine learning applications. ANN demonstrated to be particularly effective in classifying different types of land use/land cover and has been used in various studies for mapping land use/cover, including the use of remotely sensed data (Alshari et al., 2023; Mishra et al., 2018; Kadavi and Lee, 2018; Rojas et al., 2020; Dede et al., 2022). A study by Kadavi and Lee (2018) found that ANN outperformed SVM. The same observation was supported by Ge et al (2020), where ANN had the highest accuracies compared to other three machine learning algorithms (RF, KNN, SVM).

To date, there has been no significant study done yet to quantify and model the expansion of solar farms using machine learning, RS and GIS techniques. As such, this study aims to use ANN-Cellular Automata (CA) via the GIS platform and remote sensing data to assess the transition of croplands to solar farms and other LULC. Results of this study can provide scientific information to relevant stakeholders in both the agriculture and energy sectors with regards to formulating policies and plans to achieve food security and energy security.

2. STUDY AREA

The study area covers the city of Tarlac, and municipalities of Capas, Concepcion, Gerona, La Paz, Pura and Victoria (Fig. 1). Two large solar power plants are currently operating in the area namely, the PetroSolar/Tarlac City and Concepcion Solar Power Plants. Tarlac province is home to more than 107,000 hectares of agricultural farms as of 2002 (NSO, 2012).

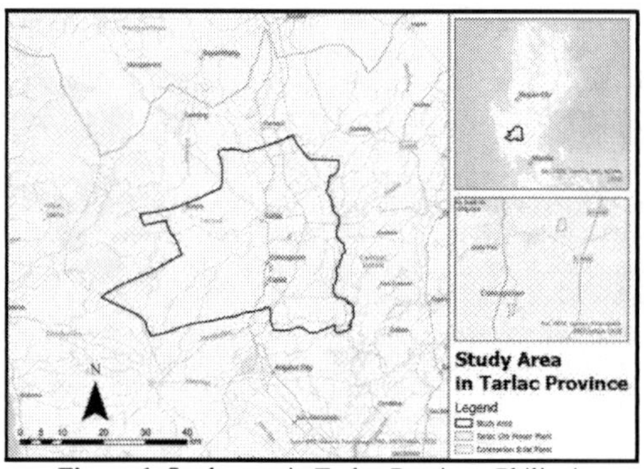

Figure 1. Study area in Tarlac Province, Philippines.

3. METHODOLOGY

3.1 Data

Three satellite products are used in this study summarized in Table 1.

Data	Description	Resolution	Year
MCD12Q1	Land cover types	500m	2015, 2019
VNL (v2.1)	Nighttime lights	500m	2019
AHI-SWR	Shortwave radiation	5 km	2019
LandScan	Population density	1 km	2019

Table 1. Satellite data used in this study.

MCD12Q1 Land Cover Type Product is a global land cover dataset at 500-m spatial resolution generated by performing a supervised classification on Moderate Resolution Imaging Spectroradiometer (MODIS) reflectance data. The said product was downloaded from the Land Processes Distributed Active Archive Center (LP DAAC) of NASA's Earth Observing System Data and Information System (EOSDIS) (L.P. DAAC, 2019). Five LULC classes are defined in this study: built-up/water (no solar PV installation areas), forest, savanna, cropland, grassland, and solar power plants (SPP).

Meanwhile, the version 2.1 of the annual global Visible and Infrared Imaging Suite (VIIRS) nighttime lights (VNL) was an upgrade of Version 2, corrected for a bug in calculating for the annual average radiance filtering technique (Elvidge et al, 2021). VNL data was downloaded from the Earth Observation Group of Payne Institute for Public Policy (https://eogdata.mines.edu/products/vnl).

Lastly, the shortwave radiation (SWR) is a product from Advanced Himawari Imager (AHI) 8/9 (AHI-SWR) and downloaded via the JAXA Himawari Monitor (eorc.jaxa.jp/ptree/index.html). The AHI-SWR has a spatial resolution of 5 km and temporal resolution ranging from 10-minutes to 1-month for the full disk extent. Each raster contains multiple bands which includes data on the total atmosphere optical thickness, total atmosphere angstrom exponent, photosynthetically active radiation, shortwave radiation, ultraviolet-A radiation, ultraviolet-B radiation (JAXA-EORC, 2015; Bessho et al., 2016). Level 3 monthly data of the AHI-8 SWR/PAR product is derived from the hourly data at each pixel. Raster Calculator from the Raster analysis tools in QGIS was then used in this study to process the annual SWR data of the Philippines from the AHI-8 L3 SWR monthly NetCDF rasters. The resolution, spatial extent and output coordinate reference system were copied and calculated from the existing raster data.

3.2 Methods

The general methodology used in this study is shown in Fig. 2. Satellite data were downloaded and resampled to 500-m spatial resolution and clipped to the study area's boundary. Pixels in the 2019 land cover dataset were manually edited to reflect the actual location of solar farms via the Serval plugin in QGIS and later assigned a separate pixel value (i.e., "5"). The MOLUSCE plugin in QGIS was then launched. The Modules for Land Use Change Evaluation (MOLUSCE) is an open-source plugin for QGIS 2.0 and above, developed by Asia Air Survey Co., Ltd. designed to analyse, model, and simulate LULC changes (Gismondi, M. et al., 2014). The 2015 and 2019 LC data were assigned as "Initial" and "Final" LULC dataset, respectively, while data on nighttime lights, population density and shortwave radiation were assigned as "Spatial variables". Next is the evaluation of correlation among the input variables, computation for area changes, transition potential modelling using ANN (multi-layer perceptron) method, and CA simulation.

Cellular automata (CA) is a mathematical model consisting of a network of cells, each with a state that changes over time according to a set of rules (Wolfram, 1983). CA is used in land use/land cover studies to simulate changes in territory cover over time due to variables like population increase, urban development, and climate change (Guzman et al., 2020; Pratomoatmojo, 2018; Pinto et al., 2021). Moreover, studies combined cellular automata-artificial neural networks, termed as the CA-ANN method, with the ANN predicting the possible outcomes of transitions in the CA model (Norizah et al., 2023;

This contribution has been peer-reviewed.
https://doi.org/10.5194/isprs-archives-XLVIII-M-1-2023-255-2023

Yang at al., 2016; Sajan et al., 2022). This method is known to be useful for forecasting changes in land use/land cover and can aid in long-term management and planning efforts.

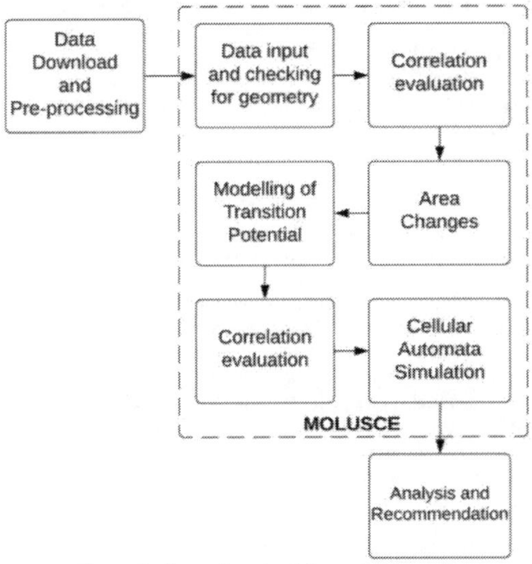

Figure 2. General methodology of this study.

The calibrated CA-ANN model was run to simulate the LULC distribution for the year 2023 using the input data for years 2015 and 2019. Since there is no available MCD12Q1 data yet for year 2023, the simulated LULC map was compared with latest high resolution satellite image from Google Earth with emphasis on the possible expansion of solar farms. The model was rerun to simulate a LULC map for the study area for year 2027 this time using input data for years 2019 and 2023.

Percent change in land use/cover for a particular year compared to the base year (i.e., 2015) was computed per LULC class using Eq. (1):

$$\Delta LULC_{i,j,k} = \frac{\sum Pixel_{k,j} - \sum Pixel_{k,i}}{\sum Pixel_{k,i}} \times 100\% \quad (1)$$

where $\Delta LULC_{i,j,k}$ is the percentage change in a LULC class k for a proceeding year j as compared to the base year i, $Pixel_{k,i}$ and $Pixel_{k,j}$ are the pixel values assigned to class k for years i and j, respectively.

Moreover, Eq. (2) is used to compute for the estimated monetary loss from agricultural produce due to solar farm conversion:

$$P_{loss} = A_{loss} \times Y \times P_{farmgate} \times 1000 \quad (2)$$

where P_{loss} is the monetary equivalent of agricultural produce lost due to solar farm conversion, Y is the palay yield per hectare (tons/ha), and $P_{farmgate}$ is the average farmgate price of palay (dry) (Php/kg), and 1000 is the unit conversion factor for tons to kg.

Lastly, further analyses were done from which recommendations would be based.

4. RESULTS AND DISCUSSIONS

4.1 Correlation of data inputs and area changes

Among the three spatial variables, the nighttime lights (VNL) and population density (PD) had the highest Pearson's correlation value of 0.54, followed by shortwave radiation (SWR) and PD (0.21); while SWR and VNL had the lowest correlation (0.16).

Fig. 3 shows the cropland conversions to grassland, savanna and SPP corresponding to 0.9%, 5.5%, and 0.4% LC changes, respectively.

(a) (b) (c)

Figure 3. Cropland conversions (2015-2019) to (a) grassland, (b) savanna, and (c) solar power plants.

4.2 Transition potential mapping and simulation

The transition potential modelling was done via the ANN multi-layer perceptron which was optimized with the following parameter values: neighbourhood = 1 px; learning rate = 0.010; maximum iterations = 100; hidden layers = 5; momentum = 0.060; current validation kappa = 0.82. The neural network learning curve is shown in Fig. 4a. Meanwhile, Fig. 4b shows the LULC transition for 2015-2019 (actual) and 2019-2023 (simulated). This simulation showed a further decrease in cropland areas to about 0.28% from 2019 or -4.39% from 2015.

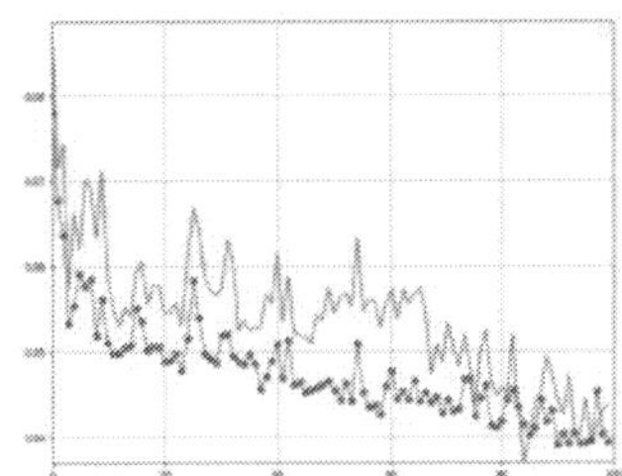

(a)

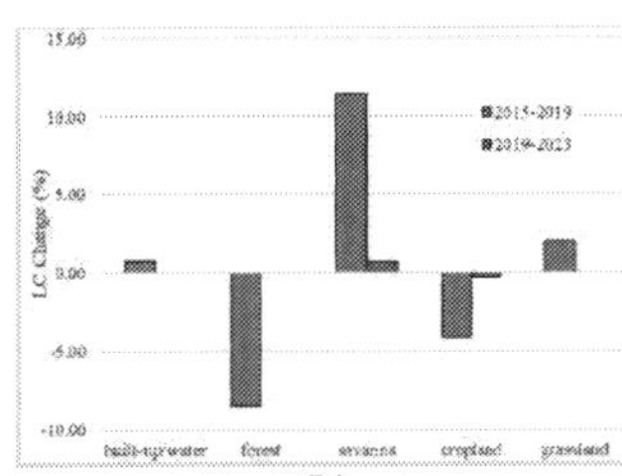

(b)

Figure 4. Result of transition potential modelling in MOLUSCE: **(a)** Neural network learning curve, **(b)** LULC transitions.

The LULC map for 2019 is shown in Fig. 5 together with the resulting maps from the model runs for the years 2023 and 2027. The said maps show a seemingly unchanged distribution of LULC classes through the time periods considered. However, upon closer inspection, new four pixels solar farms can be seen as having been added in year 2023 in a span of four years (i.e., 2019-2023). These new solar farms pixels can be translated to about 100 hectares of land which are previously croplands. Based on the latest satellite image from Google Earth Pro, these pixels are indeed located in the new solar farms called Concepcion Solar and Sta. Rosa Solar being operated by Solar Philippines Tarlac Corporation and Terasu Energy Inc which have been in commercial operation since February 2020 and January 2022, respectively (DOE, 2021; DOE, 2022).

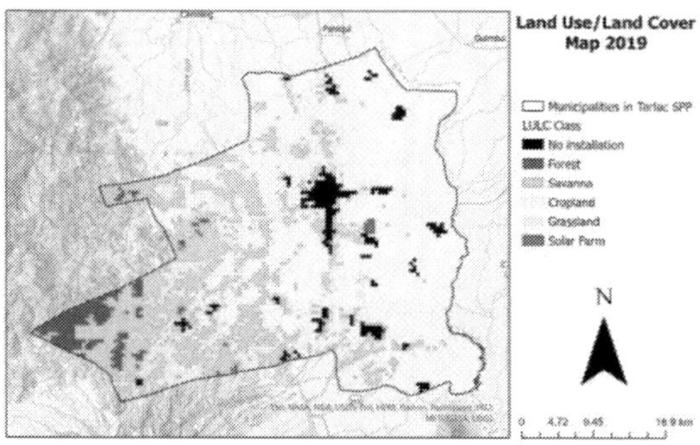

(a)

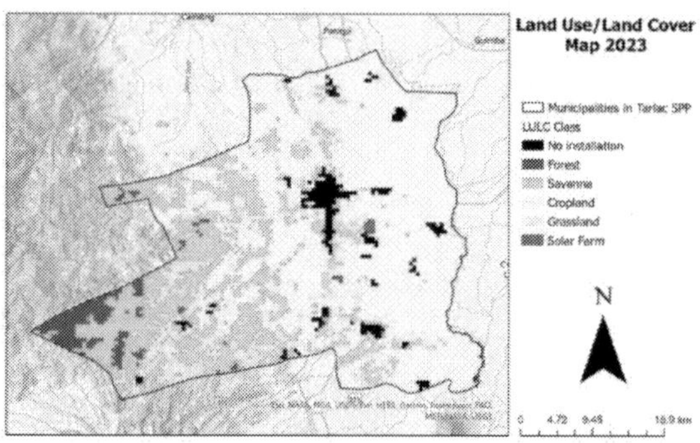

(b)

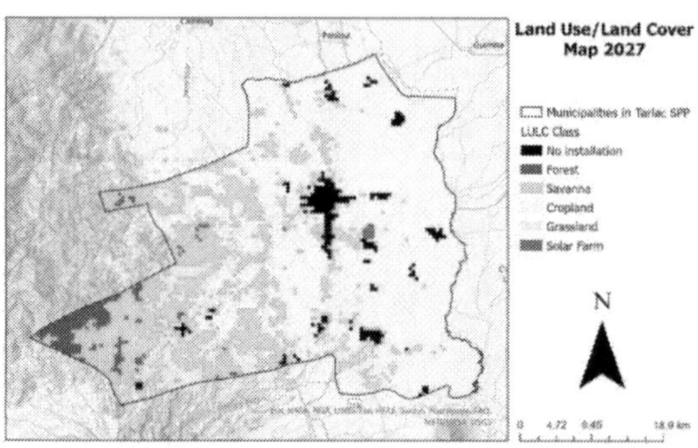

(c)

Figure 5. Land use/land cover map for the years **(a)** 2019, **(b)** 2023 and **(c)** 2027. The last two maps were simulated using the Cellular Automata in MOLUSCE plugin.

According to the Philippine Statistics Authority, the palay yield in Tarlac is 4.29 metric tons/ha as of the third quarter of 2022 while the average farmgate price of palay (dry) is at Php 17.68/kg as of December 2022 (PSA, 2022; PSA 2023). Assuming that the estimated 100-ha area that were previously planted with rice, Tarlac being one of the country's major rice-producing provinces (Mamiit et al, 2021), the estimated monetary loss from agricultural produce due to solar farm conversion is P_{loss} = Php 7,584,720.00 which is equivalent to the total average annual income of about 67 families in Tarlac province based on statistical data (NSO, 2008).

In the 2027 simulation of LULC, it is apparent that some of the solar farm pixels remain as is, but some were replaced by other LULC class including cropland (Fig. 6). Although it is possible, this conversion type (i.e., solar farm to cropland) is highly unlikely as evidenced by the latest satellite imagery of the same area. With this observation, a refinement of the model is deemed needed. It is recommended to add more spatial data as inputs to the CA-ANN model including road network, digital elevation model, energy transmission/distribution line network, among others. If historical and most recent LULC maps in shapefile format are available from the concerned government agency (e.g., such as the Philippine National Mapping and Resource Information Authority), these can be used as initial and final LULC maps as opposed to the course spatial resolution of MCD12Q1 raster data.

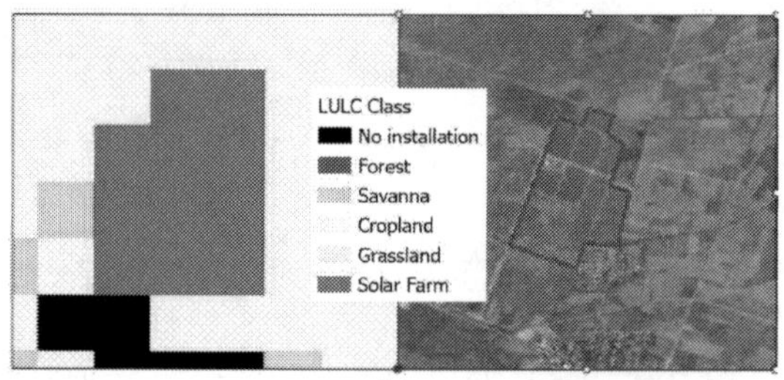

(a)

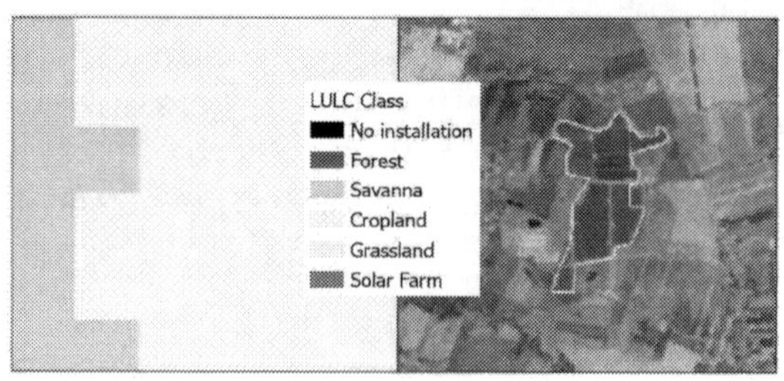

(b)

Figure 6. Predicted of LULC map for year 2027: **(a)** realistic prediction since solar farm remains as is; **(b)** unrealistic prediction since solar farm is converted back to cropland.

Fig. 7 shows the percentage of LULC changes for years 2019, 2023 and 2027 with 2015 as the base year. In general, there is an observed increasing trend for savannas and grassland with ΔLULC values of +11.4% to +15.1% and +0.2% to 3.5%, respectively. Meanwhile, a decreasing trend is observed for built-up/water, forest, and cropland with ΔLULC values of -3.0% to -6.3%, -8.5% to -21.1%, and -3.9% to -4.2%, respectively. As there are no savannas in the Philippines, the two classes—savannas and woody savannas—as defined in the

https://doi.org/10.5194/isprs-archives-XLVIII-M-1-2023-255-2023 | © Author(s) 2023. CC BY 4.0 License.

IGBP system can be attributed to the percent of canopy/vegetation cover with the latter having higher percentage than the former. As such, in dealing with these classes, their percentage of vegetation canopy cover can be used to indicate possible conversion to solar farms, with the decrease in canopy cover suggesting a high possibility of conversion from vegetation to solar farm.

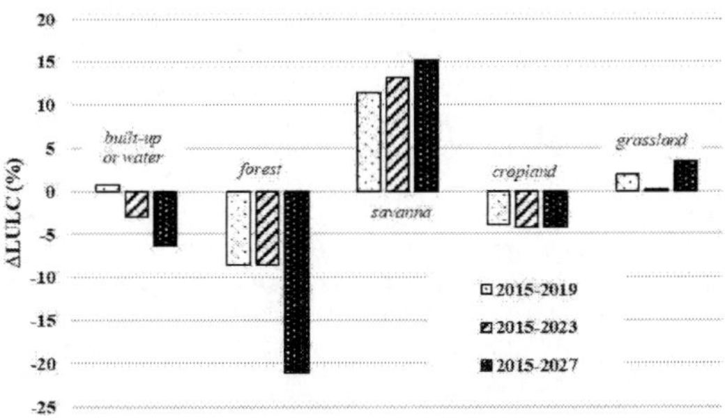

Figure 7. Percentage of LULC changes from 2015.

5. CONCLUSION AND RECOMMENDATION

This study utilized various remotely sensed products, GIS and machine learning techniques to assess the impact of cropland conversion to solar farms. Specifically, satellite-derived data on land cover types (MCD12Q1), nighttime lights (VNL v2.1), shortwave radiation (AHI-SWR), and population density (LandScan) were inputs to the MOLUSCE plugin in QGIS to simulate via the CA-ANN algorithm the LULC changes in Tarlac Province, Philippines. In terms of LULC change rate, it is observed that there is an increasing trend for savannas and grassland with ΔLULC values of +11.4% to +15.1% and +0.2% to 3.5%, respectively. Meanwhile, a decreasing trend is observed for built-up/water, forest, and cropland with ΔLULC values of -3.0% to -6.3%, -8.5% to -21.1%, and -3.9% to -4.2%, respectively. Results also showed a conversion of a 100-ha area of croplands to solar farm from year 2019 to 2023 which translates to an estimated monetary loss from agricultural produce due to solar farm conversion amounting to Php 7,584,720.00 (~USD 138,000) which is equivalent to the total average annual income of about 67 families in Tarlac. Lastly, the simulated 2027 LULC Map showed pixels with unrealistic conversions from solar farm (year 2023) to cropland (year 2027). It is therefore recommended to add more spatial data to effectively factors that may contribute to the expansion of solar farms in the future. Moreover, high resolution LULC maps (vector maps if available) can be used instead of a course resolution satellite-derived raster data. Nonetheless, this study has demonstrated the use of RS, GIS and machine learning techniques to model cropland conversion to solar farms. Results from this study can provide scientific data to policy makers, solar industry players and other relevant stakeholders of both agriculture and solar energy in doing technoeconomic assessment of solar farm development and expansion considering its effect on energy security and food security towards national sustainable development.

ACKNOWLEDGMENTS

This study was done through the FF Cruz Prof. Chair in Geodetic Engineering. The authors wish to thank the NASA's Earth Observing System Data and Information System (EOSDIS) for the MODIS Land Cover (MCD12Q1) product, the Earth Observation Group of Payne Institute for Public Policy for the VNL data, the Japan Aerospace Exploration Agency (JAXA) for the SWR data, and the University of the Philippines for the license to use ArcGIS Pro. Lastly, the DOST-Science Education Institute (DOST-SEI) Engineering Research and Development for Technology (ERDT) Program for the research dissemination grant.

REFERENCES

Alshari, E. A., Abdulkareem, M. B., and Gawali, B. W., 2023: Classification of land use/land cover using artificial intelligence (ANN-RF), *Front. Artif. Intell.*, vol. 5. https://www.frontiersin.org/articles/10.3389/frai.2022.964279 (Feb. 27, 2023).

Bessho, K., Date, K., Hayashi, M., Ikeda, A., Imai, T., Inoue, H., Kumagai, Y., Miyakawa, T., Murata, H., Ohno, T., Okuyama, A., Oyama, R., Sasaki, Y., Shimazu, Y., Shimoji, K., Sumida, Y., Suzuki, M., Taniguchi, H., Tsuchiyama, H., Uesawa, D., Yokota, H., and Yoshida, R., 2016: An Introduction to Himawari-8/9— Japan's New-Generation Geostationary Meteorological Satellites. *Journal of the Meteorological Society of Japan*, 94, pp. 151-183.

Dede, M., Asdak, C. and Setiawan, I., 2022: Spatial dynamics model of land use and land cover changes: A comparison of CA, ANN, and ANN-CA, *Regist. J. Ilm. Teknol. Sist. Inf.*, vol. 8, no. 1, Art. no. 1. doi: 10.26594/register.v8i1.2339.

de Luna, A. D., Pascual, C. E. B., Principe, J. A., and Ang, M. R. C. O. (2021). "COST-BENEFIT ANALYSIS OF CONVERTING AGRICULTURAL LAND INTO SOLAR FARM USING RS & GIS: CASE OF TARLAC PROVINCE," Int. Arch. Photogramm. Remote Sens. Spatial Inf. Sci., XLVI-4/W6-2021, pp. 133–140; https://doi.org/10.5194/isprs-archives-XLVI-4-W6-2021-133-2021.

DOE, 2021. LIST OF EXISTING POWER PLANTS (GRID-CONNECTED) AS OF DECEMBER 2020. Department of Energy. https://www.doe.gov.ph/list-existing-power-plants (25 February 2023).

DOE, 2022. LIST OF EXISTING POWER PLANTS (GRID-CONNECTED) AS OF NOVEMBER 2022. Department of Energy. https://www.doe.gov.ph/list-existing-power-plants (25 February 2023).

Elvidge, C.D, Zhizhin, M., Ghosh, T., Hsu, FC, and Taneja, J., 2021: Annual time series of global VIIRS nighttime lights derived from monthly averages: 2012 to 2019. *Remote Sensing* 13(5), pp. 1-14.

Friedl, M. A., and Brodley, C. E., 1997: Decision tree classification of land cover from remotely sensed data, *Remote Sens. Environ.*, vol. 61, no. 3, pp. 399–409. doi: 10.1016/S0034-4257(97)00049-7.

Frouin, R. and Murakami, H., 2007: Photosynthetically Available Radiation Algorithm: Estimating photosynthetically available radiation at the ocean surface from ADEOS-II global imager data. *J. Oceanography*, 63, pp. 493-503.

Ge, G., Shi, Z., Zhu, Y., Yang, X., and Hao, Y., 2020: Land

use/cover classification in an arid desert-oasis mosaic landscape of China using remote sensed imagery: Performance assessment of four machine learning algorithms, *Glob. Ecol. Conserv.*, vol. 22, p. e00971. doi: 10.1016/j.gecco.2020.e00971.

Gislason, P. O., Benediktsson, J. A. and Sveinsson, J. R., 2006: Random Forests for land cover classification. *Pattern Recognit. Lett.*, vol. 27, no. 4, pp. 294–300. doi: 10.1016/j.patrec.2005.08.011.

Gismondi, M., Kamusoko, C., Furuya, T. et al., 2014: MOLUSCE an open source land use change analyst for QGIS. Asia Air Survey Co., Ltd, Tokyo.

Gomez, E.J., 2019. Land conversion memo out next month–DAR.https://www.manilatimes.net/2019/02/21/business/business-top/land-conversion-memo-out-next-month-dar/514634/(23 Feb. 2021).

Guo, L., Xi, X., Yang, W., Liang, L., 2021: Monitoring Land Use/Cover Change Using Remotely Sensed Data in Guangzhou of China. *Sustainability* 13, 2944. https://doi.org/10.3390/su13052944

Guzman, L. A., Escobar, F., Peña, J. and Cardona, R., 2020: A cellular automata-based land-use model as an integrated spatial decision support system for urban planning in developing cities: The case of the Bogotá region, *Land Use Policy*, vol. 92, p. 104445. doi: 10.1016/j.landusepol.2019.104445.

Hassan, Z., Shabbir, R., Ahmad, S.S. et al., 2016: Dynamics of land use and land cover change (LULCC) using geospatial techniques: a case study of Islamabad Pakistan. *SpringerPlus* 5(812). https://doi.org/10.1186/s40064-016-2414-z

Hussain, S., Mubeen, M., Akram, W. et al., 2020: Study of land cover/land use changes using RS and GIS: a case study of Multan district, Pakistan. *Environ Monit Assess* 192(2). https://doi.org/10.1007/s10661-019-7959-1.

JAXA-EORC, 2015. JAXA Himawari Monitor P-Tree System User Guide. Japan Aerospace Exploration Agency, Earth Observation Research Center. https://www.eorc.jaxa.jp/ptree/userguide.html (27 February 2023).

Kadavi, P. R. and Lee, C.W., 2018: Land cover classification analysis of volcanic island in Aleutian Arc using an artificial neural network (ANN) and a support vector machine (SVM) from Landsat imagery, *Geosci. J.*, vol. 22, no. 4, pp. 653–665. doi: 10.1007/s12303-018-0023-2.

Kamaraj, M. and Rangarajan, S., 2022: Predicting the future land use and land cover changes for Bhavani basin, Tamil Nadu, India, using QGIS MOLUSCE plugin. *Environ Sci Pollut Res.* https://doi.org/10.1007/s11356-021-17904-6.

Kulkarni, A. D., and Lowe, B., 2016: Random Forest Algorithm for Land Cover Classification," *Int. J. Recent Innov. Trends Comput. Commun.*, vol. 4, no. 3.

L.P. DAAC, 2019. MCD12Q1 MODIS/Terra+Aqua Land Cover Type Yearly L3 Global 500m SIN Grid V006 [Data set]. https://doi.org/10.5067/MODIS/MCD12Q1.006.

Mamiit, R. J., Yanagida, J., and Miura, T., 2021: Productivity Hot Spots and Cold Spots: Setting Geographic Priorities for Achieving Food Production Targets. *Front. Sustain. Food Syst., 5*, pp.1-15. doi: https://doi.org/10.3389/fsufs.2021.727484.

Megahed, Y., Cabral, P., Silva, J., and Caetano, M., 2015: Land Cover Mapping Analysis and Urban Growth Modelling Using Remote Sensing Techniques in Greater Cairo Region—Egypt. *ISPRS Int. J. Geo-Inf.*, 4, 1750-1769. doi:10.3390/ijgi4031750.

Mishra, V. N., Rai, P. K., Prasad, R., Punia, M., and Nistor, M. M., 2018: Prediction of spatio-temporal land use/land cover dynamics in rapidly developing Varanasi district of Uttar Pradesh, India, using geospatial approach: a comparison of hybrid models, *Appl. Geomat.*, vol. 10, no. 3, pp. 257–276. doi: 10.1007/s12518-018-0223-5.

Muhammad, R., Zhang,W., Abbas, Z., Guo, F., Gwiazdzinski, L., 2022: Spatiotemporal Change Analysis and Prediction of Future Land Use and Land Cover Changes Using QGIS MOLUSCE Plugin and Remote Sensing Big Data: A Case Study of Linyi, China. *Land* 11 (419). https://doi.org/10.3390/land11030419.

Norizah, K., Jamhuri, J., Balqis, M., Mohd Hasmadi, I. and Nor Akmar, A. A., 2023: Land Use and Land Cover Change Prediction Using ANN-CA Model," in *Tropical Forest Ecosystem Services in Improving Livelihoods For Local Communities*, Z. Samdin, N. Kamaruddin, and S. M. Razali, Eds. Singapore: Springer Nature, pp. 107–125. doi: 10.1007/978-981-19-3342-4_7.

Norovsuren, B., Tseveen, B., Batomunkuev, V., Renchin, T., Natsagdorj, E., Yangiv, A., and Mart, Z., 2019: Land cover classification using maximum likelihood method (2000 and 2019) at Khandgait valley in Mongolia," *IOP Conf. Ser. Earth Environ. Sci.*, vol. 381, no. 1, p. 012054. doi: 10.1088/1755-1315/381/1/012054.

NSO, 2008. TARLAC QUICKSTAT. Databank and Information Services Division, NSO, Quezon City. https://www.psa.gov.ph/sites/default/files/attachments/ird/quickstat/Decemebr.pdf (01 Mar. 2023).

NSO, 2012. TARLAC QUICKSTAT. Databank and Information Services Division, NSO, Quezon City. https://psa.gov.ph/sites/default/files/attachments/ird/quickstat/tarlac_may12.pdf (30 Sept. 2022).

Pinto, N., Antunes, A. P., and Roca, J., 2021: A Cellular Automata Model for Integrated Simulation of Land Use and Transport Interactions, *ISPRS Int. J. Geo-Inf.*, vol. 10, no. 3, Art. no. 3. doi: 10.3390/ijgi10030149.

Pratomoatmojo, N. A., 2018: LanduseSim Algorithm: Land use change modelling by means of Cellular Automata and Geographic Information System, *IOP Conf. Ser. Earth Environ. Sci.*, vol. 202, no. 1, p. 012020. doi: 10.1088/1755-1315/202/1/012020.

PSA, 2022. Palay and Corn Situation in Tarlac. http://rsso03.psa.gov.ph/tags/corn-and-palay-situation-tarlac (01 Mar. 2023).
PSA, 2023. Updates on Farmgate Prices of Palay (Dry) December 2022. https://psa.gov.ph/farmgate-prices-palay/node/176400 (01 Mar. 2023).

Rojas, F., Rubio, C., Rizzo, M., Bernabeu, M., Akil, N., and Martín, F., 2020: Land Use and Land Cover in Irrigated

Drylands: a Long-Term Analysis of Changes in the Mendoza and Tunuyán River Basins, Argentina (1986–2018), *Appl. Spat. Anal. Policy*, vol. 13, no. 4, pp. 875–899. doi: 10.1007/s12061-020-09335-6.

Sajan, B., Mishra, V. N., Kanga, S., Meraj, G., Singh, S. K. and Kumar, P., 2022: Cellular Automata-Based Artificial Neural Network Model for Assessing Past, Present, and Future Land Use/Land Cover Dynamics, *Agronomy*, vol. 12, no. 11, Art. no. 11. doi: 10.3390/agronomy12112772.

Srimani, D. P. K., and Prasad, S. N., 2012: DECISION TREE CLASSIFICATION MODEL FOR LAND USE AND LAND COVER MAPPING- A CASE STUDY, *Int. J. Curr. Res.*, vol. 4.

Storie, C. D. and Henry, C. J., 2018: "Deep Learning Neural Networks for Land Use Land Cover Mapping," in *IGARSS 2018 - 2018 IEEE International Geoscience and Remote Sensing Symposium*, pp. 3445–3448. doi: 10.1109/IGARSS.2018.8518619.

Svoboda, J., Štych, P., Laštovička, J., Paluba, D., and Kobliuk, N., 2022: "Random Forest Classification of Land Use, Land-Use Change and Forestry (LULUCF) Using Sentinel-2 Data—A Case Study of Czechia," *Remote Sens.*, vol. 14, no. 5, Art. no. 5. doi: 10.3390/rs14051189.

Taati, A., Sarmadian, F., Mousavi, A., Pour, C. T. H., and Shahir, A. H. E., 2015: Land Use Classification using Support Vector Machine and Maximum Likelihood Algorithms by Landsat 5 TM Images, Walailak J Sci & Tech.

Takenaka, H., Nakajima, T. Y., Higurashi, A., Higuchi, A., Takamura, T., Pinker, R. T., and Nakajima, T., 2011: Estimation of solar radiation using a neural network based on radiative transfer. *J. Geophys. Res.*, 116, D08215. https://doi.org/10.1029/2009JD013337.

United Nations, 2015: Transforming our World: The 2030 Agenda for Sustainable Development. United Nations, NY.

Upadhyay, A., Shetty, A., Kumar Singh, S., and Siddiqui, Z., 2016: Land use and land cover classification of LISS-III satellite image using KNN and decision tree. In *2016 3rd International Conference on Computing for Sustainable Global Development (INDIACom)*, Mar. 2016, pp. 1277–1280.

Wolfram, S., 1983: Statistical mechanics of cellular automata," *Rev. Mod. Phys.*, vol. 55, no. 3, pp. 601–644. doi: 10.1103/RevModPhys.55.601.

Yang, X., Chen, R., and Zheng, X. Q., 2016: Simulating land use change by integrating ANN-CA model and landscape pattern indices, *Geomat. Nat. Hazards Risk*, vol. 7, no. 3, pp. 918–932. doi: 10.1080/19475705.2014.1001797.

Yu, Y., J. Shi, T. Wang, H. Letu, P. Yuan, W. Zhou, and L. Hu., 2018: Evaluation of the Himawari-8 Shortwave Downward Radiation (SWDR) Product and its Comparison With the CERES-SYN, MERRA-2, and ERA-Interim Datasets. *IEEE Journal of Selected Topics in Applied Earth Observations and Remote Sensing,* 12 (2), pp. 1–14.

Zhang, Y. & Chen, L. & Liu, J., 2022: ESTIMATION OF DAILY GLOBAL SOLAR IRRADIANCE FROM HIMAWARI-8 PRODUCTS OVER CHINA. *ISPRS Annals of the Photogrammetry, Remote Sensing and Spatial Information Sciences*, V-3-2022, pp. 493-500. 10.5194/isprs-annals-V-3-2022-493-2022.

This contribution has been peer-reviewed.
https://doi.org/10.5194/isprs-archives-XLVIII-M-1-2023-255-2023

The importance of Earth Observation for monitoring cultural heritage sites affected by fire events: The Case Study of Arakapas, Cyprus using Sentinel 2 data

Maria Prodromou[1,2], Daniele Cerra[3], Kyriacos Themistocleous[1,2], Gunter Schreier[3], Thomas Krauss[3], Diofantos Hadjimitsis[2,1]

[1] Eratosthenes Centre of Excellence, Franklinou Roosvelt 82 Limassol, Cyprus. (maria.prodromou, k.themistocleous) @eratosthenes.org.cy

[2] Cyprus University of Technology (CUT), 30 Arch. Kyprianos Str. 3036 Limassol, Cyprus. (k.themistocleous, d.hadjimitsis)@cut.ac.cy

[3] DLR, Germany. (Gunter.Schreier, Daniele.Cerra, Thomas.Krauss)@dlr.de

KEY WORDS: Sentinel 2, Fires, Cultural Heritage, Earth observation

ABSTRACT:

This paper presents the importance of using Earth observation to assess the impact of fire events on monitoring cultural heritage sites. The use of sensors such as Sentinel 2 can detect burnt areas in order to determine the extent of fire events on cultural heritage sites, using the fire event in Arakapas, Cyprus as a case study. Sentinel-2 is a multispectral optical sensor acquiring information in a range extending from the visible up to the short-wave infrared, which is the most sensitive spectral range for the detection of damages caused by fire. Sentinel-2 can support cultural heritage monitoring and assessment allowing rapid revisit of any site of interest using pre- and post-event images, where burned areas can be detected with change detection techniques coupled with suitable, spectral indices. Such derived products assist in quickly raising awareness on the endangerment of cultural and natural heritage sites and landscapes, emphasising the importance of Earth observation data for monitoring natural hazards for the protection of valuable cultural heritage sites.

1. INTRODUCTION

Forest fires have significant impact on both natural and anthropogenic ecosystems. Fire is an integral part of many ecosystems; however, in recent decades, there has been a significant increase in the Mediterranean area in the number of fires and the extent of the surface burned because of its distinctive climate and vegetation features. (Chergui et al., 2018). Wildland fire is a major and potentially catastrophic disturbance affecting ecosystems and their interface with urban areas, which results in loss of property, deaths, and significant environmental degradation.

Earth observation and remote sensing are a valuable and effective tool which includes a large number of systems that focus on using data from spectroradiometers, satellites, airborne and UAV platforms using visible, near and thermal infrared, microwave and other wavelengths (Santos et al., 2021). Such tools can be effectively used for fire prevention projects, assessment and monitoring purposes, as well as detecting areas affected by wildfires, estimating fire severity and burnt severity/ratio (GÜLCİ et al., 2021; Hu et al., 2021). The Sentinel 2 satellite has been used in several studies to detect burnt areas in order to determine the location and extent of fire events and to monitor environmental recovery (GÜLCİ et al., 2021; Hu et al., 2021; Pádua et al., 2020; Schroeder et al., 2016). According to the wider literature (Morresi et al., 2022) (Dindaroglu et al., 2021) (Zidane et al., 2021)(Abdikan et al., 2022) (Luo and Wu, 2022) (Putra et al., 2022)(Saulino et al., 2020), the analysis of the damage that an area suffers after a fire can be investigated using spectral indices. The most widely used index for mapping the forest fire disturbance is the Normalised Burn Ratio (NBR) (García and Caselles, 1991; Key and Benson, 2006)

The majority of cultural heritage and archaeological sites, especially in the Mediterranean region, are covered with vegetation - increasing the risk of fires – and are located near to forest regions or in abandoned areas covered with vegetation (Grammalidis et al., 2011). Several factors contribute to the increased risk of forest fires, such as prolonged drought, hot summers, strong winds, large slopes of forests and flammable dry vegetation (Prodromou et al, 2022), which have resulted in a dramatic increase in the number of wildfires in many forested areas that have become disastrous for many cultural heritage sites.

The use of Earth Observation (EO) data and remote sensing techniques has been successfully utilize for several archaeological applications, mostly for the systematic detection of CH sites (Cerra et al., 2016) as well for the Cultural Heritage Disaster Risk Management (Agapiou et al., 2020). A recent study by (Themistocleous, 2023) was focus on the monitoring cultural heritage sites affected by geohazards in Cyprus using earth observation based on the development of long-term low-impact monitoring systems as well as indirect analysis of environmental contexts to investigate changes and decay of structure, material and landscape. Also,(Cerra and Plank, 2020) in their study shows a comprehensive case of study for damage detection and monitoring in time through a series of satellite images acquired over the city of Palmyra, Syria, which suffered huge losses related to its cultural heritage in a time span longer than a year.

In this paper, a first assessment for the case of the Arakapas fire in Cyprus is presented. The fire event began on Saturday, the 3rd of July 2021 in the Limassol district near the village of Arakapas and was controlled after approximately 24 hours. The area affected by the fire is designated as an area of special aesthetic value of the Troodos Mountain range and is included in the Troodos UNESCO global geo-park. According to the Department of Antiquities, there were 13 cultural heritage sites in the extended region of the fire. Indeed, several churches of significant cultural value were in danger, as they were located close to the fire.

https://doi.org/10.5194/isprs-archives-XLVIII-M-1-2023-263-2023

2. METHODOLOGY

2.1 Study Area

The fire event began on Saturday, the 3rd of July 2021 in the Limassol district near the village of Arakapas in Cyprus and was finally controlled after less than 24 hours. Cyprus is located in Eastern Mediterranean, which is an area where forest fires frequently occur, especially during the summer period. Forest fires are considered as a major and persistent threat for the forests of Cyprus. Cyprus has an increased risk of forest fires especially during the summer period, because of the high temperatures, the prolonged drought periods, the strong winds, the configuration of the ground and the extremely flammable vegetation.

The area affected by the fire is designated as an area of special aesthetic value of the Troodos Mountain range to the Southwest Shores and is included in the Troodos UNESCO global geo-park, which characterizes it as a natural heritage landscape.

The cultural heritage buildings and monuments located in the fire-affected areas comprise churches, ancient settlements as well as water and olive mills. According to the Department of Antiquities, there are 22 cultural heritage sites in the extended region of the fire which the point of them are presented in Figure 1 and some of them are accompanied with their photo and the names of them are presented on the Table 1. Although the flames reached very close to some monuments, fortunately no monuments were damaged by fire.

Figure 1 The 22 cultural heritage sites in the extended region of the fire event.

Table 1 The names of the CH sites near to the burned area

CH site - Village
Agia Marina - Eptagoneia
Agia Marina - Odou
Agia Marina - Ora
Agie Anargiri - Agioi Vavatsinias
Agios Fotios & Anikitos - Eptagoneia
Agios Georgios - Akapnou
Agios Georgios - Eptagoneia
Agios Georgios - Vavatsinia
Agios Nicolaos - Klonari
Akapnou Bridge - Akapnou
Church of Archangelou Michael - Dierona
Church of Panayia - Vavatsinia
Dierona water mill - Dierona
Kimiseos tis Theodokou - Lageia
Monasteri of Panayias Iamatikis - Arakapas
Panayia Chryseleoussa - Melini
Panayia Iamatikis - Arakapas
Panayia tou Kampou - Akapnou
Panayis tis Agapis - Vavla
Profitis Elias - Odou
Profitis Elias - Sykopetra
Timiou Stavrou - Arakapas

2.2 Materials and Methods

The image processing was performed using custom scripts in the Google Earth Engine (GEE) platform using the JavaScript programming interface and the spatial analysis was performed using the ArcGIS Pro. The GEE is a planetary-scale platform for scientific analysis and visualization of geospatial datasets. In this platform, the open-source images acquired by several satellites, including Sentinel-2, Landsat-7, and Landsat-8 etc are accessible and can be efficiently imported and processed in the cloud without the necessity of downloading the data to local computers. Moreover, several image-driven products and many remote sensing algorithms, including classification algorithms and cloud masking methods, are available in this platform (Gorelick et al, 2017; Kumar & Mutanga 2018). In the GEE, the Sentinel-2 SR data which are orthorectified atmospherically corrected (surface reflectance) can be directly called.

The Sentinel-2 satellite images were selected due to the free access and the highest resolution compared to other free access data such as MODIS and Landsat satellite images. The S2 satellite's MSI (Multi-Spectral Instrument) sensor has 13 spectral bands with 10–60 m spatial resolution.

The Sentinel-2 collection in reflectance was filtered according to cloud cover, time interval, and region. Specifically, for the estimation of the fire severity, two Sentinel-2 images taken prior the fire event (2/7/2021) as shown in Figure 2 and after the fire event (27/7/2021) as shown in Figure 3 the fire incident were used due to the fact that were cloud-free and the dates were close to the fire event.

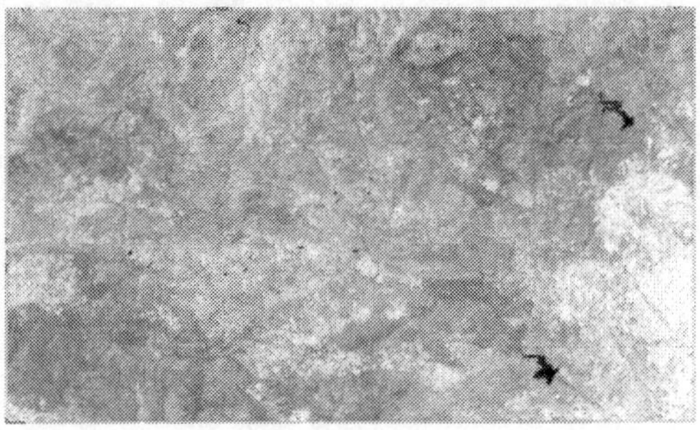

Figure 2 Sentinel-2 pseudo-color imagery (B12, B11 and B8) before the fire event 2/7/2021

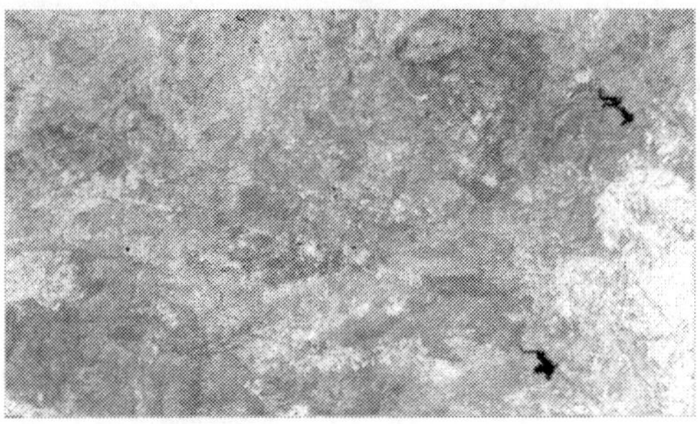

Figure 3 Sentinel-2 pseudo-color imagery (B12, B11 and B8) after the fire event 27/7/2021.

To explore and compare the spectral signal between pre- and post-fire images the reflectance values of B8 and B12 were extracted and used to create the histogram data plots for each spectral band, for each image for our study area as shown in Figure 4 and Figure 5. The histogram in green corresponds to the pre-fire satellite imagery and depicts the spectral signal of burned surfaces in B8 and B12, while the histogram in blue corresponds to the post-fire satellite imagery. Figure 4 and Figure 5 compares the histogram data plots of B8 and B12 for pre- and post-fire imagery respectively and shows that the distribution of the reflectance values of the used bands for the calculation of the dNBR spectral index is almost identical, as shown in the histograms which present a relevant overlap.

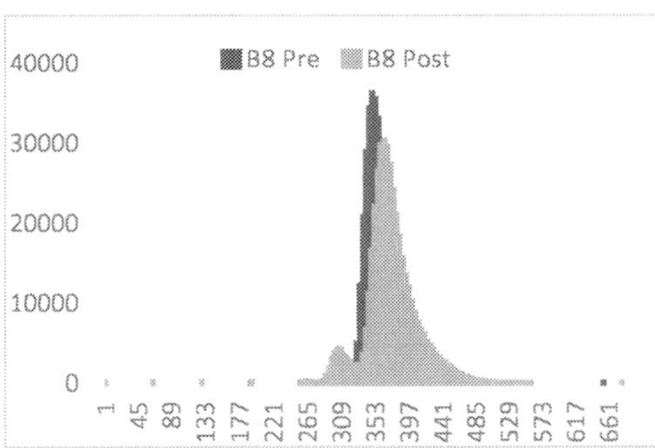

Figure 4 Histogram data plot for B8 before and after the fire event. The units on y-axes the frequency of the pixel values and the x-axis the pixel value of the reflectance scaled in 1e4.

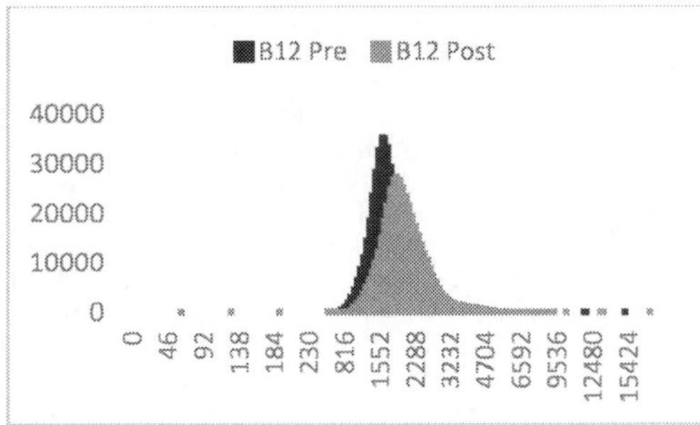

Figure 5 Histogram data plot for B12 before and after the fire event. The units on y-axes the frequency of the pixel values and the x-axis the pixel value of the reflectance scaled in 1e4.

Additionally, a cloud-masking technique relying on the assessment of the quality of each sensor's pixel was applied to minimize the chance of clouds in the analyzed region, even though images free of clouds were already selected. In this study, the Normalized Burn Ratio (NBR) indices were estimated before and after the fire event in order to estimate the burn severity using the difference of the NBR (dNBR). The Normalised Burn Ratio (NBR) was selected because is the most widely used index for mapping the forest fire disturbance and for detecting fire scars is the (García and Caselles, 1991; Key and Benson, 2006) which is given by the ratio of the difference between the near-infrared (NIR) and short-wave infrared (SWIR) bands of electromagnetic radiation and the sum of NIR and SWIR, where the NIR and SWIR refers to band 8 and band 12 of the Sentinel-2 data, respectively. The NBR spectral index takes values from -1 to +1 where the low values represent bare ground and recently burnt areas and the high values represents healthy or unburned vegetation. Also, the differenced Normalised Burn Ratio (dNBR) which is calculated by the difference of the NBR before and after the fire is an effective method for measuring burnt severity in various plant species. The low values of dNBR index refer to less burnt severity and unburned areas in contrast with high values which indicate severe damage Key & Benson (2006). The NBR and dNBR spectral indices obtained by the equations (1) and (2) respectively.

$$NBR = \frac{\text{NIR}-\text{SWIR}}{\text{NIR}+\text{SWIR}} \quad (1)$$

$$dNBR = NBR_{preFire} - NBR_{postFire} \quad (2)$$

Where:

NIR: near-infrared
SWIR: short-wave infrared

Moreover, in order to identify the Cultural Heritage (CH) sites that were at risk of being destroyed by fire, we considered the spatial overlap between the dNBR map derived from the Sentinel-2 satellite data and the cultural heritage sites. The CH sites were acquired from data.gov.cy, which is the central information portal of the public sector which provides access to datasets of all Cypriot government. Also, the nearest distance

This contribution has been peer-reviewed.
https://doi.org/10.5194/isprs-archives-XLVIII-M-1-2023-263-2023 | © Author(s) 2023. CC BY 4.0 License.

from each severity class was estimated in order to identify the riskiest CH site.

3. RESULTS AND DISCUSSION

The area affected by the fire was identified by the differenced Normalized Burn Ratio (dNBR) index based on Sentinel-2 satellite data where the pixel values were classified based on the thresholding technique according to the severity levels Table 2proposed by the (Key and Benson, 2006). The total burned area derived from Sentinel-2 data based on dNBR was 44,25Km2. For each fire severity class, the total area has been estimated and presented in Table 2.

Table 2: Fire severity classes and the burned area per class

Severity level	dNBR range (scaled by 1000)	Area (Km^2)
Enhanced Regrowth, High	-500 to -251	0.08726
Enhanced Regrowth, Low	-250 to -101	0.51229
Unburned	-100 to +99	387.32837
Low	+100 to +269	9.86474
Moderate-Low	+270 to +439	17.6953
Moderate-High	+440 to +659	15.72674
High	+660 to +1300	0.96959

The map regarding the spatial distribution of the dNBR based on the abovementioned thresholds, for the fire event in Arakapas village in Cyprus, is shown in the Figure 6.

Figure 6 Map of the differenced Normalized Burn Ratio for the study area.

A total of ten villages was evacuated because of the fire: Ora, Eptagoneia, Melini, Odou, Agioi Vavatsinias, Arakapas, Vavatsinia, Akapnou, Lageia and Dierona. Furthermore, based Corine Land Cover 2018 provided by Copernicus Land Service Figure 7, in this area, the vegetation is characterized by shrubs and other types of Mediterranean vegetation, pines and olive trees.

The estimation of total burned area and the fire severity shown that the process can be done rapidly thanks to the power of the Google Earth Engine.

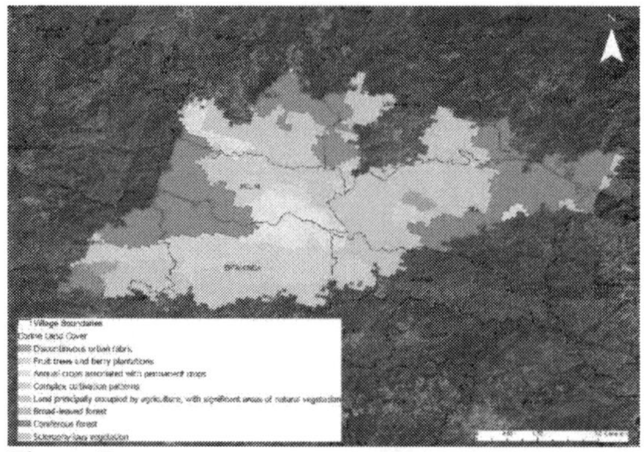

Figure 7 Vegetation type based on Corine land Cover data (2018)

In order to define and geographically locate cultural heritage assets with which to perform the assessment, we used https://www.data.gov.cy/ , that is the central information portal of the public sector which provides access to datasets of all Cypriot government.

Based on a 3Km buffer zone out of the burned area as shown in Figure 8, there are 22 cultural heritage sites inside and outside of the burned area. The list of cultural heritage sites potentially affected by the fire event in July 2021 near to Arakapas village was obtained by performing a spatial overlap between the dNBR map derived from the Sentinel-2 satellite data and the cultural heritage sites as shown in Table 3. The points which are overlapping with the burned area maps were selected.

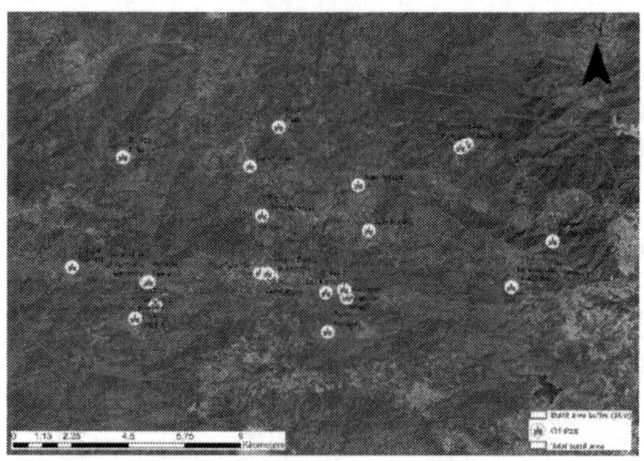

Figure 8 3Km buffer zone around the burnt area- and the included CH sites

Despite the great destruction that occurred in the area of Arakapa, which burned approximately ~42 Km^2, the 22 points of interest fall into the "unburned" according to the Figure 9 and the dNBR pixel values.

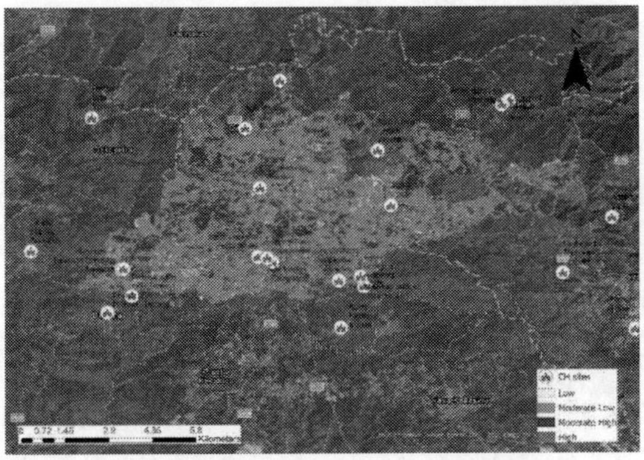

Figure 9 Map of the differenced Normalized Burn Ratio and the distribution of cultural heritage sites

Table 3 CH sites and the average distance from the burned area

CH site - Village	Average Distance from the Burned area (Km)
Agios Georgios - Eptagoneia	0.16
Agios Fotios & Anikitos - Eptagoneia	0.18
Panayia tou Kampou - Akapnou	0.23
Agia Marina - Eptagoneia	0.23
Profitis Elias - Odou	0.25
Agia Marina - Odou	0.29
Panayia Iamatikis - Arakapas	0.30
Monasteri of Panayias Iamatikis - Arakapas	0.31
Dierona water mill - Dierona	0.38
Akapnou Bridge - Akapnou	0.48
Agie Anargiri - Agioi Vavatsinias	0.52
Panayia Chryseleoussa - Melini	0.62
Agia Marina - Ora	0.64
Agios Georgios - Akapnou	0.67
Church of Archangelou Michael - Dierona	0.96
Timiou Stavrou - Arakapas	0.96
Church of Panayia - Vavatsinia	1.09
Agios Georgios - Vavatsinia	1.31
Agios Nicolaos - Klonari	1.38
Panayis tis Agapis - Vavla	1.73
Kimiseos tis Theodokou - Lageia	1.91
Profitis Elias - Sykopetra	2.56

Also buffers zones for CH site were created at 0,5Km, 1Km and 2Km in order to identify the nearest severity classes an example for the buffer zones was presented in the Figure 10.

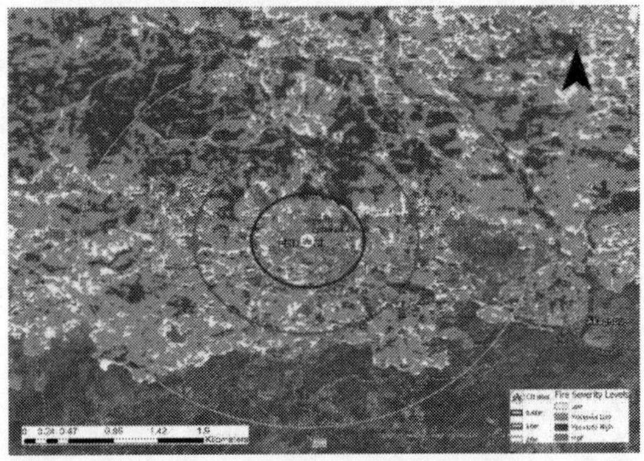

Figure 10 An example at Agia Marina (Eptagoneia) for buffer zones in 0.5Km, 1Km and 2Km for the identification of the neighbour severity classes.

Furthermore, in APPENDIX the Figure 11 presents the nearest distance from each severity class by CH site. Based on the CH site's average distance from the total burned area in the range of the 0 to 0,5Km radius the nearest CH sites which has the highest risk to be destroyed were the Agios Georgios – Eptagoneia (0,16Km), Agios Fotios & Anikitos – Eptagoneia (0,18Km), Panayia tou Kampou – Akapnou (0.23Km), Agia Marina – Eptagoneia (0.23Km), Profitis Elias – Odou (0.25Km), Agia Marina – Odou (0.29Km), Panayia Iamatikis – Arakapas (0.30Km), Monastery of Panayias Iamatikis – Arakapas (0.31Km), Dierona water mill – Dierona (0.38Km), Akapnou Bridge – Akapnou (0.48Km), Agie Anargiri - Agioi Vavatsinias (0.52Km). Furthermore, in the range of 0.5 to 1Km radius included the Panayia Chryseleoussa- Melini (0.62Km), Agia Marina – Ora (0.64 Km), Agios Georgios – Akapnou (0.67Km), Church of Archangelou Michael – Dierona (0.96 Km), Timiou Stavrou – Arakapas (0.96Km) and in the radius of the 1 to 2Km included the Church of Panayia – Vavatsinia (1.09 Km), Agios Georgios – Vavatsinia (1.31Km), Agios Nicolaos – Klonari (1.38Km), Panayis tis Agapis – Vavla (1.73Km), Kimiseos tis Theodokou – Lageia (1.91Km) and Profitis Elias – Sykopetra (2.56Km). In the range of 1 to2Km based on the distance values the CH sites are nearest to the lower severity classes and are further away from the higher severity classes.
This approach shows the risk of CH sites based on their proximity to the burned area as well as based on the degree of severity. In our future plans in order to have a more integrated estimation of the destruction risk of the CH sites, a risk map will be created which will combine other parameters, such as vegetation conditions (distribution, type, density), elevation and slope, soil moisture, fuel map, wind speed and direction statistics.

4. CONCLUSIONS

This paper provided the methodology of the monitoring and assessment of the fire event in Arakapas village. This study identified a set of potentially affected cultural heritage assets based on remote sensing techniques, using maps derived from Sentinel-2 satellite data in order to estimate the total burned area and the fire severity. The methodology was able to determine that the fire event occurred in an area with 22 cultural heritage sites, several of which were located near the site of the fire and in

https://doi.org/10.5194/isprs-archives-XLVIII-M-1-2023-263-2023 | © Author(s) 2023. CC BY 4.0 License.

danger of being damaged. Decision-makers do not recognize heritage as a priority. The global Disaster Risk Reduction sector is currently not concerned with cultural heritage. To guide the decision-making process where safety and resource values are evaluated in terms of fire risk and appropriate fire management, response strategies are identified for wildland fires. There is a need to provide a framework for fire management strategies through the use of specific fire mitigation actions as well as a platform to cooperate more fully in planning and implementing wildland fire programs across the natural and cultural monuments. The proposed methodology shows very promising results for the assessment of fire severity and the estimation of the burned area to assess the impacts on the environment and especially the cultural heritage sites with a very quick way thanks to the computational power of the Google Earth Engine platform.

ACKNOWLEDGEMENTS

This work has been supported by the project 'ERATOSTHENES: Excellence Research Centre for Earth Surveillance and Space-Based Monitoring of the Environment-EXCELSIOR' (https://excelsior2020.eu/) that has received funding from the European Union's Horizon 2020 research and innovation programme under grant agreement No 857510 (Call: WIDESPREAD-01-2018-2019 Teaming Phase 2) and the Government of the Republic of Cyprus through the Directorate General for European Programmes, Coordination and Development.

REFERENCES

Abdikan, S., Bayik, C., Sekertekin, A., Bektas Balcik, F., Karimzadeh, S., Matsuoka, M., Balik Sanli, F., 2022. Burned Area Detection Using Multi-Sensor SAR, Optical, and Thermal Data in Mediterranean Pine Forest. Forests 13, 347. https://doi.org/10.3390/f13020347

Agapiou, A., Lysandrou, V., Hadjimitsis, D.G., 2020. Earth Observation Contribution to Cultural Heritage Disaster Risk Management: Case Study of Eastern Mediterranean Open Air Archaeological Monuments and Sites. Remote Sens. 12, 1330. https://doi.org/10.3390/rs12081330

Cerra, D., Plank, S., 2020. Correction to: Towards Early Warning for Damages to Cultural Heritage Sites: The Case of Palmyra. pp. C1–C1. https://doi.org/10.1007/978-3-030-10979-0_17

Cerra, D., Plank, S., Lysandrou, V., Tian, J., 2016. Cultural Heritage Sites in Danger—Towards Automatic Damage Detection from Space. Remote Sens. 8, 781. https://doi.org/10.3390/rs8090781

Chergui, B., Fahd, S., Santos, X., Pausas, J.G., 2018. Socioeconomic Factors Drive Fire-Regime Variability in the Mediterranean Basin. Ecosystems 21, 619–628. https://doi.org/10.1007/s10021-017-0172-6

Dindaroglu, T., Babur, E., Yakupoglu, T., Rodrigo-Comino, J., Cerdà, A., 2021. Evaluation of geomorphometric characteristics and soil properties after a wildfire using Sentinel-2 MSI imagery for future fire-safe forest. Fire Saf. J. 122. https://doi.org/10.1016/j.firesaf.2021.103318

García, M.J.L., Caselles, V., 1991. Mapping burns and natural reforestation using thematic Mapper data. Geocarto Int. 6, 31–37. https://doi.org/10.1080/10106049109354290

Grammalidis, N., Çetin, E., Dimitropoulos, K., Tsalakanidou, F., Kose, K., Gunay, O., Gouverneur, B., Torri, D., Kuruoglu, E., Tozzi, S., Benazza, A., Chaabane, F., Kosucu, B., Ersoy, C., 2011. A multi-sensor network for the protection of cultural heritage. Eur. Signal Process. Conf. 889–893.

GÜLCİ, S., YÜKSEL, K., GÜMÜŞ, S., WİNG, M., 2021. Mapping Wildfires Using Sentinel 2 MSI and Landsat 8 Imagery: Spatial Data Generation for Forestry. Eur. J. For. Eng. https://doi.org/10.33904/ejfe.1031090

Hu, X., Ban, Y., Nascetti, A., 2021. Sentinel-2 MSI data for active fire detection in major fire-prone biomes: A multi-criteria approach. Int. J. Appl. Earth Obs. Geoinf. 101, 102347. https://doi.org/10.1016/j.jag.2021.102347

Key, C.H., Benson, N.C., 2006. Landscape Assessment (LA) sampling and analysis methods. USDA For. Serv. - Gen. Tech. Rep. RMRS-GTR.

Luo, H., Wu, J., 2022. An Assessment of the Suitability of Sentinel-2 Data for Identifying Burn Severity in Areas of Low Vegetation. J. Indian Soc. Remote Sens. 7. https://doi.org/10.1007/s12524-022-01518-7

Morresi, D., Marzano, R., Lingua, E., Motta, R., Garbarino, M., 2022. Mapping burn severity in the western Italian Alps through phenologically coherent reflectance composites derived from Sentinel-2 imagery. Remote Sens. Environ. 269, 112800. https://doi.org/10.1016/j.rse.2021.112800

Pádua, L., Guimarães, N., Adão, T., Sousa, A., Peres, E., Sousa, J.J., 2020. Effectiveness of Sentinel-2 in Multi-Temporal Post-Fire Monitoring When Compared with UAV Imagery. ISPRS Int. J. Geo-Information 9, 225. https://doi.org/10.3390/ijgi9040225

Putra, E.I., Ramadhi, A., Shadiqin, M.F., Saad, A., Setianto, E., Nurhayati, A.D., Saharjo, B.H., Cochrane, M.A., 2022. Assessing the severity of forest fire in Sungai Buluh Protected Peat Forest, Jambi. IOP Conf. Ser. Earth Environ. Sci. 959. https://doi.org/10.1088/1755-1315/959/1/012059

Santos, S.M.B. dos, Bento-Gonçalves, A., Vieira, A., 2021. Research on Wildfires and Remote Sensing in the Last Three Decades: A Bibliometric Analysis. Forests 12, 604. https://doi.org/10.3390/f12050604

Saulino, L., Rita, A., Migliozzi, A., Maffei, C., Allevato, E., Garonna, A. Pietro, Saracino, A., 2020. Detecting Burn Severity across Mediterranean Forest Types by Coupling Medium-Spatial Resolution Satellite Imagery and Field Data. Remote Sens. 12, 1–21. https://doi.org/10.3390/rs12040741

Schroeder, W., Oliva, P., Giglio, L., Quayle, B., Lorenz, E., Morelli, F., 2016. Active fire detection using Landsat-8/OLI data. Remote Sens. Environ. 185, 210–220. https://doi.org/10.1016/j.rse.2015.08.032

Themistocleous, K., 2023. Monitoring Cultural Heritage Sites Affected by Geohazards in Cyprus Using Earth Observation. pp. 359–377. https://doi.org/10.1007/978-3-031-13810-2_19

Zidane, I.E., Lhissou, R., Ismaili, M., Manyari, Y., Bouli, A., Mabrouki, M., 2021. Characterization of Fire Severity in the Moroccan Rif Using Landsat-8 and Sentinel-2 Satellite Images. Int. J. Adv. Sci. Eng. Inf. Technol. 11, 72–83. https://doi.org/10.18517/ijaseit.11.1.10342

APPENDIX

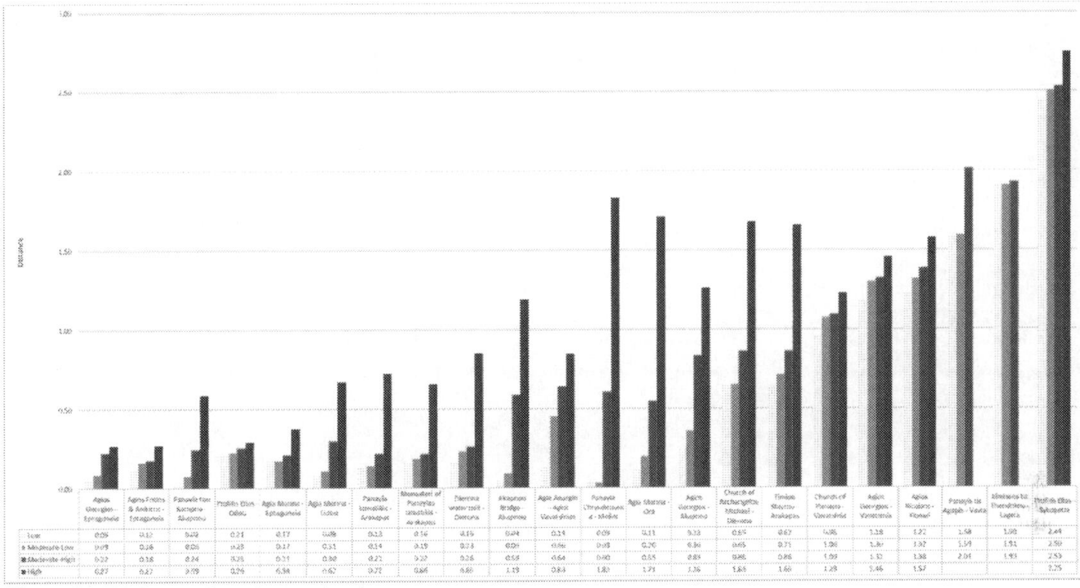

Figure 11 CH distance from the severity classes

https://doi.org/10.5194/isprs-archives-XLVIII-M-1-2023-263-2023 | © Author(s) 2023. CC BY 4.0 License.

ASSESSMENT OF LAND DEFORMATION AND LAND USE RELATIONSHIP IN GARMSAR ALLUVIAL FAN USING SENTINEL-1 DATA

F. Rafiei [1], S. Gharechelou [2]*, M. Hosseinpoor [1]

[1] Dept. of Water and Environment, Faculty of Civil Engineering, Shahrood University of Technology, Iran – Fateme_rafiei@outlook.com, Hosseinpoor.masood@yahoo.com
[2] Dept. of Geotechnics-Transportation and Surveying, Faculty of Civil Engineering, Shahrood University of Technology, Iran - sgharachelo@shahroodut.ac.ir

KEY WORDS: DInSAR, Land Deformation, Land Use, Garmsar, Alluvial Fan, Sentinel-1.

ABSTRACT:

Differential Synthetic Aperture Radar Interferometry (DInSAR) allows displacements to be detected with millimeter accuracy as well as more advanced methods, such as Persistent Scatterer Interferometry (PSI). Sentinel-1 data have been collected systematically under the COPERNICUS program at a high temporal resolution with global coverage, helping us to build a wide user community and develop miscellaneous SAR-based applications. In the Garmsar alluvial fan, the long-term groundwater overexploitation due to agricultural and urban demands, the utilization of urban space, and erosion have led to land deformation. In this study, the analysis of land subsidence in Garmsar fan was assessed by using the Differential Interferometric Synthetic Aperture Radar (DInSAR) technique based on 20 Sentinel-1 SAR images from January 2019 to June 2022. Distinct variations of land subsidence were found in the study regions however, it can be seen in most land use types. The maximum annual land subsidence rate has occurred in urban areas with an average rate of 95.2 mm/year from 2019 to 2022. Analysis showed that serious land subsidence mainly occurred in the following land use types: urban areas, agricultural lands, and bare lands.

1. INTRODUCTION

Differential Interferometric Synthetic Aperture Radar (DInSAR) is a satellite-based remote sensing technique that can be used to measure small displacements of the Earth's surface by removing or diminishing all the contributors to the interferometric phase except the displacement (Zhou et al., 2017). Differential SAR Interferometry (DInSAR) was initially used to measure the deformation of the land surface through the interpretation of Interferograms, with a combination of digital elevation models (DEM) to remove the topographic contributions. DEMs were obtained with the SAR Interferometry by using one or two additional SAR images and it was called three, or four-pass interferometry (Mancini et al., 2021).

Land subsidence is a gradual, long-term geological catastrophe that can be brought on by both natural and human-caused compaction (Zhang et al., 2014). In actuality, excessive groundwater extraction is the most frequent cause of land subsidence (Dehghani Bidgoli et al., 2021). Measuring ground deformation is key to understanding the ground response to groundwater overexploitation, land use change, natural hazards, etc. Land subsidence has become a global issue threatening our environment.

Like any other remote-sensing method, Differential Interferometric Synthetic Aperture Radar (DInSAR) has drawn a lot of interest in the field of monitoring subsidence since it can assess ground deformation. InSAR, on the other hand, detects deformation between a satellite and an object in the line of sight in a single dimension (LOS). SAR must be used with other methods, such as GNNS or leveling, in order to obtain 3D deformation components.

Several approaches have been used to detect and measure land subsidence caused by groundwater over-extraction such as traditional levelling survey, extensometer, numerical modelling of the aquifer, and global positioning system (GPS). The conventional methods usually measure displacement point by point. Therefore, they are inadequate to map the spatial variability of land subsidence which is largely driven by different land cover types and the spatial heterogeneity of the aquifer (Ahmad et al., 2019).

Compared with other techniques, traditional levelling, and GPS, Interferometric Synthetic Aperture Radar (InSAR) techniques have been successfully applied to study land deformation using multiple SAR images with a wide spatial range and millimeter-scale precision (Zhou et al., 2017).

The generated deformation maps can provide crucial information about the magnitude and spatial distribution of subsidence, and therefore the critical zones can be monitored easily to implement the remedial measures. Land subsidence due to groundwater abstraction is mainly controlled by the spatial pattern of groundwater consumption revealed by the land use types and the physical properties of the hydrogeological formation of the area (Ahmad et al., 2019).

Land subsidence characteristics in various land use types were investigated in Beijing Plain, China using InSAR techniques based on TerraSAR-X images over a 5-year period. The analysis showed that the land subsidence mainly occurred in

* Corresponding author

This contribution has been peer-reviewed.
https://doi.org/10.5194/isprs-archives-XLVIII-M-1-2023-271-2023

water areas and wetlands, paddy fields, upland soils, vegetable lands, and peasant-inhabited lands. The land subsidence in this area increased with the decrease in groundwater levels (Zhou et al., 2017). Land use change in the Vietnamese Mekong delta and its relationship with subsidence based on a 20-year time series was studied using Landsat-5 and InSAR-derived subsidence rates by previous studies. Results for the land use change sequences agreed well with the InSAR-derived subsidence rates, confirming the strong relation between land-use history and subsidence (Minderhoud et al., 2018). Quetta valley, Pakistan, has experienced land subsidence due to excessive groundwater extraction for irrigation and domestic supply. Ahmad et al., utilized Sentinel-1 and Sentinel-2 data to categorize the spatial distribution of land subsidence in 2014-2016. The results revealed that during the study period, the entire area suffered from varying degrees of accumulated land subsidence. The land subsidence was linearly proportional to the depth and depletion of groundwater observed at four piezometers located in the study area. The land subsidence over the study area showed a high correlation with land covers; e.g., the urban area and orchard vegetation with a large amount of groundwater abstraction showed an excessive amount of cumulative land subsidence (Ahmad et al., 2019).

Many of the major deltas in the world have gone through significant changes during the past decades. As as example, in the Southeast Asia, the Vietnamese Mekong delta stands out as hotspot of anthropogenic land-use change. These deltas can experience change in irrigation system, progressively becoming urbanized, with expanding networks which lead to fast-growing settlements (Minderhoud et al., 2018).

The land subsidence phenomenon is happening in most delta systems because deltaic sediments are highly compressible and susceptible to significant natural compaction during deposition and subsequent soil formation. The land subsidence is mainly due to human activities are widely recognized in deltas. Subsidence can be amlified by direct loading of the delta surface, both by natural material, such as water and sediment, and by manmade artifacts, like buildings and infrastructure. Also, the drainage of wetlands to prepare for agricultural use leads to a lowering of the phreatic water table, causing compaction in the subsoil. The organic material decomposition (oxidation) causes additional volume reduction. Also, the extraction of groundwater from deeper aquifers, to meet the increasing freshwater demands of rapidly urbanizing areas and agriculture, can lead to aquifer-system compaction.

Water Scarcity has affected the Garmsar alluvial fan due to its dry climate. However, it is one of the most important plains in the country with traditional agriculture has been modernized in recent years. Garmsar has a special economic and industrial landscape due to its proximity to Tehran as the capital city. In the past decades, rapid industrial and agricultural development in the region accompanied by rapid population growth and an increase in water demand, have led to an increase in the withdrawal of groundwater resources annually dropping groundwater aquifers level (Dehghani Bidgoli et al., 2021).

In this research, the DInSAR method was applied to obtain land deformation information in the Garmsar alluvial fan based on Sentinel-1 data from 2019 to 2022. The land use data of Garmsar alluvial fans have been assessed to examine the relationship between land use and subsidence in this area.

2. METHODOLOGY

2.1 Study Area

The study area (Figure 1) is a typical area in the northern part of the central desert of Iran with arid and semi-arid climates in hot summers and cold winters (Gharechelou et al., 2020). Garmsar plain has a total area of 320km^2 lays on the southern margin of Alborz and in the west of Semnan province (Dehghani Bidgoli et al., 2021).

The annual evapotranspiration in this area is 3200 mm, while the average annual rainfall is 140 mm (Gharechelou et al., 2020). The elevation of this area varies between 800 and 900 m above sea level. The groundwater in this area is mainly exploited by deep wells for agricultural lands and household purposes.

After leaving the southern Alborz Mountains, the Hablehrood River forms this plain at the foot of the mountain and in the north of the desert. This alluvial plain has always been affected by water scarcity due to its dry climate. However, it is one of the most important plains in the country in which traditional agriculture has been modernized in recent years.

Over the past decades, rapid industrial and agricultural development in this region, accompanied by rapid population growth which has led to an increasing water demand and groundwater resources extraction, have caused an annual dropping of groundwater level in aquifers 1.3 m (Dehghani Bidgoli et al., 2021)

Excessive groundwater abstraction, low rainfall, and high evaporation are the most important factors of desertification in this region.

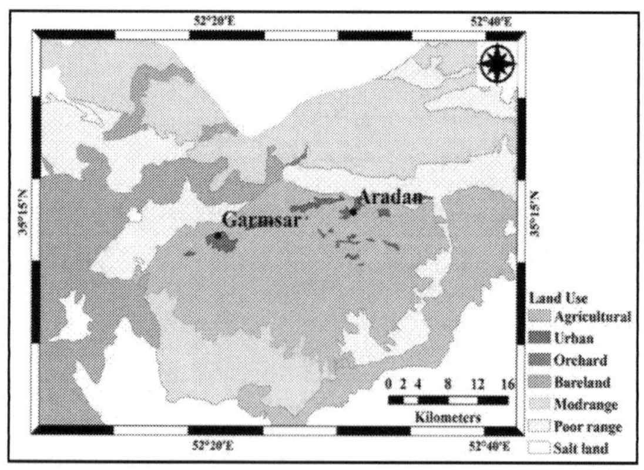

Figure 1. Location of the study area in Iran, based on land use map

2.2 SAR Data

The Sentinel-1 satellite was launched in 2014 by ESA based on Sentinel's satellites family program. The Sentinel–1 constellation of two identical SAR satellites (Sentinel-1 A and Sentinel-1 B) is the only operational mission providing global SAR data in a systematic and near real time. It also has a number of merits. First, it can image the same target location every 6 days with the two satellites. This is a significant improvement in the satellite revisit time. Secondly, it covers the globe with predefined observational scenarios periodically published and archived by the ESA. Furthermore, it has enhanced spatial and radiometric resolution. Finally, the time interval between the image acquisition and its dissemination to the public can be even 1 day (Ahmad et al., 2019).

It has a revisiting time of 12 days and operates in the C-band of HH, VV, HV, and VH polarizations. The satellite carries a C-band Synthetic Aperture Radar which will provide images in all light and weather conditions. The land deformation in this study is analyzed using 20 Sentinel-1 SAR images acquired in Single Look Complex (SLC) Level-1 from descending track between the years 2019 and 2022. The specification of acquisition SAR data is presented in Table (1).

Acquisition date	Image Type	Polarization	Direction
20190110	SLC - L1	VV	Descending
20190323	SLC - L1	VV	Descending
20190814	SLC - L1	VV	Descending
20191001	SLC - L1	VV	Descending
20191224	SLC - L1	VV	Descending
20200222	SLC - L1	VV	Descending
20200422	SLC - L1	VV	Descending
20200621	SLC - L1	VV	Descending
20200820	SLC - L1	VV	Descending
20201019	SLC - L1	VV	Descending
20201230	SLC - L1	VV	Descending
20210204	SLC - L1	VV	Descending
20210417	SLC - L1	VV	Descending
20210616	SLC - L1	VV	Descending
20210815	SLC - L1	VV	Descending
20211014	SLC - L1	VV	Descending
20211213	SLC - L1	VV	Descending
20220130	SLC - L1	VV	Descending
20220223	SLC - L1	VV	Descending
20220424	SLC - L1	VV	Descending
20220611	SLC - L1	VV	Descending

Table 1. The characteristics of data sets

The selected IW-SLC images used in the present study were processed using the Sentinel Application Platform Sentinel-1 toolbox (SNAP). The software was used to perform the Interferometric process including master image selection, topographic phase removal, and Interferogram generation. A 30-m SRTM DEM (available at https://earthexplorer.usgs.gov/) was used to remove the topographic phase. After producing the Line of Sight (LOS) displacement, the cumulative displacement map was produced to investigate the land deformation and achieve a better comprehension of the data and further analysis.

2.3 Data Processing

In DInSAR, relative distances are precisely measured by comparing the phases of two photos obtained over the same region at different periods. The DInSAR technique can be divided into two groups, namely cumulative DInSAR and consecutive DInSAR. The former uses a fixed master image and the selection of a slave image (such as 1-2, 1-3, 1-4,..., n) afterward describes the cumulative DInSAR technique. This method is particularly susceptible to temporal decorrelation, which typically gets worse in the slave images that follow, while the latter utilizes differential interferograms of nearby SAR acquisitions that are computed and added together to yield full time-series interferometric results (e.g., 1-2, 2-3, 3-4,..., n-1,n). This method minimizes temporal decorrelation by using short temporal baselines (6 days when Sentinel-1A and Sentinel- 1B are combined). This technique, however, have certain drawbacks. If any errors (residual terrain, atmospheric delay, and other phases error) are found in any of the calculated displacements in the interferograms, they propagate in the ensuing time-series deformation outcomes in the accumulation process.

In this study, consecutive DInSAR is used to overcome the mentioned drawbacks of conventional methods. The processing of data was performed using SNAP. The subswath IW3 of the Sentinel–1 IW–SLC VV polarized images acquired in relative orbit number 137 are selected because this subswath covers the study area. The precise orbit files, that provide accurate information about the location of the satellite are applied to the images to determine the accurate satellite position at the time of image acquisition (Ahmad et al., 2019).

The master and slave images are co-registered before the interferogram formation. It involves resampling the slave images with respect to the grid system of the master image so that each interferogram constitutes the same geolocation precisely. In data processing, co-registered paired images are used to create a differential Interferogram, which is an image displaying the phase change brought on by deformation. Following this, a total number of 19 interferograms were generated.

Data coherence is then calculated. Coherence is an important metric of data quality that assesses how similar the signals are in the paired images. After co-registration, the interferograms were generated and the simulated phase corresponding to the topography was removed from the SRTM DEM with 30-m pixel resolution. Coherence is measured on a scale from zero to one, with zero denoting no similarity and poor data quality and one denoting complete similarity and excellent data quality. Absolute interferometric phases are established after the Interferogram is spatially filtered to remove noise. The phase unwrapping was performed using the Statistical-cost, Network–flow Algorithm for Phase Unwrapping (SNAPHU).

The absolute phases are used to estimate displacements. The line of sight (LOS) vector, which contains both horizontal and vertical displacement, is used to quantify displacements. The features of the sensor and satellite, as well as the surface and deformation properties, determine how effectively DInSAR may be used to measure subsidence (Wempen et al, 2017).

Data decorrelation and phase saturation may reduce the DInSAR's ability to accurately quantify subsidence. Significant changes in the surface properties or the imaging characteristics over time are what generate data decorrelation. Phase saturation can be brought on by large deformation gradients. DInSAR cannot directly measure the absolute displacements when the phases are saturated.

In this investigation, data were matched across the shortest time frames possible to reduce the impact of temporal variations and phase saturation. The differential interferograms were filtered to reduce atmospheric artifacts, and those that had obvious

https://doi.org/10.5194/isprs-archives-XLVIII-M-1-2023-271-2023 |

atmospheric errors were taken out of the stack. The data were then added together to estimate the subsidence over larger time frames. The DInSAR process can be seen in Figure 2.

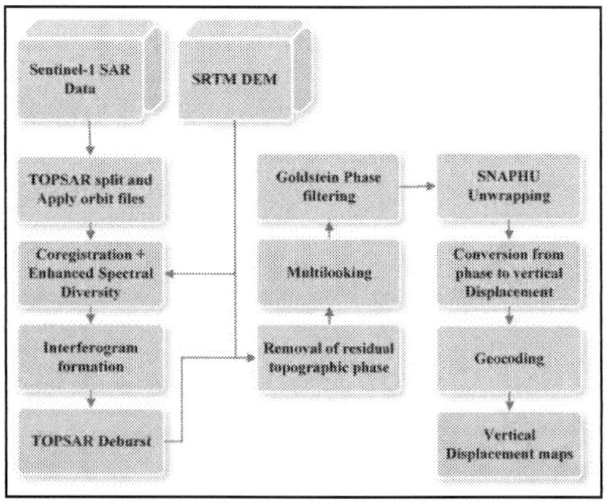

Figure 2. The conventional DInSAR workflow

2.4 Land Use Data

To investigate the relationship between land subsidence and land use types, the Land use map prepared by the Iranian Institute of Water and Soil (http://www.swri.ir/) using Landsat images. These data were used for the extraction of land deformation values in various land use types.

3. RESULT AND DISCUSSION

A total of 19 Interferograms of the study area were generated to monitor land subsidence during the 2.5 years. Figure (2) shows the displacement map of the Garmsar alluvial fan using the DInSAR method. The deformation information is along the SAR's line-of-sight (LOS) direction, therefore, the subsidence results were projected along with a vertical direction with respect to the corresponding incidence angles. The colour scale bar covers a range of 101 mm uplift and maximum subsidence of 416 mm occurring during the period in the study area.

The study was focused on four main land use types: urban areas, agricultural lands and orchards, bare lands, and rangelands. While comparing the land use (Figure 1) and the land subsidence map of Figure 3, it can be clearly seen that the land use types have a strong impact on land subsidence. The land subsidence in the agricultural areas and urban areas seems different than that of the other land use types (the most accumulated deformation and the most subsidence in the areas, respectively).

Although subsidence has occurred in most parts of the region, it can be clearly seen that the majority of subsidence has occurred in the southern and western parts of the study area. There has been a linear area indicating moving away from the satellite, which is the erosion that happened in the river bed. The accumulated subsidence resulting in different subsidence rates across the area due to different land use types and during the study period is further compared, which shows that land use type is a clear indicator of the amount of groundwater use in the study area.

According to previous studies, moving from the northwest towards the southeast of the Garmsar plain, the decline in groundwater decreases, which is aligned with the same pattern as subsidence in the area (Dehghani Bidgoli et al., 2020). The gradual spatial expansion of the red regions shows that the land subsidence is steadily propagating to the adjoining urban areas, especially, Garmsar city.

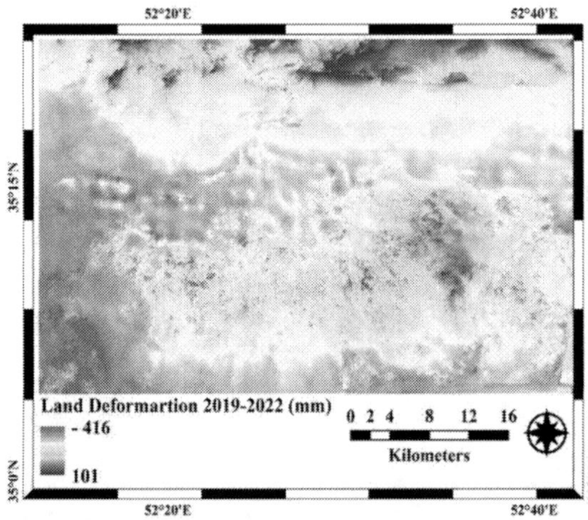

Figure 3. The spatial distribution and critical zone of Cumulative displacement detected by DInSAR analysis

While comparing the land use map (Figure 1) and the land subsidence map in Figure 3, it can be seen that the land use types influence land subsidence. The accumulated subsidence resulting in different subsidence rates across the area due to different land use types during the study period is further analyzed to investigate the relationship between land use and land deformation.

The urban area and orchards which is analogous to the alluvial plain show a more significant and monotonic tendency of subsidence with a mean rate of 95.2 and 81.2 mm/year; the deformation in these areas indicates that land subsidence is mostly caused by groundwater consumption, and is more consistent than range lands, with a recorded deformation rate of 68.8 mm/year (Table 2).

According to the table, the subsidence only between bare land and the urban area land use is not significantly different from each other at less than 5% significance level (4%), while the subsidence between all other land use types is significantly different from each other.

	Bare land	Urban area	Range land	Agricultural area
Accumulated deformation (mm)	-237	-238	-172	-203
Subsidence rate (mm/year)	-94.8	-95.2	-68.8	-81.2
Maximum value of Deformation (mm)	-75	-141	55	-62
Minimum value of Deformation (mm)	-336	-179	-320	-333

This contribution has been peer-reviewed.
https://doi.org/10.5194/isprs-archives-XLVIII-M-1-2023-271-2023 | © Author(s) 2023. CC BY 4.0 License.

Table 2. Characteristics of land use-based land deformation

The bare lands do not receive any regular irrigation water, which makes the magnitude of soil shrink significant. The deformation in range land areas is sought to be caused by erosion, and a small area showing uplift is due to vegetation growth.

It is a known fact that subsidence is mainly caused by groundwater overexploitation, and the volume of water extracted for agricultural use is more than for urban use. Also, the agricultural area in the fan experienced significant subsidence.

Unfortunately, there is not any substantial source of surface water in the study area. Furthermore, evapotranspiration is significantly higher than precipitation; therefore, it is a valid justification that the anthropogenic activities are largely supplied by pumping groundwater which leads to land subsidence, and this tendency of land subsidence is detected very well by the SAR analysis. As can be seen, the maximum subsidence that has occurred in the agricultural areas is approximately double than that of urban areas.

To achieve a better estimation, 100 random points were generated in each land use type. Then, using extract values to points, the figure for deformation in each point was extracted from the displacement map. Then, the Nan values were removed, and finally, 98 selected points in each area were used to generate a box and whisker plot according to Figure 4. As can be seen, urban and agricultural areas were affected more than two others by land subsidence and their absolute value was more than that of rangelands and bare lands.

Regarding rangelands, the deformation in these areas varied more than in others due to differences in the intensity of erosion and vegetation. One of the random points in this class indicates a little uplift in the class. Looking at bare lands, it can be seen that the land subsidence varied between approximately 75mm and around 200mm. Figure 3 is a confirmation of the influence of groundwater consumption and land use on land deformation.

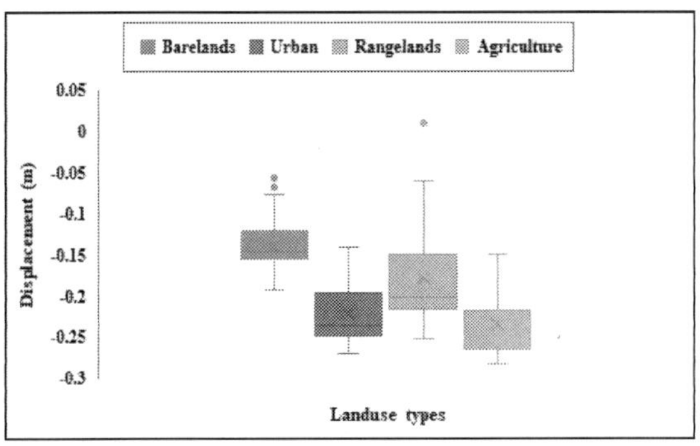

Figure 4. Box and Whisker plot for the variation of random points in the deformational areas according to the four selected land use types; Bare lands, Urban areas, Rangelands, and Agricultural areas

4. CONCLUSION

The present study indicates that the Sentinel–1 satellite data can be employed as a valuable tool to identify the spatiotemporal trend of land subsidence in the study area. DInSAR can generate data with a high spatial and temporal resolution for subsidence monitoring. It has the potential to characterize geographical characteristics of subsidence as well as give rapid and reliable reports on the amount, scope and pace of subsidence. The DInSAR method was employed to assess the relationship between land deformation and land use types in Garmsar alluvial fan. During the 2.5 years, the study area experienced a varying level of land subsidence with a minimum mean rate of 172 mm and maximum mean rate of 238 in range lands and urban areas, respectively. Results showed that land subsidence mostly occurred in urban area, which is affected by human activities, and the subsidence increases moving from east towards the western parts of the plain. Measuring and mapping the land deformation varying with different land use types could be challenging for the traditional approaches based on point measurements or topographic levelling.

The present study also shows that the Sentinel–1 satellite data can be freely used as a valuable tool for identifying the spatio-temporal land subsidence trend in the study area. Furthermore, quantifying and mapping the land subsidence according to different land use types could be a challenging job for the traditional, based on point measurements, approaches or topographic levelling. The satellite data and the data processing software is provided free of cost by the ESA and no financial cost was invested in this regard. We expect our methodology can be used to study similar risks in other locations across the world and to better manage the groundwater resources to avoid land subsidence. We would also recommend using Sentinel-2 data to investigate the land cover maps in the study area which would be updated and could also be pre-processed in open-source SNAP software.

ACKNOWLEDGMENTS

The authors would like to acknowledge the ESA Copernicus data hub for the provision of Sentinel–1 data. Also, we are thankful to the reviewers whose reviews and comments greatly improve the quality of this work.

REFERENCES

Ahmad, W., Choi, M., Kim, S. and Kim, D., 2019: Detection of land subsidence and its relationship with land cover types using ESA Sentinel satellite data: a case study of Quetta Valley, Pakistan. *Int. J. Remote Sens.*, 40(24), 9572-9603. doi.org/10.1080/01431161.2019.1633704

Dehghani Bidgoli, R. and Yazdani, M.R., 2021: Survey and visualization of land subsistence caused by groundwater depletion using Sentinel-1A IW TOPS Interferometry. *Nat. resour. environ,* 9(1), pp.107-116.doi.org/10.1016/j.jhydrol.2021.127329

Gharechelou, S., Tateishi, R. and Johnson, B.A., 2020: Mineral Soil Texture–Land Cover Dependency on Microwave Dielectric Models in an Arid Environment. *Land*, 9(2), 39. doi.org/10.3390/land9020039

Mancini, F., Grassi, F. and Cenni, N., 2021: A workflow based on SNAP–StaMPS open-source tools and GNSS data for PSI-Based ground deformation using dual-orbit sentinel-1 data: Accuracy assessment with error propagation analysis. *Remote Sens.*, 13(4), p.753. doi.org/10.3390/rs13040753

Minderhoud, P.S.J., Coumou, L., Erban, L.E., Middelkoop, H., Stouthamer, E. and Addink, E.A., 2018: The relation between land use and subsidence in the Vietnamese Mekong delta. *Sci. Total Environ.* 634, pp.715-726. doi.org/10.1016/j.scitotenv.2018.03.372

Wempen, J.M., and McCarter, M.K., 2017: Effective application of synthetic aperture radar interferometry for monitoring mine subsidence in the mountain west United States. *SME Preprint No. 17-049. Englewood, CO: SME.* doi.org/10.1016/j.ijmst.2019.12.011

Zhang, Y. Q., H. L. Gong, Z. Q. Gu, R. Wang, X. J. Li, and W. J. Zhao. 2014. Characterization of Land Subsidence Induced by Groundwater Withdrawals in the Plain of Beijing City, China. *Hydrogeol. J,* 22 (2): 397–409. doi:10.1007/s10040-013-1069-x.

Zhou, C., Gong, H., Chen, B., Li, J., Gao, M., Zhu, F., Chen, W. and Liang, Y., 2017: InSAR time-series analysis of land subsidence under different land use types in the Eastern Beijing Plain, China. *Int. J. Remote Sens.*, 9(4), 380. doi.org/10.3390/rs9040380

TREE COVER AREA ESTIMATION IN EUROPE BASED ON THE COMBINATION OF IN SITU REFERENCE DATA AND THE COPERNICUS HIGH RESOLUTION LAYER ON TREE COVER DENSITY

Christophe Sannier[1], Javier Gallego[2], Tobias Langanke[3], Usue Donezar[3], & Alexandre Pennec[4]

[1]IGN FI, 7 rue Biscornet, 75012 Paris, France

[2]Inpdependent Consultant, formerly JRC Ispra, Italy

[3]EEA, Kongens Nytorv 6, 1050 Copenhagen, Denmark

[4]CLS Lille, 61 rue de la Cimaise, 59650 Villeneuve d'Ascq, France

KEY WORDS : activity data, change detection, EEA, statistical inference, forest cover

ABSTRACT:

There is a natural tendency from the remote sensing community to extract area statistics (i.e. "Pixel counting") from EO based geospatial products to produce statistical indicators for various purposes. However, geospatial map products suffer from misclassification errors and "pixel counting" can only be justified when the accuracy of such map products reaches a level when these misclassification errors can be considered negligible, but this is possible only in very specific circumstances. Nevertheless, there has been some effort in the Remote Sensing community to assess the accuracy of map products against some form of reference data to ensure that the maps could reach a sufficient level of accuracy. However, there is generally a lack of standards and guidelines and how to perform rigorous map accuracy assessment and rigorous methods for assessing map accuracies and extracting statistics are still lacking as highlighted by McRoberts (2011). Despite substantial advances in this topic in the scientific literature in recent years notably with the paper from Olofsson et al (2014), this has yet to be fully implemented in operational projects. In addition, even if map accuracy assessment is performed correctly, high accuracy does not necessarily mean that area statistics can be directly extracted from a map.

This study is focused on developing the rigorous and appropriate use (i) of geospatial map products from satellite imagery and (ii) statistically sound methods for reporting area estimates and their associated uncertainty.

1 INTRODUCTION

There is a natural tendency from the remote sensing community to extract area statistics (i.e. "Pixel counting") from EO based geospatial products to produce statistical indicators for various purposes. However, geospatial map products suffer from misclassification errors and "pixel counting" can only be justified when the accuracy of such map products reaches a level when these misclassification errors can be considered negligible, but this is possible only in very specific circumstances. Nevertheless, there has been some effort in the Remote Sensing community to assess the accuracy of map products against some form of reference data to ensure that the maps could reach a sufficient level of accuracy. However, there is generally a lack of standards and guidelines and how to perform rigorous map accuracy assessment and rigorous methods for assessing map accuracies and extracting statistics are still lacking as highlighted by McRoberts (2011). Despite substantial advances in this topic in the scientific literature in recent years notably with the paper from Olofsson et al (2014), this has yet to be fully implemented in operational projects. In addition, even if map accuracy assessment is performed correctly, high accuracy does not necessarily mean that area statistics can be directly extracted from a map.

To illustrate this, the Table below shows a confusion or error matrix that was produced from the comparison of a set of reference data collected from the visual interpretation of sample units and a corresponding map product of Forest (F), Non-Forest (NF) and Change (C). The error matrix below is based on a probability sample (i.e. the probability of inclusion of each sample unit is known and unequal sampling intensity from different strata was corrected to ensure that each sample unit carries an equal weight in the error matrix) and different accuracy metrics can be extracted from the matrix such as overall accuracy for the entire matrix / map and user and producer accuracies for individual thematic classes which are indicative respectively of commission (i.e. over-classification) and omission (i.e. under-classification) errors. Confidence intervals can also be added to assess whether the differences between the different metrics are statistically significant at a certain confidence level.

Table 1. Error Matrix of a Forest (F), Non-Forest (NF) and Change map

		Reference				
		Stable NF	Stable F	Change	Total	User's accuracy
Map	Stable NF	276	6		**283**	0,977
	Stable F	20	691	1	**712**	0,970
	Change	1	2	2	**5**	0,427
Total		**297**	**699**	**3**	**1000**	
Producer's accuracy		0,930	0,988	0,651	Overall Accuracy	**0.969**

In the example shown, the overall accuracy is relatively high with a value close to 97%. However, the overall accuracy can be misleading in that it can hide some imbalance for individual map classes. This is particularly true for small classes such as the change class for which the accuracy is often lower when in fact changes are the most important indicators to be generated from a policy perspective. In this particular case changes are overestimated by a factor of 2/3 as compared to the reference data considering that the proportion of each classes in the matrix is properly respected. In other words, if the total area considered is 100,000km², the area of change from the reference data would represent 300km² whereas it would represent 500km² from the map.

The issue of the impact of map inaccuracies or rather imbalance between omission and commission errors has been known for some time when satellite-based products were used in combination with reference data for crop area statistics first in the USA during the 1980's (Allen 1990) and then in Europe as part of the Monitoring of Agriculture with Remote Sensing programme during the late 1980's and 1990's (Taylor et al 1997). More recently, similar approaches were developed and adapted for tropical forest monitoring to support reporting of activity data for the Forest sector as part of the United Nations Framework Convention on Climate Change (UNFCCC). Under UNFCC, there are strict requirements that estimates should neither over or underestimate reality and that their uncertainty should be known. In this context, "pixel counting" violate this statement in that there is no way of knowing (i) if the estimate is realistic (i.e. neither over or under estimation) and (ii) the associated uncertainty. The approaches from Olofsson et al (2014), but also Sannier et al (2014) were identified as suitable methods to estimate activity data and its associated uncertainty (GFOI, 2016). Methods are therefore well known and documented, but there is still a lack of widespread implementation among the remote sensing community. On the other hand, remotely sensed based products are still not very widely used as part of traditional statistical reporting systems whereas there could potentially be strong synergies between traditional purely sample based methods and the use of remotely sensed based products. Notably map products produced from satellite imagery can be:

- A cost-efficient basis for stratification thus contributing to reducing the uncertainty
- Combined with reference data through a model assisted approach to further reduce the uncertainty of the estimate

Therefore, this study is focused on developing the rigorous and appropriate use (i) of geospatial map products from satellite imagery and (ii) statistically sound methods for reporting area estimates and their associated uncertainty.

2 TECHNICAL APPROACH AND METHODOLOGY

The sections below provide a detailed description of the approach that was applied based on the extensive experience of the project team in the validation of Copernicus HRL products.

2.1 Area estimation and map validation

Both problems may seem different to each other at first sight, but they are deeply linked to each other. Most map validation indicators are computed from the confusion matrix (Table 1). Each term A_{gc} of a confusion matrix should be extrapolated to the area that belongs to the ground class g and has been classified as c. However, for the area as well as map accuracy estimates calculated form a confusion matrix, each sample unit included in the matrix must be based on a probability sample for which the sampling intensity will need to be accounted for in case of unequal sampling intensities, which may occur if a stratification has been applied. Therefore, estimating a confusion matrix is a problem of area estimation as outlined by Olofsson et al. (2014). More details are provided in the following sections.

2.2 Sample design

2.2.1 Choosing sampling units: clusters or unclustered points?

Samples for assessing the accuracy of map products and area estimation are usually drawn from area frames as opposed to list frames to provide a better representation of the population (Gallego, 2004a). In an area frame, sample units can be points, lines (often referred to as transects) or areas (often referred to as segments, described by Gallego, 1995). The first step is to define the geographical area for which the accuracy assessment is to be reported and the type of sample units. Point samples are very often used, but clusters or segments have advantages in many cases: they reduce the distortion generated by co-registration and may improve the outcome when not only thematic accuracy need to be reported, but also the geometry of mapped objects such as for CLC.

A two-stage sampling approach is implemented by further selecting Primary sampling units (PSUs) and secondary sampling units (SSU) within PSUs in the

https://doi.org/10.5194/isprs-archives-XLVIII-M-1-2023-277-2023 | © Author(s) 2023. CC BY 4.0 License.

second stage. Two-stage sampling is considered suitable for accuracy assessment of land cover maps or area estimation (Stehman, 2009) and can be adopted in certain cases because it represents a good compromise between the ease of data collection and a good geographic distribution.

In the case of the validation of HRL density product for EEA a 2-stage sampling approach was applied as shown below in which 1ha square PSUs were selected and a grid of 5x5 or 10x10 SSUs was applied to facilitate the data collection as shown in Figure 1. The same approach is suggested to be adopted here although the implementation or not of the second stage is discussed in the response design section.

Figure 1: Example of SSUs organised in a 5x5 20m grid

2.2.2 Stratification

A probability sampling design is essential for map validation and area estimation (Stehman and Foody, 2008, Olofsson et al., 2014), but this is less clear for collecting training data. In exchange, a purposive selection of locations for training data can be an acceptable choice, for example collecting data along roads (Gallego et al, 2014).

Simple random, stratified random, clustered random and systematic designs are all examples of probability sampling designs. In simple random designs, classes covering a small portion of the population may not be adequately sampled. Cluster sampling is often used to reduce the costs of the collection of reference data, but does not resolve geographic distribution problems. Stratified approaches overcome this drawback; therefore, the stratified random sampling of points is one of the most common approaches to assess map accuracy. Stratified systematic with random origin also overcome this drawback and has the advantage of enhancing traceability. The main limitation of systematic sampling is that there are no unbiased estimators of the variance, although the estimators of the target variance is unbiased (Bellhouse, 1988). In practice, the random sampling estimator of variance is used even though it may slightly overestimate the true variance. In addition, there is also a strong advantage of a systematic approach in that there is better traceability as compared with a random approach: a sample drawn systematically cannot be changed since its location is known a priori. For example, LUCAS is based on a stratified systematic sample.

In many cases, the land cover map to be assessed or validated is directly used as stratification (Lowell and Jaton, 1999). This is a good approach to estimate commission errors in a binary classification, but may be weak for other cases. If we have some information that quantifies the likely proportion of errors, for example a measure of landscape complexity, it can be a more efficient stratification for all types of errors: we shall choose a higher sampling rate in more difficult areas, where both the errors and their variances are higher.

Experience has shown that a too complex stratification will not bring major improvement and that there needs to be a clear case for stratifying the Area of Interest (AoI). A clear case is when there are marked geographical differences. Experience in Europe from the Regional Inventory programme of the MARS project in the 1990s from (Taylor, 1997) showed that the more complex the analysis on which the stratification is based, the less efficient it tends to be. In addition, there should not be too many strata because even if the stratification will make it possible to reduce the number of sample unit needed for a single stratum, the benefits will be lost if too many strata are present. (Cochran, 1977, p. 134) recommends no more than 6-8 strata in total.

If we are looking at commission errors in a binary map (impervious-non impervious, or forest-non forest) and we have decided to use both classes as strata and unclustered points as sampling units, , it is possible to estimate a suitable sample size for each stratum based on the expected acceptable error rate. The standard error of the error rate can be calculated as follows: $\sigma_h = \sqrt{\frac{p_h(1-p_h)}{n_h}}$ where n_h is the sample size for stratum h and p_h is the expected error rate. This can be reworked to express the sample size n_h as a function of p_h and desired standard error σ_h: $n_h = \frac{p_h(1-p_h)}{\sigma_h^2}$.

2.2.3 Sample allocation

When using stratified sampling, the main issue to maximise the efficiency of the stratification is to optimize the sample allocation per strata. A simple way is the use of equal allocation, such as is used e.g. for the verification of the 2012 HRLs, but this is usually not very efficient. Proportional allocation is an option, but it will give disappointing results for important classes covering a small proportion (e.g. impervious land). If we have a high priority for a class , the Neyman allocation rule is a better alternative (Cochran, 1977):

$n_h = n * (N_h * \sigma_h) / [\Sigma(N_i * \sigma_i)]$ where n_h is the sample size for stratum h, n is the total sample size, N_h is the population size for stratum h, and σ_h is the standard deviation of stratum h. According to Stehman (2009), Neyman optimal allocation should be preferred for estimating area of change as well as overall accuracy, whereas equal allocation is effective for estimating user's accuracy.

https://doi.org/10.5194/isprs-archives-XLVIII-M-1-2023-277-2023 | © Author(s) 2023. CC BY 4.0 License.

In practice (Särndal et al., 1992, p. 267 and 407) recommend a minimum within-stratum sample sizes of 10-20; (Cochran, 1977, p. 134) recommends minimum within-stratum sample sizes of 20; and for temperate forest inventories (Westfall et al., 2011) recommend within-stratum sample sizes of at least 20. Therefore, for this work, if the use of a stratification approach is suitable, the maximum number of strata should not exceed 6-8 strata and the smallest number of sample units should be not be less than 20 for a single stratum.

2.2.4 Selected approach

The approach for drawing sample units as part of the validation undertaken at pan-European level by EEA was based on the LUCAS grid which makes comparison with LUCAS results easier. However, considering the different characteristics and class definition of the HRLs the thematic information of LUCAS points was not directly used even though the LUCAS photos proved particularly useful for identifying land cover / use classes. A similar approach is suggested to be adopted here using the LUCAS grid as the sampling frame corresponding to approximately 1,100,000 points throughout the European Union where land cover or land use type is observed. Using LUCAS points ensures traceability and coherence between the different layers. In the case of the HRL, the LUCAS point is then used to define the origin of the 1ha plot to be selected as described above.

LUCAS points are located every 2 km on a regular grid, as illustrated below. A set of 81 points located on an 18x18 km square constitutes a group in which every point is associated with a number comprised between 1 and 81 (the numbers do not follow each other spatially). The same pattern with the same numbers allocation is repeated all over the grid. A replicate refers to the points with the same number selected on the whole LUCAS grid.

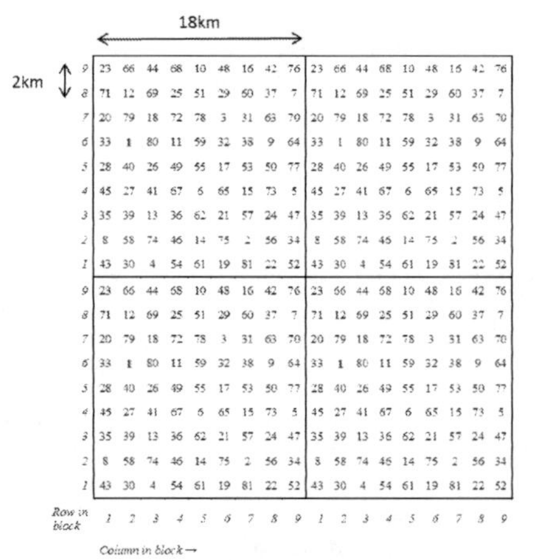

Figure 2. LUCAS points located on a regular grid

At first, the number of samples to allocate to each stratum (or thematic class) is calculated as a function of their area. In this manner the sampling design is not only systematic but also stratified. The number of sample units per stratum is to be defined to ensure sufficient level of precision at reporting level: The number of replicates to be selected for a stratum depends on its area and the number of LUCAS points intersecting the stratum.

For thematic classes covering a large proportion of the study area, 1 replicate may already exceed the defined number of samples for this class. To solve this problem, replicates are split into four sub-replicates, as illustrated by the blue numbers in the Figure below.

The opposite problem is encountered for land cover classes covering a small proportion of the study area: even by selecting 81 replicates (the maximum number), the intersecting area between the stratum and LUCAS points may be too small to reach the required number of samples. Therefore LUCAS grid can be densified by creating one point every 200 m.

Based on the lessons learnt, from previous validation exercises, the stratification procedure is simplified to only include a commission (tree cover mask 1-100) and an omission stratum (rest of the area) in order to ensure the efficiency of the approach. Stratification is based on the strata defined as follows:

- Commission: tree cover density 1-100% (minimum of 75 PSUs per country / group of countries)
- Omission: tree cover density 0% (minimum of 75 sample units per country / group of countries)
- Commission Change: all changes from 2012-2015 & 2015 -2018

If the minimum of 75 PSUs per country / group of countries is not reached at the first level of stratification, a second level is applied, per country to ensure the minimum number of PSUs.

2.3 Response design

2.3.1 Overview

Response design is the methodology used to obtain the reference data from the sample units (Stehman & Czaplewski, 1998). Stehman (2009) asserted that accuracy assessment is often made relative to some "higher quality determination of land cover". Czaplewski (2003) indicated that visual interpretation is acceptable if the spatial resolution of EO data is sufficiently better compared to the thematic classification system. For this exercise, visual interpretation of selected sample units will be used as a basis of the response design.

In a quality assessment of an EO product, the response design is a critical element: it includes all aspects of the quantification of the agreement to determine whether the map and reference data on the selected sample. The

This contribution has been peer-reviewed.
https://doi.org/10.5194/isprs-archives-XLVIII-M-1-2023-277-2023 | © Author(s) 2023. CC BY 4.0 License.

response design should closely follow the definition of the product to be assessed both in terms of geometric and thematic characteristics.

When we discuss the use of EO for statistics, the response design is still relevant especially in this case for which the use of the map products can be combined with the sampled observation in a model assisted regression estimation. Therefore, it will be important to ensure that the following characteristics are followed and closely monitored as part of the visual interpretation of selected sample units and that discrepancies between the map and sampled reference data are due to actual thematic errors:

- Minimum Mapping Unit (MMU)
- Minimum Mapping Width (MMW)
- Class definition: what is an impervious surface and what is forest?
- Ensure that the image data used is as closed as possible temporally to that used for the map production

A 2-stage sampling approach was applied:

- 1 ha square PSUs were selected based on LUCAS points as describe in section 2.2.4
- SSUs were selected in a systematic grid of 5x5 points

2.3.2 Bias Estimation and mitigation

The bias of pixel counting is the difference of the extrapolated commission and omission errors. With the notation introduced in section 2.3:

$$B = \varphi - \psi$$

The reason why classification algorithms introduce a bias is that there is no theorem than ensures that the mathematical expectation of commission and omission errors compensate each other. There is nothing similar to the *Law of Large Numbers* in sampling theory. Example of bias are provided in **Erreur ! Source du renvoi introuvable.**.

Table 2: Estimates Forest cover/change proportion forest from segments, figures in bold indicates estimates not significantly different from 0 (Source: Sannier et al., 2014).

	Forest Cover			Forest Cover Change		
	1990	2000	2010	1990-2000	2000-10	1990-2010
Direct mean	0.8568	0.8534	0.8532	-0.0034	**-0.0002**	-0.0036
Direct SE	±0.0185	±0.0186	±0.0186	±0.0009	±0.0008	±0.0012
Map estimate (pixel counts)	0.8633	0.8616	0.8613	-0.0017	-0.0003	-0.0020
Bias estimate	0.0027	0.0042	0.0049	0.0015	0.0007	0.0023
MAR Adjusted estimate	0.8606	0.8574	0.8564	-0.0032	**-0.0010**	-0.0043
MAR Adjusted SE	±0.0028	±0.0026	±0.0025	±0.0009	±0.0008	±0.0012

In some algorithms the result of pixel counting can be adjusted by tuning explicit parameters. In particular the classical maximum likelihood classifier uses the *a priori probability* of each class to push the total number of pixels of that class to higher or lower numbers. It is empirically well known that a uniform *a priori probability* usually introduces a negative bias for large classes and positive for small classes, while a proportional *a priori probability* makes the opposite.

More complex algorithms, such as random forests, neural networks and support vector machines are usually more accurate, although less robust. In these complex algorithms the parameters are hidden and cannot be tuned by the user, but the total number of pixels can be adjusted by cleaning training data or adding/removing the amount of training data in each class. Moreover, most current classification algorithms usually provide an additional probability layer which can be used to adjust the area classified by a given thematic class. These layers are not part of the products delivered for the forest layer. However, the binary layers used for the area estimation are based on density layers, the tree cover density for forest. Different threshold can be applied locally to the density values to adjust the area covered by the layers and reduce the amount of bias. Such a procedure was applied for forest cover in Gabon by Sannier al (2016b) using the global forest change dataset from Hansen et al (2013) concluding that at national level, a 70% threshold applied to the dataset was a closer depiction of the actual 30% threshold from the national forest definition, but for some areas, a more localised threshold adaptation was needed as illustrated in Figure 3 below.

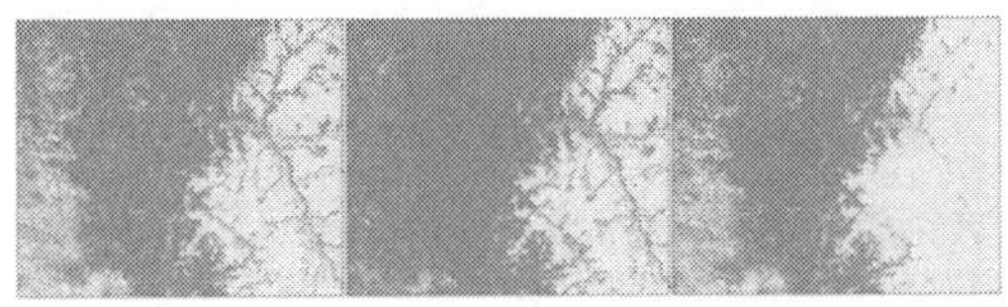

(a) (b) (c)

Figure 3: Comparison of (a) the national forest cover map, (b) the Global Forest Cover based forest cover map with a 30% tree cover threshold and (c) the Global Forest Cover based forest cover map with a 70% tree cover threshold over a 100 x 80 km extract in Eastern Gabon (Source Sannier et al, 2016)

2.3.3 Unbiased area estimates

The most usual bias correction methods are regression and calibration estimators. It should be mentioned that the "calibration estimators" used in remote sensing is not exactly the same as the "calibration estimators" as defined by Deville and Särndal (1992), more familiar to official statisticians.

Small area estimators are a larger family of more complex methods (Rao, 2003). For this study we do not intend to describe in depth small area estimators, although a brief description is foreseen for the training course.

One of the main purposes of this exercise is to produce reliable indicators of area and area change for a range of land cover/use types at selected NUTS levels based on Copernicus and map products. In addition, the efficiency of the stratification and the contribution of the map products in terms of improvements of the precision of the area and area change estimates can be evaluated by

https://doi.org/10.5194/isprs-archives-XLVIII-M-1-2023-277-2023

calculating the **stratification efficiency** (Gallego, 1995) and the **relative efficiency** of the maps combined with reference data (Taylor et al. 1997). Finally, this approach can also be used to identify the level of thematic accuracy (there is always a strong relation between thematic accuracy and precision levels) that would be required to get reliable and precise estimates of area and area change, which for the latter is always the most difficult to obtain. This level could then be used as a target for the improvement of Copernicus Land products in terms of the following:

- Minimum level of thematic accuracy to reliably detect changes for a range of Copernicus Land products and typical themes
- Minimum duration between observation periods to reliably detect change, given a specified thematic accuracy target

Area estimates can be derived from directly from the field data alone using the so-called direct expansion method as long as the data has been collected based on a probabilistic sample. The estimate of proportion (y) of class (c) and its variance are given by:

$$\bar{y}_c = \frac{1}{n} y_i \text{ and } var(\bar{y}_c) = \left(1 - \frac{n}{N}\right) \frac{1}{n(n-1)} \sum_{i=1}^{n} (y_i - \bar{y}_c)^2$$

where: y_i is the proportion of segment *i* covered by class c, *N* is total number of segments in the region, *n* is number of segments in the sample. The proportion of the study region sampled (*n/N*) is the sample fraction. The variance calculation above assumes a single stage sampling, if a 2 stage sampling is applied, this needs to be accounted for as described by Sannier et. (2014). The estimate of class area (Z) and variance in study area (D) are as follows:

$$\hat{Z}_c = D * \bar{y}_c \text{ and } var(\hat{Z}_c) = D^2 * var(\bar{y}_c)$$

where D is the area of the region. It is better to compute the estimates first as proportions rather than as absolute areas because this automatically takes account of errors resulting from small localised variations in the scale of segment maps and drawing or digitising errors. The Direct expansion estimators are calculated for each stratum present in the AOI and the total estimate just correspond to the weighted average of the proportions according to the area covered by each stratum. The standard error for the whole area is then the square root of the sum of the variance times the square of the area for each stratum:

$$\sigma_{Total} = \sqrt{\sum D_h^2 . Var_h}$$

where D_h is the stratum area. The 95% confidence interval is +/- 1.96. σ_{Total}.

However, the confidence interval of the estimate derived from direct expansion is likely to be relatively broad. To improve the precision of the estimates the sampling fraction needs to be increased or field segment data can be combined with classified satellite imagery.

A so-called Model Assisted Regression (MAR) estimator can be applied, this is more reliable than any other area estimation methodology as it provides both an area estimation per cover type together with an indication of its uncertainty.

As a summary, it relies on the combination of area estimates made at the segment level for both reference data and classified satellite imagery. The observation is paired, and a regression analysis is performed as illustrated in Figure 4.

The regression estimator methodology is fully described by Taylor et al. (1997).

The estimation of land cover type areas can be very variable from pixel counts because image classification is affected by misclassification errors affecting classes differently causing omission and commission errors. Area estimates derived from the regression estimator method are corrected from misclassification errors whilst exhibiting a more precise precision estimate than that of the direct expansion estimate thanks to the complete coverage provided by the image classification.

In summary, direct expansion estimates are unbiased, but suffer from high sampling error, pixel counts from classified satellite imagery are biased but have no sampling errors and the combination of ground data and classified imagery are unbiased and exhibit a reduced sampling error.

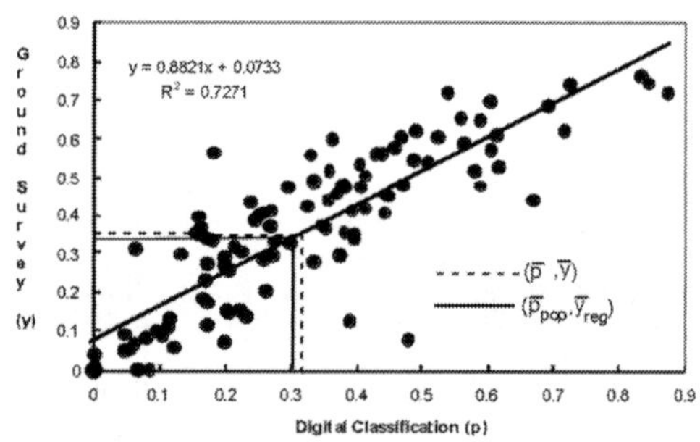

Figure 4: Relationship between digital classification and ground survey for wheat in the UK (after (Taylor, 1997).

The so-called regression estimator y_{reg} is calculated based on the regression formulae:

$$y_{reg} = \bar{y} + b * (\bar{p}_{pop} - \bar{p})$$

where $\bar{y}$ is the mean field data sample value, *b* is the slope of the regression line, $\bar{p}_{pop}$ is the proportion of pixels classified as the land cover in the whole of the region of interest and $\bar{p}$ is the classified image mean sample value. The variance of the estimate is calculated as:

$$var(\bar{y}_{reg}) = \frac{1}{n} var(y)(1 - r_{py}^2)$$

where r_{py}^2 is the regression coefficient. The variance calculation above assumes a single stage sampling, if a

2-stage sampling is applied, this needs to be accounted for as described by Sannier et. (2014).

Therefore, the higher the regression coefficient the smaller the variance and as a result the precision of the estimate. Experience for the MARS programme (Taylor, 1997) showed that high classification accuracy was correlated with high regression coefficient.

The efficiency of the regression estimator is estimated by the relative efficiency, which is the ratio of the radiance from the regression estimator method and the direct expansion estimate:

$$n_{reg} = \frac{1}{1-r_{py}^2}.$$

During the MARS project, it was shown that relative efficiency above 2 could easily be obtained, but with multi-temporal image coverage, higher classification accuracy could be achieved as part of this project resulting in reduction of variance by a greater factor. For land cover types that are very distinct such as tropical rainforest, very high relative efficiency i.e. reduction of the uncertainty can be obtained with the Regression Estimator as shown by (Sannier et al., 2016) for forest cover in Gabon for which relative efficiency close to 60 were obtained. This means that the variance of the regression is reduced by a factor of 60 as compared to the direct expansion estimate. In other words, to obtain the same level of uncertainty with the direct expansion *i.e.* without using the EO-based forest cover map, the sample size would need to be increased by a factor of 60. The 95% confidence interval of the direct expansion estimate is around 2% of the total area of Gabon when it is reduced to 0.25% for the regression estimate. It is also worth noting that this is a rare case for which the pixel count from the map is relatively close to unbiased estimates considering that it is invariably contained within the bounds of the, most precise, regression estimates.

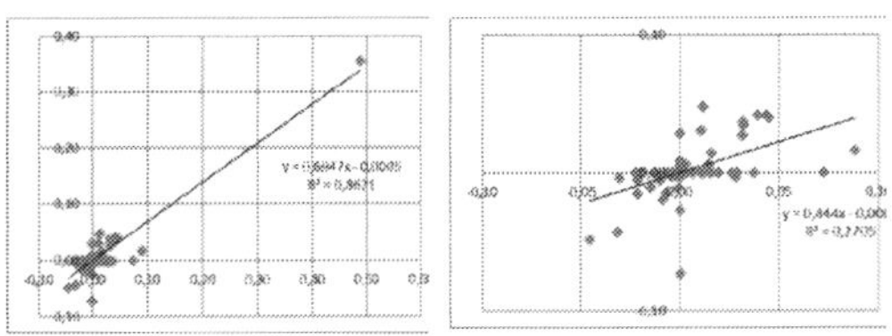

(a) (b)

Figure 5: Illustration of the importance of the quality of the linear regression with (a) all observations and (b) one observation removed.

One of the issues with the regression estimator as described above is that it is potentially very sensitive to the quality of the linear regression which is sometimes poorly reflected by the r^2 as shown in the example below (see Figure 5).

A slightly adapted estimator is also available in the form of the GREG estimator described by (Särndal et al., 1992, sec. 6.5) which focuses on the average of the differences from each observation to the mean of observations:

$$\hat{\mu}_{greg} = \frac{1}{N}\sum_{i=1}^{n} \hat{y}_i - \frac{1}{n}\sum_{i=1}^{n} (\hat{y}_i - y_i)$$

with variance:

$$\widehat{var}(\hat{\mu}_{greg}) = \frac{1}{n(n-1)}\sum_{i=1}^{n} (\varepsilon_i - \bar{\varepsilon})^2$$

where N is the number of map units, n is the reference set sample size, y_i is the observation for the ith reference set sample unit, $\hat{y}_i$ is the map class, $\varepsilon = \hat{y}_i - y_i$, and $\bar{\varepsilon} = \frac{1}{n}\sum_{i=1}^{n} \varepsilon_i$. The variance calculation above assumes a single stage sampling, if a 2-stage sampling is applied, this needs to be accounted for as described by Sannier et. (2014).

It has the advantage of not being sensitive to the quality of the linear regression but could potentially provide an estimate with a larger confidence interval compared with the direct method when the EO-based map is of poor quality for the land cover class considered. Such an approach was applied by Sannier et al. (2014) in Gabon and was applied here over Europe for Tree Cover (no forest definition was applied).

3 RESULTS AND CONCLUSIONS

The Stratified Random Sampling or Direct Estimates ad well as the GREG or Model Assisted Regression estimates were produced for the entire area of the EEA 38+UK countries covered by the Copernicus High Resolution layer on Tree Cover Density, the estimates are shown in the table below for 2012, 2015 and 2018.

Table 3. Tree cover area proportionestimates at EEA38+UK level for 2018, 2015 &nd 2012

	2018		2015		2012	
	Estimate	CI95%	Estimate	CI95%	Estimate	CI95%
Direct mean	0,305	0,004	0,290	0,004	0,273	0,005
Map (pixel counts)	0,250	n/a	0,269	n/a	0,250	n/a
MAR	0,287	0,002	0,294	0,002	0,256	0,004
Relative Efficiency		1,42		3,32		5,34

It appears that there has been a significant increase in Tree cover between 2012 and 2015. Which seems to have stabilised in 2018. It should be noted that the combination of sample and map observations provides estimates with reduced uncertainty and that the quality of the HRL TCD appears to have improved substantially between From 2012 with a relative efficiency of just 1.42 in 2012 and over 5 in 2018.

4 REFERENCES

Allen, J.D., 1990. A Look at the Remote Sensing Applications Program of the National Agricultural. *J. Off. Stat.* 6, 393–409.

Bellhouse D.R., 1988, Systematic sampling, *Handbook of Statistics*, vol. 6, ed. P.R. Krisnaiah, C.R. Rao, pp. 125-146, North-Holland, Amsterdam

Cochran W., 1977, *Sampling Techniques*. New York: John Wiley and Sons

Czaplewski, R. L. (2003). Chapter 5: accuracy assessment of maps of forest condition: statistical design and methodological considerations, pp. 115–140. In Michael A.Wulder, & Steven E. Franklin (Eds.), *Remote sensing of forest environments: concepts and case studies*. Boston: Kluwer Academic Publishers (515 pp.).

Deville J.C., Särndal C.E., 1992, Calibration estimators in survey sampling, *Journal of the American Statistical Association*, Vol 87, pp. 376-382

Gallego, J. (1995). *Sampling Frames of Square Segments*. Official Publication of the European Communities, EUR 16317, 72 pp.

Gallego, J. (2004a). Area Frames for Land Cover Estimation: Improving the European LUCAS Survey. *Third International Conference on Agricultural Statistics*. November 2-4, 2004, Mexico.

GFOI (2016), *Integration of remote-sensing and ground-based observations for estimation of emissions and removals of greenhouse gases in forests: Methods and Guidance from the Global Forest Observations Initiative*, Edition 2.0, Food and Agriculture Organization, Rome.

Hansen, M. C., P. V. Potapov, R. Moore, M. Hancher, S. A. Turubanova, A. Tyukavina, D. Thau,S. V. Stehman, S. J. Goetz, T. R. Loveland, A. Kommareddy, A. Egorov, L. Chini, C. O. Justice, and J. R. G. Townshend. (2013). High-Resolution Global Maps of 21st-Century Forest Cover Change. *Science* 342 (15 November): 850–53. Data available on-line from:http://earthenginepartners.appspot.com/science-2013-global-forest.

Lowell K., Jaton A. (ed), 1999, Spatial Accuracy assessment: Land information uncertainty in natural resources, Ann Arbor Press, Chelsea, Michigan, USA. 443 pp.

McRoberts, R. E. (2011). Satellite image-based maps: Scientific inference or pretty pictures?. *Remote Sensing of Environment*, *115*(2), 715-724.

Olofsson, P., Foody, G. M., Stehman, S. V., & Woodcock, C. E. (2013). Making better use of accuracy data in land change studies: Estimating accuracy and area and quantifying uncertainty using stratified estimation. *Remote Sensing of Environment*, 129, 122–131. doi:10.1016/j.rse.2012.10.031

Olofsson, P., Foody, G.M., Herold, M., Stehman, S.V., Woodcock, C.E., Wulder, M.A., 2014, Good practices for estimating area and assessing accuracy of land change. *Remote Sensing of Environment* 148, 42 - 57

Rao, J.N.K. (2003). *Small Area Estimation.* Wiley, New York.

Sannier, C., McRoberts, R. E., Fichet, L. -., & Makaga, E. M. K. (2014). Using the regression estimator with landsat data to estimate proportion forest cover and net proportion deforestation in gabon. *Remote Sensing of Environment, 151*, 138-148.

Sannier, C.; Gallego, J.; Dahmer, J.; Smith, G.; Dufourmont, H.; Pennec, (2016a) A. Validation of Copernicus high resolution layer on imperviousness degree for 2006, 2009 and 2012. Spatial Accuracy Assessment, Montpellier, 5–8 July 2016. http://www.spatial-accuracy.org/system/files/Accuracy%202016%20Proceedings.pdf

Sannier, C., McRoberts, R. E., & Fichet, L. V. (2016b). Suitability of Global Forest Change data to report forest cover estimates at national level in Gabon. *Remote Sensing of Environment*, *173*, 326-338.

Särndal, C.-E., Swensson, B., Wretman, J.H., 1992. *Model assisted survey sampling*. Springer, New York.

Selkowitz, D. J., & Stehman, S. V. (2011). Thematic accuracy of the National Land Cover Database (NLCD) 2001 land cover for Alaska. *Remote Sensing of Environment*, 115(6), 1401–1407. doi:10.1016/j.rse.2011.01.020.

Stehman, S. V. (2009). Sampling designs for accuracy assessment of land cover. *International Journal of Remote Sensing, 30*(20), 5243-5272.

Stehman, S. V., & Czaplewski, R. L. (1998). Design and analysis for thematic map accuracy assessment: fundamental principles. *Remote Sensing of Environment*, 64, 331–334.

Stehman, S. V., & Foody, G. M. (2008). Accuracy assessment. *The SAGE handbook of remote sensing* (pp. 297-314) doi:10.4135/9780857021052.n21

Taylor J., Sannier C., Delincé J, Gallego F.J. (1997), *Regional Crop Inventories in Europe Assisted by Remote Sensing: 1988-1993. Synthesis Report*. EUR 17319 EN, Office for Pubblications of the EC. Luxembourg. 71pp

Westfall, J.A., Patterson, P.L., Coulston, J.W., 2011. Post-stratified estimation: within-strata and total sample size recommendations. *Can. J. For. Res.* 41, 1130–1139. https://doi.org/10.1139/x11-031

This contribution has been peer-reviewed.
https://doi.org/10.5194/isprs-archives-XLVIII-M-1-2023-277-2023 | © Author(s) 2023. CC BY 4.0 License.

ENVIRONMENTAL SUSTAINABILITY ASSESSMENT OF A HIMALAYAN CATCHMENT WITH LAND COVER INDICES AND LST RELATIONSHIP USING PRINCIPAL COMPONENT ANALYSIS – A GEOSPATIAL APPROACH

[1]*M. Sathyaseelan, [1]Sanjay Kumar Ghosh, [1]Chandra Shekhar Prasad Ojha

[1]Department of Civil Engineering, Indian Institute of Technology Roorkee, Uttarakhand - 247 667, India.
*msathyaseelan@ce.iitr.ac.in, sanjay.ghosh@ce.iitr.ac.in, c.ojha@ce.iitr.ac.in

KEY WORDS: Land Cover Indices, Land Surface Temperature, Principal Component Analysis, Sustainability.

ABSTRACT:

Environmental sustainability assessment is a crucial part of the management of natural resources. Remote Sensing based environmental land cover indices such as Normalized Difference Vegetation Index (NDVI), Normalized Difference Water Index (NDWI), Normalized Difference Built-up Index (NDBI), Normalized Difference Moisture Index (NDMI), and its associated Land Surface Temperature (LST) are the major governing factors for the environmental processes that happen on the surface of the earth. These NDVI, NDWI, NDBI, NDMI, and LST are generated for 2020 using the Landsat satellite datasets. The process-based relationship among them is complex and involves various parameters but may be easily represented by multiple linear regression models. Principal Component Analysis (PCA) is one such type that efficiently handles and evaluates the contribution of each of these factors to each other based on the sampling units. The study area is the upper Ramganga catchment in the Indian Himalayas, consisting of 117 sub-catchments. These catchment units (samples) are entangled with these environmental factors. The results of the PCA reveal the relationship between each of the environmental factors and their priority. Based on the uncorrelated factors priority suggestion from the PCA, catchment units were classified as high, moderate, or low categories based on their dominance in the relationship among the factors. These spatial variations in the environmental factors can help to assess the sustainability of resources in the Himalayan catchment.

1. INTRODUCTION

1.1 Environmental Sustainability Assessment

Environmental sustainability assessment is an essential part of the conservation and management of natural resources of a region for the existence of life. Remote Sensing based land cover indices help feature extraction of land and water resources and are used for modelling various environmental and ecosystem processes concerning environmental sustainability (Robinson et al. 2017). Remote Sensing based Land Surface Temperatures (LSTs) play a vital role in monitoring and modelling the global or local Surface Energy Balances (SEBs) and water exchange processes at the land-atmosphere interactions (Zhao et al. 2019). The land cover indices and LST are useful in studying water resources management, agrobiodiversity conservation, soil health, climate change mitigation and adaptation strategies, and environmental sustainability (Singh et al., 2021). As per the United Nations (UN) Global Sustainable Development Report (GSDR, 2019) and the Sustainable Development Goals Report (SDGR, 2022), it is essential to formulate policy-making and implementation of SDGs at the local stakeholder level. The effects of climate change and land use changes need to be benchmarked in river basins (regional scale) so that the role of global trends versus local changes can be assessed and factored into decision-making (Lawford et al., 2013). Monitoring and managing natural resources at the local/regional scale as sub-catchments, administrative blocks or village levels is the current implementation scenario required for several policy-making. Science, economic viability, and community expectations is varying significantly over time. There is subsequently an increase in model complexities and difficulties in evaluating the sustainability of river basins (Horne, 2017). It is important to understand the interactions of societal factors that influence people's preferences in policy-making or decisions for sustainability (Roobavannan et al., 2020). Several studies were reported on the environmental sustainability aspects such as Eco-Hydrological sustainability (Khatun et al., 2021; Shiran et al., 2021), eco-environmental vulnerability (Nguyen et al., 2016; Liou et al., 2017; Venkatesh et al., 2020; and Kurniawan et al., 2022), vegetation dynamics (Qureshi et al., 2020), Soil moisture-LST based drought characterization (Saha et al., 2018). Several hydrological, ecological, environmental, and socio-economic problems and their solutions have been addressed by analyzing the Remote Sensing based land cover changes for environmental sustainability (Liou et al., 2017; Venkatesh et al., 2020; Kurniawan et al., 2022; Saleh et al., 2022). It includes changes in the built-up area, wetlands, forest cover, agricultural cropping patterns, and hydrological responses. The applicability and usage of those studies may vary based on the geographical prevalence of the study area and its spatial scales.

The Himalayas are notable for the richness of resources in terms of water, biodiversity, and varied ecosystems (Khare and Ghosh, 2016; Singh et al., 2021; Taloor et al., 2021; Gurjar et al., 2022; Piyoosh and Ghosh, 2022). Himalayan catchments are prone to several natural, anthropogenic, and other environmental problems associated with the hydrological cycle and the impact of climate change on the resources (Gurjar et al., 2022). In this study, using the geospatial approach, the assessment of environmental sustainability of the Upper Ramganga catchment has been carried using various land cover spectral indices and their relationship with LST. The study area has been delineated into 117 sub-catchments through the Shuttle Radar Topographic Mission (SRTM) Digital Elevation Model (DEM). Remote Sensing datasets efficiently monitor and

This contribution has been peer-reviewed.
https://doi.org/10.5194/isprs-archives-XLVIII-M-1-2023-285-2023 | © Author(s) 2023. CC BY 4.0 License.

manage these changes at variable scales. The Google Earth Engine (GEE) Platform has been used for data extraction from Landsat 8 for 2020. The land cover indices, such as the Normalized Difference Vegetation Index (NDVI), Normalized Difference Water Index (NDWI), Normalized Difference Built-up Index (NDBI), Normalized Difference Moisture Index (NDMI) and LST, are chosen as major land Remote Sensing based environmental variables that help to regulate the environmental flows, their vulnerabilities and sustainability of natural resources (Nguyen et al., 2016; Firozjaei et al., 2021; Khatun et al., 2021; Kurniawan et al., 2022). The results of these indices and LST are used to delineate the sub-catchment units as the sampling units. The threshold limits of the results are based on statistical measures such as minimum, maximum, mean, and standard deviation of the factors. The results of the indices and LST have been classified into low, moderate, high, and very high classes.

The relationship between these factors considering the catchment unit as samples is established using statistical measures such as Pearson's correlation coefficient, Regression analysis and Factor analysis, including the Principal Component Analysis (PCA). Since the number of factors is five, the PCA method is applicable and efficiently analyses the relationship among them. It is clear to identify the influencing factors using the eigenvector values and their percentage of variance among the principal component (PCs) (Qureshi et al., 2020; Abhir and Saha, 2021; Kumar et al., 2021). The results of the classification of the indices and LST are ranked based on the conceptual relation to the idea of sustainability and weighted based on the PCA results. These results are taken into the Geographical Information System (GIS) as layers, and the overlay operation is performed. The results show the variations of environmental sustainability classes such as low, moderate, high, and very high for the sub-catchments. The study is unique and deviates from other works by portraying a catchment unit-based geospatial approach for analyzing environmental sustainability results.

2. MATERIALS

2.1 Study Area

The study area is the Upper Ramganga Catchment in the Western Himalayas, which partially covers the states of Uttarakhand (53%) and Uttar Pradesh (47%) of India. It is one of the major tributaries of the Upper Ganga Basin. The geographical location of the study area lies between 28°15' 32.55" N to 30°6'32.42" N latitude and 78°15'45.37" E to 79°50'45.74"E longitude, as shown in Figure 1. The geographical area is about 18,000 km^2, and the elevation ranges from 76 m to 3000 m. As per the revised Koppen climate classification system, it falls under the categories of sub-tropical humid (cfa) and monsoonal (cwa). The mean annual rainfall varies from 2200mm to 800mm, and the average temperature ranges from 5°C to 25°C in winter and from 20°C to more than 40°C in summer. The average annual pan evaporation rate for Ramganga Basin is 4.88mm/day. The major towns such as Bareilly, Moradabad, Pilibhit, Rampur, and Shahjahanpur have populations ranging from 1 to 10 lakhs (Ramganga Basin Plan, 2020).

2.2 Datasets Used

The Digital Elevation Model (DEM) is one of the preliminary datasets for catchment or basin-level studies since any river pattern follows the topography. The catchment delineation for the study area has been obtained from the recent NASADEM_HGT version 1 an updated version of Shuttle Radar Topographic Mission (SRTM) DEM data of one arc-second resolution (~30m spatial resolution) with improved accuracy in height (Tran et al., 2023). The GEE platform has been used in this study to get the Landsat 8 Operational Land Imager (OLI) / Thermal Infrared Sensors (TIRS) Surface Reflectance (SR) collection products for 2020. Since the data is mostly available in the pre-processed format, there is no requirement for atmospheric and radiometric correction to the datasets. A known error exists in the Landsat Surface Temperature retrievals relative to clouds and possibly cloud shadows. These issues have been characterized by Cook et al., 2014. However, the presence of cloud cover in the data could be a problem. So, the cloud masking procedure has been included in the GEE program to remove 10% of the cloud cover in the data. Land cover indices such as NDVI, NDWI, NDBI, NDMI, and LST are derived from the pre-processed Landsat 8 datasets in the GEE platform.

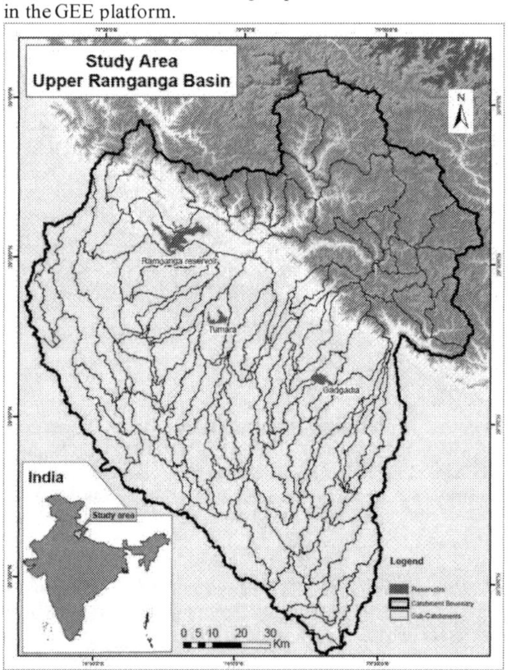

Figure 1. Study Area

Satellite Datasets Used	Band(s) used for analysis	Data Resolution
NASADEM (updated SRTM in Height measurements)	Single band raster NASADEM_HGT dataset	1-Arc Second (about 30m Ground Sampling Distance (GSD))
Landsat 8 OLI(LC08) Collection-2 Tier-1 Level-2 SR datasets	Band 4 – Red; Band 5 - Near Infrared (NIR); and Band 6 – Shortwave Infrared 1 (SWIR1)	30m X 30m GSD
Landsat 8 TIRS (LC08) Collection-1 Tier-1 Level-1 datasets accessed before 30th December 2022.	Band 10 – TIRS (this is used because USGS suggests band 10 for LST estimation rather than band 11(with uncertainty)	Resampled from 100 m to 30 m GSD data using the cubic convolution technique.

Table 1. Datasets Used

https://doi.org/10.5194/isprs-archives-XLVIII-M-1-2023-285-2023 | © Author(s) 2023. CC BY 4.0 License.

3. METHODS

Remote Sensing based degree of Environmental Sustainability has been assessed at the sub-catchment level through a workflow as shown in Figure 2.

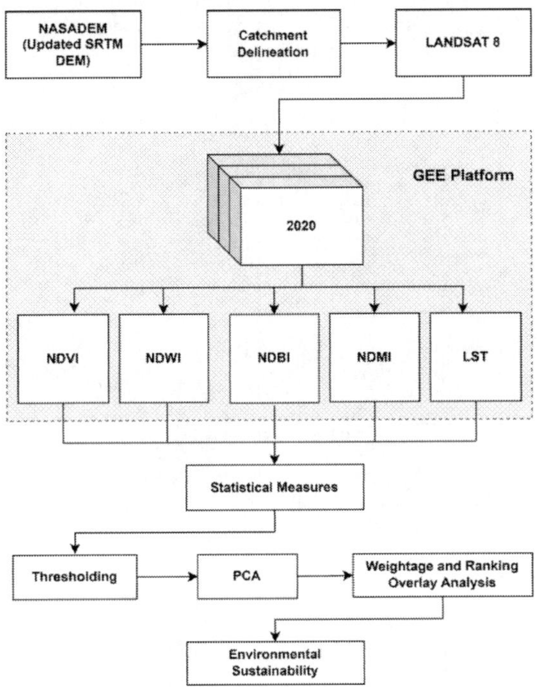

Figure 2. Flow Chart showing the methodology adopted.

3.1 Catchment delineation and Land Cover Spectral Indices

The catchment delineation has been carried on the upgraded NASA SRTM DEM with improved height measurements. Several approaches have been proposed for studying the various aspects of environmental sustainability. It is observed that the spectral indices such as NDVI, NDWI, NDBI, NDMI, and LST are used frequently, and that the same has been chosen in this study with catchment units as sampling units. Landsat 8 data has been chosen for the year 2020, with the catchment boundary as the area of interest in the GEE platform. The NDVI, NDWI, NDBI, NDMI and LST are derived from the GEE platform using the formulas suggested by Gao (1996); Weier and Herring (2000); Stathopoulou and Cartalis (2007); He et al. (2010); Khare et al. (2016); Piyoosh and Ghosh (2020) and Taloor et al. (2021) respectively.

3.2 Land Surface Temperature (LST)

The LST is calculated using the Single Channel Algorithm (SCA) (Jimenez-Munoz et al., 2014), which follows the spectral radiance of the Landsat 8 band 10. Since USGS recommends using band 10 Landsat 8 TIRS-1 datasets for the LST estimation, the same has been adopted for this study. The readily available Surface Temperature (ST) products from Collection-2 L2 ST is not used since it uses the emissivity derived from ASTER NDVI datasets which is of coarser resolution when compared with NDVI derived from Landsat 8 SR products (Table 1). The calculation of LST also exhibits known inaccuracy to pixel dimensions, clouds, and cloud shadows, which was characterized by Cook et al. (2014). So, a minimal cloud coverage of 10% has been chosen for the analysis. The prior knowledge of Land Surface Emissivity (LSE) is necessary for calculating LST (Chander et al., 2009; Valor and Caselles, 1996; Sobrino et al., 2004). So, an operational procedure to estimate LSE (i.e., ε) using the NDVI-based proportion of vegetation as proposed by Sobrino et al. (2008) and Van de Griend and Owe (1993). It has been used with the NDVI values of study area. Sobrino et al. (2004) proposed LSE calculation for the mixed pixels of vegetation. In case of unknown vegetation and soil, the emissivity values are taken as $\varepsilon_v = 0.985$ and $\varepsilon_s = 0.960$ (Valor and Caselles, 1996; Sekertekin and Bonafoni, 2020), respectively. The proportion of vegetation is also referred to as fractional vegetation cover, which is used for determining the surface emissivity of the area. In this study, the resulting emissivity is calculated by the above methods as 0.981. Based on the emissivity value, the Land Surface Temperature (LST) is calculated using the formula given by Stathopoulou and Cartalis (2007); Piyoosh and Ghosh (2020).

3.3 Thresholding

The factors contributing to the environmental sustainability are quantitatively assessed by the zonal statistical measures for the catchment unit areas. The thresholding of the statistical measures is based on the standard deviation from the mean of the histogram of data. The factors are classified as low, moderate, high, and very high based on the thresholding classes for each of the sampling catchment units in relevance to the concept of sustainability.

3.4 Principal Component Analysis (PCA)

Principal Component Analysis (PCA) is a multivariate statistical technique that applies a linear transformation to the original data to study the complex relationship between those environmental and ecological variables (Estornell et al., 2013; Jolliffe and Cadima, 2016; Firozjaei et al., 2021; Kurniawan et al., 2022; Kumar et al., 2022). PCA also reduces the dimensionality of the data matrix through Singular Value Decomposition (SVD) algorithm, and to segregate the dominant information in the data (Zhou et al., 2018; Qureshi et al., 2020). The uncorrelated linear principal components (PCs) after the orthogonal transformation gives the variance relationship between the components. The PCA is also supported by the Pearson's correlation coefficient to study the linear one-to-one relationship between the factors.

3.5 GIS Overlay

The weights of the GIS layers are obtained from PCA, and the corresponding conceptual ranking has been assigned to each of the layer classes relevant to the sustainability of environment. Later the intersection of GIS layers by overlay operation provides the final environmental sustainability with the threshold classes.

4. RESULTS

4.1 Catchment Delineation

The SRTM DEM with a spatial resolution of 1-arc second (approximately 30 m) is sufficient for this study and the catchment delineation threshold of about 30 km^2 area coverage.

This contribution has been peer-reviewed.
https://doi.org/10.5194/isprs-archives-XLVIII-M-1-2023-285-2023 | © Author(s) 2023. CC BY 4.0 License.

This size is optimal for planning and managing catchment-level environmental sustainability studies, where the shape may vary subject to the village or block boundary. Also, the variations in the size of the catchments are subjective and relevant to the type of applications and planning purposes where it is implemented. Here, the number of catchment units delineated were 117 as given in Figure 1.

4.2 Results of the Spectral Indices and LST

NDVI is the most appropriate indices for studying vegetation cover in general and its characteristics. The NDVI values of zero or less than that indicate an area with no vegetation, such as barren lands or urban or water bodies. The results of the NDVI for the study area range from -0.126 to 0.898 and shown as in Figure 3 as NDVI (A) and NDVI (B). The results also show that the vegetation is higher along the stretch of lower hilly areas where the reservoirs are seen and along some portions of the higher elevations. All other areas are devoid of waterbodies, while urban areas show moderate vegetation cover. At the sub-catchment levels, lower vegetation is seen in fewer basins.

NDWI index segregates water bodies from the remote sensing datasets. The downside of the NDWI is reflectance observed in the green band from build-up or barren areas added to the calculation of the index with an overestimation of water bodies. However, from the results, it is observed that the NDWI can classify the waterbodies efficiently, but meanwhile, it is also considering the urban areas moderately to some extent in their index range. The NDWI is values range between -0.805 and 0.378. The thresholding of the NDWI class is based on statistical measures, and the results are illustrated in Figure 3 as NDWI (A) and NDWI (B). The results show that higher values represent as waterbodies, mostly the moderate values are found in the plain regions of the basin, and the higher vegetation regions show less water index values which is a downside of NDWI trade-offs.

NDBI is the mapping index for the build-up areas. The results are shown in Figure 3. It is observed from the results that NDBI classifies the barren land into the built-up area to some extent because of the reflectance observed in the SWIR and NIR band. Users should consider the trade-offs between the efficiency and accuracy of these indices when choosing between the actual performance of the indices to the desired application. The value of the NDBI ranges between -0.433 to 0.246. The results show that the NDBI segregates the build-up area at higher values and moderately classifies barren land or less/no vegetation areas. Some higher potential build-up areas are identified at the sub-catchment level based on the statistical thresholding of NDBI results.

The NDMI is a Remote Sensing based indices for calculating the overall moisture content in vegetation, soil, and water areas/wetlands. The NDMI is calculated using the same bands of NDBI with slight changes in the numerical operation of the bands. Also, it can overcome the drawbacks of the NDWI in calculating the waterbodies/wetlands or soil moisture area by taking advantage of the Shortwave Infrared band (SWIR1). The value of the NDMI ranges between 0.602 and -0.337, as illustrated in Figure 3 as NDMI(A) and NDMI(B).

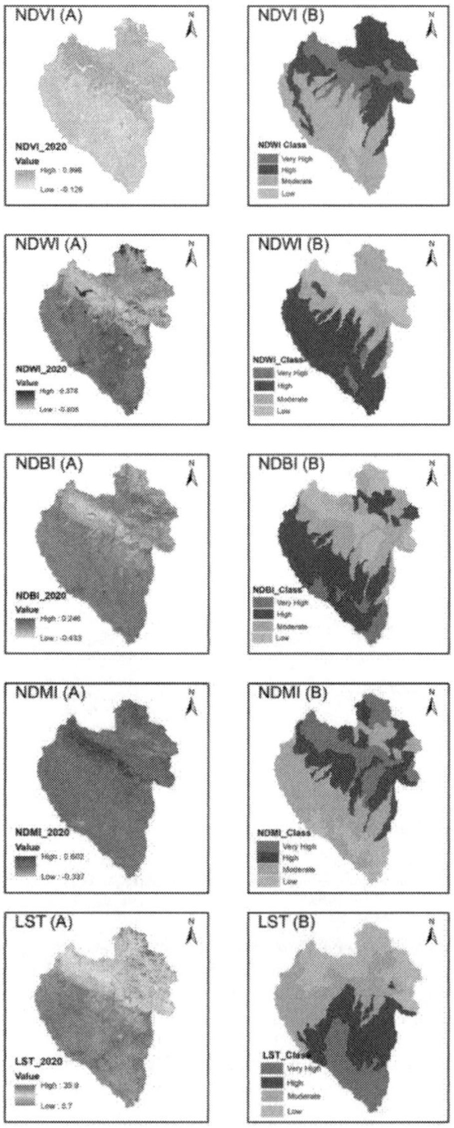

Figure 3. Land Cover Indices and LST

The Land Surface Temperature (LST) results show that it varies from 5.7°C to 35.9°C for 2020. It is calculated from the NDVI-based Land Surface Emissivity (LSE) given in Section 2. of the paper. The build-up areas show higher temperature ranges, and mostly the plain areas where the agricultural cover and waterbodies/wetlands show moderate to higher temperature ranges. The hilly areas with higher elevations show lower temperature ranges. The results of the LST are illustrated in Figure 3.

4.3 Results of the Statistical thresholds

The results of the thresholding based on the zonal statistical measures for the catchment unit corresponding to the factors is given in Table 2.

https://doi.org/10.5194/isprs-archives-XLVIII-M-1-2023-285-2023 | © Author(s) 2023. CC BY 4.0 License.

Factors	Threshold values	Category	Rank	Weight
NDVI	< 0.362	Low	IV	0.216
	0.362 <0.463	Moderate	III	
	0.463 <0.564	High	II	
	0.564 <0.720	Very High	I	
NDWI	< -0.542	Low	IV	0.188
	-0.542 < -0.452	Moderate	III	
	-0.452 < -0.362	High	II	
	>-0.362	Very High	I	
NDBI	< -0.165	Very High	I	0.203
	-0.165 < -0.118	High	II	
	-0.118 < -0.0071	Moderate	III	
	>-0.0071	Low	IV	
NDMI	< 0.122	Low	IV	0.259
	0.122 < 0.174	Moderate	III	
	0.174 < 0.226	High	II	
	> 0.226	Very High	I	
LST	< 17.416	Very High	I	0.187
	17.416 < 21.7	High	II	
	21.7 < 25.984	Moderate	III	
	25.984 < 29.502	Low	IV	

Table 2. Threshold values for factors

4.4 Results of the Correlation and the PCA analysis

The results of the correlation matrix given in Table 3. shows a strong negative correlation between the NDVI-LST; NDMI-LST and a good positive correlation between the NDWI-LST; NDBI-LST, respectively.

The NDVI has a strong negative correlation with all the factors except NDMI, which has a strong positive correlation. It is observed that the NDWI can classify the waterbodies efficiently while also classifying the urban areas moderately. It is important to note that, the NDWI has a strong negative correlation with the NDVI. In general comparison it may seems to be a wrong correlation, here, the results of the land cover indices are given as a mean of the catchment units and hence, the correlation takes the average of the pixels of NDWI apart from the water pixels, it also considers the non-water pixels areas, resulting in a strong negative correlation.

Factors	NDVI	NDWI	NDBI	NDMI	LST
NDVI	1				
NDWI	-0.95	1			
NDBI	-0.70	0.48	1		
NDMI	0.71	-0.51	-1.00	1	
LST	-0.70	0.60	0.62	-0.64	1

Table 3. Correlation Matrix of the factors

So, a positive correlation is seen among the NDWI-LST, and it must be noted that seasonal factors also play a role in defining the correlation results (Guha and Govil, 2021). However, here only the annual data have been considered for this study. The results of the NDMI show a positive correlation with the NDVI negative correlation with NDWI, NDBI, and LST. Here, the NDMI and the NDBI use the same bands with slight changes in the sign operations; hence, there is a strong negative correlation of -1 between them. Meanwhile, there is a negative correlation between the NDBI-NDVI and a positive correlation between NDBI-NDWI and NDBI-LST, respectively.

The relationship between the factors can be further studied more accurately using the PCA technique. The eigenvalues corresponding to each principal component (PCs) provides the hierarchy of the components in the relation. The eigenvalues of PC factors, their corresponding variances, and the cumulative variances are given in Table 4.

PCs	Eigenvalues	Standard deviation	Proportion of Variance	Cumulative Proportion of Variance
PC1	3.778	1.943	0.755	75.555
PC2	0.798	0.893	0.159	91.521
PC3	0.410	0.639	0.081	99.712
PC4	0.013	0.115	0.002	99.977
PC5	0.001	0.034	0.000	100.000

Table 4. Results of all major PCs and its variance

Based on these results, the PC1 with an eigenvalue greater than one with a higher variance of about 75.60% is considered for the weightage analysis.

Factors	PC 1	Component Score Coefficients	Corresponding Weights Calculated with Sum=1
NDVI	-0.936	-0.248	0.216
NDWI	0.815	0.216	0.188
NDBI	0.881	0.233	0.203
NDMI	-0.894	-0.237	0.206
LST	0.813	0.215	0.187

Table 5. Weightage of the factors using PC1 score coefficients.

The scores of the PC1, as given in Table 5., are rescaled to the weightage of the factors with sum equals to one.

4.5 Results of the Environmental Sustainability Assessment

Based on the results of PCA, the weights of the factors that contribute to the Environmental sustainability are calculated as given in Table 5. is then assigned to the corresponding factors as given in Table 2.

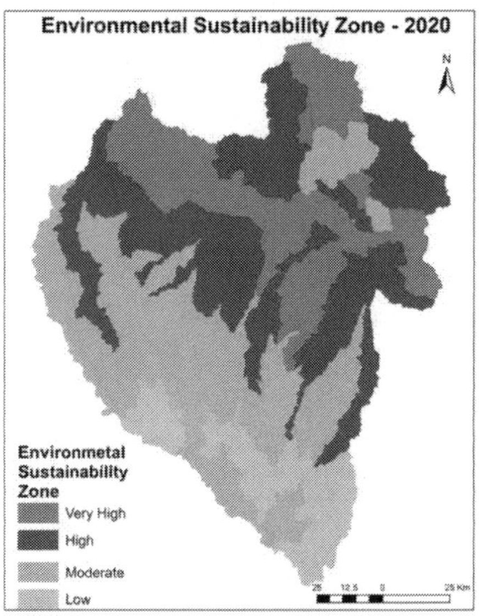

Figure 4. Environmental Sustainability Zones

All the factors that contribute to the analysis are classified as low, moderate, high, and very high classes based on the Table 2. The factors are taken into the GIS platform, where the layers are weighted, based on the PCA results and the corresponding ranks have been assigned to each class in all the layers, and the GIS overlay operation is performed. The final environmental sustainability zones for the sub-catchments were identified and classified as low, moderate, high, and very high classes based on the GIS overlay result, as shown in Figure 4. The results show that the low sustainability catchments are identified at the lower region of the basin, very high sustained zones are identified at the higher elevation regions, and along the stretch of the lower hilly regions of the Himalayas, where the reservoirs and dams are located. Higher sustainable catchment zones are found on either side of the very high sustainable zones. Most of the moderate zones of sustainable catchments are identified in the plain areas, and few are in the hilly areas.

5. DISCUSSION

The results of the land cover indices, such as the NDVI, NDWI, NDBI, NDMI, and LST, are studied for 2020 only. Their seasonal relationship with the factors also plays an important role in the seasonal assessment. However, this study has not considered it, which is one of the limitations. The results of NDWI in the zonal statistics with catchment areas yield a negative correlation with the NDVI which may leads to contradictory results because of considering the non-water areas in calculating the mean. The other limitations are that the number of factors contributing to environmental sustainability assessment is less. The limitation of the study includes the watershed prioritization based on the morphometric characteristics, several other socio-economic statuses of the region, and demographic factors, like population density and migration is not included and may be considered in future works. The correlation and the PCA technique uses the relationship the uncorrelated factors with eigenvalues and eigenvectors, where the relationship is established among of one-to-many factors has been established, and the individual factor contribution is identified. Since the number of PCs that is more than the eigenvalue of 1 is only one principal component, and hence there is no rotated component matrix exists for the same. If the number of significant components is more than 1, then we should consider the rotated component matrix for determining the weight of the factors contributing to environmental sustainability. Based on the environmental sustainability assessment results, the regions at the lower elevation or plain areas are identified as low sustainable catchment zones, where the presence of built-up and the LST are higher and favorable for the likelihood of urban agglomeration. The moderately sustainable catchments are the region where there is a need for mitigation measures to ensure or stop the degradation of environmental sustainability. The higher and very high zones are less vulnerable to the degradation of sustainability.

6. CONCLUSION

The overall study uses minimal remote sensing-based land cover indices and LST for the analysis. Despite the limitations discussed above, this study tried to assess the environmental sustainability of a catchment based on the remote sensing inputs at the sub-catchment level. The results of the PCA analysis their relationship between the factors helps in modelling the scenario and may help in prediction of future conditions using the Machine Learning (ML) algorithms. This study may help planners and decision-makers to formulate strategies and adaptation policies or measures for environmental sustainability under climate change initiatives.

ACKNOWLEDGEMENTS

The authors sincerely thank the Department of Civil Engineering, IIT Roorkee, India and the Ministry of Education, Government of India, for providing the MHRD Fellowship.

REFERENCES

Abir, F. A., & Saha, R. (2021). Assessment of land surface temperature and land cover variability during winter: a spatio-temporal analysis of Pabna municipality in Bangladesh. *Environmental Challenges*, *4*, 100167. https://doi.org/10.1016/j.envc.2021.100167

Chander, G., Markham, B. L., & Helder, D. L. (2009). Summary of current radiometric calibration coefficients for Landsat MSS, TM, ETM+, and EO-1 ALI sensors. *Remote sensing of environment*, *113*(5), 893-903. https://doi.org/10.1016/j.rse.2009.01.007

Cook, M., Schott, J. R., Mandel, J., & Raqueno, N. (2014). Development of an operational calibration methodology for the Landsat thermal data archive and initial testing of the atmospheric compensation component of a land surface temperature (LST) product from the archive. *Remote Sensing*, *6*(11), 11244-11266. https://doi.org/10.3390/rs61111244

Estornell, J., Martí-Gavilá, J. M., Sebastiá, M. T., & Mengual, J. (2013). Principal component analysis applied to remote sensing. *Modelling in Science Education and Learning*, *6*, 83-89. https://doi.org/10.4995/msel.2013.1905

Firozjaei, Mohammad Karimi, Solmaz Fathololoumi, Majid Kiavarz, Asim Biswas, Mehdi Homaee, and Seyed Kazem Alavipanah. "Land Surface Ecological Status Composition Index (LSESCI): A novel remote sensing-based technique for modeling land surface ecological status." *Ecological Indicators* 123 (2021): 107375. https://doi.org/10.1016/j.ecolind.2021.107375

Gao, B. C. (1996). NDWI—A normalized difference water index for remote sensing of vegetation liquid water from space. *Remote sensing of environment*, *58*(3), 257-266. https://bpb-us-w2.wpmucdn.com/sites.udel.edu/dist/d/1835/files/2014/06/NDWI-A-Normalized-Difference-Water-Index-for-Remote-Sensing-of-Vegetation-Liquid-Water-From-Space-1ko95nn.pdf

Guha, S., & Govil, H. (2021). An assessment on the relationship between land surface temperature and normalized difference vegetation index. *Environment, Development and Sustainability*, *23*, 1944-1963. https://doi.org/10.1007/s10668-020-00657-6

Gurjar, S. K., Shrivastava, S., Suryavanshi, S., & Tare, V. (2022). Assessment of the natural flow regime and its variability in a tributary of Ganga River: Impact of land use and land cover change. *Environmental Development*, *44*, 100756. https://doi.org/10.1016/j.envdev.2022.100756

He, C., Shi, P., Xie, D., & Zhao, Y. (2010). Improving the normalized difference built-up index to map urban built-up areas using a semiautomatic segmentation approach. *Remote Sensing Letters*, *1*(4), 213-221. https://doi.org/10.1080/01431161.2010.481681

Jiménez-Muñoz, J. C., Sobrino, J. A., Skoković, D., Mattar, C., & Cristobal, J. (2014). Land surface temperature retrieval methods from Landsat-8 thermal infrared sensor data. *IEEE Geoscience and remote sensing letters*, *11*(10), 1840-1843. https://doi.org/10.1109/LGRS.2014.2312032

Jolliffe, I. T., & Cadima, J. (2016). Principal component analysis: a review and recent developments. *Philosophical transactions of the royal society A: Mathematical, Physical and Engineering Sciences*, *374*(2065), 20150202. https://doi.org/10.1098/rsta.2015.0202

Khatun, R., Talukdar, S., Pal, S., Saha, T. K., Mahato, S., Debanshi, S., & Mandal, I. (2021). Integrating remote sensing with swarm intelligence and artificial intelligence for modelling wetland habitat vulnerability in pursuance of damming. *Ecological Informatics*, *64*, 101349. https://doi.org/10.1016/j.ecoinf.2021.101349

Khare, S., Ghosh, S. K., Latifi, H., Vijay, S., & Dahms, T. (2017). Seasonal-based analysis of vegetation response to environmental variables in the mountainous forests of Western Himalaya using Landsat 8 data. *International journal of Remote Sensing*, *38*(15), 4418-4442. https://doi.org/10.1080/01431161.2017.1320450

Kumar, A., Singh, S., Pramanik, M., Chaudhary, S., Maurya, A. K., & Kumar, M. (2021). Watershed prioritization for soil erosion mapping in the Lesser Himalayan Indian basin using PCA and WSA methods in conjunction with morphometric parameters and GIS-based approach. *Environment, Development and Sustainability*, 1-39. https://doi.org/10.1007/s10668-021-01586-8

Kurniawan, R., Saputra, A. M. W., Wijayanto, A. W., & Caesarendra, W. (2022). Eco-environment vulnerability assessment using remote sensing approach in East Kalimantan, Indonesia. *Remote Sensing Applications: Society and Environment*, *27*, 100791. https://doi.org/10.1016/j.rsase.2022.100791

Lawford, R., Bogardi, J., Marx, S., Jain, S., Wostl, C. P., Knüppe, K., ... & Meza, F. (2013). Basin perspectives on the water–energy–food security nexus. *Current Opinion in Environmental Sustainability*, *5*(6), 607-616. https://doi.org/10.1016/j.cosust.2013.11.005

Liou, Y. A., Nguyen, A. K., & Li, M. H. (2017). Assessing spatiotemporal eco-environmental vulnerability by Landsat data. *Ecological indicators*, *80*, 52-65. https://doi.org/10.1016/j.ecolind.2017.04.055

Nguyen, A. K., Liou, Y. A., Li, M. H., & Tran, T. A. (2016). Zoning eco-environmental vulnerability for environmental management and protection. *Ecological Indicators*, *69*, 100-117. https://doi.org/10.1016/j.ecolind.2016.03.026

Nguyen, B. Q., Vo, N. D., Le, M. H., Nguyen, Q. D., Lakshmi, V., & Bolten, J. D. (2023). Quantification of global Digital Elevation Model (DEM)–A case study of the newly released NASADEM for a river basin in Central Vietnam. *Journal of Hydrology: Regional Studies*, *45*, 101282. https://doi.org/10.1016/j.ejrh.2022.101282

Piyoosh, A. K., & Ghosh, S. K. (2022). Analysis of land use land cover change using a new and existing spectral indices and its impact on normalized land surface temperature. *Geocarto International*, *37*(8), 2137-2159. https://doi.org/10.1080/10106049.2020.1815863

Qureshi, S., Alavipanah, S. K., Konyushkova, M., Mijani, N., Fathololomi, S., Firozjaei, M. K., ... & Kakroodi, A. A. (2020). A remotely sensed assessment of surface ecological change over the Gomishan Wetland, Iran. *Remote Sensing*, *12*(18), 2989. https://doi.org/10.3390/rs12182989

Ramganga Basin Plan. (2020). http://www.swaraup.gov.in/BAPS%20Reports/Ramganga%20Basin%20Plan%20Final/Ramganga%20Basin%20Plan%20(March%202020)%20Vol-1.pdf

Robinson, D. A., Panagos, P., Borrelli, P., Jones, A., Montanarella, L., Tye, A., & Obst, C. G. (2017). Soil natural capital in Europe; a framework for state and change assessment. *Scientific reports*, *7*(1), 6706. https://doi.org/10.1038/s41598-017-06819-3

Roobavannan, M., Kandasamy, J., Pande, S., Vigneswaran, S., & Sivapalan, M. (2020). Sustainability of agricultural basin development under uncertain future climate and economic conditions: A socio-hydrological analysis. *Ecological Economics*, *174*, 106665. https://doi.org/10.1016/j.ecolecon.2020.106665

Saha, A., Patil, M., Goyal, V. C., & Rathore, D. S. (2018). Assessment and impact of soil moisture index in agricultural drought estimation using remote sensing and GIS techniques. *Multidisciplinary digital publishing institute proceedings*, *7*(1), 2. https://doi.org/10.3390/ECWS-3-05802

Saleh, S. K., Sanaei, A., Amoushahi, S., & Ranjbar, S. (2022). Effect of landscape pattern changes and environmental indices on land surface temperature in a fragile ecosystem in southeastern Iran. *Environmental Science and Pollution Research*, 1-17. https://doi.org/10.1007/s11356-022-24602-4

Sekertekin, A., & Bonafoni, S. (2020). Land surface temperature retrieval from Landsat 5, 7, and 8 over rural areas: Assessment of different retrieval algorithms and emissivity models and toolbox implementation. *Remote sensing*, *12*(2), 294. https://doi.org/10.3390/rs12020294

Shiran, M., Mozzi, P., Adab, H., & Asadi, M. A. Z. (2021). Remote sensing assessment of changes of surface parameters in response to prolonged drought in the arid zone of central Iran (Gavkhoni playa). *Remote Sensing Applications: Society and Environment*, *23*, 100575. https://doi.org/10.1016/j.rsase.2021.100575

Singh, R., Patel, S. K., Tiwari, A. K., & Singh, G. S. (2021). Assessment of flood recession farming for livelihood provision, food security and environmental sustainability in the Ganga River Basin. *Current Research in Environmental Sustainability*, *3*, 100038. https://doi.org/10.1016/j.crsust.2021.100038

Sobrino, J. A., Jiménez-Muñoz, J. C., & Paolini, L. (2004). Land surface temperature retrieval from LANDSAT TM 5. *Remote Sensing of environment*, *90*(4), 434-440. https://doi.org/10.1016/j.rse.2004.02.003

Sobrino, J. A., Jiménez-Muñoz, J. C., Sòria, G., Romaguera, M., Guanter, L., Moreno, J., ... & Martínez, P. (2008). Land surface emissivity retrieval from different VNIR and TIR sensors. *IEEE transactions on geoscience and Remote Sensing*, *46*(2), 316-327 https://doi.org/10.1109/TGRS.2007.904834

Stathopoulou, M., & Cartalis, C. (2007). Daytime urban heat islands from Landsat ETM+ and Corine land cover data: An application to major cities in Greece. *Solar Energy*, *81*(3), 358-368. https://doi.org/10.1016/j.solener.2006.06.014

Taloor, A. K., Manhas, D. S., & Kothyari, G. C. (2021). Retrieval of land surface temperature, normalized difference moisture index, normalized difference water index of the Ravi basin using Landsat data. *Applied Computing and Geosciences*, *9*, 100051. https://doi.org/10.1016/j.acags.2020.100051

The Global Sustainable Development Report (GSDR). (2019). The future is Now: Science for Achieving Sustainable Development. https://sdgs.un.org/sites/default/files/2020-07/24797GSDR_report_2019.pdf

The Sustainable Development Goals Report (SDGR). (2022). The United Nations. https://unstats.un.org/sdgs/report/2022/The-Sustainable-Development-Goals-Report-2022.pdf

Tran T.-N.-D, Nguyen, B. Q., Vo, N. D., Le, M. H., Nguyen, Q. D., Lakshmi, V., & Bolten, J. D. (2023). Quantification of global Digital Elevation Model (DEM)–A case study of the newly released NASADEM for a river basin in Central Vietnam. *Journal of Hydrology: Regional Studies*, *45*, 101282. https://doi.org/10.1016/j.ejrh.2022.101282

Valor, E., & Caselles, V. (1996). Mapping land surface emissivity from NDVI: Application to European, African, and South American areas. *Remote sensing of Environment*, *57*(3), 167-184. https://doi.org/10.1016/0034-4257(96)00039-9

Van de Griend, A. A., & Owe, M. (1993). On the relationship between thermal emissivity and the normalized difference vegetation index for natural surfaces. *International Journal of remote sensing*, *14*(6), 1119-1131. https://doi.org/10.1080/01431169308904400

Venkatesh, R., Abdul Rahaman, S., Jegankumar, R., & Masilamani, P. (2020). ECO-ENVIRONMENTAL VULNERABILITY ZONATION IN ESSENCE OF ENVIRONMENTAL MONITORING AND MANAGEMENT. *International Archives of the Photogrammetry, Remote Sensing & Spatial Information Sciences*, *43*. https://doi.org/10.5194/isprs-archives-XLIII-B5-2020-149-2020

Weier, J., & Herring, D. (2000). Measuring Vegetation (NDVI & EVI). NASA Earth Observatory. *Washington, DC, USA*. URL:https://earthobservatory.nasa.gov/features/MeasuringVegetation

Horne, J. (2017). The politics of water reform and environmental sustainability in the Murray–Darling Basin. *Water International*, *42*(8), 1000-1021. https://doi.org/10.1080/02508060.2017.1412201

Zhao, W., He, J., Wu, Y., Xiong, D., Wen, F., & Li, A. (2019). An analysis of land surface temperature trends in the central Himalayan region based on MODIS products. *Remote Sensing*, *11*(8), 900. http://dx.doi.org/10.3390/rs11080900

TECHNIQUES FOR COMPARING MULTI-TEMPORAL ARCHIVE AERIAL IMAGERY FOR GLACIER MONITORING WITH POOR GROUND CONTROL

Marco Scaioni [1], Arsalan Malekian [2], Davide Fugazza [3]

[1] Politecnico di Milano, Dept. of Architecture, Built Environment and Construction Engineering (DABC)
via Ponzio 31, 20133 Milano, Italy – email: marco.scaioni@polimi.it
[2] Politecnico di Milano, Lecco Campus, via G. Previati 1/c, 23900 Lecco, Italy – email: arsalan.malekian@mail.polimi.it
[3] Università degli Studi di Milano, Department of Environmental Science and Policy (DESP), Via Celoria 2, 20133 Milano, Italy
email: davide.fugazza@unimi.it

KEY WORDS: Archive Aerial images, Change detection, Glacier Monitoring, Ground Control, Multi-Temporal, Point Cloud Comparison, Structure-from-Motion

ABSTRACT:

The measurement of geometric changes in Alpine glaciers is an essential aspect to assess their reaction to climate change effects. Archive aerial images may integrate valuable information to this purpose at times when other types of remotely sensed data are not available. The application of photogrammetric techniques such as Structure-from-Motion (SfM) and Multi-View-Stereo matching allows the extraction of dense point clouds to model the glacier environment. The comparison of multiple datasets requires to setup a stable reference system, a task that in archive photos is generally carried out by means of ground control points (GCPs). This paper would like to propose and assess some techniques to cope with the lack of ground control. Multitemporal SfM (MSfM) is presented and tested on a dataset including six different aerial blocks collected by means of analogue (PAN/RGB) and digital airborne cameras from 1967 to 2006. These images have been downloaded from the IGNF online repository and cover the area of the Val Veny (Brenva and Miage glaciers) in the Mount Blanc massif, at the border between Italy and France. Coupled with other solutions (i.e., extraction of GCPs from maps and DTMs and ICP co-registration of point clouds), MSfM has revealed as a suitable technique for coregistration of multiple photogrammetric blocks of aerial photos with minimum ground control. Some tests carried out in the case study area demonstrated that the integration of MSfM and ICP coregistration for refinement may significantly improve the comparison between multiple point clouds, which is a fundamental pre-requisite for the analysis of glacier changes over time.

1. INTRODUCTION

Monitoring glaciers to assess their reaction to climate change effects includes several aspects and requires multiple technologies and sensors. The evaluation of geometric changes plays a fundamental role to detect and to measure ice-mass variations. In recent years tailored UAV (Unmanned Aerial Vehicles) missions have been organized to investigate specific glaciers (see, e.g., Fugazza et al., 2018), while high-resolution satellite imagery provides information to reconstruct the geometry in the recent decades, to be continued in the future.

Historical archive aerial photos (Poli et al., 2020) may provide valuable information to this purpose combined with up-to-date photogrammetric techniques, such as Structure-from-Motion (SfM) and Multi-View-Stereo (MVS) dense matching (Barazzetti et al., 2009). Point clouds of reconstructed 3D surface of glaciers may be used to detect thickness changes and height differences over time. This task can be accomplished by applying suitable techniques for computing distances between each pair of point clouds (Fey et al., 2017; Malekian et al., 2022).

Under the technical point-of-view, one of the crucial points to cope with the alignment of point clouds derived from archive photos is how to deal with poor ground control (Tian et al., 2022). Ground control is fundamental to setup a stable reference system for the computation of coordinates to be compared within time. In addition, a good ground control also helps compensate for possible bias in the exterior orientation (EO). This task is usually implemented by introducing enough well distributed *ground control points* (GCPs), whose coordinates are measured by using differential GNSS techniques. On the other hand, the development of IMU/GNSS sensors may directly provide the EO of a photogrammetric block, but this solution cannot be used with archive photos dating back a few decades ago. Unfortunately, GCPs are often unavailable to be used with archive photos. Due to the lower resolution and data quality of these datasets, on one side, and the changes on the terrain (new constructions, vegetation growth, soil erosion, landslides, etc.), it is often difficult to measure new GCPs to be recognizable in photos collected in the past decades.

During some research projects aiming at studying glacier changes within time, the authors have realized the need for alternative techniques to replace or to integrate the use of GCPs when comparing multitemporal datasets including photos from the archives.

These techniques have been tried during a study regarding a group of Alpine glaciers in the Mount Blanc massif in the Italian Alps at the border with France. Six datasets of digitized analogue aerial photos provided by the National Geographic and Forestry Institute of France (IGNF) have been selected, downloaded, and used for photogrammetric processing. These datasets cover a time span of approximately 40 years from 1967 to 2006 and include images acquired with either analogue (panchromatic and RGB) and digital airborne cameras. While the change in ice thickness of these glaciers was relatively small until the mid-1990s, this study revealed an increasing reduction rate at the beginning of 21st century.

The proposed paper would like to describe the methodological approach for dealing with poor ground control. More scientific results on the case study can be found in Malekian et al. (2022; 2023).

2. TECHNIQUES FOR GROUND CONTROL

2.1 Overview

On the basis of the preliminary issues illustrated in Section 1, we consider a generic set of aerial blocks, which may come either from the archives and from more recent aerial missions. In general, historical datasets of archive photos have not been collected for investigation/monitoring of glaciers but consisted in typical photogrammetric blocks at scales used for topographic mapping projects (1:7,000 - 1:30,000) and standard overlap (60%-70% along strip and 20%-30% side lap). Exception to this is the case of blocks flown over Antarctic and Arctic regions, which were collected with the purpose of scientific investigation (see, e.g., Fieber et al., 2018; Tian et al., 2022).

While for block collected in recent years the ground control can be established by a set of GCPs or by IMU/GNSS data (if available), in the past the datasets may consist in the images only.

In the following, three different techniques will be illustrated to overcome the lack of ground control:

1. Multitemporal Structure-from-Motion (MSfM);
2. Extraction of GCPs from maps and DTMs; and
3. Surface coregistration based on ICP (Iterative Closest Point).

These techniques are not intended to be only use independently from one another, but they may be also cascaded in a structured procedure.

2.2 Multitemporal Structure-from-Motion

Structure-from-Motion Photogrammetry (SfM – see Granshaw, 2018) is commonly referred to as the entire automatic procedure for computing the EO of a block of images and to output a dense point cloud derived from the application of *Multi-View-Stereo* (MVS) matching techniques (see Eltner et al., 2016; James et al., 2019). While at the beginning SfM could be only applied to close-range blocks, successive development in processing and computing techniques extended its application to any types of images (Barazzetti et al., 2020). Today SfM can be used to process a photogrammetric block by using low-cost and open-source software packages running on a standard laptop. Cloud-computing resources are also popular to process large blocks without needing local hard resources.

When photogrammetric blocks must be compared to detect changes in the surveyed area, the standard approach would be to process each dataset independently from others. Each of them is georeferenced by means of a suitable set of GCPs to be included in the bundle block adjustment BBA as weighted pseudo-observations or to apply a similarity transformation to the derived point cloud. The adopted approach depends on the options offered by each software package. After image EO and MVS matching are completed, a dense point cloud is obtained to be compared with respect to other datasets. In this approach, GCPs may not be the same at different epochs, but they should all refer to the same reference system (see Barbarella et al., 2015). Direct orientation based on INS/GNSS integrated sensors is another solution to cope with georeferencing of an aerial photogrammetric block. This solution, anyway, cannot be available when using datasets recorded before this technique entered in the state-of-the-art, i.e., before late '90s.

In many cases, GCPs cannot be used to align multitemporal datasets obtained from archive photos. This may be due to different reasons: the institution distributing the images did not collect GCPs, different reference systems and techniques adopted for their measurements, low quality. In such a case, an alternative procedure is to bundle together multitemporal blocks and to compute the image EO/camera calibration at the same time. This solution (referred to hereafter as *multitemporal SfM* – MSfM) presents the advantage that images from all epochs will be aligned together, as proposed by Feurer and Vinatier (2018) in their so-called "Time-SIFT" method. The knowledge of accurate GCPs becomes less relevant, because: (1) GCPs are not necessary for the alignment, since blocks are already aligned after SfM; (2) the availability of GCPs is useful only for global georeferencing, requiring a quality that may also be at lower accuracy (see Subsect. 2.3). Of course, if one of the adopted photogrammetric blocks is provided with accurate GCPs, this could be used for georeferencing all datasets.

Disadvantages and problems may be related to the different quality, resolution, and radiometric content (panchromatic *vs* RGB) of the aerial images, and changes in the content. This last aspect is very crucial when dealing with glaciers and, in general, the high mountain environment (Scaioni et al., 2018). For the sake of completeness, Di Rita et al. (2020) already suggested to apply MSfM to some datasets of UAV images recorded on an Alpine glacier for monitoring purpose.

2.3 Extraction of GCPs from maps and DTMs

If GCPs are not available, a possible solution is to use other types of geographic information to derive their coordinates. To do this, the planimetric component is usually split from the altimetric component. The former may be obtained from existing maps produced close to the acquisition time of the photogrammetric block to be georeferenced. In such a case, the accuracy of GCP planimetric position is usually in the order of ± 10 m. The latter may be derived from digital terrain models (DTMs), that in general are more stable over time (at least in some areas) due to the minor changes in terrain topography (see Okolie and Smit, 2022). In such a case, the accuracy of GCP height is in the order of ± 5 m.

The estimated accuracy of GCPs derived from other geographic data may seem quite poor and thus not useful. On the other hand, in many cases this accuracy is enough to provide a first alignment between different photogrammetric datasets, to be refined later (see Subsect. 2.4). An example of this application is reported in Tian et al. (2022), where data from satellite photos collected in the '60s have been compared to recent satellite data.

2.4 Coregistration of point clouds using ICP

This approach allows to align a pair of point clouds based on algorithms for point sets registration. If more point clouds need to be aligned, as in the case study presented in Section 3, one is chosen as reference ("master") and all others ("slaves") are registered to it. The criteria for "master" selection may rely on the one with its own georeferencing, or the one with the best data quality, or the one with content less dissimilar from others.

The most popular technique to apply is Iterative Closest Point (ICP), see Pomerleau et al. (2013) for a review about this subject. ICP allows to compute a rigid-body transformation between a pair of point sets by starting from an initial approximate alignment. This can be obtained from manual picking of at least three corresponding points, or by using independent georeferencing of both datasets. In fact, ICP is generally also applied to refine the coregistration already obtained by using other techniques described in Subsections 2.2 and 2.3.

Since in this study we would like to apply change-detection techniques (Lindenbergh and Pietrzyk, 2015) to understand the

This contribution has been peer-reviewed.
https://doi.org/10.5194/isprs-archives-XLVIII-M-1-2023-293-2023 | © Author(s) 2023. CC BY 4.0 License.

glacier evolution from multiple point clouds derived from photogrammetric processing of historical aerial photos, we should consider that some areas might undergo important modifications over time. The ICP-based coregistration should only include some manually stable areas to avoid biased outputs. This task is usually accomplished by manually extracting these stable regions. Otherwise, a robust registration technique may be applied to automatically discriminate between areas with and without changes (see Wujanz et al., 2016). Stable regions will be used to compute the rigid-body transformation to align the whole "slave" point cloud to the "master". After coregistration is computed, it is important to assess the quality of this result by checking residuals. If this test is positive, the same coregistration parameters obtained for stable areas can be applied to the remaining part of the "slave" point cloud.
After point cloud coregistration, the comparison can be operated (Tavakoli et al., 2021). Several techniques exist to this purpose, either on the direct basis of the point clouds, and the use of derived models such as meshes (e.g., Triangulated Irregular Networks) or grid raster digital elevation models (DEMs).
In the final analysis of changes, it is important to consider the of uncertainty in point cloud registration. This option is possible in some advance algorithms, such as M3C2 (Lague et al., 2013).

3. EXPERIMENTAL SECTION

3.1 Dataset presentation

The glaciers under investigation belong to the Mount Blanc massif in the western sector of Italian Alps at the border with France. These glaciers flow in the Val Veny, as can be seen in Figure 1, which also reports the locations of one out of six photogrammetric blocks from aerial missions operated in 1967-2006. As shown in Table 1, the oldest datasets were collected by means of analogue cameras, while the latest could already exploit digital aerial cameras (DAC) that are today the state-of-the-art technology in topographic aerial mapping. Datasets are available as panchromatic or RGB colour images (see Table 1).
Images have been downloaded in digital form through the geoportal "IGN – Remontez le temps" of the IGNF (https://remonterletemps.ign.fr/). This archive also includes other older aerial missions over the same area (i.e., in 1952 and 1958) that would make the analysis even more interesting. The image quality and the cameras adopted to collect these datasets was more difficult to be processed. For this motivation, we decided not to use them in this study, but to postpone their use in the future (see Sect. 4).
Photos recorded in analogue format have been scanned with photogrammetric scanners at pixel size of 20 μm. Some datasets are also provided with a calibration certificate containing information on the camera and lens (Kraus, 2008), and with camera locations recorded by onboard GNSS sensors (1996, 2000 and 2001). No GCPs are available. In Malekian et al. (2023) some considerations about image quality are reported.

Table 1. Properties of photogrammetric datasets from IGNF online repository (PAN : panchromatic; RGB : colour images).

Acquisition time	#Images (type)	Camera Model (focal length – [mm])	Avg. flying altitude [m a.s.l.]
2006 (Aug 23-Sept 5)	78 (RGB)	Digital	5,540
2001 (Aug 1-13)	73 (RGB)	Zeiss RMKTOP15 (150)	4,780
2000 (Jun 23-Aug 1)	25 (RGB)	Zeiss RMKTOP15 (150)	5,540
1996 (Jul 3-Aug 4)	35 (RGB)	Zeiss RMKTOP15 (150)	5,020
1988 (Jul 26)	23 (PAN)	Leica RC10 (150)	6,150
1967 (Oct 11-12)	24 (PAN)	Leica RC10 (150)	6,350

Figure 2. Satellite view of the area under investigation including: glacier outlines (see legend), ground control points (GCPs), and locations of aerial photos from the Dataset 2000 (acknowledgements to Google Earth® for the background satellite image).

This contribution has been peer-reviewed.
https://doi.org/10.5194/isprs-archives-XLVIII-M-1-2023-293-2023 | © Author(s) 2023. CC BY 4.0 License.

3.2 Application of multitemporal SfM

Multitemporal SfM (MSfM) has been applied to the image blocks listed in Table 1 to understand the effectiveness of this procedure to align aerial datasets featuring multiple characteristics (e.g., type of camera, radiometric content, flight geometry, photo scale) and different acquisition time. Both groups of differences are particularly challenging, but the changes on the ground are significant, since they account for the glacier retreat, different snow cover, soil erosion, anthropogenic modification over time.

In some previous publications (Malekian et al., 2022; 2023), all datasets listed in Table 1 were independently oriented using SfM. Some sets of GCPs were obtained as described in Subsection 2.3. The ICP alignment between point clouds derived from MVS dense matching was operated at a later stage by using the approach presented in Subsection 2.4.

Here we have followed a different approach, which is based on georeferencing only one dataset (2000) by using 7 GCPs, as presented in previous publications (Fig. 1). Once georeferenced, Dataset 2000 has been used as geographic reference to align all other photogrammetric blocks to it and to evaluate changes.

Photogrammetric processing has been carried out in Agisoft Metashape Professional® ver. 1.7.5 (in the following AMP), which represents a standard package available at low-cost but with the capability to process datasets of large aerial images as the ones employed here.

The process for the orientation of selected aerial blocks is illustrated in the following paragraphs. Before applying SfM, images have been pre-processed to improve the image quality as illustrated in Malekian et al. (2023).

3.2.1. Independent SfM of single blocks: In a first stage each block has been independently processed without considering overlap with other datasets. GCPs have been measured only in Dataset 2000 and used to setup the Italian geodetic datum (RDN 2008.0 / UTM32N) in the BBA. In other datasets, a minimum constraint datum has been used (see Luhmann et al., 2014) to compute the EO in a first stage. Image orientation has been computed based on SfM implemented in AMP: processing has always been computed at "Highest" quality level setup, that corresponds on the use of the full-resolution imagery. The image size in the considered blocks ranged from approx. 16 Mpixels to 136 Mpixels. Almost all photos in each block could be oriented apart from a few images (less than 3% of the total number of available images). Residuals in image and object space (the latter only for Dataset 2000) are reported in Malekian et al. (2023). Their values have been retained sufficient considering the type and quality of adopted images (see Aguilar et al., 2013).

Another important consideration concerns the inner orientation and camera calibration. Due to the lack of calibration certificates for some datasets, self-calibration based on standard Brown model has been included in the BBA (Luhmann et al., 2014). This option has been selected with the purpose of defining a common processing workflow for all types of aerial blocks. In addition, this solution is supported by the presence of a sufficient number of images per each block (see Luhmann et al., 2016) and a suitable block geometry in terms of strip overlap. Approximate focal lens and pixel size have been provided. The unique block acquired by DAC (Dataset 2006) has not been processed with self-calibration, since images have been delivered as distortion free and a camera calibration certificate was retrieved upon request to the IGNF.

The most datasets were collected using different high-end analogue cameras (Table 1), thus they reported the outer frame with fiducial marks and instrument recordings. According to the experiments already reported in Malekian et al. (2023), we preferred to mask the outer camera frame to exclude the corresponding area from the extraction of key-points and tie points to input in the BBA. This also avoided the recognition and measurement of fiducial marks (Forlani et al., 1999). A simple rectangular mask has been applied to all the images of the same dataset. Figure 2 shows an example of a such mask applied to an image from Dataset 1967. This image also reported tie points after (blue points) and before (blue and white points) internal blunder rejection.

Figure 2. Example of an image from Dataset 1967 (panchromatic film recorded by analogue aerial camera Leica RC10) masked to avoid the interference of the outer frame during Structure-from-Motion (SfM). Overlaid points represent extracted tie points, while blue points are the ones which passed the internal blunder rejection (acknowledgements to IGNF® for the aerial photo).

3.2.2 Multitemporal SfM: In a second stage, multiple blocks have been progressively included in a multitemporal SfM (MSfM) project in AMP. Dataset 2000 has been used to establish the datum thanks to the presence of 7 GCPs. The measurement of these points has been also extended to some images from other datasets where they could be clearly identified and manually measured (Fig. 3). This solution has led to an average number of 10 rays per GCP, which increased the reliability of corresponding observations in the BBA.

To setup the MSfM, a common block including photos from all datasets has been first created by aligning and merging individual projects in AMP. This software allows to import them as "chunks" (i.e., subprojects) in a new project. By using the extracted tie points in each block and by looking at intra-block correspondences, all "chunks" could be aligned with respect to one of them selected as reference (in this case: Dataset 2000). After merging all blocks in a unique project, they have been all aligned w.r.t. the same datum. At this point, a BBA has been run again to improve the image orientation, which has received as input 103,259 tie points (average 2.92 rays per point). Also in this case, processing has been computed at "Highest" quality level. Table 2 reports some statistics related to the BBA in MSfM.

During BBA, GCPs have been fixed as weighted observations.

https://doi.org/10.5194/isprs-archives-XLVIII-M-1-2023-293-2023 | © Author(s) 2023. CC BY 4.0 License.

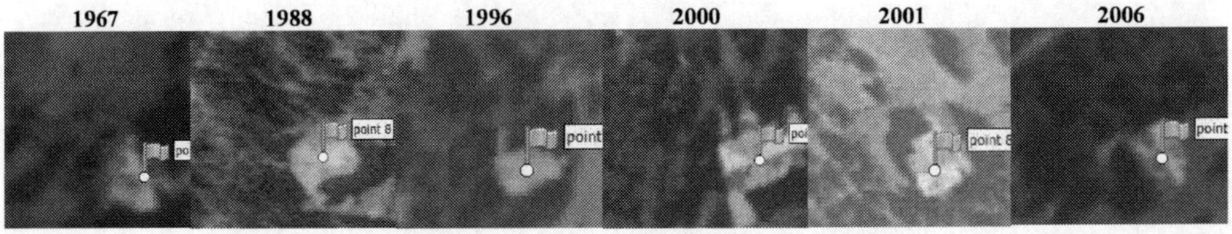

Figure 3. Images of a GCP tracked in multiple photos across different epochs.

Table 2. Root Mean Square Errors (RMSE) of GCPs and other statistical parameters after bundle block adjustment of Multitemporal SfM (MSfM); results refer to sub-blocks (one per epoch) and to the multitemporal block as a whole.

Year of the dataset	RMSE on GCPs ground coordinates			Statistics on tie points		Statistics on GCPs	
	XY [m]	Z [m]	Total [m]	Avg. residuals on image coord.s [pix]	Avg. # tie points per image	Avg. residuals on image coord.s [pix]	Avg. # GCPs per image
2006	–	–	–	0.41	3903	–	–
2001	–	–	–	1.12	2310	–	–
2000	–	–	–	0.51	2190	–	–
1996	–	–	–	0.83	2560	–	–
1988	–	–	–	1.54	1928	–	–
1967	–	–	–	0.73	3670	–	–
all	7.08	5.80	11.59	0.78	3060	3.09	0.3

According to the poor quality of GCP measurement on the images (see Fig. 3) and the source of 3D coordinates (from existing maps and a DTM), we decided to assign a low accuracy (σ_{GCP}=10 m) not to influence the intrinsic accuracy of estimated EO in the BBA, but only to guarantee a general absolute common georeferencing to all blocks. This decision is different from the one adopted in previous publications, where a higher accuracy was selected for GCP object coordinates, resulting in smaller residuals.

3.2.3 Dense cloud generation: After the EO estimation based on MSfM, a dense point cloud has been generated per each original dataset in AMP. The quality level "High" has been selected, as trade-off between final quality and processing time.
Point clouds have been cleaned and segmented by using a common Region-of-Interest (RoI). The size of dense points clouds has ranged from 26 million points (Dataset 1967) to 56 million points (Datasets 1996 and 2006).
Dense point clouds have been exported from AMP in LAS interchange format and imported in the open-source software CloudCompare (CC) Ver. 2.12.4 "Kyiv" (www. Cloudcompare.org) for successive processing operations.
In CC each point cloud has been processed first to remove duplicated points by selecting a minimum distance between closest points larger than 20 cm. Secondly, a filter to remove residual isolated points was applied.
The quality assessment has been performed in CC base on five stable areas, as presented in next paragraph. After that, the same areas have been used for computing a rigid body transformation to minimize distances between different epochs based on ICP registration. Then each set of transformation parameters has been applied to the remaining part of each "slave" point clouds to refine the alignment with respect to the "master" obtained from MSfM.
In Figure 4, point clouds obtained from Datasets 1967 and 1988 are compared by using M3C2. Differences are shown to highlight glacier changes.

3.2.4 Quality assessment: The final empirical quality assessment has been based on the comparison of dense point clouds inside the RoI. Due to the presence of several and large changes between different epochs, the analysis has been limited to five stable areas, as in previous publications. Points within five stable windows have been segmented and compared by using M3C2 (Multi-scale Model to Model – see Lague et al., 2013) technique, that was applied to detect changes between point cloud pairs (see also Tavakoli et al., 2021). M3C2 is based on the computation of local normals, which are used to tailor in each location the direction to detect changes. In addition, M3C2 is resistant to changes in point density and point cloud noise and allows to exploit information on the point clouds' accuracy (if available) to distinguish between real displacements and noise (James et al., 2017).
All points in stable areas per each epoch have been merged together and compared to other epochs. Table 3 reports the Root Mean Square Errors (RMSE) between different epochs, where the older point cloud has been used as "master" and the other as "slave" in the M3C2 comparison.
In Table 4 we reported more statistics about the comparisons between consecutive epochs only.

Table 3. RMSE after M3C2 comparison of point clouds in five stable areas from different epochs (results in [m]).

	1988	1996	2000	2001	2006
1967	1.49	2.44	0.86	0.88	1.00
1988	–	1.99	1.26	1.61	1.81
1996	–	–	2.41	2.70	2.89
2000	–	–	–	0.93	1.00
2001	–	–	–	–	0.73

This contribution has been peer-reviewed.
https://doi.org/10.5194/isprs-archives-XLVIII-M-1-2023-293-2023 | © Author(s) 2023. CC BY 4.0 License.

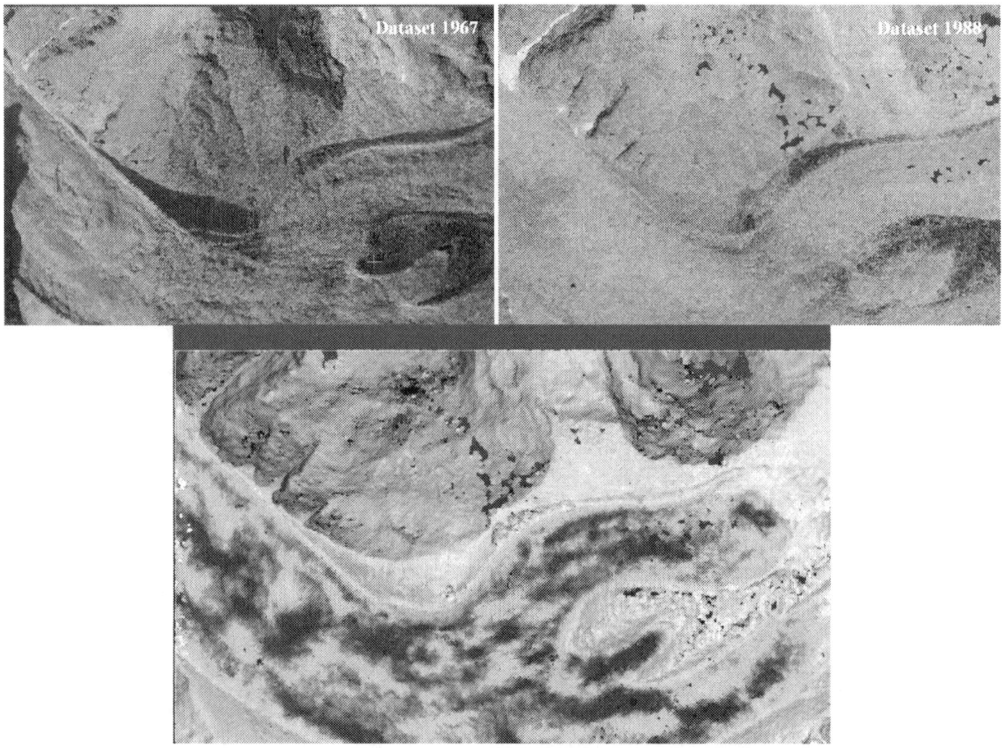

Figure 4. Result of the comparison based on M3C2 technique between Datasets 1967 and 1988 (panchromatic) over the terminus of Miage Glacier in Val Veny (distances in [m]). In the upper part the point clouds obtained from the application of MSfM are shown. After ICP refinement of coregistration, point clouds have been compared by means of M3C2 (see the map of changes in the lower part).

3.3 Discussion

The core of the methodology for the alignment of multiple photogrammetric blocks proposed here consists in the MSfM as alternative to the independent orientation of single blocks. MSfM has demonstrated to be more convenient from the operational point-of-view, requiring a unique set of GCPs with the only purpose of global georeferencing.
As far as the quality of block co-registration is concerned, the analysis of residuals on stable areas between successive epochs (Table 4) shows that, in terms of RMSE, the application of MSfM has resulted in significant improvements w.r.t. the independent SfM on single datasets. In the case of older datasets, the improvement in terms of RMSE has ranged from approx. 30% to slightly less than 50%. With more recent datasets, the improvement has reached approx. 80%. These results are motivated by the good performance of MSfM in terms of extracted tie points, as shown in Table 2.
It is important to notice that MSfM works well also for the integration of photogrammetric blocks featuring big differences in terms of sensing technology (analogue *vs* digital), radiometric content (panchromatic *vs* RGB), and relevant topographic changes.
By looking at the components contributing to RMSE in Table 4, standard deviations play the most relevant role. This demonstrates that data dispersion counts more than bias after MSfM. This is confirmed by results after the ICP refinement reported in the rightmost column of Table 4. Mean differences are reduced in terms of a few centimetres (max mean difference is -22 cm between Datasets 1996 and 2000).
Table 3 shows the RMSE after M3C2 comparison of points in stable areas for all possible combinations of epochs. General results are like the ones obtained from consecutive datasets reported in Table 4. Also in this case, larger values of RMSEs have been obtained when older datasets have been compared. This is motivated by the improved quality of aerial photos that could be obtained in the last 25 years. Dataset 1996 has revealed to be more problematic, probably due to the lower number of images and the block geometry.

4. CONCLUSIONS AND FUTURE DEVELOPMENTS

This paper presented the application of Multitemporal Structure-from-Motion (MSfM) as alternative technique for the alignment of multiple datasets of archive aerial photos collected at different times. The target application consisted in monitoring changes in Alpine glaciers as consequence of climate change. MSfM was tested in a case study (Val Veny in the Mount Blanc massif, Italy) where multiple datasets of aerial photos could be retrieved from the online archive repository of National Geographic and Forestry Institute of France (IGNF).

Table 4. Statistics after comparison of points in stable areas according to three different approaches: single SfM based on independent GCPs per each dataset (Malekian et al., 2023); MSfM with GCPs from Dataset 2000; MSfM followed by ICP refinement.

		Single SfM [m]	MSfM [m]	MSfM + ICP [m]
1967	*mean*	-1.31	-0.82	0.02
-	*std.dev.*	3.04	1.25	1.09
1988	*RMSE*	3.31	1.49	1.09
1988	*mean*	0.67	-0.11	-0.17
-	*std.dev.*	3.10	1.99	1.87
1996	*RMSE*	3.18	1.99	1.88
1996	*mean*	-1.82	-0.46	-0.22
-	*std.dev.*	3.82	2.37	2.07
2000	*RMSE*	4.47	2.41	2.08
2000	*mean*	1.61	0.18	0.05
-	*std.dev.*	4.39	0.91	0.72
2001	*RMSE*	4.48	0.93	0.72
2001	*mean*	0.31	0.19	0.08
-	*std.dev.*	3.80	0.71	0.67
2006	*RMSE*	3.81	0.73	0.67

Results shows that this technique is highly suitable for the alignment of selected datasets, which also included quite big challenges in terms of camera technology, radiometric content, and relevant topographic changes. This outcome suggests that MSfM can be generally applied to other case studies with similar performances. In addition, a further improvement could be achieved from the integration between MSfM and the refinement of point-cloud alignment based on ICP computed in stable areas. Here stable areas, which can be also exploited for quality assessment, have been manually selected. One of the possible future developments should be their automatic selection (see Wujanz et al., 2016).

Other interesting evolutions of the presented methodology is the application of new image matching techniques based on Artificial Intelligence (e.g., Deep Learning), see Morelli et al. (2022) and Mildenhall et al. (2022).

Another important aspect to develop is about the comparison of point clouds based on M3C2 technique. Recently new developments of this method were published (Winiwarter et al., 2021; Zahs et al., 2022), also considering the possibility of the contemporary analysis of multiple datasets at the same time, instead of pairwise comparison (Anders et al., 2021).

Acknowledgements

The authors would like to acknowledge the National Geographic and Forestry Institute of France (IGNF) for delivering archive photos through the geoportal "IGN - Remonter le temps" (https://remonterletemps.ign.fr/).

Acknowledgements also go to the developers of CloudCompare open-source software we used in this research.

REFERENCES

References from books:

Barazzetti, L., Gianinetto, M., Scaioni, M., 2020. Automatic Processing of Many Images for 2D/3D Modelling. In: B. Daniotti et al. (Eds.), *Digital Transformation of the Design, Construction and Management Processes of the Built Environment,* Research for Development, SpringerOpen, Cham, Switzerland, pp. 355-364, DOI: 10.1007/978-3-030-33570-0_32.

Barbarella, M., Fiani, M., Lugli, A., 2015. Multi-temporal Terrestrial Laser Scanning Survey of a Landslide. In: M. Scaioni (Ed.), *Modern Technologies for Landslide Monitoring and Prediction*, pp. 89-121, Springer Natural Hazards, Cham, Switzerland, DOI: 10.1007/978-3-662-45931-7_5.

Luhmann, T., Robson, S., Kyle, S., Boehm, J., 2014. *Close-range photogrammetry and 3D imaging*. Walter de Gruyter, pages 684, Berlin, Germany.

Kraus, K., 2008. *Photogrammetry - Geometry from Images and Laser Scans*. Walter de Gruyter, Berlin, Germany.

References from journals:

Anders, K., Winiwarter, L., Mara, H., Lindenbergh, R., Vos, S.E., Höfle, B., 2021. Fully automatic spatiotemporal segmentation of 3D LiDAR time series for the extraction of natural surface changes. *ISPRS J. Photogramm. Remote Sens.*, Vol. 173: 297-308.

Aguilar, M., Aguilar, F., Fernandez, I., Mills, J., 2013. Accuracy assessment of commercial self-calibrating bundle adjustment routines applied to archival aerial photography. *Photogramm. Rec.*, Vol. 28(141): 96-114.

Eltner, A., Kaiser, A., Castillo, C., Rock, G., Neugirg, F., Abellán, A., 2016. Image-based surface reconstruction in geomorphometry – merits, limits and developments. *Earth Surf. Dyn.*, Vol. 4: 359-389.

Feurer, D., Vinatier, F., 2018. Joining multi-epoch archival aerial images in a single SfM block allows 3-D change detection with almost exclusively image information. *ISPRS J. Photogramm. Remote Sens.*, Vol. 146: 495-506.

Fey, C., Wichmann, V., Zangerl, C., 2017. Reconstructing the evolution of a deap seated rockslide (Marzell) and its response o glacial retreat based on historic and remote sensing data. *Geomorph.*, Vol. 298: 72-85.

Fieber, K.D., Mills, J.P., Miller, P.E., Clarke, L., Ireland, L., Fox, A.J., 2018. Rigorous 3D change determination in Antarctic Peninsula glaciers from stereo WorldView-2 and archival aerial imagery. *Remote Sens. Env.*, Vol. 205: 18-31.

Fugazza, D., Scaioni, M., Corti, M., D'Agata, C., Azzoni, R.S., Cernuschi, M., Smiraglia, C., Diolaiuti, G.A., 2018. Combination of UAV and terrestrial photogrammetry to assess rapid glacier evolution and map glacier hazards. *Nat. Hazards Earth Syst. Sci.*, Vol. 18: 1055-1071.

Granshaw, S.I., 2018. Structure from motion: Origins and originality. *Photogramm. Rec.*, Vol. 33:6-10.

James, M.R., Chandler, J.H., Eltner, A., Fraser, C., Miller, P.E., Mills, J.P., Noble, T., Robson, S., Lane, S.N., 2019. Guidelines on the use of structure-from-motion photogrammetry in geomorphic research. *Earth Surf. Proc. Landforms*, Vol. 44(10): 2081-2084.

Lague, D., Brodu, N., Leroux, J., 2013. Accurate 3D comparison of complex topography with terrestrial laser scanner: Application to the Rangitikei canyon (N-Z). *ISPRS J. Photogramm. Remote Sens.*, Vol. 82: 10-26.

Lindenbergh, R., Pietrzyk, P., 2015. Change detection and deformation analysis using static and mobile laser scanning. *Appl. Geomatics*, Vol. 7: 65-74.

Luhmann, T., Fraser, C.S., Maas, H.-G., 2016. Sensor modelling and camera calibration for close range photogrammetry. *ISPRS J. Photogramm. Remote Sens.*, Vol. 115: 37-46.

Mildenhall, B., Srinivasan, P.P., Tancik, M., Barron, J.T., Ramamoorthi, R., Ng, R., 2022. NeRF: Representing Scenes as Neural Radiance Fields for View Synthesis. *Comm. ACM*, Vol. 65(1): 99-106.

Okolie, C.J., Smit, J.L., 2022. A systematic review and meta-analysis of Digital elevation model (DEM) fusion: pre-processing, methods and applications. *ISPRS J. Photogramm. Remote Sens.*, Vol. 188: 1-29.

Pomerleau, F., Colas, F., Siegwart, R., Magnenat, S., 2013. Comparing ICP variants on real-world data sets. *Autonomous Robots*, Vol. 34: 133-148.

Tian, Y.X., Xia, M.L., An, L., Scaioni, M., Li, R.X., 2022. Innovative Method of Combing Multidecade Remote Sensing Data for Detecting Precollapse Elevation Changes of Glaciers in the Larsen B Region, Antarctica. *IEEE J. Sel. Top. App. Remote Sens.*, Vol. 15: 9699-9715.

Winiwarter, L., Anders, K., Höfle, B., 2021. M3C2-EP: Pushing the limits of 3D topographic point cloud change detection by error propagation. *ISPRS J. Photogramm. Remote Sens.*, Vol. 178: 240-258.

Wujanz, D., Krueger, D., Neitzel, F., 2016. Identification of Stable Areas in Unreferenced Laser Scans for Deformation Measurement. *Photogramm. Rec.*, Vol. 31(155): 261-280.

Zahs, V., Winiwarter, L., Anders, K., Williams, J.G., Rutzinger, M., Höfle, B., 2022. Correspondence-driven plane-based M3C2 for lower uncertainty in 3D topographic change quantification. *ISPRS J. Photogramm. Remote Sens.*, Vol. 183: 541-559.

References from conference proceedings:

Barazzetti, L., Remondino, F., Scaioni, M., 2009. Combined use of Photogrammetric and Computer Vision techniques for fully automated and accurate 3D modeling of terrestrial objects. In: Proc. of Int. Conf. "Videometrics, Range Imaging, and Applications X", 2-3 Aug, San Diego (CA - USA), *Proc. of SPIE*, Vol. 7447, paper No. 74470M, 12 pp., DOI: 10.1117/12.825638.

Di Rita, M., Fugazza, D., Belloni, V., Diolaiuti, G., Scaioni, M., Crespi, M., 2020. Glacier Volume Change Monitoring from UAV Observations: Issues and Potentials of State-of-the-Art Techniques. *Int. Arch. Photogramm. Remote Sens. Spatial Inf. Sci.*, Vol. XLIII, Part B2-2020: 1041-1048, DOI: 10.5194/isprs-archives-XLIII-B2-2020-1041-2020.

Forlani, G., Pinto, L., Scaioni, M., 1999. A Procedure for Automatic Inner Orientation. *Int. Arch. Photogramm. Remote Sens. Spat. Inf. Sci.*, Vol. 32, Part 3-2W5: 131-135.

Malekian, A., Fugazza, D., Scaioni, M., 2022. 3D Surface Reconstruction and Change Detection of Miage Glacier (Italy) from Multi-date Archive Aerial Photos. In: Proc. "Computational Science and Its Applications – ICCSA 2022," 4-7 June, Malaga (Spain), *Lecture Notes in Computer Science* (LNCS), Vol. 13379, pp. 450-465, Springer Nature, Cham, Switzerland, DOI: 10.1007/978-3-031-10545-6_31.

Malekian, A., Fugazza, D., Scaioni, M., 2023. Photogrammetric Reconstruction and Multi-temporal Comparison of Brenva Glacier (Italy) from Archive Photos. *ISPRS Ann. Photogramm. Remote Sens. Spatial Inf. Sci.*, Vol. X, Part 4/W1: 459-466, DOI: 10.5194/isprs-annals-X-4-W1-2022-459-2023.

Morelli, L., Bellavia, F., Menna, F., Remondino, F., 2022. Photogrammetry Now and Then – From Hand-Crafted to Deep-Learning Tie Points. *Int. Arch. Photogramm. Remote Sens. Spatial Inf. Sci.*, Vol. XLVIII, Part 2/W1: 163-170, DOI: 10.5194/isprs-archives-XLVIII-2-W1-2022-163-2022.

Poli, D., Casarotto, C., Strudl, M., Bollmann, E., Moe, K., Legat, K., 2020. Use of Historical Aerial Images for 3D Modelling of Glaciers in the Province of Trento. *Int. Arch. Photogramm. Remote Sens. Spat. Inf. Sci.*, Vol. 43, Part B2: 1151-1158, DOI: 10.5194/isprs-archives-XLIII-B2-2020-1151-2020.

Scaioni, M., Crippa, J., Corti, M., Barazzetti, L., Fugazza, D., Azzoni, R., Cernuschi, M., Diolaiuti, G.A., 2018. Technical Aspects Related to the Application of SfM Photogrammetry in High Mountain. *Int. Arch. Photogramm. Remote Sens. Spatial Inf. Sci.*, Vol. XLII, Part 2: 1029-1036, DOI: 10.5194/isprs-archives-XLII-2-1029-2018.

Tavakoli, K., Zadehali, E., Malekian, A., Darsi., Longoni, L., Scaioni, M., 2021. Landslide Dam Failure Analysis Using Imaging and Ranging Sensors. In: Proc. "Computational Science and Its Applications – ICCSA 2021," 13-16 Sept, Cagliari (Italy), LNCS, Vol. 12955, pp. 3-17, Springer Nature, Cham, Switzerland, DOI: 10.1007/978-3-030-87007-2_1.

ADAPTATION OF DEEPLAB V3+ FOR DAMAGE DETECTION ON PORT INFRASTRUCTURE IMAGERY

M. Scherff[1], F. Hake[1], H. Alkhatib[1]

[1] Geodetic Institute, Leibniz Universität Hannover, Germany - scherff,hake,alkhatib@gih.uni-hannover.de

KEY WORDS: Image segmentation, Damage Detection, Deep Learning, Optimization, Supervised, Weakly Supervised

ABSTRACT:

Regular inspection and maintenance of infrastructure facilities are crucial to ensure their functionality and safety for users. However, current inspection methods are labor-intensive and can vary depending on the inspector. To improve this process, modern sensor systems and machine learning algorithms can be deployed to detect defects based on rapidly acquired data, resulting in lower downtime. A quality-controlled processing chain allows to provide hence informed uncertainty assessments to inspection operators. In this study, we present several Deeplab V3+ models optimized to predict corroded segments of the quay wall at JadeWeserPort, Germany, which is a dataset from the 3D HydroMapper research project. Our models achieve generally high accuracy in detecting this damage type. Therefore, we examine the use of a Region Growing-based weakly supervised approach to efficiently extend our model to other common types in the future. This approach achieves about 90 % of the results compared to corresponding fully supervised networks, of which a ResNet-50 variant peaks at 55.6 % Intersection-over-Union regarding the test set's corrosion class.

1. INTRODUCTION

Infrastructure buildings must meet high construction standards and require regular maintenance throughout their complete life cycle to ensure safe and reliable usage for its users like pedestrians, workers or even machines. Scheduled inspections are performed to detect emerging defects that could affect the facility's functionality and stability, supplemented by additional inspections as needed. National legislation, inspired by norm standards, typically regulates this process. Commissioned inspection companies compile reports detailing damaged parts or sections and providing a complete infrastructure assessment, allowing for informed maintenance decisions.

Besides commonly known infrastructure objects as bridges or tunnels, this concept has also to be applied for port facilities or similar building objects such as watergates or canals. However, the traditional approach of performing inspections by hand, with only basic tools for documentation, causes significant downtime for the facility, especially in underwater sections where divers face complicated visual conditions. This downtime leads to high costs for the facility operator. To shorten this process step, an automated inspection process should be introduced, which rely on the capabilities of today's high accurate sensors. An example of such capabilities can be found within the 3D HydroMapper research project, in which (Hesse et al., 2019) developed a swimming multi-sensor system (MSS) sensing the environment below and above the water surface. By that approach, the complete facility can be captured within several minutes or a few hours depending on its size, allowing to detect for instance regular appearing corrosion spots in the so-called splash zone more objectively afterwards. Therefore, various software systems can assist building inspectors digitally by providing data-driven predictions and optionally uncertainty information for damaged areas in order to assess the infrastructure knowledgeably.

In this study, we aim to enhance the damage detection process by adapting the Deeplab V3+ segmentation network (Chen et al., 2018) to analyze an image corrosion dataset obtained from the 3D HydroMapper project. Initially, we establish a baseline model, and explore different combinations of backbone networks and loss functions to create a robust feature-rich network. Subsequently, we conducted an extensive performance analysis and examined the impact of introducing weakly supervised ground truth images obtained through the Region Growing algorithm during the training process.

The structure of the paper is as follows: Chapter 2 gives a brief overview on related studies to damage detection of infrastructure buildings, with a focus on corrosion investigations. In Chapter 3, we describe the methodology used for the dataset acquisition and preparation. Chapter 4 presents the process of obtaining the best performing segmentation model in detail. Subsequently, we provide and discuss the evaluation results of the validation and test sets. Finally, Chapter 6 presents the conclusion and an outlook of further investigations.

2. RELATED WORK

The field of infrastructure inspection can nowadays be approached using a wide variety of sensors or tools. Among these, cameras are an excellent acquisition instrument that can effectively examine building surfaces due to their universal and simple deployment. Because of the progress achieved in image processing and computer vision, especially in the last decade, early damage detection became a feasible goal. This particular task has been investigated recently i. a. by (Duy et al., 2020), (Katsamenis et al., 2020) and (Tanveer et al., 2022). They were able to detect different damage types to a certain extent by means of Deep Learning (DL) segmentation models on concrete and metal surfaces, respectively. However, some damage types are more recognizable than others due to their colour, shape or size. While corroded segments can be unambiguously distinguished from the concrete quay wall and background in general at a certain severity level, their shape and size can vary significantly, which may lead to erroneous detections. On the other hand, for instance cracks can have more similar properties overall, but individual examples can be overlooked due to their small width

This contribution has been peer-reviewed.
https://doi.org/10.5194/isprs-archives-XLVIII-M-1-2023-301-2023 | © Author(s) 2023. CC BY 4.0 License.

or low resolution. Nevertheless, DL models like Convolutional Neural Networks (CNN) are currently the dominant objective in this research field and have a high potential to classify, detect, or even segment these damaged areas correctly due to their self-learning feature capabilities. Hence, this image processing strategy is a promising approach to digitally assist the inspection operator effectively.

In the study by (Duy et al., 2020), corroded areas on electric poles are detected using several architectural modified segmentation networks ranging from hundred thousands to ten millions trainable parameters. Three models were evaluated with and without prior background removal, achieving high Intersection-over-Union (IoU) scores for the pole and background class, but struggled to detect corrosion reliably. To overcome this limitation, the dataset was extended with a focus on the corrosion class, and the Mask R-CNN architecture (He et al., 2017) was investigated to partition these segments based on their severity level. But this network type was already examined with common CNNs by (Katsamenis et al., 2020) in the same year achieving 71 % F1 score on another corrosion dataset. In their experiments, Mask R-CNN outperforms the other candidates significantly. Their purposely developed boundary refinement approach, a data projection method dividing the model prediction in confident and fuzzy regions, effects the quantified results of this model barely. In addition, (Tanveer et al., 2022) evaluated the performance of various low parameter segmentation CNNs compared to the Deeplab V3+ architecture with ResNet as backbone network. The group trained these on a concrete dataset containing the damage types cracks, efflorescence, rebar exposure and spalling and found on average a relative maximum discrepancy of around 17 % in terms of F1 score to the reference. All these groups use Cross-entropy loss for the classification purpose in model training and no extensive optimization or evaluation was performed. Moreover, only a single model instance was learned, which limits the uncertainty investigation. Here comes (Nash et al., 2022) into play. This group trained different models of modified HRNetV2 (Sun et al., 2019) networks by Variational Inference, Monte Carlo Dropout and an Ensemble method on a further corrosion dataset. These Bayesian transformed models were evaluated especially in terms of aleatoric and epistemic uncertainty. In our study, we combine these concepts to provide an informed overview of different modified Deeplab V3+ models trained on an optimal loss function that is selected from a pool of distribution- and region-based error terms using our self-labeled port dataset. In contrast, (Hake et al., 2023b) focused only on single trained model instances and restricted architecture changes with the smaller and differently structured 3D HydroMapper dataset.

The most precise network structure is iteratively trained on simulated weakly supervised label images generated once by a Region Growing algorithm (Adams and Bischof, 1994) before any model training begins. However, (Huang et al., 2018) proposed a more sophisticated approach by using this algorithm on seed cues generated by a classification network in their segmentation pipeline. Their Deep Seeded Region Growing method works hence on image-level label information. We aim to estimate still the expected performance decline for this straight-forward approach when the manual labeling effort is significantly reduced in our investigation. These findings will guide our approach to further datasets containing other damage types that will be fed to the same model architecture to support the infrastructure inspection process. It is currently intended to mark a small region in the damage segment instead of bordering it completely by a polygon if that procedure is proven successfully. Due to the nature of the current dataset, we cannot efficiently benefit from bounding box or class activation map approaches.

3. DATASET

3.1 Data acquisition

The dataset used in this work was specifically acquired within the 3D HydroMapper project to aid the development of AI-assisted software for future port inspections based on MSS data. This image sequence shows the quay wall of the JadeWeserPort in Wilhelmshaven (Germany) in an overlapping manner. We observe similar coverage of up to 80 % between subsequent captures. Occasionally, the perspective is adjusted during the acquisition process by flying closer to the wall. This provides a wide variety of real-world information for the detection models to be trained on. The complete dataset was recorded on a single day using a controlled drone equipped with a Canon EOS 5D Mark III reflex camera, hovering or flying over the water a few meters in front of the wall. For this project the focal length was fixed at 35 mm and the resolution of pre-processed images are either 5760 x 3840 or cropped to 4608 x 3456 pixels. The dataset consists of 1300 high-resolution digital images of the infrastructure, where the object pixel size varies approximately between 2 and 5 mm depending on the individual capture. Due to this, we can find fine or arising defects and achieve accurate results in terms of position and shape of the damaged segments. This allows us to retrace the growing process over time and modify the inspection period or maintenance measures accordingly to related predictions. The labeling process neglected damages besides corrosion, such as cracks, efflorescence, or spalling due to their seldom occurrence. Therefore, our trained models cannot detect them yet. However, (Hake et al., 2023a) have already demonstrated an AI-assisted approach to detect geometrical related damage types in heightfields extracted from point cloud data. Currently, we had images of an efflorescence-damaged watergate available, which were taken by means of a tripod at several locations within this facility. Hence, these are not suitable for the current inspection concept based on a swimming or consistent moving sensor platform.

3.2 Generate ground truth images

To train the supervised segmentation models, we required labeled images indicating which pixels or segments correspond to the corrosion class. Initially, our own Matlab program was used to obtain these ground truth images, which converts manually drawn polygons successively into a 8-bit black/white (b/w) image. Later on, we used the service of Supervise.ly (Supervisely contributors, 2023) to complete our currently considered dataset consisting of 84 image-label pairs. Thereon, we can incorporate model predictions directly into the labeling process. These image pairs are uniformly distributed over the quay wall, showing a big portion of the total infrastructure, because overlapping areas were avoided for the most part. The ground truth images are based on our knowledge in this field and the gained experience over the time of this procedure, but have not been validated yet by experts. However, due to investigations in handling label noise in the DL context especially by (Nash et al., 2019), particular for corrosion detection, we expect no significantly improved corrosion detection capabilities. Nonetheless, we can not ensure that the achieved metric scores listed in chapter 5 correspond to the true values, especially with the provided number of digits.

This contribution has been peer-reviewed.
https://doi.org/10.5194/isprs-archives-XLVIII-M-1-2023-301-2023 | © Author(s) 2023. CC BY 4.0 License.

To adapt the Deeplab V3+ structure for the corrosion segmentation in the beginning by means of cross validation, and evaluate the performance of the various models methodically correct afterwards, we divide the created dataset randomly into six subsets à 14 images each. Four of them form the train set, and each of the remaining ones forms the validation and test sets, receptively. The validation set determines the optimal parameter state of any model based on the corresponding lowest epoch-wise loss value, and the test images act as independent evaluation measures. An example of the aforementioned image-label pairs is illustrated in Figure 1. As visualized by the ground truth image (black: damage-free/background, white: corrosion), corroded segments represent the minority class of this segmentation task. Because damage detection has to deal regularly with highly imbalanced datasets due to its underlying nature, models will by design favor the majority class to minimize the prediction error or maximize the total accuracy. To reorder the model's behaviour in terms of equal class detection abilities, data processing itself can be modified or the optimization process of the considered learning method has to be adjusted. In this study, we focus mainly on the second point by selecting more robust loss functions with (additionally) integrated class weights.

3.3 Simulating weakly supervised images

For the weakly supervised approach, we label the same images as before by simulating a starting region for each corroded segment. These regions are subsequently extended using the Region Growing algorithm. The quality of this process is evaluated using the original label images serving as reference. This procedure was chosen primarily because of the labor-intensive and time-consuming nature of manual labeling, which is especially true for the here investigated segmentation task. By using a brush annotation tool to mark only a small fraction or region, rather than labeling the entire segment pixel-accurately with a polygon, the process can be significantly sped up. Based on our own experience with the port dataset, labeling a single $\geq$ 16 megapixel image can take up to 30 minutes, even one with only a few damaged areas requires between five to ten minutes. Assuming that the suggested brush method reduces the manual effort by only 50 %, it can still result in time savings of 1-2 working days for the small image dataset we considered. This scales proportionally to the true labeling duration, which may vary between individual samples due to the number of damaged segments, their variable sizes and shapes. The subsequent Region Growing processing can be conducted after working hours to minimize utilization of the available computing capacity, or directly after labeling a single image to reduce the total processing time. On the other hand, this approach can also lead to a higher number of image and ground truth pairs in the same time period, resulting in a more robust trained segmentation network. However, a major disadvantage of the chosen algorithm in particular is that it is not suitable for all kinds of damage types or every segmentation class in general. For Region Growing to approximate fully supervised labels qualitatively, the true damage segments must have homogeneous colours and must clearly distinguish themselves from the background and surrounding classes. While this may be true on average for corrosion and efflorescence in this domain, it does not to concrete spalling, for example.

The version of Region Growing used here is based on standardizing each colour channel c independently to determine whether an adjacent pixel should be added to the growing start region. A pixel x_i is added if the maximum standardized absolute deviation from the expected colour is within a given threshold t. Equation 1 defines the deviation:

$$\Delta x_{max} = max\left(\left|\frac{x_i - \mu_c}{\sigma_c}\right|\right) \leq t \tag{1}$$

Here, Δx_{max} represents the deviation, μ_c and σ_c contain the channel-wise statistical quantities, and the max function selects the largest element of the standardized RGB vector $\in \mathbb{R}^{3}$. The threshold is set to a strict value of 1.25 to prevent overgrowing damage segments, especially for this pile quay wall and to provide the segmentation network with mostly correctly classified pixels, even if they represent only a small fraction of the true damaged areas. We believe that it is better to add initial missing corrosion pixels along the way than to include a large number of incorrect background pixels in the weakly supervised damage segments. To counteract wrongly generated corrosion segments during training, we use Conditional Random Field (CRF) implementation[1] by (Krähenbühl and Koltun, 2012) in each iteration. The threshold parameter is not specifically tuned to optimize certain metrics based on the original labeled data but is selected by visually comparing different outcomes for a small set of images. At the end of the algorithm, we apply equation 1 once again to every pixel within the grown region to possibly remove outliers. The starting segments for Region Growing are created by a randomly orientated ellipse that satisfies a certain overlapping condition with the reference label. The initial set semi-axes are dependent on the segment's size and are shrunken every time to the condition is not fulfilled. Using this method, the weak labels were evaluated in order to gain insight into the model's performance in the iterative training process. The initial weak labels achieved IoU scores of 92.4 and 36.2 % for the background/undamaged and corrosion class, respectively, with respect to the manual images. In Figure 1, we present the original RGB and both labeled ground truth images for a visually appealing sample of the test dataset.

4. SEGMENTATION NETWORKS

4.1 Baseline model

The Deeplab V3+ architecture is widely recognized for its impressive segmentation capabilities, and its basic modules, such as spatial pyramid pooling (SPP) and encoder-decoder structures, which are broadly applied even outside the core AI research community. The network captures object boundaries to some extent by gradually upsampling concatenated low- and high-level feature maps, and detects variable-sized objects with ease thanks to the Atrous SPP (ASPP) module, which encodes multi-scale contextual information (Chen et al., 2018). Generally, object boundaries can be refined using post-processing steps, but in this work, we orient ourselves on feature attention mechanisms ((Azad et al., 2020), (Hsu et al., 2022), (Zeng et al., 2020)) within the encoder part to achieve this. Such mechanisms recalibrate the functional relationship represented by the model to rely more on meaningful feature extractions. We aim to mimic this behaviour by modifying the dilation rates within ASPP's convolutions. By backpropagating the prediction error through the network during training, the feature maps or responsible filter blocks should be modified accordingly to boost detection capabilities of certain spatial resolutions slightly. This

[1] https://github.com/lucasb-eyer/pydensecrf

Figure 1. Example of a cropped RGB image showing the east-facing sheet pile wall of the JadeWeserPort and corresponding weakly and manually supervised ground truth images (top to bottom).

approach is expected to be beneficial due to the limited variability of occuring damaged areas in the particular port images we use, and for further related datasets that will be captured by similar systems to (Hesse et al., 2019). In other words, the damaged areas on quay walls or buildings can be closely sensed with high-resolution, allowing the ASPP module, regradless of the damage size, to segment the border area of different classes with high quality by taking optimized local context into account.

To improve segmentation results, we investigate different dilation rates of the corresponding Atrous convolution operation. The combined pyramid pooled features regain spatial context when merged with low-level features from the feature extraction network (alias backbone), primarily consisting of object boundaries. In the decoder, two 4x upsampling operations reconstruct the original image size, and contiguous segments are produced. To prevent information loss in the network structure, we apply some convolutions to combat the simplicity of linear interpolation. We examine whether adding additional blocks (0, 1 and 2) in between the Upsampling layers can significantly improve the segmentation result. For the dilation rates we consider the single to triple amount of 3, 6 (default), 12 and 24 for the Atrous convolutions. These hyperparameters are tested with three different models: VGG-16 (Simonyan and Zisserman, 2014) (Batch normalizations included), ResNet-50 (He et al., 2015) (version 1.5), and EfficientNet V2 (Tan and Le, 2021) (Small). VGG-16 has the most basic structure due to its early development in 2014 and consists of around 16 million of the lowest number of trainable parameters. The other two models are built with more advanced concepts in mind, such as Residual blocks, Neural Architecture Search optimized model structure, and Stochastic Depth, but can still be considered as rather basic or small DL models, with approximately 23 and 20 million parameters, respectively. We individually modify the models such that the output stride is set to 16 by removing the fifth MaxPooling layer or setting the step size of the last strided convolution(s) to one. Furthermore, we freeze the initial layers up to Deeplab's low-level feature output to preserve the extensively learned parameters from the ImageNet database (used weight version 1 according to the torchvision (Marcel and Rodriguez, 2010) documentation). While the purpose of this work is to assist inspection operators, fast inference times are not necessarily required. However, to prevent overfitting or long training duration for new acquired datasets in the future, we have not examined significantly larger or more complex backbones.

To train our models, we initially divide the image-label pairs in (non-)overlapping patches of 512x512 pixels. This approach allowed the models to be fed with rich semantic context that exceeded their output layer's receptive field, enabling them to learn from different perspectives of cut-off corrosion segments located near the edges. We use a 25 % overlap for the train dataset but nothing is utilized otherwise. The available 24 GB VRAM of the GeForce RTX 3090 and later 4090 is fully utilized by setting the batch size to 8 (for the EfficientNet) or 12. For an efficient training advancement, the initial learning rate is determined to be 5e-4 and 1e-3, respectively, by the most negative slope of a learning rate finder curve.

To counteract the imbalance between undamaged and corroded areas, we use image augmentation methods such as affine transformation, colour jitter and Gaussian blur, with individual associated probabilities and class weights (1:1 for baseline determination to emphasize architecture changes effectively) within the different considered loss functions. To optimize the network architecture robustly, we employ k-fold Cross-Validation with the four aforementioned training subsets extensively. Each combination of dilation rates and number of additional convolutions is trained on three of these subsets and subsequently evaluated on the validation set to determine the optimal model parameter state based on the loss value. Through this procedure, the maximum number of epochs has never exceeded 30 for all types of model trainings.

In our experiments, we employ the combination of distribution- and region-based loss terms in the form of Categorical Cross-Entropy and IoU loss. To identify the optimal model structure, we calculate the total uncertainty (entropy of softmax output), accuracy, F1 and IoU scores based on the validation patches. The average uncertainty of this subset ranges from 0.06 to 0.12, while the total accuracy varies only between 95.2 and 96.1 %. The class-specific F1 and IoU scores reveal the details in the detection capabilities on the other hand. Overall, undamaged areas are recognized with more than 97 and 94 %, respectively, with deviations of around 1 % across all experiments. The most significant performance differences occur in the corrosion class. The F1 and IoU scores for this class vary by 8 % between the best and worst performing Deeplab network, with upper bound of 59 and 42 %, respectively. By counting the best performing models for both parameter sets individually, we determined that no additional convolution leads to a performance boost, and the standard dilation rates of 3 and 12 are almost equally useful for the segmentation task. Therefore, we changed the kernel spaces within the Atrous convolutions to 5, 10 and 20 to combine these ASPP blocks roughly. Thereby, we focus primarily on a local field of view in order to detect segment boundaries accurately, but find more deviating and rich features at the same time due to the variable dilation rates. To further improve the baseline model, we consider a wide range of modified loss functions with class weights, and evaluate all three backbone networks.

This contribution has been peer-reviewed.
https://doi.org/10.5194/isprs-archives-XLVIII-M-1-2023-301-2023

4.2 Training process of (weakly) supervised models

In the first stage of finetuning, we trained the modified Deeplab V3+ architecture with several pretrained backbones and overall five different loss terms on the complete training dataset. We evaluated them equally to the previous step to determine the optimal loss function for each feature extraction network. After that, we adjusted the hyperparameters related to the cost function to improve the model performance. We then used the separate testing set to independently assess the best performing Deeplab V3+ models among the feature extraction networks (s. chapter 5). The top architecture-loss pairs are then used to train the Deeplab model on the weakly supervised dataset in an iterative process, whereby the training and validation sets are created by the model predictions post-processed with a Potts model. We changed the standard deviations regarding pixel distance and RGB difference slightly compared to the default values listed at Pydense package's Github page (s. footnote 1) to account for the properties of our dataset. In particular, we set s_{xy} and s_{rgb} of the bilateral term to 40 and 10, respectively. This enables predicted corrosion segments to be extended more strictly based on the colour difference to nearby pixels, which is necessary for the corrosion segments onto the specific pile quay wall. We stooped when the IoU scores of the validation set do not improve compared to the last iteration. The initially trained models change the datasets extensively, leading to little overlapping segments. For instance, when the complete weakly supervised dataset is deployed, the first model achieves around 30 % IoU for the corrosion class. Subsequently, the other models benefit from more coherent samples post-processed by a CRF approach, gradually refining the damage segments. This entire procedure restricts the final model from learning self-produced label noise. We provided a detailed visualization of this segmentation approach with respect to the validation and test set in subsection 5.2.

All the models are in general trained using Stochastic Gradient Descent with Nesterov Momentum and L2 regularization, with associated parameters μ and λ set to 0.9 and 1e-5, respectively, in the Pytorch framework (Paszke et al., 2019). The learning rate is halved when the validation loss does not decrease for three epochs in a row, and Batch Normalization layer momentum starts at 0.2 and gradually decreases to help the non-frozen network section adapt quickly to the task. The explored loss functions include Categorical Cross-Entropy, IoU loss, their combination (compound loss), and the base variants modified by the Focal term or parameter $\gamma = 2$ that shifts the model's attention to hard-to-classify pixels (Lin et al., 2017). Additionally, class weights of 1:3 were added to each error term. We observed significant performance improvements with the Focal IoU loss networks. Further Deeplab models were examined by incrementally increasing the parameter up to 5 for each backbone individually. The aforementioned weakly supervised model was finally trained with ResNet-50 and Focal IoU loss with $\gamma = 4$, and the impact of uncertain ground truth images was investigated by varying their fraction within the dataset between 60 to 100 %. Manually labeled images remained unchanged during each of these training runs to simulate real conditions.

5. EVALUATION AND DISCUSSION

5.1 Supervised models

The performance of the baseline Deeplab V3+ model is evaluated on the dedicated validation dataset after training it with moderate basic backbones on various distribution- and region-based loss functions. Each image patch, which lacks surrounding context, is processed by the individual segmentation networks and analyzed quantitatively using metrics related to total uncertainty and confusion matrix. In this supervised context, its evident that VGG-16 performs significantly worse than the other feature extraction networks on average due to the lower output mapping capabilities. However, the default Focal IoU loss variant ($\gamma = 2$) is marginally ahead of the corresponding EfficientNet V2 network. Additionally, we observe that region-based exclusive cost functions exhibit up to one magnitude superior uncertainty scores across all models, indicating their superiority for segmentation tasks with imbalanced datasets. We report only the corrosion class-related F1 and IoU scores in detail in this step of model optimization, as the background detection rates are comparable to those in subsection 4.1 with differences of around 3 %. Given the imbalance of this particular dataset, the total accuracy values can be neglected completely. Overall, the remaining quantities for the different Deeplab models are visualized in Figure 2.

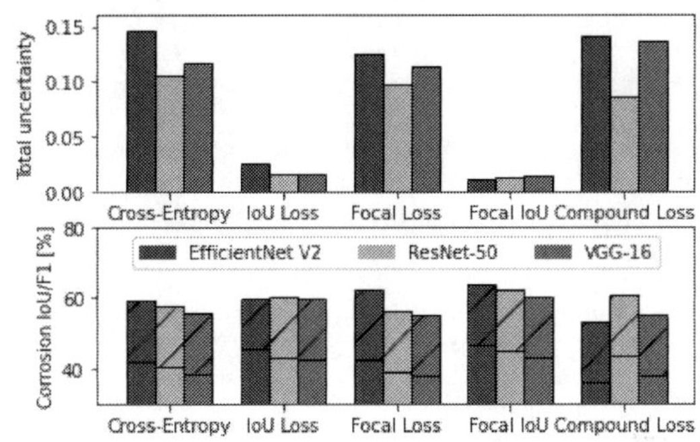

Figure 2. Total uncertainties and corrosion class related metrics of Deeplab V3+ models with respect to validation set consisting of different backbones and trained on various loss functions. The hatched bar sections indicate the difference between IoU and F1 score.

Due to the excellent predictive performance achieved using the Focal IoU loss, we conducted further experiments to optimize the γ parameter for each backbone network separately. The best-performing parameter value was then applied for the first time on the testing set. To obtain complete images instead of smaller patches for visualization or later applications, we used a blending method described in (Chevallier, 2017). This method constructs smooth and continuous segments across patch edges by means of a window function that combines several orientated model predictions. These original sized images were than analyzed using a metric designed to assess agreement with complete corroded segments in the ground truth data. This is particularly helpful because inspection operators are primarily interested in the presence of damaged areas at a certain location, rather than their exact shape and/or size. The resulting scores originate from False Positive and Negative corrosion pixels, and can therefore be understood as supportive for the detection task. To make it more comprehensible: the remaining percentages to 100 represent fully made-up or overlooked damage segments. These metrics are presented in Table 1, alongside common quality measures. In total, we trained five different segmentation network instances by randomly reinitializing the parameters (except the pretrained backbone) to determine the aleatoric and epistemic uncertainties individually. The total uncertainty is derived by the entropy of the composed Deep Ensemble, while epistemic uncertainty is derived by mutual information, and aleatoric uncertainty is equivalent to their difference. Therefore, the computed result(s) may deviate slightly from the results the single model shown in Figure 2.

Table 1. Results of best Deeplab V3+ segmentation network with respect to test set

Metric	Score
Total uncertainty (aleatoric, epistemic)	0.03492 (0.03491, 1e-5)
Total accuracy [%]	95.7
F1 Score [%] (backg., corr.)	97.7, 71.5
Pixel-wise IoU [%] (backg., corr.)	95.5, 55.6
Corrosion supporting FP and FN [%]	80.6, 86.3

The determined uncertainties are in-line with the other Deep Ensembles in terms of their relative relation. On average, the individual Deeplab models among the frameworks vary by 1e-6 to 1e-5, while the images themselves are responsible for almost the complete amount. This implies that the pretrained backbones, combined with the available but general small dataset, do not hinder the segmentation networks from learning corrosion damage appropriately, as aleatoric uncertainty remains unaffected by more training images. However, this information does not provide a detailed overview of the corrosion predictions in the images themselves. Looking at the other scores, we see that the non-damaged or background areas are classified with an accuracy of 95 % or more. The model struggles with precisely identifying the corrosion areas, but the value is as good as (Katsamenis et al., 2020) found on their available dataset. Due to the domain difference, we conclude that the optimization steps were necessary to achieve this level of performance with the restricted network architecture and overall available resources.

To get insight into the prediction results and their uncertainties in relation to the manually labeled images, we provide colour-coded corrosion segments and greyscale images below for the same scene as shown in Figure 1. In the top image, green, blue and red colors represent True Positive, False Positive and False Negative detections, respectively, while the brightness level indicates regions with higher uncertainty in the bottom images. The individual greyscale values are computed by converting an image-level scaled float value to an 8-bit unsigned integer, ensuring the full spectrum is always occupied despite the significant differences indicated by the tabular values. This example confirms that the trained models usually differ within a narrow buffer around the manually labeled damage segments.

The trained model often misses or wrongly detects corrosion near the areas that we identified as damaged with high confidence manually. Our integrated metrics reflect this by showing values above 80 %. Inspection operators dealing with manual classification on a daily basis can benefit from this information, despite other metrics indicating lower performance. If we also provide them additionally for example with the uncertainty images, they can adjust the segment boundaries using a prospect software package as part of a digitally assisted procedure.

Figure 3. ResNet-50 Deeplab V3+ model's cropped blending prediction with post-processed colour-coding according to corrosion segments from ground truth data (green: True Positive, blue: False Positive, red: False Negative). Aleatoric and epistemic uncertainty calculated from a related five instance Deep Ensemble network.

5.2 Weakly Supervised model

The best-performing segmentation network, Deeplab V3+ ResNet-50, in the supervision context is then demonstrated in our simple designed training pipeline. This pipeline starts with ground truth images generated from a strict Region Growing algorithm applied to the challenging JadeWeserPort dataset. Due to the lack of accurate label images over the complete iterative training process, performance degradation is inevitable. However, the time-consuming labeling effort is significantly reduced. This gives researchers the possibility to obtain a larger dataset, if necessary, in order to improve model's generalization. In the following Figure 4, we show the performance development for different fractions of weakly label images used within the training process.

We draw here for the first time conclusions based on the agreement of blended patch predictions to the test set regarding the individual model training procedures. Especially the initial training phase can be highlighted, because the corresponding Region Growing labeled images achieved only 24.3 and 39.1 % in terms of corrosion IoU and F1 score, respectively. The final model of each training pipeline was determined by the highest IoU damage value of the individual evolved validation set, requiring 3, 2, 3, 3 and 2 iterations in descending order of the weakly label fraction. According to these results, the models iteratively trained on the complete weakly supervised dataset were the most successful, while the optimal Deeplab networks learnt from 70-90 % weak labels had a 5-10 % lower performance in terms of corrosion IoU. The dataset with the largest fraction of manual labels performed worst, with an additional 10 % drop in performance. This outcome was expected, as the strategy of finding the best model relys on monotonically increasing metric scores, which is equivalent to efficient conver-

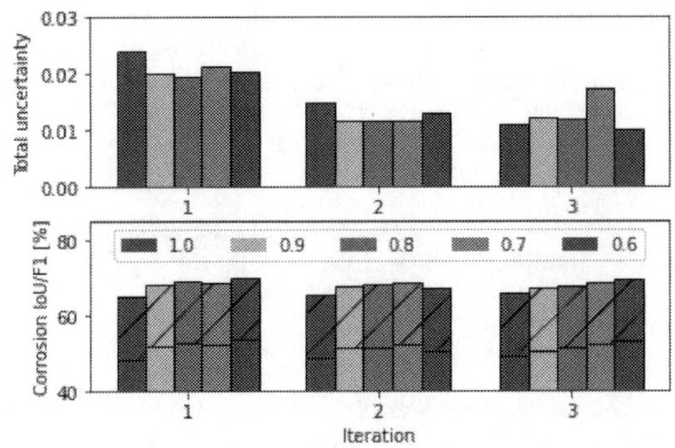

Figure 4. Total uncertainties and corrosion class related metrics of optimal Deeplab V3+ ResNet-50 with respect to test set trained on weakly supervised dataset. Different colours stand for the fraction of deployed weak labels. The hatched bar sections indicate the difference between IoU and F1 score.

gence. When an individual dataset contains inaccurate but obvious corrosion labels, the model can learn the underlying complexity more effectively. However, when a significant amount of the dataset consists of contradictory image pairs in terms of weak labels, there is no fast and straightforward solution. But the iteratively refined images may approximate the true damage segments step by step at different speeds. Hence, we can achieve supervised-like performance after several model trainings when a reasonable amount of accurate labels is available. However, we cannot measure such an advancement in general due to a lack of reference data. Nevertheless, we found a significant improvement between different models (with weak supervised fractions between 0.7 and 1.0) in terms of the labeled test set after a few iterations, as demonstrated in Figure 4. But how much labeling effort can be saved to achieve that? If we continue to calculate with the 50 % example from subsection 3.3, we would reduce the labeling effort for the complete training and validation dataset (70 images in total) by roughly 12 hours. For every accurately produced segmentation ground truth image taking approximately 10 - 15 more minutes, the model gains about 0.15 % IoU corrosion performance within an initial linear growing phase. Afterwards, the training may take more iterations than 3 to convergence, resulting in unpredictable performance capabilities, followed by a period in which a single model recognize weak labels confidently as outliers and take this into account for its parameter optimization automatically. Additionally, the determined total uncertainty over the test data is barely affected for the final models by the relative portion of manual labeled images. Therefore, we suggest that one should only label up to one third of small, imbalanced image datasets precisely to deploy the corresponding model in a short period of time. This will have a noticeable effect within the process chain and guarantees, based on our experiments, at least 90 % of the maximum achievable performance. This gap might be further closed by applying certain image augmentation methods and/or domain-specific loss functions.

In this paragraph, we present a detailed analysis of the performance of our best model on the test set, which provides insights into the potential of our method for real-world damage detection applications. We present the results using the same format as in the supervised setting, with Table 2 and Figure 5. Although most of the quantities in Deeplab model have

Table 2. Results of Deeplab V3+ segmentation network trained on 70 % weakly supervised images with respect to test set

Metric	Score
Total uncertainty (aleatoric, epistemic)	0.02248 (0.02248, 2e-6)
Total accuracy [%]	95.2
F1 Score [%] (backg., corr.)	97.4, 68.5
Pixel-wise IoU [%] (backg., corr.)	94.9, 52.1
Corrosion supporting FP and FN [%]	81.6, 75.0

only slightly changed compared to the fully supervised model, it tends to miss corrosion segments more often, resulting in a higher number of False Negative pixels (75.0 compared to 86.3 %). Conversely, the number of type 1 error for damaged segments decreases. These factors lead to relatively minor changes in the metric scores, but has a more visible impact on the prediction images themselves. Figure 5 illustrates the behaviour explicitly for the centrally located corrosion segments. In contrast, we observe finer, uncertain boundaries around the predicted corrosion segments, especially in aleatoric case, which is reflected in the lower scores in Table 2 compared to the fully supervised results. This difference can be explained by the iteratively trained models that learn from individual refined label images and build up confidence over time.

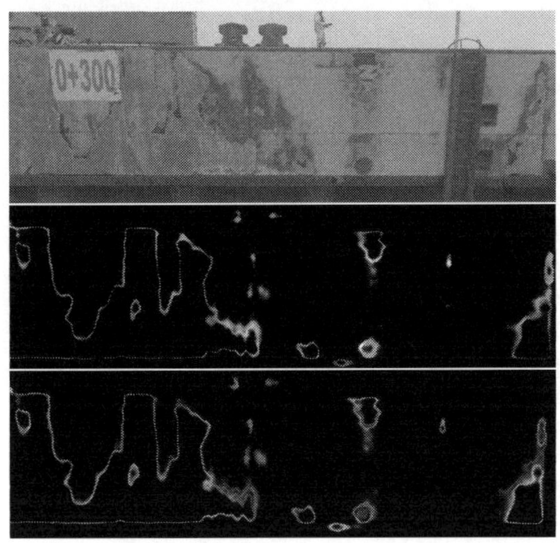

Figure 5. Cropped blending prediction by best ResNet-50 Deeplab V3+ model trained on 70 % weakly supervised images. Colour-coded in post-processing according to corrosion segments from ground truth data (green: True Positive, blue: False Positive, red: False Negative). Aleatoric and epistemic uncertainty calculated from an affiliated five instance Deep Ensemble network.

6. CONCLUSION

In this study, we developed a modified Deeplab V3+ model for segmentation of corrosion in infrastructure inspections. The model was optimized using 4-fold Cross-Validation on a corrosion dataset, and we found that the combination of ResNet-50 backbone and Focal IoU loss with $\gamma = 4$ (with integrated

This contribution has been peer-reviewed.
https://doi.org/10.5194/isprs-archives-XLVIII-M-1-2023-301-2023 | © Author(s) 2023. CC BY 4.0 License.

class weights of 1:3) achieved optimal results for our imbalanced dataset. We restricted our optimization to moderate large feature extraction networks for efficient training on consumer-level hardware. Our DL model is designed to assist operators visually in future inspections, and we plan to retrain it on further datasets that contain also other types of damage. We also investigated an iterative weakly supervised approach based on Region Growing, which can reduce the manual effort required for ground truth labeling. We found that roughly one third of manual labels are sufficient to achieve around 90 % of the fully supervised detection capabilities. The produced Deeplab V3+ models are part of the automation process of infrastructure inspections to reduce on-site downtime and subjective assessments (for example under stress). In future research, we plan to evaluate the model on a wide range of infrastructure datasets and refine the weakly supervised approach possibly with existing concepts to train a more robust segmentation network. We also plan to provide possibly model predicted uncertainty maps to inspection operators for a justified decision.

Funding This research was funded by the Deutsche Forschungsgemeinschaft (DFG, German Research Foundation) - NE 1453/5-1 and as part of the Research Training Group i.c.sens [RTG 2159].

REFERENCES

Adams, R., Bischof, L., 1994. Seeded region growing. *IEEE Transactions on Pattern Analysis and Machine Intelligence*, 16(6), 641-647.

Azad, R., Asadi-Aghbolaghi, M., Fathy, M., Escalera, S., 2020. Attention deeplabv3+: Multi-level context attention mechanism for skin lesion segmentation. *ECCV 2020 Workshop on BioImage Computing*.

Chen, L.-C., Zhu, Y., Papandreou, G., Schroff, F., Adam, H., 2018. Encoder-decoder with atrous separable convolution for semantic image segmentation.

Chevallier, G., 2017. Using a u-net for image segmentation, blending predicted patches smoothly is a must to please the human eye. `https://github.com/-Vooban/Smoothly-Blend-Image-Patches`.

Duy, L. D., Anh, N. T., Son, N. T., Tung, N. V., Duong, N. B., Khan, M. H. R., 2020. Deep learning in semantic segmentation of rust in images. *Proceedings of the 2020 9th International Conference on Software and Computer Applications*, ICSCA 2020, Association for Computing Machinery, New York, NY, USA, 129–132.

Hake, F., Lippmann, P., Alkhatib, H., Oettel, V., Neumann, I., 2023a. Automated damage detection for port structures using machine learning algorithms in heightfields.

Hake, F., Scherff, M., Neumann, I., Alkhatib, H., 2023b. Using semantic segmentation for the damage detection of port and marine infrastructures. Herbert Wichmann Verlag, accepted.

He, K., Gkioxari, G., Dollár, P., Girshick, R., 2017. Mask r-cnn.

He, K., Zhang, X., Ren, S., Sun, J., 2015. Deep residual learning for image recognition.

Hesse, C., Holste, K., Neumann, I., Hake, F., Alkhatib, H., Geist, M., Knaack, L., Scharr, C., 2019. 3D HydroMapper: Automatisierte 3D-Bauwerksaufnahme und Schadenserkennung unter Wasser für die Bauwerksinspektion und das Building Information Modelling. *Hydrographische Nachrichten*, 113, 26-29.

Hsu, C.-Y., Hu, R., Xiang, Y., Long, X., Li, Z., 2022. Improving the Deeplabv3+ Model with Attention Mechanisms Applied to Eye Detection and Segmentation. *Mathematics*, 10(15). https://www.mdpi.com/2227-7390/10/15/2597.

Huang, Z., Wang, X., Wang, J., Liu, W., Wang, J., 2018. Weakly-supervised semantic segmentation network with deep seeded region growing. *2018 IEEE/CVF Conference on Computer Vision and Pattern Recognition*, 7014–7023.

Katsamenis, I., Protopapadakis, E., Doulamis, A., Doulamis, N., Voulodimos, A., 2020. Pixel-level corrosion detection on metal constructions by fusion of deep learning semantic and contour segmentation.

Krähenbühl, P., Koltun, V., 2012. Efficient Inference in Fully Connected CRFs with Gaussian Edge Potentials. https://arxiv.org/abs/1210.5644.

Lin, T.-Y., Goyal, P., Girshick, R., He, K., Dollár, P., 2017. Focal loss for dense object detection.

Marcel, S., Rodriguez, Y., 2010. Torchvision the machine-vision package of torch. *Proceedings of the 18th ACM International Conference on Multimedia*, 1485-1488.

Nash, W. T., Powell, C. J., Drummond, T., Birbilis, N., 2019. Automated corrosion detection using crowd sourced training for deep learning.

Nash, W., Zheng, L., Birbilis, N., 2022. Deep learning corrosion detection with confidence. *npj Materials Degradation*, 6.

Paszke, A., Gross, S., Massa, F., Lerer, A., Bradbury, J., Chanan, G., Killeen, T., Lin, Z., Gimelshein, N., Antiga, L., Desmaison, A., Köpf, A., Yang, E., DeVito, Z., Raison, M., Tejani, A., Chilamkurthy, S., Steiner, B., Fang, L., Bai, J., Chintala, S., 2019. Pytorch: An imperative style, high-performance deep learning library.

Simonyan, K., Zisserman, A., 2014. Very deep convolutional networks for large-scale image recognition.

Sun, K., Zhao, Y., Jiang, B., Cheng, T., Xiao, B., Liu, D., Mu, Y., Wang, X., Liu, W., Wang, J., 2019. High-resolution representations for labeling pixels and regions.

Supervisely contributors, 2023. Supervisely. `https://github.com/supervisely/supervisely`.

Tan, M., Le, Q. V., 2021. EfficientNetV2: Smaller Models and Faster Training. https://arxiv.org/abs/2104.00298.

Tanveer, M., Kim, B., Hong, J., Sim, S.-H., Cho, S., 2022. Comparative Study of Lightweight Deep Semantic Segmentation Models for Concrete Damage Detection. *Applied Sciences*, 12(24). https://www.mdpi.com/2076-3417/12/24/12786.

Zeng, H., Peng, S., Li, D., 2020. Deeplabv3+ semantic segmentation model based on feature cross attention mechanism. *Journal of Physics: Conference Series*, 1678, 012106.

This contribution has been peer-reviewed.
https://doi.org/10.5194/isprs-archives-XLVIII-M-1-2023-301-2023

The International Archives of the Photogrammetry, Remote Sensing and Spatial Information Sciences, Volume XLVIII-M-1-2023
39th International Symposium on Remote Sensing of Environment (ISRSE-39) "From Human Needs to SDGs", 24–28 April 2023, Antalya, Türkiye

SELF-SUPERVISED LEARNING FOR CROP CLASSIFICATION USING PLANET FUSION

Caglar Senaras[1*], Piers Holden[2], Timothy Davis[2], Akhil Singh Rana[2], Maddie Grady[2], Annett Wania[2], Richard de Jeu[1]

[1] Planet Labs Haarlem, Netherlands - (caglar.senaras, ries)@planet.com
[2] Planet Labs Germany GmbH, Berlin, Germany - (piers, akhil.singh.rana, tim.davis, maddie.grady, annett)@planet.com

KEY WORDS: Self-supervised learning, Deep Learning, RapidAI4EO, Crop Classification, Planet Fusion

ABSTRACT:

Benefiting from the high cadence and spatial resolution of the new generation of Earth observation satellites, remote sensing technology is allowing us to derive more valuable information for the agricultural sector. Crop classification is one of the fundamental information derivatives from Earth observation data researchers used for food security, crop monitoring, and economic assessment. The robustness of a crop classification model to variations in environmental and management conditions due to time and location is one of the crucial requirements. To achieve this, we developed a novel self-supervised method using the advantage of unlabeled samples and transformer architectures. We used six different areas in Germany and four years to evaluate the robustness of the model. Our experiments showed that self-supervised deep learning methods could provide a significant advantage in handling these variations. In some cases, we observed around 30 percentage points improvements in F1-score performance compared to a Random Forest based model.

1. INTRODUCTION

Developing agricultural policies and sustainable farm management is essential for handling the increasing global demand for food. Accurate and timely information about the types and extent of crops present in an area is necessary for evaluating crop health and productivity, forecasting crop yields, and optimizing farming practices. For that reason, developing a robust machine learning-based crop classification method has become an important research topic.

Most state-of-the-art crop classification approaches use the advantages of supervised learning methods such as Random Forest (RF), Support Vector Machines (SVM) (Saini, 2018). As with other remote sensing applications, supervised deep learning-based methods have also become popular for crop analysis in recent years (He, 2019), (Zhao, 2021). These deep learning approaches have demonstrated strong performance on datasets with limited geographical and temporal extent (Metzger, 2021). However, large datasets with crop type labels covering multiple years and countries have not been readily available until recently (eg. EuroCrops (Schneider, 2021)). Due to the data-intensive nature of training deep methods, this historical lack of data has limited efforts to comprehensively evaluate the potential of deep learning methods to perform at continental scale, across the full range of climate conditions that require many years of data to be fully expressed.

Self-supervised learning (SSL) represents a paradigm-shift in the machine learning domain, enabling the pre-training of deep learning models using sets of unlabeled samples. Many studies demonstrated that models pre-trained with SSL have better classification performance than fully supervised models when the number of labelled samples is limited (Wang, 2022). They showed that self-supervision outperforms supervision when reducing the number of labels using BigEarthNet (Sumbul, 2019) land cover classification dataset. In one of the recent studies, Marszalek pretrained a deep learning-based crop classifier with SSL on a limited number of unlabeled data (Marszalek, 2022). They trained the crop classifiers for a given region by using multiple years and measured the performance. According to them, it is essential to introduce a few labels for the predicted year during the training.

In this study, we adapt one of the state-of-the-art SSL architectures, called SimSiam (Chen, 2021), to pre-train our attention-based crop classifier. Since we need a large number of unlabeled samples to train our model in a self-supervised manner effectively, we combine the advantages of two datasets: RapidAI4EO (Marchisio, 2021) and EuroCrops. EuroCrops provides the boundary of the fields around Europe. On the other hand, RapidAI4EO satisfies the need for high-resolution and high-cadence time series for some of these fields. To provide a reliable crop classification approach, spatial and temporal robustness become essential, where spatial robustness refers to the ability of an algorithm to classify crops in different locations accurately, and temporal robustness refers to its ability to classify crops accurately in different years. Therefore, we design a set of experiments to compare the ability of a deep learning model pre-trained with SSL to generalize to unseen spatial and temporal contexts, against a baseline of deep learning models (without SSL pre-training) and Random Forest models.

The following Section 2 will describe the proposed method, including the sample selection strategy using an unlabeled SSL dataset and the deep learning architecture. Section 3 introduces the dataset used for the crop classification downstream task. Section 4 presents the results of our experiments, including data analysis to observe the impact of these (temporal and spatial) variability on crops and quantitative evaluations of the classifiers' performance. Finally, in Section 5, we conclude the paper by summarizing our main findings and discussing their implications for future work.

* Corresponding author

This contribution has been peer-reviewed.
https://doi.org/10.5194/isprs-archives-XLVIII-M-1-2023-309-2023

2. PROPOSED METHOD

The proposed method includes two important steps. In the first one, we combine publicly available datasets to prepare a large number of positive pairs, which will be used during the pretraining of the model in an SSL fashion without any crop type information. After that, we pretrain our model using SSL and apply supervised training to our model to handle crop classification downstream tasks.

2.1 Self-Supervised Learning Dataset

As mentioned in the introduction, we integrate the advantages of two datasets: RapidAI4EO and EuroCrops.
The RapidAI4EO corpus, which was produced under Horizon2020 Project RapidAI4EO (https://rapidai4eo.eu/), is one of the most extensive remote sensing datasets, and it is publicly available to the machine learning and remote sensing community. It contains two data sources, Sentinel-2 and Planet Fusion (Planet Fusion, 2021), and covers 500,000 patches of 600m x 600m, distributed across Europe (Figure 1).
In this study, we use Planet Fusion data, which provides a sensor-harmonized, gap-filled, and cloud-free time series. It represents the highest spatial and temporal resolution data in the corpus with a ground resolution of three meters, five-day cadence, and four spectral bands (VNIR) for 2018 and 2019. A major advantage of using Planet Fusion imagery was that it provided regular, complete and high-quality time series that required almost no pre-processing to prepare them for use in our machine learning pipelines, highlighting its suitability for these types of workflows.

The EuroCrops dataset, an initiative of the Technical University of Munich (https://www.eurocrops.tum.de/), aims to combine all publicly available self-declared crop reporting datasets from countries of the European Union. It provides the boundaries of the fields and the crop type for a given year.

We observed that 91,944 RapidAI4EO locations intersect with 862,691 EuroCrops fields in total. However, for most countries, EuroCrops includes the crop type information after 2019. Since the temporal range of the RapidAI4EO corpus does not cover this period, we use only the field boundary information. We assume that farmers planted only one type of crop at a given time. As a result, we can assume that randomly selected samples in a field will represent the same crop type.

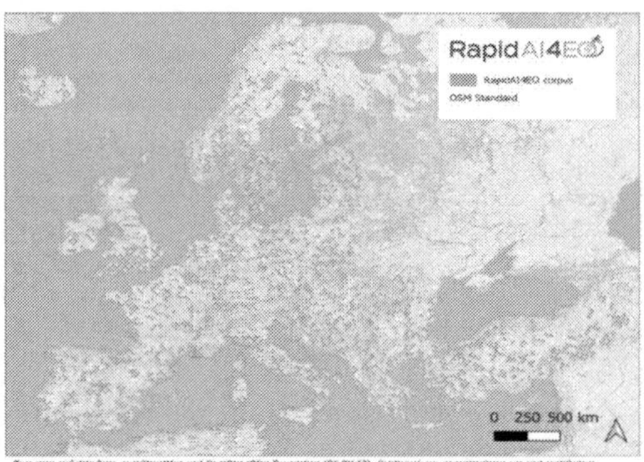

Figure 1. The approximate location of the 500,000 RapidAI4EO patches

To prepare our unlabeled dataset, we select 24,407 RapidAI4EO locations and use the largest two fields for each location. For each field, we randomly pick ten samples (Figure 2). During the training of the SSL model, two of these samples are randomly selected as positive pairs and fed to our model as described in Section 2.2.

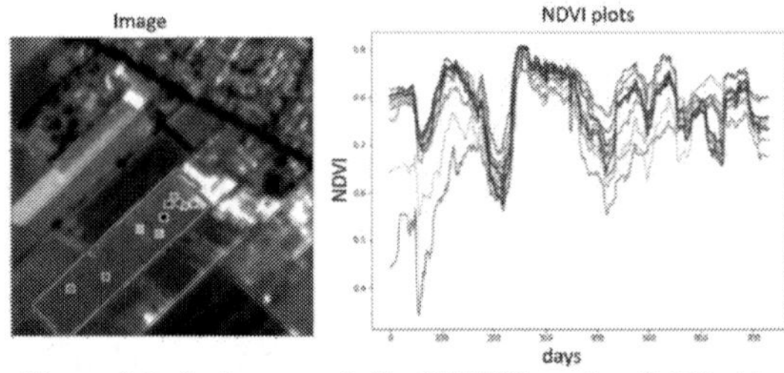

Figure 2. Left: An example RapidAI4EO patch, a field inside that patch (red polygon in the left image), and ten sample points inside the field. Right: 2-year NDVI time-series (right) of the collected ten samples for the given field.

2.2 SSL Crop Classification Architecture

In this study, we adapted SimSiam to introduce a simple and effective framework for unsupervised representation learning based on the Siamese network architecture without requiring any negative pairs. In SimSiam, the model is trained to predict the parameters of a fixed encoder network, given the inputs of an augmented view of the same data. Thus the model learns to identify essential characteristics in the data, creating a valuable representation that can be utilized for the downstream tasks.

After the positive pairs are generated, as explained in Section 2.1, we use them for self-supervised learning. For each pair, we obtain daily time series of the surface reflectance (SR) values and feed these time series to SimSiam to train our attention-based encoder. The encoder is inspired by (Garnot, 2020). The encoder combines the Pixel-Set Encoder (PSE) and the Transformer architecture to extract features from the input images. We adapt the PSE to use a single SR value for each field by using fully connected layers. After the pretraining, we modify our encoder to handle crop classification downstream by adding fully connected and soft-max layers.

3. CROP CLASSIFICATION DATASET

For the crop classification downstream task, we create a new dataset that will allow us to evaluate the temporal and spatial robustness of the models across a single-nation context. To achieve our goal, we selected six areas in Germany (Figure 3).

https://doi.org/10.5194/isprs-archives-XLVIII-M-1-2023-309-2023 | © Author(s) 2023. CC BY 4.0 License.

Area-5 and Area-6 were previously used in the AI4FoodSecurity[2] challenge. The selected 4 new areas were longitudinally stratified, to ensure variability in climate, planting practices and environmental conditions, each being part of a distinct region. Each area has a size of 24 km x 24 km. All areas except Area-5 and Area-6 have a daily Planet Fusion time series between 2019 and 2021. On the other hand, Area-5 and Area-6 have only the daily samples for 2019 and 2018, respectively.

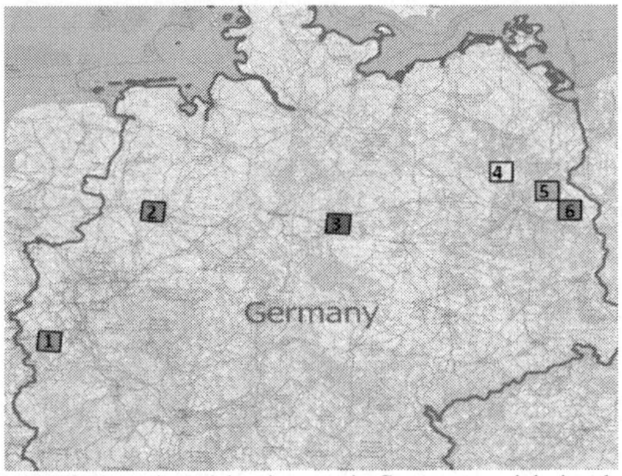

Figure 3. Location of the six areas in Germany used during the crop classification experiments. Red and Purple tiles are used for testing and the remaining four are used for the training phase.

Crop planting information and field boundaries for these areas were sourced from publicly available user submitted datasets that form the basis of official EU reporting (Integrated Administration and Control System). The datasets were accessed through the region specific data portals[3]. After consideration, we decide to focus on six crop types: winter wheat, winter barley, winter rye, maize, winter oilseed rape, and sugar beet. These crops were well-represented across all tiles in the dataset and with adequate numbers to avoid wildly imbalanced classes. Figure 4 shows the NDVI time series of the selected crop types.

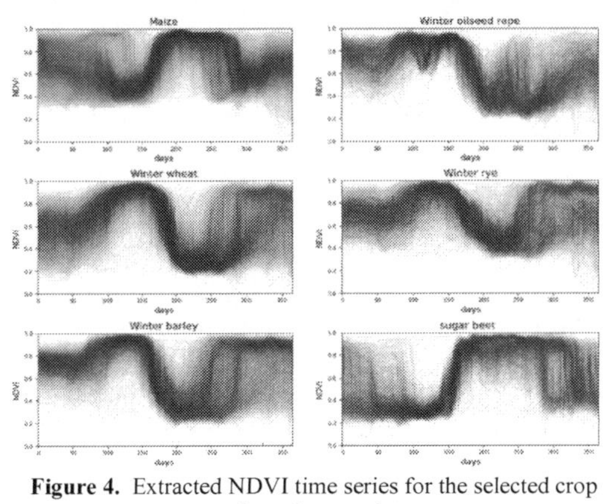

Figure 4. Extracted NDVI time series for the selected crop types by using our dataset.

Finally, we split the dataset into training and test sets by considering the class distributions and geolocations of the areas. The Area-3 and Area-6 are used for testing, and the rest are used for training. Ten percent of the training samples are randomly selected for validation purposes. Table 1 summarizes the number of samples for each crop class used for training, validation and testing purposes.

Crop Type	Training	Validation	Testing
Maize	11,618	1,295	1,190
Winter barley	5,158	602	1,072
Winter oilseed rape	1,418	155	1,218
Winter rye	2,304	275	946
Winter wheat	7,079	744	4,144
Sugar beet	1,766	190	1,160

Table 1. The number of samples for each crop class.

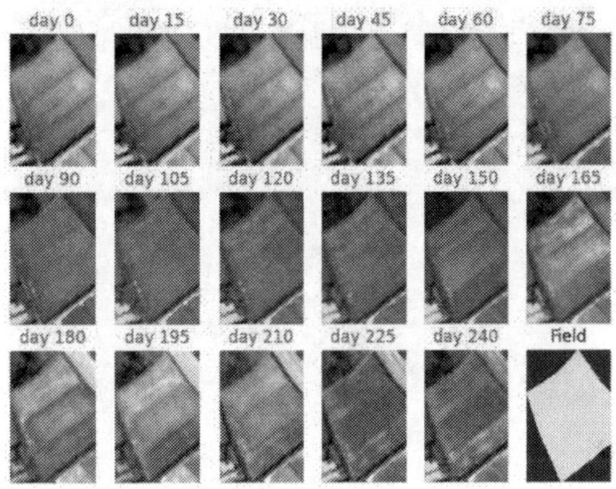

Figure 5. False-color images of a field in our dataset with 15 days cadence.

4. EXPERIMENTS

4.1 Data Analysis

Before starting to measure the classification performance of different models, we prefer to visually observe the impact of the temporal and spatial variances on crops. For that reason, we analyze the Normalized Difference Vegetation Index (NDVI) time series of selected crops under different scenarios.

First, we aim to focus on temporal impact. Therefore, we focus on a single area, Area4 and plot the NDVI time series of winter barley fields for three sequential years (Figure 6). Although the harvesting time for the three years are close to each other, the earlier phases show significant differences. Similar pattern is observed also for different areas.

[2] https://platform.ai4eo.eu/ai4food-security-germany
[3] Brandenburg: https://geobroker.geobasis-bb.de
Lower Saxony:
https://sla.niedersachsen.de/landentwicklung/LEA/
North Rhine-Westphalia: https://ckan.open.nrw.de/

This contribution has been peer-reviewed.
https://doi.org/10.5194/isprs-archives-XLVIII-M-1-2023-309-2023 | © Author(s) 2023. CC BY 4.0 License.

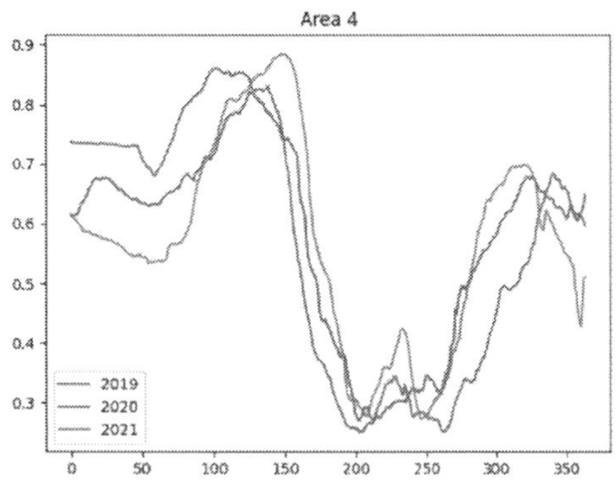

Figure 6. The average NDVI times series of winter barley in Area 4 for three years

Then we analyze the impact of spatial differences on crops. In Figure 7, we plot the NDVI time series of winter barley fields for a single year (2020) in four different areas. Since we focus on a single country, we believe that the spatial variance is not as problematic as the temporal variance for this dataset. However, when we need to solve temporal and spatial variability simultaneously, the intra-class dissimilarity and inter-class similarity may create some problems for some of the classifiers. Figure 8 displays the NDVI plots of two different cereals, which are obtained from four different areas and three years. A similar issue is observed between sugar beet and maize.

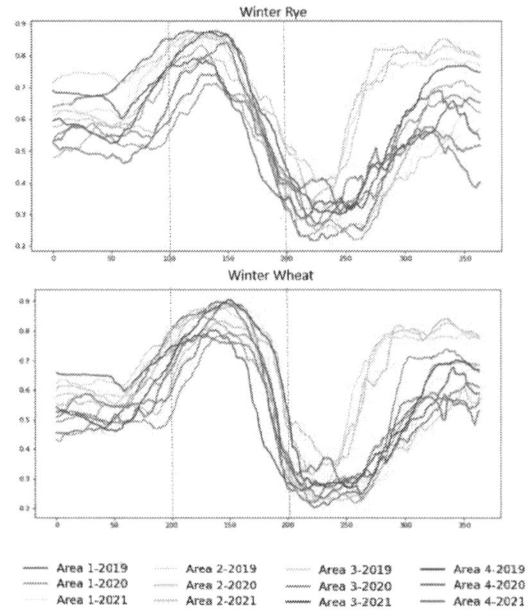

Figure 8. The average NDVI times series of winter barley and winter wheat for each different area and different years.

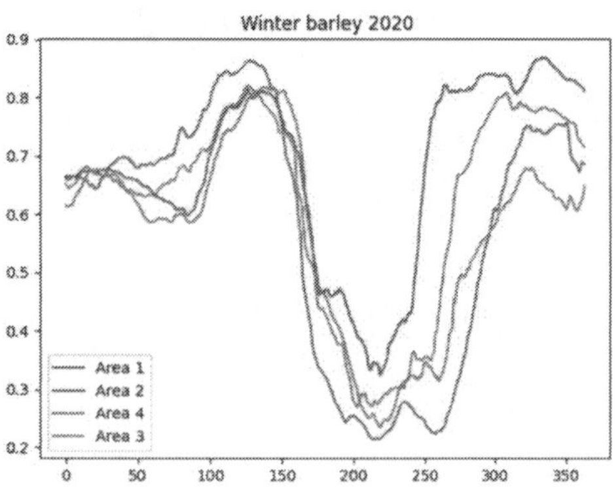

Figure 7. The average NDVI times series of winter barley for our 4 areas.

4.2 Experiment Methodology

In order to evaluate the robustness of our approach, we design two experimental setups. For each setup, we train our proposed model for ten epochs. Similarly, to observe the impact of SSL, we trained the same encoder without any SSL pretraining for ten epochs. In addition to deep learning-based methods, we train two different Random Forest (RF) classifiers.

As a feature set of the first RF classifier, we use NDVI, Normalized Difference Water Index (NDWI), and Green-Blue NDVI (GBNDVI) indices. The second RF classifier uses only surface reflectance values obtained from Planet Fusion. We use grid search to optimize the hyper-parameters of both RF classifiers.

To evaluate the effectiveness of the classifiers, we employ three commonly used evaluation metrics: F1-Macro, F1-Micro, and F1-weighted average. F1-Macro calculates the F1 score for each class and takes the average of all the classes. F1-Micro calculates the F1 score by considering the total number of true positives, false negatives, and false positives across all classes. F1-weighted average calculates the F1 score for each class and takes a weighted average of all the classes.

4.3 Experiment 1- Evaluate Spatial Robustness

The spatial robustness of a crop classifier is a crucial aspect of its accuracy and reliability. To evaluate the spatial robustness of a crop classifier, it is essential to train and test it in different locations with varying environmental and management conditions. In the first experiment, we aim to understand the spatial robustness of the proposed model. Therefore we use the same years (2019, 2020 and 2021) for training and testing. Since Area-6 has only samples from 2018, we exclude it from our test set. Table 2 summarizes the selected areas for this experiment. In total, we used 29,343 fields during the training and tested on 8,668 fields.

https://doi.org/10.5194/isprs-archives-XLVIII-M-1-2023-309-2023

Tile	Selected Year(s)	Subset
Area-1	2019, 2020, 2021	Train
Area-2	2019, 2020, 2021	Train
Area-4	2019, 2020, 2021	Train
Area-5	2019	Train
Area-3	2019, 2020, 2021	Test

Table 2. The selected areas for the Experiment-1

We compare four models' performances based on metrics described in section 4.2 (Table 3). The proposed model with SSL pretraining achieves the highest performance values. The second best model is the same encoder without any pretraining. We observe that SSL pretraining helped us to distinguish winter wheat and rye crop types in a better way for this experiment. The crop-based precision values of the winter rye are 63% and 91% without and with SSL, respectively.

Interestingly, we observe that RF performs significantly better when we use surface reflectance values instead of indices. The indices-based RF had a lower classification performance, especially for sugar beets and maize classes.

Model Name	F1-Micro	F1-Macro	F1-Weight.
Random Forest (indices)	87.64%	83.99%	87.43%
Random Forest (SR)	92.52%	90.39%	92.38%
Our model without SSL	93.62%	92.58%	93.95%
Our model	**96.09%**	**95.41%**	**96.06%**

Table 3. Performance Comparison under spatial variance

4.4 Experiment 2- Evaluate Spatial and Temporal Robustness

The second experiment is an extension of the first, and in addition to spatial variability, it involves temporal variability by training and testing samples from different years. The goal is to evaluate the classifiers' ability to handle variations in environmental and management conditions over time. During the experiment, we trained the models using daily PF data of 20,012 fields captured in 2019 and 2020 from Areas 1, 2, 4, and 5. After the training, we evaluated the models on 3,865 fields captured in 2018 and 2021 from Area-6 and Area-3, respectively. Table 4 shows the classification performance of four different models.

In this experiment, we observe that deep-learning models have better performances. Again the proposed model with an SSL pretraining achieves the highest performance values. On the other hand, RF-based classifiers face a significant performance decrease under the spatial and temporal variability between train and test sets. After including the temporal variability, the performance decrease is more than 20 percent for the indices-based RF classifier. When we compare two different RF-based classifiers, we observe that the indices-based one has a lower capability to distinguish cereals from each other (i.e., wheat, barley, and rye). Similarly, it confuses sugar beets and maize classes.

Model Name	F1-Micro	F1-Macro	F1-Weight.
Random Forest (indices)	69.39%	61.73%	67.30%
Random Forest (SR)	86.18%	83.14%	85.79%
Our model without SSL	94.15%	93.19%	94.13%
Our model	**95.21%**	**94.77%**	**95.20%**

Table 4. Performance Comparison under temporal and spatial variance

These results are consistent with our preliminary data analysis explained in Section 4.1. It is possible to observe the NDVI similarity within cereals and the NDVI similarity of maize-sugar beet pairs in Figure 4. The generated feature space for indices-based RF is not enough to handle intra-class dissimilarity and inter-class similarity with high accuracy.

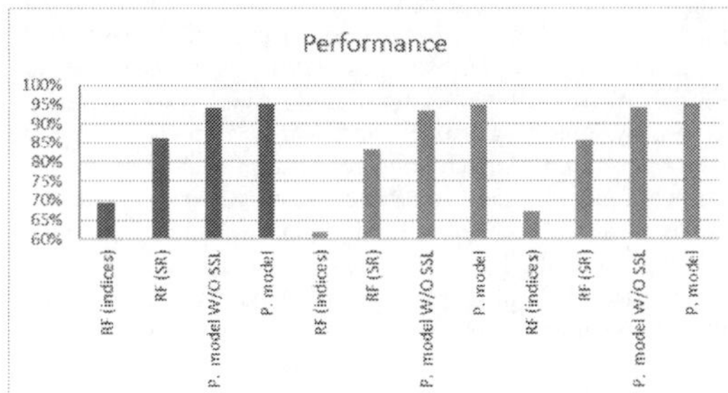

Figure 8. The performance Comparison under temporal and spatial variance. The blue, orange and green colors represent F1-Micro, F1-Macro and F1-Weighted respectively.

When we analyze the proposed model, we observe that it achieves very satisfying results for all crop types. Table 5 displays each crop type's class-based precision, recall, and f-score values. Although our dataset has some imbalance issues and the number of sugar beet fields used for training is 10x less than maize fields, the model was able to classify sugar beets with very high accuracy.

Crop Type	Precision	Recall	F-score
Maize	97.72%	98.41%	98.07%
Winter barley	86.05%	94.13%	89.91%
Winter oilseed rape	99.84%	99.52%	99.68%
Winter rye	92.11%	83.60%	87.65%
Winter wheat	95.16%	95.29%	95.23%
Sugar beet	98.63%	97.55%	98.09%

Table 5. Crop based classification performance for DL with SSL pretraining under temporal and spatial variance

Winter barley and rye had the lowest classification performances, with 89.91% and 87.65% F-scores, respectively. To understand the reason, we generated the confusion matrix (Table 6), and as expected, we observed that distinguishing between cereal types is relatively more challenging than classifying other crops.

https://doi.org/10.5194/isprs-archives-XLVIII-M-1-2023-309-2023 |

		Prediction					
Actual		Maize	W. Barley	W. Oil. R.	W. Rye	W. Wheat	S. Beets
	Maize	558	1	0	3	2	3
	W. Barley	1	401	0	5	18	1
	W. Oil. R.	1	1	623	0	1	0
	W. Rye	1	22	0	362	48	0
	W. Wheat	4	40	1	22	1377	1
	S. Beet	6	1	0	1	1	359

Table 6. Confusion matrix for DL with SSL pretraining under temporal and spatial variance

5. CONCLUSION

In this study, we focus on crop classification, an important research area for remote sensing in agriculture. We observe that the characteristics of crops vary significantly in terms of both time and location. For a reliable crop classification solution, it's essential to satisfy both temporal and spatial robustness. We successfully apply self-supervised learning methods to our transformer-based deep learning techniques to achieve our goal. We test our model under two different scenarios to see the impact of temporal as well as temporal and spatial variability. The results show that our method provides the best results in both scenarios, with a performance difference of over 30 points compared to a traditional random forest classifier based on indices.

Our findings show that:

- Our Deep Learning architecture outperforms RF on our single-nation, multi-year dataset up to 30 percentage points.
- SSL pre-training delivers a performance improvement over standard supervised training when testing on unseen spatial and temporal contexts
- The total performance gain from SSL pre-training is variable. The gain was largest when evaluating spatial generalization, which is also the dimension in which the pre-training dataset expressed the greatest variation. The gain was smaller when evaluating generalization to new years, which may be partly explained by the fact that the pre-training dataset only covered two years.

Based on our results, we would expect SSL pre-training to deliver a more substantial performance improvement over the other methods when:

- limited training labels are available and for only a limited spatial or temporal extent; or
- the unlabeled dataset covers larger regions of time and space that are not covered in the labelled data, but are relevant to the domain.

For our dataset, the deep learning-based method without SSL pretraining already has very high F1-micro and F1-weighted values above 94%. Therefore it is unclear whether SSL pre-training delivers enough of a performance advantage to warrant the additional computational effort required. We believe it is important to validate the performance impact of SSL on more challenging problems.

In future work, we plan to use more labelled samples from around Europe to increase the number of classes and to test for higher spatial variability. As another exciting topic, the high-frequency data from Planet Fusion may be beneficial for early-season crop classification which would be a valuable building-block for real-time food security and supply monitoring across field, regional and national scales .

6. ACKNOWLEDGEMENTS

This work was done in the frame of the AgriCaptureCO2 project which has received funding from the European Union's Horizon 2020 research and innovation programme under grant agreement number 101004282 (https://agricaptureco2.eu/).

7. REFERENCES

Chen, X., & He, K. (2021). Exploring simple siamese representation learning. In Proceedings of the IEEE/CVF Conference on Computer Vision and Pattern Recognition (pp. 15750-15758).

He, T., Xie, C., Liu, Q., Guan, S., & Liu, G. (2019). Evaluation and comparison of random forest and A-LSTM networks for large-scale winter wheat identification. Remote Sensing, 11(14), 1665.

Garnot, V. S. F., & Landrieu, L. (2020, September). Lightweight temporal self-attention for classifying satellite images time series. In International Workshop on Advanced Analytics and Learning on Temporal Data (pp. 171-181). Springer, Cham.

Guo, D., Xia, Y., & Luo, X. (2021). Self-supervised GANs with similarity loss for remote sensing image scene classification. IEEE Journal of Selected Topics in Applied Earth Observations and Remote Sensing, 14, 2508-2521.

Marchisio, G., Helber, P., Bischke, B., Davis, T., Senaras, C., Zanaga, D., ... & Wania, A. (2021, July). RapidAI4EO: A Corpus for Higher Spatial and Temporal Reasoning. In 2021 IEEE International Geoscience and Remote Sensing Symposium IGARSS (pp. 1161-1164). IEEE.

Marszalek, M. L., Saux, B. L., Mathieu, P. P., Nowakowski, A., & Springer, D. (2022). Self-supervised learning--A way to minimize time and effort for precision agriculture?. arXiv preprint arXiv:2204.02100.

Metzger, N., Turkoglu, M.O., D'Aronco, S., Wegner, J.D., Schindler, K. (2021). Crop Classification under Varying Cloud Cover with Neural Ordinary Differential Equations. arXiv preprint arXiv:2012.02542

Saini, R., & Ghosh, S. K. (2018). Crop classification on single date sentinel-2 imagery using random forest and support vector machine. The International Archives of Photogrammetry, Remote Sensing and Spatial Information Sciences, 42, 683-688.

Seydi, S. T., Amani, M., & Ghorbanian, A. (2022). A Dual Attention Convolutional Neural Network for Crop Classification Using Time-Series Sentinel-2 Imagery. Remote Sensing, 14(3), 498.

Schneider, M., Broszeit, A., & Körner, M. (2021). Eurocrops: A pan-european dataset for time series crop type classification. arXiv preprint arXiv:2106.08151.

Sumbul, G., Charfuelan, M., Demir, B., & Markl, V. (2019, July). Bigearthnet: A large-scale benchmark archive for remote sensing image understanding. In IGARSS 2019-2019 IEEE International Geoscience and Remote Sensing Symposium (pp. 5901-5904). IEEE.

Wang, Y., Albrecht, C. M., Braham, N. A. A., Mou, L., & Zhu, X. X. (2022). Self-supervised learning in remote sensing: A review. arXiv preprint arXiv:2206.13188.

Zhao, H., Duan, S., Liu, J., Sun, L., & Reymondin, L. (2021). Evaluation of five deep learning models for crop type mapping using sentinel-2 time series images with missing information. Remote Sensing, 13(14), 2790.

Planet Fusion 2022, https://assets.planet.com/docs/Fusion-Tech-Spec_v1.0.0.pdf

This contribution has been peer-reviewed.
https://doi.org/10.5194/isprs-archives-XLVIII-M-1-2023-309-2023

The International Archives of the Photogrammetry, Remote Sensing and Spatial Information Sciences, Volume XLVIII-M-1-2023
39th International Symposium on Remote Sensing of Environment (ISRSE-39) "From Human Needs to SDGs", 24–28 April 2023, Antalya, Türkiye

A PROPOSED FRAMEWORK FOR SURVEILLANCE OF DENGUE DISEASE AND PREDICTION

V. Sharma[1*(0000-0002-3812-4151)], **S. K. Ghosh**[1(0000-0001-7849-9313)], **S. Khare**[1(0000-0002-6996-4583)]

[1] Civil Engineering Department, Indian Institute of Technology Roorkee, Uttrakhand-247667

Keywords: Dengue; framework; Dengue Disease Monitoring; DDM; GIS; Remote Sensing, Risk mapping; Surveillance; Prediction; Modelling

ABSTRACT:

Recurring outbreaks of dengue during past decades have affected public health and burdened resource constraint health systems across the world. Transmission of such diseases is a conjugation of various complex factors including vector dynamics, transmission mechanism, environmental conditions, cultural behaviours, and public health policies. Modelling and predicting early outbreaks is the key to an effective response to control the spread of disease. In this study, a comprehensive framework has been proposed to model dengue disease by integrating significant factors using different inputs, such as remote sensing, epidemiological data, and health infrastructure inputs. This framework for Dengue Disease Monitoring (DDM) model provides a conceptual architecture for integrating different data sources, visualization and assessment of disease status, and prediction analysis. The developed model will help forewarn the public health administration about the outbreak for planning interventions to limit the spread of dengue. Further, this forecasting model may be applied to manage the existing public health resources for medical and health infrastructure, also to determine the efficacy of vector surveillance and intervention programmes.

1. INTRODUCTION

1.1 BACKGROUND

Dengue is the fastest-spreading vector-borne diseases with a growth rate of 8 times in the last two decades (WHO, 2021). The highest burden is observed in tropical and sub-tropical regions (Fig 1). With about 128 countries currently at risk of infection, 70% of the burden is contributed by Asia for 2019. For dengue prevention is the only measure to reduce the health burden as no specific therapeutics are defined (WHO,2021). Therefore, there is a need to undertake a holistic approach and adopt adequate measures to control the onset of such a disease. Developing a disease surveillance system that has the capability to detect disease outbreaks may provide proactive and effective control measures during the onset of the outbreak. Further, it should be able to monitor the trends of incidences including temporal and geographic distribution of cases, but also be able to assess and confirm the possibility of an outbreak considering various aspects of the surrounding environment and demographic factors, geo-spatial physiographic features, entomological, epidemiological and serological evidence by aiding an Early warning system (EWS).

1.2 DETERMINANTS OF DENGUE TRANSMISSION

The ecology of the dengue vector has been widely studied and modelled using weather dependency as its occurrence is highly seasonal (Jayaraj et al., 2019; Nuraini et al., 2021). In addition, complex dynamics of dengue is governed by the interaction of multiple agents i.e. humans (host), mosquitoes (vector) and dengue virus over time and space. Despite complexities, an analysis involving identified dengue cases linked with significant factors, analysing and modelling their dependence for predicting future cases, may be used to analyse the impact of public health and vector intervention measures opted for disease control. For efficient planning, implementation and evaluation of the interventions to reduce dengue transmission, equal emphasis on the virus, vector, and their interaction with humans along with environmental factors is required. Well-timed and reliable estimates of dengue emergence, including location, time, and intensity, may enhance proactive disease management.

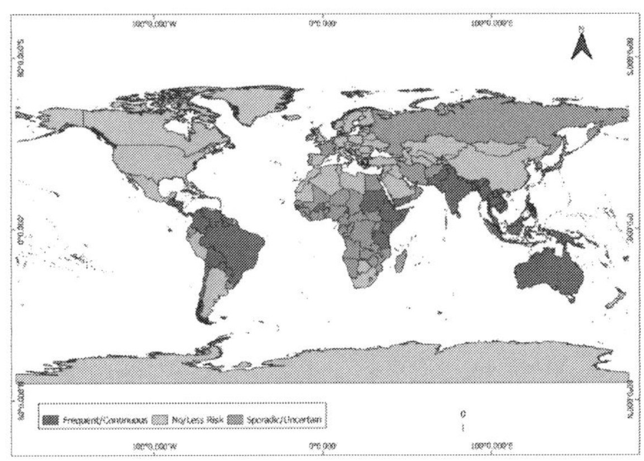

Fig 1: Geographic distribution of dengue cases (Data Source: CDC, 2021)

Dengue is extremely sensitive to weather conditions as the vectors hence its highly seasonal. However, this seasonality is not uniform across the globe. It varies with geographical location, macro and micro climatic patterns like draught cycle, and trend of rainfall. The geographical range in which vectors' movement occurs along with their habitat preferences increases the exposure of humans to vectors(Vanwambeke et al., 2007). In addition, the growth, and development of the viruses in vectors and hosts occur under specific environmental conditions. The geographical shift in the distribution of dengue cases across the globe owing to climate change is being widely studied. It has been observed under warmer conditions, the extrinsic incubation period of the dengue virus may be shortened, which may lead to increased efficiency of viral transmission by mosquitoes. In contrast, extreme conditions like very high temperatures and low temperatures may hinder mosquito growth as it may lead to high mosquito mortality or hinder the development of the virus, therefore decreasing the risk(Sirisena et al., 2017). Rainfall, on the other hand, is necessary to create and maintain breeding sites for Aedes mosquitoes and therefore has a strong influence on vector distribution. The major governing factors for dengue transmission identified in various studies are summarised in Table 1.

Table 1: Major contributing factors for dengue occurrences

Determinant Factor	Author	Impact/Magnitude
Built-up	Machault et al., 2014	Maximum dengue cases were found in the built-up area with a 300-m radius buffer.
	(Dom et al., 2016)	High levels of dengue cases were reported within a buffer of 600 m around residential areas.
	(Zahouli et al., 2017)	Development time of larvae in urban areas is significantly shorter than that in rural and sub-urban areas due to comparatively warmer temperatures.
	(Ferraguti et al., 2016)	Vector abundance were higher in vegetated and rural areas as compared to urban areas.
Water Bodies	(Ferraguti et al., 2016)	High mosquito distribution is found in wetland areas.
Vegetation Density	(Sarfraz et al., 2012)	Land cover consisting of orchards, rangeland and deciduous forests contributed more towards vector growth. Standing water in paddy fields and swamp forests upto 2.5 cm to 30 cm in depth are observed as most common habitats for the dengue vector
Temperature	Wu et al., 2007	Every 1 °C increase in average monthly temperature above 18°C increased the risk for dengue transmission by 1.95 times.
	(Xiao et al., 2018)	It was observed the biting activities of mosquitoes were restricted to temperatures between 15 °C and 35 °C.
Season	(Appawu et al., 2006)	A high number of cases in the dry season is due to the increase in the storage of water.
	(Sirisena et al., 2017; Zahouli et al., 2017)	Bimodal rainfall patterns corresponded to two peaks in dengue distribution noted on an annual scale.
	(Captain-Esoah et al., 2020)	Unimodal rainfall pattern corresponded to a single peak in dengue case distribution
Elevation	(Gyawali et al., 2020)	The suspected elevation limit for Aedes spp. was found to be up to 2100 m. However, an increased incidence of dengue was found to be in cities located above 1300 m of elevation compared to cities in the same elevation due to urban heat island effects.

The major risk components of dengue transmission are classified into exposure, sustainability and adaptive capacity (Fig 2). Exposure determines the vector-human interaction in an environment while susceptibility governs the proneness of a community to be infected by dengue. The adaptive capacity provides the ability to the community to deal with the disease. It may be enhanced by enhancing the quality of healthcare facilities, vector control measures, and spreading awareness among the public.

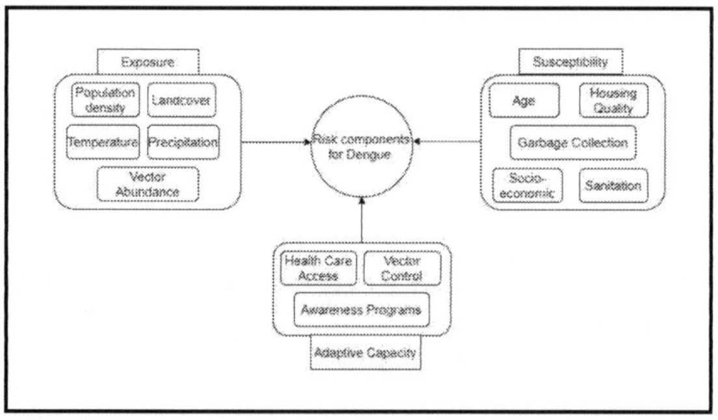

Fig 2: Factors affecting dengue risk and susceptibility

1.3 DENGUE INCIDENCE MODELLING

The high global burden of dengue led to much research across the globe depending upon the severity of the problem and varying from studying the distribution of cases, assessing contributing factors, determining the population at risk and predicting future incidences. A variety of statistical and machine learning methods are applied to model dengue transmission and produce risk maps. While plotted disease incidence may facilitate the allocation of public health resources, but also there is a need for methods that allows an assessment of interventions over time and space. Disease modelling utilizes existing estimates of disease incidence to predict impact depending on the expected changes in population demographics, environmental and climatic conditions. It is crucial to plan the resources, create emergency

SOPs, task allocation and most important restrict the disease in case of occurrence. While most of the studies undertaken in the past decade used different statistical models to analyse contributing factors to dengue or to carry out prediction. However, the latest researches are more focused on using advanced methods like the use of machine learning or deep learning tools in order to predict disease occurrences. The various significant modelling techniques along with contributing variables for dengue used across the globe are shown in Table 3.

Table 3: Modelling Techniques used for dengue prediction

Model Name	Developed By	Parameters Used
Logistic Regression	Joseph Berkson,1944 (Cramer, 2002)	Land use, vegetation density, slope, aspect, distance from open areas and gold mining sites along with malaria incidence data.
Generalised Linear Model	John Nelder and Robert Wedderburn (Nelder et al., 1972)	Temperature, humidity, and the total number of dengue cases per week. Remote sensing-derived products like NDVI, NDWI, Daytime Temp, Night-time Temp, and Precipitation in both urban and rural areas are used as a proxy for environmental conditions, mosquito population
Analytical Hierarchy Process	T.L. Saaty (1971-1975) (Saaty, 1987)	Dengue data, land use, housing type, residential buffering, elevation, population density, Land Surface Temperature
Support Vector Machine	Cortes and Vapnik (Cortes and Vapnik, 1995)	Dengue incidence data, mean temperature, rainfall, relative humidity, vector density, elevation, remote sensing derived products like NDVI and NDBI as a proxy for vegetation and urban density
Poisson Regression	Simon Poisson, 1837	Weekly dengue cases, Oceanic Ni˜no Index (ONI), temperature, rainfall, humidity, sunshine hours, wind speed, built-up and vegetation density.
Auto-Regressive Integrated Moving Average	Box and Jenkins (Wilson, 2016)	Temperature, rainfall, humidity, dengue cases, vector density, wind speed and direction
Gradient Boosting	Jerome H. Friedman (Friedman, 1999)	Confirmed dengue incidence data, Land surface temperature, precipitation, LULC, population density, urban accessibility, wind speed and direction
Random Forest	Leo Breiman (Breiman, 2001)	Minimum temperature, maximum temperature, precipitation, NDVI, relative humidity, urbanicity, population density, urban accessibility, vector data, land surface temperature
Generalised Additive Model	Trevor Hastie and R. Tibshirani (Hastie and Tibshirani, 2005)	Dengue incidence data mean temperature, average relative humidity and rainfall, and population density.

The advantage of using a statistical model against a machine learning model is that a statistical model may be easily interpreted, and the governing factors may be assessed and may be understood by public health managers (Mudele et al., 2021). Moreover, it may be further integrated into real-life health systems and aid robust decision-making. Another advantage is the ease of processing and visualization which in contrast, for machine learning methods specialized libraries are required, followed by parameter tuning and long training cycles. However, machine learning algorithms may help by determining hidden patterns by combining different variables, which is not possible in statistical analysis. Overall it is observed that ARIMA, GAM, regression, RF and BRT have got maximum accuracy while modelling dengue transmission. However, these models have not been simultaneously applied to the same datasets for a region. Therefore, it is proposed to verify the suitability of the such model for a given region after a comparative analysis.

2. DENGUE DISEASE MONITORING MODEL

2.1 CONCEPTUAL FRAMEWORK

Lack of data availability in many under-developed countries led to the development of frameworks like the Water Associated Disease Index, Analytical Hierarchical Process, and Spatial Multi-Criteria Evaluation. Such frameworks worked on determining highly vulnerable areas based on a set of criteria like risk factors, and computing pair-wise comparisons and lacked maintaining consistency itself. Very few studies incorporated the socio-environmental aspects of dengue transmission including their spatial association and relative risk concerning neighbourhood regions. Moreover, the risk associated with neighbourhood regions and the non-availability of the medical facility on dengue occurrences has not been studied. Thus, this study aims to overcome these limitations and aims to bring together a conceptual framework that may assist in modelling and predicting dengue cases and addresses how the dynamics of DENV transmission are modulated by significantly identified factors. Therefore, this study proposes a framework model which provides a comprehensive support for exploratory analysis, spatiotemporal analysis, risk analysis and prediction depending upon the level of analysis required at the time of the outbreak. The framework aims at (i) identifying hotspots and rate of disease spread, (ii) determining risk zones that may be prone to future outbreaks and (iii) predicting the occurrences of dengue in time and space. Based on the findings from the various literature, the conceptual framework is presented in Fig 3.

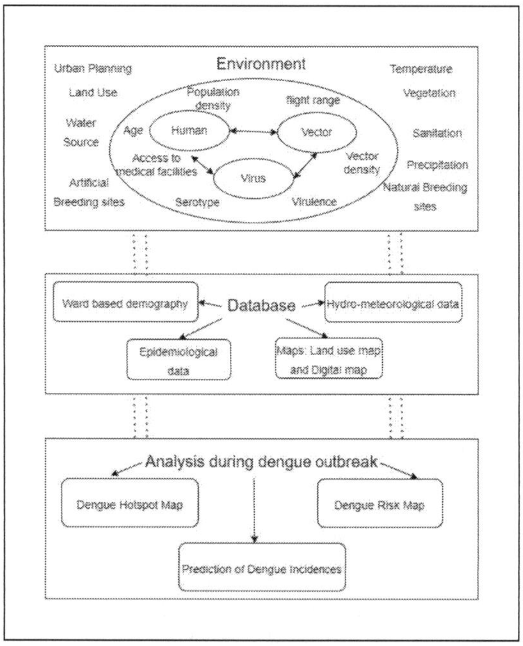

Fig 3. Conceptual framework of the proposed model for dengue disease monitoring

This framework integrates the three important aspects of dengue transmission i.e. environmental component, the database containing disease-related information and analysing of the dengue outbreak. The base layer consists of integrating all the factors associated with dengue incidence. Thereafter, the database consisting of dengue illness, the demography of the study region and its geographic information is integrated to model the existing situation of the outbreak. The final layer consists of the outputs based on the type of support required depending on stages of the disease outbreak viz disease occurrence, spread, and fall. For robust prediction, the such model will be trained to predict dengue occurrences in the region over time. The analysis may provide graphical output in terms of hotspot maps, risk maps and prediction of dengue incidence. The advantages of having graphical output are interpretation, ease of data sharing, and communication across various sectors and levels of administration including public communication. DDM may be further used for developing an early warning system which may alarm public health administrators to take preventative measures to minimize the further spread of the outbreaks by enforcing required interventions.

2.2 PROPOSED WORKFLOW:

The workflow followed for the DDM model is summarised in Fig 4. Overall it consists of input, processing and output blocks. The input block includes data harvesting processes from different sources like remote sensing, station data, ground-truth data, and surveying. A database will be generated containing geo-coded reported cases with supplementary information about the patient details, demographic information and the area/ward from which the patient belongs. The standard and homogeneous form of such data are required such database will contain epidemiological data of at least 5 years or the latest available data, whichever is earliest. The datasets for environmental parameters like vegetation density, built-up, and water bodies and hydro-meteorological parameters like temperature, rainfall, and humidity will be collected. The preliminary data processing carried out at the input block includes data aggregation, homogenisation, storing, filtering and visualisation. The data aggregation process as a part of data governance will integrate structured, semi-structured and unstructured data coming from various sources after homogenisation based on the spatial resolution of acquired data. Thereafter the dataset will be processed and analysed in a processing block using a set of algorithms and equations to determine the trend, regions of dengue risk, and predicted dengue occurrences. The processing block of the DDM model is further divided into three distinct modules, with further descriptions of these modules provided in subsequent subsections.

Module I: Dengue Spread Model

Spatio-temporal patterns of DF cases will be determined to study the variation of DF cases over time and geographical regions. Surface Trend Analysis will be used to determine the rate and direction of outbreak spread. Such analysis is useful for Public Health administration by planning preventative/control measures to limit the further spread by enforcing interventions. The geo-coded epidemiological data will be first converted from a daily scale to a weekly scale in the respective wards of the municipal boundary. The rate of disease spread will be estimated using polynomial regression models (Tutt-gu et al., 2021; Zinszer et al., 2015). The rate of change (in weeks/km) of dengue incidences will be obtained based on the best fit linear model, using partial derivatives with respect to x and y. The vectors will then be converted for determination of magnitude and direction.

Cluster Analysis is used to determine significant clusters of DF outbreaks. Such analysis is useful for planning a medical facility or for additional resource allocation as a mitigation measure to control an outbreak. The discrete Poisson Model will be used to identify overlapping spatiotemporal and spatial clusters. The temporal unit may be selected depending on the incidence of DF cases in the region. For example, in case the incidence is low, an annual DF occurrence per municipality per year will be selected (Tutt-gu et al. 2021) and the case it is highly seasonal variation per municipality may be considered (Henry and Mendonça 2020).

Relative Risk (RR) will be calculated for each significant cluster providing risk within that cluster relative to that of the outside cluster and calculated as per equation (1)

$$RR = c/E(c)(C-c)/(C-E(c)) \quad \ldots\ldots\ldots \quad (1)$$

This contribution has been peer-reviewed.
https://doi.org/10.5194/isprs-archives-XLVIII-M-1-2023-317-2023 | © Author(s) 2023. CC BY 4.0 License.

Where, c= observed cases within the cluster, E(c) expected number of cases, C= total number of cases in the dataset.

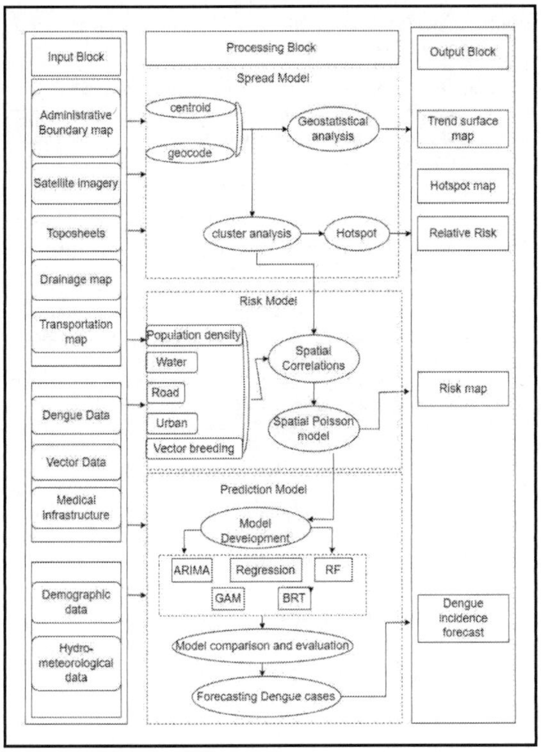

Fig 4 Workflow of the DDM model

Module 2: Dengue Risk Model

A multivariate temporal approach will be used to identify vulnerable regions prone to DF outbreaks. A risk model along with a spatial Poisson point process model will be generated using individual layers of epidemiological data consisting of patient locations, land-use maps, roads, and total buildings in high-risk areas determined in the first objective. Raster layers will be developed separately for patient data, breeding places and road map layers using the kernel density function and the euclidean distance function will be used to calculate distances of roads, water bodies, and buildings. Spatial correlations between DF case locations in each high-risk area and other layers will be determined using Pearson's product correlation coefficient (Mala and Jat, 2019; Withanage et al., 2021). The risk values will be ranked for each layer depending upon their contribution to DF occurrence in the range of 0(minimum) to 10(maximum). The resulting risk map may help healthcare workers and decision-makers in taking preventative measures in vulnerable areas.

Mathematical modelling of patient cases with developed layers: Spatial Poisson point process model will be used to determine high-risk localities using patient locations as a function of raster layers created from breeding places, road maps, water bodies and buildings.

$$\ln(i) = \alpha + \beta 1 \text{ Breeding Places-} + \beta 2(\text{Roads}) + \beta 3 (\text{water bodies}) + \beta 4(\text{buildings})\ldots \quad (2)$$

where (i) is the modelled point pattern intensity for dengue incidences at location "i", α is the base intensity derived from the multivariate model and β_1 to β_4 are the estimated coefficients for each respective variable. Thereafter, model outputs will be used to plot the predicted intensity of dengue in the areas to identify high-risk localities.

Module III: Dengue Prediction Model

The disease data will be split into 2:1 for training and testing. Different statistical- machine learning models will be developed and compared to forecast the dengue outbreaks based on the epidemiological data and predictive variables such as monthly rainfall, rainy days and temperature. Different models will be generated using ARIMA, GAM, regression, Random Forest and Boosted Regression Trees. The best model will be selected based on the lowest Akaike's information criterion (AIC), Bayesian Information criterion (BIC), R-squared (R^2), root mean square errors (RMSE), mean absolute errors(MAE) and mean absolute percentage error (MAPE) of prediction and will be used for analysing dengue disease data for prediction.

The processed output will be presented to the users in the form of a graphical representation using a web application. By using a graphical user interface the user experience may be enhanced and allow for eased data communication and sharing. Based on a detailed literature review on dengue surveillance and modelling, this proposed framework provides an overview and foundation for a new approach to handling public health problems.

3. CONCLUSION

This framework provides different stages of analysis depending upon the severity of the situation which may provide information about the location, timing, and intensity of infectious diseases that will help public health stakeholders in taking proactive disease containment and management efforts. This framework includes comprehensive, end-to-end support for exploratory analysis, spatiotemporal analysis, disease spread modelling and prediction. It provides an architecture for epidemiologists to break up disease management into sub-problems and solve the sub-problems with appropriate modelling approaches. The framework supports the integration of structured data (disease incidents, environmental conditions, demographics, health status, and other data. The study will provide support to the different stages of the disease viz disease occurrence, spread, and fall which will allow public health administrators to better planning and preparedness.

https://doi.org/10.5194/isprs-archives-XLVIII-M-1-2023-317-2023 | © Author(s) 2023. CC BY 4.0 License.

References:

Appawu, M., Dadzie, S., Abdul, H., Asmah, H., Boakye, D., Wilson, M., Ofori-adjei, D., 2006. Surveillance of viral haemorrhagic fevers in Ghana: entomological assessment of the risk of transmission in the northern regions. Ghana Med. J. 40, 137–141. https://doi.org/10.4314/gmj.v40i3.55269

Breiman, L., 2001. Random Forests. Mach. Learn. 45, 5–32. https://doi.org/10.1023/A:1010933404324

Captain-Esoah, M., Baidoo, P.K., Frempong, K.K., Adabie-Gomez, D., Chabi, J., Obuobi, D., Amlalo, G.K., Veriegh, F.B., Donkor, M., Asoala, V., Behene, E., Boakye, D.A., Dadzie, S.K., 2020. Biting behavior and molecular identification of Aedes aegypti (Diptera: Culicidae) subspecies in some selected recent yellow fever outbreak communities in Northern Ghana. J. Med. Entomol. 57, 1239–1245. https://doi.org/10.1093/jme/tjaa024

Cortes, C., Vapnik, V., 1995. Support-Vector Networks. Mach. Learn. 20, 273–297. https://doi.org/10.1109/64.163674

Cramer, J., 2002. The origins of Logistic Regression.

Dom, N.C., Ahmad, A.H., Latif, Z.A., Ismail, R., 2016. Application of geographical information system-based analytical hierarchy process as a tool for dengue risk assessment. Asian Pacific J. Trop. Dis. 6, 928–935. https://doi.org/10.1016/S2222-1808(16)61158-1

Ferraguti, M., Martínez-De La Puente, J., Roiz, D., Ruiz, S., Soriguer, R., Figuerola, J., 2016. Effects of landscape anthropization on mosquito community composition and abundance. Sci. Rep. 6, 1–9. https://doi.org/10.1038/srep29002

Friedman, J.H., 1999. Stochastic Gradient Boosting 1–10.

Gyawali, N., Johnson, B.J., Dixit, S.M., Devine, G.J., 2020. Patterns of dengue in Nepal from 2010–2019 in relation to elevation and climate. Trans. R. Soc. Trop. Med. Hyg. https://doi.org/10.1093/trstmh/traa131

Hastie, T., Tibshirani, R., 2005. Generalized Additive Model. Encycl. Biostat. https://doi.org/10.1002/0470011815.b2a09018

Jayaraj, V.J., Avoi, R., Gopalakrishnan, N., Raja, D.B., Umasa, Y., 2019. Developing a dengue prediction model based on climate in Tawau, Malaysia. Acta Trop. 197, 105055. https://doi.org/10.1016/j.actatropica.2019.105055

Machault, V., Yébakima, A., Etienne, M., Vignolles, C., Palany, P., Tourre, Y.M., Guérécheau, M., Lacaux, J., 2014. Mapping Entomological Dengue Risk Levels in Martinique Using High-Resolution Remote-Sensing Environmental Data. Int. J. Geo-Information 3, 1352–1371. https://doi.org/10.3390/ijgi3041352

Mala, S., Jat, M.K., 2019. Implications of meteorological and physiographical parameters on dengue fever occurrences in Delhi. Sci. Total Environ. 650, 2267–2283. https://doi.org/10.1016/j.scitotenv.2018.09.357

Mudele, O., Frery, A.C., Zanandrez, L.F.R., Eiras, A.E., Gamba, P., 2021. Modeling dengue vector population with earth observation data and a generalized linear model. Acta Trop. 215, 105809. https://doi.org/10.1016/j.actatropica.2020.105809

Nelder, J.A., Wedderburn, M., R.W., 1972. Generalized Linear Models. J. R. Stat. Soc. 135, 370–384.

Nuraini, N., Fauzi, I.S., Fakhruddin, M., Sopaheluwakan, A., Soewono, E., 2021. Climate-based dengue model in Semarang, Indonesia: Predictions and descriptive analysis. Infect. Dis. Model. 6, 598–611. https://doi.org/10.1016/j.idm.2021.03.005

Saaty, R.W., 1987. The analytic hierarchy process-what it is and how it is used. Math. Model. 9, 161–176. https://doi.org/10.1016/0270-0255(87)90473-8

Sarfraz, M.S., Tripathi, N.K., Tipdecho, T., Thongbu, T., Kerdthong, P., Souris, M., 2012. Analyzing the spatio-temporal relationship between dengue vector larval density and land-use using factor analysis and spatial ring mapping. BMC Public Health 12, 1–19. https://doi.org/10.1186/1471-2458-12-853

Shen, J.C., Luo, L., Li, L., Jing, Q.L., Ou, C.Q., Yang, Z.C., Chen, X.G., 2015. The impacts of mosquito density and meteorological factors on dengue fever epidemics in Guangzhou, China, 2006-2014: A time-series analysis. Biomed. Environ. Sci. 28, 321–329. https://doi.org/10.3967/bes2015.046

Sirisena, P., Noordeen, F., Kurukulasuriya, H., Romesh, T.A., Fernando, L.K., 2017. Effect of Climatic Factors and Population Density on the Distribution of Dengue in Sri Lanka: A GIS Based Evaluation for Prediction of Outbreaks. PLoS One 12. https://doi.org/10.1371/journal.pone.0166806

Tutt-gu, M., Yuan, M., Szaroz, D., Mckinnon, B., Kestens, Y., Guillot, C., Leighton, P., Zinszer, K., 2021. Modelling Spatiotemporal Patterns of Lyme Disease Emergence in Québec. Int. J. Environ. Res. Public Health 18, 9669. https://doi.org/10.3390/ijerph18189669

Vanwambeke, S.O., Lambin, E.F., Eichhorn, M.P., Flasse, S.P., Harbach, R.E., Oskam, L., Somboon, P., Van Beers, S., Van Benthem, B.H.B., Walton, C., Butlin, R.K., 2007. Impact of land-use change on dengue and malaria in northern Thailand. Ecohealth 4, 37–51. https://doi.org/10.1007/s10393-007-0085-5

WHO, 2021. Dengue and severe dengue [WWW Document]. URL https://www.who.int/news-room/fact-sheets/detail/dengue-and-severe-dengue (accessed 6.19.21).

Wilson, G.T., 2016. Time Series Analysis: Forecasting and Control, 5th Edition, by George E. P. Box, Gwilym M. Jenkins, Gregory C. Reinsel and Greta M. Ljung, 2015. Published by John Wiley and Sons Inc., Hoboken, New Jersey, pp. 712. ISBN: 978-1-118-67502-1. J. Time Ser. Anal. 37, 709–711. https://doi.org/10.1111/jtsa.12194

Withanage, G.P., Gunawardana, M., Viswakula, S.D., 2021. Multivariate spatio - temporal approach to identify vulnerable localities in dengue risk areas using Geographic Information System (GIS). Sci. Rep. 11, 1–

This contribution has been peer-reviewed.
https://doi.org/10.5194/isprs-archives-XLVIII-M-1-2023-317-2023 | © Author(s) 2023. CC BY 4.0 License.

11. https://doi.org/10.1038/s41598-021-83204-1

Wu, P.C., Guo, H.R., Lung, S.C., Lin, C.Y., Su, H.J., 2007. Weather as an effective predictor for occurrence of dengue fever in Taiwan. Acta Trop. 103, 50–57. https://doi.org/10.1016/j.actatropica.2007.05.014

Xiao, J., Liu, T., Lin, H., Zhu, G., Zeng, W., Li, X., Zhang, B., Song, T., Deng, A., Zhang, M., Zhong, H., Lin, S., Rutherford, S., Meng, X., Zhang, Y., Ma, W., 2018. Weather variables and the El Niño Southern Oscillation may drive the epidemics of dengue in Guangdong Province, China. Sci. Total Environ. 624, 926–934. https://doi.org/10.1016/j.scitotenv.2017.12.200

Zahouli, J.B.Z., Koudou, B.G., Müller, P., Malone, D., Tano, Y., Utzinger, J., 2017. Urbanization is a main driver for the larval ecology of Aedes mosquitoes in arbovirus-endemic settings in south-eastern Côte d'Ivoire. PLoS Negl. Trop. Dis. 11, 1–23. https://doi.org/10.1371/journal.pntd.0005751

Zinszer, K., Morrison, K., Anema, A., Majumder, M.S., Brownstein, J.S., 2015. The velocity of Ebola spread in parts of west Africa. Lancet Infect. Dis. 15, 1005–1007. https://doi.org/10.1016/S1473-3099(15)00234-0

EVALUATING THE POTENTIAL OF 8 BAND PLANETSCOPE DATASET FOR CROP CLASSIFICATION USING RANDOM FOREST AND GRADIENT TREE BOOSTING BY GOOGLE EARTH ENGINE

Vishal Sharma[1], Sanjay Kumar Ghosh[2]

[1] Department of Civil Engineering, Indian Institute of Technology, Roorkee, India - vsharma1@ce.iitr.ac.in
[2] Department of Civil Engineering, Indian Institute of Technology, Roorkee, India - sanjay.ghosh@ce.iitr.ac.in
Orcid: 0000-0003-0252-1726[1]; 0000-0001-7849-931[2]

KEY WORDS: PlanetScope, Crop Classification, Machine Learning, Random Forest, Gradient Tree Boosting, Google Earth Engine.

ABSTRACT:

One of the challenging task in agriculture is mapping of crops using satellite images as spectral properties of the crops looks similar to each other, and there are many complexities which are there in the field such as small land holdings, heterogeneous and often distinct field patterns. In past, mapping with high resolution satellite images was not possible due to the non-availability of data and even they were costlier. However, with the free availability of PlanetScope dataset having 8 spectral bands, with daily revisit cycle and resolution of 3m, it is now possible to carry out mapping and monitoring of agriculture crops. The objective of the study is to classify major crops of rabi (December-April) season using single date PlanetScope imagery in the Haridwar district, Uttarakhand, India. Classification of crops has been carried out using an ensemble based machine learning algorithms within Google earth engine. Off late, Google Earth Engine (GEE), a cloud based platform has attracted the attention of remote sensing analyst since it expediates the classification yielding good results of high quality. In this study, Random Forest (RF) and Gradient Tree Boosting algorithm (GTB) have been used after performing hyper-parameter tuning and results shows that overall accuracy obtained by RF and GTB are 88% and 86.5% respectively. According to the results, both classifier performed well but RF achieved 1.5% high accuracy over GTB. Analysis of the results show that highest accuracy was attained by agriculture class (wheat) while other crops class exhibited lower accuracy. In this study, RF was found to be more competent as compared to GTB in classifying the crops and PlanetScope 8 band dataset has also proved its potential in classifying crops in heterogeneous fields.

1. INTRODUCTION

Mapping of crops is considered an important task in many different fields as it provides countless advantage for understanding the food system and yield prediction. It is also helpful during the challenging time caused due to climate change and also provides sustainable agriculture practice. Classification of crop provides useful information's for various agencies such as insurance agencies, geoportal, regional and national agricultural boards for decision making. Therefore, timely and accurate crop mapping is required which otherwise is often expensive, labour intensive and destructive process if performed manually.

Therefore, Remote Sensing data is considered a useful source of data for timely and accurately mapping and monitoring of crops. In the past years, freely available MODIS data with high temporal resolution, Landsat and Sentinel 2 data with 30m and 10m spatial resolution have been used for crop mapping (Zheng et al. 2015; Saini and Ghosh 2018). One of the biggest disadvantage with MODIS and Landsat data was spectral mixing due to low and coarse resolution data. In past, mapping with high resolution data was not possible because of its availability and cost, but now with free available PlanetScope dataset having a spatial resolution of 3m and daily revisit cycle is being used widely. Initially, PlanetScope provided 4 band surface reflectance data (Blue, Green, Red and NIR), but from 2022 onwards, it has started to provide a 8-band atmospheric corrected surface reflectance dataset which is data. There are many studies which reveals the advantage of 4 band PlanetScope dataset. A study has been carried out to compare PlanetScope dataset with Sentinel-2 dataset by (Mudereri et al. 2019) for mapping Striga weed in Kenya and results shows that PlanetScope dataset is more accurate in mapping as compared to Sentinel-2 dataset. Further, PlanetScope dataset has a great potential for producing the crops intensity maps at detailed resolutions (Rafif et al. 2021). The study used the 4 band data for mapping four different types of crop classes and found that NIR band to be most extensive for analysing and evaluating the cropping intensity. Acharki, (2022) deployed Random Forest machine learning algorithm for LULC classification using 4 band PlanetScope dataset, Landsat-8 and Sentinel-2 and the result show the potential of PlanetScope data which give an overall accuracy of 97%.

With the increase availability of free remote sensing data and improvement in the image resolution (radiometric, spectral, spatial, temporal), there is lot of scope for the researchers to adopt exhaustive techniques for data analysis. However, with improved spatial, spectral, radiometric and temporal resolutions, the volume of data has increased manifolds. This brought a shift in the working environment of traditional workstations to cloud- based platforms. Google Earth Engine (GEE) is one of the widely utilized cloud-based geospatial analytic platform which allow the researchers and users to tackle the difficulties faced while managing the huge dataset

This contribution has been peer-reviewed.
https://doi.org/10.5194/isprs-archives-XLVIII-M-1-2023-325-2023

and processing. It provides the platform where user can analyse the available remote sensing images using Integrated Development Environment (IDE) code editor without downloading the dataset. In addition to this, GEE has large number of options of several machine learning algorithms. The use of GEE engine from 2013 is been increased drastically (Tamiminia et. al, 2020; Sharma et.al, 2022).

One of the advantage of 8-band and 4-band PlanetScope data is the introduction of Red-edge band (697-713 nm) between the red band and NIR band. This Red-edge band has huge potential in the area of vegetation mapping especially agriculture. Thus, one of the prime objective of the study is to explore the potential of 8-band PlanetScope data for classifying the major crops using machine learning algorithms. Here Random Forest (RF) and Gradient Tree Boosting (GTB) have selected for analysis using Google Earth Engine. The efficacy of RF and GTB algorithms are evaluated in the study for identifying different crops in the study area. The structure of the paper can be defined as follows: Section 2 highlights the characteristics of the study area followed by the satellite data-set utilized; Section 3 showcases the framework of methodological description of machine learning algorithms; Section 4 reveals and portrays the results and corresponding discussion and at last Section 5 presents the conclusive remarks and the future scope of the work.

2. STUDY AREA AND DATA USED

In this study, Hardiwar district is selected as a study area which is bounded between 29.58° N latitude and 78.13° E longitude covering an area of 2310.76 Km^2. The region lies within the close proximity of River Ganges on the east side and River Yamuna on the west side. The region has sandy loam soil with high porosity. It is located at the foothill of Himalaya and is mainly dependent on tourism, animal husbandry and agriculture. The major agricultural crops grown here are sugarcane (32%), wheat (18%) and rice (5%) and rest is cash crops such as groundnut, maize, mustard, potato and lentil etc. (https://haridwar.kvk4.in/district-profile.html). Thus, it is important to have accurate information regarding the crops being sown.
PlanetScope satellite data for 19 February 2022 has been used in the study as it corresponds to the peak winter season of crop growth. The 8-bands dataset has been used for classification to assess the utility of newly added bands. The satellite data was downloaded from https://www.planet.com/explorer/ and detailed specification of the PlanetScope dataset such as band name, spatial resolution and wavelength is given in Table1.

Ground truth data is collected using the Trimble Handheld Global Positioning System during the growing season (shown in Figure 2). According to the survey, the study area is split up into 12 different Land Use and Land Cover (LULC) classes such as: built-up, water, barren, sand, fallow-land, forest, shrub, orchards, wheat, sugarcane, mustard and other crops (wheat with trees, mustard with trees, vegetables etc.). Here, major crops in the study area are Wheat, Mustard and Sugarcane. Other Crop classes mainly consist of double crops, vegetable and many more. Orchard class mainly consists of a group of trees mainly guava trees during winter season and trees which are planted on the boundary of agricultural field. Figure 2. shows the False Color Composite (FCC) of the study area which was created using PlanetScope imagery.

Table 1. PlanetScope 8 Band Dataset Detailed Specification

Band No	Band Name	Resolution (m)	Wavelength(nm)
b1	Coastal Blue	3	431-452
b2	Blue	3	465-515
b3	Green	3	513-549
b4	Green	3	547-583
b5	Yellow	3	600-620
b6	Red	3	650-680
b7	Red-Edge	3	697-713
b8	NIR	3	845-885

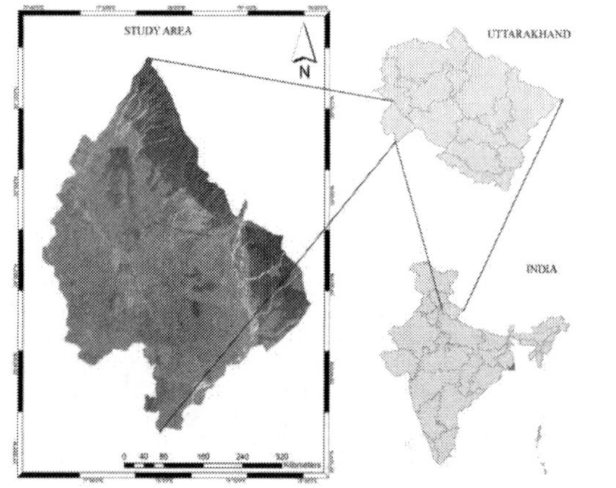

Figure 1. False colour composite (FCC) of the study area using PlanetScope Image

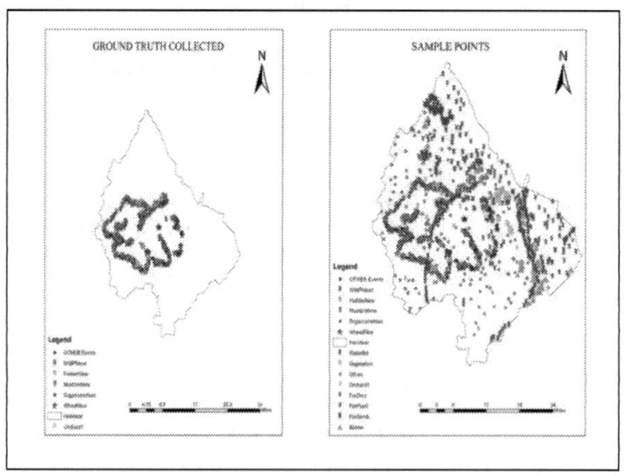

Figure 2. a) Ground Truth Data Collected for Major Crops and Sample Reference points generated for other classes using high resolution data.

3. METHODOLOGY ADOPTED

The methodology incorporated in the research work for crop classification is schematically presented in Figure 3. The main

This contribution has been peer-reviewed.
https://doi.org/10.5194/isprs-archives-XLVIII-M-1-2023-325-2023

concentration of the study is for crop classification; therefore, Single date imagery of Planet scope was acquired during the growing season of winter crop. The study area is covering around 10 to 14 tile of planet data thus there is a need to mosaic the data before any processing. Mosaicking of dataset is performed in PCI geomatica outside the Google Earth Engine (GEE). Once the mosaicked image is generated the image is uploaded into GEE using its code editor.

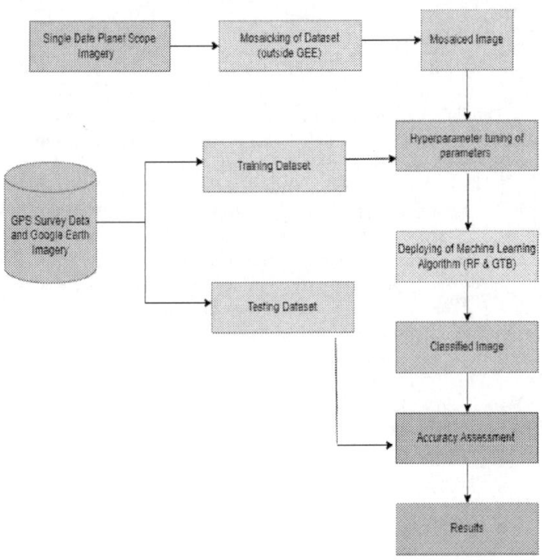

Figure 3. Methodology Adopted in Study

Table 2: No of Points used in Training and Testing of classifier for different classes

S. No	Class Name	Total Points	Training Points	Testing Points
1.	Built-Up	193	123	70
2.	Water	154	116	38
3.	Barren	115	80	35
4.	Sandline	253	195	58
5.	Fallow Land	85	60	25
6.	Deciduous Forest	71	50	21
7.	Shrub	113	79	54
8.	Orchards	76	53	23
9.	Wheat	169	118	51
10.	Sugarcane	145	102	43
11.	Mustard	110	77	33
12.	Other Crops	55	32	23

GEE is used in the study for further processing and analysis. Ground truth data was collected during the field survey using the Trimble GPS and sample for some classes were collected using high resolution google earth imagery which collectively creates a reference dataset (shown in Figure 2). Now with the help of GEE functions, the reference dataset is spited into 70% of training and 30% of testing data. RF and GTB machine learning algorithm have been used for classification. Both Algorithms are powerful and popular supervised machine learning algorithms with some pros and cons. The training dataset is further used for optimizing the parameters of RF and GTB to attain high accuracy. Now these optimized parameters (describe in Table 3) are used along with training dataset to train the models to predict the various classes in the input image. This helps in generating the classified map of the study area. Thereafter, testing dataset is used to test the accuracy of the models. In this study, for accuracy assessment F1- score has been used. It helps in accessing the accuracy of specific class, by combining the precision and recall of a classifier into a single metric by taking their harmonic mean of that particular class.

Table 3. Parameters used for Tuning by RF and GTB

Algorithm	Hyper Parameter	Tested values	Optimal Values
RF	numberOfTrees	[50,100,150,200,250,300]	250
	variablesPerSplit	[2,3,4,5]	3
	bagFraction	[05,0.6,0.7,0.8,0.9]	0.7
GTB	numberOfTrees	[50,100,150,200,250,300]	200
	Learning rate	[0.05,0.06,0.07,0.08,0.09]	0.08

3.1 Classification Algorithms

In the present study, two powerful classifiers such as Random Forest and Gradient Tree boosting are used for classifying the crops.

3.1.1 Random Forest Classifier (RF)

Random Forest (RF) algorithm (Breiman,2001) is a supervised Stastical classification approach which is based on the concept of learning strategy. It is an ensemble classifier and non-parametric algorithm, which combines the decision tree approach and aggregation approach. This classifier is capable of determining the variable importance and has revealed a high accuracy as compared to other tradition classifiers such as SVM, kNN and Decision tree (Saini & Ghosh, 2018; Lopes et al., 2020). RF classifier randomly selects the set of training data from the training dataset using the replacement policy which improves the accuracy by reducing the variance from the input data. It consists of a collection of tree structured classifiers and majority voting scheme is used to determine the output. RF classifier has three main tuning parameters which needs to be optimized. These parameters are: ntree (number of trees to grow), variablesPerSplit (number of variables per split) and bagFraction (fraction of input to bag per tree). From the literature, variablesPerSplit or Mtry and ntree is considered as the most critical parameters which had a great influence on classification results(Saini & Ghosh, 2018 ; Acharki, 2022). RF has come advantages (Breiman,2001):

i. RF performs well in many cases when compared to Adaboost.
ii. RF can handle huge amount of data easily.
iii. RF is robust to outliers as well as noise.
iv. RF can provide useful internal estimates of error, strength, correlation and variable importance.
v. RF can easily run on large datasets.

3.1.2 Gradient Tree Boosting (GTB)

The Gradient Tree Boosting (GTB) model (Friedman, 2001) comprises of an integrated decision-tree and boosting algorithms. The integrated algorithmic approach computes the residuals between predicted values and the actual value. Furthermore, in order to decipher the classification problem, the approach utilizes gradient, boosting as well as the decision trees. Additionally, this also assists in performing the process of

https://doi.org/10.5194/isprs-archives-XLVIII-M-1-2023-325-2023 | © Author(s) 2023. CC BY 4.0 License.

regression for predictions. In general, boosting corresponds to numerous weak classifiers combined together for creating a prominent classifier. On the other hand, when the model executes and estimates the loss function, incremented flexibility corresponds to the gradient function. The GTB model contains less number of model parameters yielding higher computational speed and robust stability when compared to the conventional support vector machine model.

3.1.3 Accuracy Assessment

Confusion Matrix (CM) technique is used in the study for accuracy assessment (Foody et. al, 1992). In this technique, various statistics parameter is evaluated for all classified maps. Statistics parameter are as follows:

Overall Accuracy (OA):
Overall accuracy can be defined as "the ratio of correctly classified samples to that of the total sample space".

$$OA = \frac{\textit{Number of correctly classified samples}}{\textit{total number of samples}} * 100 \quad \ldots (1)$$

Precision (P): Precision can be defined as "the ratio of correctly classified pixels of particular class to the total number of classified sites". It mainly calculates the presence of correctly predicted training data.

$$P = \frac{\textit{Correctly classified pixel of class}}{\textit{total number of classified sites}} \quad \ldots\ldots (2)$$

Recall (R): Recall can be defined as "the ratio of number of correctly predicted training data points of a class to the total training data points provided".

$$R = \frac{\textit{Correctly classified GroundTruth sites}}{\textit{total number of GroundTruth sites}} \quad \ldots\ldots (3)$$

F1-Score: It can be defined as the harmonic mean of precision and recall.

$$F1 = \frac{2*(P*R)}{(P+R)} \quad \ldots\ldots (4)$$

Where P is precision and R is recall

4. RESULTS AND DISCUSSION

In the study, Planet Scope images acquired in the growing season were mosaicked and the resultant image is used for the crop classification. All bands (except 'b1: coastal blue') at 3m resolution are used for crop mapping. The training and testing data used described in table 3 from which it can be observed that 70% of each class is been utilized by classifier for training purpose and 30% of data is utilized as a testing dataset. The total of 1539 data points is used from which 1072 points are used for training purpose while 492 points used for testing. This portioning of dataset is done using split function in GEE. In remote sensing, the quality of the classified map can be assessed with the help of accuracy assessment, the accuracy only defines whether the generated map can be used for the particular application (Foody, 2002). According to literature, for evaluating the process, confusion matrix and kappa coefficients is widely used. In this study, for evaluating the performance of both classifiers overall accuracy along with F1 score were evaluated. All these accuracy parameters were evaluated by using the optimal value of tuning parameters obtained from respective classifiers.

Table 4. Overall Accuracy by GTB and RF

Dataset	GTB	RF
OA (%)	81.99%	83.68%

The list of tuning parameters along with tested and optimal values are shown in table 3. In RF, numberOfTrees, variables per split and bagFraction are the three parameters which are optimized and can affect the classification accuracy. The testing value for numberOfTrees ranges from 50 to 300 with an interval of 50; variablesPerSplit ranges from 2 to 5 with an interval of 1 and bagFraction ranges from 0.5 to 0.9 with an interval of 0.1. The optimized value for the parameters are 250, 3 and 0.7 respectively for RF classifier. Similarly, for GTB classifier numberOfTrees and learning rate (shrinkage) were the tuning parameters and testing value for numberOfTrees range from 50 to 300 and learning rate or shrinkage value range from 0.05 to 0.09 with an interval of 0.01. The optimal values for GTB classifier are 200 and 0.08 respectively. With these optimal value classification is performed by both classifier and it was observed that GTB achieved 81.99% as the overall accuracy and 79.66% as the kappa coefficient while RF attained 81.99% as overall accuracy and 81.56% as kappa coefficient. The classified image for RF and GTB classifier is shown in Figure 4 and Figure 5.

It has been observed that both classifier performed well in the field of agriculture i.e. for crop classification. But from the result it can be seen that RF yields better result as compared to GTB. This can be because of the property of Random Forest that it is robust to noise and outliers while GTB is not. Table 4. shows the class specific accuracy, from which it can be inferred that in crops, wheat attains the high accuracy by RF classifier followed by Sugarcane and Mustard. This can be because during 19th February wheat crop must be at flowering stage attaining high level of chlorophyll content while mustard must be at ripening stage because during the mid of march harvesting of mustard starts in the study area and sugarcane is either harvested or if retains for next year. So, during this period it is easy to discriminate wheat from other crops and this can be observed from table 5 also that wheat is attaining high accuracy among all crops.

While Sugarcane has achieved 84.55% of accuracy by RF classifier and 89.56% by GTB. From the error matrix, it has been observed that maximum intermixing of Sugarcane pixels can be seen in shrub and other crop class due to the similar spectral characteristics. In other crops sample from double crops are also included which shows similar spectral behaviour and hamper the accuracy. Mustard crop is achieving 85.95 %of accuracy with GTB classifier and 88.08% accuracy with RF classifier. It has been observed that other crops (include double crops) attain lowest accuracy by both classifiers. While this can be a reason because few samples were collected for this class and due to double crops maximum pixels of this class is been misclassified into wheat or mustard class.

This contribution has been peer-reviewed.
https://doi.org/10.5194/isprs-archives-XLVIII-M-1-2023-325-2023 | © Author(s) 2023. CC BY 4.0 License.

Apart from crops from Table 5, it is also observed that Urban is not attaining a specific accuracy of 85% as it can be noticed that maximum pixels of urban is misclassified as Sandline and vice versa. This is because sand and urban have similar spectral reflectance and according to literature, SWIR band can help in discriminating Sandline from Urban but in PlanetScope data SWIR band is not present. This mainly reducing the accuracy of Urban as well as Sand line. However, other classes show nearly same accuracy with both classifiers but water attains high accuracy with RF Classifier as compared to GTB Classifier.

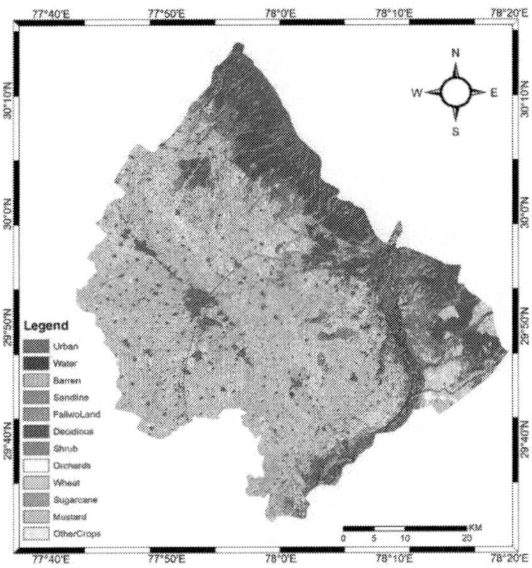

Figure 4. Classified Map generated using Random Forest (RF)

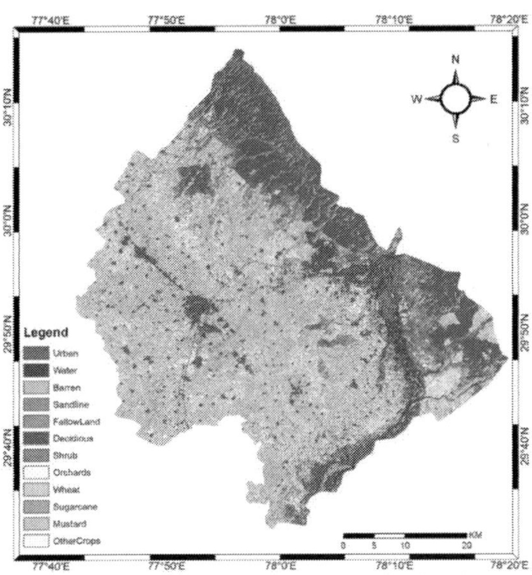

Figure 5. Classified Map generated using Gradient Tree Boosting (GTB)

Table 5. F1 score of all classes by RF and GTB

Class Name	GTB	RF
Urban	76.37	78.57
Water	82.66	92.10
Barren	95.64	95.89
Sand line	78.29	81.35
Fallow Land	84.30	87.66
Deciduous	72.96	76.97
Shrub	83.01	80.88
Orchards	87.09	88.19
Wheat	91.15	93.20
Sugarcane	89.56	84.55
Mustard	85.95	88.08
Other Crops	28.04	32.21

Table 6: Total No of Points Correctly Classified by RF and GTB

S. No	Class Name	Correctly Identified by RF	Correctly Identified by GTB	Total Points
1.	Built-Up	55	55	58
2.	Water	35	31	46
3.	Barren	35	33	36
4.	Sandline	46	44	76
5.	Fallow Land	21	21	25
6.	Deciduous Forest	16	16	21
7.	Shrub	45	46	54
8.	Orchards	20	20	23
9.	Wheat	48	46	51
10.	Sugarcane	36	39	43
11.	Mustard	31	29	33
12.	Other Crops	8	8	23

Table 6. clearly shows the correctly identified pixel by both classifiers. When the values of correctly classified pixel by RF and GTB is observed, it is seen that nearly equal no of pixels were correctly identified by both classifiers. But RF yield better accuracy as compared to GTB Classifier for nearly all classes especially for crops.

5. CONCLUSION

In this study, machine learning based two powerful classifiers RF and GTB were used for crop classification for Haridwar district. Hyper parameter tuning of important parameters were performed for both classifiers to attain highest accuracy. Planet Scope dataset consist of 8 bands from which 7 bands were utilized in the study leaving band 1 which is a coastal blue. The results of the classification reveals that RF classifier attains high accuracy as compared to GTB classifier. According to F1 score of specific classes it can be observed that wheat attains maximum accuracy in the crop classes followed by sugarcane and mustard. This is due to the reason of chlorophyll level being high in wheat during the February period as it attains its flowering stage while mustard must have shifted from reproductive stage to ripening stage. Other crops in both classifier attains minimum accuracy due to the presence of double crops. Barren attains high accuracy using both classifier while water attains high accuracy by RF classifier. Both classifiers perform well while RF attains high (+1.5) accuracy as compared to GTB. The results of the study can conclude that

This contribution has been peer-reviewed.
https://doi.org/10.5194/isprs-archives-XLVIII-M-1-2023-325-2023

PlanetScope 8 band data has a great potential to classify crops by using RF classifier. But if we need to improve accuracy of some specific class than some spectral indices related to the application has to be integrated along with PlanetScope bands.

REFERENCES

Acharki, Siham. 2022. "PlanetScope Contributions Compared to Sentinel-2, and Landsat-8 for LULC Mapping." *Remote Sensing Applications: Society and Environment* 27(April):100774.https://doi.org/10.1016/j.rsase.2022.100774.

Breiman, L. (2001). Random forests. Machine learning, 45(1), 5-32.

Foody, G. M., Campbell, N. A., Trodd, N. M., & Wood, T. F. (1992). Derivation and applications of probabilistic measures of class membership from the maximum-likelihood classification. Photogrammetric engineering and remote sensing, 58(9), 1335-1341.

Foody, G. M. (2002). Status of land cover classification accuracy assessment. Remote sensing of environment, 80(1), 185-201.

Friedman, J. H. (2001). Greedy function approximation: a gradient boosting machine. Annals of statistics, 1189-1232.

Lopes, Mailys et al. 2020. "Improving the Accuracy of Land Cover Classification in Cloud Persistent Areas Using Optical and Radar Satellite Image Time Series." *Methods in Ecology and Evolution* 11(4): 532–41.https ://doi. org/10.1111/2041-210X.13359.

Mudereri, B. T. et al. 2019. "A Comparative Analysis of Planetscope and Sentinel Sentinel-2 Space-Borne Sensors in Mapping Striga Weed Using Guided Regularised Random Forest Classification Ensemble." *International Archives of the Photogrammetry, Remote Sensing and Spatial Information Sciences - ISPRS Archives* 42(2/W13): 701–8.https://doi.org/10.5194/isprs-archives-XLII-2-W13-701-2019

Rafif, Raihan et al. 2021. "Crop Intensity Mapping Using Dynamic Time Warping and Machine Learning from Multi-Temporal Planetscope Data." *Land* 10(12). https:// doi.org/10.3390/land10121384

Saini, R., and S. K. Ghosh. 2018. "Crop Classification on Single Date Sentinel-2 Imagery Using Random Forest and Suppor Vector Machine." *The International Archives of the Photogrammetry, Remote Sensing and Spatial Information Sciences* XLII–5(November): 683–88. https://doi.org/10.5194/isprs-archives-XLII-5-683-2018.

Tamiminia, H.; Salehi, B.; Mahdianpari, M.; Quackenbush, L.; Adeli, S.; Brisco, B. Google Earth Engine for geo-big data applications: A meta-analysis and systematic review. ISPRS J. Photogramm. Remote Sens. 2020, 164, 152–170.

Sharma, V., & Ghosh, S. K. (2022). IMPACT OF CLIMATE ON VEGETATION INDICES OVER RAINFED DISTRICTS OF UTTARAKHAND, INDIA. The International Archives of Photogrammetry, Remote Sensing and Spatial Information Sciences, 43, 989-996.

Zheng, Baojuan, Soe W. Myint, Prasad S. Thenkabail, and Rimjhim M. Aggarwal. 2015. "A Support Vector Machine to Identify Irrigated Crop Types Using Time-Series Landsat NDVI Data." *International Journal of Applied Earth Observation and Geoinformation*34(1):103–12.http://dx.doi.org/0.1016/j.jag.2014.07.002.

CHARACTERISING THE SLUM ENVIRONMENT FROM SPACE FOR ACHIEVING SDGs

Sulochana Shekhar and Kannan Ravi

Central University of Tamil Nadu, Thiruvarur, India

KEYWORDS: Slum development, SDG Index, Chennai, Remote sensing, Inclusive urban development

ABSTRACT:

With less than eight years left to achieve the Global Goals, nations are accelerating the design and implementation of sustainable solutions to the world's biggest challenges ranging from poverty to climate change. India has assessed its status in achieving SDGs through SDG Index, an aggregate measure that computes goal-wise scores on the 16 SDGs. For SDG 11, the 2020-21 assessment showed that all other 32 states scored above 65 as front runners except for seven. One such criterion to assess the progress of SDG11 is the "percentage of urban households living in Katcha houses". In the present study, to understand the slum development activities to attain sustainability at a city level, Chennai has been chosen as a case study. Using VHR data from Google Earth Pro, the slums of Chennai are mapped, and the recent changes in the slum status are identified.

The slums are categorised into four categories: existing, emerged, expanded and evicted slums, based on their characteristics. The study shows that between 2001 and 2011 there are 110 new slums emerged in the study area, which accounts for 1.43 sq. km. There are only a few slums that emerged between 2011 and 2022 (0.35 sq. km), but there was a considerable slum expansion (1.07 sq. km.) during that period, and some slums were evicted from their previous locations. The results based on satellite data have been verified and validated at the field level. The ongoing slum development programs in Chennai give optimistic hope for achieving inclusive urban development by 2030.

1. INTRODUCTION

Globally, Rapid urbanisation has been observed in many regions since the latter half of the 20th century. In 2008, the urban population surpassed the rural population for the first time, and predictions suggest it could reach 68% of the world's total by 2050 (United Nations, 2019). The majority of urban expansion occurs in developing countries, where there is a rapid migration from rural to urban areas. Rural unemployment, natural population growth and other factors drive migration from rural to urban (Jedwab et al., 2017). Cities often struggle to provide enough job opportunities for their rapidly growing populations, resulting in a high prevalence of insecure informal employment that pays low wages. The rural population migrates to cities for work/better living. Still, it often ends up in unsanitary housing clusters due to a lack of affordable housing, resulting in slums which provide substandard living conditions. The proliferation of slums in numerous Indian cities is causing difficulties in ensuring significant infrastructure and essential services. Thus, the foremost challenge facing urban areas is striking a balance between economic progress and sustainable development (Shekhar, 2021).

Urban development necessary for sustainable development. Sustainable cities drive it by providing job/business opportunities, safe and affordable housing, and robust societies/economies through investment in public transport, green spaces, and participatory planning/management (Thomas et al., 2021). Goal 11, "Sustainable Cities and Communities," is part of the 17 SDGs adopted by UN members to end poverty, protect environment, and achieve peace/prosperity by 2030, as defined by UNDP. Goal 11 (Sustainable Cities and Communities) is either directly or indirectly tied to Goals 1, 2, 3, 6, 7, 8, 9, and 10. Progress in any of these goals benefits sustainable cities and communities. India aligns its growth plan with the sustainable development goals as a global partner (Shekhar, 2021).

Chennai city, the capital of Tamil Nadu, has been selected as a case study for this research in order to comprehend the slum development efforts required for inclusive and sustainable urbanisation at the city level. The main objective of the study is to identify the present status of slums for achieving sustainable development goals and estimate the percentage of urban households living in Katcha houses for the year 2022.

2. MATERIALS AND METHODS

2.1 Study area

The study area is Chennai city (Figure 1), capital of Tamil Nadu, India. The city located between longitude 80° 08'3.9411″ E and 80° 20′ 2.398″ E and latitude 13° 06′ 26″ N and13° 14′ 3.924″N. The Chennai city covers an area of 426 sq. km, which is divided into 15 zones and 200 divisions for administrative purposes. The Cooum and Adyar river, along with several minor canals, flow through this region. According to the 2011 census, 5.41% of India's population, which is 65.5 million people, live in slums. In Tamil Nadu, 8% of the urban population or 5.8 million people, reside in slums (Census of India, 2011). In Tamil Nadu, Chennai has the largest slum population, estimated at around 1.3 million.

This contribution has been peer-reviewed.
https://doi.org/10.5194/isprs-archives-XLVIII-M-1-2023-331-2023 | © Author(s) 2023. CC BY 4.0 License.

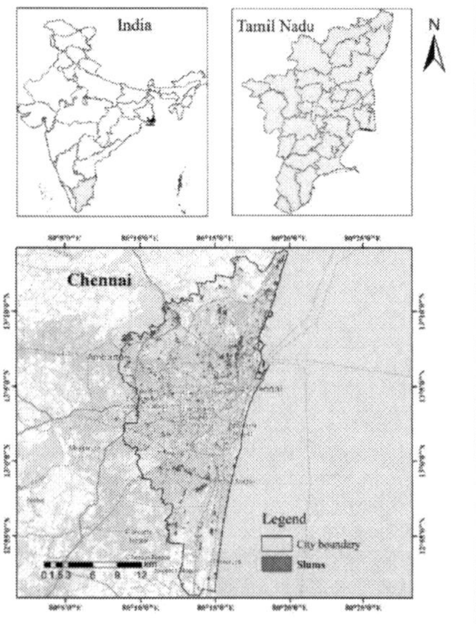

Figure 1. Study area Map

2.2 Dataset

The base data for Chennai slums was collected from Tamil Nadu Urban Habitat Development Board (TNUHDB). The base data contains the information related to slums and rehabilitated settlements (relocated settlements). This includes slum names (of 1131slums), ward numbers, zone numbers, the total number of households, the total population, and the spatial boundary (of the year 2014) of each slum. Google Earth time series imagery for the years 2001, 2011, and 2022 have been used in this study to find the spatial-temporal variation of slums.

2.3 Methodology

The following flow chart (Figure 2) shows the methodology adopted in identifying the characteristics of slums using spatial data.

Initially, an ESRI shapefile was created for the study area. The shapefile encompasses 15 zones, 200 ward boundaries of the Chennai, and delineated slum boundaries. It comes in seven distinct file formats, which include supplementary information such as spatial boundaries, attribute data, projections, and coordinate systems (Gruebner et al., 2014). The 2014 slum boundary was obtained from the TNUHDB, and based on that, slum boundary layers were created for the years 2001, 2011, and 2022 using Google Earth Pro.

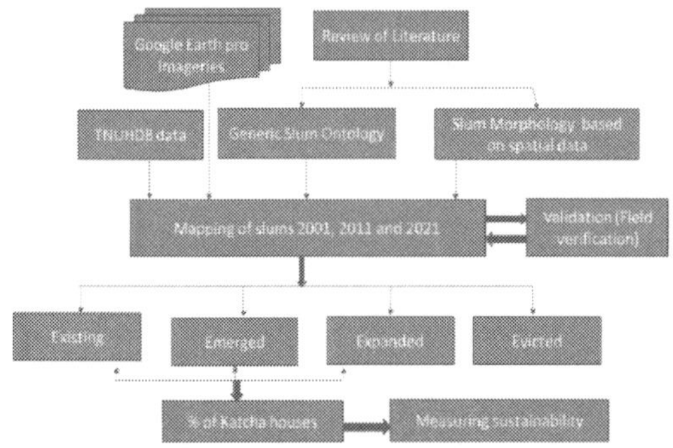

Figure 2. Methodology flowchart

2.3.1 Slum Mapping

This methodology for mapping slums is explained in this section. As previously mentioned in the dataset session, spatial-temporal analysis was conducted using historical images of Google Earth Pro from 2001, 2011, and 2022. The study was limited by the availability of high-resolution satellite imagery from Google Earth for the study area and the quality of the images, which could be impacted by clouds and shadows. The availability of historical imagery from Google Earth was also restricted by the amount of commercially available VHR data (Duque et al., 2017). Since 2001, historical VHR imagery for the Chennai district has been accessible. The years 2001, 2011, and 2022 were digitized from Google Earth's historical imagery, utilising the slum base data from TNUHDB as a reference.

Slums that were not previously included in the database have been digitized using a generic slum ontology and the characteristics of slum morphology. The term "Generic Slum Ontology (GSO)" refers to the worldwide understating of slums. GSO aims to bridge the gap by offering a thorough description of slums (Pratomo et al., 2017). The GSO categorises slums into three distinct spatial levels: the object level, which includes building and access network characteristics; the settlement level, which encompasses settlement shape and density; and the environs level, which refers to the location and neighbourhood features (Kohli et al., 2012, 2019; Shekhar, 2013, 2019). The morphology features used to identify slums include building size, height, density, orientation, and the diversity of patterns in the slum's spatial layout (Taubenböck, 2018). The Google Earth Historical imageries of 2001, 2011, and 2022 were used to create slum layers.

2001: No of Slums = Existing Slums --------(1)

2011: No of Slums = (Existing slums + Spatial Expansion of slums+ Emerged slums*) – Evicted slums -------------(2)

2022: No of Slums = (Existing slums + Spatial Expansion of slums+ Emerged slums**) – Evicted slums -------------(3)

* Slums formed between 2001-2011

** Slums formed between 2011-2021

The above-said methodology has been used to determine the number of slums in 2001, 2011 and 2021 and to quantify the spatial variation or growth of slums between 2001-11 and 2011-22.

https://doi.org/10.5194/isprs-archives-XLVIII-M-1-2023-331-2023 | © Author(s) 2023. CC BY 4.0 License.

Characteristics	Informal (Slum)	Formal Built-up areas
Size	Small (substandard) building sizes and varied	Generally larger building sizes and varied
Density	High to Very high roof density Lack of open and green spaces within or in the proximity of slum areas.	Low to moderate density areas Presence of open and green spaces within or nearby areas
Pattern	Organic layout structure No planned street patten	Regular layout pattern with planned regular road
Site characteristics	Often at hazardous locations and vulnerable Proximity to transportation lines and livelihood opportunities	Land has the basic suitability for being built-up with necessary infrastructure

Table 1. Morphological features typical for slum areas (Adapted from Kuffer et al., 2016; Aminipouri et al., 2009)

2.3.2 Field validation

The digitized slums require validation and it was done through field verification. The slums in the Chennai city had been divided into four categories, and samples were obtained from Google Earth Pro imageries for each category. The samples were taken across the study area from 2001, 2011, and 2022 imageries and verified in the field to confirm the presence and the growth of slums.

2.3.3 Estimation of Katcha houses

To determine the "percentage of urban households living in Katcha houses", a hybrid method was employed. Which involves the use of secondary and primary data i.e., using existing slum database and field visits. The process of calculating this percentage started with the delineation of new slums/emerged slums and expanded slums, which are not available in the existing slum database. The number of Katcha houses in these slums were then determined through a combination of field visits and the use of Google Earth Pro.

A total of 95 slums were selected from the emerged slums, and out of those, fifty slums were visited to gather data on the number of Katcha houses in person. During these visits, the number of houses made of materials like mud, bamboo, or thatch, which are considered to be less permanent or sturdy, was counted. The remaining 45 slums were analysed using Google Earth Pro, which allowed us to calculate the number of Katcha houses remotely.

The following method was used to calculate the number of slum households living in Katcha houses, considering the field visits and the data from Google Earth Pro. Finally, the results were cross-verified using the 2011 census data to determine if the number of Katcha houses had increased or decreased. This hybrid approach helped to ensure the accuracy and reliability of the findings and provided a comprehensive picture of the housing situation in urban areas.

Total number of Katcha houses = No of Katcha houses in (existing Slums + emerged slums + expanded slums) --------------------- (4)

3 RESULTS

This section elucidates the outcome of mapping informal settlements and their characteristics for achieving sustainable development.

3.1 Slum growth from 2001 to 2011

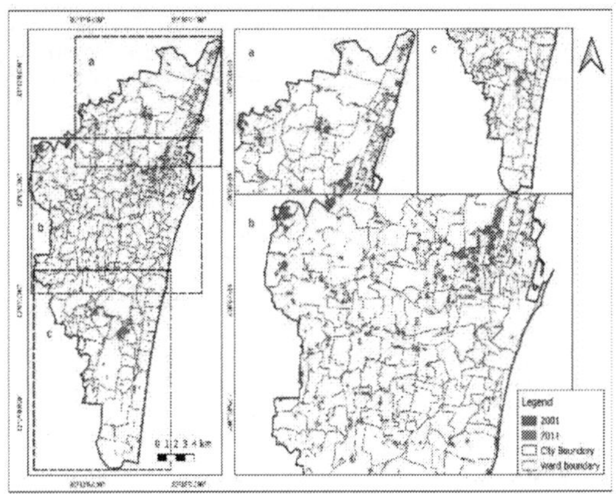

Figure 3. Spatial and temporal analysis of slums (year 2001 – 2011)

Figure 3 above shows the spatiotemporal variation of slums. The map was divided into three parts for better visualisation. A total of 110 slums underwent changes between 2001 and 2011. These slums are found in 50 wards out of 200 and spread across the fifteen zones of Chennai corporation (Table 2). A maximum of seven wards experienced the formation of new slums, and slum expansion comes under zone 3, Madhavaram and is followed by zone 9, Teynampet, where 6 wards underwent these changes. These are all the wards that show the invasion of slum settlements into areas previously not having slums. It has been identified that the slums which emerged in the following wards, numbers 8,14,28,39,63,77, 98,100,107,111,116,124, 125,176, and 194 were through the rehabilitation procedure. The rehabilitation process is part of TNUHDB activities which involve the rehabilitation and resettlement schemes to improve the environs of the slums and the living standards of the urban slum families (https://tnuhdb.tn.gov.in/).

https://doi.org/10.5194/isprs-archives-XLVIII-M-1-2023-331-2023 | © Author(s) 2023. CC BY 4.0 License.

	2001 - 2011				2011-2022			
Zone number	**Ward Number of Emerged and Expanded Slums (A)**	**No of the Wards experienced (A)**	**Ward number of evicted slums (B)**	**No of the Wards experienced (B)**	**Ward Number of Emerged and Expanded Slums (A)**	**No of the Wards experienced (A)**	**Ward number of evicted slums (B)**	**No of the Wards experienced (B)**
1. Thiruvotriyur	4, 5, 7, 9	4	8, 14	2	-	-	14	1
2. Manali	16, 17, 18, 19, 20	5	-	-	-	-	17	1
3. Madhavaram	22, 24, 25, 27, 28, 29, 33	7	28	1	23, 24, 25, 28	4	-	-
4. Tondiarpet	37	1	39	1	34	1	42, 45	2
5.Royapuram	56, 57, 58, 59, 60	5	63	1	56	1	58, 59, 61, 63	4
6. Thiru. Vi. Ka. Nagar	70	1	77	1	65	1		
7. Ambattur	82, 83, 85	3	-	-	-	-	82, 90, 93	3
8. Anna Nagar	102	1	98, 100, 107	3	-	-	100, 106, 107	3
9. Teynampet	114, 117, 118, 122, 124, 126	6	111, 116 124, 125	4	-	-	110, 111, 117, 120, 122, 124	6
10. Kodambakkam	138, 139, 142	3	-	-	128,138,142	3	138, 142	2
11. Valasaravakkam	143	1	-	-	-	-	143,144	2
12. Alandur	156, 157, 160, 161	4	-	-	-	-	157	1
13. Adyar	171, 172, 173, 175	4	176	1	-	-	171, 172	2
14. Perungudi	183, 189, 191	3	-	-	-	-	187, 191	2
15. Sholinganallur	198, 200	2	194	1	-	-	196	1
Total number of wards		**50**		**15**		**10**		**30**

Table 2. Zone wise distribution of emerged and existing slums

3.2 Slum growth from 2011 to 2022

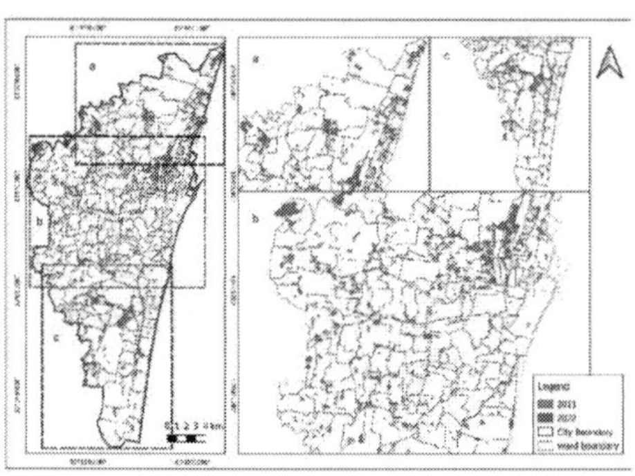

Figure 4. Spatial and temporal analysis of slums for the year 2011 – 2022

Figure 4 shows the spatiotemporal variation of slums from 2011 to 2022. The map was divided into three parts for better visualization. The new slums have emerged in wards 23, 24, 25, 28,34,56, 65, 128, 138, and 142. Again, the maximum of 4 wards in Madhavaram zone experienced the expansion of slum settlements in this decade. Slums were evicted from 30 wards distributed across the zones (Table2) during this period. Teynampet zone experienced more eviction than other zones. The output shows that compared to 2001 to 2011, eviction happened more frequently in the 2011 – 2022 period.

Year of Mapping	Based on the TNUHDB base layer	Based on slum ontology and slum morphology	Total number of slums@	The total area of slums km^2
2001	995	97	1092	18
2011	1103	99	1202	20.68
2022	1106	95	1201	22.1

Table 3. Growth of slums

@ Total number of slums do not include the evicted slums

Table 3 shows the growth of slums for 2001, 2011, and 2022. It also represents that the year between 2001 and 2011 there

https://doi.org/10.5194/isprs-archives-XLVIII-M-1-2023-331-2023 | © Author(s) 2023. CC BY 4.0 License.

are 110 new slums emerged in the study area. Since 2011, there were few slums formed between 2011 and 2022, and the total area showed an increase due to expansion of slum boundaries. Table 4 shows that northern Chennai has more slum growth than the central and south Chennai..

Region	Zone number	2001 – 2011		2011 - 2022	
		Total number of wards (Formation of new slums and Slum expansion)	Total number of wards (Slum eviction)	Total number of wards (Formation of new slums and Slum expansion)	Total number of wards (Slum eviction)
North Region	1, 2, 3, 4, 5	22	5	6	8
Central Region	6, 7, 8, 9, 10	14	8	4	14
South Region	11, 12, 13, 14, 15	14	2	-	8

Table 4. Distribution of slums at regional level

Being elongated in shape, Chennai is divided into three regions namely Northern Chennai, Central Chennai and South Chennai. The number of wards have slum population is high in northern Chennai compared to other regions. Presence of industrial and commercial activities along with more open space during 2000s also reasons for major slum growth and slum expansion in this region.

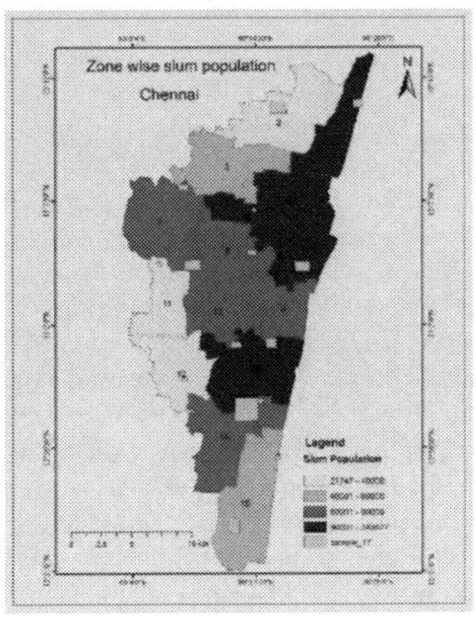

Figure 5. Zone wise slum population of Chennai district

Figure 5 represents the zone-wise slum population of the Chennai. The result of the study mainly reveals that 28% of Chennai's urban population lives on 5.16% of the land. The slum population has been divided into four categories, and four samples have been taken from each category (4x4=16) of slums. The samples were collected for the years 2001, 2011, and 2022 from Google Earth Pro. The samples were used for the ground validation and to confirm the growth of slums.

	200 1	2011	2022
A	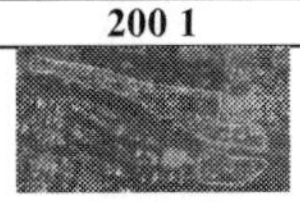 4.71 hectare	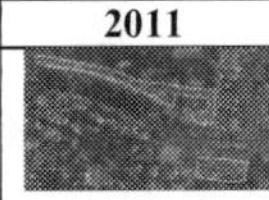 2.18 hectare	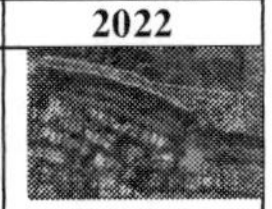 2.18 hectare
B	 7.48 hectare	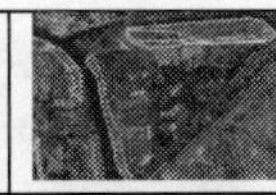 8.37 hectare	 3.81 hectare
C	 9.47 hectare	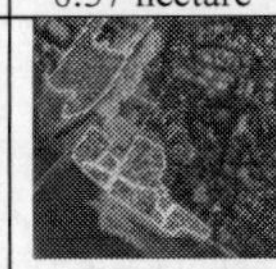 8.25 hectare	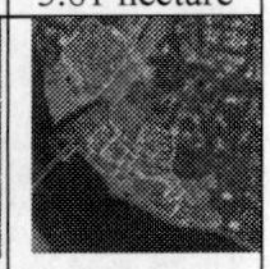 7.24 hectare
D	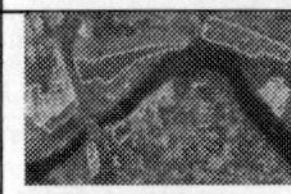 4.54 hectare	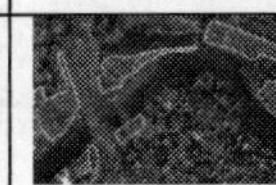 4.05 hectare	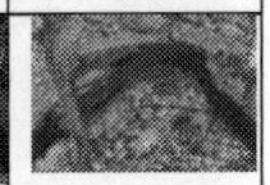 2.35 hectare
E	 NIL	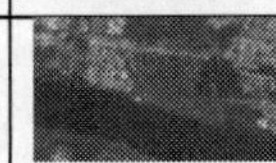 0.09 hectare	 0.19 hectare
F	 77.66 hectare	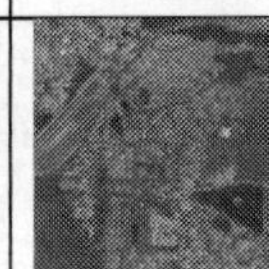 90.15 hectare	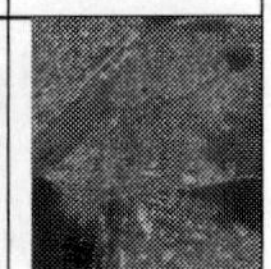 55.57 hectare
G	 NIL	 0.82 hectare	 0.89 hectare
H	 5 hectare	 13.27hectare	 16.76hectar e
I	 4.53 hectare	 6.43 hectare	 8.73 hectare
J	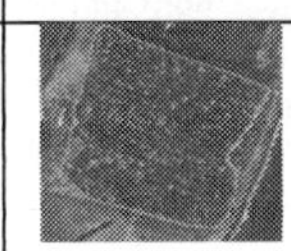 21.89 hectare	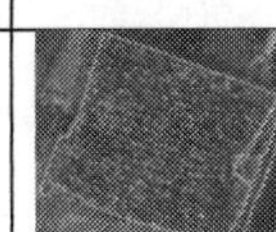 22.24 hectare	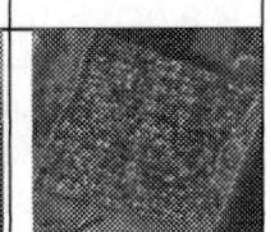 23.3 hectare

https://doi.org/10.5194/isprs-archives-XLVIII-M-1-2023-331-2023 | © Author(s) 2023. CC BY 4.0 License.

K	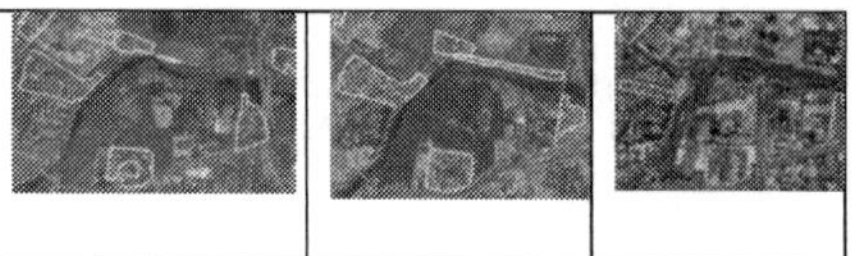		
	11.72 hectare	13.29 hectare	12.98 hectare
L			
	0.48 hectare	1.23 hectare	1.44 hectare
M			
	1.05 hectare	2.32 hectare	2.48 hectare
N	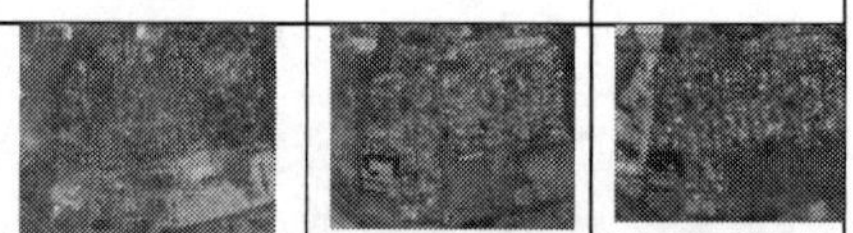		
	3.56 hectare	4.45 hectare	4.75 hectare
O			
	2 hectare	2.59 hectare	3.65 hectare
P			
	10.89 hectare	11.59 hectare	13.2 hectare
	New slums and slum expansion Slum Eviction Slums		

Table 5. Samples show the Changes in slum areas

Table 5 represents the growth of slums, and columns show that times 2001, 2011, and 2022 and rows show the 16 samples by alphabet. The table represents the expansion of existing slums, formation of new slums, and eviction of existing slums that had taken place in the study area.

3.3 Katcha houses

In the SDG index, for SDG 11, "Sustainable cities and communities", the year 2020-21 assessment showed that except for seven states in India, all other 32 states scored above 65 as front runners. One such criterion to assess the progress of SDG11 is the "percentage of urban households living in Katcha houses". Tamil Nadu state has scored 79 out of 100 for SDG11, with only 0.8 per cent of households live in Katcha houses. In the present study, to understand the slum development activities to attain inclusive and sustainable urbanisation at a city level, Chennai, the capital of Tamil Nadu, has been chosen as a case study. Using high-resolution spatial data from Google Earth Pro, the slums of Chennai are mapped, and the recent changes in the slum status are identified.

The 2011 census showed that Chennai had a total of 12,60,298 households. Out of these households, 3.7% or 47,446 were identified as Katcha houses (Office of Registrar General of India, 2011). This term refers to houses that are made of materials like mud, bamboo, wood, thatch or plastic (polythene), which are not permanent or sturdy. These houses are considered to be less secure and have poor living conditions compared to houses made of more durable materials. Thus, they represent the slum households.

In 2022, the result of this study shows that the percentage of urban households living in Katcha houses in Chennai had increased to 4.58% or 57,771. This represents a significant increase from the previous census and highlights a growing concern for adequate housing in the city.

4. DISCUSSION

The slums are grown rapidly in Chennai since 2001. Various factors have contributed to slum growth, such as rapid population growth and rural-urban migration. In recent decades, developing extensive industrial facilities in Chennai and its suburbs has led to a significant population increase (Viswanatha and Tharkar, 2010). Migration occurs due to push factors such as limited job prospects, inadequate wages, drought, unavailability of essential amenities, lack of land ownership, and social issues in rural areas. Conversely, pull factors such as increased job opportunities, higher wages, improved infrastructure, and access to better facilities attract migration from rural to urban areas (Zhang 2016). Urban areas can't provide impoverished migrants with enough jobs, infrastructure, and secure tenure. Migrant labourers choose settlement sites depending on job prospects, leading to slum expansion (Shekhar, 2021). In 1971, Chennai's migrant population was 0.78 million and increased to 0.98 million in 2001(Krishnamurthy and Desouza, 2015). According to the 2011 census, 1.32 million people migrated; within that, 0.96 million were intra-state migration, and 0.33 million were inter-state migration (Census of India, 2011). Consequently, the disparity between the requirement and availability of essential services and infrastructure in metropolitan areas increases, negatively affecting the standard of living (Shekhar, 2019). In 2001, Chennai's urban population lived in slums was 18.9%, which increased to 28% in 2011 (CMDA 2008; Office of Registrar General of India, 2011). This figure shows the rapid growth in the slum population in the study area.

Since 1901, Chennai's spatial size or area has expanded from 68 km^2 to 426 km^2(Krishnamurthy and Desouza, 2015). The city has had to expand its spatial boundary due to its natural growth and immigration. As per the survey conducted in

https://doi.org/10.5194/isprs-archives-XLVIII-M-1-2023-331-2023 | © Author(s) 2023. CC BY 4.0 License.

1986, there were 996 slums in the city, and in 2014, 1131 slums were surveyed by TNUHDB. The results show that between the years 2001 to 2011, 125 new slums emerged, and from 2011 to 2022, 37 new slums emerged, and spatial expansion of slums took place.

Though all the zones underwent changes in their slum characteristics, some of them experienced notable changes. The northern Chennai had a significant growth of slums and the Madhavaram zone had more emerged slums during 2001-11 and 2011-22 due to its large informal sector and vacant land. Textile-based business activities in the Teynampet zone also attracted new slums during that period. But the Teynampet zone also experienced more slum evictions during 2001-11 and 2011-22; because of its high commercial land value and high density, the slums were resettled in various other zones.

Due to the Millennium Development Goal developmental activities and the Slum-free India mission of the Government of India, Slum development program initiatives from the State Government, the emergence of new slums was reduced between 2011 and 2021. At the same time, many slums underwent spatial expansion during this period, along with rehabilitation and in-situ development.

But the increase in the number of Katcha houses from 2011 to 2022 in Chennai is worrying and it may be due to various factors such as population growth, urbanization, and limited availability of affordable housing options, etc. This data underscores the need for action to address the housing challenges faced by the residents of Chennai, including improving the living conditions of those living in Katcha houses.

5.CONCLUSION

The result of the study shows that Chennai experienced significant slum growth during 2001- 2011 and slum expansion during 2011-2021. In this study, Google Earth Pro open-source software was predominantly used for slum mapping and validation. The findings revealed the emergence of slums and indicated that 28% of urban dwellers live on 5.16% of the land. As part of Slum development programs of both Central and State Government schemes, slums have been evicted from their original habitation and resettlement or rehabilitation taken place in many wards of Chennai. As a result of this study, an updated dataset was prepared for 2022. The result also shows that 57,771(4.58%) Katcha houses exist in Chennai. Compared to the 2011 census, the present study showed an increase in the number of households living in Katcha houses. This demands an urgent need for affordable housing and the implementation of central and state Government schemes effectively. The eleventh goal of Sustainable development is exclusively for inclusive slum development. This is possible with inclusive urban planning involving the slum community, Public and private partnerships, Community-Based Organizations or Non-Government Organizations and all other stakeholders. The present study is an earnest attempt in this direction using openly available spatial data. Urban planners and urban local bodies can use this dataset or spatial database and can make informed decisions for slum development to achieve sustainable urban development. The decisions we make and actions we take today will have momentous consequences for future generations.

ACKNOWLEDGEMENTS

The first author sincerely thanks DST-SERB for providing a Travel grant to attend the ISRSE39 conference in Antalya, Turkey. The second author thanks University Grants Commission for providing a fellowship for his doctoral research.

REFERENCES

Census (2011). Census, 2011, Office of the Registrar General of India, 2011. https://www.census2011.co.in/slums.php

Duque, J. C., Patino, J. E., & Betancourt, A. (2017). Exploring the potential of machine learning for automatic slum identification from VHR imagery. Remote Sensing, 9(9), 895.

Gruebner, O., Sachs, J., Nockert, A., Frings, M., Khan, M., Hossain, M., ... & Hostert, P. (2014). Mapping the Slums of Dhaka from 2006 to 2010. Dataset Papers in Science, 2014.

Jedwab, R., Christiaensen, L., & Gindelsky, M. (2017). Demography, urbanization and development: Rural push, urban pull and... urban push? Journal of Urban Economics, 98, 6-16.

Krishnamurthy, R., & Desouza, K. C. (2015). Chennai, India. Cities, 42, 118-129.

Kohli, D., Kuffer, M., and Gevaert, C. M. (2019). The Generic Slum Ontology : Can a Global Slum Repository be created ? 2019 Joint Urban Remote Sensing Event (JURSE), 1–4.

Kohli, D., Sliuzas, R., Kerle, N., and Stein, A. (2012). An ontology of slums for image-based classification. Computers, Environment and Urban Systems, 36(2):, 154–163. http://doi.org/10.1016/j.compenvurbsys.2011.11.001

Kuffer, M., Pfeffer, K., and Sliuzas, R. (2016). Slums from Space — 15 Years of Slum Mapping Using Remote Sensing. Remote Sensing. http://doi.org/10.3390/rs8060455

Pratomo, J., Kuffer, M., Martinez, J., & Kohli, D. (2017). Coupling uncertainties with accuracy assessment in object-based slum detections, case study: Jakarta, Indonesia. Remote sensing, 9(11), 1164.

Shekhar, S. (2013). Slum modelling by using Ontology and Geoinformatics : case study of Gulbarga, (June 2013)

Shekhar, S. (2019). Effective management of slums- Case study of Kalaburagi city, Karnataka, India. Journal of Urban Management, (February 2019):, 0–1. http://doi.org/10.1016/j.jum.2019.09.001

Shekhar, S (2021). Urbanization in India. Slum Development in India: A Study of Slums in Kalaburagi, 1-20.

Shekhar, S (2021). Slums in India (pp. 21-43). Springer International Publishing.

Taubenböck, H., Kraff, N. J., & Wurm, M. (2018). The morphology of the Arrival City-A global categorization based on literature surveys and remotely sensed data. Applied Geography, 92, 150-167.

Thomas, R., Hsu, A., & Weinfurter, A. (2021). Sustainable and inclusive–Evaluating urban sustainability indicators' suitability for measuring progress towards SDG-11. Environment and Planning B: Urban Analytics and City Science, 48(8), 2346-2362.

Viswanathan, V., & Tharkar, S. (2010). Can the divide be bridged: Overview of life in urban slums in India. Indian Journal of Community Medicine: Official Publication of Indian Association of Preventive & Social Medicine, 35(1), 198–199. http://dx.doi.org/10.4103/0970-0218.62562.

World Urbanization Prospects 2018 - Highlights. (2019). Department of Economic and Social Affairs. World Population Prospects 2018. Retrieved from https://population.un.org/wup/

Zhang, X. Q. (2016). The trends, promises and challenges of urbanisation in the world. Habitat international, 54, 241-252.

This contribution has been peer-reviewed.
https://doi.org/10.5194/isprs-archives-XLVIII-M-1-2023-331-2023 | © Author(s) 2023. CC BY 4.0 License.

REMOTE SENSING FOR FLOOD INUNDATION MAPPING USING VARIOUS PROCESSING METHODS WITH SENTINEL-1 AND SENTINEL-2

E. Stoyanova[1*]

[1]University of Architecture, Civil Engineering and Geodesy (UACEG), Faculty of Geodesy, Dept. of Photogrammetry and Cartography, Sofia, Bulgaria – stoyanova.elia@gmail.com

KEY WORDS: Sentinel-1, Sentinel-2, spectral indices, SAR, flood mapping, change detection

ABSTRACT:

The following paper discusses different remote sensing methodologies for flood mapping. The chosen study area is nearby Karlovo, Bulgaria where torrential rains in August and September 2022 caused devastating damages in the area. The location is still under critical status and damages are not completely evaluated, yet.
The aim of this study is to investigate how open satellite data can be applied in the mapping the results from the floods. The research team worked with radar and optical data from Sentinel-1 and Sentinel-2. Different technological approaches were applied to classify the land cover changes. Given the fact that the flood mass consisted mainly of mud an innovative approach was considered in order for the damages area to be computed.
Climate change predictions and deforestation suggest increasingly frequent floods in the future. For these reasons, rapid and timely flood mapping under specific conditions is very important in flood modelling, hazard and risk analysis. In order to investigate the flooded area in the region of the selected study area, different post-processing methods were used. The research team focused on using satellite imagery from both Sentinel-1 and Sentinel-2. Various spectral indices were computed as well as unsupervised and supervised classifications were applied. The results of the study were presented in the form of statistical analyses and comparison between the computational methods.

1. INTRODUCTION

A flood is a natural disaster in which a land area is temporarily flooded with a huge amount of water. It is obtained as a result of an increase in the level of water basins and riverbeds. The reasons for this are different, and most often these are intense rainfall, snowmelt, dam wall failure, soil erosion, human activity and others. Flooding can occur very slowly or occur in a matter of minutes.

They are also some of the most catastrophic natural disasters, as they cause significant damage to settlements, buildings, cars, bridges, roads and transport infrastructure. According to the Organization for Economic Co-operation and Development, floods cause more than $40 billion in damage worldwide annually. Millions of people and animals around the world are affected by floods every year, unfortunately sometimes losing their lives, with the number of deaths increasing in recent decades to over 100 people per year.

Often the worst part is not the flood itself, but the time after it. After the waters recede, the affected areas remain covered with mud and silt. Drinking water and the environment are contaminated with pesticides, various debris, fuel and untreated sewage. And in structures soaked with water, mold often forms, which can be very dangerous. Other problems for people after the floods include a lack of electricity and clean drinking water, which in turn can lead to disease outbreaks from the unsanitary conditions.

Precisely for these reasons, accurate and timely information about the size of the affected territory is needed. In this way, the damage will be able to be calculated and mapped, which will help with crisis management and recovery processes. This information is important because there is often a lack of timely, up-to-date and publicly available data on flood damage.

On August 22, 2022 the municipality of Karlovo declares a partial state of emergency due to heavy rains, as a large part of the streets and buildings are flooded. The amount of rain that fell is many times more than the monthly rainfall, but the nearby river did not overflow and people managed to cope without major damage. But the rains do not stop and on September 2, 2022, the dyke of the Stryama River broke and flooded several villages in the municipality of Karlovo, with the worst affected being the village of Bogdan and the village of Karavelovo, which are the subject of this study (Figure 1).

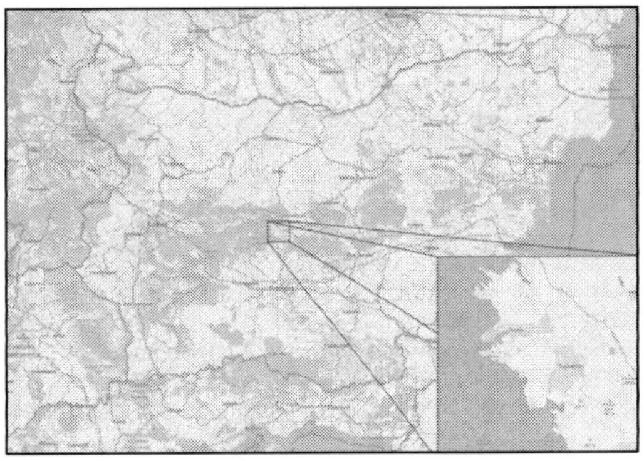

Figure 1. Location of case study: Bogdan and Karavelovo, Karlovo municipality, Bulgaria

The study is focused on the use of SNAP (Sentinel Application Platform), QGIS, Google Earth Pro and The Copernicus Open Access Hub. This contributes to the application of these studies by a wide range of interested specialists, as open access can accelerate getting good things done, helping spread the value more widely and reducing costs yet improving innovation.

This contribution has been peer-reviewed.
https://doi.org/10.5194/isprs-archives-XLVIII-M-1-2023-339-2023

2. STUDY AREA AND DATASET

2.1 Study area

As mentioned, the study focused mainly on two villages - Bogdan and Karavelovo in Karlovo municipality. They were chosen because the flood that hit them was one of the biggest disasters of 2022 in Bulgaria (Figure 2). The level of the river that floods the villages has risen by 3 meters in 8 hours, and the amount of rain that has fallen is over 250 litres per square meter.

Figure 2. The damage caused by the flood after September 2, 2022. Karavelovo village the day after the flood (*top left*); Multiple cars swept away in affected areas *(top right*); Home in the village of Karavelovo, it can be clearly seen how far the mud has reached (*bottom left*); The stadium of the village of Karavelovo covered with trees brought down from the river (*bottom right*). © Анастас Търпанов/ Anastas Turpanov (credit:https://www.dnevnik.bg/photos/2022/09/07/4388097_fotogaleriia_sled_potopa/ – accessed on 21 February 2023.)

The affected area is smaller in size. This is precisely why the research aims to explore the possibilities for remote sensing of smaller areas based on satellite images from the Copernicus program.

2.2 Dataset

For this paper, images from the Sentinel-1 and Sentinel-2 are used. They are provided free of charge by ESA as part of the Copernicus program. All images used in the study were obtained through the Copernicus Open Access Hub.

Two Sentinel-1A SAR datasets were used in the study. The first image was recorded before the heavy rains started and the river overflowed. The second image was recorded days after the floods. The main specifications of the datasets are presented in Table 1.

Mission	Date	Format-Mode	Polarization	More Information
Sentinel-1A	13 August 2022	GRD-IW	VV+VH	before the flood
Sentinel-1A	06 September 2022	GRD-IW	VV+VH	after the flood

Table 1. Specifications of Sentinel-1 images

A total of 3 multispectral images from Sentinel-2 were used. The first image was taken again before the heavy rains started and the river overflowed. The second image was again taken days after the floods. And the third was made 4 months after the disaster, the purpose of which is to see what the condition of the affected territory is. The main features of the images are presented in Table 2.

Mission	Date	Type	More Information
Sentinel-2A	18 August 2022	S2MSI2A	before the flood
Sentinel-2A	07 September 2022	S2MSI2A	after the flood
Sentinel-2B	08 January 2023	S2MSI2A	after the flood

Table 2. Specifications of Sentinel-2 images

Each of the 3 images consists of 13 bands and the features with their names, resolutions, bandwidths and central wavelengths are presented in Table 3 (Sentinel-2 User Handbook, 2015).

Sentinel-2 Bands	Central wavelength (nm)	Bandwidth (nm)	Resolution (m)
Band 1 - Coastal aerosol	443	20	60
Band 2 - Blue	490	65	10
Band 3 - Green	560	35	10
Band 4 - Red	665	30	10
Band 5 - Vegetation Red Edge	705	15	20
Band 6 - Vegetation Red Edge	740	15	20
Band 7 - Vegetation Red Edge	783	20	20
Band 8 - NIR	842	115	10
Band 8A - Vegetation Red Edge	865	20	20
Band 9 - Water vapour	945	20	60
Band 10 - SWIR - Cirrus	1375	30	60
Band 11 - SWIR	1610	90	20
Band 12 - SWIR	2190	180	20

Table 3. Features of Sentinel-2 images

3. METHODS

3.1 Sentinel-1 SAR data

To begin the practical part, the two Sentinel-1 SAR images were used. The processing was carried out according to the methodology specified in TRAINING KIT HAZA01 of RUS Copernicus.

Since the Area of Interest (AOI) is quite small and there is no need to process the whole image, subsets are made. Figure 3 shows the area covered by the study. Since the orbit state vectors provided in SAR product metadata are usually not exact, a precise exact orbit is applied. The orbital file provides accurate information about the satellite's position and velocity. Based on this information, the orbit state vectors in the abstract product metadata are updated. (SNAP Help)

Figure 3. Area of Interest for Sentinel-1 images

This contribution has been peer-reviewed.
https://doi.org/10.5194/isprs-archives-XLVIII-M-1-2023-339-2023 | © Author(s) 2023. CC BY 4.0 License.

The next stage of processing is to remove the thermal noise. It is the background energy that is generated by the receiver itself. It skews the radar reflectivity to towards higher values and hampers the precision of radar reflectivity estimates. Level-1 products provide a noise LUT for each measurement dataset, provided in linear power, which can be used to remove the noise from the product (TRAINING KIT HAZA01).

The next step is calibration. The objective is to provide imagery in which the pixel values can be directly related to the radar backscatter of the scene. Though uncalibrated SAR imagery is sufficient for qualitative use, calibrated SAR images are essential to quantitative use of SAR data. The radiometric correction is also necessary for the comparison of SAR images acquired with different sensors, or acquired from the same sensor but at different times, in different modes, or processed by different processors. (SNAP Help)

Speckle Filtering is performed since SAR images have speckles which degrade the quality of the image and make interpretation of features more difficult (Figure 4). They are caused by random constructive and destructive interference of the de-phased but coherent return waves scattered by the elementary scatters within each resolution cell. (SNAP Help)

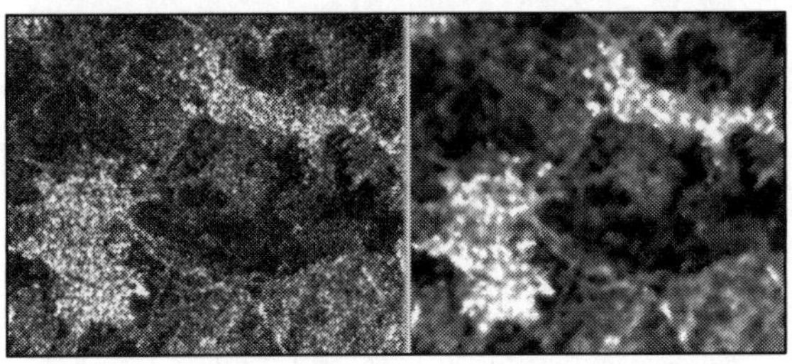

Figure 4. Speckle Filtering. View of the two villages before filtration (*left*), View after filtration (*right*).

The same threshold was chosen for both images so that the water could be separated from the other objects in the environment. The filtered backscatter coefficient histogram is used, where low values correspond to water and high values to other non-water objects. By using the color manipulation methods, the water is visualized in black, and the rest of the objects in the white-gray color range.

Topographic variations in the scene and tilt of the satellite sensor can lead to distortions of distances in SAR images. The Range Doppler orthorectification method for geocoding SAR images from single 2D raster radar geometry was used. It uses available orbit state vector information in the metadata, the radar timing annotations, the slant to ground range conversion parameters together with the reference DEM data to derive the precise geolocation information (SNAP Help). Terrain correction is used to compensate for these distortions and make the image as close as possible to its real appearance.

In both images, the two water bodies (red boxes, Figure 5) can be clearly seen. But no other indications of flooded areas are observed. Even before the flood occurs, there are more areas that can be identified as flooded. The blue arrows in Figure 5 show the areas that have undergone changes since the disaster. It is there that the river passes, leaving its bed and flooding the surrounding areas with mud.

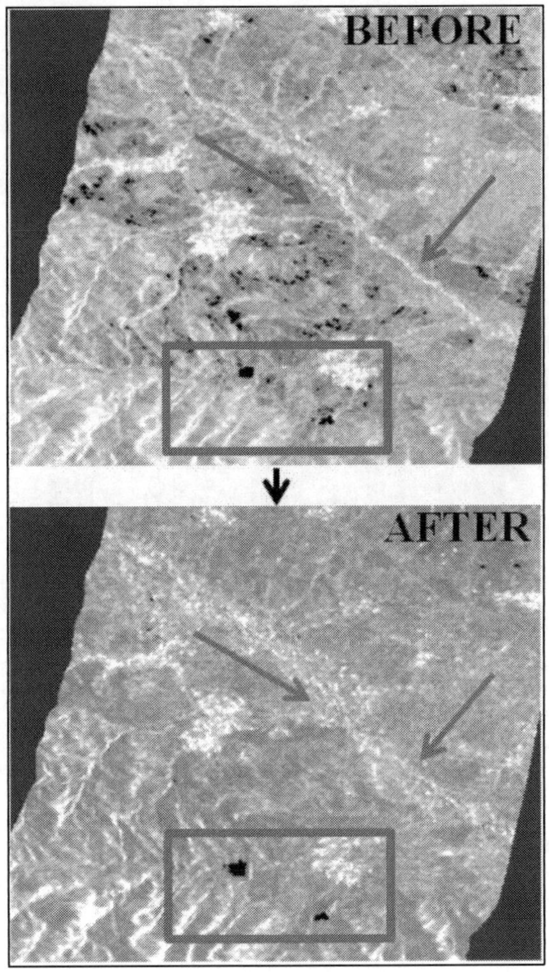

Figure 5. Results of radar data processing. Affected area before the flood (*top*), affected area after the flood (*bottom*).

So changes are observed after the processing, but since the flood mainly consists of mud, this type of treatment does not give the desired results to determine the affected area.

3.2 Sentinel-2 Optical data

Two different approaches were used in processing the optical data. One applied vegetation indices and the other emphasized supervised and unsupervised classification.

Because the same optical images from Sentinel-2 are used in both methods, the initial processing steps are identical. Processing begins with the creation of subsets, since the area of interest covers a significantly smaller area than that in the available images. This operation is applied to all three images using the same coordinate values of the endpoints of the geographic polygon (Figure 6).

Used satellite images are multi-size product since their bands are of different resolutions. The essence of processing begins with resampling the multi-size product to a single-size product. In these cases B2 band is used as reference band since it has the

This contribution has been peer-reviewed.
https://doi.org/10.5194/isprs-archives-XLVIII-M-1-2023-339-2023 | © Author(s) 2023. CC BY 4.0 License.

biggest resolution of 10m. All other nodes will be resampled to match it's size. "Nearest" is the chosen upsampling method where every pixel value in the output product is set to the nearest input pixel value.

Figure 6. Area of Interest for Sentinel-2 images

3.2.1 Vegetation indices

The method for establishing post-flood changes is based on the analysis of the chosen area of interest through differences in terrain reflectivity and the use of different types of vegetation indices. A total of four such indices were used - NDVI, NDMI, MSI and NDWI.

The Normalized Difference Vegetation Index (NDVI) is an indicator that uses the red and near-infrared spectral bands (Equation (1)). It is one of the most well-known and used vegetation indices and it is highly associated with vegetation content. High NDVI values correspond to areas that reflect more in the near-infrared spectrum. Higher reflectance in the near-infrared corresponds to denser and healthier vegetation (GU, 2019).

$$\mathrm{NDVI} = \frac{(\mathrm{NIR-Red})}{(\mathrm{NIR+Red})} = \frac{(\mathrm{B8-B4})}{(\mathrm{B8+B4})}, \tag{1}$$

where NIR = Near-infrared spectral band
Red = Red spectral band
B4 = Red band in Sentinel-2 image
B8 = NIR band in Sentinel-2 image

The post-flood image was subtracted from the pre-flood image to generate a new difference image to serve as a measure of absolute change between captured images (Equation (2)). In the new raster, higher values represent areas of greatest change (high flood severity) and low or negative numbers representing non-flooded areas.

$$\mathrm{dNDVI} = \mathrm{NDVI_{pre}} - \mathrm{NDVI_{post}}, \tag{2}$$

where dNDVI = Differenced Normalized Difference Vegetation Index
$\mathrm{NDVI_{pre}}$ = NDVI applied to a pre-flood image
$\mathrm{NDVI_{post}}$ = NDVI applied to a post-flood image

Figure 7 shows the results of the NDVI analysis. For better understanding, only the damaged/changed areas after the flood are depicted.

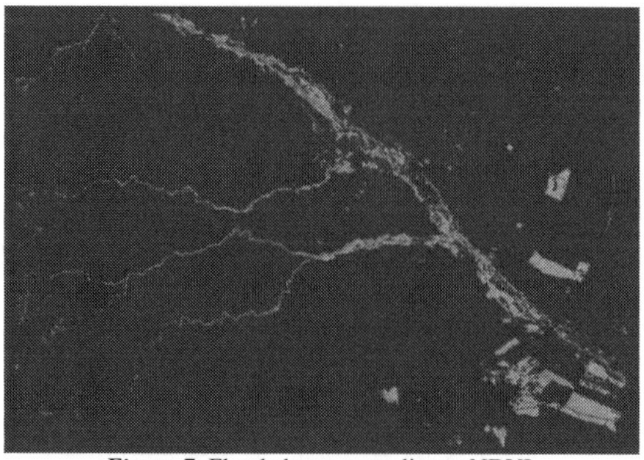

Figure 7. Flooded area according to NDVI

Normalized Difference Moisture Index (NDMI) is used to determine vegetation water content (Equation (3)). It is calculated as a ratio between the NIR and SWIR values (USGS, 2019). Generally higher value of NDMI specifies the presence of more soil moisture under massive aquatic bodies. Vegetation water content and soil moisture content are determined by it

$$\mathrm{NDMI} = \frac{(\mathrm{NIR-SWIR})}{(\mathrm{NIR+SWIR})} = \frac{(\mathrm{B8-B11})}{(\mathrm{B8+11})}, \tag{3}$$

where NIR = Near-infrared spectral band
SWIR = Short-wave infrared spectral band
B8 = NIR band in Sentinel-2 image
B11 = SWIR band in Sentinel-2 image

The same processing method was used after obtaining the results of this index. The post-flood image was subtracted from the pre-flood image to generate a new difference image to serve as a measure of absolute change between captured images (Equation (4)).

$$\mathrm{dNDMI} = \mathrm{NDMI_{pre}} - \mathrm{NDMI_{post}}, \tag{4}$$

where dNDMI = Differenced Normalized Difference Moisture Index
$\mathrm{NDMI_{pre}}$ = NDMI applied to a pre-flood image
$\mathrm{NDMI_{post}}$ = NDMI applied to a post-flood image

Figure 8 presents the results of the NDMI analysis. For better understanding, only the damaged/changed areas after the flood are depicted.

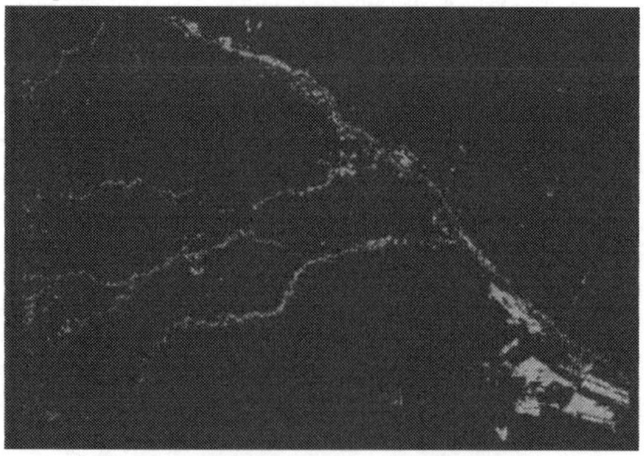

Figure 8. Flooded area according to NDMI

This contribution has been peer-reviewed.
https://doi.org/10.5194/isprs-archives-XLVIII-M-1-2023-339-2023

Moisture Stress Index (MSI) is used for canopy stress analysis, productivity prediction and biophysical modeling (Equation (5)). Interpretation of the MSI is inverted relative to other water vegetation indices. Higher values of the index indicate greater plant water stress and in inference, less soil moisture content (Welikhe et al., 2017).

$$MSI = \frac{SWIR}{NIR} = \frac{B11}{B8}, \quad (5)$$

where NIR = Near-infrared spectral band
SWIR = Short-wave infrared spectral band
B8 = NIR band in Sentinel-2 image
B11 = SWIR band in Sentinel-2 image

Again after obtaining the results of this index the post-flood image was subtracted from the pre-flood image to generate a new difference image to serve as a measure of absolute change between captured images (Equation (6)).

$$dMSI = MSI_{pre} - MSI_{post}, \quad (6)$$

where dMSI = Differenced Moisture Stress Index
MSI_{pre} = MSI applied to a pre-flood image
MSI_{post} = MSI applied to a post-flood image

Figure 9 shows the results of the MSI analysis. For better understanding, only the damaged/changed areas after the flood are depicted.

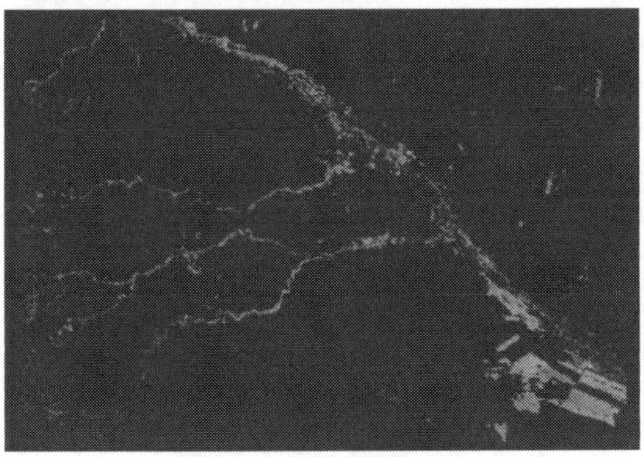

Figure 9. Flooded area according to MSI

Normalize Difference Water Index (NDWI) is used for the water bodies analysis. The index uses Green and Near-infrared bands and can enhance water information efficiently in most cases (Equation (7)). It is sensitive to build-up land and result in over-estimated water bodies (Bahadur, 2018).

$$NDWI = \frac{(Green-NIR)}{(Green+NIR)} = \frac{(B3-B8)}{(B3+B8)}, \quad (7)$$

where NIR = Near-infrared spectral band
Green = Green spectral band
B8 = NIR band in Sentinel-2 image
B3 = Green band in Sentinel-2 image

Again the above mentioned option was used where the post-flood image was subtracted from the pre-flood image to generate a new difference image to serve as a measure of absolute change between captured images (Equation (8)).

$$dNDWI = NDWI_{pre} - NDWI_{post}, \quad (8)$$

where dNDWI = Differenced Normalized Difference Water Index
$NDWI_{pre}$ = NDWI applied to a pre-flood image
$NDWI_{post}$ = NDWI applied to a post-flood image

Figure 10 presents the results of NDWI processing. For better understanding, only the damaged/altered areas after the flood are depicted.

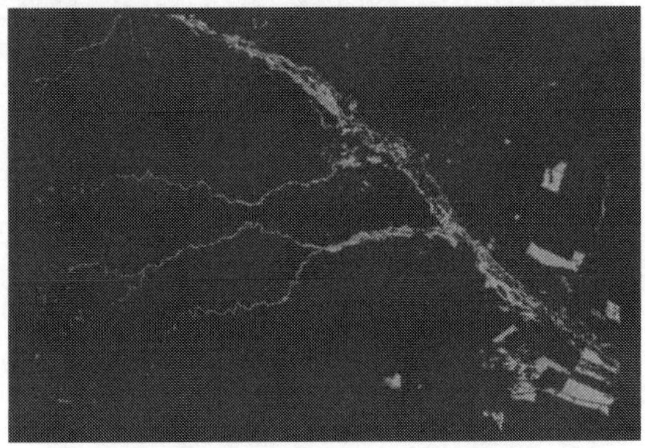

Figure 10. Flooded area according to NDWI

3.2.2 Classification methods

In order to make better use of the information contained in the spectral zones, procedures based on developed algorithms and recognizing spectral homogeneity are used. Classifications of multispectral Sentinel-2 images obtained from the real world are used to identify the individual land cover elements. Through the classifications used, different classes are recognized on the selected image based on its spectral characteristics.

Classifications are divided into two types:
- Unsupervised – in learning mode, they are automatic and do not depend on the knowledge of terrain data and do not require additional information on the assignment of pixels to the reduced classes.
- Supervised – in self-learning mode, they are semi-automatic and more accurate compared to unsupervised classes. They require prior knowledge of the classes for a sufficient number of pixels to be recognized by the analyzer, or a field search to be performed.

For each of these types, one method is applied - K-Means (KM) Cluster Analysis and Maximum Likelihood Classification.

K-Means Cluster Analysis /Unsupervised classification/ is the classification of objects into different groups, or more precisely, the division of a data set into clusters so that the data in each share some common feature - often proximity according to a certain distance measure (Filipova, 2020). Data clustering is a common statistical data analysis technique that is used in many fields, including machine learning, data mining, pattern recognition, image analysis, and bioinformatics. In this particular case, 6 clusters and 30 iterations were selected as parameters of this classification. Figure 11 presents the classification by this method and more information about the obtained results.

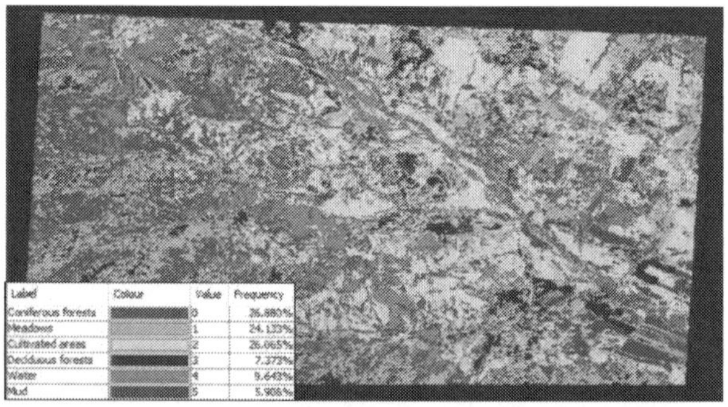

Figure 11. K-Means Cluster Analysis

The **Maximum likelihood classifier** /Supervised classification/ is one of the most popular classification methods, in which a pixel with the maximum likelihood is classified into the corresponding class. The likelihood is defined as the posterior probability of a pixel belonging to a given class (SNAP Help). A total of 6 vector data containers were created, and a different number of standards were added to them, corresponding to different objects of the terrain cover: coniferous forests, deciduous forests, meadows, water, mud and cultivated areas. Figure 12 presents the classification by this method and more information about the obtained results.

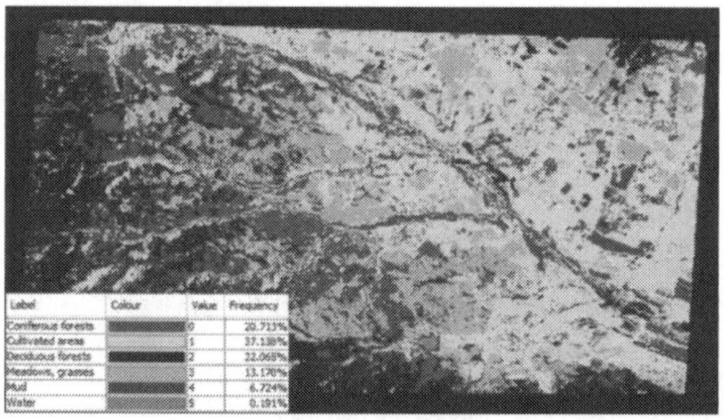

Figure 12. Maximum likelihood classifier

3.2.3 What happens after 4 months?

To understand what happens to the affected territory a few months later, an image from Sentinel-2 from January 8, 2023 was used. For this purpose, only visual interpretation was done. No vegetation indices or classifications were used, as trees and plants are already at a different stage of their phenological cycle and the results would not be comparable, which could lead to misinterpretation of the results.

Figure 13 presents images of the village of Karavelovo before and after the flood, as well as four months after it. The mud deposits that have been brought down from the river can be clearly seen. In the second series of photos, it can be established what the situation is now. It can be seen that almost all the mud is still there and that only a small part of the centre of the village has been cleaned.

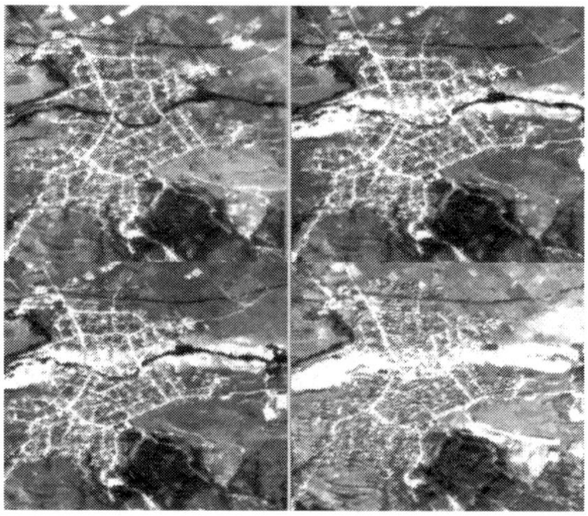

Figure 13. Karavelovo village. Photo from August 18, 2022 before the flood (*top left*); Photo from September 7, 2022 after the flood (*top right and bottom left*); photo from January 8, 2023 - 4 months after the disaster (*bottom right*).

4. RESULTS

Visually, the results of processing the SAR images from Sentinel-1 are presented in Figure 5. All the water bodies /dams/ are well represented, but since the flood mainly consists of mud, the change before and after the disaster cannot be established in good quality and quantity.

The results of the processing based on the vegetation indices are shown in Figures 7 to 10. They visualize the affected flooded areas more clearly. Figure 13 presents a visual comparison of the four indices. It can be seen that the best results are given by NDVI and NDWI, since they have few residual pixels that do not carry qualitative information for interpreting the results. Also, with these indices, the more strongly affected territories are distinguished clearly and densely with well-defined contours.

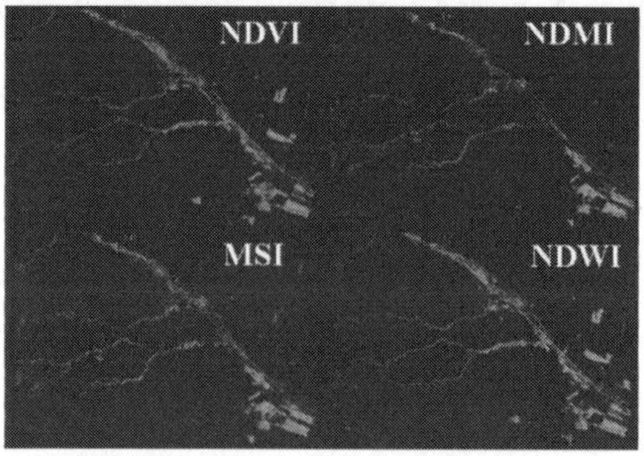

Figure 14. Comparison of flooded areas calculated from vegetation indices

The final processing products of the classification methods are shown in Figures 11 and 12, where it is obvious that the one in the training mode /Maximum likelihood classifier/ gave better results. While processing by the method without training /K-Means Cluster Analysis/ has given too many water areas. The results of the quantitative evaluation by the two optical methods of processing are presented in Table 4.

Method	Name	Flooded area [km^2]
Vegetation indicies	NDVI	8.431
	NDMI	5.336
	MSI	7.446
	NDWI	8.586
Unsupervised classification	K-Means Cluster Analysis	9.888
Supervised classification	Maximum likelihood classifier	11.178

Table 4. Size of the flooded area obtained by the different methods

5. CONCLUSION

The specifics of this case study include free-to-use images from the Copernicus program, the use of different methodologies, a small affected area, an urbanized and forested area, and a flood mainly consisting of mud. Based on the obtained results of the processing by the different methods and subsequent visual interpretation, it was found that for this case the best results are obtained by using two of the vegetation indices Normalize Difference Water Index and Normalized Difference Vegetation Index. With them, the results are best visible, the processing time is short and fast, and an affected area can be easily calculated in a desired projection.
Acceptable results are obtained with supervised classification, but there the assigned classes also have a great influence. Errors can be obtained by setting these benchmarks incorrectly. The worst results are observed when using SAR images, because the affected area is not very large, and in addition, the main part of the flood consists of mud.

According to the Bulgaria's Flood Risk Management Plans, so far the affected areas have not been identified as areas of significant potential high or medium flood risk. The results obtained from the research would support the future analyses related to this risk for the future periods.

Since the post-flood situation is not yet fully under control, the results can easily be integrated into a geographic information system and used by a wide range of specialists. This can facilitate the faster implementation of activities related to the restoration of normal life in populated areas and the reduction of damage after the disaster.

Figure 15 presents the flood affected areas obtained from NDVI results and visualized in Google Earth Pro.

Figure 15. Flooded areas

The timely use of remote and other flood monitoring methods is extremely important. But it would be best if such disasters did not happen at all. Therefore, in order to limit the risk of floods, the Ministry of the Interior of the Republic of Bulgaria advises:

- Each person should become aware of the risk of flooding for the locality in which he lives or works;
- Each person should know about the nearest safe place, the shortest and safest route to evacuate;
- Do not throw construction and household waste into the riverbeds, this reduces the conductivity of the riverbed;
- Do not block river currents for the purpose of taking water for watering or bathing animals, as these barriers multiply the destructive power of water during floods;
- Not to destroy and pierce the dams of the rivers for the purpose of taking water for irrigation or other activities;
- To report to the local municipality, town hall or regional administration, if the riverbed and the spaces under the bridges are not cleaned of waste, construction materials, trees and others;
- Do not build residential and agricultural buildings and other structures in unprotected river floodplains.

REFERENCES

Anusha, N. , Bharathi, B., 2019: Flood detection and flood mapping using multi-temporal synthetic aperture radar and optical data. The Egyptian Journal of Remote Sensing and Space Science, Volume 23, Issue 2, August 2020, Pages 207-219. https://doi.org/10.1016/j.ejrs.2019.01.001.

Bhangale, U., More, S., Shaikh, T., Patil S., More, N., 2020: Analysis of Surface Water Resources Using Sentinel-2 Imagery. Procedia Computer Science 171 (2020) 2645–2654.

Bhattacharya, S., Halder, S., Nag, S., Roy, P., Roy, M., 2021; Advances in Water Resources Management for Sustainable Use, DOI: 10.1007/978-981-33-6412-7_18.

Elhag, M., Bahrawi, J., 2016: Soil Salinity Mapping and Hydrological Drought Indices Assessment in Arid Environments Based on Remote Sensing Techniques. Geosci. Instrum. Method. Data Syst. Discuss., doi:10.5194/gi-2016-39.

Filipova S., Filipov D., Raeva P., SUSTAINABLE MANAGEMENT OF WATER BODIES BASED ON MULTITEMPORAL SATELLITE IMAGES, Proceedings Vol. 1, 8th International Conference on Cartography and GIS, 2020, Nessebar, Bulgaria ISSN: 1314-0604, pp. 693-701

Foroughnia, F.; Alfieri, S.M.; Menenti, M.; Lindenbergh, R. Evaluation of SAR and Optical Data for Flood Delineation Using Supervised and Unsupervised Classification. Remote Sens. 2022, 14, 3718. https://doi.org/10.3390/rs14153718.

Gilabert, M., Gonzalez-Piqueras, J., Martinez, B., 2011: Theory and application of vegetation indices. Remote Sensing Optical Observations of Vegetation Properties, 2010: 1- 43, ISBN: 9/8-87 -308-0421-7.

Güvela, Ş., Akgül, M., Aksu, H., Flood inundation maps using Sentinel-2: a case study in Berdan Plain. Water Supply Vol 22 No 4, 4098 doi: 10.2166/ws.2022.039.

https://giscrack.com/list-of-spectral-indices-for-sentinel-and-landsat/

https://www.mvr.bg/gdpbzn/info-center/pravila-povedenie/pri-navodnenia

https://www.nationalgeographic.com/environment/article/floods

https://pro.arcgis.com/en/pro-app/latest/help/data/imagery/indices-gallery.htm

https://scihub.copernicus.eu/

https://www.usgs.gov/landsat-missions/landsat-normalized-difference-vegetation-index

Kamenova, K., Dimitrov, P. 2020: Evaluation of Sentinel-2 vegetation indices for prediction of LAI, fAPAR and fCover of winter wheat in Bulgaria. European Journal of Remote Sensing. DOI: 10.1080/22797254.2020.1839359.

Kriegler, F.J., Malila, W.A., Nalepka, R.F. and Richardson, W., 1969, Preprocessing transformations and their effect on multispectral recognition, in: Proceedings of the sixth International Symposium on Remote Sensing of Environment, University of Michigan, Ann Arbor, MI, pp. 97-131

Marinova, S., Thematic Mapping and Visualization for Early Warning and Crisis Management, UACEG, Sofia, 2018, pp. 154, ISBN 978-954-724-108-4.

McKenna, P., Phinn, S., Erskine, P., 2018: Fire Severity and Vegetation Recovery on Mine Site Rehabilitation Using WorldView-3 Imagery. Fire 2018, 1, 22; doi:10.3390/fire1020022.

Medinal, J., Blanco, A., Candido, C., 2020: Land cover monitoring of Laguna lake watershed using MODIS NDVI data. The International Archives of the Photogrammetry, Remote Sensing and Spatial Information Sciences, Volume XLII-3/W11.

Meyer, R., Søgaard, M., Schødt, M., Horion, S., Prishchepov, A. 2022: Emergency flood mapping in Australia with Sentinel-1 and Sentinel-2 satellite imagery. BSGLg, 78, 2022, 123-138. https://doi.org/10.25518/0770-7576.6653.

Na Sun, Cailin Li, Baoyun Guo, Xiaokai Sun, Yukai Yao, Yue Wang, 2022: Research status and development trend of photogrammetry and remote sensing in urban flood disasters. ISPRS Annals of the Photogrammetry, Remote Sensing and Spatial Information Sciences, Volume X-3/W1-2022

Nhangumbe, M., Nascetti, A., Ban, Y. 2023: Multi-Temporal Sentinel-1 SAR and Sentinel-2 MSI Data for Flood Mapping and Damage Assessment in Mozambique. ISPRS Int. J. Geo-Inf. 2023, 12, 53. https://doi.org/10.3390/ijgi12020053.

RUS-Copernicus. Training Kit - HAZA01 - Flood monitoring with Sentinel-1 using s-1 toolbox – January 2015, MALAWI.

SNAP Help

Sentinel-1 SAR User Guide

Sentinel-2 User Handbook, ESA. 2015

Tarpanelli, A., Mondini, A., Camici, S., 2022: Effectiveness of Sentinel-1 and Sentinel-2 for flood detection assessment in Europe. Nat. Hazards Earth Syst. Sci., 22, 2473–2489, 2022 https://doi.org/10.5194/nhess-22-2473-2022.

Tavus, B., Kocaman, S., Nefeslioglu, H., Gokceoglu, C., 2020: A fusion approach for flood mapping using SENTINEL-1 and SENTINEL-2 datasets. The International Archives of the Photogrammetry, Remote Sensing and Spatial Information Sciences, Volume XLIII-B3-2020,https://doi.org/10.5194/isprs-archives-XLIII-B3-2020-641-2020.

Tavus, B., Kocaman, S., Nefeslioglu, H.A., Gokceoglu, C., 2019: FLOOD MAPPING USING SENTINEL-1 SAR DATA: A CASE STUDY OF ORDU 8 AUGUST 2018 FLOOD. International Journal of Environment and Geoinformatics 6(3):333-337. http://dx.doi.org/10.30897/ijegeo.666212.

Welikhe, P., Quansah, JE., Fall, S., Elhenney, WMc., 2017; Estimation of Soil Moisture Percentage Using LANDSAT-based Moisture Stress, Index. J Remote Sensing & GIS 6: 200. doi: 10.4172/2469-4134.1000200.

Zhang, M., Chen, F., Liang, D., Tian, B., Yang, A., 2020: Use of Sentinel-1 GRD SAR Images to Delineate Flood Extent in Pakistan. Sustainability 2020, 12(14), 5784; https://doi.org/10.3390/su12145784.

Road Infrastructure Mapping by Using iPhone 14 Pro: An Accuracy Assessment

B. Suleymanoglu[1], R. Tamimi[2], Y. Yilmaz[1], M. Soycan[1], C. Toth[2]

[1]Civil Engineering Faculty, Yildiz Technical University, 34220 Esenler Istanbul, Turkey – (bariss, soycan, yilmazy)@yildiz.edu.tr
[2]SPIN Laboratory, Department of Civil, Environmental and Geodetic Engineering, The Ohio State University,
470 Hitchcock Hall, 2070 Neil Ave., Columbus, OH 43210 – (tamimi.12, toth.2)@osu.edu

KEY WORDS: LiDAR, photogrammetry, point cloud analysis, feature detection, smart devices, GNSS RTK

ABSTRACT:

Vital aspects of transportation networks, such as the extraction of road information and analysis of road conditions, have become increasingly important research topics as they outline the foundation of many applications such as high-precision mapping, infrastructure planning and maintenance, intelligent transportation, or road design analysis. Therefore, regularly obtaining accurate high-density point cloud data of infrastructures supports many transportation-based applications and provides up-to-date information for smart cities or digital twins. Low-cost smartphone platforms equipped with a variety of sensors provide new and powerful data acquisition capabilities that can be exploited in the geospatial field. For example, mobile phones are now capable of collecting valuable data to generate accurate models to support digital reconstruction of infrastructures. These platforms can provide simple and effective data acquisition, while offering useful geospatial data that can be an alternative to traditional measurement techniques. However, the sensor performance with respect to spatial accuracy of point clouds generated in different applications have not yet been fully investigated. Thus, this paper evaluates the feasibility of using the point clouds generated by the built-in camera and LiDAR sensors integrated into iPhone 14 Pro for extracting road-related information. Additionally, the use of the viDoc RTK Rover on the iPhone 14 Pro increases the platform positioning accuracy, consequently improving the georeferencing accuracy of the point clouds. To validate the performance of the point clouds obtained by the iPhone 14 Pro, a reference dataset of the road features was obtained by measuring with a single-point RTK-GNSS receiver, receiving corrections from the Turkish CORS network (TUSAGA-Aktif) which provides two to three centimetres of accuracy. In addition, reference point cloud data over the same area was obtained from different platforms such as Mobile LiDAR and UAS, and the road features were extracted from these dataset and performance validated. The data acquired by the iPhone 14 Pro was processed and evaluated with respect to the reference datasets. The advantages and disadvantages of using iPhone 14 Pro are analysed in detail and the findings are reported.

1. INTRODUCTION

The extraction of road information from high accuracy point clouds has become an important tool in the geospatial field in recent years, thanks to the development of a variety of sensors that provide powerful data acquisition capabilities. Some of these methods include Mobile Mapping Systems on vehicles like automobiles or aircrafts to collect data with sensors to generate point clouds. Point clouds are three-dimensional data sets that consist of many points in space, each of which has a set of coordinates and frequently other attributes such as colour or intensity. Point clouds can be generated using a variety of sensors, including lasers and cameras. The use of the LiDAR (Light Detection and Ranging) sensor significantly improves point cloud creation by capturing more points in areas where camera sensors may struggle to see, as well as the speed at which they can be collected. Using a camera sensor, images can be taken at a set interval and stitched together utilizing photogrammetry to reconstruct the entire site. The accuracy of data collected could be improved with the use of RTK corrections, which would be essential for point clouds generated , as VRS provides the best solution currently(Puente, I., et al. 2013).

Smartphones offer a convenient and accessible means of data acquisition for various mapping applications, including road infrastructure mapping. With the advancements in sensors and technologies, such as the integration of LiDAR and RTK systems, mobile phones can provide accurate and precise data for mapping applications, making them potentially a useful and efficient tool for road infrastructure mapping.
Early testing of the iPhone shows the framework of research that indicated the potential use of mobile phones for collecting data for mapping purposes. (Sirmacek, B. et al. 2014) conducted a study to assess the accuracy of building point clouds generated from iPhone 3GS's images. They used the iPhone's camera sensor to capture images of building facades, and then used specialized software to process the images and generate point clouds. The accuracy of the building point clouds generated from the iPhone images was compared to the reference point clouds obtained using a total station, a precision surveying instrument. The results showed that the building point clouds had an average accuracy of 3-5 cm, with a maximum error of 15 cm. The authors also analysed the distribution of the errors and found that most of the errors were concentrated near the edges of the building facades, where the texture and features were more complex. The accuracy of the point clouds generated from iPhone images was concluded to be sufficient for many mapping applications, and that the technology had the potential to be a useful tool for quickly and efficiently generating building point clouds in the future.

More recent studies with iPhones include a LiDAR sensor. (Luetzenburg. et al. 2021) carried out a series of experiments to evaluate the performance of the iPhone 12 Pro LiDAR for engineering applications. They used the iPhone to measure a range of objects, including trees, buildings, and terrain, and compared the results to those obtained using other LiDAR systems. The authors found that the iPhone was able to provide accurate and precise measurements for these objects, with an average accuracy of 0.5-1.5 cm and an average precision of 0.2-0.5 cm. The iPhone performed particularly well when measuring flat surfaces and objects with clear features, such as buildings and trees. However, it performed less well when measuring objects with complex shapes or rough surfaces, such as rocks and terrain.

This contribution has been peer-reviewed.
https://doi.org/10.5194/isprs-archives-XLVIII-M-1-2023-347-2023

This case study investigates the feasibility of using a mobile phone like the iPhone 14 Pro with its integrated camera and LiDAR sensor, along with a viDoc RTK Rover to capture data to generate a point cloud and assess the accuracy for road infrastructure mapping. LiDAR-based systems have been extensively utilized in road-related applications, including the extraction of geometric characteristics, road surface analysis, traffic sign detection, and inventory mapping, as demonstrated in previous literature (Gargoum and El-Basyouny 2017; Ma et al., 2018;). The present study aimed to compare the advantages and disadvantages of iPhone 14 Pro data with respect to the commonly used mobile LiDAR and unmanned aerial vehicle (UAV) LiDAR in such applications. The study area was previously surveyed and mapped using a GNSS receiver with a Virtual References System (VRS) on the Turkish Continuously Operating Reference System (CORS) network (TUSAGA-Aktif) to generate a reference dataset for result comparison. The purpose of this paper is to evaluate the performance and accuracy of the mobile phone geospatial sensor technology, including performance validation in a controlled environment. This will provide insight into the capabilities of using an iPhone 14 Pro and viDoc RTK Rover for road infrastructure mapping, and how it compares to traditional methods.

2. MATERIALS AND METHODS

2.1 Study Areas, Data Acquisition Systems and Datasets

2.1.1 iPhone 14 Pro + viDoc RTK Rover

The viDoc RTK Rover was an essential component of the mapping process, providing real-time kinematic (RTK) positioning to improve the accuracy of the mapping results with respect to the stock GNSS receiver built in the iPhone. The CORS stations continuously monitor signals from satellites and transmit RTK correction signals to the viDoc RTK Rover, which uses this information to provide centimetre-level accuracy positioning. The viDoc RTK Rover was connected to the iPhone 14 Pro via Bluetooth, allowing the iPhone 14 Pro to receive real-time corrections, and improve the positioning of the system. The iPhone 14 Pro was equipped with a high-resolution camera and LiDAR (light detection and ranging) sensor, which were used to capture imagery and depth data, respectively. The camera captured images of the road infrastructure, while the LiDAR sensor generated a point cloud of the area, providing a comprehensive 3D representation of the object space that allows us to interpret road infrastructure. The Pix4Dcatch app was utilized for image processing and automatic alignment between the images and LiDAR data, improving the accuracy and efficiency of the mapping process. The Pix4Dcatch app was used to process the data collected by the iPhone 14 Pro and RTK Rover to create a 3D reconstruction of the road infrastructure. The app automatically aligned the images and LiDAR and used the RTK data to improve the geolocation accuracy of the mapping results. The Bluetooth connectivity between the iPhone 14 Pro and the RTK Rover allowed for real-time corrections to be received by the iPhone 14 Pro, further improving the accuracy of the mapping results.

2.1.2 UAV-LiDAR

Another survey of the study area was performed using a UAV-LiDAR system. The specific system used in this study was the Phoenix Alpha-32 LiDAR system (see Figure 1). The system consisted of several components, including a Velodyne laser scanner, a high-precision inertial measurement unit (IMU) from OEM-ADIS16488, NovAtel OEM6 dual-frequency GNSS receivers and a microcomputer.

Figure 1. UAV-LiDAR System (Phoneix, 2023)

2.1.3 Mobile LiDAR

Another measurement system used in the study area was a Mobile LiDAR system. The system consisted of several components, including a Velodyne laser scanner, a Ladybug5 camera, two dual-frequency GNSS receivers a high-precision inertial measurement unit (IMU), and an odometer (refer to Figure 2).

Figure 2. Mobile LiDAR System (AnkaGeo, 2023)

2.1.4 Study Areas and Datasets

The experiment took place on the Davutpasa Campus of Yildiz Technical University in Istanbul. The study area was a 90-meter-long and 7-meter-wide corridor with a 5.5% slope, as shown in Figure 3. This corridor is a two-lane road that includes a change in slope both across and along the road, as well as sidewalks.

(a) (b)

Figure 3. Study Area:
3D view of the point cloud (a), 2D image (b)

https://doi.org/10.5194/isprs-archives-XLVIII-M-1-2023-347-2023 | © Author(s) 2023. CC BY 4.0 License.

2.2 Methodology

2.2.1 Point Cloud Generation

The combination of the iPhone 14 Pro and the Pix4Dcatch app allowed for real-time data collection and processing of the LIDAR point cloud data in the field. The Pix4Dcatch app processed the data from the LiDAR sensor to generate a point cloud. To improve the geolocation accuracy of the mapping results, it also utilized data from the viDoc RTK rover received via the VRS network. The LiDAR point cloud data captured by the Pix4Dcatch app on the iPhone 14 Pro provided a preliminary representation of the point cloud, but further processing of the images was required to create a high-resolution point cloud by providing more data captured by the camera sensor. The data collected on the iPhone 14 Pro was then transferred to the Pix4Dmatic software on a computer for further processing. The Pix4Dmatic software utilized the LIDAR point cloud data in combination with the imagery data captured by the iPhone 14 Pro by fusing both point clouds to generate a more detailed and accurate representation of the road infrastructure. The software also employed advanced algorithms and computer vision techniques to further improve the accuracy of the mapping results. Besides using RTK GNSS data, the Pix4Dmatic software incorporated georeferencing data, control points gathered by a GNSS receiver in the field. This data was used to reference the positions in the point cloud that were marked and easily identifiable by the user. Photogrammetric image processing used in Pix4Dmatic allowed for the creation of a highly detailed and accurate 3D representation of the object space. The software automatically aligned the LIDAR point cloud data and imagery data and used the RTK GNSS data and georeferencing data to improve the accuracy of the mapping results. .

2.2.2 Accuracy Assessment of Point Cloud

The absolute accuracy of the point cloud data was assessed by comparing manually measured RTK-GNSS ground control points (GCPs) with digital elevation models (DEMs) created from the point cloud data. Point clouds generated by systems, such as UAV-Lidar, MLS, and iPhone, needs some common reference from to perform comparisons. Here we used a regular DEM for our evaluation and
selected interpolation method of inverse distance weighting (IDW), which is widely used for interpolating irregularly spaced data sets, such as point clouds (Shams et al., 2018; Barbarella et al., 2019). The optimal grid resolution for this study was determined using Equation 1, which was suggested by Hengl (2006) and it can accurately calculate the minimum grid resolution (p) based on the data density. Different DEM surfaces were generated at various grid resolution calculations for each dataset.

$$p = 0.5 * \sqrt{\frac{1}{D}} \qquad (1)$$

where D is the average point density (number of point/dm^2)

The elevation difference was calculated between each RTK-GNSS GCP (Z_{GCP}) and the elevation of the point at the same position (Z_{DEM}) in DEM. Furthermore, the root mean square error (RMSE) and standard deviation (SD were computed by utilizing the vertical discrepancies between the observed RTK-GNSS control points (Z_{GCP}) and the points on the DEM surface at corresponding positions (Smith et al., 2014, Tamminga et al., 2015). These points are independent of the point cloud and DEM generation and spaced throughout the study area. Figure 4. illustrates the study area's as shaded relief DEM, derived from point cloud data along with the spatial distribution of 170 ground control points. The RMSE and SD were calculated as follows:

$$\text{Root-Mean-Square-Error } (RMSE) = \sqrt{\frac{\sum_{i=1}^{n}(Z_{GCP(X_i-Y_i)} - Z_{DEM(X_i-Y_i)})}{n}} \qquad (2)$$

$$\text{Standard Deviation } (SD) = \sqrt{\frac{\sum_{i=1}^{n}(\left(Z_{GCP(X_i-Y_i)} - Z_{DEM(X_i-Y_i)}\right) - \mu)}{n-1}} \qquad (3)$$

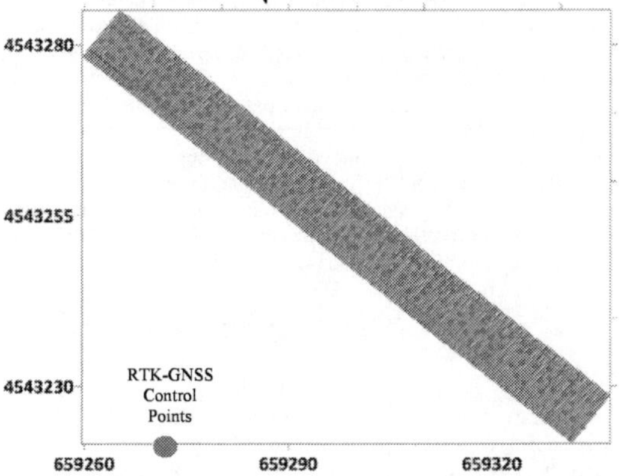

Figure 4. Example of shaded relief map of DEM where red dots shows distribution of RTK-GNSS ground control points

2.2.3 Trajectory based Partitioning and Road Edge Detection

To reduce processing time and increase the accuracy of the road infrastructure assessment, the images captured by the iPhone 14 Pro's camera sensor position at each measurement point during the test run was used as trajectory data. The point cloud data was partitioned perpendicular to the trajectory data from the iPhone using a predefined length. Subsequently, numerous cross sections were formed from each data block with a predefined width. Each cross section thus contains data belonging to the road surface, curbs, sidewalks, and various objects including trees, traffic lights, and vehicles. As seen in Figure 5(a), the study area is divided into rectangular blocks, and cross-sections extracted from each block are projected perpendicular to the plane, as shown in Figure 5(b). The red dots indicate the trajectory data for iPhone 14 Pro camera sensor. Since all three datasets were obtained from the same area, the same trajectory data was also used to divide the MLS and UAV-LiDAR data into blocks.

Based on the extracted cross-sections from each data block, curb points were determined from the raw point cloud data using our proposed methodology, which is based on the slope and elevation difference parameters. In general, road surfaces consist of smooth surfaces with low slope and height differences, and there are serious variations in slope and height in the transition to sidewalks. However, as there are points belonging to different objects (such as trees and streetlights) in each section, these points must first be identified and removed from the dataset. Therefore, the detection and filtering of these points were performed using the elevation difference in each cross-section. For this, the elevation difference between points with maximum and minimum height was calculated and compared with the height of the curb in the study area. The point is classified as an object point if its elevation differences are higher than the curb

https://doi.org/10.5194/isprs-archives-XLVIII-M-1-2023-347-2023 | © Author(s) 2023. CC BY 4.0 License.

elevation. This iterative process continued until there were no more points than the curb height in each cross-section.

$$\Delta Z_i = \max(Z_i) - \min(Z_i)$$
$$\text{if } Z_{curb} < \Delta Z_i \rightarrow \text{object points}$$
$$\text{otherwise } p_i \text{non-object points} \quad (4)$$

Where, ΔZ_i maximum elevation difference, curb height, which is 20 cm in the study area. After detecting the object points, our slope-based methodology was applied to detect curb points. Thus, the search algorithm was implemented from the scan centre to two different directions in each cross-section, and slope values were calculated between consecutive points using the formulas in equation 5. As stated above, the elevation of the pavement points is higher than the road surfaces, and in general, there are abrupt slope changes in the transition from road surfaces to sidewalks. Accordingly, the curb points were determined by comparing the calculated slope values with the threshold value.

$$Sp_i^{i+1} = \frac{Z_{i+1}-Z_i}{\sqrt{(X_{i+1}-X_i)^2-(Y_{i+1}-Y_i)^2}} \quad (5)$$

Where, $(X_{i+1,i}, Y_{i+1,i}, Z_{i+1,i})$ express the coordinate values of consecutive points, Sp_i^{i+1} denote calculated slope values. If the calculated slope value is greater than the predetermined slope threshold value ($S_{threshold}$), it is classified as a curb point.

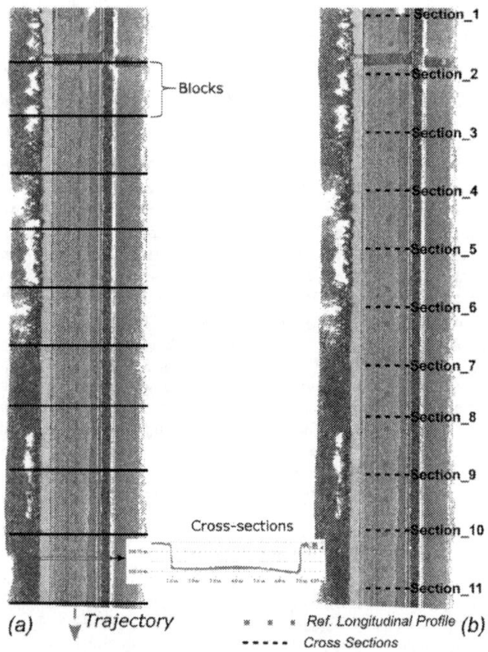

Figure 5, a) Splitting the study area into rectangular blocks and forming cross-sections, b) red dots show reference longitudinal profile points in the study area, black dashed lines indicate reference cross-sections.

Following the detection of candidate curb points, the road boundaries were segmented into left and right road boundaries based on trajectory information. Subsequently, falsely detected curb points were identified and eliminated through the implementation of the RANSAC algorithm (Fischler and Bolles, 1981). This methodology enabled the removal of erroneous data points and improved the accuracy of the curb detection process. Consequently, the boundaries of the road were extracted using detected curb points with the polynomial curve fitting method.

2.2.4 Extraction and Comparison of Road Geometry Parameters (road width, road centerline, Longitudinal Profile and if possible cross-slope)

The data obtained from the previous step, which involved determining the road surface and boundaries, was utilized to determine the geometric parameters of the road. Thus, many important geometric parameters such as road width, elevation, and slope for longitudinal and cross-sectional profiles of the road surface were calculated quickly and accurately together with the road centreline.

The methodology proposed by Yadav et al. (2018) was employed to determine the road width and road centreline. Firstly, refined curb points identified in the previous section were selected and segmented by sequentially aligning them along the road boundary. Next, a polynomial line was fitted to the points within this segment, and the slope values (s_i) of the line were calculated. The slope of the perpendicular line (s_j) with respect to this line was determined using equation 6. Lastly, the points on the other side of the road boundary were substituted into equation7, and the points with the minimum residuals were chosen (Yadav et al., 2018).

$$s_i * s_j = -1 \quad (6)$$
$$y - y_1 = s_j * (x - x_1) \quad (7)$$

As mentioned earlier, one of the significant geometric parameters that were obtained is the longitudinal profile information. To obtain this information, the planimetric coordinates and height values of the reference longitudinal profile points were measured using the RTK-GNSS technique, with intervals of approximately 8 meters on average as seen in Figure 5(b). To compare the reference longitudinal data with point clouds data obtained by MLS, UAV-LiDAR and iPhone-viDoc RTK, the points closest to the reference longitudinal points within these point clouds were determined using a k-nearest neighbour search (k is selected as 1).

The following step involved identifying eleven transverse sections along the road line, as shown in Figure 5(b), and measuring and calculating the cross-slope values of these sections. To evaluate performance accuracy, the cross-slope values derived from the point clouds were compared to the reference cross-slope values of the identified sections. The reference cross-slope values were calculated by measuring the pavement points or lane lines using the RTK-GNSS technique providing 2-3 centimetre accuracy (Badescu et al.,2011; Garrido et al.,2011).. The cross-slope value was obtained by applying linear regression to the extracted data of the point clouds within specified cross-sections, and the cross-slope values were determined as the slope of the regression line. All the procedures were implemented in the MATLAB environment.

3. RESULTS

3.1 Accuracy Assessment of Point Clouds

Absolute accuracy was evaluated based on manually measured RTK-GNSS measurements. The calculated error statistics for DEM surfaces with respect to each RTK-GNSS point are given in table 1. As seen in the table, all statistical values are below the centimetre level. Upon analysing the error statistics of both systems, it was determined that the point cloud data obtained from all systems exhibited minimal discrepancies. Furthermore, the results indicated that the vertical accuracy of all point cloud data was notably high, and the datasets were georeferenced with satisfactory precision at the centimetre scale, which was deemed suitable for the purpose of this study.

	Min (m)	Max (m)	STN (m)	RMSE (m)
UAV-LiDAR	-0.03	0.10	0.02	0.03
MLS	-0.05	0.10	0.03	0.03
iPhone-viDoc	-0.06	0.01	0.02	0.04

Table 1. Error statistics for point cloud data

Figure 6 presents a comparative analysis of longitudinal profiles between the reference data and the data derived from UAV-LiDAR, MLS, and iPhone-viDoc. The dashed black line, blue circle, and orange crosshair indicate the height differences between the reference longitudinal profiles and those derived from UAV-LiDAR, MLS, and iPhone-viDoc, respectively. Statistical values for these height differences are given in Table 2. Upon examining the height differences obtained from all three systems, deviations were observed in certain areas. However, further investigation revealed that the error values remained within the centimetre level, indicating highly consistent results across all three systems. In addition, the variances of the height differences are slightly higher with MLS and UAV-LiDAR, compared to the iPhone. As a result, the georeferencing of iPhone data may exhibit greater consistency in this regard.

Reference Profiles Comparison	Min (m)	Max (m)	STN (m)	RMSE (m)
UAV-LiDAR	-0.03	0.02	0.01	0.01
MLS	-0.04	0.02	0.02	0.02
iPhone-viDoc	-0.02	-0.01	0.01	0.01

Table 2. Error statistics values of longitudinal profile differences

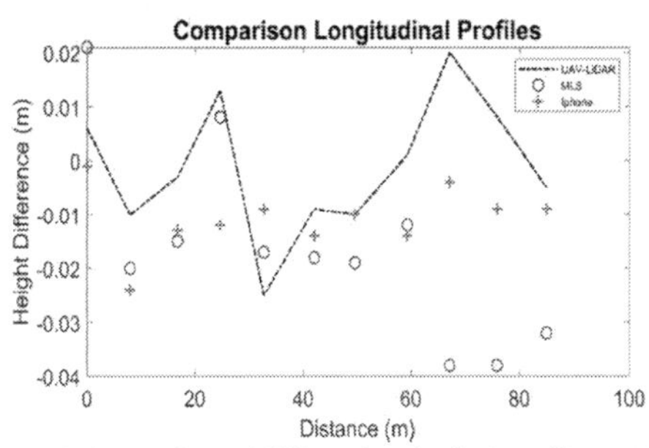

Figure 6. Comparison of different longitudinal profiles: UAV-LiDAR, MLS, and iPhone

The result of our road boundary extraction is depicted as a set of red points and a vectorized road boundary represented by a green line. These representations have been overlaid onto the point clouds obtained from unmanned aerial vehicles (UAVs), mobile laser scanning (MLS), and iPhone sources, respectively, as presented in Figure 7.

We visually inspected several parts of typical road scenes that have been enlarged using our method. The road boundaries are generally extracted correctly, but there may be instances where our method fails to draw boundaries due to the nature of the point cloud data or because of differences in pavement structures. Thus, as illustrated in Figure 7 (c, d), it has been determined that road boundaries are unable to be extracted due to insufficient data on the curb points using the UAV LiDAR system in those road sections. To alleviate the limitations of this system, several scans should be acquired from different viewpoints to generate comprehensive point clouds of a scene. As shown in Figure 7 (g-k-l), deviations from the curb boundary were observed in certain sections due to the incorrect detection of some points as pavement points from MLS and iPhone-based point clouds. Both systems have the ability to acquire a more comprehensive and complete representation of the road surface, primarily due to sensors scanning angles and positions. This feature ensures that the proposed methodology is capable of obtaining appropriate data. However, it should be noted that the scanning capacity of the iPhone is restricted to shorter routes, limiting its application in certain scenarios.

In addition, the average road width was calculated from the obtained refined boundary points. The calculated road width values were compared with field measurements to determine any deviations from the reference values. The results indicated that the road width values obtained from MLS, UAV-LiDAR, and iPhone deviated from the reference values by 2.7%, 0.5%, and 3.4%, respectively. Additionally, The UAV-LiDAR performed better than the MLS and the iPhone because, as shown in figure 7, both systems had faulty boundaries detected in some parts of the road.

In the Table 3, reference cross-slope values were determined using RTK-GNSS data and subsequently compared against values obtained from UAV-LiDAR, MLS, and iPhone. The cross-slope difference between reference and UAV-LiDAR ranged from 0.01% to 0.65%, with an average difference of 0.25%. Similarly, the difference between reference and MLS cross-slope values ranged from 0.03% to 0.44%, with an average difference of 0.20%. For the iPhone, the cross-slope difference ranged from 0.02% to 0.69%, with an average difference of 0.19%. The acceptable accuracy for cross-slope differences, as defined by relevant literature and technical guidelines, is generally considered to be 0.2% (Shams et al.,2018; Hunt et al., 2013). Based on the obtained results, it can be concluded that the average difference values obtained by all three methods were close to this acceptable limit. Therefore, the systems evaluated in this study show promising results for cross-slope evaluation.

This contribution has been peer-reviewed.
https://doi.org/10.5194/isprs-archives-XLVIII-M-1-2023-347-2023

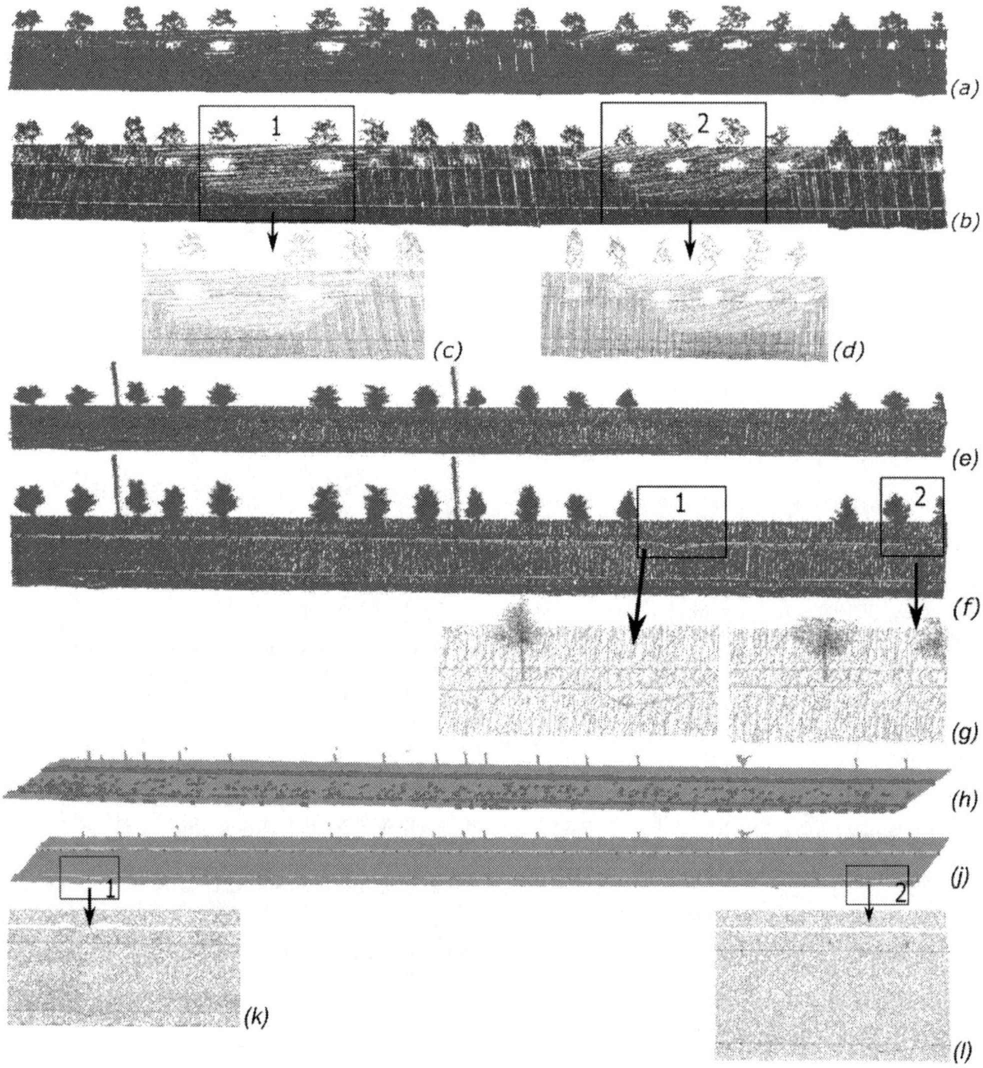

Figure 7: Road boundary extraction results for UAV-LiDAR (a-b-c-d), MLS (e-f-g), and iPhone (h-j-k-l). (a-e-h) shows the initially detected curb points, (b-f-j) shows the refined curb points (in red), and the vectorized road boundary is shown in green. Enlarged images in (c-d-g-k-l) show missing road parts and false curb boundaries.

4. DISCUSSION AND CONCLUSIONS

The study aimed to evaluate the accuracy of road infrastructure mapping and cross-slope evaluation using an iPhone 14 Pro with the viDoc RTK Rover. The absolute accuracy of the point cloud data was evaluated based on manual measurements taken from a GNSS receiver. The iPhone-viDoc point cloud was also compared to MLS and UAV-LiDAR data obtained in the same study area. The results indicate that the vertical accuracy of the iPhone-viDoc point cloud data was noticeably higher, with error values around the centimetre level. The dataset was georeferenced with satisfactory precision at the centimetre scale. This suggests that all three methods can be used for road boundary extraction and cross-slope evaluation with a high level of accuracy.

The road boundary extraction results for all three systems indicated that road boundaries were generally extracted correctly. However, there were instances where the method failed to draw boundaries due to the nature of the point cloud data or differences in pavement structures. Deviations from the curb boundary were also observed in certain sections due to the incorrect detection of some points as pavement points from MLS and iPhone-based point clouds. The longitudinal profiles derived from all three systems showed highly consistent results with deviation values within the centimetre level.

This contribution has been peer-reviewed.
https://doi.org/10.5194/isprs-archives-XLVIII-M-1-2023-347-2023 | © Author(s) 2023. CC BY 4.0 License.

		Differences at Ground Truth		
Sec.	**Ground Truth (%)**	**UAV LiDAR (%)**	**MLS (%)**	**iPhone (%)**
1	0.74	0.04	0.34	0.69
2	0.02	0.10	0.12	0.17
3	0.29	0.58	0.29	0.22
4	0.27	0.65	0.27	0.02
5	0.31	0.19	0.13	0.09
6	0.23	0.26	0.03	0.12
7	0.40	0.17	0.09	0.00
8	1.01	0.31	0.44	0.28
9	0.85	0.16	0.33	0.19
10	0.04	0.01	0.10	0.04
11	0.25	0.25	0.10	0.30
Mean		**0.25**	**0.20**	**0.19**

Table 3. Cross-slope comparison between ground truth and UAV-LiDAR, MLS, and iPhone

The cross-slope evaluation results indicate that the average difference values obtained by all three methods were close to the acceptable limit of 0.2%, including the data obtained by the iPhone 14 Pro with the viDoc RTK Rover.

In conclusion, the study demonstrates the potential of using an iPhone 14 Pro with the viDoc RTK Rover for road infrastructure mapping and cross-slope evaluation with high accuracy. The results also show that the iPhone-viDoc point cloud data can be used interchangeably with MLS and UAV-LiDAR data for these purposes. To further improve the experiments, future studies could consider using PPK on the viDoc RAW data to achieve even higher accuracy in georeferencing. Additionally, incorporating machine learning techniques could help improve the road boundary extraction results by automatically identifying and removing pavement points from the point cloud data. The study provides valuable insights for the use of mobile mapping systems in road infrastructure mapping and evaluation.

ACKNOWLEDGEMENT

We would like to express our gratitude to Pix4D for generously supplying us with a viDoc RTK rover, which was instrumental in improving the accuracy of our mapping results. We would also like to thank AnkaGeo for providing us with mobile LiDAR data that was crucial to the success of our project. Without the support and contributions of both companies, this project would not have been possible.

REFERENCES

Shams A., Sarasua W.A., Famili A., Davis W.J., Ogle J.H., Cassule L., Mammadrahimli A. Highway Cross-Slope Measurement using Mobile LiDAR. *Transp. Res. Rec.*, Vol.2(02), 2018, p.p 175-199.

Barbarella M., De Blasiis M.R., Fiani M. Terrestrial laser scanner for the analysis of airport pavement geometry. Int. J. Pavement Eng., Vol.20, No.4, 2019, p.p 466-480.

Hengl T. Finding the right pixel size. Comput. Geosci. Vol.32, No.9, 2006, p.p 1283-1298

Tamminga A.D., Eaton B.C., Hugenholtz C.H. UAS-based remote sensing of fluvial change following an extreme flood event. Earth Surf. Process. Landforms Vol.40, No.11, 2015, p.p 1464-1476.

Smith M.W., Carrivick J.L., Hooke J., Kirkby M.J. Reconstructing flash flood magnitudes using "Structure-from-Motion": A rapid assessment tool. J. Hydrol. Vol.519, 2014, p.p 1914-1927

Fischler, M.A.; Bolles, R.C. Random sample consensus: A paradigm for model fitting with applications to image analysis and automated cartography. Commun. ACM 1981, 24, 381–395.

Yadav, M., Singh, A. K., & Lohani, B. (2018). Computation of road geometry parameters using mobile LiDAR system. Remote Sensing Applications: Society and Environment, 10, 18-23.

Shams A., Sarasua W.A., Famili A., Davis W.J., Ogle J.H., Cassule L., Mammadrahimli A. Highway Cross-Slope Measurement using Mobile LiDAR. *Transp. Res. Rec.*, Vol.2(02), 2018, p.p 175-199.

Hunt J. E., Vandervalk A., and Snyder D. SHRP2 Report S2-S03-RW-01: Roadway Measurement System Evaluation. Transportation Research Board of the National Academies, Washington, D.C., 2013

Phoneix LiDAR Systems. https://www.phoenixlidar.com/wp-content/uploads/2019/11/PLS-AL3-32-Spec-Sheet_180406.pdf, Accessed Time 16/02/2023

Ankageo Geopgrahic Information Technologies. https://ankageo.com/urunler/ Accessed Time 16/02/2023

Badescu, G., Stefan, O., Badescu, R., Ortelecan, M. & Veres, S. I. (2011). Positioning System GPS and RTK VRS Type, Using the Internet as a Base, a Network of Multiple Stations. FIG Working Week 2011, TS03G - GNSS CORS Networks Case Studies (Flash), 5247 / Marrakech, Morocco, 18-22 May 2011

Garrido, M. S., Gimenez, E., Lacy, M. C. & Gil, A. J. (2011). Testing precise positioning using RTK and NRTK corrections provided by MAC and VRS approaches in SE Spain. Journal of Spatial Science, 56(2), 169-184.

Gargoum, S., & El-Basyouny, K. (2017, August). Automated extraction of road features using LiDAR data: A review of LiDAR applications in transportation. In 2017 4th International Conference on Transportation Information and Safety (ICTIS) (pp. 563-574). IEEE.

Ma, L., Li, Y., Li, J., Wang, C., Wang, R., & Chapman, M. A. (2018). Mobile laser scanned point-clouds for road object detection and extraction: A review. Remote Sensing, 10(10), 1531.

MONITORING THE ENVIRONMENT IN SMART CITIES: THE IMPORTANCE OF GEOSPATIAL LOCATION REFERENCING

B. Suleymanoglu[1*], C. Toth[2], A. Masiero[3], A. Ladai[4]

[1]Civil Engineering Faculty, Department of Geomatics Engineering,
Yildiz Technical University, 34220 Esenler Istanbul, Turkey – bariss@yildiz.edu.tr
[2]Department of Civil, Environmental and Geodetic Engineering, The Ohio State University,
470 Hitchcock Hall, 2070 Neil Ave., Columbus, OH 43210 - toth.2@osu.edu
[3]Department of Civil and Environmental Engineering (DICEA), University of Florence,
50139 Florence, Italy - andrea.masiero@unifi.it
[4]Department of Manufacturing and Civil Engineering, Norwegian University of Science and Technology (NTNU) in Gjøvik,
Teknologiveien 22, 2815 Gjøvik, Norway - andras.ladai@ntnu.no

KEY WORDS: Georeferencing, Collaborative Navigation, Mobility, Smart City

ABSTRACT:
The Smart City concept is taking momentum recently as big metropolises as well as mid-size cities are intensifying their efforts to improve the life of people living in dense urban environment. Local governments are eager to have up-to-date information of every aspect of city life, including environmental data, such as air and water quality parameters; mobility data, such as traffic flow, including vehicles, transit passengers; crowd control, such as public events, mobility in hospitals; life quality data, such as social status, education level, health records; etc. Monitoring all these very different data streams in space and time is a formidable challenge. While on the data acquisition side, tremendous progress has been achieved, as sensors have been deployed in increasingly large numbers on both mobile and static platforms, there is a lack of creating accurate geotags, as the quality of georeferencing varies over a large scale. It is important to note that the data acquisition is becoming largely customer-based, as smart devices are efficient sensor systems and with advancing communication technologies, crowdsourcing is quickly becoming the dominant data source on mobile platforms. In this paper, we investigate the potential to exploit the ranging capabilities of imaging and communication sensors and use the strength of the spatial network formed by the sensors to improve the georeferencing of a group of platforms operating in a close environment, such as UAS swarm or a platoon of autonomous vehicles. Transportation in cities and in general mobility are of great interest to Smart Cities, they represent one of the most significant components of the activities, so having an optimized transportation system is essential to reduce carbon footprint, decrease commute time, and just improve the quality of life. To assess the performance of collaborative navigation based accurate georeferencing, data was acquired at a simulated intersection area at The Ohio State University, where multiple vehicles, pedestrians and cyclists were moving around. In addition, drones were flying above the area. Here we report about our initial results.

1. INTRODUCTION

Smart Cities can be defined in various ways but what makes these cities different from the others is practically the level of monitoring and data sharing capabilities (Ahvenniemi *et al.*, 2017). Sensors continuously acquire data from the built-up environment as well as from all the dynamic objects, such as people and all kind of moving platforms. Digital twins may provide a virtual representation of the city environment at a variety of levels, including buildings, vehicles, man-made objects, and then human beings, vegetation, birds, etc. These systems are usually live, i.e., their content is continuously updated as new data comes in. Besides monitoring the state of the city, they provide an optimal way to control and manage every aspect of the city operations; for example, visualization, traffic control, decision making, planning, etc.

Location information, including, at minimum, coordinates and time is a very fundamental metadat of any piece of information acquired and processed in a Smart City environment. Most of the data currently come from dedicated sensor networks, but crowdsourced data is rapidly increasing and expected to become the main source of information (Toth and Jozkow, 2015). Typical smartphone has about 30 sensors and there are about 7 billion users worldwide. These numbers dwarf the industrial sensor market, which is estimated to be about $25 billion worldwide by 2023.

The intricate and multifaceted nature of urban environments poses challenges to attaining a precise, stable, and robust localization. The uncertainty and lack of GNSS in cities makes positioning especially hard, and thus, there is a necessity for new methodologies that integrate different sensors and data sources to improve the robustness of the localization processes; mainly, to assist mobility efforts (Benevolo *et al.*, 2016). Mobility is an essential element of Smart Cities, the main challenge is the georeferencing of moving objects, such as people, vehicles, UAS, UAV, etc. There are many technologies to achieve this as well as several methods. Since in most situation the mobile platforms closely share the object space, their navigation systems can collaborate by sharing location and benefit from it. Knowing the location in real-time is critical for optimizing the flow of people, vehicles, goods, etc.

In previous works, we have directed efforts towards collaborative localization techniques that incorporate Ultra-Wideband (UWB) technology (Masiero *et al.*, 2022), vision and LiDAR-based positioning (Masiero *et al.*, 2021b, Hosseinyalamdary *et al.*, 2015) as well as crowdsourced vehicle data (Toth *et al.*, 2018) in combination with GNSS/IMU sensors to explore various scenarios and environments, including those where GNSS signals are challenged.

https://doi.org/10.5194/isprs-archives-XLVIII-M-1-2023-355-2023 | © Author(s) 2023. CC BY 4.0 License.

This study presents a brief overview of the test scenarios and experimental sensor configurations employed during a data collection campaign executed at OSU West Campus, including preliminary findings. In contrast to an earlier effort, which considered a 2D road environment (Retscher *et al.*, 2020), the navigation space was extended to 3D by including four UAS during the experiments. Thus, the focus here is on jointly positioning a group of vehicles, swarms of robots, such as UAS and UGV, etc., in urban environment.

2. COLLABORATIVE NAVIGATION

The basic concept of collaborative navigation is that platforms navigating in close proximity can share their sensor data and thus a joint navigation solution for all the platforms should be better than the individual platform solutions. A key condition is that platforms should have inter-platform range/distance measurements, which allow to geometrically constrain the platform locations. In simple terms, for example, good GPS/GNSS position information can be propagated to other platforms which may have compromised GPS/GNSS. Fig. 1 shows a simple four-car arrangement, where assuming 2D positioning, knowing all the six inter-vehicle ranges, positioning of all platforms is feasible if at least two platforms have good GPS/GNSS positions. Obviously, the spatial relationship conditions the collaborative solutions. For example, if the four cars are in a line, then there is no observation in the lateral direction, so only compromised solutions can be obtained. Similarly, if only one platform has GPS/GNSS data, then relative positioning is possible, but the heading is undefined for all the vehicles; unless, aiding from other sources is available.

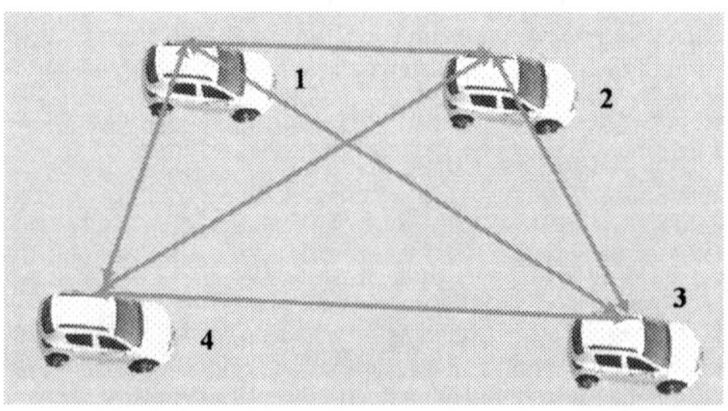

Figure 1. Collaborative navigation concept

The concept of collaborative navigation is not really new, but lack of affordable technologies prevented implementations until recently. Early works focused on integrating the inter-nodal range/bearing measurements or locally generated maps (Roumeliotis and Beke, 2002; Bryson and Sukkarieh, 2009; Grejner-Brzezinska et al., 2009). Also, a similar concept, community relative navigation, was studied before GPS was introduced, and can be traced back at least to the 1970s (e.g., Rome and Stambaugh, 1977; Widnall and Gobbini, 1982; Schneider, 1985).

Recent developments in communication and sensors have made the implementation of collaborative navigation solutions affordable as well as attractive. In the RF domain, for example, inter-platform range measurements can be obtained in a wireless sensor network through RSS (Received Signal Strength), TOA (Time of Arrival), TDOA (Time Difference of Arrival), and AOA (Angle of Arrival) techniques. Dedicated technologies, such as Ultra-Wideband (UWB) ranging, using a broad bandwidth and thus providing for time transfer, communication and centimeter level ranging capability, is of particular interest to positioning and navigation applications (MacGougan et al., 2009). In addition, the UWB transceivers can self-localize and form an ad-hoc network of beacons (Sluis and toth, 2021; Ladai and Toth, 2022). Furthermore, optical sensors, such as cameras and LiDAR can also provide range and AOA information that can be used alone or combined with RF sensor-based observations (Masiero *et al.*, 2021a). Finally, vehicle motion modeling and IMU data can further improve the efficiency of collaborative navigation implementations.

When clear LOS (Line of Sight) is available, obtaining range and angular data (AOA) is generally feasible. But when obstacles exist in the space where the platforms are deployed, there may be not enough information to estimate the network. Therefore, we selected a generally open area with a few buildings where reference points could be easily established and surveyed at high accuracy.

3. DATA ACQUISITION

A massive field test, including multiple ground and UAS platforms was organized and executed on OSU West Campus, May 8-16, 2022. The main objective was to simultaneously acquire a rich dataset from multiple transportation modalities that can support research on collaborative navigation. The ground platforms included four vehicles, two cyclists and two pedestrians, and four simultaneously flying UAS in the air. There was a ground control network established and UWB sensors were deployed. The overall effort included more than 22 people and the data collection lasted over eight days. Fig. 2 shows a session when all platforms were utilized; note that the bird's eye view image was taken from the UAS platform flying at the lowest altitude.

Figure 2. Test are with ground platforms

The data acquisition campaign included ground control deployment and surveying, ground vehicle data collection sessions, UAS rehearsals, and combined ground/air data acquisitions as well as a few repeat tests in each category. A key aspect of the campaign was to obtain highly accurate platform trajectory data, which can be used as reference for performance evaluation as well as for various simulation scenarios. The ground platforms and drones primarily moved/flew predefined trajectories and rich sensor data was collected. There was no intention to run collaborative navigation software in real-time due to lack of communication capabilities, and thus, everything was logged for postprocessing.

A remote parking lot of OSU West Campus was selected for all the experiments as it was practically unused during the tests and allowed to fly UAS without significant restrictions. Another

This contribution has been peer-reviewed.
https://doi.org/10.5194/isprs-archives-XLVIII-M-1-2023-355-2023 | © Author(s) 2023. CC BY 4.0 License.

consideration was that the selected location was in an open-sky and RF friendly area to assure that as accurate GPS/GNSS as possible was acquired to obtain high-quality refence solutions. A simulated intersection was defined by ground control, as this represents one of the most important traffic situations in Smart Cities.

From the 50 m by 50 m area used for the main tests, the central area of 30 m by 30 m was populated by 13 signalized ground targets. During the data acquisition sessions, tripods were installed over the target locations and used to mount the UWB transmitters. A total of 12 UWB units formed the ground network that was used to provide local positioning capabilities; note that it was totally independent from GNSS. Fig. 3 shows the ground control layout in the area; note that the inter-node ranges are marked by green. Fig. 4 shows a ground control point and an UWB transceiver mounted on a tripod.

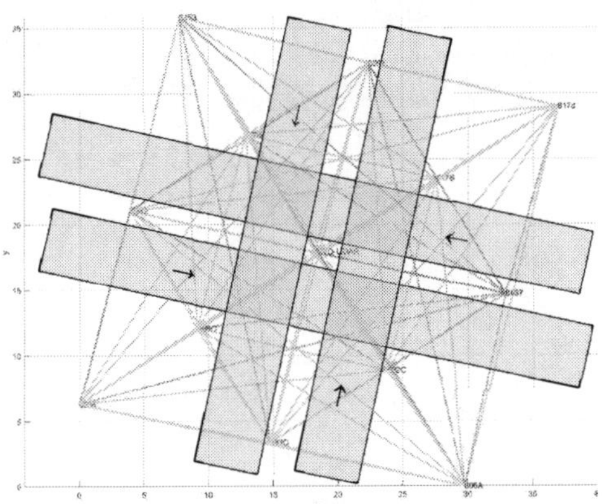

Figure 3. Ground control network deployed at the simulated intersection area

Figure 4. Ground target to support image-based georeferencing and UWB transceiver installed on tripod area

4. VALIDATION OF GPS/GNSS PNT SOLUTIONS

Creating a reference navigation solution is essential for any subsequent investigation, such as validating other sensor data-based solutions, introducing GPS/GNSS gaps to assess the impact on integrated navigation solutions, simulating spoofing (Wang *et al.*, 2022), introducing hardware malfunctioning, etc. In our tests, most platforms carried geodetic-grade GNSS receivers, which provided excellent data in the open-sky environment. The Inertial Explorer was primarily used for reference solution creation, though RtkLib as well as in-house tools were also utilized for validation. The Inertial Explorer provides an internal estimate of the precision, which is generally viewed as somewhat optimistic, yet it is a good relative measure of performance. Table 1 lists the results for all the platforms where GNSS data was available. As expected, the component values are in 1-2 cm range.

Table 1. Precision of post-processed GNSS/INS and GNSS solutions

Platform	Solution Type	Position [m]		
		East	North	Height
GPSVan	GNSS/INS	0.001	0.001	0.003
Pilot	GNSS/INS	0.001	0.001	0.003
CRV	GNSS	0.007	0.006	0.013
CyberCar	GNSS	0.007	0.006	0.013
Cyclist 1	GNSS	0.007	0.006	0.012
Cyclist 2	GNSS	0.007	0.006	0.013
Pedestrian 1	GNSS	-	-	-
Pedestrian 2	GNSS	-	-	-
UAS1	GNSS	0.007	0.006	0.013
UAS2	GNSS	0.007	0.007	0.014
UAS3	GNSS	0.007	0.007	0.014
UAS4	-	-	-	-

For two vehicles, there were high-end IMU data streams available, so attitude data was also obtained, 5-10 arcsec and 20-30 arcsec range for pitch/roll and heading, respectively. While it is not independently validated, they provide a realistic reference for performance validation; consider the UWB ranging accuracy to be in the 10-30 cm range in open areas (clear LOS).

Reference trajectories computed for ground vehicles are shown in Fig. 5. Note the different patterns, such as a car, cyclist and then a pedestrian all make turns at different curvature, and obviously, the speed varies too; though not shown in the figures.

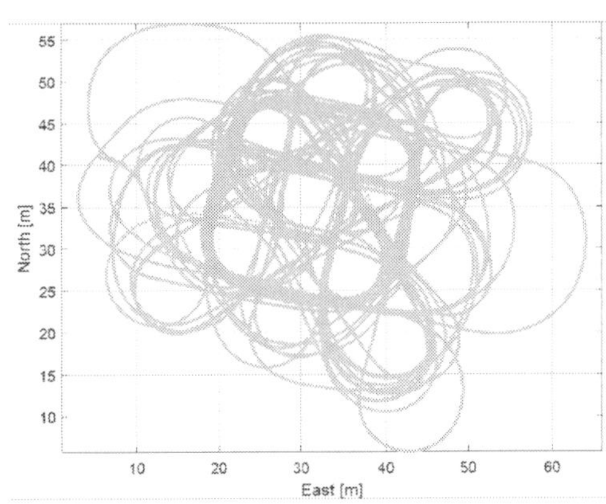

(a) Typical vehicle trajectory of the GPSVan

This contribution has been peer-reviewed.
https://doi.org/10.5194/isprs-archives-XLVIII-M-1-2023-355-2023 | © Author(s) 2023. CC BY 4.0 License.

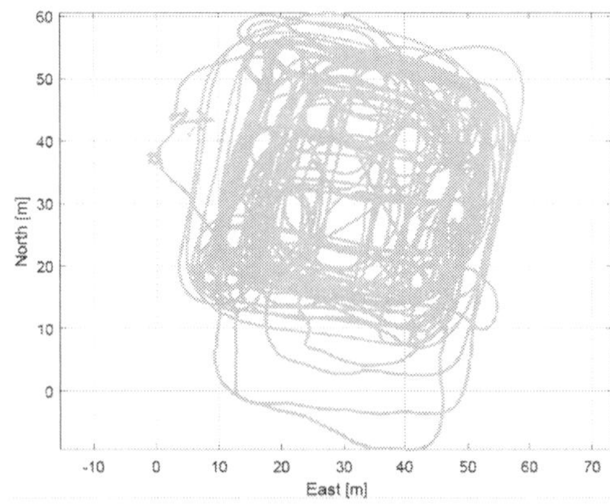

(b) Typical trajectory of a cyclist

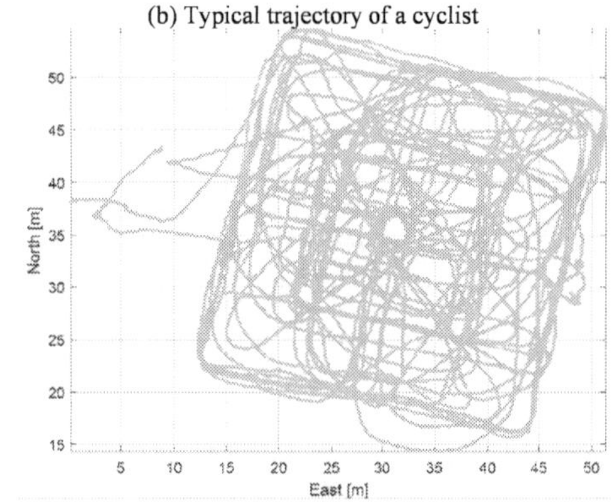

(c) Typical trajectory of a pedestrian

Figure 5. Typical platform trajectories from a test run

5. UWB PERFORMANCE EVALUATION

The UWB transceivers played an important role in our experiments as they provided the range measurements between the various platforms in the first phase. 12 units, as anchors formed the ground network, see Fig. 3, and then 8 units were deployed on four vehicles and four UAS platforms. Cyclists and pedestrians had only GNSS receivers installed on helmets, and thus, range data could be estimated from GNSS solutions.

Since UWB transceivers are restricted to operate at a very low power level, comparable to GNSS signals, the measurements are repeated several times to obtain acceptable results. Obviously, the distance between two transceivers as well obstacles, such as other vehicles or pedestrians blocking LOS, further decrease the performance. Therefore, the UWB acquired range data was analyzed first, and then cleaned before it was used for positioning. Fig. 6 shows two examples for range data acquired from mobile platforms: the GPSVan and UAS4. The horizontal lines show periods when the platforms were not moving. Note that the number of outliers, ranges that are likely incorrect is larger when the platforms are moving. Knowing the dynamics of the motion those outliers can be filtered, and then the data can be used for collaborative network formation. Also note that the time duration is different for the measurements, as the drone has a limited flight time, while the GPSVan continued running for a while.

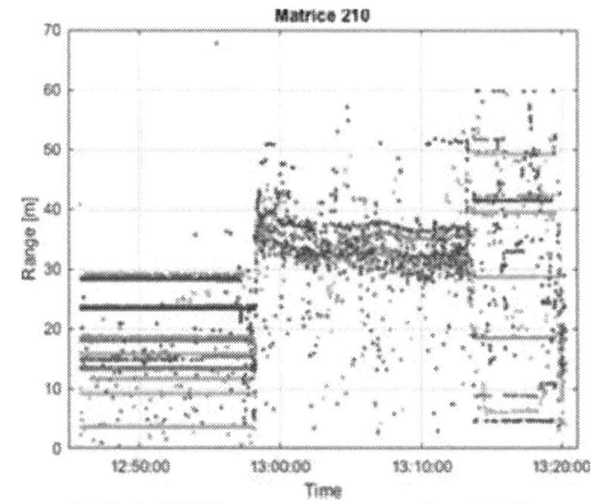

(a) Typical UWB range measurements from UAS4

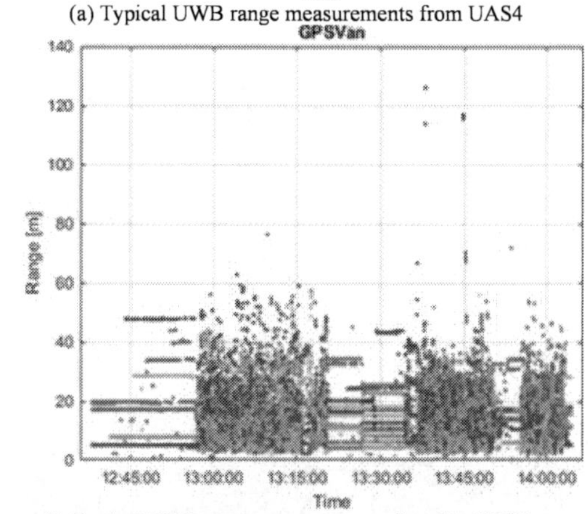

(b) Typical UWB range measurements from the GPSVan

Figure 6. Typical UWB range data acquired by two platforms

First, the self-localization of the ground UWB network was analyzed, using 13 transceivers. The data was acquired in a static mode, and then ranges were used to estimate the relative position of UWB units; i.e., a 2D geodetic network in a local coordinate system. Note that not all possible measurements, $\binom{n}{2}$ were acquired due to environmental limitations. Only about 85% of the ranges were available, which still represented a highly overdetermined system. Using standard least squares adjustment, a 2D solution was computed, as the area is reasonably flat. The internally estimated precision was about 5 cm (1σ). The relative positions of the local network were then compared to the surveyed ground control points. The RMSE obtained was about 10 cm (2D), which is larger than expected but can be explained by two things. First, the network formed by UWB ranges assumed a planar distribution of the locations, while the ground control points were surveyed in 3D with a few-cm accuracy. Second, the tripod setup over the surveyed points and the mounting of the UWB transceivers introduced some mounting bias too.

This contribution has been peer-reviewed.
https://doi.org/10.5194/isprs-archives-XLVIII-M-1-2023-355-2023 | © Author(s) 2023. CC BY 4.0 License.

In the next step, a vehicle platform trajectory was computed based on using the static UWB network. The data rate of acquiring UWB ranges varies due to the environmental conditions. On an average for a single vehicle data collection, we typically obtained an about 20 Hz, which represents about 0.25-0.5 m spatial sampling at the typical moving platform speed of 5-10 m/s. Fig. 7 shows the estimated positioning precision of an about 1-minute run; the horizontal axis is in epochs and the vertical is in m. This decimeter-level positioning performance is clearly adequate for transportation applications; even for lane-following it is suitable.

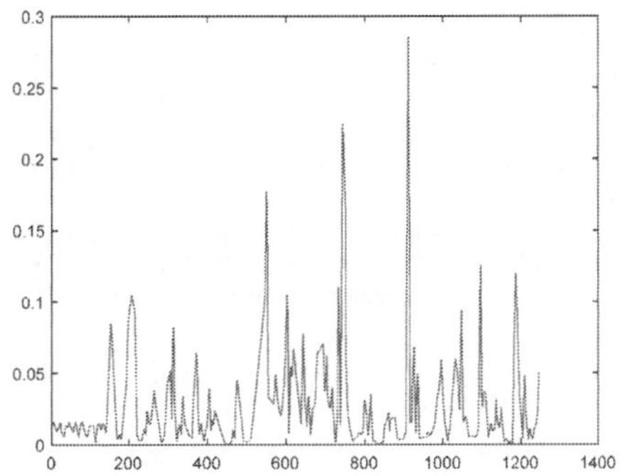

Figure 7. Internally estimated positioning precision in 2D

6. COLLABORATIVE NAVIGATION RESULTS

In earlier research efforts, we tested our collaborative network-based approach only on 2D data, such as four vehicles moving on road or in an intersection area. Now for the first time, we have real 3D data, as the drones were simultaneously flying over the ground vehicles. The approximate size of the envelop, the bounding box that includes all the various platform trajectories was about 100 m x 100 m x 80 m. Initial trajectory solutions are shown in Figure 8. The top image (a) presents the 4-drone only solution while the one below (b) provides the joint 4-drone and 4-vehcile solution. Note that the 12-unit ground UWB network was also used in obtaining collaborative navigation solutions.

The quantitative evaluation is still in the works, as obtaining the optimal solution and comparing it to the reference solution is a formidable task; say, in contrast to the static network-based positioning of a single mobile platform. A major challenge is the timing, time synchronization, i.e., using the correct time tags for all sensor data streams, which is, obviously, essential to achieve the highest feasible accuracy for the given specification of the sensors.

All sensors were GPS time-synchronized locally, either built-in GPS or via an Android interface, at an about 5-10 msec or better accuracy. Therefore, given the platform dynamic, the error introduced by timing is negligible. Having eight mobile UWB transceivers constantly polling each others as well as the 12-unit ground network, however, increases the chances for interference; despite the fact that UWB works with very short pulses and high repetition rate. Our data shows a varying rate of range data acquisition of about 2 – 5 Hz. Considering the vehicle motion of 5-10 m/s, a platform may move a few meters at that rate. Therefore, motion compensation should be considered, such as for straight line trajectories, projecting the ranges to a virtual location (epoch) to compute a consistent network solution. Turning and 3D motion, such as the case for UAS, may need better modeling. The local motion modeling of platforms is generally available from an EKF-based navigation filter; though, the quality may fluctuate over a large range. The UWB range-based network solutions can be considered as GNSS fixes in the standard GNSS/INS integration scheme, i.e., they provided fixes when new measurements become available.

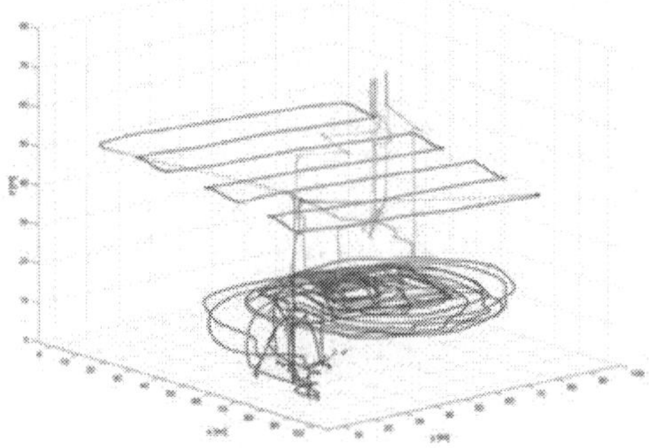

(a) Solutions for four drones

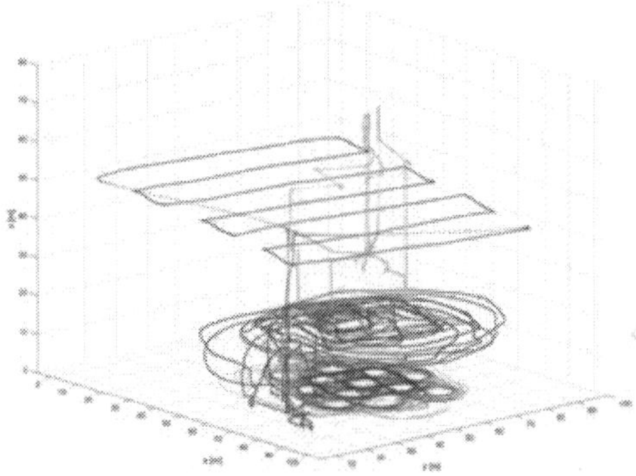

(b) Solutions for four drones and four vehicles

Figure 8. Collaborative network-based positioning

Another aspect of having platform solution is that UWB measurements are not always reliable, as there could be outliers and no measurements at all. An outlier detection algorithm for collaborative navigation introduced in (Xiong *et al.*, 2021) models the GNSS measurements and inter-platform range measurements into common and specific parts and tries to exclude the faulty measurements with a greedy search strategy. Test results show the algorithm has a better detection of GNSS faults than tradition Receiver Autonomous Integrity Monitoring (RAIM) and good sensitivity to the faulty UWB range measurements.

7. CONCLUSION

With the proliferation of sensors, inter-platform range measurements are becoming available, which, combined with increasing communication capabilities, provide a good bases for implementing collaborative navigation. The key idea is to exploit the strength of the geometry, defined by a geodetic network, formed by the platforms, to provide independent constraints for a group of platforms, and thus, achieving better overall navigation solution for all the platforms.

In this study, an UWB sensor network, installed on both static and mobile platforms provided the range measurements, but, of course, any other sensor measurement can be used. The initial evaluations show that fairly good ranging accuracy could be obtained in an intersection size area, resulting in performance that meets the requirements of most transportation activities in Smart City environment.

In summary, the results obtained so far in a 3D scenario clearly demonstrate the potential of collaborative navigation. It is important to emphasize that using this approach requires not more than utilizing already available sensor measurements on smart devices. Then, while inter-platform communication is available, such as DSRC (Dedicated Short-Range Communications), which is being phased out, cloud-based peer-to-peer communication is becoming widely available.

REFERENCES

Ahvenniemi, H., Huovila, A., Pinto-Seppä, I., Airaksinen, M. (2017), What are the differences between sustainable and smart cities? *Cities*, Volume 60, Part A, February 2017, Pages 234-245 https://doi.org/10.1016/j.cities.2016.09.009

Benevolo, C., Dameri, R.P., D'Auria, B. (2016). Smart Mobility in Smart City. In: Torre, T., Braccini, A., Spinelli, R. (eds) Empowering Organizations. *Lecture Notes in Information Systems and Organisation*, vol 11. Springer, Cham. https://doi.org/10.1007/978-3-319-23784-8_2

Bryson, M. and Sukkarieh, S. (2009). Architectures for cooperative airborne simultaneous localization and mapping. *Journal of Intelligent and Robotic Systems*, 55(4-5), 267-297.

Gabela, J., Retscher, R., Goel, S., Perakis, H., Masiero, A., Toth, Ch, Gikas, V., Kealy, A., Koppanyi, Z., Błaszczak-Bąk, V., Li, Yan., Grejner Brzezinska, D., 2019 Experimental Evaluation of a UWB-Based Cooperative Positioning System for Pedestrians in GNSS-Denied Environment, *Sensors* 2019, 19, 5274; doi:10.3390/s19235274

Hosseinyalamdary, S., Balazadegan, Y., & Toth, C. (2015). Tracking 3D moving objects based on GPS/IMU navigation solution, laser scanner point cloud and GIS data. *ISPRS Int. J. Geo-Inf.* 2015, 4(3), 1301-1316; *https://doi.org/10.3390/ijgi4031301*

Ladai, A., C. Toth (2022), Range-based Network Formation to Support Georeferencing Of Mapping Swarm Platforms, *Int. Arch. Photogramm. Remote Sens. Spatial Inf. Sci.*, XLVI-M-2-2022, 147–152, 2022

Masiero, A., Toth, C., Gabela, J., & Retscher, G. (2022). Analysis of the Performance of AN Uwb-Based Cooperative Positioning for Different CAR Platoon Configurations. *The International Archives of Photogrammetry, Remote Sensing and Spatial Information Sciences*, 43, 467-472.

Masiero, A.; Toth, C.; Gabela, J.; Retscher, G.; Kealy, A.; Perakis, H.; Gikas, V.; Grejner-Brzezinska, D. (2021) Experimental Assessment of UWB and Vision-Based Car Cooperative Positioning System. *Remote Sens.* 2021, 13, 4858. https://doi.org/10.3390/rs13234858

Masiero, A., Dabove, P., Di Pietra, V., Piragnolo, M., Vettore, A., Cucchiaro, S., ... & Gabela, J. (2021b). A case study of pedestrian positioning with UWB and UAV cameras. *Int. Arch. Photogramm. Remote Sens. Spatial Inf. Sci.*, XLIII-B1-2021, 111–116, 2021 https://doi.org/10.5194/isprs-archives-XLIII-B1-2021-111-2021

MacGougan, G., O'Keefe, K., and Klukas, R. (2009). Ultra-wideband ranging precision and accuracy. *Measurement Science and Technology*, 20(9), 095105.

Matlab Computer Vision Toolbox™ https://www.mathworks.com/products/computer-vision.html

Retscher, G., Kealy, A., Gikas, V., Gabela, J., Goel, S., Li, Y., ... & Grejner-Brzezinska, D. (2020). A Benchmarking Measurement Campaign to Support Ubiquitous Localization in GNSS Denied and Indoor Environments. In Beyond 100: The Next Century in Geodesy: *Proceedings of the IAG General Assembly*, Montreal, Canada, July 8-18, 2019 (pp. 123-128). Cham: Springer International Publishing.

Rome, H. and Stambaugh, J. (1977). Evaluation of the accuracy of community relative navigation organization concepts. *Navigation*, 24(2), 168-180.

Roumeliotis, S. and Beke, G. (2002). Distributed multi-robot localization. *IEEE Transactions on Robotics and Automation*, 18(5), 781{795. doi:https://doi.org/10. 1109/TRA.2002.803461.

Sluis, B., Toth, Ch. (2021) 2-Dimensional Geometric Analysis of a Simple Free Network, *ISPRS Annals of the Photogrammetry, Remote Sensing and Spatial Information Sciences*, Volume V-4-2021, pp.147-152.

Toth, C., Jozkow, G. (2015) Remote Sensing Platforms and Sensors: A Survey, *ISPRS Journal of Photogrammetry & Remote Sensing*, 115 (2016), pp. 22-36.

Toth, C. K., Koppanyi, Z., & Lenzano, M. G. (2018). New Source of Geospatial Data: Crowdsensing by Assisted and Autonomous Vehicle Technologies. Int. Arch. Photogramm. Remote Sens. Spatial Inf. Sci., XLII-4/W8, 211–216, 2018, https://doi.org/10.5194/isprs-archives-XLII-4-W8-211-2018

Wang, X., Ch. Toth, D. Grejner-Brzezinska, A. Masiero (2022), Collaborative Navigation: Supporting PNT System Operational Anomaly Detection, IFAC-PapersOnLine, Volume 55, Issue 24, 2022, Pages 285-291, https://doi.org/10.1016/j.ifacol.2022.10.298

Widnall, W. and Gobbini, G. (1982). Community relative navigation based on measurement sharing. In Proceedings of the IEEE Position, Location, and Navigation Symposium (PLANS 82).

Xiong, J., Cheong, J., Xiong, Z., Dempster, A., Tian, S., and Wang, R. (2021). Integrity for multi-sensor cooperative positioning. IEEE Transactions on Intelligent Transportation Systems, 22(2), 792-807. https://doi:https://doi.org/10.1109/TITS.2019.2956936

THE SPATIAL DISTRIBUTION OF SELECTED OPTICAL ACTIVE COMPONENTS IN THE GULF OF IZMIT USING BIVARIATE/MULTIVARIATE REGRESSION ANALYSIS

F. Sunar[1], A. Dervişoğlu[1], N. Yağmur[2], E. Aslan[3], H. Atabay[3]

[1] ITU, Civil Engineering Faculty, Geomatics Engineering Department, 80626 Maslak Istanbul, Turkiye
(fsunar, adervisoglu)@itu.edu.tr
[2] Gebze Technical University, Engineering Faculty, 41400 Kocaeli, Turkiye – nyagmur@gtu.edu.tr
[3] TUBITAK MRC Marine and Coastal Research Group, Kocaeli, Turkiye
(ertugrul.aslan, hakan.atabay)@tubitak.gov.tr

KEY WORDS: Water Quality, Gulf of Izmit, Sentinel-2, Bivariate/Multivariate Regression Models, Spatial Mapping

ABSTRACT

Compared to traditional field-based (in-situ sampling) measurements, satellite-based remote sensing is an accurate, timely and cost-effective approach to monitor the dynamics of water bodies using images at different spatial and temporal scales. With satellite-based remote sensing techniques, qualitative measurements obtained directly from satellite images are integrated with in-situ measurements, enabling the establishment of spectral statistical relationships between satellite data and water quality physical indicators such as suspended solids, turbidity and chlorophyll-a. In this study, the spatial distribution of three water quality parameters (Chlorophyll-a (Chl-a), Secchi disk and Conductivity (EC)) which are optical active components (OAC) in the Gulf of Izmit were evaluated using in-situ water quality measurements, together with both field-spectroradiometer measurements and Sentinel-2 satellite imagery. In-situ water quality and field-spectroradiometer measurements were collected at the same date with the satellite overpass. Bivariate and multivariate regression models were established to analyse the correlation of in-situ water quality measurements with two different measurement datasets (i.e. satellite and spectroradiometer), and then the results were evaluated with two accuracy metrics Root Mean Square Error (RMSE) and Mean Absolute Error (MAE); and compared visually with the spatial distribution maps of the three water quality parameters generated by the ordinary Kriging interpolation method.

1. INTRODUCTION

Coastal areas, where approximately 40% of the world's population live, are among the most sensitive environments, as any change in these ecosystems resulting from anthropogenic activities could endanger the entire marine ecology (Gholizadeh et.al., 2016). Therefore the quality of water bodies is an important issue for the health of coastal ecosystems and is under the influence of human activities and climate change. As a result of the deterioration of water quality, water pollution events such as harmful algal blooms (Karlson et al., 2021), hypoxia (Lai et al., 2022) and mucilage (Sunar et al., 2022) occur. Field-based (in-situ sampling) measurements that rely solely on frequent and/or long-term samples as a traditional approach cannot fully address the spatio-temporal variation of water quality in any watershed. Consequently, an integrated water conservation and reclamation approach is required for effective water quality monitoring, in which long-term field-based measurements can be used in conjunction with space technologies such as remote sensing. In other words, integrating traditional data collection with remote sensing technology today provides an effective monitoring approach for the dynamics of water bodies using images at different spatial and temporal scales (Hu et al., 2022; Zhu et al., 2022).

As known, the optically active constituents (OACs) (such as suspended sediments, colored dissolved organic matter (CDOM), chlorophyll-a (Chl-a) etc.) present in natural water bodies have inherent optical properties (IOPs) that affect the water's ability to absorb light. In order to determine underwater light dispersion, the photosynthetic process, and remote sensing reflectance, the IOPs of the individual OACs in the area are crucial (Shi et al., 2017). Many studies showed that water quality parameters can be retrieved with varying accuracy, as the spectral properties of water change depending on the interaction between solar radiation and OAC (Pizani et al., 2020). However, other water quality constituents, such as nutrient concentrations, dissolved oxygen levels, and microorganisms/pathogens, cannot be directly estimated by remote sensing data as they are not optically active; i.e. they may be estimated indirectly but still remains a big challenge (Sagan et al., 2020). In general, statistical regression models (bivariate and multivariate) that are often considered site-specific, are established to quantify measured reflectance – OAC relationships.

For monitoring the pollution and its effects on seas and coastal waters in Turkey, the Ministry of Environment, Urbanisation and Climate Change has conducted an Integrated Marine Pollution Monitoring Program (DEN-IZ) in cooperation with the Scientific and Technological Research Council of Turkey - Marmara Research Center (TUBITAK-MAM) since 2014. The scope of this program is to carry out in-situ measurements and analyses of water quality parameters at the designated stations in all seas (i.e. Black Sea, Marmara Sea and the Straits, Mediterranean and Aegean Sea), and then to report the results and evaluations periodically (Url-1). Gulf of Izmit, known as one of the most polluted marine ecosystems of Turkey and located in the east of the Marmara Sea, is one of the fragile coastal areas evaluated within the scope of the mentioned DEN-IZ program. In this study, the spatial distribution of the three OACs (Chl-a, Secchi disk and Conductivity) was evaluated using in-situ water quality measurements, together with handheld field spectroradiometer measurements and Sentinel-2 satellite imagery obtained simultaneously. For this purpose, the correlation of in-situ water quality measurements with two other different measurement datasets (satellite and spectroradiometer) was established with regression models and the results were evaluated with accuracy metrics.

https://doi.org/10.5194/isprs-archives-XLVIII-M-1-2023-361-2023

2. STUDY AREA

The Gulf of Izmit, which is a semi-enclosed coastal ecosystem located in the east of the Marmara Sea, chosen as the study area, has an area of approximately 310 km^2, 60 km long and 2-10 km wide (Tugrul et al. 1989) (Figure 1). Due to salinity differences, stratification in the water column occurs throughout the year (Ergul et al., 2018).

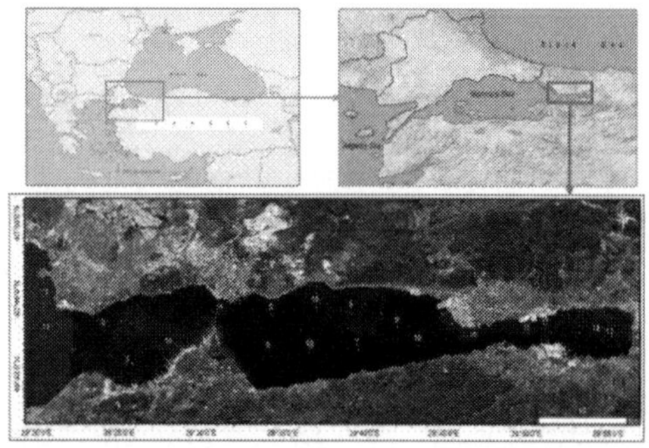

Figure 1. The map of Gulf of Izmit and location of 17 in-situ sampling points in Sentinel-2 satellite image (© Copernicus).

Kocaeli province, located in the Gulf of Izmit, is known as a city with a giant natural harbor. The fact that the region has different natural potentials that are promising and that can attract many investors has led to the intense use of these coastal areas under many factors such as natural environment characteristics, human structure, industrial activities and transportation routes (Garipagaoglu and Uzun, 2014; Ergül et al., 2018). Hence, in the province, which has a current population of approximately 2 million people, a significant amount of uncontrolled sewage, industrial and agricultural wastes are discharged to the Gulf as a result of the increasing population and industrial activities since the 1980s (Figure 2).

Figure 2. Locations of main freshwater input source points and wastewater treatment plants in the eastern, central and western basins of the Gulf of Izmit (adapted from Ergül, 2016; Tolun et al., 2012).

3. MATERIALS AND METHODOLOGY

3.1 Materials

Sentinel-2 satellite image acquired on October 21, 2021 was used in the study. Sentinel-2 satellite images can be obtained free of charge by the European Space Agency in different processing levels such as Top-Of-Atmosphere (TOA) and Bottom-Of-Atmosphere (BOA). Its spatial resolution varies in different wavelength ranges (VNIR - SWIR spectral range) as 10 m, 20 m and 60 m. The spectral resolution of Sentinel-2 data, which has an open data access policy, is given in Figure 3.

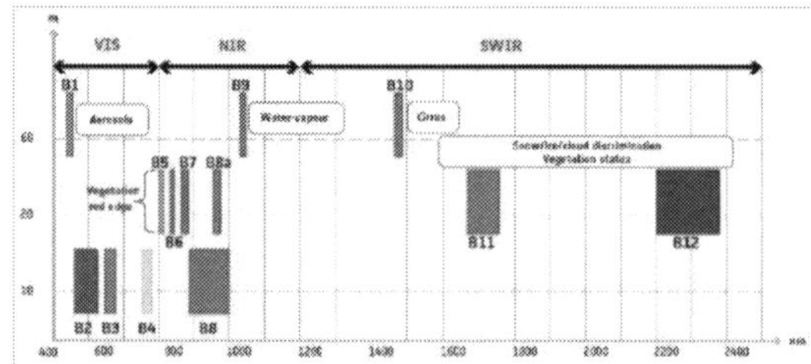

Figure 3. The spectral resolution of Sentinel-2 data (©ESA).

In the study, BOA satellite image data was used together with 17 simultaneous spectroradiometer measurements and in-situ water quality measurements. The in-situ measurements are homogeneously distributed along the Gulf (Figure 1). The spectra of the points were measured by the hand-held ASD FieldSpec 4 with a spectral range of 325-1075 nm (Figure 4). 13 of the measurements were used in the regression models and 4 samples were reserved for accuracy assessment. The spectral values exhibit a similar tendency, with the exception of a few sampling locations.

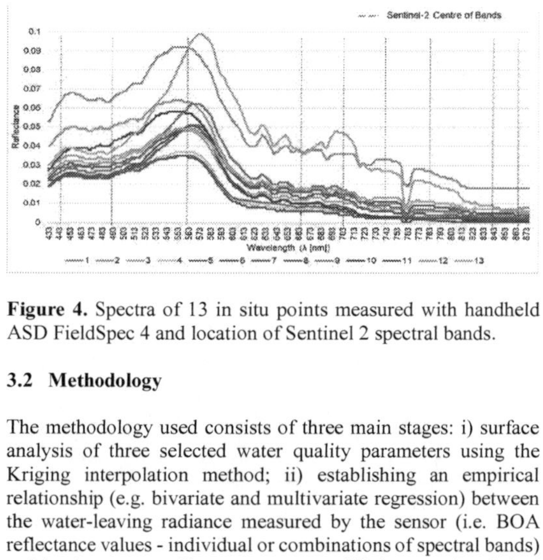

Figure 4. Spectra of 13 in situ points measured with handheld ASD FieldSpec 4 and location of Sentinel 2 spectral bands.

3.2 Methodology

The methodology used consists of three main stages: i) surface analysis of three selected water quality parameters using the Kriging interpolation method; ii) establishing an empirical relationship (e.g. bivariate and multivariate regression) between the water-leaving radiance measured by the sensor (i.e. BOA reflectance values - individual or combinations of spectral bands) and in-situ water quality measurements; iii) the use of statistical accuracy metrics to evaluate the accuracy of the models accepted.

3.2.1 Kriging interpolation method: One of the geostatistical interpolation techniques used for modeling spatial data is the Kriging approach (Prusova et.al., 2012). This method is widely used in a water quality analysis to estimate the values of unmeasured water quality parameters and to analyze spatial distribution patterns of the parameters of interest. In this study, the ordinary Kriging interpolation method was applied for the surface analysis of the selected three water quality parameters and then compared with the results of the regression models (produced by satellite data and field spectroradiometer measurements) used to estimate these parameters.

3.2.2 Regression analysis: In this study, bivariate and multivariate linear regression analyses were performed for the aforementioned three OAC components. As a common statistical method, regression models are an efficient tool for investigating the relationship between variables, even when small sample size

This contribution has been peer-reviewed.
https://doi.org/10.5194/isprs-archives-XLVIII-M-1-2023-361-2023 | © Author(s) 2023. CC BY 4.0 License.

variables are used (Razi and Athappilly, 2005). Linear regression models, also called simple bivariate regression, evaluate the linear relationship between two variables (i.e. dependent and independent variables) based on the least squares, and the best model has minimum square error between the observed and predicted parameters (Zare Abyaneh, 2014).

Multivariate linear regression (MLR), an extension of bivariate linear regression, is a common statistical modeling technique to explain a dependent variable with more than one independent variable. More clearly, MLR is used to predict a response variable (y) from multiple explanatory variables ($x_1, x_2, ..., x_n$) to represent the linear relationship between many independent predictor variables (spectral bands) and a single dependent variable (a water quality parameter such as Chl-a) as a single functional formula (Shi, 2017).

As noted in the literature, correlations between in-situ water quality measurements and spectral reflectance values can be complex and nonlinear, especially for Case 2 water bodies (such as Gulf of Izmit), as these parameters respond differently to various spectral wavelengths (Gholizadeh et al., 2016; Hafeez et al., 2019; Topp et al., 2020). Therefore multivariate regression was also used in this study, assuming that there is no strong correlation between each independent variable, as the estimation of each OAC may depend on more than one independent variable (i.e. more than different bands or combinations). As a result, bivariate and multivariate linear regression analyzes were performed by using two different datasets; i) in-situ water quality measurements and satellite imagery and ii) in-situ water quality measurements and field spectroradiometer measurements. The best models in these two scenarios were evaluated using Root Mean Square Error (RMSE) and Mean Absolute Error (MAE).

3.2.3 Model performance evaluation: After the regression models are built, the goodness of fit of the model and the statistical significance of the estimated parameters need to be confirmed. Most commonly used p-value is the most important step to accept or reject a null hypothesis. More specifically, if the p-value is less than a common threshold (e.g. 0.05), the null hypothesis is rejected, indicating a relationship between the variables.

In order to evaluate the performance of the models used in regression analysis, R-squared (a measure to determine how well the model fits the dependent variables), Root Mean Square Error (RMSE - popular metric that shows how close the estimate is on average to the actual value, but also indicates the effect of large errors) and Mean Absolute Error (MAE - similar to the RMSE, the MAE is an indicator of how close the estimates are, on average, to the actual model) given in the Equations below were used and the accepted models for the selected parameters were tested using with in-situ measurements allocated as test data.

$$R^2 = 1 - \frac{\sum_i (y_i - \hat{y}_i)^2}{\sum_i (y_i - \overline{y})^2} \quad (1)$$

$$RMSE = \sqrt{\frac{\sum_{i=1}^{n} (y_i - \hat{y})^2}{n}} \quad (2)$$

$$MAE = \frac{1}{n} \sum_{i\ 1}^{n} |y_i - \hat{y}_i| \quad (3)$$

where y_i is the measured value,
$\hat{y}_i$ is the estimated value and
n is the number of measurements.

In general, the higher the R-squared, the better the model fits the data. On the other hand, the lower values of RMSE and MAE indicate higher accuracy of a regression model (i.e. the model estimates correctly).

4. RESULTS

First, the ordinary Kriging interpolation method was applied for the surface analysis of the selected three water quality parameters and spatial distribution maps produced are given in Figure 5. Four unused samples were considered for the accuracy assessment of each Kriging model and the results are given in Table 1.

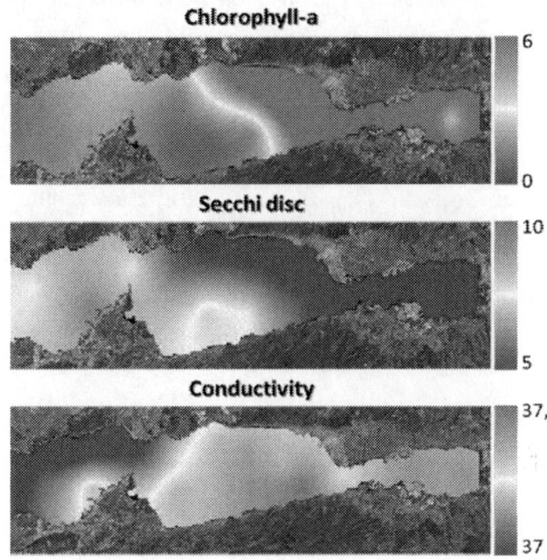

Figure 5. The spatial distribution patterns of three selected water quality parameters produced by the ordinary Kriging interpolation method.

In Figure 5, it is clearly seen that the eastern basin (i.e. interior) and some parts of the central basin close to the eastern basin are more polluted (high Chl-a and low Sechi disc values) than other parts of the basin. The decrease in Chlorophyll-a values (primary production indicator) from the eastern basin to other regions is clearly evident and realistic in the Kriging model, because the interior basin has high pollutant load and low water circulation (Tuğrul and Morkoc, 1990). The eastern basin is influenced by industrial and domestic loads, particularly from the Kiraz (Kullar) and Kumla (Sarı) streams, and nutrient and chlorophyll-a levels confirm the findings (Figure 6) (Tan and Aslan, 2020).

Figure 6. Locations of Kiraz and Kumla streams connected to the interior basin.

This contribution has been peer-reviewed.
https://doi.org/10.5194/isprs-archives-XLVIII-M-1-2023-361-2023

Secchi disc visibility is likewise lowered where chlorophyll-a levels are high. In Figure 5, a sharp line in the Kriging interpolation method, especially in chlorophyll-a, is caused by the low number of in-situ measurements used in the model, so it does not fully reflect the real situation and a distribution as in Figure 7 is found to be more appropriate.

Although conductivity is considered an indirect indicator of pollution, it is a good parameter for determining water quality, as it presents a close relationship with the dissolved salt content in the water column of coastal water bodies, which is often associated with sewage discharge (Tolun et al., 2012; de Sousa et al., 2014; Eyuboglu and Eyüboğlu, 2020). Therefore, high EC values, which can generally be associated with the presence of domestic wastewater, confirm the high conductivity values, especially in the eastern and central basins (Figure 2).

In order to find the best estimator of the 3 selected water quality parameters in the Gulf of Izmit, bivariate and multivariate linear regression models were used. The best regression models selected for Chl-a, Secchi disk and Conductivity parameters in two different measurement datasets (i.e. Sentinel-2 Level 2A and field spectroradiometer) are given in Table 1. Again, four unused in-situ measurements were considered for the accuracy assessment of each model developed. The best regression models selected and the spatial distribution of the 3 parameters produced using these selected models are given in Figure 7.

Parameter	Variables	Data	Model	t	Significance (p-value)	F	Standard Error	R	R^2	RMSE	MAE
Chl-a	Kriging									4.237	2.328
	B4/B2	Spectroradiometer	Bivariate	5.40	0.0002	29.20	5.83	0.85	0.73	8.605	7.822
	B3 B4	Spectroradiometer	Multivariate	-0.03 0.79	0.125	2.57	9.49	0.58	0.34	3.863	3.313
	B3/B2	Sentinel-2	Bivariate	4.15	0.002	17.19	6.96	0.78	0.61	4.671	3.394
	B2 **B3** **B4**	**Sentinel-2**	**Multivariate**	-3.03 1.14 0.14	**0.028**	4.90	7.59	**0.79**	**0.62**	**3.068**	**2.353**
Secci disk	Kriging									0.436	0.126
	B4	Spectroradiometer	Bivariate	-1.38	0.20	1.89	2.13	0.38	0.15	1.102	0.967
	B3 B4	Spectroradiometer	Multivariate	0.18 -0.61	0.44	0.88	2.23	0.39	0.15	0.744	0.545
	B4/B2	Sentinel-2	Bivariate	-5.79	0.0001	33.49	1.14	0.87	0.75	4.118	3.356
	B2 **B3**	**Sentinel-2**	**Multivariate**	6.34 -7.51	**0.0001**	30.65	0.90	**0.93**	**0.86**	**1.373**	**1.262**
Conductivity	Kriging									0.265	0.188
	B2+ B8A	**Spectroradiometer**	**Bivariate**	-3.25	**0.008**	10.58	0.36	**0.70**	**0.49**	**0.285**	**0.227**
	B2 B8A	Spectroradiometer	Multivariate	-1.06 -0.58	0.034	4.82	0.38	0.70	0.49	0.260	0.198
	(B3-B4)/B2	Sentinel-2	Bivariate	3.31	0.007	10.97	0.36	0.71	0.50	0.340	0.308
	B2 B3 B4	Sentinel-2	Multivariate	-2.26 1.64 -1.15	0.067	3.40	0.38	0.73	0.53	1.771	1.765

Table 1. The bivariate and multivariate regression models used for three water quality parameters on two different measurement datasets (the best predictive models selected are marked in bold).

As seen in Table 1, the multivariate regression model constructed using Sentinel-2 image (with bands B2, B3, and B4) estimated Chl-a concentrations with an R^2 value of over 60%. Also, this model has the lowest RMSE and MAE errors, i.e. 3.07 and 2.35, respectively. On the other hand, the spatial distribution of this model was found to be highly compatible with the Chl-a model produced by the Kriging interpolation method. Although the estimation of the spectroradiometer dataset for this parameter was higher with the bivariate model (i.e. $R^2 \approx 73\%$), however, both accuracy measures were found to be high and their spatial distribution did not appear to be compatible with the surface model produced by the Kriging interpolation method (Figure 6).

For Secchi disk parameter retrieval, the multivariate model using the two spectral bands of the Sentinel-2 image (i.e. B2 and B3 bands) outperformed (i.e. high R^2 value; 86%) all models built with the spectroradiometer dataset. Although the multivariate model created with the spectroradiometer data set had the lowest RMSE and MAE accuracies, the R^2 values were found to be quite low (i.e. 15%).

Spectroradiometer dataset with bivariate model provided greater RMSE and MAE accuracies for the last parameter, Conductivity parameter retrieval, compared to Sentinel-2 image data, despite R^2 values being in almost the same range (i.e. 49% $\approx$ 50/53%).

This contribution has been peer-reviewed.
https://doi.org/10.5194/isprs-archives-XLVIII-M-1-2023-361-2023 | © Author(s) 2023. CC BY 4.0 License.

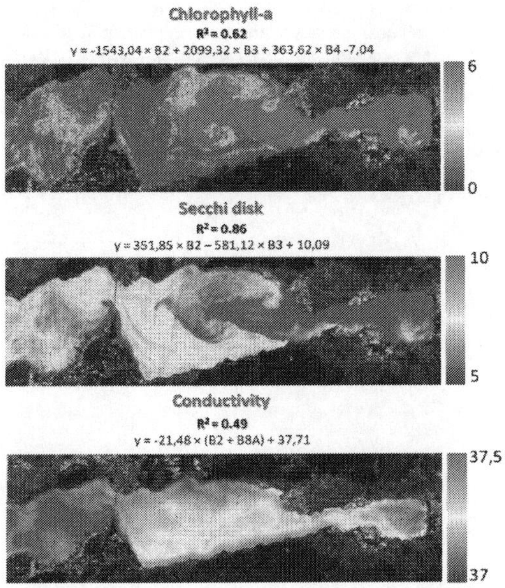

Figure 7. The spatial distribution patterns of three selected water quality parameters produced by the best regression models selected.

When two figures (Figure 5 and Figure 6) showing the spatial distribution patterns of three selected water quality parameters were compared, it was observed that the results for Chl-a and Secchi disk characteristics were quite comparable. On the other hand, it was determined that the models' compatibility and comparability for the conductivity parameter were quite low, therefore further analysis is still required.

The results of this study demonstrate that improvements are still required for the generalization and transferability of the established models for other basins in the Marmara Sea with regard to the accuracy of the estimation of these three OACs. Additionally, it is thought that the variations seen for all parameters may also result from stratified formations that alter the distributions of water quality parameter along the water column of the Gulf of Izmit and from meteorological conditions that affect oceanographic features (Ergul, 2016).

5. CONCLUSIONS

In this study, bivariate and multivariate linear model accuracies of three OACs in the Gulf of Izmit were evaluated using satellite data. The findings show that pollutants in wastewater from residential and industrial regions, particularly in the inner and middle parts of the Gulf, have a negative impact on the water quality. The results of the established models were found to be relatively consistent with in-situ measurements of water quality parameters, but it is thought that more extensive field measurements and different satellite datasets across the Gulf should be used to improve the results.

Despite its potential, there are several issues in water quality monitoring using space-borne remote sensing that need to be addressed before the developed prediction models can be adapted for operational use in the Marmara Sea; i) since the Gulf of Izmit is a coastal inland body of water, it contains optically active components at various concentrations, making it difficult to utilize, compare, and apply the prediction models created in other basins in the Marmara Sea; ii) since the optical properties of these waters are complex and change over time, the seasonal and annual fluctuations of these parameters need to be evaluated using a wider range of in-situ data collection.

As a continuation of this study, the models created in October 2021 will be tested with additional in-situ measurements and spectroradiometer data performed in October 2022, and their utility and compatibility of the models will be evaluated in order to reduce the aforementioned problems.

ACKNOWLEDGEMENTS

This work has been supported by the "Integrated Marine Pollution Monitoring 2020-2022 Programme" carried out by the Ministry of Environment, Urbanization and Climate Change/General Directorate of EIA, Permit and Inspection/ Department of Laboratory, Measurement and coordinated by TUBITAK- MRC ECPI.

REFERENCES

de Sousa, D.N.R., Mozeto, A.A., Carneiro, R.L., Fadini, P.S., 2014. Electrical conductivity and emerging contaminant as markers of surface freshwater contamination by wastewater. *Science of the Total Environment*, 484, 19-26.

Ergul, H.A., 2016. Evaluation of seasonal physicochemical conditions and chlorophyll-a concentrations in Izmit Bay, Marmara Sea. *Journal of the Black Sea/Mediterranean Environment*, 22(3).

Ergul, H.A., Aksan, S., Ipsiroglu, M., 2018. Assessment of the consecutive harmful dinoflagellate blooms during 2015 in the Izmit Bay (the Marmara Sea). *Acta Oceanologica Sinica*, 37(8), 91-101.

Eyuboglu, H., Eyuboglu, O., 2020. Effects of the environmental pressures in Izmit Bay (In Turkish: İzmit Körfezi' nde Çevresel Baskıların Etkileri), *Journal of Anatolian Environmental and Animal Sciences*, 5 (1), 25-37.

Garipagaoglu, N., Uzun, S., 2014. İzmit Körfezi kıyılarında kıyı alanı kullanımı. Türk Coğrafya Dergisi, (63), 9-22.

Gholizadeh, M.H., Melesse, A.M., Reddi, L., 2016. A comprehensive review on water quality parameters estimation using remote sensing techniques. *Sensors*, 16(8), 1298.

Hafeez, S., Wong, M.S., Ho, H.C., Nazeer, M., Nichol, J., Abbas, S., Tang, D., Lee, K.H., Pun, L., 2019. Comparison of machine learning algorithms for retrieval of water quality indicators in case-II waters: A case study of Hong Kong. *Remote Sensing*, 11(6), 617.

Hu, M., Ma, R., Xiong, J., Wang, M., Cao, Z., Xue, K., 2022. Eutrophication state in the Eastern China based on Landsat 35-year observations. *Remote Sensing of Environment*, 277, 113057.

Karlson, B., Andersen, P., Arneborg, L., Cembella, A., Eikrem, W., John, U., West, J.J., Klemm, K., Kobos, J., Lehtinen, S., Lundholm, N., 2021. Harmful algal blooms and their effects in coastal seas of Northern Europe. *Harmful Algae*, 102, p.101989.

https://doi.org/10.5194/isprs-archives-XLVIII-M-1-2023-361-2023

Lai, Y., Jia, Z., Xie, Z., Li, S., Hu, J., 2022. Water quality changes and shift in mechanisms controlling hypoxia in response to pollutant load reductions: A case study for Shiziyang Bay, Southern China. *Science of The Total Environment*, 842, 156774.

Pizani, F.M., Maillard, P., Ferreira, A.F., de Amorim, C.C., 2020. Estimation of water quality in a reservoir from Sentinel-2 MSI and Landsat-8 OLI sensors. ISPRS Annals of Photogrammetry, Remote Sensing & Spatial Information Sciences, 5(3).

Prusova, M., Orlikova, L., Hanzlova, M., 2012. Comparison of geostatistical approaches dealing with the distribution of snow. *Open Geosci*, 4, 603–613.

Razi, M.A., Athappilly, K., 2005. A comparative predictive analysis of neural networks (NNs), nonlinear regression and classification and regression tree (CART) models. *Expert Systems with Applications*, 29(1), 65-74.

Sagan, V., Peterson, K.T., Maimaitijiang, M., Sidike, P., Sloan, J., Greeling, B.A., Maalouf, S., Adams, C., 2020. Monitoring inland water quality using remote sensing: Potential and limitations of spectral indices, bio-optical simulations, machine learning, and cloud computing. *Earth-Science Reviews*, 205, p.103187.

Shi, L., Mao, Z., Wu, J., Liu, M., Zhang, Y., Wang, Z., 2017. Variations in spectral absorption properties of phytoplankton, non-algal particles and chromophoric dissolved organic matter in Lake Qiandaohu. *Water*, 9(5), 352.

Sunar, F., Dervisoglu, A., Yagmur, N., Colak, E., Kuzyaka, E., Mutlu, S., 2022. How Efficient can Sentinel-2 Data Help Spatial Mapping of Mucilage Event in the Marmara Sea?. International Archives of the Photogrammetry, Remote Sensing & Spatial Information Sciences, 43, 181-186.

Tan, I., Aslan, E., 2020. Metal pollution status and ecological risk assessment in marine sediments of the inner Izmit Bay. *Regional Studies in Marine Science*, *33*, p.100850.

Tolun, L.G., Ergenekon, S., Hocaoglu, S.M., Donertas, A.S., Cokacar, T., Husrevoglu, S., Beken, C.P., Baban, A., 2012. Socioeconomic response to water quality: a first experience in science and policy integration for the Izmit Bay coastal system. *Ecology and Society*, 17(3).

Topp, S.N., Pavelsky, T.M., Jensen, D., Simard, M., Ross, M.R.V., 2020. Research trends in the use of remote sensing for inland water quality science: Moving towards multidisciplinary applications. *Water*, 12, 169.

Tugrul, S., Morkoc, E., Okay, O., 1989. The determination of oceanographic characteristics and assimilation capacity of the Izmit Bay. In: Kalafatoglu E (ed) Wastewater treatment and disposal studies: NATO-TU WATERS. Final Report, TUBITAK Marmara Scientific and Industrial Research Center, Departmentof Chemical Engineering, Gebze, Istanbul, 193–259.

Tuğrul, S., Morkoc, E., 1990. Transport and water quality modeling in the Bay of Izmit. NATO TU-WATERS Project, Technical Report. TÜBİTAK-MRC Publ. Kocaeli, Turkey.

Zare Abyaneh, H., 2014. Evaluation of multivariate linear regression and artificial neural networks in prediction of water quality parameters. *Journal of Environmental Health Science and Engineering*, 12(1), 1-8.

Zhu, X., Guo, H., Huang, J.J., Tian, S., Xu, W., Mai, Y., 2022. An ensemble machine learning model for water quality estimation in coastal area based on remote sensing imagery. *Journal of Environmental Management*, 323, 116187.

Url-1: Integrated Marine Pollution Monitoring Program, https://ced.csb.gov.tr/denizlerde-butunlesik-kirlilik-izleme-programi-i-84266, (in Turkish), Accessed January 23, 2023.

https://doi.org/10.5194/isprs-archives-XLVIII-M-1-2023-361-2023 | © Author(s) 2023. CC BY 4.0 License.

DROUGHT MONITORING FROM 2001-2019 IN NORTHEAST THAILAND USING MODIS NDVI IMAGE TIME SERIES AND Savitzky-Golay APPROACH

Savittri Ratanopad Suwanlee [1], Nudthawud Homtong [2], Jaturong Som-ard [1]*

[1] Department of Geography, Faculty of Humanities and Social Sciences, Mahasarakham University, Maha Sarakham Province, 44150, Thailand - (jaturong.s, savittri.s)@msu.ac.th
[2] Department of Geotechnology, Faculty of Technology, Khon Kaen University, Khon Kaen Province, 40002, Thailand - nudth@kku.ac.th

KEY WORDS: Drought, Earth Observation, Remote Sensing, MODIS, NDVI, VCI, Savitzky-Golay.

ABSTRACT:

Drought directly threatens food security and livelihoods, thereby increasing socioeconomic risks and remains a challenge for natural resource management, particularly in frequently affected regions. Earth observation (EO) satellites provide extensive spectral and temporal data for long-term drought monitoring. This study monitored droughts in Northeast Thailand from 2001 to 2019 using the MODIS normalised difference vegetation index (NDVI) image time series. The Savitzky-Golay (S-G) method was used to remove noise and fill gaps in the image datasets. Optimal indicators as the vegetation condition index (VCI) and the standard vegetation index (SVI) were used to monitor drought distribution patterns over the previous 19 years. S-G filtering effectively reduced the impact of undetected clouds and water vapour, while VCI had the highest accuracy coefficient of determination (R^2) for rainfall data at 0.85. Long-term droughts occurred frequently in 2005, 2004, 2007, and 2001 with the northern and central regions most severely affected. Severe drought primarily impacted agricultural land, forest and miscellaneous areas. Inter-annual drought variability for one and three time steps was clearly demonstrated in May and April to June from 2001 to 2019. Overall, the VCI provided a high level of satisfaction for drought monitoring in this region and clearly displayed the spatial distribution of long-term drought regions. Our findings provide a valuable resource for drought mitigation planning and warning systems.

1. INTRODUCTION

Drought is an ubiquitous and recurring natural phenomenon worldwide, resulting in a lack of water supplies, decreased agricultural productivity, and increased socioeconomic risks (Ding et al., 2011). Drought occurrences have become more severe, prolonged, and frequent in recent decades as a result of climate change and population development, particularly in tropical and subtropical regions (AghaKouchak et al., 2015; Ullah et al., 2022). With this, drought management and monitoring offer valuable historical information.

Northeast Thailand is in the tropics, and the soil is primarily sandy with low water-holding capacity. Droughts have become more severe and frequent, wreaking havoc on agricultural and economic sectors, with farmers suffering from low crop production (Som-ard, 2020; Thavorntam et al., 2015).

Earth Observation (EO) data can be used to monitor the spatial distribution of drought occurrences and provide crucial information for response operations aimed at mitigating the most severely affected areas.

EO data has evolved into a vital tool for drought mapping, providing near-real-time global coverage as well as long-term service (Klisch & Atzberger, 2016). EO data have been collected by numerous sensors such as Landsat-5, the Thematic Mapper (L5 TM); Landsat-7, the Enhanced Thematic Mapper (L7 ETM+); Landsat-8, the Operational Land Imager (L8 OLI); and Moderate Resolution Imaging Spectroradiometer (MODIS). The EO data is increasingly being used for drought monitoring (Ghaleb et al., 2015; Klisch & Atzberger, 2016; Ullah et al., 2022). These mapping results have proven to be highly satisfactory for monitoring drought events, particularly when detailed temporal EO data is used.

The Terra and Aqua platforms' comprehensive archives of long-term MODIS satellite data provide highly efficient observation frequencies with large area coverage to capture rapidly changing land dynamics. All sensor data is open source and has been widely used in natural resource management (Justice et al., 2002; Wang et al., 2018). Numerous studies have demonstrated the high potential of MODIS dense temporal observations and large land coverage time series to track drought events under frequently cloudy conditions (Cammalleri et al., 2019; Klisch & Atzberger, 2016; Kumar et al., 2021; Thavorntam et al., 2015). MODIS time series data can be used to track drought episodes in harsh climate zones.

The advantages of using MODIS time series data together with several vegetation indices such as the normalized difference vegetation index (NDVI), vegetation condition index (VCI), temperature condition index (TCI), temperature-vegetation drought index (TVDI), vegetation health index (VHI), standardized vegetation index (SVI) for drought mapping have been shown in several studies. For example, Zhao et al. (2021) detected drought areas in the Yellow River Basin in China from 2003 to 2019 using MODIS data, while Kumar et al. (2021) developed a tool to monitor agricultural drought in the Tamil Nadu State of India. Klisch and Atzberger (2016) evaluated the optimal monitoring systems for drought events in Kenya using MODIS NDVI time series, while Rotjanakusol and Laosuwan (2019) tracked drought using Terra-MODIS from 2014 to 2016 in the lower northeast of Thailand. Their results showed highly

* Corresponding author

https://doi.org/10.5194/isprs-archives-XLVIII-M-1-2023-367-2023

temporal and spatio-temporal drought patterns from various vegetation indices. Long-term drought monitoring across wide regions has been neglected in developing countries including Thailand. Monthly drought monitoring in Thailand remains a challenge due to highly cloudy conditions and inclement weather regions.

Therefore, the main objectives of this study were to:

- analyze the optimal vegetation index in Northeast Thailand from 2001 to 2019 using MODIS NDVI time series.
- identify the most severe drought distribution patterns between 2001 and 2019 in Northeast Thailand.

2. MATERIAL AND METHODS

The implemented methods included filtering image time series, analyzing the optimal drought index and mapping drought distribution patterns (Figure 1).

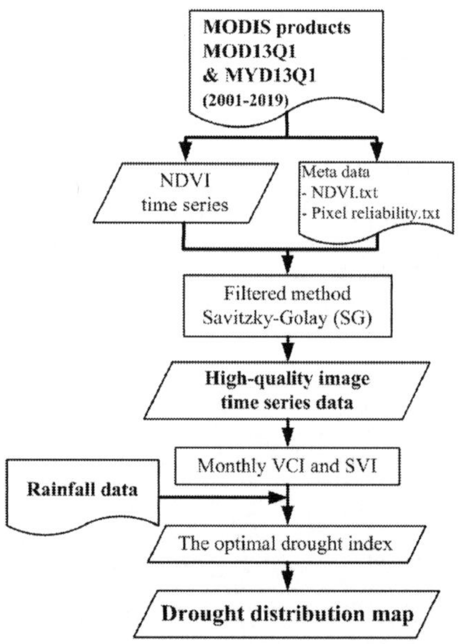

Figure 1. Illustration of the implemented workflow diagram for drought monitoring from 2001 to 2019 in Northeast Thailand.

2.1 Study Area

Northeast Thailand was selected as the study covers 14°00′ to 18°27′ and 101°00′ to 105°35′ with an area of 168,854 km^2 (Figure 2). The 30-year average temperature is 26.9 °C and the average annual rainfall is 1,446 mm. This region is categorized as having a tropical semi-humid dry-savannah climate (Köppen climate, classification: Aw) with three seasons of summer, rainy season and winter.

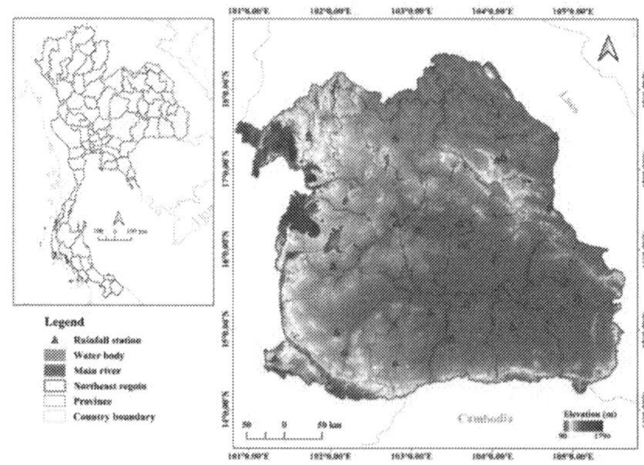

Figure 2. Study region (Northeast Thailand): background shows elevation with a height of 90-1,790 mm. The red triangle points are the rainfall stations of the Thai Meteorological Department.

2.2 MODIS Data Sets

2.2.1 Data Acquisition: MOD13Q1 and MYD13Q1 NDVI with five MODIS Terra and Aqua satellite products, were obtained through the University of Natural Resources and Life Sciences (BOKU) online platforms. These products were mosaicked, multi-tiled, and re-projected to geographic coordinates (datum WGS84) by the BOKU Geomatics server (http://ivfl-info.boku.ac.at/) (Vuolo et al., 2012). NDVI products were selected from both sensor data acquired every 8 days (temporal resolution) at 250 m spatial resolution. A total of 874 images were collected, taken between 2001 and 2019 across the study region.

2.2.2 Filtered Image Time Series: MODIS NDVI time series data are affected by environmental conditions such as clouds, water vapor and smoke, resulting in poor image data quality. This study adapted the Savitzky-Golay (S-G) filtering approach to eliminate noise and fill gaps for the time series data. S-G filtering can be employed to smooth, reconstruct and fit spectral values using the least square algorithm (Savitzky & Golay, 1964). The S-G filtering was conducted in TIMESAT software with two setting parameters: the half-width of the smoothing window (*m*) and the number of degrees of the smoothing polynomial (*n*). The S-G filtering was utilized to derive the smoothed image time series data using equation 1. This analysis fitted the optimal filter parameters of 8 (*m*) and 3 (*n*), and used 10 iterations to determine the best fitting effect for processing. High-quality time series datasets were then applied to the study region as the image output.

$$Y_j^* = \sum_{i=-m}^{i=m} \frac{C_i Y_{j+1}}{n} , \quad (1)$$

where Y^*_j is the filtered image time series data, Y_j is the input data, C_i is the coefficient of filter fitting, i.e., the weight of input data, n is the filtered processing data and the smoothing window size ($2m+1$), and m is the half-size of the smoothing window.

2.3 Calculated Drought Indices

2.3.1 The VCI is ideal for assessing the impact of weather on vegetation by enhancing inter-annual vegetation index variations based on EO satellite data (Kogan et al., 2003). The maximum VCI value is 100% with minimum of zero. For this study the VCI was used to map drought anomalies following equation 2, by utilizing the *raster* package in R software (version 3.6.2) (Hijmans et al., 2015) for this study.

$$VCI_i = 100 \text{ x } (NDVI_i - NDVI_{min,i})/(NDVImax,i - NDVImin,i), \quad (2)$$

where VCI_i is the updated vegetation condition index (VCI) at time step i, $NDVI_i$ is the filtered NDVI image at time i and $NDVImin,i$ or $NDVImax,i$ are the lowest/highest values of filtered NDVI pixels from 2001 to 2019 during eight days.

The VCI categorized the thresholds for drought anomalies compared to rainfall data, with drought ranks shown in Table 1.

VCI (%)	Drought category
≤ 15	Extreme drought
16 to 35	Severe drought
36 to 50	Moderate drought
51 to 65	Normal/ No drought
> 65	Wet

Table 1. Thresholds used to update the vegetation condition index (VCI) and related drought categories based on rainfall data.

2.3.2 The SVI measures: the probability of vegetation condition deviation from normal preferred to NDVI value in a year. The performed SVI was summarized by Peters et al. (2002). SVI values between 0 and 1 were applied for assessing drought, as calculated in equation 3.

$$SVI_i = (Z_{ijk} - Z_{ijMIN})/(Z_{ijMAX} - Z_{ijMIN}), \quad (3)$$

where SVI_i is the updated standardized vegetation index (SVI) at time step j, Z_{ijk} is z-value of the filtered NDVI pixel i at time j for year k and Z_{ijMIN} or Z_{ijMAX} are the lowest/highest filtered NDVI pixels i from 2001 to 2019 at time j during eight days.

The SVI was defined as the threshold of drought anomalies in comparison to rainfall data. The vegetation anomalies were classified and shown in Table 2.

SVI	Drought category
≤ 0.05	Extreme drought
0.06 to 0.25	Severe drought
0.26 to 0.75	Moderate drought
0.76 to 0.95	Normal/ No drought
≥ 0.96	Wet

Table 2. Thresholds related to vegetation anomalies using the standardized vegetation index (SVI) based on rainfall data.

2.4 Drought Indicator Assessment

The VCI and SVI were compared to rainfall data using the coefficient of determination (R^2) to determine the best drought index for this research region. R^2 was calculated by collecting VCI and SVI pixel values at a Sisaket rainfall station and utilizing monthly rainfall data from 2007 to 2017. The optimal indicator was then used to monitor the spatial distribution of drought patterns from 2001 to 2019.

2.5 Mapping Drought Distribution Patterns

To monitor drought affected regions, the monthly indicator was temporally and spatially aggregated to map drought occurrences. One month demonstrated short-term drought from 2001 to 2019, with three months of VCI (VCI3M) aggregated to map drought distribution patterns. Annual maps showed long-term trends for the most severe drought years between 2001 and 2019.

3. RESULTS

3.1 MODIS NDVI Image Time Series Dataset

The NDVI profile of a randomly selected pixel in the center of an arid region (row 1228, column 946) is shown in Figure 3. The filtered result was consolidated (smoothed) as the brown line compared to the original observation data.

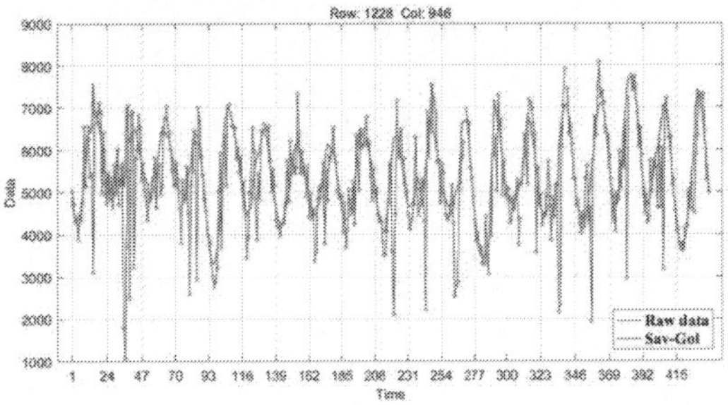

Figure 3. MODIS NDVI pixels of raw data (blue line) and filtered information (brown line) of filtered NDVI image time series data from 2001 to 2019 using the Savitzky-Golay (S-G) method.

A comparison between the original and filtered MODIS NDVI image time series data (exemplary output in August 2019) is demonstrated in Figure 4. Filtered map values were smoothed, with noise pixel time series values masked to improve the image datasets. The map results presented a high-quality dataset by fitting the best parameters of the S-G method.

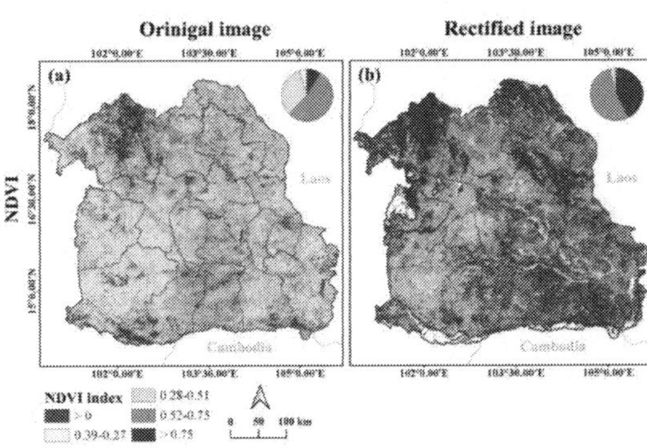

Figure 4. Comparison of MODIS NDVI image time series data with exemplary of the observed images in August 2019: (a) original and (b) filtered image dataset.

3.3 Drought Indicator

Figure 5 illustrates the R^2 values of the VCI and SVI monthly vegetation indices, as well as rainfall data, from 2007 to 2017. In 2017, the R^2 values were 0.85 (VCI) and 0.74 SVI. The VCI value was slightly greater than the SVI. The VCI indicator demonstrated remarkable sensitivity for extremely variable monthly rainfall levels, such as those found in Thailand.

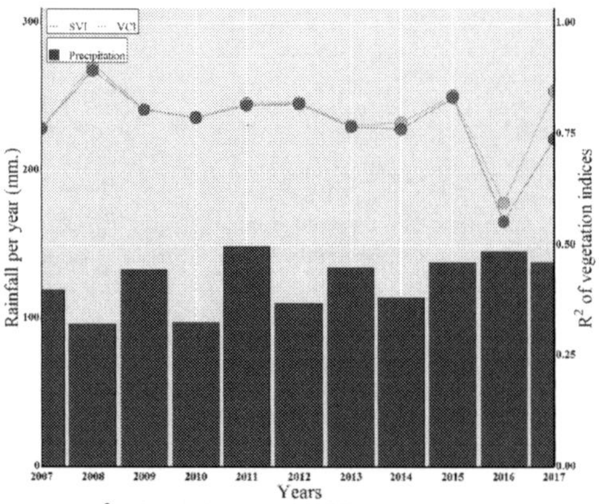

Figure 5. R^2 values of SVI and VCI monthly vegetation indices and rainfall data from 2007 to 2017 in Northeast Thailand.

3.4 Spatial Drought Distribution Pattern

Figure 6 shows a boxplot of annual VCI (%) from 2001 to 2019. The presence of mean VCI (%) values below 50% (red line) indicated a drought anomaly. Drought events were detected in the seven years ranked in regard to their severity as 2005, 2004, 2007, 2001, 2010, 2002, and 2006.

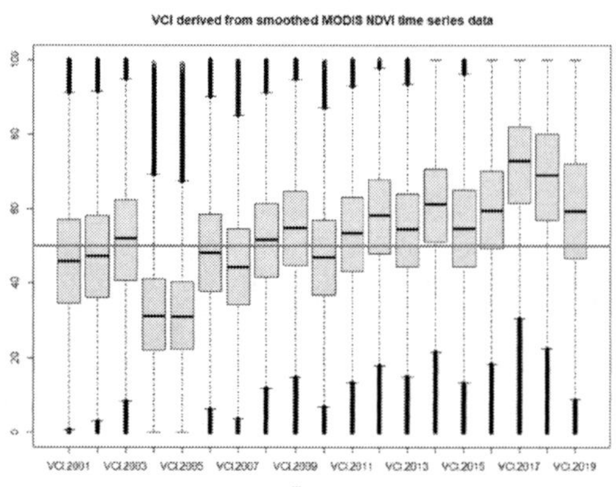

Figure 6. Boxplot of mean annual vegetation condition index (VCI) (%) from 2001 to 2019 in Northeast Thailand. Years under the red line indicate drought anomalies.

Long-term drought monitoring from 2001 to 2019 is shown in Figure 7. Drought distribution patterns are based on VCI (%), with the worst drought areas located in north and central areas.

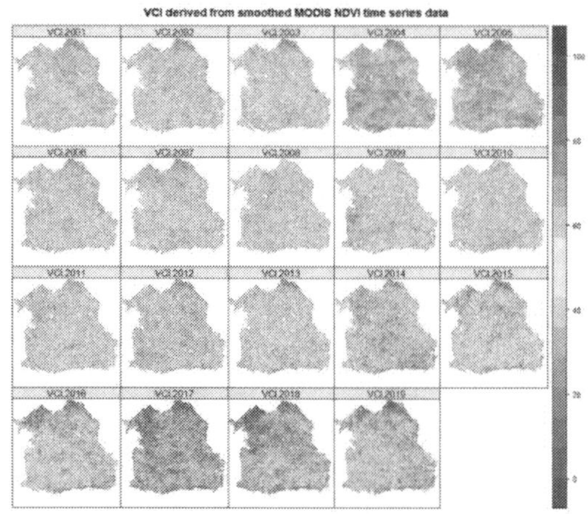

Figure 7. Long-term drought monitoring in Northeast Thailand from 2001 to 2019 derived from the mean annual vegetation condition index (VCI) (%).

Temporal and spatial VCI (%) for the three months April to June were aggregated to show spatial distribution patterns of the worst drought years (Figure 8), with mean VCI3M (%) values of 27.24 (2005), 45.30 (2004), 45.50 (2007), and 46.16 (2001). The VCI3M of 2005 shows the greatest drought anomaly over the four years. Droughts were most severe in the north, center, and south, where they predominantly affected agricultural areas.

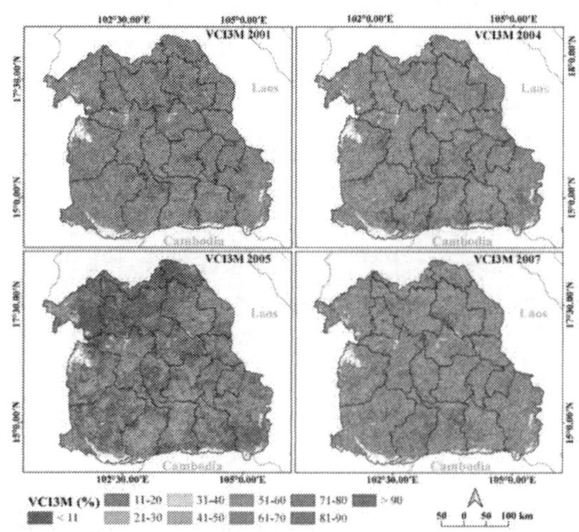

Figure 8. Inter-annual variability of three time steps (April to June) for drought in Northeast Thailand in 2001, 2004, 2005, and 2007 based on the vegetation condition index (VCI) (%).

The inter-annual variability for a single time-step in May for the years of the most severe drought occurrences (2001, 2004, 2005, and 2007) is illustrated in Figure 9. This month is generally the beginning of the rainy season in Thailand and is highly effective in presenting different vegetation growth due to rainfall. The map result showed the high frequency and extreme intensity of drought event patterns, which mostly appeared in the central, south, and southeast. The drought mostly impacted agricultural land and miscellaneous areas. Results suggested repeated drought events across Northeast Thailand.

This contribution has been peer-reviewed.
https://doi.org/10.5194/isprs-archives-XLVIII-M-1-2023-367-2023 | © Author(s) 2023. CC BY 4.0 License.

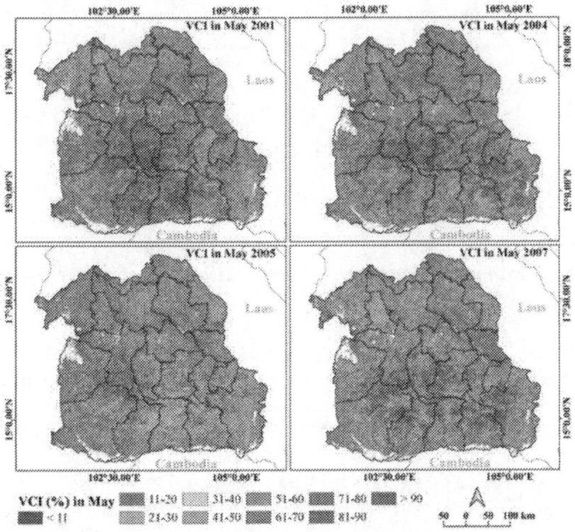

Figure 9. Inter-annual variability for single-time step (May) with drought in Northeast Thailand in 2001, 2004, 2005 and 2007 based on the vegetation condition index (VCI) (%).

4. DISCUSSION

The S-G method showed high potential to remove noise and fill gaps in the MODIS NDVI image time series data. Our findings agreed with those of Cai et al., (2017), emphasizing the ability to predict average phenology across the entire region. This study also achieved highly smoothed NDVI image time series data for estimating drought events in high cloud coverage areas like Northeastern Thailand, highlighting the performance for predicting average phenology over the entire region. This study also achieved highly smoothed NDVI image time series data for estimating drought events in high cloud coverage areas like Northeastern Thailand.

The VCI was the optimal indicator for monitoring drought events in this region. The VCI value is highly related to the R^2 value when compared to monthly rainfall data, concurring with Klisch and Atzberger (2016). Their findings demonstrated that it is more efficient to monitor drought dynamics. AghaKouchak et al. (2015), Thavorntam et al. (2015), Kumar et al. (2021) and Zhao et al. (2021) demonstrated VCI as a powerful tool for monitoring long-term vegetation conditions and drought events. The SVI showed promising results, but accuracy decreased with high differences in monthly rainfall (Luetkemeier et al., 2017). Our findings confirmed that VCI derived from the smoothed NDVI image time series datasets provided high efficiency for monitoring drought events in this region from 2001 to 2019.

Over the period of 19 years, long-term spatial drought monitoring revealed anomalous distribution patterns. Our data revealed drought anomalies in seven years, with the greatest variation occurring in 2005, 2004, 2007, 2001, 2010, 2002, and 2006. These drought occurrences were caused by variability in rainfall data, which resulted in insufficient chlorophyll for vegetation phenology (Thavorntam et al., 2015). According to Limsakul and Singhruck (2016), the mean VCI3M (April to June) map in 2005 exhibited Thailand's greatest drought due to the El Niño event (2016). The monthly drought maps produced excellent outcomes for this region's drought distribution patterns, with drought occurrences primarily occurring in agricultural and forest areas. Over a 9-year period, Thavorntam et al. (2015) observed widespread and intense drought occurrences, mainly in the north, central, and southern areas.

Droughts are common in the Northeast Thailand, and their effects on agriculture are significant, as plant growth is heavily reliant on rainwater, irrigation, and groundwater during seasons of low rainfall. However, the region has continued to experience drought occurrences, which cause rice harvests to perish, cows to go without grazing, and natural water levels to plummet. The main cause of drought is a lack of rainfall, as well as the absence of seasonal rain, which usually happens from late June to the end of July. An extended period of no rainfall is frequently encountered in the region's middle and lower reaches. Because these two areas are not impacted by the Southeast Asian Monsoon, there is less water in natural sources and groundwater bodies. As a result, moisture in the soil is drained, impacting plant growth (Sumpong et al., 2021), particularly in the paddy fields of Thung Kula Rong Hai district, which is a key area for jasmine rice cultivation and received first prize in the 2020 World's Best Rice Award (Jirapornvaree et al., 2021; Chouichom, 2021).

As a result, the government has assessed which regions are ideal for rice farming and which should be used instead for the growth of other crops that require minimal water during the dry season. Because water is limited, certain places may be converted into something, such as rice fields to crop farms cultivating sugarcane, cassava, maize, or rubber-all of which have now become the country's principal economic crops, as well as important exported items. Finally, drought is clearly a crucial problem that causes economic, social, and environmental damage, particularly in agricultural sectors; hence, solid planning as well as rapid effective technologies, including quick accurate forecast, are essential to prevent and mitigate the effects of drought. We utilized high temporal MODIS time series datasets to track the spatial spread of drought from 2001 to 2019. Drought maps with one and three time increments can also be used to properly evaluate and monitor the spatiotemporal changes of agricultural drought occurrences. The results demonstrated great efficiency as a tool for providing critical information to relevant stakeholders for well-organized drought management and planning.

Our study highlighted anomalies in drought distribution patterns from 2001 to 2019 across Northeast Thailand. Future studies should apply climatic and dynamic factors for large-scale drought monitoring, following the recent study of Ullah et al. (2022), to further improve the accuracy of drought monitoring.

5. CONCLUSIONS AND OUTLOOK

High temporal data from the MODIS time series datasets can be used to monitor drought. This study tracked long-term drought events over cloudy and rainy days in Thailand. Results showed high efficiency as an essential tool for drought management.

Key findings and further recommendations are presented below:

- Map results showed drought distribution patterns from 2001 to 2019 that can be used as essential land management and planting tools.

https://doi.org/10.5194/isprs-archives-XLVIII-M-1-2023-367-2023 |

- To provide high-quality NDVI time series datasets for inclement weather regions, the Savitzky-Golay (S-G) method effectively reduced pixel error in time series by smoothing and removing. Future studies can use our proposed method.

- The VCI had high efficiency. However, other drought indices (i.e., temperature condition index (TCI), temperature-vegetation drought index (TVDI), and vegetation health index (VHI)) together with other climatic data and powerful machine learning methods such as random forest (RF), support vector machines (SVM), and artificial neural networks (ANN) should also be used to improve drought monitoring in this highly adverse region.

- Our results are useful for drought mitigation planning and warning systems. Long-term Earth Observation (EO) and dense temporal data provide highly accurate drought mapping in areas where drought events repeatedly occur to improve land management and agricultural cultivation.

ACKNOWLEDGEMENTS

This research project was financially supported by Mahasarakham University Department Fund. The authors would like to acknowledge the Institute of Geomatics, University of Natural Resources and Life Sciences (BOKU) who provided the platform to access the MODIS NDVI time series data.

REFERENCES

AghaKouchak, A., Farahmand, A., Melton, F., Teixeira, J., Anderson, M., Wardlow, B. D., Hain, C., 2015. Remote sensing of drought: Progress, challenges and opportunities. *Rev. Geophys.*, *53*(2), 452-480.

Cai, Z., Jönsson, P., Jin, H., Eklundh, L., 2017. Performance of smoothing methods for reconstructing NDVI time-series and estimating vegetation phenology from MODIS data. *Remote Sens.*, *9*(12), 1271.

Cammalleri, C., Verger, A., Lacaze, R., Vogt, J., 2019. Harmonization of GEOV2 FAPAR time series through MODIS data for global drought monitoring. *Int J Appl Earth Obs and Geoinf.*, *80*, 1-12.

Chouichom, S., 2021. Readiness of entrepreneurs towards group performance development of OTOP product: A case study in Northeastern Thailand. In: Behnassi, M., Barjees Baig, M., El Haiba, M., Reed, M.R. (eds) *Emerging Challenges to Food Production and Security in Asia, Middle East, and Africa*. Springer, Cham.

Ding, Y., Hayes, M. J., Widhalm, M., 2011. Measuring economic impacts of drought: a review and discussion. *Disaster Prev. Manag.*, *20*(4),434-446.

Ghaleb, F., Mario, M., Sandra, A. N., 2015. Regional landsat-based drought monitoring from 1982 to 2014. *JCLI*, *3*(3),563-577.

Hijmans, R.J., Etten, J., Cheng, J., Mattiuzzi, M., Sumner, M., Greenberg, J.A., Lamigueiro, O.P., Bevan, A., Racine, E.B., Shortridge, A. and Ghosh, A., 2014. Geographic data analysis and modeling. R package version 2.3-12. https://cran.r-project.org/web/packages/raster/index.html (1 October 2022).

Jirapornvaree, I., Suppadit, T., Kumar, V., 2021. Assessing the economic and environmental impact of jasmine rice production: Life cycle assessment and Life Cycle Costs analysis. *J. Clean. Prod.*, *303*, 127079.

Justice, C., Townshend, J., Vermote, E., Masuoka, E., Wolfe, R., Saleous, N., Roy, D., Morisette, J., 2002. An overview of MODIS Land data processing and product status. *Remote Sens. Environ*, *83*(1-2), 3-15.

Klisch, A., Atzberger, C., 2016. Operational drought monitoring in Kenya using MODIS NDVI time series. *Remote Sens.*, *8*(4), 267.

Kogan, F., Gitelson, A., Zakarin, E., Spivak, L., Lebed, L., 2003. AVHRR-based spectral vegetation index for quantitative assessment of vegetation state and productivity. *Photogramm Eng Rem S.*, *69*(8), 899-906.

Kumar, K. A., Reddy, G. O., Masilamani, P., Turkar, S. Y., Sandeep, P., 2021. Integrated drought monitoring index: A tool to monitor agricultural drought by using time-series datasets of space-based earth observation satellites. *Adv. Space Res.*, *67*(1), 298-315.

Limsakul, A., Singhruck, P., 2016. Long-term trends and variability of total and extreme precipitation in Thailand. *Atmos Res.*, *169*, 301-317.

Luetkemeier, R., Stein, L., Drees, L., Liehr, S., 2017. Blended drought index: integrated drought hazard assessment in the Cuvelai-Basin. *JCLI*, *5*(3), 51.

Peters, A. J., Walter-Shea, E. A., Ji, L., Vina, A., Hayes, M., Svoboda, M. D., 2002. Drought monitoring with NDVI-based standardized vegetation index. *Photogramm Eng Rem S.*, *68*(1), 71-75.

Rotjanakusol, T., Laosuwan, T., 2019. Drought evaluation with NDVI-based standardized vegetation index in lower northeastern region of Thailand. *Geogr. Tech.*, *14*(1).

Savitzky, A., Golay, M. J., 1964. Smoothing and differentiation of data by simplified least squares procedures. *Anal. Chem.*, *36*(8), 1627-1639.

Som-ard, J., 2020. Rice security assessment using geo-spatial analysis. *Int. J. Geoinformatics*, *16*(1), 21-38.

Sumpong S., Kaensombat, J., Boonanak, A., 2021. Water with drought in the Northeast. *Humanit J. (Graduate School), Ramkhamhaeng University*, *10*(2), 67-80.

Thavorntam, W., Tantemsapya, N., Armstrong, L., 2015. A combination of meteorological and satellite-based drought indices in a better drought assessment and forecasting in Northeast Thailand. *Nat Hazards* (Dordr), *77*(3), 1453-1474.

Ullah, I., Ma, X., Ren, G., Yin, J., Iyakaremye, V., Syed, S., Kaidong, Lu., Xing, Y. Singh, V. P., 2022. Recent Changes in Drought Events over South Asia and Their Possible Linkages with Climatic and Dynamic Factors. *Remote Sens.*, *14*(13), 3219.

This contribution has been peer-reviewed.
https://doi.org/10.5194/isprs-archives-XLVIII-M-1-2023-367-2023 | © Author(s) 2023. CC BY 4.0 License.

Vuolo, F., Mattiuzzi, M., Klisch, A., Atzberger, C., 2012. Data service platform for MODIS Vegetation Indices time series processing at BOKU Vienna: current status and future perspectives. *Proc. SPIE Earth Resour. Environ. Remote Sens./GIS Appl. III*, *8538*, 83-92.

Wang, F., Wang, Z., Yang, H., Zhao, Y., Li, Z., Wu, J., 2018. Capability of remotely sensed drought indices for representing the spatio–temporal variations of the meteorological droughts in the Yellow river basin. *Remote Sens.*, *10*(11), 1834.

Zhao, X., Xia, H., Pan, L., Song, H., Niu, W., Wang, R., Li, R., Bian, X., Guo, Y., Qin, Y., 2021. Drought monitoring over Yellow river basin from 2003–2019 using reconstructed MODIS land surface temperature in Google Earth Engine. *Remote Sens.*, *13*(18), 3748.

LAND COVER DYNAMIC CHANGE MODES ANALYSIS BASED ON EMPIRICAL MODE DECOMPOSITION

Liang Tang[1], Haijun Yang[1*], Ping Tang[2], Zhuang Li[1], Yuhao Tang[2]

[1]Hainan Tropical Ocean University, Sanya, China - tyh9977@163.com,hyang@hntou.edu.cn, hitlz@163.com
[2]Aerospace Information Research Institute, Chinese Academy of Sciences, Beijing, China - tangping@radi.ac.cn, tangyh0707@163.com

KEY WORDS: Time-series, Land Cover, Dynamic Analysis, Empirical Mode Decomposition.

ABSTRACT:

The dynamic change mode of land cover not only reflects the change characteristics of the land cover itself, but also reflects the influence exerted by external factors. Empirical mode decomposition is an effective time-frequency analysis method for nonlinear and nonstationary signals. In this paper, empirical mode decomposition firstly uses to each pixel time series to get its intrinsic mode functions, each intrinsic mode function relates to change mode with some time scale, clustering those intrinsic mode functions over the image, and different land cover change modes are obtained. The experimental dataset used is MODIS subset from the Jan 2011 to Dec 2020, located in the Ningxia Hui Autonomous Region, China. The clustering results of half-annual cycle and one-year cycle of intrinsic mode functions show the influence of local climate and policy factors on land cover change.

1. INTRODUCTION

Land cover is one of the core topics in the study of global change. It not only affects the natural basis of human survival and development, but also affects the structure and function of the Earth's biochemical layer and the circulation of energy and material in the Earth system. On the global, national or regional scale, land use change always drives land cover change. The change information plays an important role in global climate change research, ecological environment monitoring and other fields. The Terra Moderate Resolution Imaging Spectrometer (MODIS) Normalized Difference Vegetation Index data products are employed for land cover change detection (Tucker, 1979). The vegetation in terrestrial ecosystem plays an important role in the global cycle of material and energy and is becoming more and more crucial to the sustainability in the context of global climate change and intensifying human activities. Vegetation changes can create feedback to climate system via the biogeophysical effects, i.e., increased leaf cover, tundra shrubification, and tree line advance can mediate land–atmosphere interactions (Zhang, 2013, 2018). To study non-stationary vegetation response, we selected Ningxia Hui Autonomous Region(NHAR) as a study area.

The normalized difference vegetation index (NDVI) is currently widely used to estimate vegetation cover, biomass and leaf area index and thus its changes can be used as a proxy for greening or browning (Alcaraz-Segura, 2010). In recent decades, several scholars have used NDVI data to conduct in-depth research on the vegetation change (Cihiar, 1991; Nicholson, 1994; Eklundh, 1998). Their results showed that the change of vegetation is closely related to not only climate factors but also human activities (Vicente-Serrano et al 2005).

NDVI time series are usually non-stationary, i.e., they present different frequency components, such as periodic or seasonal variations, long-term and short-term fluctuations, which may not be limited only to the mean of the series but also affects its overall variance structure. Typically, such series are characterized by patterns like periodicity, trends and localized abrupt changes or discontinuities resulting from disturbance events (de Beurs, 2005).

In this study, we selected Ningxia Hui Autonomous Region as the study area. We first decomposed the frequency components using the empirical mode decomposition (EMD) in the NDVI from Jan 2011 to Dec 2020. And then classified the half-annual cycle and one-year cycle of intrinsic mode functions using K-means clustering algorithm. The objective of this study is to develop a spatio-temporal coupling method for multi-scale change information extraction and analysis. These findings will promote the development and application of remote sensing time series image information mining technology.

2. DATA AND METHODS

2.1 Data and Preprocessing

The NDVI time series as the study area come from MOD13A2 product data of the Moderate-resolution imaging spectroradiometer (MODIS) during the period from January 2011 to December 2020. The product downloaded from website http://ladsweb.nascom.nasa.gov/data/search.html, 1km spatial resolution and 16day NDVI Maximum Value Composite (MVC) Product. The data has 23 phases per year and a total of 230 phases in 10 years. The MOD13A2 products are converted format from HDF to TIFF and re-projected by MODIS Reprojection Tools (MRT). Due to atmospheric variations, cloud contamination or multidirectional effects, MOD13A2 product still contain a lot of residual noises and outliers by MVC processed. These residual noises outliers hinder further analysis and have a risk to produce erroneous results, and be eliminated by further processing. The optimum-length Savitzky-Golay filter was selected as reprocessing to reconstruct high-quality NDVI time-series by approaching the upper NDVI envelope via an iteration process (Tang, 2019).

2.2 Empirical Mode Decomposition

EMD is proposed as a time-frequency signal analysis method by Huang et al. in 1998(N.E.Huang, 1998). It can decompose time-frequency signal into a finite number of intrinsic mode

* Corresponding author

This contribution has been peer-reviewed.
https://doi.org/10.5194/isprs-archives-XLVIII-M-1-2023-375-2023

function(IMF) according to the characteristics of the data signal itself. Each IMF represents the characteristic components of the signal on different time scales and does not interfere with each other. Any intrinsic modal function should satisfy the following two conditions: the difference between the number of extreme points and the number of zeros of the eigenmodal function is not more than 1; at any time, the mean value of the upper and lower envelope of the intrinsic modal function is zero. Unlike Fourier and wavelet transform, the decomposed IMF components is derived from the data itself, and contain local characteristic signals of different time scales of the original signal, which is especially suitable for the analysis of non-stationary and non-linear signals(Yu, 2005). Therefore, EMD method has been effectively applied in different engineering fields since it was proposed(Gloersen, 2003; Huang, 1998; Huang, 1999; N.E.Huang, 2003; Liang, 2000; Kaplan, 2013).
The essence of the EMD is to get the intrinsic mode pattern through data time scale features and perform data decomposition. The EMD of the one-dimensional signal $x(t)$ can be expressed as after equations:

$$x(t) = \sum_{i=1}^{n} imf_i(t) + r(t) \tag{1}$$

where $imf_i(t)$ = ith IMF
$r(t)$ = residual function
n = IMF component number

Figure 1 shows the results of a 10-year NDVI time series decomposed by EMD at a certain point.

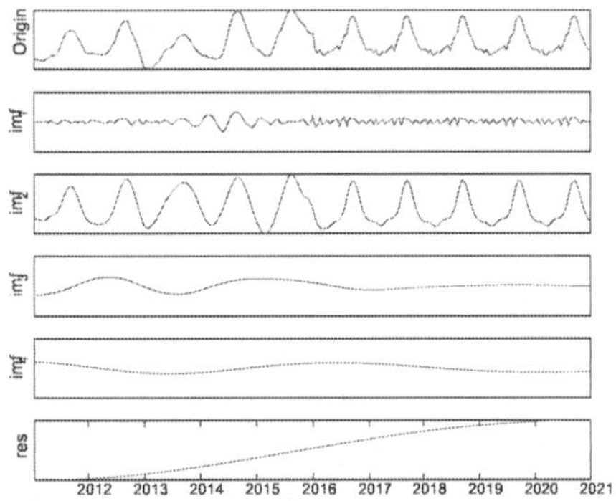

Figure 1. Decomposition results by EMD method

In figure 1, the frequency of signal from the component imf_1 to imf_4 decreased gradually, and the residual was denoted as res, which is a monotonic function and indicated an average trend. In res curve, NDVI value increased slowly from 2012 to 2013, and increased relatively rapidly between 2014 and 2018, and remains almost unchanged from 2011 to 2012, and until 2019, there was a slow increase.

2.3 K-means Clustering Algorithm

K-means clustering algorithm is an unsupervised clustering algorithm, which widely used in field of data mining across different disciplines in the past fifty years. The clustering results are not fixed, but have good convergence. In particular, the algorithm has low complexity and can quickly cluster chaos into several classes (Kumar, 2017). The basic idea of K-means algorithm is that for a given number of clusters k, an initial partition is created randomly, that is, some representative data points are randomly selected as the initial cluster centers. According to the distance between the remaining data points and each cluster center, it is divided into each class, and then the new cluster center is determined again, and so on. The iterative method is used to move the cluster center continuously to try to further improve the division, until the cluster center is not Change again.

We begin by providing a formal description of the k-means problem and two serial algorithms for finding an approximate solution. Let X = $\{x_1, \ldots, x_n\}$ be a set of N data points, each with dimension d. The K-means algorithm aims to partition the observations into K sets: C=$\{c_1,c_2,\ldots,c_k\}$f g such that $K \leqq N$:The k-means problem seeks to find a set of k means M = $\{\mu_1, \ldots, \mu_k\}$ which minimizes the objective function

$$J(k) = \sum_{i=1}^{N}\sum_{k=1}^{K} z_{ik} \| X_i - \mu_k \|^2 \tag{2}$$

Where $d_{x_i} = \| x_i - \mu_k \|^2$ is the minimum distance measured between observation X_i and the cluster center μ_k and z_{ik} is an indicator variable belonging/not to the kth cluster.

3. EXPERIMENT AND DISCUSSIONS

3.1 Study Region

Study region is NHAR of China. It is situated in the upper and middle reaches of the Yellow River, north-western China, and covers an area of approximately 66,400 km^2, extending from 104°17′ E to 109°39′ E, and from 35°14′ N to 39°14′ N. It is a transitional zone between desert and the Loess Plateau and has a temperate continental arid and semi-arid climate. There are three major deserts surrounding it and desertification is a challenging threat there. Several decades of efforts have been made therefore to protect and restore vegetation for desertification combatting, including the implementation of the Grain for Green Program. That program is the largest ecological restoration project in central and western China initiated by the Chinese Central Government in 1999. It was reported that the desertification there has been significantly alleviated. Therefore, many scholars are of great interest to understand it from the perspective of vegetation cover change in this area(Tang, 2020). Figure 2 shows a true-color-composite image of Landsat TM.

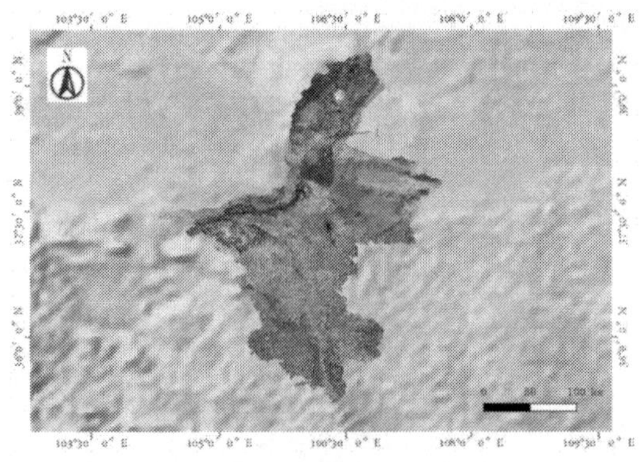

Figure 2. Location of the NHAR

3.2 Experiment procedure

We validated our decomposition model by a 16-day composition NDVI time series of forest type. Satellite image time series were collected from January 2011 to December 2020 (denoted as yellow block 1 in Figure 2), which was used to illustrate and discuss the time series decomposition procedure.

Figure 3 illustrates the time series of block 1, shown in Figure 2. The EMD decomposition result of this time series is further illustrated in Figure 4, and it was decomposed into five IMFs and one residue.

In figure 3, the NDVI value of this time series was at a relatively low value from 2011 to 2016, with little change. But from June 2016 to 2020, the value of NDVI had a relatively large increase, and then entered a period of equilibrium.

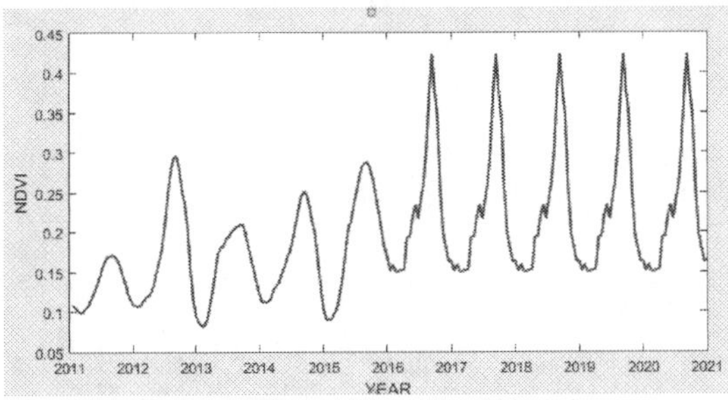

Figure 3. NDVI time series of block 1

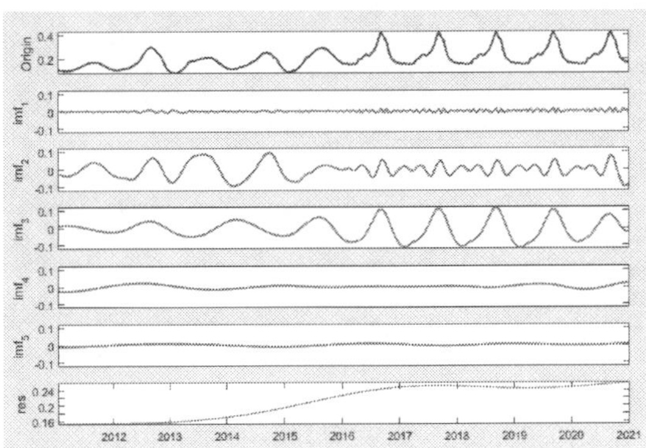

Figure 4. NDVI time series decomposed results using EMD

In figure 4, the time series (blue line) is decomposed into 5 IMFs (red line) and a residue (green line). The *imf₁*, which was close to white noise, was subtracted from the original. Two components, *imf₂* and *imf₃*, represent high-frequency component of the time series. which shown seasonal, weekly changes in vegetation. It shown period changes of vegetation. *Imf₄* and *imf₅* were low frequency components.

Components *imf₂* and *imf₃* represent the high-frequency component of the signal, which shows seasonal, periodic changes in vegetation. *Imf₄* and *imf₅* represent the low frequency components, which together with the res component constitute the trend term, representing the slow, trended change of vegetation. In res curve, NDVI value increased slowly from 2011 to 2013, and increased relatively rapidly between 2014 and 2017, and the trend of vegetation has a process of decreasing first and then increasing from 2017 to 2021.

To verify the validity of seasonal components, we analyzed the correlation between the biophysical characteristics of vegetation and the IMF2 and IMF3, taking into account that the NDVI is derived from the red and near-infrared reflectance ratio. It is easy to observe that the oscillation of the near infrared and red spectral reflectance time series is closely related to the IMF2 and IMF3 obtained by decomposing the NDVI time series. It is noted that the spectral reflectance time series in the near infrared (red) band have similar wave properties. In particular, the semi-annual wave of IMF2 is caused by the semi-yearly wave of reflectivity in near-infrared band, while the annual wave in IMF3 is the joint result of near-infrared and red band(Kong, 2015).

The average IMF period after EMD decomposition is calculated, which is equal to the ratio of the number of samples of the IMF to its maximum value points (or minimum value points). We first obtain the IMF components corresponding to the semi-annual period and one-year period of each time series, and then carry out K-means clustering analysis on these two components(Kong, 2015). According to the characteristics of land cover types in the study area, the land features are divided into 10 categories. The color configuration table is shown in Table 1.

INDEX	TYPE	VOLUATION	COLOR
1	No data	0	
2	Cultivated land	10	
3	Forest	20	
4	Grassland	30	
5	Shrubland	40	
6	Wetland	50	
7	WaterBodies	60	
8	Artificial surface	80	
9	Bareland	90	
10	Permanent snow and ice	100	

Table 1. Figure Color Configuration Table

The K-means clustering results and the clustering centers of the semi-annual time series are shown in Figure 5 and 6.

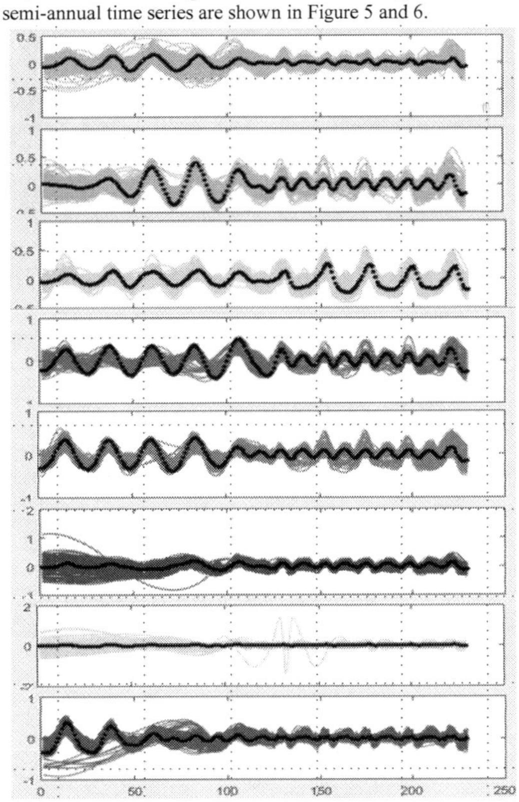

Figure 5. semi-annual time series k-means clustering results

In Figure 5 and 6. "*" represents the cluster center, and show the approximate proportion of each category. Because the isolated points and noise points will have a large impact on the k-means clustering results, the interpolation of missing values and the processing of noise such as cloud pollution are the key points. In this experiment, the results of time series pretreatment are not ideal, and need to be further strengthened.
The results of K-means clustering for the NDVI time series of the study area are shown in Figure 7 and Figure 8.

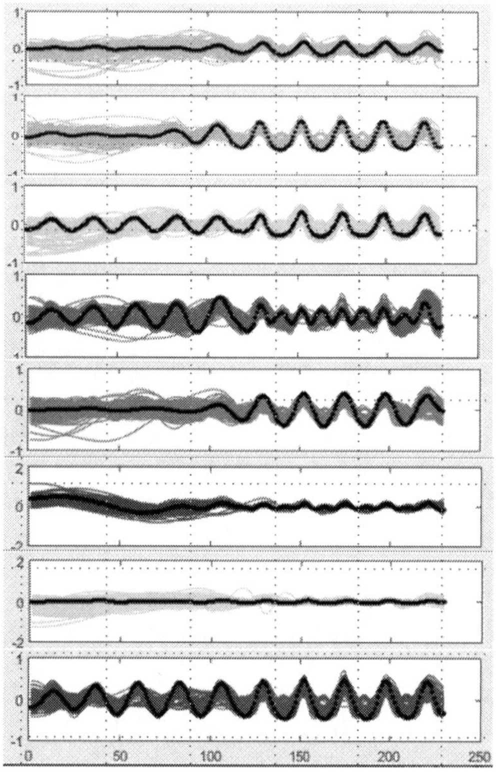

Figure 6. annual time series k-means clustering results

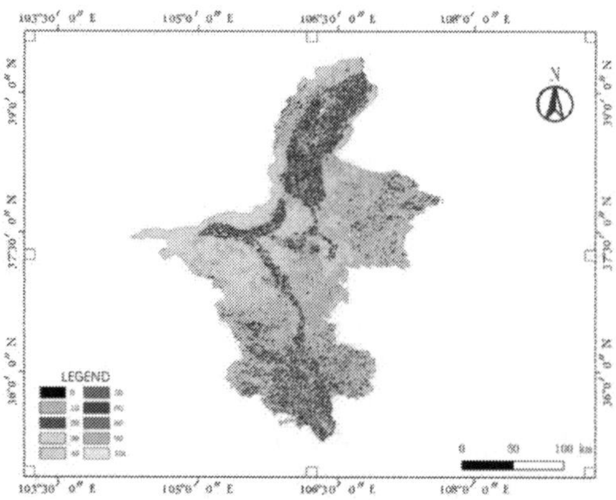

Figure 7. k-means clustering results for semi-annual time series

Figure 7 shows the k-means clustering results for semi-annual time series.

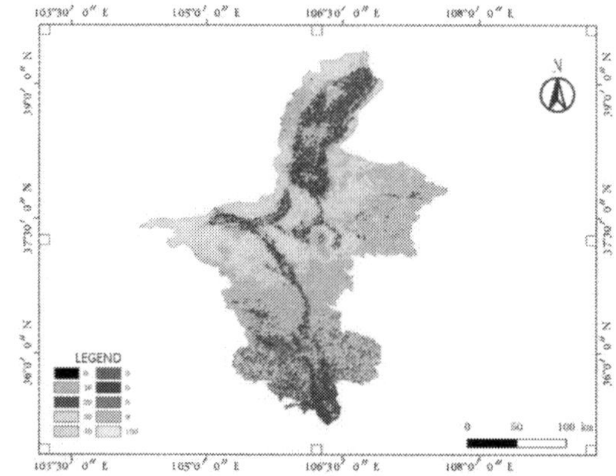

Figure 8. k-means clustering results for a one-year period time series

Figure 8 shows the k-means clustering results for a one-year period time series. From the two clustering results, the results of the two clusters are very similar, and the change of the clusters reflects the influence of local climate and policy factors on land cover change.
From the results of Figure 7 and Figure 8, the fragmentation of classification results is serious. The reasons for this problem may be as follows: Firstly, there may be some problems in the temporal sequence preprocessing method, resulting in outliers and noise points affecting the EMD decomposition results; Secondly, K-means clustering algorithm is sensitive to mutation point, which makes the clustering results scattered. Third, there is no post-processing, and part of the clustering results should be classified and merged.

4. CONCLUSIONS

Due to the influence of vegetation phenology, the surface of a certain area usually presents seasonal changes with a period of half a year or a year, which causes periodic fluctuations in the satellite image sequence of the area. This fluctuation pattern is usually stable and is determined by the local land cover type. When the land cover changes, the fluctuation pattern of the satellite image series will change, from one kind of land to another.
In this paper, remote sensing time series has the characteristics of seasonality, instability, regionality, multiscale, spatiotemporal autocorrelation, The semi-annual and one-year cycle clustering results of K-means clustering based on empirical mode decomposition are introduced, and the results are analyzed, as well as the relationship with local climate and policy factors.
Based on the current relevant research status, the detection of dynamic change information in remote sensing time series images mainly reflects the following development characteristics:
(1) A large amount of accumulated historical data and the newly launched high temporal resolution remote sensing sensors provide basic data guarantee for the research in this field.
(2) Sustained change is a relatively slow change in land use and cover type.
Land surface change includes some specific change patterns, such as land use / cover of cultivated land / unused land change to land of construction, grassland, forest, bare land and sandy land. Using the time series data, we can get the characteristics

of the surface changes with time, which provides an effective data source for the information mining of the surface changes.

(3) Multi-feature temporal sequence decomposition of multidimensional EMD. Traditional EMD decomposition for a single feature, but the single feature is not enough to reflect the characteristics of land cover change, In order to give full play to the advantages of remote sensing data and help to extract spatio-temporal dynamic features of remote sensing time series, it is necessary to try multi-feature and multi-dimension time series EMD decomposition, which can not only simplify the analysis method, but also can verify each other and improve the accuracy of detection.

ACKNOWLEDGEMENTS

This research was funded by the National Natural Science Foundation of China (Grant No. 42061064), the Hainan Natural Science Foundation of China (Grant No. 2019RC235) and the National key R&D Program of China(Grant No.2021YFF0704004).

REFERENCES

Chan, K.L., Qin K., 2017: Biomass burning related pollution and their contributions to the local air quality in Hong Kong. *Int. Arch. Photogramm. Remote Sens. Spatial Inf. Sci.*, XLII-2/W7, 29-36. doi.org/10.5194/isprs-archives-XLII-2-W7-29-2017.

Alcaraz-Segura, D., Chuvieco, E., Epstein, H. E., Kasischke, E. S., Trishchenko, A., 2010. Debating the greening vs. browning of the north American boreal forest: differences between satellite datasets. Global Change Biology, 16(2), 760-770.

Cihiar, J., Laurent St, et al., 1991. Relation between the NDVI and ecological variables. *Remote Sensing of Environment*, 35(3), 279-298.

de Beurs, K. M., & Henebry, G. M., 2005. A statistical framework for the analysis of long image time series. International Journal of Remote Sensing, 26, 1551−1573.

Eklundh, L., 1998. Estimating relations between AVHRR NDV I and rainfall in East Africa at 102 day and monthly time scales. International Journal of Remote Sensing, 19(3), 563-568.

Gloersen, P., Huang, N.,2003. Comparison of interannual intrinsic modes in hemispheric sea ice covers and other geophysical parameters. *IEEE Trans. Geosci. Remote Sens.* **2003**, *41*, 1062–1074. 19.

Huang, W., Shen, Z., N.E.Huang., et al, 1998. Use of intrinsic modes in biology: examples of indicial response of pulmonary blood pressure to step hypoxia. Proc. Natl Acad. Sci. USA, 1998, 95: 12766-12771.

Huang, W., Shen, Z., N.E.Huang, et al, 1999. Nonlinear indicial response of complex nonstationary oscillations as pulmonary hypertension responding to step hypoxia. Proc. Natl Acad. Sci. USA, 1999, 96: 1834-1839.

Kaplan, N., Erer, I., 2013. An additive empirical mode decomposition based method for the fusion of remote sensing images. In Proceedings of the 6th International Conference on Recent Advances in Space Technologies (RAST), Istanbul, Turkey, 12–14 June 2013; pp. 165–169.

Kong, Y. L., 2015. Anomaly Detection in Remote Sensed Imagery Time Series Based on EMD. Beijing: Institute of Remote Sensing and Digital Earth Chinese Academy of Sciences.

Kong, Y. L., Meng, Y., Li, W., Yue, A. Z., and Yuan, Y., 2015. Satellite image time series decomposition based on EEMD. Remote Sensing, 7(11): 15583–15604 [DOI: 10.3390/rs71115583]

Kumar, K., Reddy, A.R.M., 2017. An efficient k-means clustering filtering algorithm using density based initial cluster centers, Information Sciences, Volumes 418–419,2017, Pages 286-301, https://doi.org/10.1016/j.ins.2017.07.036.

Liang, H., Lin, Z., Mc, Callum RW, 2000. Artifact reduction in electrogastrogram based on the empirical mode decomposition method. Medical and Biological Computing, 2000, 38(1): 35-41.

N.E.Huang, Shen, Z., Long, S. R., Wu, M. C., Shih, H. H., Zheng, Q., Yen, N. C., Tung, C. C., and Liu, H. H.,1998. The empirical mode decomposition and the Hilbert spectrum for nonlinear and non-stationary time series analysis, in Proceedings of the Royal Society of London A: mathematical, physical and engineering sciences, 454(1971), The Royal Society, 903-995.

N.E.Huang, Wu, M. L., Qu, W. D., et al, 2003. Applications of Hilbert-Huang transform to non-stationary financial time series analysis. Applied Stochastic Models in Business and industry, 2003, 19: 245-268.

N.E.Huang, Shen, S. S., 2005. *Hilbert-Huang Transform and Its Applications*, World Scientific: Singapore, 2005; Volume 5. 20.

Nicholson, SE., Farrar, TJ., 1994. The influence of soil type on the relationship between NDVI, rainfall, and soil moisture in semiarid Botswana. *Remote Sensing of Environment*, 50(2), 107-120.

Tang, L., Zhao, Z., Tang, P., et al., 2019. SURE-based optimum-length SG filter to reconstruct NDVI time series iteratively with outliers removal[J]. International Journal of Wavelets, Multiresolution and Information Processing, 2019,18(2): 20500010.

Tang, L., Yang, H., J., 2020. Monotonic Vegetation Trend Detection Based on Ensemble Empirical Mode Decomposition in Ningxia, China. Conference Series: Earth and Environmental Science 502 (2020) 012042.

Tucker, C.J., 1979. Red and photographic infrared linear combinations for monitoring vegetation. *Remote Sens. Environ.* **1979**, *8*, 127–150.

Yu, D., Cheng, J., and Yang Y., 2005. Application of EMD method and Hilbert spectrum to the fault diagnosis of roller bearings. Mechanical systems and signal processing, 19(2), 259-270.

Zhang, W., Miller, P.A., Smith, B. et al., 2013. Tundra shrubification and tree-line advance amplify arctic climate warming: results from an individual-based dynamic vegetation model. Environ. Res. Lett., 8 (2013), Article 034023.

Zhang, W., Miller, P.A., Jansson, C., et al., 2018, Self-amplifying feedbacks accelerate greening and warming of the Arctic. Geophys. Res. Lett., 45 (2018), pp. 7102-7111.

Vicente-Serrano, SM., Lasanta, T., Romo, Alfredo., 2005. Analysis of spatial and temporal evolution of vegetation cover in the Spanish central Pyrenees: Role of human management. Environment Management, 34(6), 802-818。

This contribution has been peer-reviewed.
https://doi.org/10.5194/isprs-archives-XLVIII-M-1-2023-375-2023

ESTIMATING TREE HEIGHT USING LOW-COST UAV

Giuseppina Vacca [1]

[1] DICAAR, Dept. of Civil-Environmental Engineering and Architecture, University of Cagliari, Cagliari, Italy - vaccag@unica.it

KEY WORDS: UAV, SfM, Precision agriculture, photogrammetry.

ABSTRACT:

Precision agriculture (PA) is defined as an agricultural management based on the observation, measurement, and response of a set of quantitative and qualitative variables that affects agricultural production. The first step in the approach to precision agriculture is, therefore, the acquisition and collection of data from optical, multispectral, geophysical sensors, etc. aimed at obtaining knowledge and monitoring of the crop. In recent years UAVs are assuming an important role in precision agriculture. The possibility of mounting RGB, multispectral, LiDAR sensors on them make these systems fast, accurate, and usability compared other geomatic methods. In this study, we focus on a very low-cost UAV system to assess individual tree height and generate 3D Model and orthophoto of the study area. We used the DJI Mini2, a low-cost UAV that can be used without a flight rating and without restrictions due to its light weight of only 250 grams. The case study where the survey was performed is an agricultural area of about 1 hectare, where are some fruit trees and a small vineyard. The area was selected because it contained both tall and small trees. The study concerned the influence of the relative flight altitude and therefore of the GSD of the images on the extraction of the dimensional data of the trees. From the results obtained, it can be stated that the flight altitude has certainly more influence on the measurement of small tree (around one meter tall) compared to tall ones (around 4 m).

1. INTRODUCTION

Precision agriculture (PA) is defined as an agricultural management based on the observation, measurement, and response of a set of quantitative and qualitative variables that affects agricultural production (Ammoniaci, 2021). The first step in the approach to precision agriculture is, therefore, the acquisition and collection of data from optical, multispectral, geophysical sensors, etc. aimed at obtaining knowledge and monitoring of the crop (Pagliai, 2022).
Precision agriculture (PA) is a relatively new discipline that was developed in the middle of the 1980s, and that has been listed among the top ten developments in agriculture in the last decades (Crookston, 2006).
PA includes a wide range of techniques and methodologies for the management and monitoring of agriculture which aim to develop an increasingly sustainable agriculture.
The increase in the world population has led the FAO to estimate an increase in the need for products and food needs of 60% compared to the annual average analysed from 2005 to 2007, in relation to the growth forecast of the world population established at around 9 billion by the year 2050 (Alexandratos, 2012). The Commission for Agriculture and Rural Development of the European Parliament has also provided the same estimate regarding the increase in the world population, underlining how the demand for healthy food and optimal nutrition constitutes one of the major future challenges worldwide. The new challenge for agriculture will consequently be to produce more in a more sustainable way. A challenge that presupposes a key concept for agriculture, namely innovation.
The geomatics techniques then become a tool of considerable importance to contribute to this challenge, just remember in the early 1990s the first agriculture machines with on-board GNSS receivers spread to map production yields and satellite images were used to discriminate areas with different soil characteristics (Casa, 2018). In addition to these methodologies, the first optical sensors were used, mounted on agricultural machines to monitor the vegetative vigour (Casa, 2018). In the early 2000s, the new major players in precision agriculture were drones, which made it possible to provide, at low cost, very high-resolution images both in the visible and in the multispectral, becoming decision support systems in the prevention of diseases and water stress and in the determination of some tree parameters (Hobart, 2020).
Usually, the UAV's are developed with an automated drone system with sensors and cameras in order to monitor the condition and height of the crops (Velusamy, 2022).
The role of UAV is taken care by the captured images. The images can be RGB or multispectral so to help the farmers to take appropriate measures at the right time to protect the crops from diseases.
The advantage of using UAVs is certainly related to fast performance, high resolution and low cost compared to other geomatic methods (Zhang, 2021).
Among the data that are required in precision agriculture and that find an answer in the use of UAVs, are the determination of the width of the canopy and the height of the plants. In particular, the height of the plants is important for assessing the economic value of the plant and its health condition (Krause; Vaglio Laurin, 2019). The classic methods used to determine plant heights are not suitable for large-scale use (Dempewolf, 2017).
The use of multimages UAV photogrammetry with the Structure from Motion (SfM) approach allows to reduce costs and acquisition times by producing images and 3D models that allow the extraction of accurate metric information. There are various research that, in recent years, are testing the potential of UAVs for measuring the size of plants and their growth.
In (Kameyama, 2020) the research aimed to study the effects of the flight and shooting conditions of a UAV on the accuracy of the height and volume measurement of trees.

This contribution has been peer-reviewed.
https://doi.org/10.5194/isprs-archives-XLVIII-M-1-2023-381-2023

In (Alin, 2017) a low-cost UAV was used to obtain the canopy height model, by extracting the digital surface model from the digital terrain model, and filtered it locally based on the pixel-based window size using the provided algorithm.
In this study, we focus on a very low-cost UAV system to assess individual tree height and generate 3d Model and orthophoto of the study area. We used the DJI Mini2, a low-cost UAV that can be used without a flight rating and without restrictions due to its light weight of only 250 grams.
The case study where the survey was performed is an agricultural area of about 1 hectare, where are some fruit trees and a small vineyard. The area was selected because it contained both tall and small trees.

2. MATERIALS AND METHODS

The aim of this research was to evaluate whether the use of a very low-cost UAV can contribute to providing useful parameters for precision agriculture.
For this purpose, a low-cost and open category drone was used, i.e. usable without particular requests for piloting qualifications. This was the DJI Mini 2, a quadcopter weighing less than 250 grams. The research involved agricultural land planted in part with fruit trees, with an average height of about 4 meters, and a vineyard with vines with an average height of about 1.3 m.
To this end, 3 flights were performed at different relative flight heights: 30 m, 50 m and 80 m, as we wanted to study the accuracy of the measurements with such a low-cost instrumentation as well as the influence of the relative flight altitude and therefore of the resolution of the images in determining the trees heights.
The validation of the measurements was performed with the measurements of the trees height obtained with classical topographic methodology using a total station.

2.1 Case study

The area used as a case study is in the countryside of Furtei (Sardinia - Italy). It is an agricultural land of approximately 1.2 ha, within which there are 23 plants, mainly olive and fruit trees, arranged along the edges of the land. At the center of the area there is a vineyard made up of 5 rows of about 80 m long each. In figure 1 the area on which the experimentation was carried out.

Figure 1. Case study

2.2 UAV survey

A DJI quadcopter, the MINI 2, was used for the research. The main feature of this drone, in addition to its low cost, is its weight, which is less than 250 grams; it is classified as a light UAV, low-cost and easy to use.
The technical specifications guarantee a maximum battery life of 30 minutes and a wind resistance of 29-38 km/h. The small quadcopter supports HD video transmissions up to 10 km thanks to the OcuSync 2.0 video transmission system which gives greater safety in flight. The DJI Mini2 (figure 2) is equipped with a GNSS satellite positioning system and a digital camera with a 3-axis stabilizer, with an inclination of -90°. The digital camera allows to make 4k 30fps videos and 12MP images thanks to a 1/2.3" CMOS sensor with a pixel size of 1.5 mm and a lens with an f/2.8 aperture.

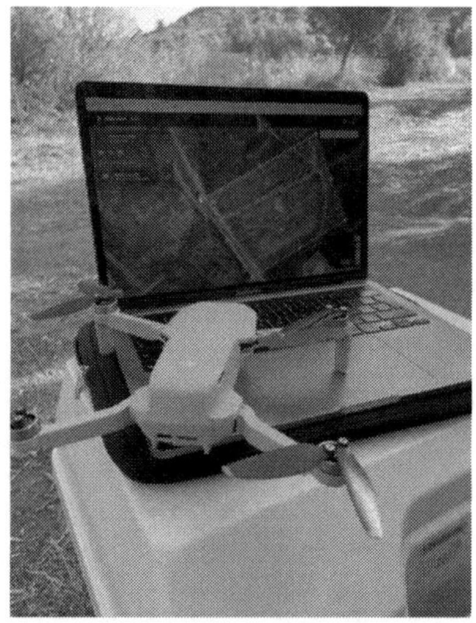

Figure 2: DJI Mini2

In the research, 3 flights were carried out at different flight relative quote: 30 m, 50 m and 80 m, with the aim of studying the influence of the flight quote on the data collected. Table 1 shows the flight parameters of the 3 measurements.

	Flight 1	Flight 2	Flight 3
Relative quote	30 m	50 m	80 m
Overlap	80%		
Sidelap	40%		
GSD	1.03 cm	1,72 cm	2,75 cm
Flight time	3' 48''	2' 03''	1' 16''
Max velocity	11 km/h	12 km/h	15 km/h
n. images	83	27	8
n. flight lines	4	2	1

Table 1. UAV Flight Parameters

As can be seen from table 1, as the flight relative quote increased, the number of images decreased but at the same time the Ground Sample Distance (GSD) increased. We went from 83 images for the 30 m flight to only 8 images for the 80 m flight, while the GSD went from 10.3 cm for the 30 m flight to 2.75 cm for the 80 m flight. Having flights with few images means reducing both acquisition and processing times.

The flights were carried out in August 2022 on a windless day to ensure the stability of the UAV.
In order to georeference the point clouds, 10 Ground Control Points (GCPs) distributed throughout the agricultural area were used. For their arrangement the prescriptions of (Santos Santana, 2021) were followed. In figure 3 the disposition of the GCPs.
Another 4 points were materialized and measured on the ground and used as Check Points (CPS).

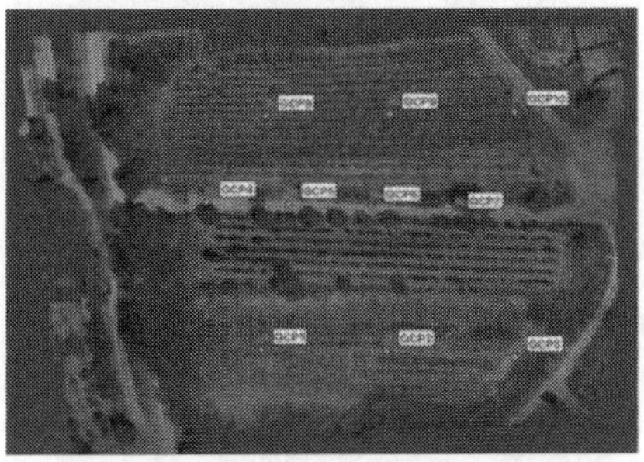

Figure 3. GCPs position

The GCPs and the CPs were materialized on the ground through red and white circular targets of 20 cm in diameter (figure 4). The shape and size of the targets have been planning to be visible on images acquired from a height of about 80 m. The points were surveyed using a GNSS receiver in RTK mode, with the ITALPOS Permanent Stations Network, determining their coordinates in the ETRF2000 datum with geodetic height.

Figure 4. Target

The images from 3 surveys were processed using the Metashape by Agisoft software, that implements the Structure from Motion (SfM) algorithm. The SfM is a low-cost photogrammetric method for high-resolution topographic reconstructions. The SfM operates under the same basic tenets of the stereoscopic photogrammetry. However, they fundamentally differ because in SfM the geometry of the scene, camera positions and orientation are solved automatically without points known. In the other method, instead, points are solved simultaneously using a highly redundant, iterative bundle adjustment procedure, based on a database of features automatically extracted from a set of multiple images with a high degree of overlap (Szeliski, 2010).
The Metashape workflows consist in the following main steps: image import, image alignment, generation of the sparse point cloud, optimization of image alignment, georeferenced, generation of the dense point cloud, generation DEM and generation orthophoto (Vacca, 2018; Giannattasio, 2013).

2.3 Trees height measurements

All the plants were measured with classic geomatic methods, in particular the plants over 1 m in height were measured with the Total Station, while the smaller ones were measured with a metric pole (figure 5).

Figure 5. Tree height measurements

The measurements were used to validate the heights extracted from the UAV surveys.

3. RESULTS

The images of the 3 flights were processed with the Metashape by Agisoft software. The processing of the dense cloud points of the 3 flights was performed with High quality parameter. The dense point clouds were georeferenced using 10 GCPs, while 4 are the CPs. Table 2 shows the Root Mean Square (r.m.s.), in centimeters, on the GCPs and CPs for the 3 flights.

	30 m (cm)			50 m (cm)			80 m (cm)		
	E	N	Q	E	N	Q	E	N	Q
r.m.s GCPs	1.2	1.1	0.5	0.5	0.5	1.1	0.2	0.2	0.1
r.m.s CPs	2.3	1.7	2.0	1.0	1.4	5.0	0.6	2.2	4.0

Table 2. GCPs and CPs r.m.s.

Figures 5, 6 and 7 show the values of the r.m.s. over the GCPs.

This contribution has been peer-reviewed.
https://doi.org/10.5194/isprs-archives-XLVIII-M-1-2023-381-2023

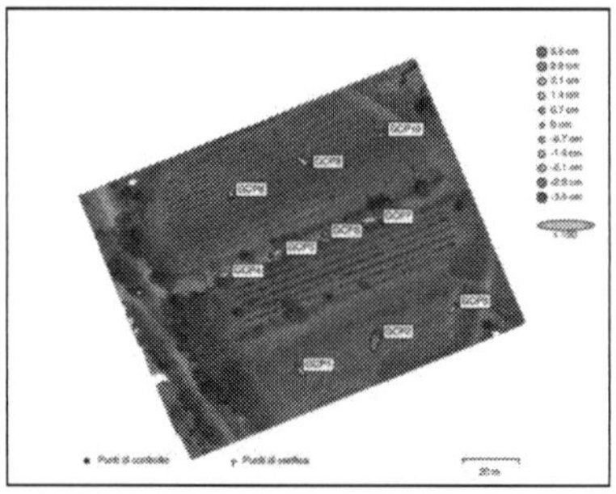

Figure 6. r.m.s. GCPs – 30 m flight

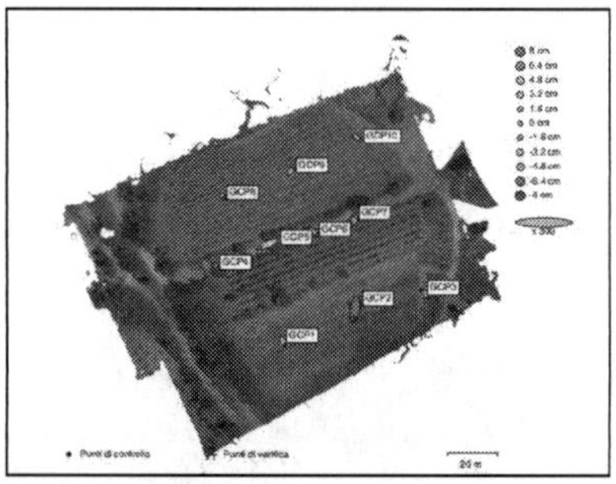

Figure 7. r.m.s. GCPs – 50 m flight

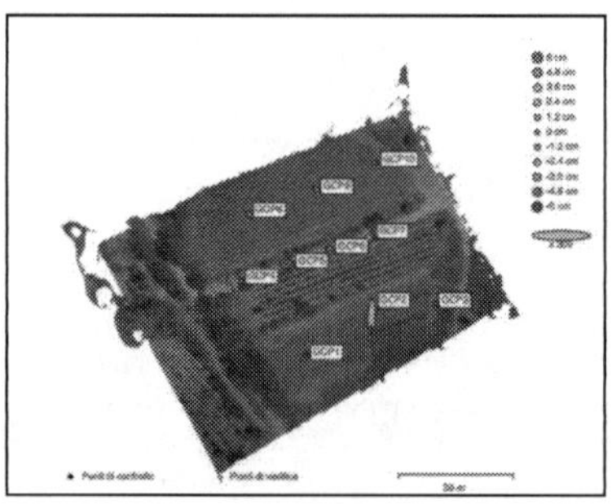

Figure 8. r.m.s. GCPs – 80 m flight

Table 3 reports the number of points in the dense point cloud for each flight and the size of the corresponding file.

	30 m	50 m	80 m
N. points	191.032.802	65.456.200	24.695.518
File Size	2.68 GB	1.10 GB	421 MB

Table 3. N. points and dimension of the dense point cloud

The figures 8, 9 and 10 show the dense point cloud for each flight.

Figure 9. Dense Point cloud – 30 m

Figure 10. Dense point cloud - 50 m

Figure 11. Dense Point Cloud - 80 m

For each flight the mesh and orthophotograph were also produced.

3.1 Trees height measurement

After processing the dense point clouds, the heights of the plants were extracted from them. This operation has required particular attention as, for each plant, both the height of the land and the height of the highest part of the plant had to be determined. The measurements were made within the Metashape software with the help of some of its tools.
To identify the points of interest in the dense cloud, we initially highlighted the crown of the plant using the "circular selection" command, then the highest altitude point among the selected points of the crown was extracted.

Figure 12 shows the procedure followed.

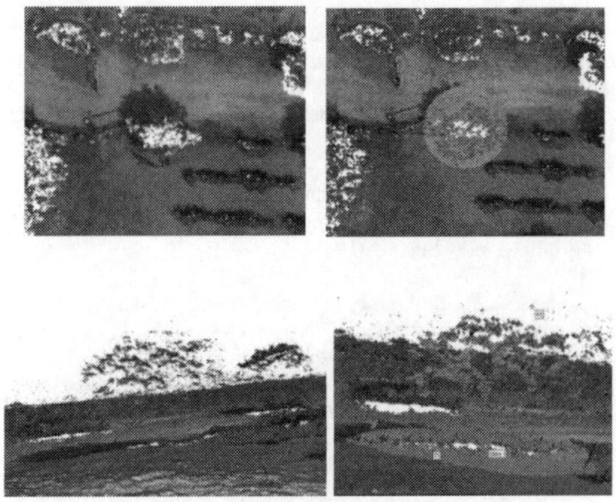

Figure 12. Tree height measurement

The tree hight was therefore estimated from the difference between the height of the trees and the height of the ground.
After determining the height of all the trees, these were compared with direct measurements. Table 4 shows the statistical results of these differences, while figure 13 shows the normal distribution of the differences.

	30 m	50 m	80 m
media (m)	0.276	0.256	0.091
r.m.s. (m)	0.391	0.372	0.326

Table 4. Statistical results of trees

	30 m	50 m	80 m
media (m)	0.112	0.214	0.253
r.m.s. (m)	0.08	0.183	0.198

Table 5. Statistical results of vines

	30 m	50 m	80 m
media (m)	0.228	0.244	0.137
r.m.s. (m)	0.334	0.366	0.325

Table 6. Statistical results of trees+vines

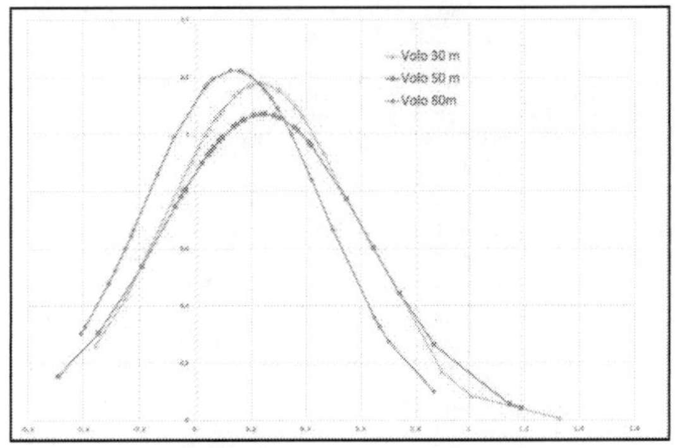

Figure 13. Normal distribution of differences

4. DISCUSSION

The research involved the use of a low-cost UAV for determining the heights of trees and vines. In particular, the study concerned the influence of the flight relative quote and therefore of the GSD of the images on the extraction of trees metrical parameters.
From the results obtained it would seem that the flight relative quote does not decisively influence the measurement of the height of tall trees. The flight at 80 m would seem to give better results with a rms of 33 cm compared to 39 cm of the flight at 30 m. For small trees, such as vines, flying at lower relative quote returns better values. For 30 m flight the rms is 8 cm, for 50 m flight it is 18 cm, and for 80 m flight it is 20 cm.
Considering the set of tall and small plants, the rms of the differences are 33 cm for the flight at 30 m, 37 cm for the flight at 50 m and 32 cm for the flight at 80 m.
We can therefore say that the influence of the flight altitude is certainly more incident in the measure of small plants (around the meter) than the high ones (around 4 m).

5. CONCLUSION

The research had the objective of studying the potential and accuracy of a very low-cost drone for measuring the heights of trees. The possibility of using this methodology, as an alternative to the normal measurement procedures, puts us in front of interesting scenarios. The procedure performed would allow overcoming the limits associated with traditional geomatic measurement techniques which are more slow and costly both in terms of time and man-hours.
The results obtained are comforting and in line with those obtained from other research, further improvements can certainly be obtained by integrating nadiral flights UAV with oblique flights UAV. This would allow to obtain more complete point clouds of the foliage avoiding holes and thus improving the process of calculating heights.

ACKNOWLEDGEMENTS

We thank Alessandro Aramu for his contribution to the DJI Mini2 UAV surveys.

REFERENCES

Ammoniaci, M., Kartsiotis, S.-P., Perria, R., Storchi, P., 2021: State of the Art of Monitoring Technologies and Data Processing for Precision Viticulture. Agriculture 2021, 11, 201. https://doi.org/10.3390/agriculture11030201

Pagliai, A., Ammoniaci, M., Sarri, D., Lisci, R., Perria, R., Vieri, M., D'Arcangelo, M.E.M., Storchi, P., Kartsiotis, S.-P., 2022: Comparison of Aerial and Ground 3D Point Clouds for Canopy Size Assessment in Precision Viticulture. Remote Sens. 2022, 14, 1145. https://doi.org/10.3390/rs14051145

Crookston, R.K., 2006: A top 10 list of developments and issue impacting crop management and ecology during the past 50 years. Crop Science, 46 (5) (2006), pp. 2253-2262

Casa, R., 2018: Agricoltura di precisione. Metodi e tecnologie per migliorare l'efficienza e la sostenibilità dei sistemi colturali. Edagricole New Bus. Media 2018.

Hobart, M., Pflanz, M., Weltzien, C., Schirrmann, M. Growth, 2020: Height Determination of Tree Walls for Precise Monitoring in Apple Fruit Production Using UAV Photogrammetry. Remote Sens. 2020, 12, 1656. https://doi.org/10.3390/rs12101656

Alexandratos, N. and J. Bruinsma. 2012: World agriculture towards 2030/2050: the 2012 revision. ESA Working paper No. 12-03. Rome, FAO.

Velusamy, P., Rajendran, S., Mahendran, R.K., Naseer, S., Shafiq, M., Choi, J.-G., 2012: Unmanned Aerial Vehicles (UAV) in Precision Agriculture: Applications and Challenges. Energies 2022, 15, 217. https://doi.org/10.3390/en15010217

Zhang, H.L., Tian, W.T., Yin, J., 2021: A Review of Unmanned Aerial Vehicle Low-Altitude Remote Sensing (UAV-LARS) Use in Agricultural Monitoring in China. Remote Sens. 2021, 13, 1221

Krause, S., Sanders, T. G. M., Mund, J. P., & Greve, K., 2019: UAV-based photogrammetric tree height measurement for intensive forest monitoring. Remote Sensing, 11(7), 1–18. https://doi.org/10.3390/rs11070758

Vaglio Laurin, G., Ding, J., Disney, M., Bartholomeus, H., Herold, M., Papale, D., & Valentini, R., 2019: Tree height in tropical forest as measured by different ground, proximal, and remote sensing instruments, and impacts on above ground biomass estimates. International Journal of Applied Earth Obser vation and Geoinformation, 82 (June), 101899. https://doi.org/10.1016/j.jag.2019.101899

Dempewolf, J., Nagol, J., Hein, S., Thiel, C., & Zimmermann, R., 2017: Measurement of with in-season tree height growth in a mixed forest stand using UAV imagery. Forests, 8(7), 1–15. https://doi.org/10.3390/f8070231

Kameyama, S., Sugiura, K. Estimating Tree Height and Volume Using Unmanned Aerial Vehicle Photography and SfM Technology, with Verification of Result Accuracy. Drones 2020, 4, 19. https://doi.org/10.3390/drones4020019

Anıl Can Birdal, Uğur & Tarık Türk, 2017: Estimating tree heights with images from an unmanned aerial vehicle, Geomatics, Natural Hazards and Risk, 8:2, 1144-1156

Huseyin Yurtseven, Mustafa Akgul, Suleyman Coban, Sercan Gulci, Determination and accuracy analysis of individual tree crown parameters using UAV based imagery and OBIA techniques, Measurement, Volume 145, 2019, Pages 651-664, ISSN 0263-2241

Krause, S., Sanders, T.G.M., Mund, J., Greve, K. UAV-Based Photogrammetric Tree Height Measurement for Intensive Forest Monitoring. Remote Sens. 2019, 11, 758

Santos Santana, L., E. Araújo, G. Silva Ferraz, D. Bedin Marin, B. Dienevam Souza Barbosa, L. Mendes Dos Santos, P. Ferreira Ponciano Ferraz, L. Conti, S. Camiciottoli, and G. Rossi., 2021: Influence of Flight Altitude and Control Points in the Georeferencing of Images Obtained by Unmanned Aerial Vehicle. European Journal of Remote Sensing 54 (1): 59–71. doi:10.1080/22797254.2020.1845104.

Szeliski, R., 2010: Computer vision: algorithms and applications. Springer

Vacca, G., Furfaro, G., Dessì, A., 2018: The use of the UAV images for the building 3D model generation. Int. Arch. Photogramm. Remote Sens. Spatial Inf. Sci., XLII-4/W8, 217–223, https://doi.org/10.5194/isprs-archives-XLII-4-W8-217-2018

Giannattasio, C., Grillo, S.M., Vacca, G., 2013: Interdisciplinary study for knowledge and dating of the San Francesco convent in Stampace, Cagliari – Italy (XIII-XXI Century) ISPRS Annals of the Photogrammetry, Remote Sensing and Spatial Information Sciences, 2 (5/W1), pp. 139-144. Cited 10 times. www.isprs.org/publications/annals.aspx doi:10.5194/isprsannals-II-5-W1-139-2013

http://it.smartnet-eu.com/ Accessed August, 2022

http://www.agisoft.com/ Accessed August, 2022

CHALLENGES OF ARTIFICIAL INTELLIGENCE APPLICATION FOR DISASTER RISK MANAGEMENT

D. Velev [1] *, P. Zlateva [1, 2]

[1] University of National and World Economy, Sofia, Bulgaria – (dgvelev, plamzlateva)@unwe.bg
[2] Institute of Robotics, Bulgarian Academy of Sciences, Sofia, Bulgaria

KEY WORDS: Natural disasters, Disaster risk management, Disaster risk reduction, Artificial Intelligence

ABSTRACT:

Artificial Intelligence (AI) has the potential to play a significant role in disaster risk management - from predicting disasters to optimizing response efforts. However, its application in disaster management also presents significant challenges. These challenges include the need for high-quality and diverse data, compatibility with existing systems and technologies, ethical and social implications, and ongoing research and development. In addition, data privacy and security is a critical issue as the use of AI in disaster management often involves the collection and analysis of sensitive information. Addressing these challenges is crucial to ensuring that AI is developed and used in ways that are fair, equitable, and effective in reducing the impact of disasters. The aim of the paper is to provide an analysis of challenges of artificial intelligence application for disaster risk management.

1. INTRODUCTION

In recent years, there has been a steady increase in natural disasters worldwide (Marshak, Rauber, Johnson, 2022). There is also an increase in their negative impact on societies, economies and ecology. Disasters are events that can cause significant damage, destruction or loss of life and property, and they usually occur unexpectedly or with little warning. Disasters can be natural or they can be caused by human actions. Disasters can have far-reaching consequences on the affected populations, including physical injuries, emotional trauma, economic disruption and social unrest (Cunningham et al., 2023; WEF, 2021). Effective disaster management and preparedness are critical for minimizing the impact of disasters and for helping communities recover and rebuild after such events occur (Bankoff et al., 2004; Castillo et al., 2021).

The most severe natural disasters can be classified as (Murray, et. al., 2021; Relyea, Ricklefs, 2018; Watts, 2022):

- Earthquakes - caused by the movement of tectonic plates and they can lead to widespread destruction and loss of life, especially in densely populated areas.
- Hurricanes and Typhoons - powerful storms that form over oceans and bring high winds, heavy rainfall and flooding to coastal areas.
- Volcanic Eruptions - releases ash, gas, and lava that can destroy homes, crops, and cities, and they can cause widespread health problems from ash inhalation
- Tsunamis - giant ocean waves caused by underwater earthquakes, volcanic eruptions, or typhoons, can cause massive coastal flooding and damage.
- Tornadoes - powerful swirling air columns that can cause extensive damage to buildings and infrastructure, as well as injure or kill people.
- Floods - occur when water overflows the normal riverbeds or other body of water, causing damage to homes, businesses, and infrastructure.
- Wildfires - fast-moving fires that can spread quickly and destroy homes, wildlife habitats, and other property.
- Droughts - prolonged periods of dry weather that can lead to crop failure and water shortages, causing widespread food and water insecurity.

It is important to note that the severity of a natural disaster can be affected by factors such as the level of preparedness of the affected communities and their infrastructure, as well as the response of emergency services and relief organizations (Sandler, Schwab, 2022). Due to that unfavourable trend, the risk management represents an important area field of research for disaster risk reduction and mitigation of the potential damage to people, infrastructures, business, environment, etc. (Sharma, Quah, 2018).

Nowadays, the timely and effective disaster risk management should rely on the processing of real data and information collected from different available resources, which data comes in vast volumes (Eslamian, Eslamian, 2021).

During a disaster, vast volumes of data are generated from different sources, such as mobile communications, social networking sites (Facebook, Twitter, Instagram, YouTube, etc.), sensors performing in Internet of Things environments, images from satellites, etc. (Tatano, Collins, 2019). Although this data acquisition should be considered as a first major step in the preparation for the following disaster risk management, here comes the question what should be the most suitable ICT that will handle effectively such situations. Fortunately enough, artificial intelligence (AI) has a steadily increasing utilization, and its applications gradually cover all types of disasters.

The aim of the paper is to provide an analysis of challenges of artificial intelligence application for disaster risk management. Artificial Intelligence (AI) has the potential to play a significant role in disaster risk management.

* Corresponding author

https://doi.org/10.5194/isprs-archives-XLVIII-M-1-2023-387-2023 | © Author(s) 2023. CC BY 4.0 License.

2. ARTIFICIAL INTELLIGENCE BRIEF

Artificial Intelligence (AI) is one of the fastest developing areas of the information and communication technology. Artificial intelligence refers to the simulation of human intelligence in machines that are programmed to think like humans and imitative their actions. The ideal characteristic of artificial intelligence is its ability to rationalize and take actions that have the best chance of achieving a specific goal (Russel, Norwig, 2021).

A subset of artificial intelligence is Machine Learning (ML), which refers to the concept that computer programs can automatically learn from and adapt to new data without being assisted by humans (Serrano, 2021). Deep Learning (DL) enable automatic learning through the processing of huge amounts of unstructured data such as text, images, or video (Goodfellow, Bengio, Courville, 2016).

AI requires a specialized hardware and software for writing and training its algorithms. AI systems process vast amounts of labeled training data, analyzing the data for correlations and patterns, and using these patterns to make predictions about future states. AI programming focuses on three cognitive skills - learning, reasoning and self-correction.

Recent advances in AI allow scientists and researchers to access and analyze new and bigger data sources than ever before. Studies show that the application of the artificial intelligence for disaster risk management is a promise. With the help of artificial intelligence, it is possible to predict where a natural disaster may occur, what the negative consequences would be, as well as to propose adequate measures for effective disaster risk management (Abid, et al. 2021; Arfan, et al. 2019).

By combining AI and ML with data analytics technologies are creating new AI systems for disaster risk management. These systems enable their users to ask specific and targeted questions and to receive useful answers real-world datasets.

AI techniques such as image recognition and classification can be quite suitable in assessing the damage as they can observe and analyze images and identify objects and features such as damaged buildings, flooding areas, blocked roads, etc. AI and sentiment analysis on social media data (Facebook, Twitter, Instagram, etc.) about natural disaster risk can to provide early warnings, ground-level location data, and real-time reports.

ML approaches such as predictive analytics can also analyze past events to identify and extract patterns and populations vulnerable to natural calamities. A large number of learning approaches are used to identify risk areas and improve predictions of future natural disasters.

3. ESSENCE OF THE ARTIFICIAL INTELLIGENCE APPLICATION FOR DISASTER RISK MANAGEMENT

Natural disasters have a significant impact on the generation and collection of data. During a disaster, the priority is often on immediate response and recovery efforts, which can lead to disruptions in the collection and reporting of data. This can lead to a lack of accurate information on the extent of the disaster, its impact on people and infrastructure, and the resources needed for recovery. The effective use of data in response to natural disasters is critical to ensure that resources are directed to where they are needed most, and that recovery efforts are based on the most up-to-date information. The integration of data from various sources can also help provide a more complete picture of the disaster and its impact. There are successfully developed and implemented information technologies for data acquisition:

Internet of Things (IoT) plays a significant role in improving disaster response and recovery efforts (Azrour, Irshad, Chaganti, 2022). IoT devices can provide real-time data on the conditions in affected areas, allowing responders to make more informed decisions and respond more quickly to the needs of communities (Solanki, Kumar, Nayyar, 2021). IoT can play a role in mitigating the impact of natural disasters by providing early warning systems that can alert communities to potential dangers. IoT-enabled sensors can be used to monitor environment conditions and to trigger automatic alerts to communities in the path of a hurricane or flood. This can provide them with critical time to evacuate or to take other necessary precautions to protect themselves and their property. IoT can also be used to monitor critical infrastructure in real-time, allowing responders to quickly identify and address potential hazards. the integration of IoT devices with other data sources, such as satellite imagery and drone footage, can provide a more comprehensive picture of the disaster and its impact, helping to inform decision-making and response efforts (Taser, 2021).

Edge computing can play a crucial role in the response to natural disasters by providing local processing and data analysis capabilities even in the absence of a stable network connection (Gao, Li, Zhuang, 2021). This can help organizations respond more effectively to the needs of affected communities, especially in remote or hard-to-reach areas. Edge computing supports real-time data analysis, which can be critical in the aftermath of a disaster when time is of the essence. By processing data locally, organizations can quickly analyze information and make decisions, even in the absence of a stable network connection, since it can reduce the amount of data that needs to be transmitted over the network, which can reduce latency and improve the performance of critical systems. Edge computing can support real-time collaboration and information sharing among response teams, even in remote or disconnected areas. This can help organizations coordinate their efforts, pool resources, and make informed decisions (Misra et al., 2022).

Remote sensing can play a critical role in the response to natural disasters by providing real-time information and data about the extent and impact of a disaster (Ahmed, Alam, 2022). This information can help organizations assess the situation, prioritize response efforts, allocate resources, and make informed decisions. Remote sensing can provide real-time information about weather conditions and potential hazards, allowing organizations to proactively prepare for and respond to natural disasters. allowing organizations to focus their response efforts in these areas (Chang, Bai, 2018). Remote sensing can provide information about the location and extent of damage to critical infrastructure, such as roads, bridges, and communication networks, allowing organizations to prioritize their efforts to restore these essential services. In addition, remote sensing can provide information about the extent and impact of environmental degradation, such as soil erosion, landslides, and changes in water quality, allowing organizations to prioritize their efforts to address these impacts (Lavender, S., Lavender, A., 2023). It is also important to consider the integration of remote sensing data with other data sources, such as ground-based observations and social media, to provide a

This contribution has been peer-reviewed.
https://doi.org/10.5194/isprs-archives-XLVIII-M-1-2023-387-2023 | © Author(s) 2023. CC BY 4.0 License.

comprehensive and accurate picture of the situation. Remote sensing data can be integrated with other data sources to provide a more complete and accurate picture of the impact of a disaster, allowing organizations to respond more effectively and efficiently (Rosso, Sebastianelli, Ullo, 2021).

Social Media has become an increasingly important tool in the response to natural disasters (Light, Moody, 2020). It allows people to share real-time information on the extent and impact of a disaster, providing valuable data to response teams (Galety et al., 2022). It can also be used to coordinate response efforts, mobilize volunteers, and provide information to affected communities. It could be regarded as a popular means for crowdsourcing information from people on the ground, providing a more comprehensive picture of the disaster and its impac (Gupta, Sahoo, 2021). Social media can be used to monitor the response to a disaster, providing valuable feedback on the effectiveness of response efforts and helping to identify areas for improvement.

Big data plays a significant role in the response to natural disasters by providing real-time information and insights about the extent and impact of a disaster (Rani et al., 2023). This information can support decision-making, prioritize response efforts, and inform long-term recovery planning. Big data can be used to analyze social media and other online sources to provide real-time information about the extent and impact of a disaster (Schintler, McNeely, 2020). This information can be used to identify areas in need of immediate assistance, prioritize response efforts, and support the allocation of resources. Big data can also be used to analyze large amounts of satellite imagery and other remote sensing data to provide a comprehensive picture of the extent and impact of a disaster. This information can be used to support the assessment of the situation, prioritize response efforts, and inform recovery planning. Big data can be used to analyze historical data about past disasters, allowing organizations to identify patterns and make informed decisions about how to respond to future disasters and big data can be used to identify areas that are particularly vulnerable to natural disasters and to support the development of early warning systems and risk mitigation strategies.

Cloud computing can play a critical role in the response to natural disasters by providing on-demand access to computing resources, data storage, and applications (Hurson, Wu, 2021). This can help organizations respond more quickly and effectively to the needs of affected communities. cloud computing can provide organizations with scalable, flexible, and secure data storage, which can be critical in the aftermath of a disaster when large amounts of data need to be collected, processed, and analyzed (Sharma, Sharma, Elhoseny, 2021). This data can be used to inform decision-making, prioritize response efforts, and track progress. cloud computing can provide organizations with access to powerful computing resources, such as high-performance computing and artificial intelligence, which can be used to analyze complex data sets and model potential scenarios in real-time. This can help organizations make more informed decisions and respond more effectively to the needs of affected communities. Another advantage of cloud computing in the context of natural disasters is the ability to collaborate and share information in real-time. This can help organizations coordinate their efforts, pool resources, and make informed decisions.

All the above technologies generate large volumes of data. Data quality and quantity play a crucial role in disaster response and recovery. In the context of natural disasters, the accuracy and timeliness of data can have a significant impact on the effectiveness of response and recovery efforts.

Data quality refers to the accuracy and completeness of the data, and its relevance to the situation (Vluymans, 2018). High-quality data can support effective decision-making, prioritize response efforts, and inform long-term recovery planning. Data quality is particularly important because it can have a direct impact on the safety of affected communities and the effectiveness of response efforts.

Data quantity refers to the amount of data available. Large amounts of data can be generated from a variety of sources, including social media, satellite imagery, and remote sensing data. This data can provide a comprehensive picture of the extent and impact of a disaster, and can support effective decision-making and resource allocation.

To ensure data quality and quantity, it is important to have a well-defined data management and processing strategy in place. This should include protocols for collecting, storing, and processing data, as well as for verifying the accuracy of the data. Data sources should be carefully selected and monitored to ensure that the data is relevant, accurate, and up-to-date (Srivastav et al., 2020; Tagarev, 2019). In the context of natural disasters, data can be sourced from a variety of sources, including government agencies, non-profit organizations, and individuals. It is important to consider the reliability and credibility of each data source and to ensure that data is collected and processed in a consistent and transparent manner. It is important to have a robust system in place for data analysis and interpretation. This could include advanced analytics and machine learning algorithms to process and analyze large amounts of data, and to support informed decision-making and resource allocation.

Data and AI are closely interconnected (Sipola, Kokkonen, Karjalainen, 2023). AI relies on data to learn and make decisions. AI models cannot be trained or improve over time without data. Data is the fuel to the AI algorithms, and the more data that is available, the more accurate and effective AI systems become (Sutton, 2021). The process of training an AI model involves feeding it large amounts of data and allowing it to learn from that data by identifying patterns and making predictions. Data plays a critical role in the development of AI models. Data scientists and engineers use data to train and refine AI models, and to validate their accuracy and performance.

Combining with the processing of large volumes of generated data, there is an increasing number of AI solutions that can be used for disaster risk management:

Machine learning algorithms can be used to analyze large amounts of data from a variety of sources, such as satellite imagery, weather data, and social media, to identify patterns and make predictions about the impact of a disaster (Joshi, 2023). The machine learning algorithms can be used to predict the spread of a fire, the likelihood of flooding in a certain area, or the extent of damage caused by an earthquake (Marques, González-Briones, López, 2022).

https://doi.org/10.5194/isprs-archives-XLVIII-M-1-2023-387-2023 | © Author(s) 2023. CC BY 4.0 License.

Deep learning algorithms can be used to process large amounts of data, such as satellite imagery and sensor data, to support disaster response and recovery efforts (Hong et al., 2020). The deep learning algorithms can be used to identify objects, such as buildings and roads, in satellite imagery, to support damage assessment efforts.

Artificial neural networks can be used to process large amounts of data to make predictions about the impact of a disaster. The algorithms can be used to predict the spread of a fire or the likelihood of flooding in a certain area (Asadnia et al., 2022).

Reinforcement learning algorithms can be used to support decision-making in real-time during a disaster. The algorithms can be used to optimize the deployment of resources, such as emergency supplies and personnel, to ensure that they are delivered to the areas that need them most (Abdel-Basset et al., 2022).

Decision tree algorithms can be used to support decision-making during a disaster (Agrawal et al., 2022). The algorithms can be used to determine the most effective response to a disaster based on a range of factors, such as the size of the affected area, the number of people affected, and the availability of resources.

Natural language processing (NLP) can be used to analyze unstructured data sources, such as social media, news articles, and online forums, to identify areas that have been affected by a disaster, to monitor the spread of false information, and to support effective communication and information sharing among response and recovery organizations (Campesato, 2020).

Predictive modeling can be used to forecast the impact of a disaster before it occurs. AI can be used to predict the path of a hurricane or the potential impact of an earthquake on a particular region. This information can be used to prepare for and respond to disasters more effectively (Subramanian et al., 2023).

Predictive maintenance can be used to monitor critical infrastructure, such as power plants and communication networks, to identify potential failures and take proactive measures to prevent or mitigate their impact (Sakurai, Shaw, 2021).

Image and video analysis be used to analyze satellite imagery, drone footage, and other visual data sources to support disaster response and recovery efforts. The algorithms can be used to identify areas that have been affected by a disaster, to assess the extent of damage, and to support search and rescue operations.

Optimization algorithms can be used to optimize the deployment of resources, such as emergency supplies and personnel, to ensure that they are delivered to the areas that need them most. These algorithms can also be used to support logistics and resource allocation during a disaster response.

Early warning systems - AI can be used to develop early warning systems that alert people to the approach of a disaster. AI can be used to analyze data from sensors and other sources to detect the early signs of a landslide, a flash flood, or a tornado.

Decision support systems - AI can be used to support decision-making during a disaster by providing real-time information about the situation on the ground (Sumathi et al., 2022). AI can be used to analyze data from sensors, drones, and other sources to provide up-to-date information about the extent of the disaster, the number of people affected, and the availability of resources.

Risk assessment - AI can be used to assess the risk of a disaster occurring in a particular region (Müller, 2016). AI can be used to analyze data about past disasters, as well as data about factors such as weather patterns, geology, and infrastructure, to identify areas that are at high risk of future disasters (Roberts, Tonna, 2022; Tamboli, 2019).

Post-disaster analysis - AI can be used to analyze data after a disaster has occurred to identify patterns, determine causes, and support recovery efforts. AI can be used to analyze satellite imagery and other data to identify areas that have been affected by a disaster and to support damage assessment efforts.

Robotics - AI algorithms can be used to control robots and drones that can be deployed in disaster zones to provide real-time information and support the response effort. Robots and drones can be used to inspect damaged buildings, assess damage, and transport supplies.

The specific algorithms used may depend on the type of disaster, the data and resources available, and the goals and objectives of the response and recovery effort. It should be noted that these AI methods and algorithms are not mutually exclusive, and that they can be combined and integrated in various ways to support disaster response and recovery efforts.

4. ARTIFICIAL INTELLIGENCE LIMITATIONS FOR DISASTER RISK MANAGEMENT

Although AI is advanced enough to help manage natural disaster risk, its limitations prevent its widespread application in real world. To reduce AI limitations, it is necessary to: employment of experienced researchers and technical experts; collect good quality data for training the AI-powered applications; conducting multiple trainings with the data and adaptation of the obtained results, etc.

There is still much research to be done in order to fully leverage the potential of AI in disaster risk management. Some of the key challenges and limitations could include:

Data quality and quantity - One of the biggest challenges in applying AI to disasters risk management is the quality and quantity of available data. In many disaster scenarios data can be scarce, unreliable or inconsistent, which can make it difficult to train AI models and to make accurate predictions (Özyer, Bakshi, Alhajj, 2019).

Privacy and security - The use of AI in disaster risk management often involves processing large amounts of personal data, which raises privacy and security concerns (Douglass et al., 2023). It is crucial to ensure that the data is securely stored and processed in accordance with privacy regulations and that the rights of individuals are protected (Gerunov, 2023; Montasari, 2022).

This contribution has been peer-reviewed.
https://doi.org/10.5194/isprs-archives-XLVIII-M-1-2023-387-2023 | © Author(s) 2023. CC BY 4.0 License.

Computational resources - Applying AI to disasters can require significant computational resources, especially for complex models such as deep learning algorithms (Akerkar, 2020). In some disaster scenarios, such as in remote or isolated areas, there may not be adequate computational resources available to support AI-based solutions.

Integration with existing systems - AI solutions for disasters must be able to integrate with existing systems, such as command and control centers, emergency response networks, and sensor networks (Wang, Moriarty, 2018). This can be challenging, especially when dealing with heterogeneous systems and data formats.

Real-time performance - real-time performance is critical in many disaster scenarios. AI-based solutions must be able to process and analyze large amounts of data in real-time in order to support decision-making and to provide real-time situation awareness.

Explainability and transparency - AI-based solutions must be transparent and explainable, especially in high-stakes scenarios such as disasters (Masakowski, 2022). Decision-makers must understand the decisions made by AI models and the reasoning behind those decisions.

Bias and fairness - AI models can perpetuate biases and unfairness if the data used to train them is biased or if the algorithms are designed to perpetuate certain biases. This can be especially problematic in disaster scenarios, where decisions made by AI models can have ambiguous consequences.

Responsibility and accountability - The deployment of AI in disaster scenarios can raise questions about the responsible and accountable persons for the decisions made by AI algorithms. It's essential to establish clear rules of responsibility and accountability for AI decision-making in disaster scenarios.

Transparency - The deployment of AI in disaster scenarios can raise questions about the transparency and interpretability of AI algorithms. Hence, it is essential to ensure that AI algorithms are transparent and their decision-making processes can be understood and scrutinized.

Ethical and legal considerations - AI solutions for disaster management must be designed and deployed in an ethical and legally compliant manner, which should include considerations such as privacy, data protection, and liability.

Lack of human interaction - AI algorithms can lack the human interaction and intuition that is often necessary in disaster scenarios, especially in complex and unpredictable situations.

Limited adaptability - AI algorithms may not be able to adapt to new and unexpected scenarios, especially in rapidly evolving disaster situations.

Algorithm specificity - AI algorithms may be designed for specific types of disasters or scenarios and may not be applicable to other types of disasters.

Overreliance on technology - Overreliance on AI technology can lead to a decrease in human decision-making skills and a lack of situational awareness (Forney, Sadar, 2021).

Cost - Implementing AI-based solutions for disasters can be expensive, especially for developing countries that may lack the resources and infrastructure to support such solutions.

Technical expertise - Developing and deploying AI-based solutions for disasters requires a high level of technical expertise, which may not be readily available in many disaster-affected areas (Forsyth, 2019).

Despite these challenges, there is a growing consensus among experts in the field that AI has the potential to significantly improve disaster response and recovery efforts. By leveraging AI algorithms to process and analyze large and complex datasets, organizations and governments can make more informed decisions, allocate resources more effectively, and respond to disasters more quickly and efficiently.

In the future, AI is expected to play an increasingly important role in disaster management and response. Some of the key trends in the use of AI in disasters include:

Improved prediction and early warning systems - AI algorithms can be used to analyze large amounts of data from a variety of sources, including weather patterns, satellite imagery, and social media, to make more accurate predictions about natural disasters, such as hurricanes, earthquakes, and tsunamis (Prakash et al., 2021).

Automated damage assessment - AI algorithms can be used to analyze satellite imagery and other data sources to quickly assess the extent of damage caused by a disaster and prioritize response efforts.

Integration with IoT and 5G technology - AI algorithms can be integrated with IoT devices and 5G networks to provide real-time information and facilitate rapid response in disaster scenarios (Hussain, Sheng, Peng, 2022).

Increased use of unmanned aerial vehicles (UAVs) - AI algorithms can be integrated with UAVs to quickly assess damage and prioritize response efforts, as well as to deliver supplies and aid to affected communities (Koubaa, Azar, 2021; Yamagata, 2020).

Personalized and targeted response - AI algorithms can be used to analyze data from a variety of sources to personalize and target disaster response efforts based on the specific needs and vulnerabilities of affected communities.

Improved resource allocation - AI algorithms can be used to analyze data on resource needs, supply chain disruptions, and other factors to optimize resource allocation and logistics in disaster scenarios (Pani et al., 2021).

Enhanced collaboration and communication - AI technologies, such as chatbots and voice assistants, can be used to facilitate communication and collaboration between disaster response organizations and affected communities (Saroliya et al., 2023).

Human augmentation - AI can be used to augment the abilities of human responders in disaster scenarios, by providing real-time information, automating critical tasks, and reducing response times (Priyadarshini et al., 2022).

This contribution has been peer-reviewed.
https://doi.org/10.5194/isprs-archives-XLVIII-M-1-2023-387-2023

Scalability - The use of AI in disaster risk management needs to be scalable to accommodate different types and sizes of disasters. This requires the development of AI systems that can be easily adapted to different disaster scenarios and that can be deployed quickly and effectively in response to a disaster.

Cost-effectiveness - AI systems for disaster risk management need to be cost-effective as resources for disaster response and recovery efforts are often limited. This requires the development of AI systems that are both efficient and effective in their use of resources, while still providing the necessary information and support to decision-makers.

Evaluation and assessment - The impact and effectiveness of AI systems for disaster risk management need to be regularly evaluated and assessed. This requires the development of appropriate metrics and methods for evaluating the performance of AI systems, and the collection and analysis of data to understand their impact on disaster response and recovery efforts.

Public awareness and acceptance - there is a need to increase public awareness and acceptance of AI in disaster risk management. This requires effective communication and education strategies to build trust in AI systems and to ensure that they are used in ways that are ethical, responsible, and in the best interest of disaster-affected communities.

5. CONCLUSIONS

The use of AI holds promise in helping avert, mitigate and manage natural disaster risks by analysing large pools of data. However, more efforts are required to ensure that AI technologies are deployed in a responsible manner. All interested institutions to make informed decisions on natural disaster risk reduction can successfully use the proposed analysis of AI application challenges. Despite these challenges, the benefits of AI in disaster risk management are too great to ignore. Investment in research, innovation, and collaboration among stakeholders will be necessary to overcome the challenges and unleash the full potential of AI in disaster risk management. By addressing these challenges, AI has the potential to transform disaster risk management, improving the speed and accuracy of disaster prediction, enhancing early warning systems, and facilitating more effective response and recovery efforts.

ACKNOWLEDGEMENT

The first author has been financially supported by National Science Fund of Bulgaria under the Grant №KP-06-H55/5 from 15.11.2021, titled "Smart risk management for business from adverse events and natural disasters". The second author has been financially supported by the Science fund of the University of National and World Economy, Sofia, Bulgaria under the Grant № NID NI 1/2021 titled "Development and use of artificial intelligence in education and economy".

REFERENCES

Abdel-Basset, M., et al., (Eds.), 2022. Deep Learning Approaches for Security Threats in IoT Environments, Wiley.

Abid, S.K.; Sulaiman, N.; Chan, S.W.; Nazir, U.; Abid, M.; Han,H.; Ariza-Montes, A.; Vega-Muñoz, A. Toward an Integrated Disaster Management Approach: How Artificial Intelligence Can Boost Disaster Management. Sustainability, 2021, 13, 12560. https://doi.org/, 10.3390/su132212560

Agrawal, P., et al., (Eds.), 2022. Machine Learning and Data Science - Fundamentals and Applications, Wiley.

Ahmed, B., Alam, A., (Eds.), 2022. Remote Sensing of Natural Hazards, MDPI.

Akerkar, R., (Ed.), 2020. Big Data in Emergency Management: Exploitation Techniques for Social and Mobile Data, Springer.

Arfan, M., Khan, Z., Qadri, N., Hameed, M. H. & Amir, A. R. (2019). Role of Artificial Intelligence (AI) in combined disaster management, 3(2), 97–121.

Asadnia, M., et al., (Eds.), 2022. Artificial Intelligence and Data Science in Environmental Sensing, Academic Press.

Azrour, M., Irshad, A., Chaganti, R., (Eds.), 2022. IoT and Smart Devices for Sustainable Environment, Springer.

Bankoff, G., et al., (Eds.), 2004. Mapping Vulnerability: Disasters, Development and People, Earthscan.

Campesato, O., 2020. Artificial Intelligence, Machine Learning, and Deep Learning, Mercury Learning and Information.

Castillo, F., Wehner, M., Stone, D.A., (Eds.), 2021. Extreme Events and Climate Change - A Multidisciplinary Approach, Wiley.

Chang, N.B., Bai, K., 2018. Multisensor Data Fusion and Machine Learning for Environmental Remote Sensing, CRC Press.

Cunningham, W. P., et al., 2023. Environmental SCIENCE - A Global Concern, 16th Edition, McGraw Hill LLC.

Douglass, R., et al., (Eds.), 2023. IoT for Defense and National Security, Wiley.

Eslamian, S., Eslamian, F., (Eds.), 2021. Handbook of Disaster Risk Reduction for Resilience - New Frameworks for Building Resilience to Disasters, Springer.

Forney, S.Z., Sadar, A.J., 2021. Environmental Risk Communication - Principles and Practices for Industry, CRC Press.

Forsyth, D., 2019. Applied Machine Learning, Springer.

Galety, M.G., et al., (Eds.), 2022. Social Network Analysis - Theory and Applications, Wiley.

Gao, J., Li. M., Zhuang, W., 2021. Connectivity and Edge Computing in IoT: Customized Designs and AI-based Solutions, Springer.

Gerunov, A., 2023. Risk Analysis for the Digital Age, Springer.

Goodfellow, I., Bengio, Y., Courville, A., 2016. Deep Learning, MIT Press.

Gupta, B.B., Sahoo, S.R., 2021. Online Social Networks - Security Principles, Algorithm, Applications, and Perspectives, CRC Press.

Hong. T., et al., (Eds.), 2020. Deep Learning for Social Media Data Analytics, Springer.

Hurson. A.R., Wu, S., (Eds.), 2021. Advances in Computers - AI and Cloud Computing, Volume 120, Academic Press.

Hussain, A., Sheng, G.T., Peng, L., (Eds.). 2022. IoT and AI Technologies for Sustainable Living - A Practical Handbook, CRC Press.

Joshi, A. V., 2023. Machine Learning and Artificial Intelligence, 2nd Edition, Springer.

Kanbara, S., Shaw, R., Kato, N., Miyazaki, H., Morita, A., (Eds.), 2022. Society 5.0, Digital Transformation and Disasters - Past, Present and Future, Springer.

Koubaa, A., Azar, A.T., (Eds.), 2021. Deep Learning for Unmanned Systems, Springer.

Lavender, S., Lavender, A., 2023. Practical Handbook of Remote Sensing, 2nd. Edition, CRC Press.

Light. R., Moody, J., (Eds.), 2020. The Oxford Handbook of Social Networks, Oxford Academic.

Marques, G., González-Briones, A., López, J.M.M., (Eds.), 2022. Machine Learning for Smart Environments/Cities - An IoT Approach, Springer.

Marshak, S., Rauber, R., Johnson, N., 2022. Natural Disasters, W. W. Norton & Company.

Masakowski, Y. R., (Ed.), 2022. Artificial Intelligence and Global Security: Future Trends, Threats and Considerations, Emerald Publishing.

Misra. S., et al., (Eds.), 2022. Artificial Intelligence for Cloud and Edge Computing, Springer.

Montasari, R., (Ed.), 2022. Artificial Intelligence and National Security, Springer.

Müller. V., (Ed.), 2016. Risks of Artificial Intelligence, CRC Press.

Murray, V., et. al., 2021. Hazard Information Profiles - Supplement to : UNDRR-ISC Hazard Definition & Classification Review - Technical Report, United Nations.

Özyer, T., Bakshi, S., Alhajj. R., (Eds.), 2019. Social Networks and Surveillance for Society, Springer.

Pani. S.K., (Eds.), 2021. Applications of Machine Learning in Big-Data Analytics and Cloud Computing, River Publ.

Prakash, K. B., et al., (Eds.), 2021. Advanced Deep Learning for Engineers and Scientists - A Practical Approach, Springer.

Priyadarshini, P. et al. (Eds.), 2022. Artificial Intelligence - Applications and Innovations, CRC Press.

Rani, S., et al., (Eds.), 2023. Big Data, Cloud Computing and IoT - Tools and Applications, CRC Press.

Relyea, R., Ricklefs, R., 2018. Ecology: The Economy of Nature, 8th Edition, W. H. Freeman and Company.

Roberts, T., Tonna, S. J., 2022. Risk Modeling - Practical Applications of Artificial Intelligence, Machine Learning, and Deep Learning, Wiley.

Rosso, M.P., Sebastianelli. A., Ullo, S.L., (Eds.), 2021. Artificial Intelligence Applied to Satellite-based Remote Sensing Data for Earth Observation, Institution of Engineering and Technology.

Russel, S., Norwig, P., 2021. Artificial Intelligence – A Modern Approach, 4th Edition, Pearson.

Sakurai, M., Shaw, R., (Eds.), 2021. Emerging Technologies for Disaster Resilience Practical Cases and Theories, Springer, 262.

Sandler, D., Schwab, A.K., 2022. Hazard Mitigation and Preparedness, 3rd. Edition, Routledge.

Saroliya, A., et al., (Eds.), 2023. Internet of Things and Fog Computing- Enabled Solutions for Real- Life Challenges, CRC Press.

Schintler. L.A., McNeely, C.L., (Eds.), 2020. Encyclopedia of Big Data, Springer.

Serrano, L. G., 2021. Grokking Machine Learning, Manning Publications.

Sharma, P., Sharma, M., Elhoseny. M., (Eds.), 2021. Applications of Cloud Computing Approaches and Practices, CRC Press.

Sharma, S.K., Quah. E., (Eds.), 2018. Economics of Natural Disasters, World Scientific Publishing.

Sipola, T., Kokkonen, T., Karjalainen, M., (Eds.), 2023. Artificial Intelligence and Cybersecurity - Theory and Applications, Springer.

Solanki, A., Kumar, A., Nayyar, A., (Eds.), 2021. Digital Cities Roadmap - IoT-Based Architecture and Sustainable Buildings, Wiley.

Srivastav, A.L., et al., 2020. Visualization Techniques for Climate Change with Machine Learning and Artificial Intelligence, Elsevier.

Subramanian, T., et al., (Eds.), 2023. Artificial Intelligence, Machine Learning and Blockchain in Quantum Satellite, Drone and Network, CRC Press.

Sumathi, S., et al., (Eds.), 2022. Machine Learning for Decision Sciences with Case Studies in Python, CRC Press.

Sutton, D., 2021. Information Risk Management - A practitioner's guide, 2nd. Edition, 2021, BCS Learning and Development Ltd.

This contribution has been peer-reviewed.
https://doi.org/10.5194/isprs-archives-XLVIII-M-1-2023-387-2023 | © Author(s) 2023. CC BY 4.0 License.

Tagarev, N., 2019. Role of Case Study Analyses in Education of Cybersecurity Management. 13th International Multi-Conference on Society, Cybernetics and Informatics (IMSCI 2019), vol. 1, 61-64.

Tamboli, A., 2019. Keeping Your AI Under Control: A Pragmatic Guide to Identifying, Evaluating, and Quantifying Risks, Apress.

Taser, P.Y., 2021. Emerging Trends in IoT and Integration With Data Science, Cloud Computing, and Big Data Analytics, IGI Global.

Tatano, H., Collins, A., (Eds.), 2019. Proceedings of the 4th Global Summit of Research Institutes for Disaster Risk Reduction, Springer.

Vluymans, S., 2018. Dealing with Imbalanced and Weakly Labelled Data in Machine Learning using Fuzzy and Rough Set Methods, Springer.

Wang, S.J., Moriarty, P., 2018. Big Data for Urban Sustainability - A Human-Centered Perspective, Springer.

Watts, C., 2022. Natural Disasters, New Edition, DK Publ. Inc.

WEF, 2021.The Global Risks Report 2021, 16th Edition, World Economic Forum.

Yamagata, Y., (Ed.), 2020. Spacial Analysis Using Big Data - Methods and Urban Applications, Academic Press.

This contribution has been peer-reviewed.
https://doi.org/10.5194/isprs-archives-XLVIII-M-1-2023-387-2023 | © Author(s) 2023. CC BY 4.0 License.

MACHINE LEARNING BASED BIAS CORRECTION FOR MODIS AEROSOL OPTICAL DEPTH IN BEIJING

Mengjuan Wang [1], Meng Fan [2], Zhibao Wang [1,3*], Liangfu Chen [2], Lu Bai [4], Yuanlin Chen [1], Mei Wang [1]

[1] School of Computer and Information Technology, Northeast Petroleum University, Daqing 163318, China
[2] State Key Laboratory of Remote Sensing Science, Aerospace Information Research Institute of Chinese Academy of Sciences, Beijing 100101, China
[3] Bohai-Rim Energy Research Institute, Northeast Petroleum University, Qinhuangdao 066004, China
[4] School of Computing, Ulster University, Belfast BT15 1ED, UK

KEY WORDS: Aerosol optical depth, bias correction, machine learning, artificial neural network, support vector regression.

ABSTRACT:

Aerosol refers to suspensions of small solid and liquid particles in the atmosphere. Although the content of aerosol in the atmosphere is small, it plays a crucial role in atmospheric and the climatic processes, making it essential to monitor. In areas with poor aerosol characteristics, satellite-based aerosol optical depth (AOD) values often differ from ground-based AOD values measured by instruments like AERONET. The use of 3km DT, 10km DT and 10km DTB algorithms in Beijing area has led to significant overestimation of AOD values, highlighting the need for improvement. This paper proposes the use of machine learning techniques, specifically support vector regression (SVR) and artificial neural network (ANN), to correct the deviation of AOD data. Our approach leverages ground-based monitoring data, meteorological reanalysis data and satellite products to train the models. Our results show that the ANN model outperforms the SVR model achieving R2, RMSE and Slope values of 0.88, 0.12 and 0.97, respectively, when applied to nearly two decades of data from 2001 to 2019. This study significantly improves the accuracy of MODIS AOD values, reducing overestimation and bringing them closer to ground-based AOD values measured by AERONET. Our findings have important applications in climate research and environmental monitoring.

1. INTRODUCTION

1.1 Introduction

Aerosols are tiny particles that can be solid, liquid, or a mixture of both, with an aerodynamic diameter of fewer than 100 μm. When carried by the atmosphere, they are referred to as atmospheric aerosols. Despite their small quantity, aerosols have a significant impact on the physical and chemical processes in the atmosphere, as well as a crucial role in the climate system (Luo et al., 2014). Aerosols particles change the solar radiation reaching the Earth's surface by scattering and absorbing solar short-wave and Earth long-wave radiation, thereby affecting the climate. Additionally, aerosol particles can indirectly influence the climate by acting as cloud particles, which can that and alter the concentration of various chemical components in the atmosphere and contribute to the cloud formation. Aerosols have substantial impacts on the Earth's radiation budget balance, cloud and precipitation formation, regional air pollution visibility, human health (Colbeck and Lazaridis, 2010). The occurrence of winter haze weather significantly reduces the visibility of cities, causing inconvenience in people's daily lives. Recent studies have indicated that Particulate Matter (PM) in polluted air, particularly inhalable fine particles like PM2.5, is linked to heart and respiratory diseases (T. Li et al., 2018; Ni et al., 2015; Tecer et al., 2008). These particles can pose a threat to human health. Research shows that aerosol and PM2.5 have strong correlation (Guo et al., 2017; Li et al., 2015). Therefore, monitoring and controlling the atmospheric environment are critical.

Aerosol optical depth (AOD) is an important parameter for characterizing the optical properties of atmospheric aerosols. It represents the integrated extinction coefficient over the entire atmosphere in the vertical direction and is the primary optical parameter that can be retrieved using ground-based and satellite remote sensing methods. Satellite detection technology is increasingly being used to monitor and assess environmental changes (Li et al., 2020; Shi et al., 2022; Yang et al., 2013; J. Zhang et al., 2021). Satellites equipped with various sensors and cameras can capture detailed images and data of the Earth's surface, atmosphere, and oceans (Bovensmann et al., 2010; Yu et al., 2021; X. Zhang et al., 2021a, 2021b; Zhu et al., 2021). While ground-based observation provides high temporal resolution data on aerosol characteristics at a specific location, it has limited spatial coverage and cannot monitor large areas (Z. Li et al., 2018). In contrast, satellite-based monitoring of aerosols is advantageous due to its wide range, low cost and lack of geographical constraints (Dubovik et al., 2019). As satellite detection technology and capabilities continue to improve, the role of satellite remote sensing in atmospheric aerosol research is also increasing. Thus, maximising the potential of satellite remote sensing for AOD-related research real-life applications is of paramount importance.

Satellite remote sensing has become increasing valuable for filling gaps in AOD data where the characteristics are not readily apparent. However, the accuracy of satellite products is limited by assumptions regarding aerosol and surface properties, and AOD adjustment through topography and canopy (Weber et al., 2010). However, ground measurements are not subject to these limitations, which is why satellite

* Corresponding author

This contribution has been peer-reviewed.
https://doi.org/10.5194/isprs-archives-XLVIII-M-1-2023-395-2023 | © Author(s) 2023. CC BY 4.0 License.

derived AOD values are often verified by AERONET AOD values (Estellés et al., 2012). Currently, Moderate Resolution Imaging Spectroradiometer (MODIS) is among the most used sensors for aerosol detection. MODIS is one of the primary sensors on the Terra and Aqua satellites, which collaborate to repeatedly observe the Earth's surface every 1-2 days.

Studies indicate that C6.1 dark target algorithm (3km, 10km DT) and a combination of the dark target and deep blue algorithm (10km DTB) AOD products in Beijing are often overestimated(Wei and Sun, 2016). Nevertheless, there is still room for improvement in the accuracy of MODIS AOD (AODM). An effective approach is to use satellite observations as input and ground observations as output to build models based on both sources. These models can correct AOD values obtained from satellite observations, with sufficient training data. The models can be adapted to different retrieval scenarios and outperform deterministic algorithms. Machine learning methods satisfy these requirements and have been widely adopted in the aerosol science. For instance, (Lanzaco et al., 2016a) used machine learning techniques to correct the AOD value at 550nm, resulting in a significant improvement in the AOD value obtained from satellite data. (Palancar et al., 2016) used a similar method to obtain AOD values of 340 nm from MODIS and calculated the aerosol radiative forcing in the UV-B region, and the obtained values were consistent with those obtained using AERONET AOD (AODA) as input. These studies use AODA to correct AODM biases but are not applicable in areas with low AERONET sites density or for different underlying surfaces, which limits their reliability on a global scale.

In this paper, a method to improve AODM is proposed using AODA as reference. AOD products of Three different algorithms of AOD products, including 3km DT,10km DT, 10km DTB from Terra and Aqua satellites in Beijing area, are employed in this method. The AERONET site data in Beijing area is used as the ground truth for bias correction. The method leverages machine learning technology to improve the correlation between AODM and AODA and trains the model using three types of AOD inversion products from Beijing area to correct the overestimation phenomenon. The method preserves abnormal data in the dataset to make it applicable to original data. Moreover, the method requires only annual average trend data of meteorological variables, meteorological conditions as possible inputs to keep the requirements minimal.

2. DATASET

2.1 AERONET data

AERONET ("AERONET," n.d.) is a federation of ground-based remote sensing aerosol networks jointly established by NASA and LOA - PHOTONS (CNRS), provides high-quality measurements of aerosol optical properties (Holben et al., 1998). With around 500 sites worldwide, it represents a key resource for aerosol observation. The aerosol products derived from AERONET observations are classified into three levels: Level 1.0, Level 1.5, and Level 2.0. In this study, the Level 2.0 data from four aerosol sites in Beijing were used. The detailed information of these sites is presented in Table 1.

Table 1. Statistics of AERONET ground stations in China

Site name	longitude, latitude	observation time	MODIS Terra/Aqua data volume
Beijing	116.381, 39.977	2001.05-2018.11	438/361
Beijing_PKU	116.31, 39.992	2016.07-2019.5	134/107
Beijing_RADI	116.379, 40.005	2010.05-2019.05	67/166
Beijing-CAMS	116.317, 39.933	2012.08-2019.05	261/189

AERONET does not provide the AOD observation value at 550 nm band. However, the 550 nm Ångström index can be obtained by interpolation of the Ångström index at 500 nm and 675nm bands. The estimation of AOD at 550 nm can be obtained by using the equations (1) and (2), as proposed by (Eck et al., 1999).

$$\alpha_{500-675} = -\frac{\ln(\tau_{500}/\tau_{675})}{\ln(500/675)}, \quad (1)$$

$$\tau_{550} = \tau_{675}(500/675)^{-\alpha_{500-675}}, \quad (2)$$

In the above equation: τ_{550} and τ_{675} are the AOD values of 500nm and 657nm respectively, $\alpha_{500-675}$ is the Ångström index of 500-657nm, $\tau_{550}$is the AOD value of 550nm obtained by interpolation.

2.2 MODIS data

The United States National Aeronautics and Space Administration (NASA) launched a series of comprehensive Earth observation satellites in 1991. Terra and Aqua are two solar polar orbit synchronous satellites that were launched in 1999 and 2002 respectively. They were known as "morning star" and "afternoon star" because of their transit times of 10:30 a.m. and 1:30 p.m., respectively. These satellites can acquire daily global coverage data using a spectrometer with a measurement range from 0.4μm to 14.4μm. Since aerosol absorption levels in the mid-infrared region are low, surface reflectance can be determined, from which AOD can be derived (Lanzaco et al., 2017). However, different inversion algorithms may yield different AOD results.

To correct for bias in the AOD products, this work uses the AOD products of the Terra (from March 2001 to May 2019) and Aqua satellite (from July 2002 to May 2019) MODIS 3km DT,10km DT, 10km DTB. The corrected data is verified using the AOD data from AERONET stations in Beijing to ensure accuracy.

2.3 Meteorological data

Meteorological conditions play an essential role in predicting AOD, and they have a strong influence on the concentration distribution, chemical composition, and optical properties of AOD. Variations in AOD have been found to be significantly linked to meteorological factors. Therefore, in this study, the fifth-generation global climate reanalysis dataset ERA5 ("Climate Data Store," n.d.), was published by the European Centre for Medium-Range Weather Forecasts (ECMWF). ERA5 provides hourly atmospheric, terrestrial, and oceanic climate variables, terrestrial and oceanic climate variables at a spatial resolution of 0.25° × 0.25° (Jiang et al., 2021). In this study, meteorological parameters extracted include the U wind

This contribution has been peer-reviewed.
https://doi.org/10.5194/isprs-archives-XLVIII-M-1-2023-395-2023 | © Author(s) 2023. CC BY 4.0 License.

component (u10) and V wind component vectors (v10), 2m dew point temperature (d2m), 2m air temperature (t2m), surface temperature (skt), and atmospheric pressure (sp). Uv10 is calculated based on u10 and v10.

To capture the annual trend of meteorological conditions, we use the day of the year as a parameter. However, since January 1 and December 31 have the same average, this parameter is modified to create a more representative parameter. Specifically, we set the variable to 1 on January 1 and increase it to a maximum of 183 on July 1 and 2, then decrease it again to a minimum on December 31 to better reflect the seasonality of meteorological conditions (Lanzaco et al., 2016b).

2.4 Data pre-processing

The feature sample set was constructed from data gathered between 2001 to 2019, and a range of different features were used for model training. These parameters are listed in Table 2. Time data features were specified by selecting the year, month, and day. The transit times for Terra and Aqua satellites are 10:30 AM and 1:30 PM, respectively. In order to maintain consistency in timing, the EAR5 data was selected are 10:00 and 13:00 because the time resolution of the EAR5 data is in hours.

Table 2. Feature information table

Type	Data	Temporal resolution	Spatial resolution	Data origin
Ground station data	AERONET _AOD	15 minutes	——	AEROSOL ROBOTIC NETWORK
Satellite data	3km DT AOD	1 day	3km	MODIS
	10 km DT AOD	1 day	10km	MODIS
	10km DTB AOD	1 day	10km	MODIS
Meteorological data	uv10	1 hour	0.25°×0.25°	ERA5
	d2m	1 hour	0.25°×0.25°	ERA5
	t2m	1hour	0.25°×0.25°	ERA5
	skt	1 hour	0.25°×0.25°	ERA5
	sp	1hour	0.25°×0.25°	ERA5
	days	1day	——	——
Time	year	——	——	——
	month	——	——	——
	day	——	——	——
Geographic data	lons	——	——	——
	lats	——	——	——

3. METHODOLOGY

Machine learning is a subfield of artificial intelligence that focuses on learning from data to improve predictions. In this study, two types of machine learning algorithms, Artificial Neural Network (ANN) and Support Vector Regression (SVR) were used to correct the bias between AODM and AODA. The datasets used in this study were MODIS Terra and MODIS Aqua, which were divided into training and test sets. For each machine learning algorithm, the training set was split into "input" and "output" variables, and the algorithm searched for the relationships between them.

The goal of the study was to obtain AODM that closely matched AODA. The inputs used for the machine learning models included the three AODM products, meteorological variables, time (year, month, and day), latitude and longitude, and the annual average trend of meteorological conditions, while AODA was used as the output variable.

3.1 ANN

ANN is a machine learning method inspired by biological systems (Basheer and Hajmeer, 2000). It is a complex network comprising interconnected neurons that work in parallel. In a typical ANN structure, the input is connected to one or more layers of neurons, which are connected to the output. The connection weights between the neurons are parameters that are adjusted during the training process. The training process is iterative, where the Root Mean Square Error (RMSE) is calculated in each cycle and the connection weights are adjusted based on the results until the network can correctly predict the output. This type of process is called supervised learning because it provides realistic outputs.

When training an ANN model, the dataset is randomly split into a training set and a test set. The training set is used to train the weights of the neural network while the test set is used to evaluate the performance of the trained model on new, unseen data. The convergence of the training process is evaluated using the RMSE and the number of training iterations. Once the model is trained, a scatter plot is generated from the test set data, and the regression line is determined to evaluate the accuracy of the predicted outputs. This validation is rigorous and independent since the test set is randomly selected from all available temporal and spatial datasets.

3.2 SVR

Support Vector Machine (SVM) was first introduced by Cortes and Vapnik and developed into classification and regression problems (Sain, 1996) . SVM is widely accepted in the field of learning because of its ability to generalize. One of its important features is its capability to project linearly inseparable data into a high dimensional feature space, making it linearly separable. SVMs were later extended to regression problems, named SVR (Schölkopf et al., 2000; Smola and Schölkopf, 2004). In this study we used the SVR method proposed by LIBSVM (Chen, Fan and Lin 2006, Fan et al. 2005).

The goal of the SVR is to find a function that has a minimum deviation from the output of the training dataset, while also maintaining a flatness to control the complexity of the system and the training error. The penalty factor of the model is continuously adjusted through the training set, and the coefficient of determination (R2) and RMSE of the current model are evaluated in each iteration of training. The convergence of the training is determined by using the coefficient of R2, RMSE, and the penalty factor training iterations.

3.3 Technical flow chart

In this study, two machine learning algorithms: ANN and SVR were used to correct the bias between AODM and AODA. This study aims to approximate the AODM as close as possible to the AODA. Two dataset - MODIS Terra and MODIS Aqua were used together with meteorological data. The sample set is created by spatial temporal matching the features from satellite data with ground station (AERONET) data. As shown in Figure 1, machine learning algorithms were used to map the features inputs to correct AOD.

3.4 Correlation analysis

Feature selection was performed based on the correlation analysis, as shown in Figure 2. Pearson correlation coefficients were calculated between ground stations and all characteristic parameters. The AODA was positively correlated with 3km DT, 10km DT and 10km DTB, and the correlation exceeded 0.9. Temperature affects the Brownian motion of aerosol and the vertical distribution of aerosol, so that the change of its mass concentration satisfies the negative exponential law. All meteorological characteristic parameters were positively correlated. Among them, the strongest correlation was d2m with a correlation coefficient of 0.4, and the AODA was positively correlated with longitude (lons) and latitude (lats). Moreover, according to Figure 2, the correlation between year, month, sp and AODA was negative, and they were removed from the sample. Based on the results from correlation analysis, the final selected features train the ANN and SVR models are 3km DT, 10km DT, 10km DTB, day, lons, lats, uv10, d2m, t2m, skt, and days. The data was split into a training set of 70% and a test set of 30%.

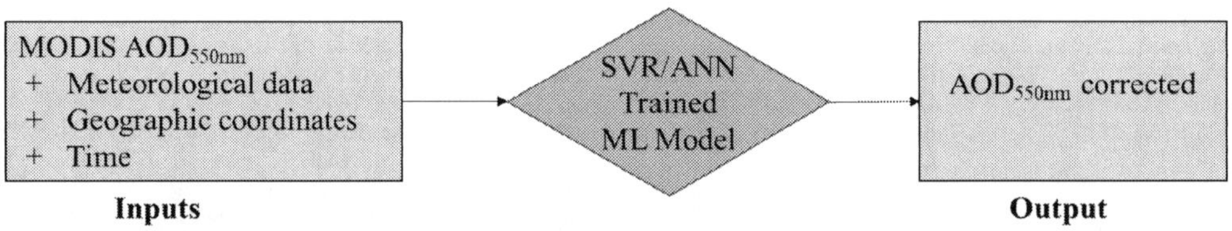

Figure 1. Flowchart of the proposed bias correction method

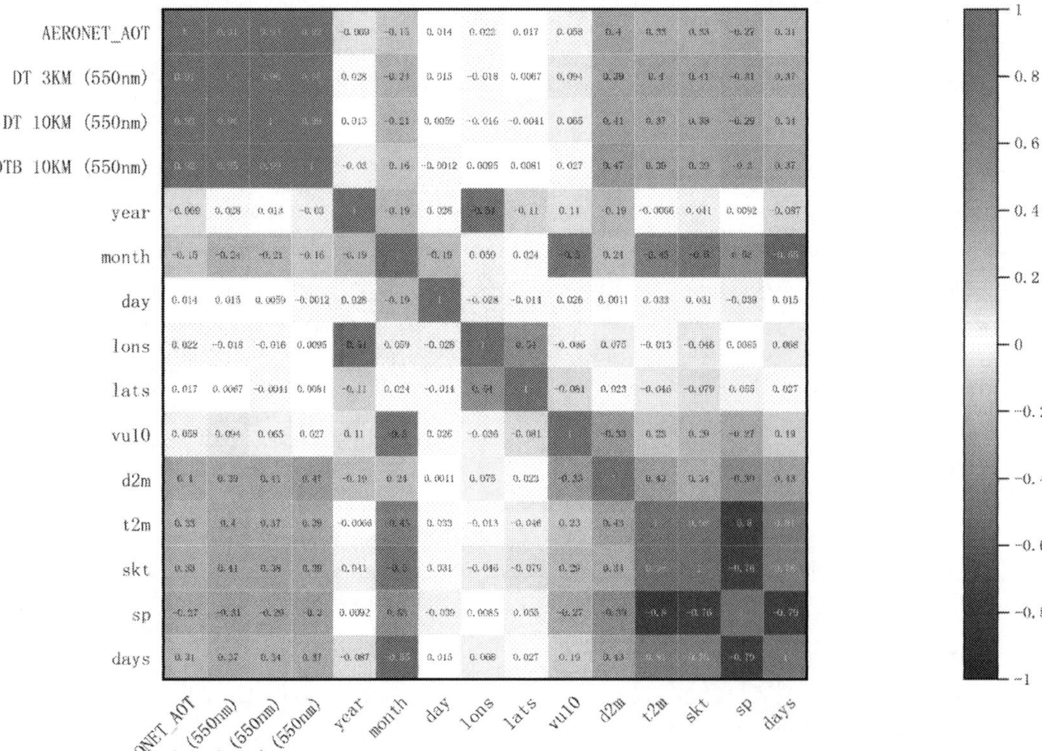

Figure 2. Correlation matrix of AOD Features

This contribution has been peer-reviewed.
https://doi.org/10.5194/isprs-archives-XLVIII-M-1-2023-395-2023 | © Author(s) 2023. CC BY 4.0 License.

4. RESULTS AND DISCUSSIONS

4.1 Comparison results of AOD values from MODIS data and AERONET

We first compare the AODM with AODA. Three MODIS AOD products that were used in this study, including 3km DT, 10km DT, and 10km DTB in Beijing region. The results are shown in Figure 3, R2 of the 3km DT Terra dataset is 0.578, and that of the Aqua dataset is 0.167. R2 of the 10km DT Terra dataset is 0.639, and that of Aqua is 0.585. R2 of the 10km DTB Terra dataset is 0.722, and that of Aqua is 0.657. Compared with the AODA, AODM is overestimated and needs to be improved. Therefore, it is necessary to propose a method to improve the accuracy of AODM.

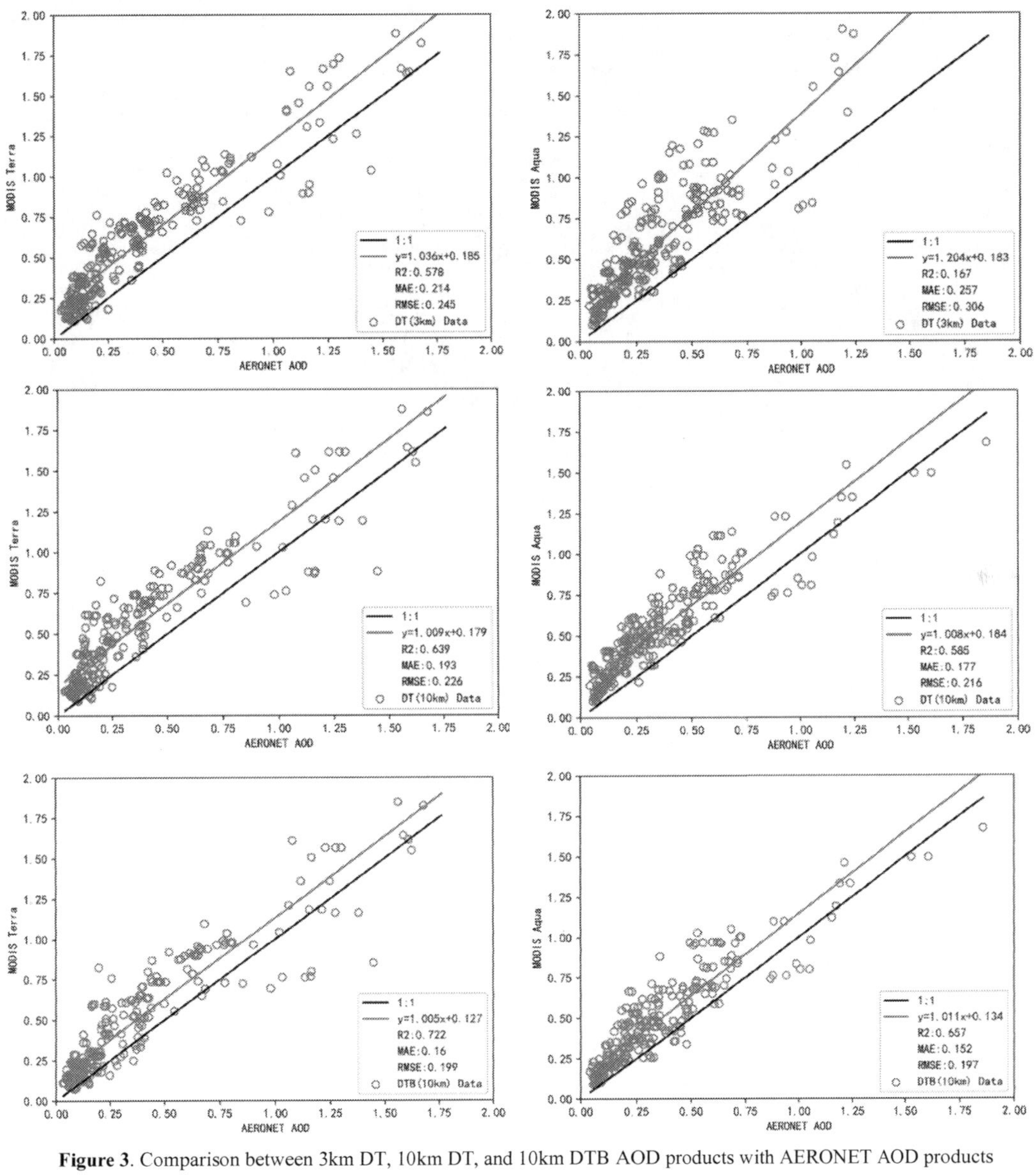

Figure 3. Comparison between 3km DT, 10km DT, and 10km DTB AOD products with AERONET AOD products

4.2 Determine the optimal parameters

To evaluate the bias correction effect of ANN and SVR model in AOD model, the training set and test set are used, and the accuracy is verified by 10-fold cross validation method. In the process of model training, the parameters of the model are fine-tuned to finally determine the optimal parameters. The specific parameters are shown in Table 3.

This contribution has been peer-reviewed.
https://doi.org/10.5194/isprs-archives-XLVIII-M-1-2023-395-2023 | © Author(s) 2023. CC BY 4.0 License.

Table 3. The model optimizes the relevant parameters

Algorithm name	Name of parameter	Value
ANN	Number of input layer neurons	40
	Number of hidden layer neurons	20
	Number of output layer neurons	1
	Number of iterations	1000
SVR	Penalty factor	5000

4.3 Bias correction results

After applying our proposed machine learning based bias correction method, the overestimation has been significantly improved. The results from SVR and ANN models are shown in Figures 4 and 5 respectively. After correction, R2 of the Terra dataset is 0.843 (SVR) and 0.876 (ANN), MAE is 0.11 and 0.083, and RMSE is 0.148 and 0.125. R2 of the Aqua dataset is 0.858 and 0.88, MAE is 0.09 and 0.088, and RMSE is 0.12 and 0.12. Based on the above analysis, it is noted that the deviation correction effect of ANN on AOD is better than that of the SVR method.

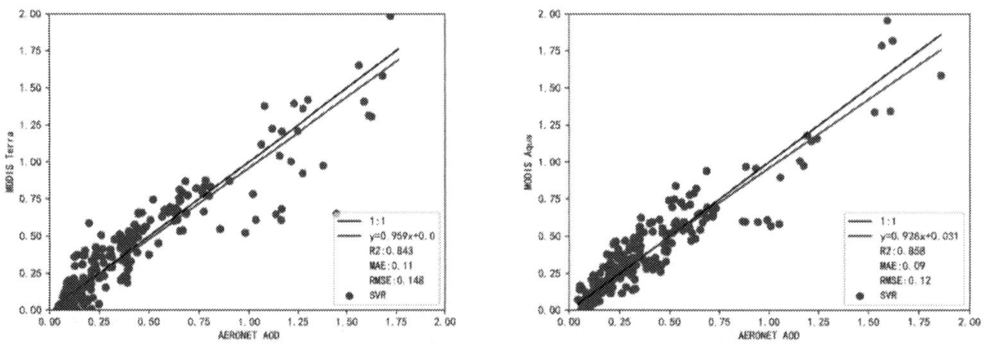

Figure 4. SVR model deviation correction results

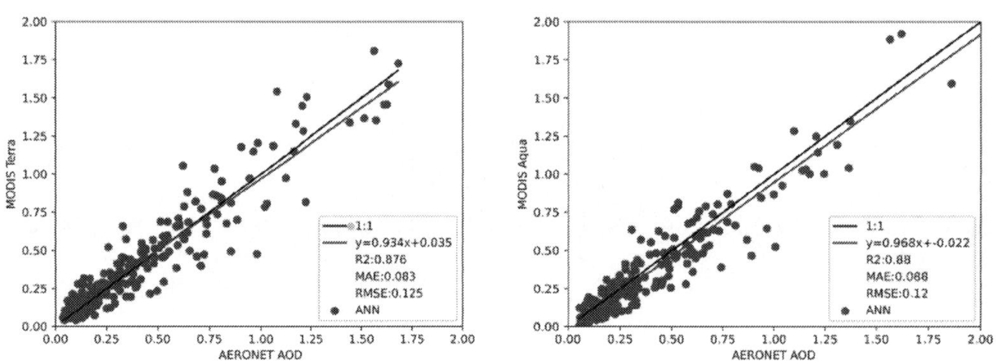

Figure 5. ANN model deviation correction results

5. CONCLUSIONS

In this study, we propose a machine learning based method to correct MODIS DT AOD at 3km, 10km, and MODIS DTB at 10km. MODIS AOD data and meteorological data are used as input to the proposed machine learning models, including SVR and ANN. Four ground stations from AERONET located in Beijing are used as ground truth. The results show that our proposed methods significantly improve the overestimation from the MODIS AOD and ANN performance is better than SVR. In future, this machine learning based method can also be applied to other parts of the world, especially in areas with poor aerosol characteristics, such as Africa and Australia, where the reliability of MODIS algorithm is relatively poor. Extending the practicability of this method to define a region around each site and train the model for each region separately can improve this research to a certain extent. Estimation of PM2.5 and PM10 based on satellite measurements is a topic very relevant to atmospheric science, and there is a strong correlation between aerosol and PM. By using the measurement data of particulate matter in a certain area and combining with the method used in this study, it is possible to obtain better PM2.5 and PM10 from satellite measurement data.

REFERENCES

AERONET [WWW Document], n.d. URL https://aeronet.gsfc.nasa.gov/ (accessed 2.20.23).

Basheer, I.A., Hajmeer, M., 2000. Artificial neural networks: fundamentals, computing, design, and application. J. Microbiol. Methods 43, 3–31.

This contribution has been peer-reviewed.
https://doi.org/10.5194/isprs-archives-XLVIII-M-1-2023-395-2023 | © Author(s) 2023. CC BY 4.0 License.

Bovensmann, H., Buchwitz, M., Burrows, J.P., Reuter, M., Krings, T., Gerilowski, K., Schneising, O., Heymann, J., Tretner, A., Erzinger, J., 2010. A remote sensing technique for global monitoring of power plant CO2 emissions from space and related applications. Atmospheric Meas. Tech. 3, 781–811. https://doi.org/10.5194/amt-3-781-2010

Climate Data Store [WWW Document], n.d. URL https://cds.climate.copernicus.eu/#!/home

Colbeck, I., Lazaridis, M., 2010. Aerosols and environmental pollution. Naturwissenschaften 97, 117–131.

Dubovik, O., Li, Z., Mishchenko, M.I., Tanré, D., Karol, Y., Bojkov, B., Cairns, B., Diner, D.J., Espinosa, W.R., Goloub, P., 2019. Polarimetric remote sensing of atmospheric aerosols: Instruments, methodologies, results, and perspectives. J. Quant. Spectrosc. Radiat. Transf. 224, 474–511.

Eck, T.F., Holben, B., Reid, J., Dubovik, O., Smirnov, A., O'neill, N., Slutsker, I., Kinne, S., 1999. Wavelength dependence of the optical depth of biomass burning, urban, and desert dust aerosols. J. Geophys. Res. Atmospheres 104, 31333–31349.

Estellés, V., Campanelli, M., Utrillas, M., Expósito, F., Martínez-Lozano, J., 2012. Comparison of AERONET and SKYRAD4. 2 inversion products retrieved from a Cimel CE318 sunphotometer. Atmospheric Meas. Tech. 5, 569–579.

Guo, J., Xia, F., Zhang, Y., Liu, H., Li, J., Lou, M., He, J., Yan, Y., Wang, F., Min, M., Zhai, P., 2017. Impact of diurnal variability and meteorological factors on the PM2.5 - AOD relationship: Implications for PM2.5 remote sensing. Environ. Pollut. 221, 94–104. https://doi.org/10.1016/j.envpol.2016.11.043

Holben, B.N., Eck, T.F., Slutsker, I. al, Tanré, D., Buis, J., Setzer, A., Vermote, E., Reagan, J.A., Kaufman, Y., Nakajima, T., 1998. AERONET—A federated instrument network and data archive for aerosol characterization. Remote Sens. Environ. 66, 1–16.

Jiang, T., Chen, B., Nie, Z., Ren, Z., Xu, B., Tang, S., 2021. Estimation of hourly full-coverage PM2. 5 concentrations at 1-km resolution in China using a two-stage random forest model. Atmospheric Res. 248, 105146.

Lanzaco, B.L., Olcese, L.E., Palancar, G.G., Toselli, B.M., 2017. An improved aerosol optical depth map based on machine-learning and MODIS data: Development and application in South America.

Lanzaco, B.L., Olcese, L.E., Palancar, G.G., Toselli, B.M., 2016a. A method to improve MODIS AOD values: Application to South America.

Lanzaco, B.L., Olcese, L.E., Palancar, G.G., Toselli, B.M., 2016b. A method to improve MODIS AOD values: Application to South America.

Li, J., Carlson, B.E., Lacis, A.A., 2015. How well do satellite AOD observations represent the spatial and temporal variability of PM 2.5 concentration for the United States? Atmos. Environ. 102, 260–273. https://doi.org/10.1016/j.atmosenv.2014.12.010

Li, J., Pei, Y., Zhao, S., Xiao, R., Sang, X., Zhang, C., 2020. A Review of Remote Sensing for Environmental Monitoring in China. Remote Sens. 12, 1130. https://doi.org/10.3390/rs12071130

Li, T., Hu, R., Chen, Z., Li, Q., Huang, S., Zhu, Z., Zhou, L.-F., 2018. Fine particulate matter (PM2. 5): The culprit for chronic lung diseases in China. Chronic Dis. Transl. Med. 4, 176–186.

Li, Z., Xu, H., Li, K., Li, D., Xie, Y., Li, L., Zhang, Y., Gu, X., Zhao, W., Tian, Q., 2018. Comprehensive study of optical, physical, chemical, and radiative properties of total columnar atmospheric aerosols over China: An overview of Sun–Sky Radiometer Observation Network (SONET) measurements. Bull. Am. Meteorol. Soc. 99, 739–755.

Luo, Y., Zheng, X., Zhao, T., Chen, J., 2014. A climatology of aerosol optical depth over China from recent 10 years of MODIS remote sensing data. Int. J. Climatol. 34, 863–870.

Ni, L., Chuang, C.-C., Zuo, L., 2015. Fine particulate matter in acute exacerbation of COPD. Front. Physiol. 6, 294.

Palancar, G.G., Olcese, L.E., Lanzaco, B.L., Achad, M., López, M.L., Toselli, B.M., 2016. Aerosol radiative forcing efficiency in the UV-B region over central Argentina. Atmospheric Res. 176, 1–9.

Sain, S.R., 1996. The nature of statistical learning theory.

Schölkopf, B., Smola, A.J., Williamson, R.C., Bartlett, P.L., 2000. New support vector algorithms. Neural Comput. 12, 1207–1245.

Shi, K., Bai, L., Wang, Z., Tong, X., Mulvenna, M.D., Bond, R.R., 2022. Photovoltaic Installations Change Detection from Remote Sensing Images Using Deep Learning, in: IGARSS 2022 - 2022 IEEE International Geoscience and Remote Sensing Symposium. Presented at the IGARSS 2022 - 2022 IEEE International Geoscience and Remote Sensing Symposium, IEEE, Kuala Lumpur, Malaysia, pp. 3231–3234. https://doi.org/10.1109/IGARSS46834.2022.9883738

Smola, A.J., Schölkopf, B., 2004. A tutorial on support vector regression. Stat. Comput. 14, 199–222.

Tecer, L.H., Alagha, O., Karaca, F., Tuncel, G., Eldes, N., 2008. Particulate matter (PM2. 5, PM10-2.5, and PM10) and children's hospital admissions for asthma and respiratory diseases: a bidirectional case-crossover study. J. Toxicol. Environ. Health A 71, 512–520.

Weber, S.A., Engel-Cox, J.A., Hoff, R.M., Prados, A.I., Zhang, H., 2010. An improved method for estimating surface fine particle concentrations using seasonally adjusted satellite aerosol optical depth. J. Air Waste Manag. Assoc. 60, 574–585.

Wei, J., Sun, L., 2016. Comparison and evaluation of different MODIS aerosol optical depth products over the Beijing-Tianjin-Hebei region in China. IEEE J. Sel. Top. Appl. Earth Obs. Remote Sens. 10, 835–844.

Yang, J., Gong, P., Fu, R., Zhang, M., Chen, J., Liang, S., Xu, B., Shi, J., Dickinson, R., 2013. The role of satellite remote sensing in climate change studies. Nat. Clim. Change 3, 875–883.

Yu, Z., Wang, Z., Bai, L., Chen, L., Tao, J., 2021. Remote Sensing Inversion of PM10 Based on Spark Platform, in: 2021 IEEE International Geoscience and Remote Sensing Symposium IGARSS. Presented at the IGARSS 2021 - 2021 IEEE International Geoscience and Remote Sensing Symposium, IEEE, Brussels, Belgium, pp. 1685–1688. https://doi.org/10.1109/IGARSS47720.2021.9554323

Zhang, J., Wang, Z., Bai, L., Song, G., Tao, J., Chen, L., 2021. Deforestation Detection Based on U-Net and LSTM in Optical Satellite Remote Sensing Images, in: 2021 IEEE International Geoscience and Remote Sensing Symposium IGARSS. Presented at the IGARSS 2021 - 2021 IEEE International Geoscience and Remote Sensing Symposium, IEEE, Brussels, Belgium, pp. 3753–3756. https://doi.org/10.1109/IGARSS47720.2021.9554689

Zhang, X., Zhang, Y., Bai, L., Tao, J., Chen, L., Zou, M., Han, Z., Wang, Z., 2021a. Retrieval of Carbon Dioxide Using Cross-Track Infrared Sounder (CrIS) on S-NPP. Remote Sens. 13, 1163.

Zhang, X., Zhang, Y., Lu, X., Bai, L., Chen, L., Tao, J., Wang, Z., Zhu, L., 2021b. Estimation of lower-stratosphere-to-troposphere ozone profile using long short-term memory (LSTM). Remote Sens. 13, 1374.

Zhu, S., Zhu, H., Xu, J., Zeng, Q., Zhang, D., Liu, X., 2021. Satellite remote sensing of daily surface ozone in a mountainous area. IEEE Geosci. Remote Sens. Lett. 19, 1–5.

This contribution has been peer-reviewed.
https://doi.org/10.5194/isprs-archives-XLVIII-M-1-2023-395-2023

The International Archives of the Photogrammetry, Remote Sensing and Spatial Information Sciences, Volume XLVIII-M-1-2023
39th International Symposium on Remote Sensing of Environment (ISRSE-39) "From Human Needs to SDGs", 24–28 April 2023, Antalya, Türkiye

AN APPROACH TO ANALYSIS OF THE IMPACT OF NATURAL DISASTERS ON THE ECONOMIC EFFICIENCY AND PROFITABILITY OF BUSINESS

P. Zlateva [1, 2] *, M. Galabov [1], D. Velev [1]

[1] University of National and World Economy, Sofia, Bulgaria – (plamzlateva, m.galabov, dgvelev)@unwe.bg
[2] Institute of Robotics, Bulgarian Academy of Sciences, Sofia, Bulgaria

KEY WORDS: Natural disasters, Business indicators, Economic efficiency, Profitability, Business sustainability.

ABSTRACT:

The paper purpose is to propose an approach to the analysis of the impact of various natural disasters (intense rainfall, extreme high or low temperatures, droughts, strong storms, floods, earthquakes and others) on the economic efficiency and profitability of the business. In this study, more specifically, the business companies in the sphere of production, trade and services are considered. These companies generate net income from sales of production, goods and services. In view of the successful development and sustainability of these companies, it is necessary to analyse the variations in specific ratios of two of the main business indicators (economic efficiency and profitability) under the influence of different natural disasters. The ratios related to the indicator of economic efficiency and profitability are calculated as the corresponding net revenues or profits from sales (of production, goods and services) are divided by one of the following variables: sum of the average balance values of current or fixed tangible assets, incurred full costs of the activity or other. Numerical results of several applications of the proposed approach with particular values of the variables that are included in the expressions describing the two business indicators are presented. The results of the analysis of the calculated values of the ratios related to the two business indicators (economic efficiency and profitability) before and after the possible occurrence of a given natural disaster can successfully support decision-making financial managers to take adequate measures to reduce the potential negative consequences, recover the normal operative activity or increasing resilience of companies.

1. INTRODUCTION

Nowadays, due to climate changes, natural disasters have a significant negative impact on the sustainability and development of business (IPCC-Working group II, 2022; UNISDR, 2019). These impacts can be direct or indirect, and can affect companies with different characteristics from all economic sectors (Climate-ADAPT_EEA, 2022; Advisory team-World Bank, 2018). Furthermore, the extent of the impact can vary depending on the type and severity of disaster (intense rainfall, extreme high or low temperatures, droughts, strong storms, floods, earthquakes and others), as well as the geographic location of the company and the economic sector in which it operates (EEA, 2021; Commission Notice-EC, 2021; Dimitrov, Penchev, Bogomilova, 2019).

The size of a business can also affect its ability to adapt to and recover from natural disasters. Larger businesses can have more resources and contingency plans to deal with disturbances of their sustainability, while smaller businesses can be more vulnerable to the negative economic effects of natural disasters.

Some of the ways in which natural disasters can affect business sustainability and development are as follows (Bragg, 2022):

- **Physical damages**: Natural disasters such as floods, hurricanes, earthquakes, and others can cause damages to buildings, equipment, and other physical assets of a business. These damages can result in significant repair or replacement costs, as well as reduce the sustainability of the business.
- **Supply chain disruption**: Natural disasters can disrupt supply chains, which can affect a business's ability to produce and sell its products. For example, if a business relies on a supplier that is located in an area affected by a natural disaster, it may be unable to obtain the necessary raw materials or components to continue production. This can lead to delays, increased costs, and lost revenue.
- **Interruption of Operations**: Natural disasters can cause business operations to be interrupted for a time period due to difficulty in accessing the business location (road closures, difficulties for employees to get to work), power outages, several infrastructure damages or other reasons, which can result in loss of customers, suppliers, and profits.
- **Increase in prices of raw materials**: Natural disasters can cause an increase in the prices of raw materials that are needed to production. For example, a drought can reduce the amount of wheat and both lead to an increase in its price, as well as the prices of flour and bakery products.
- **Higher Costs:** Business can incur higher costs to recover from natural disasters, such as the cost of repairs and replacement of damaged equipment, loss of revenue due to downtime, increased insurance premiums, etc.
- **Need to adapt to changing climate conditions:** Natural disasters can be caused by climate change and this requires businesses to adapt to these new conditions. This can include investing in more sustainable own infrastructure, using environmentally

* Corresponding author

friendly technologies, improving energy efficiency and other similar measures.
- **Need to improve the quality of products and services**: Natural disasters can force the businesses to invest in more efficient technologies and innovations to improve the quality of their products and services. This can increase the competitiveness of the business, as well as make it more resistant to future crises.

In this study, more specifically, business companies from sectors dealing with production, trade and services are considered. These companies generate net income from sales of products, goods and services. In view of the successful development and sustainability of these companies under the influence of different natural disasters, it is necessary to analyse the variations of two of the main business indicators (economic efficiency and profitability).

The paper purpose is to propose an approach to analysis of the impact of various natural disasters on the economic efficiency and profitability of the business. The idea is to perform the analysis based on the change in the values of the specific ratios that describe the two business indicators.

2. ESSENCE OF BUSINESS INDICATORS - ECONOMIC EFFICIENCY AND PROFITABILITY

2.1 Essence of Economic Efficiency

The efficiency, in economics and organizational analysis, is defined as a measure of the input a system requires to achieve a specified output. A system that uses few resources to achieve its goals is efficient, in contrast to one that wastes much of its input (Encyclopedia Britannica, 2023).

In general, economic efficiency in an economy indicates that the available resources are being utilised in the best possible way. It is known when an economy is economically efficient, any changes made to assist one entity would harm another. From a microeconomic point of view, business's pursuit of economic efficiency means increasing output and reducing costs.

In the production case, goods are produced at their lowest possible cost, as are the variable inputs of production. For consumers, economic efficiency leads to lower prices for goods and services. For the government, more efficient companies and higher levels of productivity and economic activity increase economic growth (StudySmarter, 2023).

There are various types of economic efficiency: Productive efficiency, Dynamic efficiency, Static efficiency, Social efficiency, Allocational efficiency, Pareto efficiency, etc. (Barnier and Bellucco-Chatham, 2020; Gordon, 2022; StudySmarter, 2023):

- Productive economic efficiency exists when output is fully maximised from the available inputs. In this situation is impossible to produce more of one good without producing less of another. For a company, productive efficiency occurs when the average total cost of production is minimised.
- Dynamic efficiency is explained as the productive efficiency of a company over a long period of time. This can be accomplished if the company reduces its costs by implementing new production processes and innovations. Dynamic efficiency aids in reducing long-run average cost (LRAC).
  Dynamic efficiency is affected by the following factors: Investment - Investing in technology and more capital can lower future costs; Technology - Improved technology in a company can help reduce costs; Finance - Accessibility to finance can help the company invest more capital to improve production, which in turn leads to a reduction in costs; Motivating the workforce - Encouraging and motivating workers and managers can enable a company to reduce costs.
- Static economic efficiency is concerned with the best combination of existing resources at a particular time. It is concerned with producing outputs at a specific time by allocating resources differently.
- Pareto efficiency is when every economic good is optimally allocated across production and consumption so that no change to the arrangement can be made to make anyone better off without making someone else off.

2.2 Essence of Profitability

Profitability is the main goal of any business. Without profitability the business will not survive in the long time. Therefore, measuring current and past profitability and forecasting future profitability is very important (Hofstrand, Johanns, 2021).

Profitability refers to the extent to which a company earns a profit. Also, in many financial analyzes of various businesses, profitability and profit are used interchangeably, but the two terms are not equivalent. Both are accounting metrics that are used to analyze the financial success of a company, but there are distinct differences between the two. Although a company can realize a profit, this does not necessarily mean that the company is profitable. Profitability is a financial metric that companies use to determine how successful they are. This is a relative measure and is usually expressed as ratios. While profit is a specific number and is considered as an absolute measure (Horton, Potters, Munichiello, 2021)

In this regard, the value of a business cannot be estimated only on the basis of its profit. It is necessary to perform a profitability analysis to understand whether the company is using its resources and capital effectively

Profitability can also be defined as a measure of a company's profit relative to its expenses. Businesses that are more efficient can make more profit as a percentage of its expenses than a less-efficient business, which must spend more to generate the same profit (Gartner Inc. 2023).

It is generally accepted that a business achieves profitability if the total amount of its income is greater than the total amount of expenses for the accounting period. However, it is also necessary to consider if a company is recording its business transactions under the accrual basis of accounting, it is quite possible that the profitability condition will not be matched by the cash flows generated by the organization, since some accrual-basis transactions (such as depreciation) do not involve cash flows (Bragg, 2022).

The profitability of a company is predetermined by various factors, such as expenses/costs, demand, productivity, competition, etc. (Horton, Potters, Munichiello, 2021).

This contribution has been peer-reviewed.
https://doi.org/10.5194/isprs-archives-XLVIII-M-1-2023-403-2023 | © Author(s) 2023. CC BY 4.0 License.

Profitability is commonly expressed by various financial ratios, which are created by the income statement. The values of these profitability ratios can be considered tools for making decisions about the financial state of the given business. Also, these ratios can help management, analysts, and investors better understand how the company is able to earn the money necessary to cover its expenses and other company-related costs.

3. DESCRIPTION OF THE APPROACH TO ANALYSIS OF THE IMPACT OF NATURAL DISASTERS ON THE ECONOMIC EFFICIENCY AND PROFITABILITY OF BUSINESS

3.1 Essence of the Approach

The main idea of the proposed approach is that the financial manager needs to be well aware of the impact that a natural disaster can have on the economic and sustainable development of the company in order to be able to make informed and effective decisions to deal with potential financial problems.

In this study, it is assumed that the economic efficiency is measured by comparing the economic effect with the resources or costs invested in its achievement, i.e. the ratio is economic effect/resources and economic effect/costs. The profitability is measured as comparing the financial result with the resources or costs or revenues through which this result is realized, i.e. the ratio is financial result/resources, financial result/costs and financial result/revenues.

The essence of the proposed approach is to calculate and compare various coefficients related to the economic efficiency and profitability of the company, taking into account the impact of natural disasters on the business sustainability. In particular, the coefficients related to these two business indicators (economic efficiency and profitability) can be calculated as the corresponding net revenues/costs or profits from sales (of production, goods and services) are divided by one of the following variables: sum of the average balance values of fixed or current tangible assets, incurred total costs of the activity or other elements.

3.2 Essence of the Proposed Approach

The essence of the proposed approach is to carry out an analysis of the impact of various natural disasters on business based on the variation in the values of several specific ratios determining the levels of economic efficiency and profitability of the monitored company. In particular, the companies from real sectors dealing with production, trade and services are considered. Knowledge of these ratios and the impact of natural disasters on their values is essential for the financial management of companies. It is important for managers to know how a natural disaster can affect the business sustainability so that they can make adequate recommendations to reduce the negative consequences and increase the company resilience.

3.2.1. Ratios of economic efficiency:

Efficiency ratio of fixed tangible assets in production sector =

$$= \frac{\textit{Net income from sales of production}}{\textit{Sum of the average balance sheet values of fixed tangible assets in the production sector}} \quad (1)$$

Efficiency ratio of fixed tangible assets in the trade sector =

$$= \frac{\textit{Net income from sales of goods}}{\textit{Sum of the average balance sheet values of fixed tangible assets in the trade sector}} \quad (2)$$

Efficiency ratio of fixed tangible assets in the service sector =

$$= \frac{\textit{Net income from sales of services}}{\textit{Sum of the average balance sheet values of fixed tangible assets in the service sector}} \quad (3)$$

Efficiency ratio of current tangible assets in production sector=

$$= \frac{\textit{Net income from sales of production}}{\textit{Average balance sheet values of current tangible assets in the production sector}} \quad (4)$$

Efficiency ratio of current tangible assets in the trade sector =

$$= \frac{\textit{Net income from sales of goods}}{\textit{Average balance sheet values of current tangible assets in the trade sector}} \quad (5)$$

Efficiency ratio of current tangible assets in the service sector =

$$= \frac{\textit{Net income from sales of services}}{\textit{Average balance sheet values of current tangible assets in the service sector}} \quad (6)$$

Efficiency ratio of total costs invested in the sold production =

$$= \frac{\textit{Net income from sales of production}}{\textit{Total costs invested in the sold production}} \quad (7)$$

Efficiency ratio of total costs invested in the sold production =

$$= \frac{\textit{Net income from sales of goods}}{\textit{Total costs invested in the sold goods}} \quad (8)$$

Efficiency ratio of total costs invested in the sold production =

$$= \frac{\textit{Net income from sales of services}}{\textit{Total costs invested in the sold services}} \quad (9)$$

3.2.2. Ratios of profitability:

Profitability ratio of fixed tangible assets in production sector =

$$= \frac{\textit{Profit from sales of production}}{\textit{Sum of the average balance sheet values of fixed tangible assets in the production sector}} \quad (10)$$

Profitability ratio of fixed tangible assets in the trade sector =

$$= \frac{\textit{Profit from sales of goods}}{\textit{Sum of the average balance sheet values of fixed tangible assets in the trade sector}} \quad (11)$$

This contribution has been peer-reviewed.
https://doi.org/10.5194/isprs-archives-XLVIII-M-1-2023-403-2023 | © Author(s) 2023. CC BY 4.0 License.

Profitability ratio of fixed tangible assets in the service sector =

$$= \frac{\textit{Profit from sales of services}}{\textit{Sum of the average balance sheet values of fixed tangible assets in the service sector}} \quad (12)$$

Profitability ratio of current tangible assets in production =

$$= \frac{\textit{Profit from sales of production}}{\textit{Average balance sheet values of current tangible assets in the production sector}} \quad (13)$$

Profitability ratio of current tangible assets in trade sector =

$$= \frac{\textit{Profit from sales of goods}}{\textit{Average balance sheet values of current tangible assets in the trade sector}} \quad (14)$$

Profitability ratio of current tangible assets in service sector =

$$= \frac{\textit{Profit from sales of services}}{\textit{Average balance sheet values of current tangible assets in the service sector}} \quad (15)$$

Profitability ratio of total costs invested in sold production =

$$= \frac{\textit{Profit from sales of production}}{\textit{Total costs invested in the sold production}} \quad (16)$$

Profitability ratio of total costs invested in sold production =

$$= \frac{\textit{Profit from sales of goods}}{\textit{Total costs invested in the sold goods}} \quad (17)$$

Profitability ratio of total costs invested in sold production =

$$= \frac{\textit{Profit from sales of services}}{\textit{Total costs invested in the sold services}} \quad (18)$$

4. APPLICATIONOF THE APPROACH TO ANALYSIS OF THE IMPACT OF NATURAL DISASTERS ON THE ECONOMIC EFFICIENCY AND PROFITABILITY OF BUSINESS

4.1 Applications of the Approach to Analysis of the Impact of Natural Disasters on the Economic Efficiency of Business

If it is assumed that a natural disaster occurred at the end of the previous year, which caused significant damage to the production tangible assets of the company, and then it is logical to assume that in the current year the net sales revenue will be smaller compared to the previous year.

Let's analyze the impact of a natural disaster on the economic efficiency of the business based on variations in the values of the efficiency ratio of the total costs invested in the sold production (7).

One of the main components of total costs is prime cost of sold production. Let's assume that due to a natural disaster production has decreased, and hence the prime cost of sold production also decreases, for example by 15%. Other components of total costs are selling and administrative costs. Let's assume that due to a natural disaster they have decreased, for example by 18% and by 10% respectively. Let's assume that net income from sales of production has also decreased, for example by 20%.

Here, let's denote prime cost of sold production by A, cost of sell by B, administrative cost by C, and Net income from sales of production by V. In this case, the efficiency ratio (7) is presented as follows:

Efficiency ratio of total costs invested in the sold production =

$$= \frac{V}{A + B + C} \quad (19)$$

The efficiency ratio (19) after taking into account the reductions in the variables due to a natural disaster is rewritten as follows:

Efficiency ratio of total costs invested in the sold production =

$$= \frac{(1\text{-}0.20)\,V}{(1\text{-}0.15)A+(1\text{-}0.18)B+(1\text{–}0.1)C} = \frac{0.8V}{0.85A+0.82B+0.9C} \quad (20)$$

If, for example, $V = 200$, $A = 120$, $B = 15$, and $C = 30$, then the efficiency ratio (19) has the following value:

Efficiency ratio of total costs invested in the sold production =

$$= \frac{200}{120 + 15 + 30} = \frac{200}{165} = 1.21 \quad (21)$$

If a natural disaster occurs that damages the production activity of the company, then taking into account the indicated reductions in the values of the considered variables for the efficiency ratio (20), the following value is obtained:

Efficiency ratio of total costs invested in the sold production =

$$= \frac{0.8 \times 200}{0.85\times120+0.82\times15+0.9\times30} = \frac{160}{102+12.3+27} = \frac{160}{141.3} = 1.13 \quad (22)$$

As can be seen as a result of the damage caused by the natural disaster to the business, the value of this ratio decreased from 1.21 to 1.13 for the same duration periods, respectively before and after the disaster.

In the given example, the economic effect decreases by 20%, (from 200 to 160) and the costs invested in its realization decrease by 14%, (from 165 to 141.3) i.e. the rate of reduction of the effect is higher than the rate of reduction of costs, as a result of which the value of the ratio decreases.

It is also possible that after the natural disaster the rate of reduction in the economic effect is equal to the rate of decrease in costs in general. In this case, the value of the economic ratio (7) does not change, i.e. it remains equal to 1.21 (21).

Let's consider the third scenario, in which the rate of reduction of the economic effect is less than the rate of reduction of the costs invested in this effect. Let's take the same sample data again: the net income from sales of production (the economic effect) is 200 and the total costs invested in the sold production are 165.

Let's assume that after a natural disaster, the economic effect decreases by 20% and the total costs by 30%. In this case, the economic coefficient (7) can be expressed numerically as follows:

Efficiency ratio of total costs invested in the sold production =

$$= \frac{(1-0.20) \times 200}{(1-0.30) \times 165} = \frac{0.8 \times 200}{0.7 \times 165} = \frac{160}{115.5} = 1.39 \quad (23)$$

Here there is a reduction of net income from sales of production (the economic effect) from 200 to 160 and of total costs invested in the sold production from 165 to 115.5 for the corresponding equal periods before and after the disaster. In this case, due to the changes in the variables mentioned above, the efficiency ratio (7) increases from 1.21 to 1.39 for the same size post-disaster period.

The increase, retention and decrease of the value of the efficiency ratio of total costs invested in the sold production is relative to the period before the disaster. The pre-disaster and post-disaster periods are of equal length in all analyses.

4.2 Applications of the Approach to Analysis of the Impact of Natural Disasters on the Profitability of Business

In the denominators of the expressions for the various ratios, for profitability (10)-(18) are included the values of the resources, expense/costs and revenues, through which the financial results (profits) are obtained.

If it is assumed that a natural disaster occurred at the end of the previous year, which caused significant damage to the company's property, then it is logical that in the current year after the natural disaster, the profits for the business will be lower compared to the previous year.

If for the current year after the occurrence of a natural disaster the rates of decrease in profits are greater than the rates of decrease of the variables that are present in the denominators of the expressions, then the values of corresponding ratios are smaller for the current year compared to the previous year. This means that a unit of resource is engaged in realizing less profit from sales in the current year after the occurrence of the natural disaster compared to the previous year.

If the rates of decrease in profits are equal to the rates of decrease in the variables that are present in the denominators, then there will be no change in the the values of corresponding ratios for the current year compared to the previous year. However, this does not mean that the natural disaster did not have an adverse impact on the company, as the profits in the current year are lower compared to the previous year, and this decrease affects the financial result of the business for the current year.

If for the current year the rates of decrease in profits are smaller than the rates of decrease of the variables that are present in the denominators of the expressions, then the corresponding ratios will be greater for the current year after the occurrence of the natural disaster compared to the previous year. This means that a unit of resource is engaged in making more profit in the current year than in the previous year. However, for the current year, the profits are smaller compared to the previous one, and this has an impact in the direction of reducing the financial result of the company.

The three analyzed scenarios for the potential impact of natural disasters on the company's profitability ratios can be successfully verified through different numerical examples.

Let's consider a company that sells production with prime cost $A = 120$, costs of sales are $B = 15$, and administrative expenses are $C = 30$. Let the net income from sales of production is V=200. Then the profit from sales of production, denoted by variable P, is calculated as follows:

$$P = V-(A+B+C) = 200-(120+15+30) = 200-165 = 35 \quad (24)$$

Let's calculate the value of the *Profitability ratio of total costs invested in sold production* (16) with the above numerical values of the variables as follows:

Profitability ratio of total costs invested in sold production =

$$= \frac{P}{A+B+C} = \frac{35}{120+15+30} = \frac{35}{165} = 0.21 \quad (25)$$

Let's consider the following scenario: It is assumed that a natural disaster has occurred and caused damage to the company's property. For this reason, prime cost of sold production decreased by 15% from 120 to 102), costs of sales decreased by 18% (from 15 to 12), and administrative expenses decreased by 10% (from 30 to 27). In addition, the net income from sales of production also decreases by 20% (from 200 to 160). Therefore, in this case the profit on sales is calculated as:

$$P = V-(A+B+C) = 160-(102+12+27) = 160-141 = 19 \quad (26)$$

In this example, after comparing the values of (24) and (26), it can be seen that the *profit from sales of production, P,* is reduced from 35 to 19. This reduction is about 46% and is due to negative impact the natural disaster on the business activity.

In this case, the value of the *Profitability ratio of total costs invested in sold production* (16) is calculated as follows:

Profitability ratio of total costs invested in sold production =

$$= \frac{P}{A+B+C} = \frac{19}{102+12+27} = \frac{19}{141} = 0.13 \quad (27)$$

It can be seen from (25) and (27) that there is a decrease in the value of this profitability ratio from 0.21 to 0.13. This is due to the impact on business of the natural disaster that occurred, which in this case leads to the fact that the rate of decrease in profit from sales of production is greater than the rate of decrease in the total costs invested in the sold production. In particular, the decrease in profit from sales of production is about 46% (from 35 to 19) and the decrease in total costs invested in the sold production is about 15% (from 165 to 141).

Let's consider a second scenario in which, due to the occurrence of a natural disaster, the rate of decrease in profit from sales of production equals the rate of decrease in total costs invested in the sold production (for example with 30%):

Profitability ratio of total costs invested in sold production =

$$= \frac{(1-0.30) \times 35}{(1-0.30) \times 165} = \frac{0.70 \times 35}{0.70 \times 165} = \frac{24.5}{115.5} = 0.21 \quad (28)$$

https://doi.org/10.5194/isprs-archives-XLVIII-M-1-2023-403-2023 | © Author(s) 2023. CC BY 4.0 License.

In this case, the value of the profitability ratio (28) does not change, i.e. remains equal to 0.21 (25). However, there is a decrease in profit from sales of production due to the natural disaster that occurred, and this negatively affects the overall financial sustainability of the business.

Let's consider a third scenario and assume that after occurrence of a natural disaster the rate of decrease in profit from sales of production is less (for example with 30%) than the decrease in total costs invested in the sold production (for example with 40%). In this case, the profitability ratio (16) can be expressed numerically as follows:

Profitability ratio of total costs invested in sold production =

$$= \frac{(1-0.30) \times 35}{(1-0.40) \times 165} = \frac{0.70 \times 35}{0.60 \times 165} = \frac{24.5}{99} = 0.25 \qquad (29)$$

As can be seen from (25) and (29) , if the rate of decrease in the profit from sales of production is less than the rate of decrease in the full costs invested in the realization of this profit, then the value of the profitability coefficient increases (here from 0.21 to 0.25). However, this in case the profit from sales of production is less (P=24.5) after the impact of the natural disaster compared to the profit from sales of production before the disaster (P=35). However, it should be noted that this circumstance affects the financial sustainability of the business.

It is important to point out that for the sustainability of the business it is good for the experts working in the financial management of the company to know very well the essence of these two business indicators (economic efficiency and profitability). Therefore, after the occurrence of a natural disaster that caused material damage to the company, it is necessary to calculate and analyze these business indicators. Then it is necessary to propose to the management of the company adequate measures aimed both at overcoming the adverse consequences and at restoring the values of these business indicators to pre-disaster levels. Certainly, it is better if these measures lead to an increase in the values of the economic efficiency and profitability of business.

5. CONCLUSIONS

The essence and their respective coefficients of the two business ratios are described. Numerical results of several applications of the proposed approach with particular values of the variables that are included in the expressions describing the two business indicators are presented. The results of the analysis of the calculated values of the relevant ratios related to the two business indicators (economic efficiency and profitability of the business) before and after the possible occurrence of a given natural disaster can successfully support decision-making financial managers to take adequate measures to reduce the potential negative consequences, recover the normal operative activity or increasing resilience of companies.

ACKNOWLEDGEMENT

This study is financially supported by National Science Fund of Bulgaria under the Grant №KP-06-H55/5 from 15.11.2021, titled "Smart Risk Management for Business from Adverse Events and Natural Disasters".

REFERENCES

Advisory team of the World Bank 2018 Climate Change Adaptation – Assessment of the Disaster Risk Management Sector, Bulgaria, https://www.moew.government.bg

Barnier, B., Bellucco-Chatham, A., 2020 Economic Efficiency: Definition and Examples. The Investopedia Team, Investopedia, https://www.investopedia.com/terms/e/economic_efficiency.asp

Bragg, S. M., 2022 *Cost Accounting Fundamentals – Essential Concepts and Examples*. 7th Edition, AccountingTools Inc.

Gartner Inc. 2023. Gartner Glossary - Finance – Profitability https://www.gartner.com/en/finance/glossary/profitability

Gordon, J., 2022 Economic Efficiency – Explained, https://thebusinessprofessor.com/en_US/economic-analysis-monetary-policy/economic-efficiency-definition#academic-research-on-economic-efficiency-4

Dimitrov, D., Penchev, G., Bogomilova, E., 2019 GIS Application for Economic Assessment of Direct Disaster Losses. In: Murayama, Velev, Zlateva, (eds) *IFIP Advances in Information and Communication Technology*, Springer, Cham, 516, 123–135, https://doi.org/10.1007/978-3-030-18293-9_11

Encyclopedia Britannica, 2023 Efficiency economics and organizational analysis. https://www.britannica.com/topic/efficiency-economics-and-organizational-analysis

European Commission 2021 Commission Notice - *Technical guidance on the climate proofing of infrastructure in the period 2021-2027*, Official Journal of the European Union C 373, 1-92

European Environment Agency (EEA) 2021 *Europe's changing climate hazards* — an index-based interactive EEA report 15 https://www.eea.europa.eu/publications/europes-changing-climate-hazards-1/climate-hazards-indices.

European Environment Agency (EEA) 2022 *Climate change adaptation - European Climate Adaptation Platform* (Climate-ADAPT), https://www.eea.europa.eu/themes/climate-change-adaptation/climate-adapt

Gartner Inc. 2023. Gartner Glossary - Finance Glossary – Profitability. www.gartner.com/en/finance/glossary/profitability

Hofstrand D., Johanns, A., 2021. Understanding Profitability. Ag Decision Maker, https://www.extension.iastate.edu/agdm/

Horton, M., Potters, C., Munichiello, K., 2021 *The difference between Profitability and Profit.* https://www.investopedia.com/ask/what-difference-between-profitability-and-profit.asp

Intergovernmental Panel on Climate Change (IPCC), Working Group II *Climate Change 2022: Impacts, Adaptation and Vulnerability*. 6-th Report, https://www.ipcc.ch/report/ar6/wg2

StudySmarter 2023, Economic Principles-Economic Efficiency. https://www.studysmarter.co.uk/explanations/microeconomics/economic-principles/economic-efficiency/

United Nations Office for Disaster Risk Reduction (UNISDR) 2019 *Global Assessment Report on Disaster Risk Reduction*. https://gar.unisdr.org/report-2019.

ASSESSMENT OF LAND SUITABILITY AND CAPABILITY BY INTEGRATING REMOTE SENSING AND GIS FOR AGRICULTURE IN POTWAR REGION PAKISTAN

S. Batool [1]*, S.A. Mahmood[2], M. Muhammad Ali[2]

[1] Centre for Integrated Mountain Research University of the Punjab Lahore, Pakistan. Saira.cimr@pu.edu.pk
[2] Department of Space Science University of the Punjab Lahore, Pakistan. amer.spsc@pu.edu.pk, maryamalirehan@gmail.com

KEY WORDS: Land Suitability, crop suitability, Remote Sensing, MCE, AHP, GIS, Potwar-Pakistan

ABSTRACT:

The assessment of land suitability and capability to reduce the anthropogenic pressure on is significant for the agricultural yield management in rainfed Potwar Region Pakistan. This study examines phases of wheat growth using Landsat-8 thermal bands and temperature based site maps, and demarcates suitable sites in potwar region for sustainable crops through Multi-Criteria Evaluation (MCE), Analytical Hierarchy Process (AHP), python and geospatial techniques. The total investigation area was 22168 km^2 out of which 12256 km^2 (55%) declared highly suitable, 2998 km^2 (13%) was moderately suitable, 4843 km2 (22%) was less suitable and 2069 km^2 (10%) was found not suitable for wheat crop cultivation (WCC). The results indicated that the areas having temperature ranging between 10–18◦C, pH between 6.2-6.5 and clay loom soil type with drainage level 0.85-1.1 corresponds to extremely suitable for wheat production. Various land quality factors; that is slope, flooding, soil texture, erosion, depth and coarser bits under different land-units were assessed for the crops. Consequently, all these factors were combined to make an integrated land suitability and capability (LSC) map for each landuse and suitable soils for agricultural productions. It is concluded that better landuse practices can be implemented to enhance the suitable agricultural sites.

1. INTRODUCTION

According to the World Population dataset, there was an increase of 8 billion populations in 2022 and will be around 10 billion until 2050 (Roser et al., 2013). This rapid growth has put the planet in danger and has caused a food and water deficit as a result. The world is currently facing multiple threats from this sudden rise in population, including threats to the food supply chain and the availability of fresh water as urban sprawl is quickly displacing agricultural land. Pakistan's economy is based primarily on agriculture, which accounts for 23.3% of the country's GDP. It represents 42.3% of total employed labour force in Pakistan. Agriculture's role to the broader national economy has, however, been declining (from 27.3% in 1985 to 19.8 % to date) as compared to other sectors. Lack of significant advancement in agricultural production from major crops and yield stagnation are emerging as serious challenges for the agriculture sector (Islam, 1995).. Lack of true-to-type seeds, climate variability, deteriorating water availability, progressive changes in soil nutrients, and cultivation of some crops in inopportune locations are the causes of this yield stagnation (planting rice in areas better suited to cotton). Agribusiness is the Around 70% of Pakistanis depend on agricultural produce for their daily sustenance (Rehman, 2016). While the majority of Pakistan's crop needs are satisfied by the production of wheat, other crops such as rice, sugarcane, corn, and maize are also grown in addition to wheat. Wheat significantly boosts the country's economy, but at the same time, the amount of agricultural land has been diminished by urbanisation, infrastructural projects (Motorways, industrialization, housing schemes, etc.), and other factors (Rehman et al., 2015) To meet the rising demand for food, careful resource management and decision-making are therefore required. Ecological elements, such as different soil parameters (soil, temperature, moisture, and pH of the soil), as well as soil temperature and moisture, have an impact on plant growth (Al.karaki, 2012; Porter & Gawith, 1999; Russell & Wilson, 1994) These factors affect a plant's nutritional value, but farmers are not aware of their importance in determining how they will affect the amount of wheat produced on an acre (Onwuka & Mang, 2018). Therefore, farmers need to be aware of the special significance of the soil and the related elements that affect wheat output and growth. A particular crop's production and growth are influenced by the soil's nutrients and physical characteristics. A clear strategy is proposed to increase agricultural productivity. To get the desired yield from a partial harvest, the LSC is crucial. To evaluate the LSC of a certain landcover, the soil must be divided into its subclasses. To achieve the major goals of this study, the LSC of soil is examined and thematically mapped utilising geospatial techniques (El Baroudy, 2016). The vast spatiotemporal satellite data sets can be obtained, secured, handled, and manipulated with the help of the relatively accurate GIS tool in order to examine their thematic features and perform a more correct assessment of LSC. To accurately map LSC zones, geospatial methods can be combined with MCE methodologies (Zolekar & Bhagat, 2015). The primary goal is to incorporate the MCE techniques at GIS platform to identify LSC regions by taking into account a variety of variables to produce LSC maps and wheat production. Using MCE, possible LSC sites are evaluated based on drainage parameters and soil pH. Site data are collected using AHP methodologies and weighted linear sum, and the results are then merged into a GIS environment. (Anees et al., 2014). Without taking into account the specific use, the LSC assessment describes and rates land development units (LDUs) in a broad framework. Classes I through VIII are clearly and formally defined This classification is useful because specific soil types may be suitable for particular crops while being ineffective for others (Hossain et al., 2011). As a result, accuracy in terms of land use types is crucial. It might be portrayed in terms of different crop productions, but also in terms of how specific crops are planted and harvested (Rehman et al., 2015; Sys et al., 1991). The LSC refers to a piece of land's ability to sustainably support crop production. Its analysis provides details on the land use prospects and constraints and, as a result, leads to conclusions about the best way to use the natural resources, the information from which is a crucial requirement for LUDs. Based on the qualities of the soil, the characteristics of the terrain, and an examination of the existing land use, land may be divided into spatially allocated agriculture prospective sectors (Khan et al., 2016; Bandyopadhyay et al., 2009). Only through

This contribution has been peer-reviewed.
https://doi.org/10.5194/isprs-archives-XLVIII-M-1-2023-409-2023

scientific soil surveys, assessments of the potential for a wide range of land uses, and the creation of flexible, socially acceptable, and environmentally friendly land use regulations could agri-production be met (Khan, 2018; Sathish & Niranjana, 2010). Agricultural indices, biophysical parameters, and land use and Landcover (LULC) evaluations can all be estimated using remotely sensed data (RSD), in addition to cropping strategy research (Ahmad, 2013; Rao et al., 1996 and Panigrahy et al., 2006). However, unless the datasets are combined with site-specific climatic and soil data, RSD only cannot suggest crop compatibility for a given place. The RSD can identify several physiographic units and generate auxiliary information on site characteristics, such as the investigation area's aspect and slope direction. To begin assessing crop suitability, complete information on the properties of the soil profile is necessary. The goal of the current study is to demonstrate how soil data sets and GIS approaches may be used to assess wheat yield and LSC in the Potwar Region and apply crop sustainability. Numerous academics have already confirmed the value of combining geospatial approaches to quantify the land evaluation)

2. MATERIALS AND METHODS

2.1. Experimental Site

The study area is located in Pakistan's Punjab province. With over 56% of the country's population, Punjab is the province with the most population. It has a size of roughly 205,344 Km^2 and is situated at 31.17° N and 72.70° E. About 26% of Pakistan's total land area is covered by it. Additionally, it has ties to Rajasthan and Punjab in India. The powerful Indus River and its tributaries left behind soil deposits in the Punjab during the Quaternary Period. Upper Indus Plain (UIP) is the Pakistan's primary agricultural region. It is also known as Pakistan's "Grain Basket" (Khan et al., 2015). The area set aside for wheat in the 2014–15 Rabi season was 6.98 mh (Becker et al., 1990; Khan, 2018) Punjab is classified into three zones based on its agro-ecological characteristics: I Potwar Region, which has 10% of Punjab's rain-fed agricultural land (Ahmad, 2013) (ii) the central-south region of Punjab's semi-arid-to-arid desert, which produces less agricultural output The majority of irrigated crop production is located in the Indus basin (iii). Punjab contributes greatly to the state's economic growth and provides almost 200 million people with food security by growing wheat and other necessary foods (Barsi et al., 2014). More than 68% of rural population depends on agriculture for their livelihood. In Pakistan, Punjab province is in a semi-arid region with temperatures typically fluctuate from 2 to 45°C, which can reach 50°C in the summer and 7°C in the winter. The yearly maximum temperature ranges from 28° to 32°C, while the annual minimum temperature ranges from 15° to 19°C. The trend toward continuous temperature rises that cause global warming is observed to be greater than 1°C overall. With a total variance of 228mm during the previous 50 years, there was an upward tendency in rainfall. Punjab has three main seasons according to climate classification: rainy season (April to June), when temperatures soar to 45 to 50 °C (monsoon) Between July and September, there is an average yearly rainfall of between 508 and 630 millimeters on the plains and between 1140 and 1270 millimeters in the sub-mountain region. From October to March, the temperature dips below freezing. From 53 to 62%, the humidity changes, fluctuating with the climate conditions. The length and duration of sunny days are reflected in sunlight hours; for example, May, June, and July have longer, sunnier days with more sunshine hours than December and January.

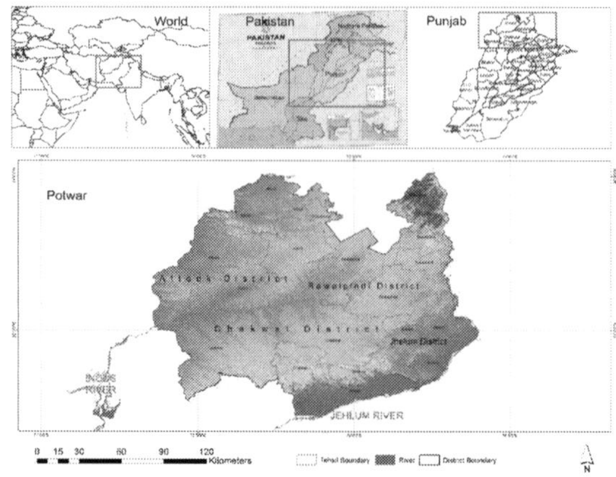

Figure 1: Location of Potwar Region

2.2. Experimental Design

Significant LSC needs for wheat production were established with the help of discussions and conversations with local agronomists, the Wheat Crop Monitoring Center, literature research, and wheat crop specialists. soil types and pH The first variables examined in this research to create adequate wheat growing areas were temperature, soil drainage, and soil electrical conductivity. The flowchart below outlines the methodical procedures needed to compute various parameters Figure 2.

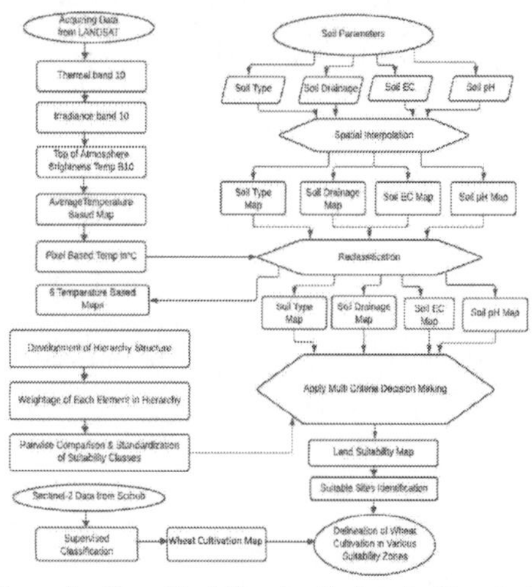

Figure 2: Flow Chart Showing Systematic Steps to Generate Various Maps

2.2.1. Python Based LST Maps from Landsat-8

According to Rajeswari and Mani (2004; Xu et al., 2004), LST is the radiative skin temperature of the land generated from infrared radiation (Weng et al., 2004). LST is a key factor in determining the behaviour of the Earth's thermal environment because it effectively regulates the radiant temperature of the planet, allowing researchers to assess a variety of aspects of life, including climatic changes and variabilities, the health of cereal crops and vegetation, agricultural and hydrological processes, and extreme events like forest fires and volcanic eruptions (Jimenez et al., 2006). On global scale, controlling Earth's surface's chemical, physical, and biological processes depends on LST evaluation using RSD (Becker et al., 1990). Numerous scholars share a same understanding of RSD for the various growth stages of wheat crop and LST computations using Python and Landsat-8 level 1 data. Since manual LST calculation is

This contribution has been peer-reviewed.
https://doi.org/10.5194/isprs-archives-XLVIII-M-1-2023-409-2023

time-consuming, chaotic, and prone to accumulation of errors, Python code was created to compute LST automatically with repeat process. This study's code was specifically designed to work with Landsat-8 data. USGS instructions state that Band-11 of the TIRS should not be used owing to calibration uncertainty, and that Band-10 should only be used for calculations. The approach that is used in the next step uses thermal bands from Landsat-8 to compute LST.

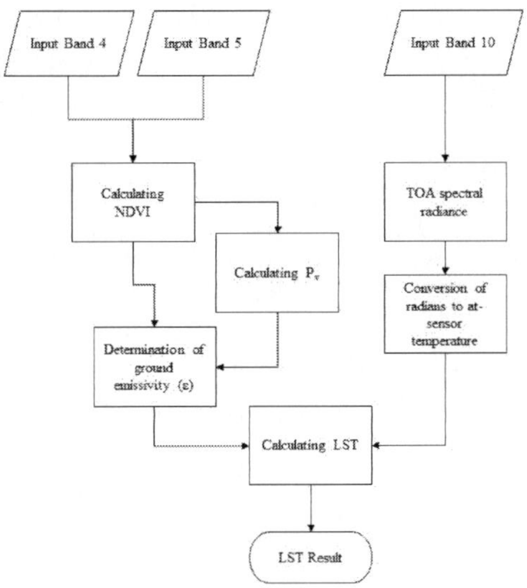

Figure 3: Flow Chart to Compute LST

Thermal Infrared band-10 was used to compute brightness temperature value (BTV) and Bands-4, 5 were used for NDVI. Metadata of satellite image used here is provided in Table 1 below.

	Thermal constant, Band-10	
K1	Rescaling factor, Band-10	1321.08
K2		777.89
ML	Correction, Band-10	0.000342
AL		0.1
Oi		0.29

Table 1: Metadata of satellite thermal bands to compute Temperature in C°.

2.2.2 Top of Atmospheric Spectral Radiance (TOA)

The TOA was estimated through the given equation below in accordance with the according to USGS,

$$TOA\ (L_\lambda) = M_L * Q_{cal} + A_L \qquad (1)$$

where, A_L, M_L and Band-10 are the additive and multiplicative band specific rescaling parameters correspondingly and L_λ is TOA (Barsi et al., 2014) To automate the technique subsequent python script was developed.

2.2.3 Conversion of TOA Recorded by Onboard Sensor

The Digital Number (DN) of every pixel was converted to spectral radiance and the radiance to temperature in degree Celsius through thermal constants derived from metadata accessible in entire package of Landsat-8 imagery by the following formula (USGS, 2013).

$$BT = \frac{K_2}{\ln[(K_1/L)+1]} - 273.15 \qquad (2)$$

Where

K_1 and K_2 = are the thermal constants with a fixed values provided in metadata.

These constants are significant to calculate temperature in degree Celsius and L is the radiance estimated earlier. The Python script was used to compute LST

2.2.4 NDVI Method for Emissivity Correction

Red and Near Infrared bands are commonly used to compute the green index called Normalized Difference Vegetation Index (NDVI) (Weng et al.,2004) NDVI estimates the amount of vegetation (P_v). The emissivity was calculated by integrating (P_v).

$$NDVI = \frac{NIR(Band5)-R(Band4)}{NIR(Band5)+R(Band4)} \qquad (3)$$

2.2.5 Computing the P_v

Pv is computed by the given formula below,

$$P_v = \frac{(NDVI-NDVI_{min})^2}{(NDVI_{max}+NDVI_{min})^2} \qquad (4)$$

Where P_v = land surface emissivity (ε).

2.2.6 Calculating Land Surface Emissivity (LSE (ε))

LSE (ε) was computed to estimate blackbody radiance (Plank's Law) to calculate the released radiance. It is much efficient and precise technique for the assessment of LST. The formula to compute emissivity is as below,

$$\varepsilon = 0.004 * P_v + 0.986 \qquad (5)$$

2.2.7 Computing LST

Finally, LST was measured by using BTV and LSE (ε) by the following equation.

$$LST = \frac{BT}{1+\left(0.00115*\frac{BT}{1.4388}\right)*\ln(\varepsilon)} \qquad (6)$$

2.2.8 Soil Suitability Parameters

The LSC data and related characteristics, such as pH, EC, soil type, and drainage factor, were provided by the Soil Survey of Pakistan (SoP), primarily in relation to the examination site for the wheat crop. Ground-controlled points with values for soil characteristics made up the LSC data. In order to trace thematic changes in the soil parameters and estimate their geographical distributions, we used inverse distance weighting (IDW) interpolation. The Analytical Hierarchy Process (AHP) was chosen to give each soil attribute a varied weight.

2.2.9 Multi-Criteria Evaluation (MCE)

The AHP is a pair-wise comparative method that assigns weights based on the significance of each soil characteristic in order to evaluate each one. LSC maps based on soil are created using AHP. All soil-related parameters were given weights in order to clarify their importance in relation to the growth rate of the wheat crop. Four LSC levels were allotted to implement and reclassify the relevant elements in accordance with the advice and recommendations of experts on their suitability for wheat growth. Clay loam and silt clay soils have a pH range of 5.5–6.5 and an EC range of 0.8–1.3, which is suitable for the growth of wheat crops. Based on the LSC values, four kinds of soil parameter were created. There were four categories of suitability: very, somewhat, less, and not suitable.

3. RESULT AND DISCUSSION

3.1 Temperature Based Potential Zones for Wheat Crop

Temperature effects the appropriateness of wheat growth in addition to other parameters. Table 1 presents the field observations and literature review. We found that whereas germination often takes place between 7 and 20 degrees Celsius, the ideal temperature range for all phases of wheat growth is 18 to 25 degrees Celsius. Temperatures in highly appropriate zones are shown in Table. 1 along with a literature evaluation. Wheat crops cannot develop at temperatures below 10°C and beyond 25°C because growth halts at these levels. At temperatures up to 20°C, the metabolic activity increases significantly, which is necessary for grain filling in the next stages. Summer wheat can grow in temperatures between 22 and 34 °C, while winter wheat may grow in temperatures between 5 and 25 °C. 0 to 5°C is the bare minimum temperature below which wheat crops cannot thrive (Aydan et al, 2016). The ideal temperature for wheat plant growth is 25°C, while above the temperature ranges of 30-32°C, wheat growth typically ends. For the whole thermal datasets listed in Table 1, LST were calculated. Based on their appropriate ranges, the maps for LST were categorized into four classes. In order to obtain temperature-based suitability maps, the average of all LST maps was taken (TBSMs).

Phonological Stage	Literature Source	Literature Temp(?C)	Observed Temp for Highly Suitable Site(?C)
Germination	(Russell & Wilson, 1994)	7.1-20	10 – 18
Tillering	(Qu & Wang, 1982)	> 8.5	10-16
Heading	(Salazar-Gutierrez, Johnson, Chaves-Cordoba, &Hoogenboom,2013)	10-18.4	14-19
Milk Dough	(Jenner, 1991)	20	16-20
Full Maturity	(Hossain, Sarker, Hakim, Lozovskaya,&Zvolinsky, 2011)	10-25	15-22

Table 2: Literature Based Optimum Temperature and Observed Temperature for Different Stages of Wheat Crop.

Scale	Soil Type	Soil Drainage	Soil PH
High Suitable	Clay Loam	0.85-1.1	6.2-6.5
Moderate Suitable	Silt Clay	1.1-1.3	5.5-6.2
Less Suitable	Saline	1.3-1.4	5.2-5.4
Not Suitable	Sandy	1.4-2.0	4.5-5.2

Table 3: Soil Suitability Parameters

Soil Parameter	Soil Type	EC	Drainage	pH	Weight
Soil Type	1				0.6000
EC	3	1			0.0857
Drainage	1/3	1/3	1		0.0286
pH	1/5	1/3	1/7	1	0.2857
Accumulated					? =1

Table 4: AHP to Compute Weight of Each Parameter

Through raster calculator in ArcMap, the soil-based maps in Figure 4 were put together and then split by their individual statistics to produce the averaged soil suitability maps (SSMs) shown in (Figure 4). The following is how the TBSMs (a-f) in (Figure 7) describe the appropriateness ranks. Due to rising temperatures, all the maps (a–f) that showed red zones were unsuitable for the growth of WCC as wheat growth is hindered in adverse climatic conditions. While the areas closest to water bodies indicated less suitability for wheat cultivation and growth, the green areas that are away from urban built-up areas appeared to be moderately to highly favorable for wheat cultivation and growth. The final TBSMs were created by averaging all the TBSMs (a-f).

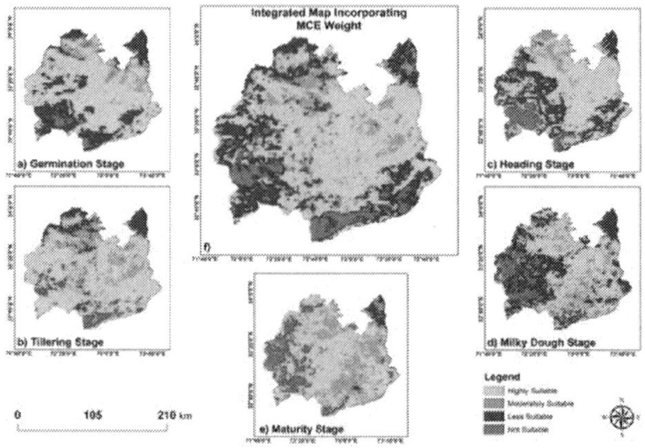

Figure 7.: Wheat Map Representing Various Growth Stages at Variable Temperatures (a), Wheat Crop in Germination Stage (b), Tillering Stage and Wheat Crop (c), Wheat Crop Heading Stage, (d), Wheat Crop in Milky Dough Stage, (e), Wheat Crop in Ripening Stage (f) Temperature Based Integrated Map incorporating all temperature-based stages.

3.2 Delineation of Potential Zones for Wheat Crop Cultivation (WCC)

Due to its propensity to behave poorly when drained, loamy clay is believed to be highly suitable (HS), whereas sand is perceived to be unsuitable (NS). The silty-clay behaves well-drained; thus it was classified moderately acceptable (MS) for the development of the wheat crop. Electrical conductivity (EC) range as 1.6 and 2.0 was NS for wheat growth and may cause the wheat crop to burn up. The EC value range between 0.85 and 1.1 was deemed extremely acceptable, while the region between 1.1 and 1.4 was shown to be less suitable (LS) (HS). For the growth of a wheat crop, a pH between 6.2 and 6.5 is considered HS, whereas a pH below 5.2 is considered NS for growing wheat. When these characteristics' AHP weights were calculated, soil type received the highest weight and soil drainage received the lowest weight (0.0286). Because the majority of wheat was grown on flat terrain, where the slop factor was small too little because it didn't alter the water distribution course, it is established that soil drainage has the lowest weight and soil types have the highest weight. Finally soil parameters were subjected to the IDW interpolation approach to examine trends and geographic variations, and the outcomes were mapped in (Figure 4.)

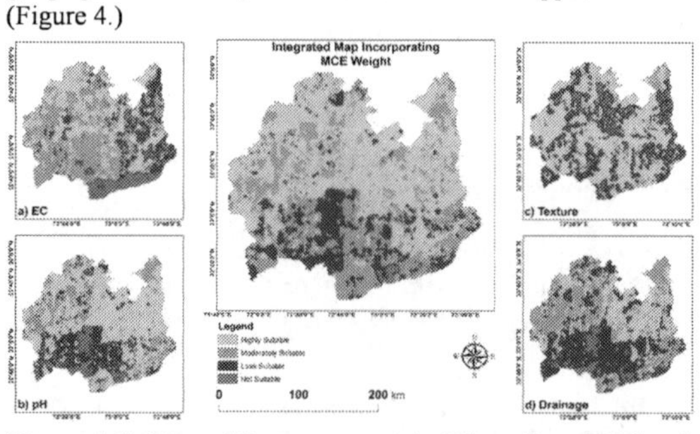

Figure 4: Soil Based Parameters and the Wheat Crop (a) Electric Conductivity Map. (b) Drainage Map. (c) pH of Soil (d) Soil Type Map. (e) Integrated Soil Map Incorporating MCE Weights

This contribution has been peer-reviewed.
https://doi.org/10.5194/isprs-archives-XLVIII-M-1-2023-409-2023 | © Author(s) 2023. CC BY 4.0 License.

3.3 MCE Applied to Both Physical Parameters and TBSMs as in Figure 8

The integrated map considering temperature and soil related parameters is mapped in (Figure 8).

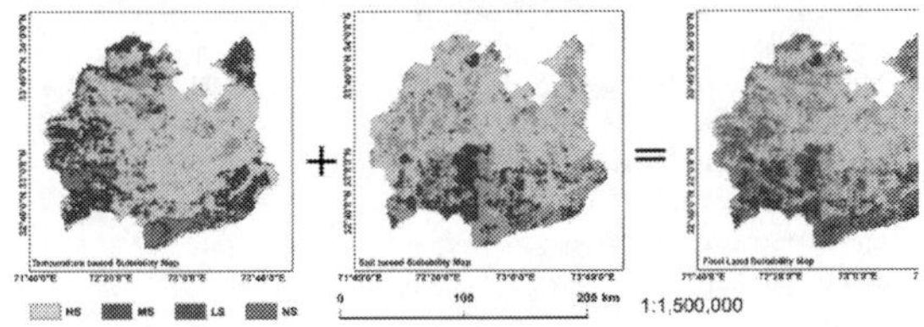

Figure 8: Study Site Mapped Incorporating Physical Parameters and Temperature Based Map.

(Figure 4) describes the MCE methods used to create the integrated map in (Figure 8) from TBSMs (for wheat crop). The overall investigation area was 22168 km^2, of which 12256 km^2 (or 55%) were deemed extremely suitable, 2998 km^2 (or 13%) were deemed moderately acceptable, 4843 km^2 (or 22%) were deemed least suitable, and around 2069 km^2 (10%) were deemed NS for WCC. Results are shown for the HS area, which has a temperature range of 8 to 24 °C, a pH level of 6.2 to 6.5, a soil type of clay loam, and a soil drainage level of 0.85 to 1.1. The NS region was distinguished by dense urban development, extremely high temperatures, and soils with a pH of less than 5.

3.4 Supervised Image Classification

Prior to the milky dough phase, we obtained satellite photographs of a wheat crop on that specific day, and we meticulously classified the images using machine learning techniques (MLTs). The classification of existing land use features using the MLTs yields encouraging results (even within the vegetation). The categorized findings show that WCC covered 16371 km2 of land. In order to cross-authenticate and determine the consistency of picture classification, along with Maali/Patwari, a ground-truthing survey was conducted, and the results were consistent up to 84%. Figure 3.4 depicts the wheat crop's size.

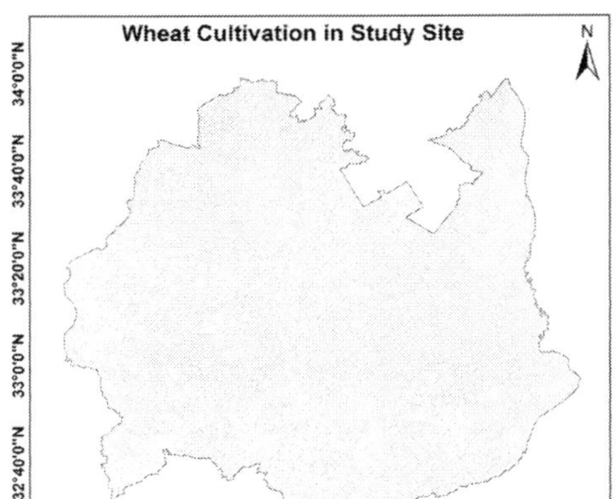

Figure 5.: Spatial Extent of Wheat Cultivation in the Study Site

To establish the actual WCC area in the LS, HS, MS, and NS LSC-zones, the categorized map was placed on the final suitability zones. The WCC areas were discovered to be as follows: Figure 3.5 shows that of the total 16371 km2, 9567 km2 were in the HS zone, 9245 km2 in the MS zone, 1203 km2 in the LS zone, and 356 km2 were in the NS zone.

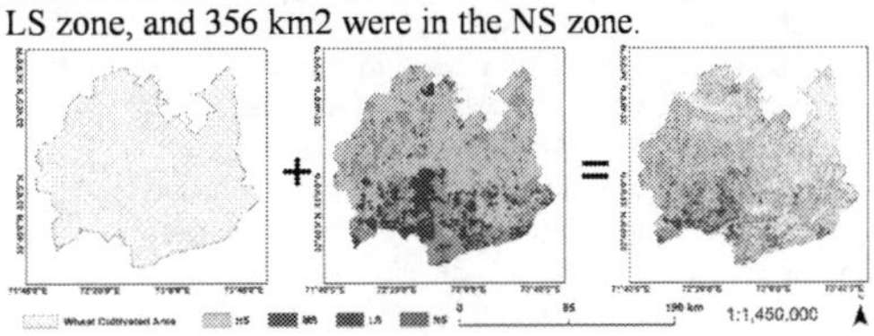

Figure 6.: Final Suitability Maps

4. DISCUSSION

Prior to the milky dough phase, we obtained satellite photographs of a wheat crop on that specific day, and we meticulously classified the images using machine learning techniques (MLTs). The classification of existing land use features using the MLTs yields encouraging results (even within the vegetation). The categorized findings show that WCC covered 16371 km2 of land. In order to cross-authenticate and determine the consistency of picture classification, along with Maali/Patwari, a ground-trothing survey was conducted, and the results were consistent up to 84%. Figure 3.4 depicts the wheat crop's size. To establish the actual WCC area in the LS, HS, MS, and NS LSC-zones, the categorized map was placed on the final suitability zones. The WCC areas were discovered to be as follows: Figure 3.5 shows that of the total 16371 km2, 9567 km2 were in the HS zone, 9245 km2 in the MS zone, 1203 km2 in the LS zone, and 356 km2 were in the NS zone. Based on the temperature ranges, a spatial distribution of the investigation site was created. LS, HS, MS, and NS were the primary classes. We saw that urban/built-up regions have developed into urban heat islands (UHI) as a result of anthropogenic activity, microclimate differences, and global warming. These activities include the production of GHG such as CO_2, SO_2, and aerosols which trap these emissions, at the local, national, and international levels. These regions were observed NS for WCC since the ideal temperature range required for WCC does not occur within this range. It is observed that the wheat crop in the nearby urban regions was short stressed and heighted. Due to the extremely low temperatures in these places, which made them LS for WCC, wheat crop growth there was substantially slowed down and required more time to reach the necessary stage of growth. Since these regions fell within the ideal range of temperatures required for wheat WCC and growth, the spatial sites within urban/built-ups and water bodies were discovered to be HS and MS for WCC and growth. Higher values are visible on the EC map, which may be caused by domestic emissions from urban homes and vehicle exhausts. The iron, magnesium, cobalt, and calcium components of these materials are directly given to the soil, which is excellent for WCC, but their weight is just 0.0857, which may not have a significant impact on WCC and growth.

Similar to this, it was discovered that the soil pH was higher close to the inner-city because of the direct interfaces of acidic wastes from both industrial and home sources that became the part of soil. The type of soil was given a relatively high weight (0.6), which indicated that it had a direct impact on the WCC. The investigation locations were split up into different types, including clayish, loamy, and silty soil. We discovered that clayish soil, which has a better capacity to retain water, was optimal for growing wheat crops, however loamy and silty soils are also somewhat beneficial. Due to greater rates of water seepage that call for frequent and reliable water supply, locations with sandy soil types were identified as NS for WCC. In relatively level places, it was found that the soil drainage weight was quite low, although the process of water distribution was unaffected by the slope factor. In contrast to other classification methods that are currently available, we used support vector machine (SVM) classification for the wheat crop. During ground

https://doi.org/10.5194/isprs-archives-XLVIII-M-1-2023-409-2023

validations and field surveys, the SVM produced significantly better results, up to 88% accuracy. In comparison to other conventional image classification techniques the total area under WCC is 65% and treated many types of vegetation e.g. wheat crops, about 45% of the investigation site was discovered to be under the WCC. The investigation's findings were determined to be extremely accurate and in line with the available data, however because our farmers do not use the most recent methods, the growth of the wheat crop has decreased in terms of the site's overall productivity.

5. CONCLUSION

The global population is increasing gradually in contrast to shrinkage of agricultural land which is decreasing rapidly as well. To address the difficulties of global food security, it is important to increase agricultural yield per acre, which is unattainable without an understanding of LSC zones. A typical farmer disregards the suitability criteria and begins agricultural techniques in the wrong locations, which decreases overall agricultural yield. To guarantee the best possible yield, the LSC zones for a specific crop must remain maintained. To map the LSC zones remote sensing and GIS spatial analytic tools can be used. The investigation area's physiographic units matched up with its appropriateness for orchards and crops. Crops were MS in the lowlands whereas slope made them MS in the uplands. The LS were caused by slope and organic carbon restrictions. Because of their low levels of organic carbon and average drainage, the midlands were MS. The study shows a strong correlation between soils and physiography. The variety in topography that results in erosion, leaching, sedimentation, and other pedogenic processes influenced by water table is responsible for the production of the diverse group of soil

REFERENCES

Ahmad, P., & Wani, M. R. (Eds.). (2013). Physiological Mechanisms and Adaptation Strategies in Plants Under Changing Environment: Volume 2 (Vol. 2). Springer Science & Business Media.

Al-Karaki, G. N. (2012). Phenological Development-Yield Relationships in Durum Wheat Cultivars under Late-Season High-Temperature Stress in a Semiarid Environment. ISRN Agronomy, 2012, 1–7. https://doi.org/10.5402/2012/456856

Anees, M. T., Javed, A., & Khanday, M. Y. (2014). Spatio-temporal land cover analysis in Makhawan Watershed (MP), India through remote sensing and GIS techniques. Journal of Geographic Information System, 2014.

Avdan, U., & Jovanovska, G. (2016). Algorithm for automated mapping of land surface temperature using LANDSAT 8 satellite data. Journal of Sensors, 2016. https://doi.org/10.1155/2016/1480307

Barsi, J. A., Schott, J. R., Hook, S. J., Raqueno, N. G., Markham, B. L., & Radocinski, R. G. (2014). Landsat-8 thermal infrared sensor (TIRS) vicarious radiometric calibration. Remote Sensing, 6(11), 11607-11626.

Becker, F., & Li, Z. L. (1990). Towards a local split window method over land surfaces. Remote Sensing, 11(3), 369-393.

Khan, A., Hansen, M. C., Potapov, P., Stehman, S. V., & Chatta, A. A. (2016). Landsat-based wheat mapping in the heterogeneous cropping system of Punjab, Pakistan. International Journal of Remote Sensing, 37(6), 1391-1410

Department Agriculture, F. and F. (2016). Production Guideline. Production Guidlines for Wheat, 1–25.El Baroudy, A. A. (2016). Mapping and evaluating land suitability using a GIS-based model. CATENA,140,96–104. https://doi.org/10.1016/J.CATENA.2015.12.010

Hossain, A., Sarker, M. A. Z., Hakim, M. A., Lozovskaya, M. V., & Zvolinsky, V. P. (2011). Effect of temperature on yield and some agronomic characters of spring wheat (Triticum aestivum L.) genotypes. International journal of agricultural research, innovation and technology, 1(1-2), 44-54. https://doi.org/10.22004/AG.ECON.305265

Islam, N. (1995). Population and Food in the Early Twenty-First Century. In Food Policy. International Food Policy Research Institute (IFPRI).

Jiménez-Muñoz, J. C., Sobrino, J. A., Gillespie, A., Sabol, D., & Gustafson, W. T. (2006). Improved land surface emissivities over agricultural areas using ASTER NDVI. Remote Sensing of Environment, 103(4), 474-487.

Khan, A., Hansen, M. C., Potapov, P., Stehman, S. V., & Chatta, A. A. (2016). Landsat-based wheatmapping in the heterogeneous cropping system of Punjab, Pakistan. International Journal of Remote Sensing, 37(6), 1391-1410 https://doi.org/10.1080/01431161.2016.1151572

Ministry of Finance, Government of Pakistan. Pakistan Economic Survey 2013–2014. (2014). http://www.finance.gov.pk/survey_1314.html

Onwuka, B. (2018). Effects of Soil Temperature on Some Soil Properties and Plant Growth. Advances in Plants & Agriculture Research, 8(1), 34–37. https://doi.org/10.15406/apar.2018.08.00288

Porter, J. R., & Gawith, M. (1999). Temperatures and the growth and development of wheat: a review. European Journal of Agronomy, 10(1), 23–36. https://doi.org/10.1016/S1161-0301(98)00047-1

Punjab CRS: Base Line Survey Agriculture Information System Building Provincial Capacity for Crop Forecasting and Estimation. (2012).
Rajeshwari A, M. N. D. (2014). Estimation of Land Surface Temperature of Dindigul District Using Landsat 8 Data. International Journal of Research in Engineering and Technology, 03(05), 122–126. https://doi.org/10.15623/ijret.2014.0305025

Rehman, A., & Hussain, I. (2016). Agricultural and Economic Development in Pakistan and Its Comparison With China, India, Japan, Russia and Bangladesh. Andamios. Revista de Investigación Social, March 2017.

Rehman, A., Jingdong, L., & Du, Y. (2015). Last Five Years Pakistan Economic Growth Rate GDP and Its Comparison with China India and Bangladesh. International Journal of Scientific & Technology Research, 4(1), 81–84.

Rehman, A., Jingdong, L., & Hussain, I. (2015). The province-wise literacy rate in Pakistan and its impact on the economy. Pacific Science Review B: Humanities and Social Sciences, 1(3), 140–144. https://doi.org/10.1016/J.PSRB.2016.09.001
Remote sensing of the urban heat island and its changes in Xiamen City of SE China - PubMed. (n.d.).

Retrieved December 5, 2021, from https://pubmed.ncbi.nlm.nih.gov/15137654/

Russell, G., & Wilson, G. W. (1994). An Agro-Pedo-Climatological Knowledge-Base of Wheat in Europe.
Joint Research Centre, European Commission, Luxembourg, 158.Using the USGS Landsat Level-1 Data Product. (n.d.). Retrieved December 5, 2021, from https://www.usgs.gov/core-science-systems/nli/landsat/using-usgs-landsat-level-1-data-product hosporterkhanonwuladem

Weng, Q., Lu, D., & Schubring, J. (2004). Estimation of land surface temperature–vegetation abundance relationship for urban heat island studies. Remote Sensing of Environment, 89(4), 467–483.
https://doi.org/10.1016/J.RSE.2003.11.005

World Population to Reach 9.9 Billion by 2050 | News | SDG Knowledge Hub | IISD. (2020). https://sdg.iisd.org/news/world-population-to-reach-9-9-billion-by-2050/

Zolekar, R. B., & Bhagat, V. S. (2015). Multi-criteria land suitability analysis for agriculture in hilly zone:
Remote sensing and GIS approach. Computers and Electronics in Agriculture, 118, 300–321.
https://doi.org/10.1016/J.COMPAG.2015.09.016

The International Archives of the Photogrammetry, Remote Sensing and Spatial Information Sciences, Volume XLVIII-M-1-2023
39th International Symposium on Remote Sensing of Environment (ISRSE-39) "From Human Needs to SDGs", 24–28 April 2023, Antalya, Türkiye

6 FEBRUARY 2023 KAHRAMANMARAŞ – TÜRKİYE EARTHQUAKES: A GENERAL OVERVIEW

C. Gokceoglu[1]

[1] Dept. of Geological Engineering, Hacettepe University, 06800 Beytepe Ankara, Turkey - cgokce@hacettepe.edu.tr

KEY WORDS: Earthquake, Kahramanmaraş, Pazarcık, Elbistan, Landslide, Rockfall, Liquefaction.

ABSTRACT:

On 6 February 2023, two major earthquakes occurred in the East Anatolian Fault Zone (EAFZ) of Türkiye. The EAFZ forms the east border of the Anatolian Plate. The magnitude 7.7 and 7.5 Kahramanmaraş – Türkiye Earthquakes that struck southern Türkiye and resulted in common destruction in 11 provinces in the region. Total 6 main fault segments of the EAFZ ruptured during the earthquake sequence on 6 February 2023, and approximately 400 km surface rupture occurred. The life losses are reported as over 50000, and approximately 300,000 buildings were collapsed or severely damaged. In addition to damages on the buildings and the infrastructures, liquefactions, landslides, rockfalls, and rock avalanches were also observed during the earthquake sequence. The purpose of this study is to present a general overview on the 6 February 2023 Earthquake sequence.

1. INTRODUCTION

The Anatolian plate is bounded in the north by the North Anatolian Fault Zone (NAFZ) and in the east by the East Anatolian Fault Zone (EAFZ). As a result of the northward movement of the Arabian Plate, the Anatolian Plate between NAFZ and EAFZ moves approximately westward. As a result of the plate movement, in the last century, devastating earthquakes have occurred along the NAFZ. However, the EAFZ had produced relatively few earthquakes in the last 200 years when compared to historical records and has therefore accumulated significant stresses along its length (Nalbant et al., 2002). For this reason, some segments of the EAFZ were seismic gaps, and hence, the region was a high seismic potential. Consequently, some severe earthquakes were expected in the region (Nalbant et al., 2002; Gokceoglu, 2022; Can et al., 2022).

In the last century, only three earthquakes produced surface rupture. These are the 1971 Bingöl Earthquake (Ms=6.8), the 2010 Karakoçan Earthquake (Mw=6.1), and the 2020 Elazığ Earthquake (Mw=6.8). In addition, it is known that the EAFZ to the east of Lake Hazar was also ruptured by the 1874 Ms 7.1, 1875 Ms 6.7, and 1866 Ms 7.2 earthquakes (Ambraseys 1988; Emre et al., 2018). In addition to these earthquakes, before the instrumental period, severe earthquakes occurred along the EAFZ (Figure 1).

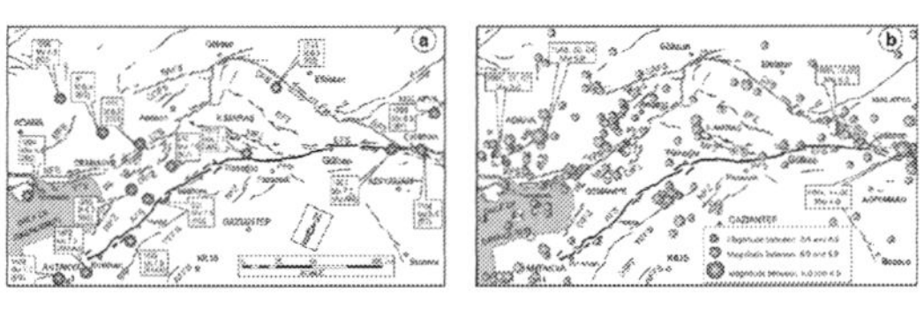

Figure 1. Fig. 2 Distribution of both historical (a) and instrumental (b) earthquakes along the southern segments of the EAFZ (taken from Duman et al., 2020)

As stated, the Erkenek, Pazarcık, Amanos, Antakya, Çardak and Sürgü segments of the EAFZ have not produced a significant earthquake but they have continued to buildup elastic strain energy, until February 6, 2023. On February 6, 2023, at 01:47 UTC, the Narlı, Erkenek, Pazarcık and Amanos segments ruptured and the Mw=7.7 Kahramanmaraş-Pazarcık earthquake occurred. Approximately 9 hours later at 10:24 UTC, the Mw=7.7 Kahramanmaraş-Elbistan earthquake occurred near Ekinözü due to rupture of the Çardak and Sürgü segments (Figure 2). As the results of the successive earthquakes, Kahramanmaraş, Adıyaman, Hatay, Osmaniye, Gaziantep, Kilis, Şanlıurfa, Diyarbakır, Malatya, Adana, and Elazığ were significantly affected. More than 50000 people died and approximately 300000 buildings were collapsed or heavily damaged. In addition to damages on the buildings and the infrastructures, surface ruptures, liquefactions, landslides, rockfalls, and rock avalanches were also observed commonly during the recent destructive earthquakes. The purpose of this study is to present a general overview on the 6 February 2023 Earthquake sequence.

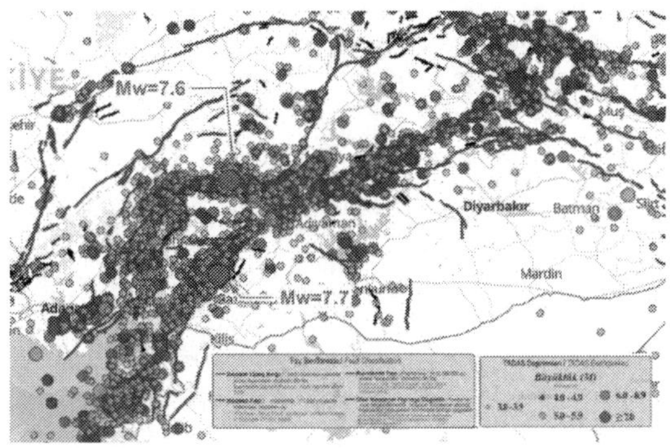

Figure 2. Epicenters of the 6 February 2023 Türkiye Earthquakes (AFAD, 2023)

2. GROUND FAILURES

2.1 Surface Ruptures

The Pazarcik earthquake was initiated on a smaller Narlı fault (326-1 in Figure 3) in the south, jumped to the north, and ruptured the Pazarcık (2-6 in Figure 3) and Erkenek (2-5 in Figure 3) segments of EAFZ towards NE and the Amanos segment (2-7 in Figure 3) towards SW. This earthquake produced left-lateral surface ruptures and the displacements reached to 4 m (Figure 4).

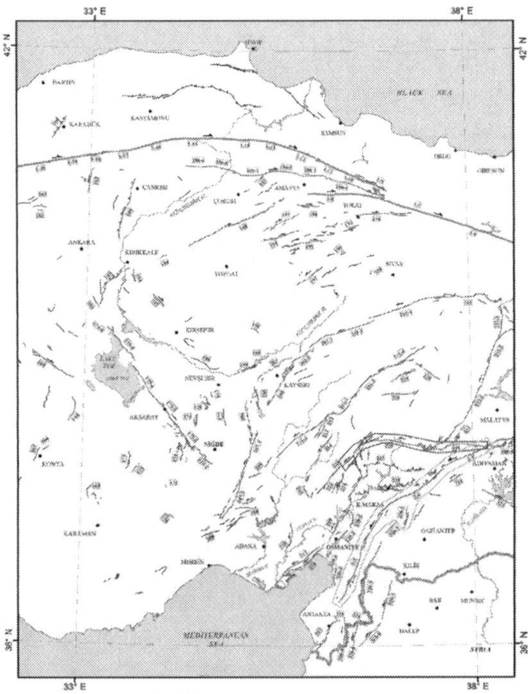

Figure 3. The ruptured segments of DAFZ (yellow frame shows the Pazarcik Earthquake while blue frame shows the Elbistan Earthquake) (modified after Emre et al., 2018).

Figure 4. Some views from the left lateral displacements occurred during the Pazarcık Earthquake.

The surface ruptures resulted in highway and railway damages and the transportation to the earthquake region was blocked. The Tarsus-Adana-Gaziantep Motorway (TAG) was cut by the displacement of the Amanos segment around Nurdaği (Figure 5).

Figure 5. The TAG Motorway cut by the surface rupture (TRT Haber, 2023)

The other important state road cut by the surface rupture is Gaziantep – Kahramanmaraş State Road (Figure 6). Access to the earthquake region was provided by secondary roads due to the highway and the state roads that were cut and damaged by the surface ruptures, which caused delays in the search and rescue efforts for the first few days after the earthquake.

Figure 6. The Gaziantep – Kahramanmaraş State Rod cut by the surface rupture

The railway system under service was constructed in the period of 1927-1937 (Nohab Saabye and Lerche Kampax, 1937). The surface ruptures were cut by various locations (Figure 7).

https://doi.org/10.5194/isprs-archives-XLVIII-M-1-2023-417-2023 | © Author(s) 2023. CC BY 4.0 License.

Figure 7. A view from a railway cut by the surface rupture

The T3 railway tunnel between Gölbaşı and Malatya was cut by surface rupture of the Erkenek segment. The fault cut the tunnel at the 80th m from the portal of Erkenek side and approximately 150 m part of the tunnel collapsed completely (Gokceoglu and Karahan, 2023) (Figure 8). The failure of this tunnel prevented the railway transportation between Malatya and Gaziantep.

Figure 8. A view from the collapsed T3 tunnel

The dam body of the Erkenek dam was cut perpendicular by the Erkenek segment (Figure 9). The dam was damaged heavily but, it is not collapsed. As soon as occurrence of the Pazarcık earthquake, the water in the reservoir was discharged by the authorities. This precaution prevented the occurrence of a secondary hazard.

Figure 9. A view from the Erkenek dam body cut by the surface rupture

In addition to the engineering structures cut by the surface ruptures, surface ruptures caused heavy damage to many buildings and caused their collapse completely (Figure 10).

Figure 10. A view from a double stories building cut by the surface rupture (Tetirlik village, Pazarcık)

During the Elbistan earthquake, the Çardak segment (226 in Figure 3) and the Sürgü segment (227 in Figure 3) ruptured. These segments also produced surface ruptures and the maximum displacement was measured as 6.7 m by Taylan Sancar (Figure 11).

This contribution has been peer-reviewed.
https://doi.org/10.5194/isprs-archives-XLVIII-M-1-2023-417-2023 | © Author(s) 2023. CC BY 4.0 License.

Figure 11. The maximum displacement measured by Taylan Sancar (modified after Özacar et al., 2023).

2.2 Landslides

During the earthquake sequence, more than 2500 landslides were triggered. The south part of the Amanos mountains are generally free from landslides but several landslides in the Amanos mountains were triggered. A landslide dam around Islahiye occurred, and the GoogleEarth images before and after the Earthquake sequence is given in Figure 12. Another large and planar landslide occurred around Tepehan (Figure 13) and such type failures were also observed around this region. In this region, the failure surface was observed clearly (Figure 14). A sakung with a volume of around 70M m^2 occurred near Ekinözü during the Elbistan earthquake (Figure 15). Especially in the northern parts of the earthquake region such as Çelikhan, Kahta, Gölbaşı, Besni, Nurhak and Tut, many landslides were triggered by the earthquakes.

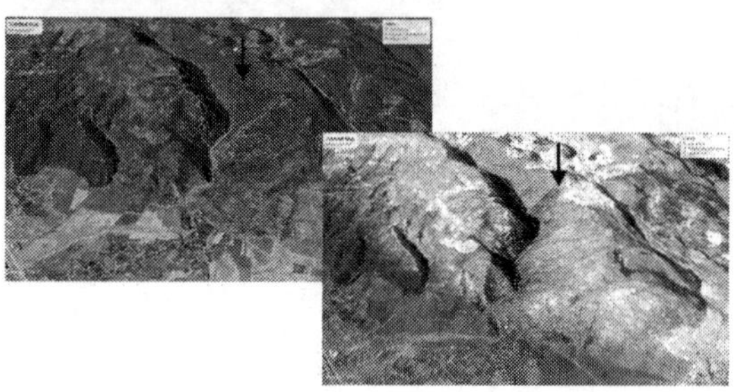

Figure 12. The GoogleEarth images before and after the earthquakes (west of Islahiye, Degirmencik)

Figure 13. A view from the Tepehan planar failure triggered by the earthquake

Figure 14. A view from the failure surface of the planar failure

2.3 Rockfall and Rock Avalanche

The rockfalls were observed commonly in the mountainous areas of the earthquake region. Especially roads and railways were affected by the rockfalls (Figures 16 and 17). In addition to roads and railways, some villages such as Bektaşlı village, Kırıkhan were damaged by rockfall (Figure 18). The rockfall in Bektaşlı village resulted in death of 50 person.

Figure 15. A view from the sakung near Ekinözü

Figure 16. A view from the rockfall on the Gölbaşı – Nurhak road

Figure 17. A view from the rockfall at the south of Fevzipaşa Railway Station

Figure 18. A view from the rockfalls in Bektaşlı village, Kırıkhan

In addition to rockfalls, the rock avalanches were observed during the 6 February 2023 Türkiye Earthquakes. Rock avalanches result from rapid fragmentation of very fast-moving, initially intact rock masses during transport (Hungr et al., 2001). Especially, along the east slopes of the Göksu river, east of Tut, Adiyaman, several rock avalanches were triggered (Figure 19). These failures are observed clearly by comparing before and after earthquake orthophotos presented on the HGM Küre platform of the General Directorate of Mapping (HGM, 2023) (Figure 20).

Rock avalanches observed near Nurhak and Tut did not cause loss of life or great damage in this earthquake, because rock avalanches triggered by the 6 February 2023 Earthquakes took place in locations far from settlements. However, this slope failures should be considered during the site selection efforts.

Figure 19. A view from the rock avalanches along the east slopes of the Göksu river

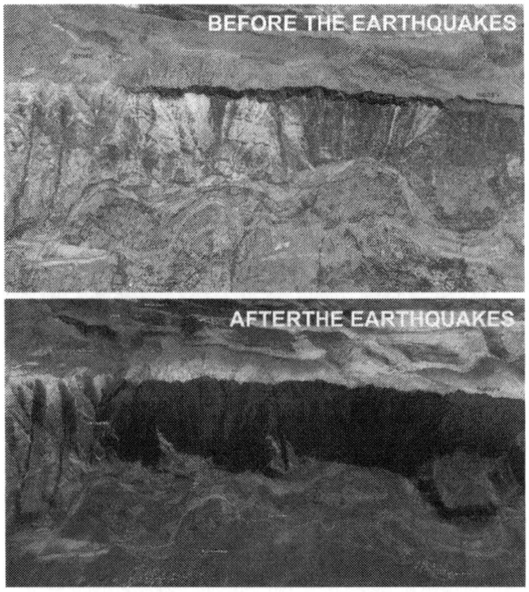

Figure 20. The HGM Küre images before and after earthquake (along the east slopes of the Göksu river)

2.4 Liquefaction

During the 6 February 2023 Earthquakes, the liquefactions occurred commonly. A liquefaction and lateral spreading map was prepared by Üstün et al. (2023) (Figure 21). As can be seen from Figure 20, severe liquefactions were observed in Elbistan, Gölbaşı, Türkoğlu, Antakya, Samandağı and İskenderun.

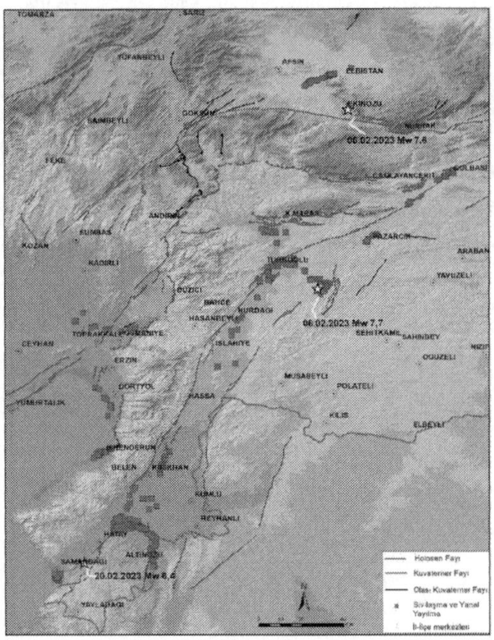

Figure 21. The liquefaction and lateral spreading map of the Earthquake region (Üstün et al., 2023).

In Gölbaşı, liquefaction damaged the buildings, roads and railways severely (Figure 22), because the main lithology of this region is lake sediments.

Figure 22. A toppled building due to liquefaction in Gölbaşı

Along the İskenderun coastline, severe liquefactions occurred, and as a result of liquefactions, the İskenderun Port was damaged heavily (Figure 23), and a serious fire was initiated. In addition, the İskenderun coastline settled more than 1.5 m (Figure 24).

Figure 23. A view from the İskenderun Port after the earthquake

This contribution has been peer-reviewed.
https://doi.org/10.5194/isprs-archives-XLVIII-M-1-2023-417-2023

Figure 24. The settled coastline of İskenderun

In the Amik Plain, severe liquefactions and lateral spreadings were also observed (Figure 25). The liquefactions in this plain damaged the Antakya – Cilvegözü road, and the runway of the Hatay Airport (Figure 26).

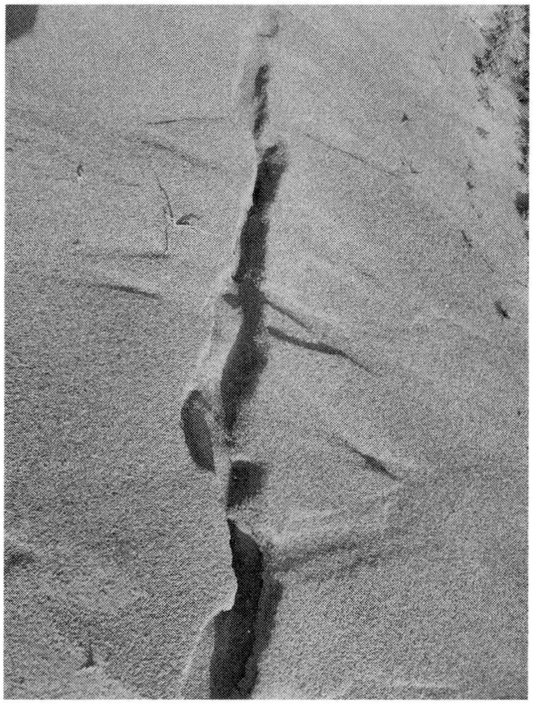

Figure 25. A typical view from liquefaction occurred in the Amik Plain

Figure 26. A view from the runway damages of the Hatay Airport

3. CONCLUSIONS

The general characteristics of the 6 February 2023 Kahramanmaraş – Türkiye Earthquakes are given in this study. The main conclusions can be drawn as follows:

(a) As a result of the rupture of the Narlı, Erkenek, Pazarcık and Amanos segments in the East Anatolian Fault Zone, a 7.7 magnitude earthquake with a Pazarcık epicenter occurred on February 6, 2023, at 04:17 local time. On the same day, another large Mw 7.6 earthquake occurred at 13:24 with the epicenter in Elbistan, as a result of the rupture of Çardak and Sürgü faults.

(b) During the successive earthquakes, approximately 100000 km^2, more than 13 million people and 11 provinces were affected directly.

(c) According to the official data, death toll is more than 50000, and approximately 300000 buildings were collapsed or heavily damaged.

(d) Approximately 400 km surface rupture occurred, and these ruptures cut railways and highways. This resulted in a serious transportation problem during the first few days.

(e) More than 2500 different types of landslides were triggered, and a landslide dam occurred in the Amanos mountains. In addition, a sakung with an approximate volume of 70M m^3 was triggered.

(f) In mountainous part of the earthquake region, rockfalls were triggered commonly, and the rockfalls contributed to the increase in damage and loss of lifes. Additionally, around Nurhak and Tut, some rock avalanches were observed.

(g) Liquefactions and lateral spreadings were observed commonly in the coastlines and plains. These soil

https://doi.org/10.5194/isprs-archives-XLVIII-M-1-2023-417-2023 | © Author(s) 2023. CC BY 4.0 License.

behaviours also contributed to increase in earthquake damages.

REFERENCES

AFAD (Afet ve Acil Durum Yönetimi Başkanlığı). 2023. 06 Şubat 2023 Pazarcık (Kahramanmaraş) Mw 7.7 ve Elbistan (Kahramanmaraş) Mw 7.6 Depremlerine İlişkin Ön Değerlendirme Raporu. (https://deprem.afad.gov.tr/ assets/pdf/ Kahramanmaras%20%20Depremleri_%20On%20Degerlendirme%20Raporu.pdf)

Ambraseys NN (1988) Engineering seismology. *J Earthq Eng Str Dyn*. 17:1–105

Can, A., Baskose, Y., and Gokceoglu, C., 2022. Stability assessments of a triple-tunnel portal with numerical analysis (south of Turkey). *Geotechnical Research*. 9 (2), 116-128, https://doi.org/10.1680/jgere.21.00028

Duman, T.Y., Elmacı, H., Özalp, S., Kürçer, A., Kara, M., Özdemir, E., Yavuzoğlu, A., Güldoğan, Ç.U., 2020. Paleoseismology of the western Sürgü–Misis fault system: East Anatolian Fault, Turkey. *Mediterranean Geoscience Reviews*. 2: 411–437. https://doi.org/10.1007/s42990-020-00041-6

Emre, Ö., Duman, T.Y., Özalp, S., Şaroğlu, F., Olgun, Ş., Elmacı, H., Çan, T., 2018. Active fault database of Turkey. *Bull Earthquake Eng*. 16, 3229-3275, https://doi.org/10.1007 /s10518-016-0041-2

Gokceoglu, C., 2022. Assessment of rate of penetration of a tunnel boring machine in the longest railway tunnel of Turkey. *SN Applied Sciences*. 4, 19, https://doi.org/10.1007/s42452-021-04903-y

Gokceoglu, C., Karahan, S., 2023. Seismic Performance Of Transportation Tunnels In The Region Affected By 6 February 2023 Türkiye Earthquake Sequence: A General Assessment. 57th U.S. Rock Mechanics / Geomechanics Symposium, ARMA 23-0960.

HGM (General Directorate of Mapping), 2023. HGM Küre Application. URL: https://hgmkure.btk.gov.tr/ (accessed on 1 May 2023).

Hungr, O., Evans, S.G., Bovis, M.J., Hutchinson, J.N., 2001. A review of the classification of landslides of the flow type. Environmental & Engineering Geosciences. 7 (3), 221-238, 10.2113/gseegeosci.7.3.221

Nalbant SS, McCloskey J, Steacy S, Barka AA (2002) Stress accumulation and increased seismic risk in eastern Turkey. *Earth Planet Sci Lett*. 195, 291-298. https://doi.org/10.1016/ S0012-821X(01)00592-1

Nohab Saabye and Lerche Kampax, 1937. Construction des lignes de chemins de fer: Irmak-Filyos and Fevzipaşa Diyarbekir. Technical Report, 258p.

Özacar, A.A., Uzel, B., Bozkurt, E., Sançar, T., Sopacı, E., Okay, H.B., Kaymakçı, N., Rojay, B., Gülerce, Z., Kıncal, C., Köksal, D., Gregory, L., 2023. Active Tectonic Setting and Seismic Source. Reconnaissance Report on February 6, 2023 Kahramanmaraş-Pazarcık (Mw=7.7) and Elbistan (Mw=7.6) Earthquakes, 485p.

TRT Haber Website, 2023. https://www.trhaber.com/gundem/ deprem-sonrasinda-hangi-yollar-tarafige-kapandi-adana-maras-gaziantep-nurdagi-h87054.html. Accessed on February 8, 2023.

Üstün, A.B., Altuntaş, G., Demirörs, U., Karayazı, O., 2023. 6 Şubat 2023 Kahramanmaraş Depremleri ve 20 Şubat 2023 Defne (Hatay) Depremi Sonucu Gelişen Sıvılaşma Yapıları Saha Gözlemleri ve Değerlendirmeler. MTA Jeoloji Etüdleri Dairesi Başkanlığı, 16s.

https://doi.org/10.5194/isprs-archives-XLVIII-M-1-2023-417-2023 | © Author(s) 2023. CC BY 4.0 License.

APPRAISAL OF CLIMATE CHANGE DISASTER AND FOOD SECURITY IN KHYBER PAKHTUNKHAW PAKISTAN USING GEOSPATIAL TECHNOLOGIES

Syed Amer Mahmood[1*], Zainab Tahir[2], Saira Batool[3], Amer Masood[4], Muhammad Haseeb[5] and Maryam Muhammad Ali[6]

[1,2,4,6] Department of Space Science, University of the Punjab, Lahore,Pakistan, amerpakistan@gmail.com, zainabtahir244@gmail.com, amerchishti_@hotmail.com, maryamalirehan@gmail.com

[3] Center For Integrated Mountain Research (CIMR), University of the Punjab,Lahore,Pakistan, sairabnaqvi5@gmail.com

[5] University of Engineering and Technology, Lahore, Pakistan, mrhaseeb223@gmail.com

KEY WORDS: Food Security, Mann-Kendall, RDI. SPI, PET

ABSTRACT

Drought is one of the most severe natural disasters. It negatively influences the crop output considerably as water resources are inadequate across large geographic areas. Food insecurity, hunger, and poverty are widespread in three Pakistani provinces: Baluchistan, Khyber Pakhtunkhwa, and Sindh. Until March/April 2022, roughly 4.66 million people, or 25% of the population examined, were severely food insecure. Historically, the severity of a drought is evaluated using several different indicators, with the selection of indices depending on the data that is easily accessible. By multiple aspects, including potential evapotranspiration and rainfall data on a standard timescale and computing monthly, seasonal, or annual data, the Reconnaissance Drought Index (RDI) has a significant advantage over the other indices in determining drought severity. When rainfall in a region is significantly lower than the amount of potential evapotranspiration, drought situations can occur on the earth. Rainfall will often occur in locations with an increase in the relative humidity of the air. The reconnaissance drought index (RDI) for Khyber Pakhtaunkhawa province in Pakistan was calculated from 1981 to 2020 for 3-, 6-, and 12-month timeframes. This research employed rainfall, minimum and maximum temperature data from KPK province to describe drought using the DrinC program, the Man Kendall test, and Sen's slope calculator to determine drought patterns and variation. All regions had moderate, severe, and extreme droughts, with RDIs between -1.0 and -3.5. RDI 3-, 6-, and 12-month evaluations highlighted 1991, 1994, 2000, 2010, 2015, and 2017. These results indicate that productivity declines in dry and subsequent years due to seasonal rainfall and soil moisture retention.Wheat production averaged 1.7 tones ha-1 year-1, with significant variations. Between October and December, it is an essential to have favorable growing conditions to increase productivity. If droughts occur in this era, the yield will be substantially more negatively impacted than if they occur later in the growing cycle. The 3-month RDI values offer a more accurate description of this variation than the six-month and twelve-months RDI Values. Geospatial drought analysis indicates a considerable variance in drought patterns across the Kpk province over the last 4 decades. Because long-term droughts sometimes affect water quality, especially groundwater, a rise in the intensity of long-term droughts could pose a risk to the water management in the regions that have been examined. Analysis of the Landsat TM 30 m data took place. At the same time, Landsat imageries were utilized to generate the Normalized Difference Vegetation Index (NDVI) and Normalized Difference Water Index (NDWI) for drought evaluation. The findings of the NDWI indicate an increase in value from 0.58 to 0.7 for the KPK region.

1. INTRODUCTION

Pakistan, a country in Southeast Asia with 225.2 million inhabitants (PBS 2021), has serious water issues (Bacha et al., 2021). In Pakistan, where there is a 52% water deficit, the severe water crisis has affected even essential services. Global climate change cannot be avoided, and Pakistan is currently in a dangerous scenario (Kirby et al., 2016; P. Kumar et al., 2021). According to the Intergovernmental Panel on Climate Change (IPCC), the average global temperature of the Earth, oceans, and atmosphere increased by 0.85 degrees Celsius between 1880 and 2012. (J. A. Khan, 2011). According to Podger et al. (2021), drought is seen as an environmental calamity that has catastrophic effects on the agriculture sector and the ecosystem. It also has long-lasting effects on ecosystems, the atmosphere, water quality, and food production. Even without the faintest comprehension of the different types of droughts, agricultural droughts are an issue. It produces a harsh climate where food production needs insufficient moisture levels, resulting in crop failure and a food shortfall (Banerjee & Bera, 2021). (Mao et al., 2017). These consequences may be observed in those areas and agricultural systems that are most vulnerable to the destruction caused by drought (for example, rain-fed agriculture) (Yu et al., 2018; Qaiser et al., 2021). Due to its severe effects on ecosystems, the atmosphere, water quality, and food production, drought is a pervasive environmental disaster that has recently piqued the interest of many researchers from a variety of fields, including agriculture, ecosystems, earth sciences, climatology, hydrology, and geography (Surendran et al., 2017). Droughts may exacerbate the country's water dilemma, which already exists in normal years. Droughts are a type of meteorological disaster that can have significant social and economic repercussions. The drought that started in 1998 had a negative impact on agricultural output in 2000–2001, which was just 2.6%. The spatial-temporal inequalities in water distribution that create water shortages are expected to worsen as the climate changes. Drought is a reoccurring issue with effects on the environment and human existence in various ways. The impacts of drought on several facets of society are substantially more severe (Haile, 2019). Droughts were predominantly felt in northwestern mid-high latitudes due to an increase in worldwide ambient temperature and a decrease in rainfall; the impact of these temperatures will continue to increase (Jiang et al., 2022). In Pakistan, 42.3% of the nation's workforce—or 67 percent of the populace—is employed in agriculture, which is primarily practised in rural areas (Bhandari et al., 2020). Rainfall in Khyber Pakhtaunkhawa varies in frequency, amount, and distribution. In the KPK region over the previous 48 years, droughts have become more common, which has reduced the amount of water available for crops (Xu et al., 2021). In regions

https://doi.org/10.5194/isprs-archives-XLVIII-M-1-2023-425-2023 | © Author(s) 2023. CC BY 4.0 License.

where arid climate causes droughts, evaluating rainfall and evapotranspiration variability is crucial (Ahmed et al., 2018; Liu et al., 2021). The DrinC (Drought Index Calculator) programme was used to calculate the drought indices. 2014 (Stagge et al.).Landsat images are used in indices like the Normalized Difference Vegetation Index (NDVI), the Vegetation Condition Index (VCI), the Scaled Drought Condition Index (SDCI), the Soil Moisture Agricultural Drought Index (SMADI)(Sánchez et al., 2016), and the Empirical Standardized Soil Moisture Index (Carro et al., 2016) to evaluate the effects of drought on vegetative cover, while also Studies on agricultural drought have been undertaken using simpler markers that were initially intended for assessing meteorological drought (Zhang et al., 2017). KPK has turned into a hotspot for all hydrometeorological and geological disasters because of its more varied topography than other Pakistani provinces. Rural areas are where KPK is most prevalent. Only 17% of the population lives in urban areas. According to the national economic framework, agriculture accounts for 30% of the region's GDP (Abid et al., 2013).

2. MATERIALS AND METHODS

2.1. Study area

The study area is the Pakistani province of Khyber Pakhtunkhawa, which is situated in the country's northwest. KPK is the smallest province in terms of territory and ranks third in terms of people behind Sindh and Punjab (Figure 1). Thecoordinates of this place are 34°57'9.44"N and 72°19'52.01"E. The Federally Administered Tribal Areas (FATA) cover 101,741 km2, or 12.8% of Pakistan's total land area, and are 27,220 km2 larger than Khyber Pakhtunkhawa (74,521 km2), (Rahman et al; 2019)

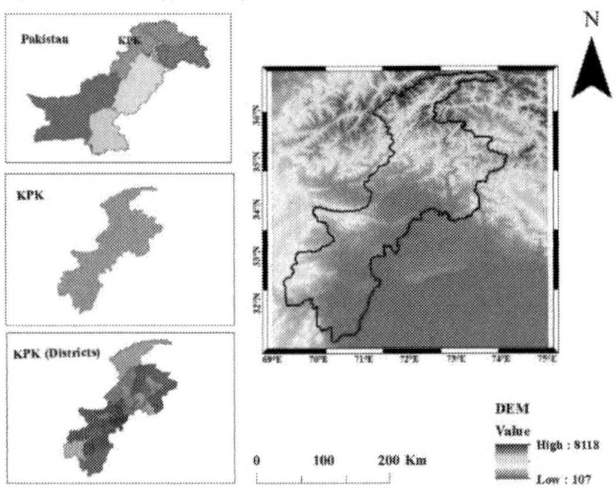

Figure 1. Map of Study Area

According to the agro-climatic classification, there is a predictable progression of climatic conditions from "dry regions" in the southern portion, through "dry semi-arid" and "wet semi-arid," to a transversal monsoon area in the middle of KPK, with a "sub-humid" and even "humid" climate in some places. (M. S. Khan, 2016) (Britannica 2012). In the province, agriculture provides a living for more than 85% of the inhabitants (Ashraf et al., 2014). The economies of agriculture and livestock are crucial to the development of the nation's economy.(Rahman et al., 2002).

3. DATA AQUISITION AND METHODOLOGY

Landsat photos from the Operational Land Imager (OLI), Enhanced Thematic Mapper Plus (ETM+), and Thematic Mapper (TM) have been extensively used in several remote sensing studies over the past three decades. According to the study area's calculations, USGS Earth Explorer's (http://earthexplorer.usgs.gov) high resolution reference data was matched. Data on temperature and precipitation from 1981 to 2020 was obtained from the TerraClimate website. Monthly rainfall statistics and the potential evapotranspiration (PET) based on minimum and maximum temperature were used to determine the annual precipitation parameters. The aridity index (AI) was determined using the UNEP approach, and the PET was calculated using Hargreave's method.

3.1. Reconnaissance Drought Index (RDI)

The standardized RDI used for our research is thought to be accurate and supported in the literature(Vangelis et al., 2013). It could be used during times when there is no rainfall. For dry and semi-arid environments, the Reconnaissance drought index is much more susceptible than the Standardized Precipitation Index (Asadi Zarch et al., 2017). The SPI has some flaws, one of which is that it ignores the impact of other variables connected to severe droughts. It is dependent on both total precipitation (P) and potential evapotranspiration (PET), which are two different factors, one measurable (P) and the other estimated (PET) (Angelidis et al., 2012). PET was estimated by applying Hargreave's approach (Equation 1) from maximum and minimum temperatures(Vangelis et al., 2013). Comparing the Thornthwaite and Blaney–Criddle is a preferable option for mountainous regions (Cai et al., 2015).

$$\mathbf{PET} = \mathbf{0.0023} \times \mathbf{Ra} \times (\mathbf{Tmean} + \mathbf{17.8}) \times \mathbf{TR}^{0.5} \quad (1)$$

Where,

Ra = Extra-terrestrial radiation

TR= Diurnal temperature variation (T min and T max)

Hargreave's approach is well-known for its efficiency (Lang et al., 2017), and it has been widely used in semi-arid and dry environments. For droughts assessment and management, the RDI can be utilize in three different ways: the initial value RDI (αk), the normalized RDI (nor), and the standardized RDI

RDI is computed using the equation 2 (assuming the log-normal distribution is used.)

$$RDIst(i) = y(i) - y / \hat{\sigma} y \quad (2)$$

Where

y(i) = is the ln (ak(i))
y = arithmetic mean
$\hat{\sigma}$ y = Standard deviation
(Asadi Zarch et al., 2017)

For estimating the specific frequency distributions of k, the gamma probability density function (pdf) was utilized (Khan et al., 2017). Compared to the normal conditions of the country, positive RDI value shows rainy seasons, and negative values show dry seasons. Drought severity is divided into four categories: extreme (2.0), severe (1.5 to 2.0), moderate (1.0 to 1.5) and mild (0.5 to 1.0).

3.2. Spatio-Temporal Analysis NDVI and NDWI

The Normalized Difference Vegetation Index, An indicator called NDVI was created to assess plant productivity using remote sensing methods (Hamrell, 2014). Due to the NDVI's

This contribution has been peer-reviewed.
https://doi.org/10.5194/isprs-archives-XLVIII-M-1-2023-425-2023 | © Author(s) 2023. CC BY 4.0 License.

accuracy as a measure of land plant productivity, it has been utilised as an indication of crop growth, land cover, crop productivity analysis, rainfall and droughts analysis, and warning systems for many years (Alhawiti & Mitsova, 2016). This technique can be used to track changes in flora, land use, and climate patterns. The NDWI is frequently used to identify changes in water body content. Water absorbs visible and infrared light, hence NDWI uses near-infrared and green electromagnetic bands to represent waterways.

$$\boldsymbol{NDVI} = NIR - \frac{RED}{NIR} + RED \quad (3)$$

$$\boldsymbol{NDWI} = \quad NIR - Green/NIR + Green \quad (4)$$

Red and near-infrared light (red reflectance) provides the basis for the NDVI (red reflectance plus near-infrared). Due to the significant chlorophyll absorption at 0.67 m, the near-IR band can detect up to eight leaf levels, but the red band can only detect one or frequently fewer (Mlenga et al., 2019). Two near-IR bands are included in the NDWI, one at 1.2 m and the other at 0.86 m. The NDWI is calculated using the Near-Infrared (NIR) and Green (G) bands. The amount of water in waterbodies is highlighted by this formula (Erban et al., 2022).

3.3 Aridity Index

The aridity index is used to gauge a region's climate patterns based on total rainfall and potential evapotranspiration (Li et al., 2017). By employing the equation, it is a powerful indication of the degree of dryness and drought identification (Qaisrani et al., 2022):

Aridity Index AI =P/PET (5)

Where,

'P' = "mean annual precipitation"

"PET" = "mean annual evapotranspiration."

PET was significant for the calculation of RDI.

4. REULTS AND DISCUSSIONS

4.1. Spatio-Temporal Analysis of Climatic variables (Interpolation by IDW Method)

Each climatic variable's interpolation performance was analysed separately. In order to estimate the values for the pixels in a raster, interpolation techniques require a minimal number of sample points (Berndt & Haberlandt, 2018). To estimate missing values, any geographic spatial information can be employed, including terrain, precipitation, pollutant levels, temperature, potential evapotranspiration, and others. IDW is an accurate interpolator, hence the maximum and minimum values in the interpolation area can only occur at specific places (Figure 2). Clusters and exceptions have an effect on the output region. IDW contends that the movement of the surfaces is caused by the neighborhood's local variability (Samanta et al., 2012).

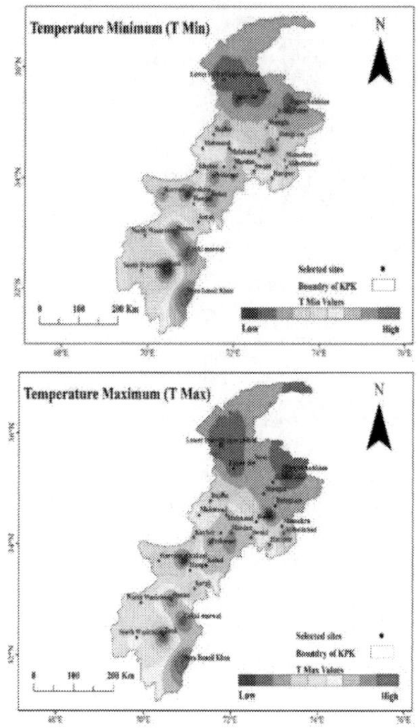

Figure 2. Interpolated Maps With Minimum/Maximum Temperatures From 1981-2020.

Recently, researchers found that temperature during droughts have been increasing more than in normal conditions in and, they relate this increase to water vapours in the atmosphere. This figure illustrates the lowest and highest minimum temperature. are 8.11, the lowest value, and 17.5 is the highest.Figure 7 displays the KPK maximum temperature, with 18.9 representing the minimum and 32 representing the maximum. This graph demonstrates that while some districts experience high temperatures, nearly 20 had low to moderate temperatures. Rainfall and PET (Figure 3) have an impact on agricultural water access in the soil, which has a clear and strong link to droughts.Using data from a 40-year average, we show the spatial-temporal variations in precipitation and PET. Changes in the drought may be influenced by a variety of causes, but changes in rainfall and ET are the most important ones (which can be reflected by PET). The spatial variance of PET generally shows a gradient lowering from south to north. The greatest and lowest values of PET are 122.1 and 88.42, respectively. When the amount of precipitation falls significantly short of the amount of potential evapotranspiration, causing the area to dry out, drought conditions arise on land. In general, rain will fall where the humidity is higher. PPT also ranges from 22.92 to 120, with 120 being the highest value. Swat and Mansehra in KPK have an increase in precipitation, whereas D.I. Khan, Bannu, and Karak experience declines. The soil will dry out if PET is higher than PPT until conditions are steady or water is supplied.

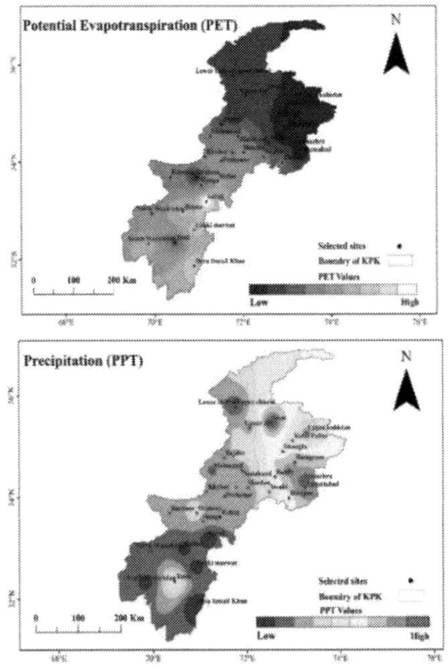

Figure 3. Maps Showing PET (Left) And PPT (Right) Variations During 1981-2020

4.2. Aridity Index Analysis

For characterising the climatic characteristics of a region, the aridity index (AI) is crucial. RDI was calculated using the UNEP method in the DrinC programme for specified locations. You may view the AI for each region and the pertinent meteorological data at each location in (Figure 5). This diagram illustrates how the aridity index is divided into categories for each of the KPK's regions. The lowest value in this case is 0.29, which occurs in a hyper-arid location, while the greatest value, 1.27, occurs in a very humid zone.

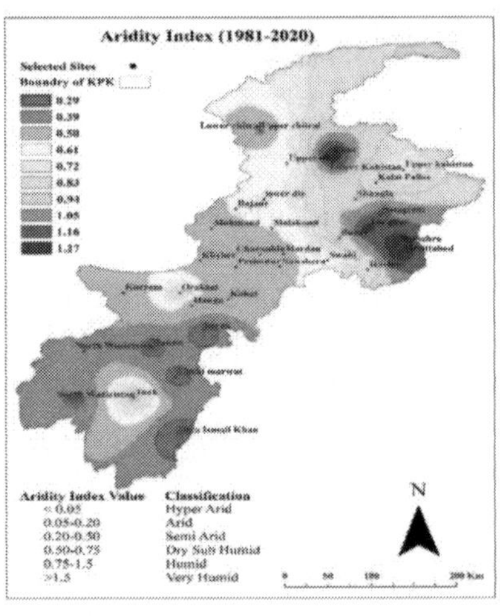

Figure 4. Spatial-Temporal Patterns of Aridity Index showing Classification for Khyber Pakhtaunkhawa.

4.3. Statistical Analysis of Climatic variables (Mann-Kendall Test)

The Mann–Kendall (M–K) technique is a nonparametric statistical test method developed by Mann and Kendall. The World Meteorological Organization (WMO) says this is the best way to study water and weather (Li et al., 2019). Drought patterns were analyzed using the nonparametric Mann-Kendall tests (Bandyopadhyay et al., 2016). Hydro-climatic time series data were analyzed using the Mann-Kendall test to look for any distinctive trends (Pablos et al., 2017). Sen presented a nonparametric method in 1968 that can determine the slopes of observed values if it has a linear trend line(Kousari et al., 2014; Yacoub & Tayfur, 2020). The most commonly used nonparametric method for analysing the trend in the time series is the Mann-Kendall (MK) test (Mann 1945, Kendall 1975). Researchers have extensively used this method for different hydro-meteorological parameters (Hirsch 1984, Aziz and Burn 2006, Mondal et al. 2012). Mann-Kendall test is an important tool for analysing the trend of climatic parameters, evaporation, reference evapotranspiration, stream flows and groundwater fluctuations (Bandyopadhyay et al. 2009, Ramazanipour and Roshani 2011). HO is a hypothesis, X1. Xn is series data, and n is the number of samples. S is calculated as(Gocic & Trajkovic, 2013);(Yue et al., 2002):

$$S = \sum_{i=1}^{n-1} \sum_{j=i+1}^{n} \operatorname{sgn}(x_j - x_i),$$

$$\operatorname{sgn}(x_j - x_i) = \begin{cases} 1 & (x_j - x_i) > 0 \\ 0 & (x_j - x_i) = 0 \\ -1 & (x_j - x_i) < 0 \end{cases}$$

Where S is a 0 mean and variance of

$$\operatorname{Var}(S) = \frac{n(n-1)(2n+5)}{18}, \quad (7)$$

Z of S formula:

$$Z = \begin{cases} \dfrac{S-1}{\sqrt{\operatorname{Var}(S)}} & S > 0 \\ 0 & S = 0 \\ \dfrac{S+1}{\sqrt{\operatorname{Var}(S)}} & S < 0 \end{cases} \quad (8)$$

If Z has a positive number, the trend is heading higher, and vice versa.

For a rank sequence:

$$S_k = \sum_{i=1}^{k} r_i \; (k = 1, 2, \ldots, n) \quad (9)$$

$$r_i = \begin{cases} 1, & x_i > x_j \\ 0, & x_i < x_j \end{cases} \; 1 \le j \le i \quad (10)$$

The UF_k statistic is defined as follows if the time series in question is genuinely random and uncorrelated:

$$UF_k = \frac{[S_k - E(S_k)]}{\sqrt{\mathrm{Var}(S_k)}} \quad (11)$$

$\mathrm{Var}(S_k)$ Indicate variance and $E(S_k)$ Indicate mean. These could be calculated as:

$$E(S_k) = \frac{k(k-1)}{4}, \mathrm{Var}(S_k) = \frac{k(k-1)(2k+5)}{72} \quad (12)$$

UF_K= Standard distribution normal. $UF_K > UF_{\alpha/2}$ is indicating trend change. $UF_K > 0$ shows a positive upward trend and vice versa.

For the lowest temperature, The time series data of minimum temperature revealed an increasing trend, as indicated by the positive values of Kendall's tau and S statistics. Zc statistics showed that the trend was statistically significant between 1981 and 2020. Positive Kendall's tau and S statistics for maximum temperatures demonstrate the upward trend in maximum temperature time series. Using time series data of climatic parameters, the effect of climate variability and its implications on potential evapotranspiration were assessed. Results revealed a decreasing trend for PET for Banuu, but an increasing trend for minimum temperature, maximum temperature, and precipitation (Figure 5). Figures 6--9 display the full graphs and their trend. The results of the analysis showed that the variability of climatic parameters can be described by the Mann-Kendall test with Sen's slope estimator. Based on the findings, it can be inferred that all climatic factors must be taken into account when assessing how climate change would affect crop water needs. be conducted in other regions of the nation.

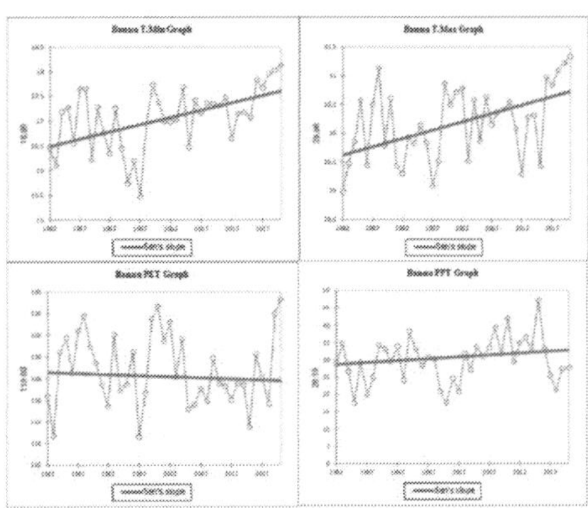

Figure 5. Sen's linear estimates of minimum, temperature, maximum temperature, PET and Precipitation for Bannu District of Khyber Pakhtaunkhawa.

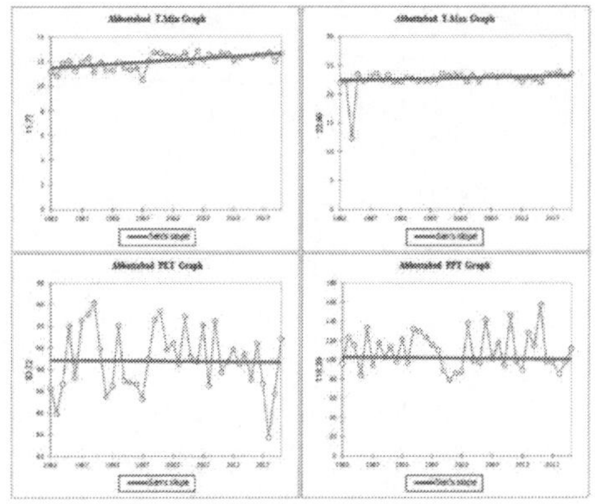

Figure 6. Sen's linear estimates of minimum, temperature, maximum temperature, PET and Precipitation for Abottabad Ditrict of Khyber Pakhtaunkhawa.

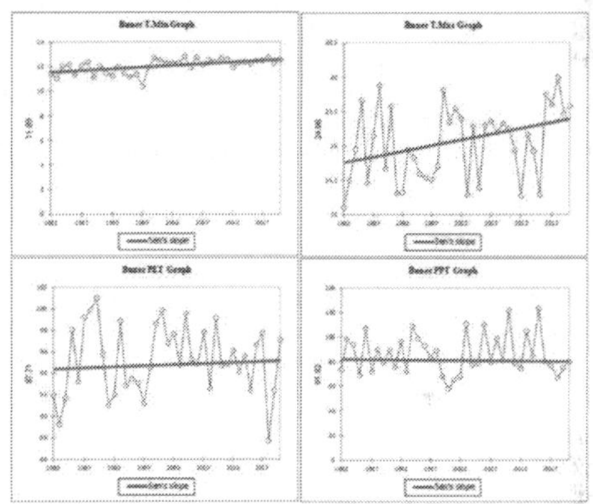

Figure 7. Sen's linear estimates of minimum, temperature, maximum temperature, PET and Precipitation for Bunner Ditrict of Khyber Pakhtaunkhawa.

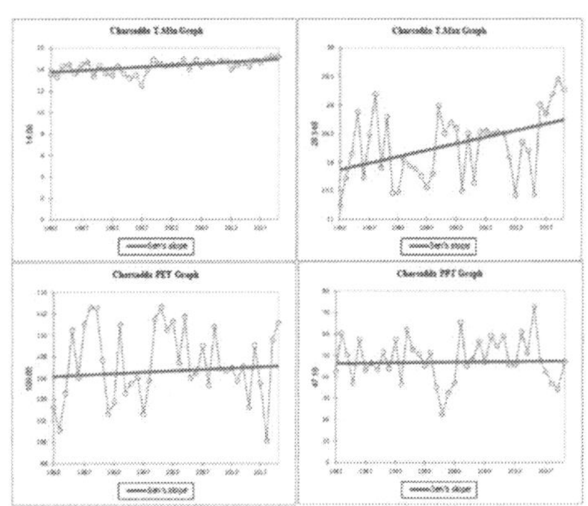

Figure 8. Sen's linear estimates of minimum, temperature, maximum temperature, PET and Precipitation for Charsadda Ditrict of Khyber Pakhtaunkhawa.

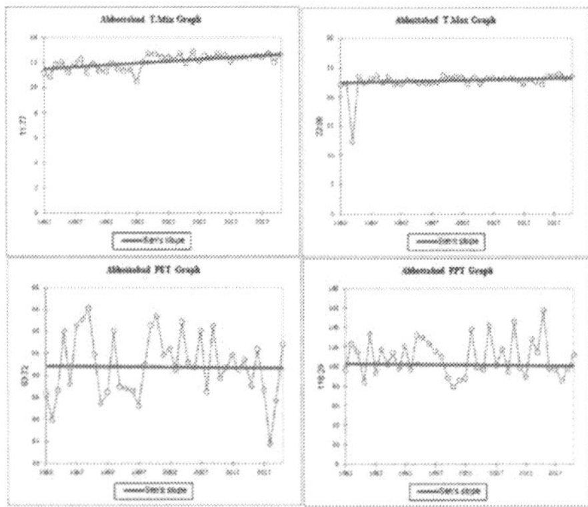

Figure 9. Sen's linear estimates of minimum, temperature, maximum temperature, PET and Precipitation D.I.Khan Ditrict of Khyber Pakhtaunkhawa

The geographical classification of droughts was developed using the inverse distance weighting (IDW) interpolation technique and 3 distinct RDI periods (3, 6, and 12 months) (Ben Abdelmalek & Nouiri, 2020). Because the data is created for a specific area, autocorrelation is not considered in the IDW technique, making it a deterministic approach that relies on calculated information (Amini et al., 2019; Cavus and Aksoy, 2019). The geographic display pictures with RDI values produced by Arc GIS 10.8 highlight the droughts season (Figure 9-11). Years that have RDI values were taken into account for the spatial analyses. The only dataset for the area that was at the time accessible was used to create the interpolation maps. These results demonstrate that the three drought classes' affected regions vary among the three RDI values. Extreme droughts were observed throughout the whole study region in 1999 and 2001. Large-scale severe droughts occurred in 1991, and severe and moderate droughts were frequent in 1981, 1998, 2010, and 2017. The other periods covered a range of geographic conditions, from mild drought to typical conditions.

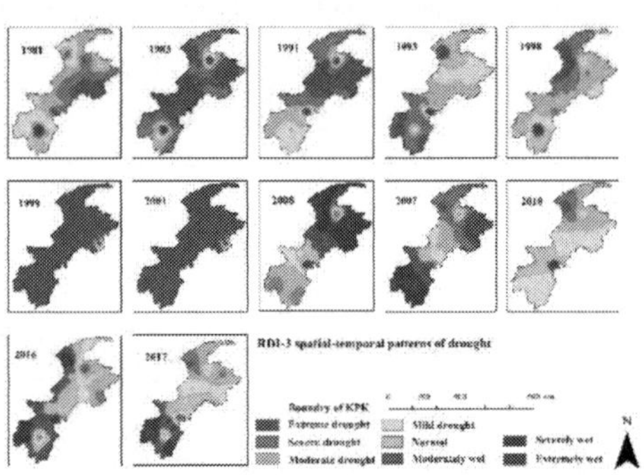

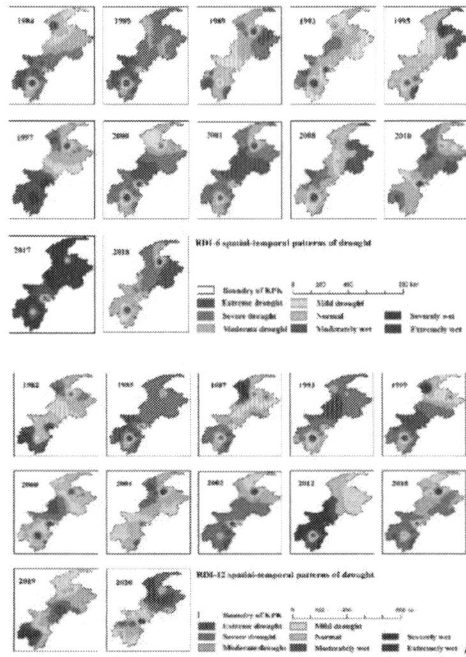

Figure 10. RDI for 3-6-12 Months showing Spatio-temporal patterns of drought in Khyber Pakhtaunkhawa for1981-2020.

4.4. NDVI and NDWI Variations

The growth phase is between October and March, when NDVI values are high, according to the average Landsat NDVI readings for KPK. The peak vegetative season lasts from January through March, and during this time, NDVI values between 0.58 and 1 indicate dense vegetation. The range of NDVI readings is typically between -1.0 and 1.0, with negative values signifying clouds and precipitation, positive values of 0 denoting dry ground, and higher positive values ranging from thin vegetation (0.1 to 0.5) to dense greenery (0.6 and above).

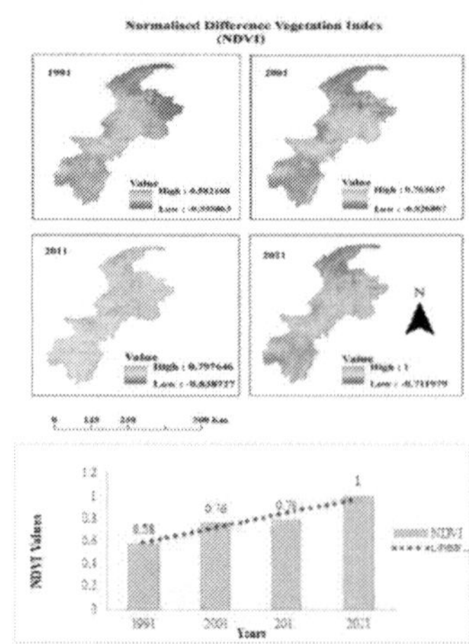

Figure 4.11. Comparative Analysis of NDVI for KPK region between 1991-2021

In 1991, there was more vegetation in the centre of KPK, such as Peshawar, Malakand, Mardan, and Swabi, than in some of the northern and southern regions. A very low NDVI value (below 0.1) this year indicates bleak terrain, snow, etc., while a very high NDVI value in 2021 indicates lush vegetation. This year's vegetation is lush, healthy, and green. Data that was gathered from a Landsat picture using the NDWI index is used to create graphs and maps that show how results vary over time as a curve. Except when the research region is a non-aqueous surface, the negative values to -1 are the apparent signs of water shortages. The positive values that reach +1 on a map usually have a bluish appearance and refer to either a high water content or a water level. The NDWI result has no dimensions and ranges from -1 to +1 depending on the type and amount of vegetation present. High NDWI readings are correlated with high vegetative cover percentage and water content (in blue). Lower NDWI values are associated with low vegetation water and low vegetation percent cover (in white).

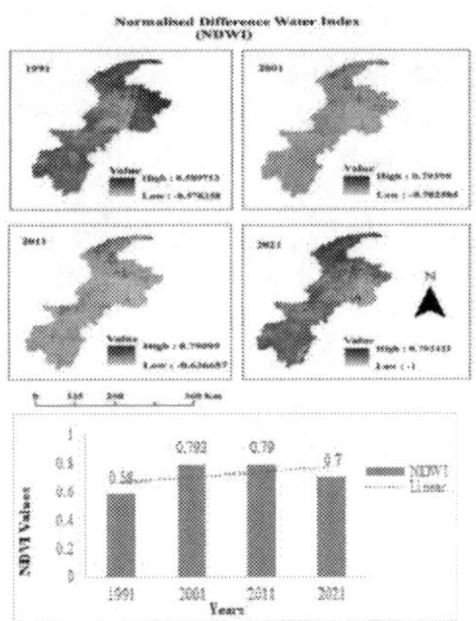

Figure **4.12.** Comparative Analysis of NDWI for KPK region between 1991-2021

NDWI levels will drop during a severe drought. The near-infrared bands increase the strong reflectance of terrestrial land use and soil properties while decreasing the low reflectivity of aquatic features. The NDWI algorithm yields positive results fo rivers while yielding negative numbers (or 0) for soils and dry vegetation.

5. CONCLUSION AND RECOMMENDATIONS

The most frequent natural calamity that has a large negative influence on crop output is drought, which occurs when there are insufficient water resources available across vast geographic areas. Depending on the data's availability, many drought indices are used to measure the intensity of the drought. This study evaluated general drought aspects in the Khyber Pakhtaunkhawa province from 1981 to 2020 using the reconnaissance drought index (RDI), which combines PPT and PET data. The area has a significant degree of rainfall variability and is located on the edge of the annual precipitation needed for wheat production. An increase in long-term drought severity could endanger the management of water resources in the studied areas because long-term droughts typically affect water quality, especially groundwater. The results indicate that these methods could contribute to the development of a calculated plan to combat the effects of droughts.Given that long-term water shortages have a significantly greater impact on water supply than short-term ones, particularly groundwater, According to the NDWI results, the value for the KPK region rises from 0.58 to 0.7. From January through March is when the vegetation is at its height; during this time, dense vegetation is reflected by NDVI values between 0.79 and 1. Examining the study month of September reveals that low NDVI values correspond to the years of drought forecast for the KPK region. The research focuses on climate-induced variability, which is crucial for understanding how the deficiency in the state of KPK emerged. These results provide national planners with references that they can use to create future policies and decisions to lessen the consequences of drought.River-based farming is prevalent due to the effective irrigation system in the nation. Water must be conserved and used judiciously for home and agricultural uses, though. To keep track of climate changes and their effects on droughts and floods, more data collection stations are needed in KPK because there aren't as many metrological stations there. As water shortage is a real-time and sustainable issue, the building of tube wells should be prioritised in drought-prone areas.

REFERENCES

Abid, S., Shah, N. A., Hassan, A., & Farooq, A. (2013). Asian Journal of Agriculture and Rural Development Growth and Trend in Area , Production and Yield of Major Crops of Khyber Pakhtunkhwa , Pakistan. 149–155.

Ahmed, K., Shahid, S., & Nawaz, N. (2018). Impacts of climate variability and change on seasonal drought characteristics of Pakistan. Atmospheric Research, 214, 364–374. https://doi.org/10.1016/j.atmosres.2018.08.020

alhawiti, r. h., & mitsova, d. (2016). using landsat-8 data to explore the correlation between urban heat island and urban land uses. 457–466.

Ashraf, M., Routray, J. K., & Saeed, M. (2014). Determinants of farmers' choice of coping and adaptation measures to the drought hazard in northwest Balochistan, Pakistan. Natural Hazards, 73(3), 1451–1473. https://doi.org/10.1007/s11069-014-1149-9

Bacha, M. S., Muhammad, M., Kılıç, Z., & Nafees, M. (2021). The dynamics of public perceptions and climate change in swat valley, khyber pakhtunkhwa, Pakistan. Sustainability (Switzerland), 13(8), 1–22. https://doi.org/10.3390/su13084464

Banerjee, A., & Bera, S. (2021). Chapter 13 - GIS-based assessment for agricultural susceptibility and sustainability to meteorological drought in West Bengal. In G. S. Bhunia, U. Chatterjee, A. Kashyap, & P. K. Shit (Eds.), Land Reclamation and Restoration Strategies for Sustainable Development (Vol. 10, pp. 265–284). Academic Press. https://doi.org/https://doi.org/10.1016/B978-0-12-823895-0.00010-5

Berndt, C., & Haberlandt, U. (2018). Spatial interpolation of climate variables in Northern Germany—Influence of temporal resolution and network density. Journal of Hydrology: Regional Studies, 15, 184–202. https://doi.org/https://doi.org/10.1016/j.ejrh.2018.02.002

Bhandari, S., Speer, J. H., Khan, A., & Ahmed, M. (2020). Drought signal in the tree rings of three conifer species from Northern Pakistan. Dendrochronologia, 63(July), 125742. https://doi.org/10.1016/j.dendro.2020.125742

Carrão, H., Russo, S., Sepulcre-Canto, G., & Barbosa, P. (2016). An empirical standardized soil moisture index for agricultural drought assessment from remotely sensed data. International

Journal of Applied Earth Observation and Geoinformation, 48, 74–84. https://doi.org/https://doi.org/10.1016/j.jag.2015.06.011

Haile, G. G. (2019). Drought : Progress in broadening its understanding Drought : Progress in broadening its understanding. January 2020. https://doi.org/10.1002/wat2.1407

Hamrell, M. R. (2014). What is an IND? FDA Regulatory Affairs: Third Edition, 266(April), 41–76. https://doi.org/10.3109/9781420073553-3

Jiang, T., Su, X., Singh, V. P., & Zhang, G. (2022). Spatio-temporal pattern of ecological droughts and their impacts on health of vegetation in Northwestern China. Journal of Environmental Management, 305, 114356. https://doi.org/https://doi.org/10.1016/j.jenvman.2021.114356

Khan, J. A. (2011). Changing climatic patterns and their impacts with special focus on pakistan. Symposium on "Changing Environmental Pattern and Its Impact with Special Focus on Pakistan," 285.

Khan, M. S. (2016). Ecological herpetology of the western Himalayas. November, 1–6.

Kirby, J. M., Mainuddin, M., & Mpelasoka, F. (2016). The impact of climate change on regional water balances in Bangladesh. 481–491. https://doi.org/10.1007/s10584-016-1597-1

Kumar, P., Shah, S. F., Uqaili, M. A., Kumar, L., & Zafar, R. F. (2021). Forecasting of drought: A case study of water-stressed region of Pakistan. Atmosphere, 12(10). https://doi.org/10.3390/atmos12101248

Li, Y., Feng, A., Liu, W., Ma, X., & Dong, G. (2017). Variation of aridity index and the role of climate variables in the Southwest China. Water (Switzerland), 9(10), 1–14. https://doi.org/10.3390/w9100743

Liu, Q., Zhang, J., Zhang, H., Yao, F., Bai, Y., Zhang, S., Meng, X., & Liu, Q. (2021). Evaluating the performance of eight drought indices for capturing soil moisture dynamics in various vegetation regions over China. Science of The Total Environment, 789, 147803. https://doi.org/https://doi.org/10.1016/j.scitotenv.2021.147803

Mao, Y., Wu, Z., He, H., Lu, G., Xu, H., & Lin, Q. (2017). Spatio-temporal analysis of drought in a typical plain region based on the soil moisture anomaly percentage index. Science of The Total Environment, 576, 752–765. https://doi.org/https://doi.org/10.1016/j.scitotenv.2016.10.116

Mlenga, D. H., Jordaan, A. J., & Mandebvu, B. (2019). Integrating standard precipitation index and normalised difference vegetation index for near-real-time drought monitoring in Eswatini. Jamba: Journal of Disaster Risk Studies,11(1),1–9. https://doi.org/10.4102/JAMBA.V11I1.917

No Title. (n.d.).

Podger, G. M., Ahmad, M., Yu, Y., Stewart, J. P., Muhammad, S., Ali, M., & Khero, Z. I. (2021). Development of the Indus River System Model to Evaluate Reservoir Sedimentation Impacts on Water Security in Pakistan.

Qaiser, G., Tariq, S., Adnan, S., & Latif, M. (2021). Evaluation of a composite drought index to identify seasonal drought and its associated atmospheric dynamics in Northern Punjab, Pakistan. Journal of Arid Environments, 185, 104332. https://doi.org/https://doi.org/10.1016/j.jaridenv.2020.104332

Qaisrani, Z. N., Nuthammachot, N., Techato, K., Jatoi, G. H., Mahmood, B., & Ahmed, R. (2022). Drought variability assessment using standardized precipitation index , reconnaissance drought index and precipitation deciles across Balochistan , Pakistan. 84, 1–12.

Rahman, G., Muhammad, R., Anwar, M., & Dawood, M. (2022). Spatio - temporal analysis of climatic variability , trend detection , and drought assessment in Khyber Pakhtunkhwa , Pakistan. Arabian Journal of Geosciences, 1–13. https://doi.org/10.1007/s12517-021-09382-4

Rahman, G., Rahman, A., Ullah, S., Miandad, M., & Collins, A. E. (2019). Spatial analysis of landslide susceptibility using failure rate approach in the Hindu Kush region, Pakistan. Journal of Earth System Science, 128(3), 59. https://doi.org/10.1007/s12040-019-1066-0

Samanta, S., Pal, D., Lohar, D., & Pal, B. (2012). Interpolation of climate variables and temperature modeling. Theoretical and Applied Climatology, 107, 35–45. https://doi.org/10.1007/s00704-011-0455-3

Sánchez, N., González-zamora, Á., Piles, M., & Martínez-fernández, J. (2016). (SMADI) Integrating MODIS and SMOS Products : A Case of Study over the Iberian Peninsula. https://doi.org/10.3390/rs8040287

Şerban, C., Maftei, C., & Dobrică, G. (2022). Surface Water Change Detection via Water Indices and Predictive Modeling Using Remote Sensing Imagery: A Case Study of Nuntasi-Tuzla Lake, Romania. Water (Switzerland), 14(4). https://doi.org/10.3390/w14040556

Stagge, J. H., Tallaksen, L. M., Xu, C., & Lanen, H. A. J. V. A. N. (2014). Standardized precipitation-evapotranspiration index (SPEI): Sensitivity to potential evapotranspiration model and parameters. 2014(October), 367–373.

Surendran, U., Kumar, V., Ramasubramoniam, S., & Raja, P. (2017). Development of Drought Indices for Semi-Arid Region Using Drought Indices Calculator (DrinC) – A Case Study from Madurai District, a Semi-Arid Region in India. Water Resources Management, 31(11), 3593–3605. https://doi.org/10.1007/s11269-017-1687-5

Xu, H., Wang, X., Zhao, C., Shan, S., & Guo, J. (2021). Seasonal and aridity influences on the relationships between drought indices and hydrological variables over China. Weather and Climate Extremes, 34, 100393. https://doi.org/https://doi.org/10.1016/j.wace.2021.100393

Yu, C., Huang, X., Chen, H., Huang, G., Ni, S., & Wright, J. S. (2018). Earth ' s Future Assessing the Impacts of Extreme Agricultural Droughts in China Under Climate and Socioeconomic Changes Earth ' s Future. 689–703. https://doi.org/10.1002/2017EF000768

Zhang, Q., Zhang, J., & Wang, C. (2017). Risk assessment of drought disaster in typical area of corn cultivation in China. Theoretical and Applied Climatology, 128(3–4), 533–540. https://doi.org/10.1007/s00704-015-1723-4

This contribution has been peer-reviewed.
https://doi.org/10.5194/isprs-archives-XLVIII-M-1-2023-425-2023 | © Author(s) 2023. CC BY 4.0 License.

STELLAR: A LARGE SATELLITE STEREO DATASET FOR DIGITAL SURFACE MODEL GENERATION

Sonali Patil[1,*] and Qi Guo[2]

[1] German Aerospace Center (DLR), Braunschweig, Germany - sonali.patil@dlr.de
[2] Purdue University, School of Electrical and Computer Engineering, West Lafayette, IN, USA - guo675@purdue.edu

KEY WORDS: Photogrammetry, digital surface model, big data challenge, digital earth, computationally intensive data processing.

ABSTRACT:

Stellar is a large, satellite stereo dataset. It contains rectified stereo pairs of the terrain captured by the satellite image sensors and corresponding true disparity maps and semantic segmentation. Unlike stereo vision in autonomous driving and mobile imaging, a satellite stereo pair is not captured simultaneously. Thus, the same object in a satellite stereo pair is more likely to have a varied visual appearance. Stellar provides flexible access to such stereo pairs to train methods to be robust to such appearance variation. We use publicly available data sources, and invented several techniques to perform data registration, rectification, and semantic segmentation on the data to build Stellar. In our preliminary experiment, we fine-tuned two deep-learning stereo methods on Stellar. The result demonstrates that most of the time, these methods generate denser and more accurate disparity maps for satellite stereo by fine-tuning on Stellar, compared to without fine-tuning on satellite stereo datasets, or fine-tuning on previous, smaller satellite stereo datasets. Stellar is available to download at `https://github.com/guo-research-group/Stellar`.

1. INTRODUCTION

Digital Surface Models (DSMs) contain 3D representations of the earth's surface. It records the shape information of terrains and human constructions of an area at specific time stamps. DSMs have broad applications in environmental studies, for example, monitoring and analyzing glaciers melting, water level changes of rivers and lakes, and human activities. Additionally, they are widely used in non-environmental applications such as urban visualization, infrastructure planning, and mixed reality.

There are two major approaches to generating DSMs nowadays. The first approach uses Lidars on UAVs to scan the landscape to produce point clouds, which are then processed to become DSMs. The DSMs from this approach have an average height error as low as 0.3 centimeters (San Diego LiDAR Report, n.d.) thanks to the high accuracy of Lidar sensors. But because the cruising altitude of UAVs is relatively low, the Lidar scanning can only cover a limited area each time. A typical full swath spacing of a UAV Lidar scan is below 1km. Thus, it is slow and expensive to build large-area DSMs via this approach. Furthermore, Lidar scanning relies on GPS for accurate geolocation, but the reliability of GPS signals can be easily affected by landscapes, atmospheric conditions, etc., in practice.

Thanks to the recent maturation of very high resolution (VHR) satellite image sensors, an alternative approach that reconstructs DSMs via stereo matching emerges. In this approach, people first perform two-view or multi-view stereo matching using VHR images of the same area captured from different satellite perspectives to generate local disparity maps. Then they fuse disparity maps of areas together to produce a DSM of a larger area. A VHR image can cover areas of hundreds of sq. km with a ground sampling resolution between 30cm-60cm depending on the viewing angle. Therefore, despite its lower accuracy compared to Lidar scanning, it is cheaper and more efficient for people to create a global DSM via satellite stereo.

*Corresponding author. This work was done at Purdue University.

The accuracy of stereo matching algorithms has improved these years significantly in street view and indoor settings thanks to the appearance of large, supervised datasets such as KITTI (Menze and Geiger, 2015) and Middlebury (Scharstein et al., 2014). These datasets provide sufficient stereo image pairs and corresponding ground truth disparity maps to train data-hungry deep-learning (DL) methods. However, in the domain of satellite stereo, developing a large-scale supervised dataset with diverse VHR images that supports quantitative evaluation is still at an initial stage in our perspective.

We present *Stellar*. It is a large satellite stereo dataset that contains stereo VHR image pairs and corresponding ground truth disparity maps generated from Lidar measurements. Stellar is so far the largest in areas of data to the best of our knowledge (Table 1). It is flexible to use. People can select stereo VHR image pairs that have specific time differences or sun angle differences. This allows users to control the appearance difference between the stereo image pairs caused by seasonal vegetation, weather, human activities, and shadows and specular reflections caused by sunlight (Figure 1). Stellar also provides semantic segmentation of the areas (*e.g.*, Figure 1), which enables quantitative analysis of stereo-matching accuracy based on types of areas, e.g., buildings, rivers, and vegetation.

Although the source data of Stellar comes from several public repositories (Bosch et al., 2016, Brown et al., 2018, Van Etten et al., 2018), there are image processing challenges that need to be overcome to build Stellar. First, the Lidar measurements and VHR images use different coordinate reference systems. This means the elevation measured from Lidar has to be calibrated to the same coordinate system as VHR images. We used a novel co-registration method to perform the alignment (Section 3.) Second, it is non-trivial to rectify stereo pairs for VHR images, as VHR sensors on satellites have nonlinear homography and non-conjugate epipolar curves. In Stellar, we developed a patch-based rectification method to overcome this challenge (Section 3.2.3.) Finally, Lidar measurements and VHR images

https://doi.org/10.5194/isprs-archives-XLVIII-M-1-2023-433-2023 | © Author(s) 2023. CC BY 4.0 License.

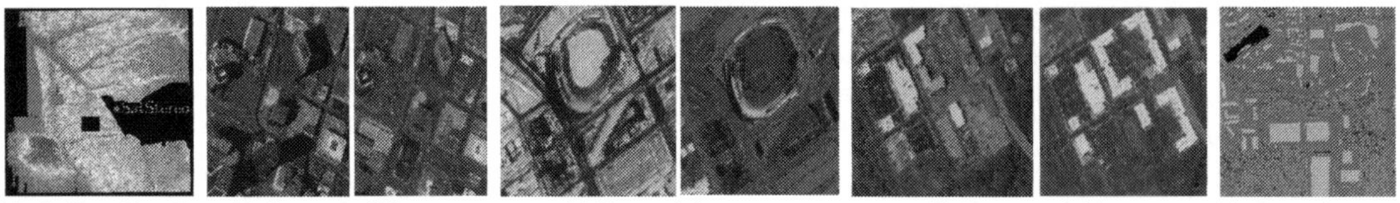

(a) Stellar Sample DSM (b) Sun angle difference. (c) Seasonal changes. (d) Human activities. (e) Semantic segmentation.

Figure 1. From left to right, (a) Stellar sample DSM as compared to a SatStereo region, (b-d) stereo pairs with varied appearances due to sun angle, season, and human activities. (e) Typical semantic segmentation. Orange for buildings, red for bridges, blue for water, green for vegetation and gray for ground.

Dataset	Area (km^2)	#City	Best Res. (m)	Unipolar Disparity	Pair Selection	Semantic Labels	Varying Stereo Image Size	Lidar DSM	True Disparity	Multi Date	Stereo Baseline Variation
MVS3DM (Bosch et al., 2016)	1.5	1	0.3	✗	✗	✗	✗	✓	✗	✓	✗
DFC (Bosch et al., 2019)	7.15	2	0.3	✗	✗	✓	✗	✓	✓	✓	✓
SatStereo (Patil et al., 2019a)	4.5	3	0.3	✗	✗	✓	✓	✗	✓	✓	✓
WHU-Stereo (Li et al., 2022b)	≈ 778	**6**	0.65	✗	✗	✗	✗	✗	✓	✗	✗
Stellar (ours)	**1,682**[1]	5	0.3	✓	✓	✓	✓	✓	✓	✓	✓

Table 1. Stellar vs. existing satellite stereo datasets. Apart from covering much larger areas than previous ones, Stellar allows users to select stereo image pairs with specific time differences and sun angle differences, etc. Stellar also supports generating stereo images with variable sizes and provides dense ground truth disparity.

of the same area may be acquired with a time difference longer than a year. The landscape may vary between the Lidar and the VHR image, and among VHR images (Figure 1). This creates noise in the ground truth disparity map, as the disparity from Lidar may not reflect the true disparity of a stereo VHR image pair. To mitigate this issue, Stellar provides semantic segmentation along with the ground truth disparity maps, which enables users to analyze methods on areas less prone to change over time (Section refsubsubsec:semantic.)

To validate the effectiveness of Stellar, we fine-tune two DL stereo matching algorithms on it and on previous satellite stereo datasets, and analyze their performance. Our experiments demonstrate that, in most cases, the DL stereo methods improve their accuracy on satellite stereo by fine-tuning on Stellar, compared to without fine-tuning or fine-tuning on previous, smaller satellite stereo datasets. The detailed description of our experiments is in Section 4.

Stellar is available to download at `https://github.com/guo-research-group/Stellar`.

[1]Lidar DSM coverage

2. RELATED WORK

2.1 Deep Learning (DL) Satellite Stereo

The classic two-stage stereo matching pipeline first computes a cost volume from a pair of rectified images, then estimates the disparity map based on the cost volume. Traditionally, semi-global matching (SGM) (Hirschmuller, 2005) is one of the most popular methods for stereo matching. In the era of deep learning, people leveraged deep neural networks to generate the cost volume (Zbontar et al., 2016), perform stereo matching (Zhang et al., 2019), or both. These DL methods clearly outperform traditional, non-DL methods on standard stereo benchmarks, such as Middlebury, KITTI, and ETH3D. Recently, more end-to-end DL methods emerged. These methods replace the classic two-stage pipeline with a single neural network architecture that takes in a rectified image pair and directly outputs the disparity map. These end-to-end DL methods have outperformed two-stage DL methods and demonstrated the top performances in computer vision benchmarks (Lipson et al., 2021, Zhao et al., 2022, Li et al., 2022a).

There have been several studies that quantitatively analyze the performance of DL stereo methods on satellite stereo (Albanwan and Qin, 2022, Gómez et al., 2022). Albanwan *et al.* (Albanwan and Qin, 2022) present an extensive comparative study

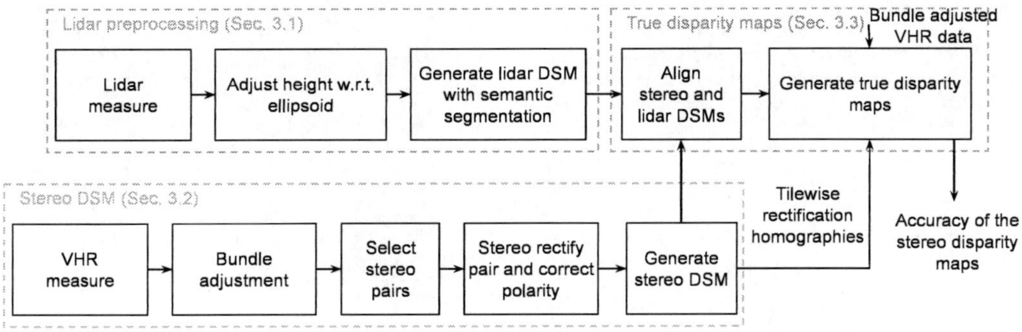

Figure 2. Stellar data generation pipeline.

of various stereo-matching approaches by fine-tuning them using a relatively small satellite stereo dataset (DFC in Table 1). They analyze two-stage approaches including SGM using a non-DL census cost volume (Zabih and Woodfill, 1994) and using a DL-based cost volume (Zbontar et al., 2016), as well as end-to-end DL methods (Kendall et al., 2017, Chang and Chen, 2018, Cheng et al., 2020). They report that the two-stage approaches, whether using DL cost volume or not, are more robust and generalizable. Meanwhile, end-to-end DL methods have the highest geometric accuracy while not generalizing well for unseen data. To increase the robustness of these end-to-end DL methods, a large satellite stereo dataset that covers more diversity in its stereo pairs seems necessary.

2.2 Satellite Stereo Datasets

Table 1 provides a comparison among previous satellite stereo datasets and Stellar. A critical challenge in creating a satellite stereo dataset is the generation of ground truth disparity. Typically, people leverage Lidar measurements to synthesize the ground truths, but the Lidar point clouds are sparse and are in different coordinate systems than satellite imagery. The Multi-View Stereo 3D Mapping dataset (MVS3DM) (Bosch et al., 2016) only contains Lidar point clouds as ground truths and the authors do not calibrate the Lidar data and the Satellite imagery into the same coordinate system, which could cause difficulty in performing supervised training using the dataset. The 2019 Data Fusion Contest dataset (DFC) (Bosch et al., 2019) includes sparse true disparity maps generated from the Lidar point clouds. Its Lidar measurements and satellite imagery are aligned by maximizing mutual information between the two sources of data. SatStereo (Patil et al., 2019a) contains dense ground truths. The authors first create true DSMs from the Lidar measurements, then transform the Lidar DSMs into dense disparity maps in the same coordinate as the satellite imagery via bundle adjustments.

A key difference between the satellite stereo datasets and stereo datasets in the computer vision community, such as Middlebury, KITTI, and ETH3D, is that the disparity maps are naturally bipolar in satellite stereo, meaning the disparity maps contain both positive and negative values. This is caused by the fundamental difference in camera model between the satellite cameras and pinhole cameras. See Sec 3.2.2 for detailed discussions. As in Table 1, all listed previous datasets contain both positive and negative disparity values. Adapting a stereo-matching algorithm to work for the both positive and negative disparity, especially those based on semi-global matching (SGM), is non-trivial. Stellar overcomes this issue. It enables all positive disparity via an innovative tile-based polarity correction (Sec 3.2.3.)

Compared to previous satellite stereo datasets, Stellar enables much higher flexibility. Compared to the recently released WHU-Stereo dataset (Li et al., 2022b), which contains a similar covered area as Stellar (Table 1), Stellar contains more than one stereo pair per region with different baseline distances and acquisition times that users can flexibly choose. This could help an algorithm learn to handle varying baseline distances and different visual appearances within stereo pairs.

3. DATA GENERATION PIPELINE

Figure 2 shows the overview of Stellar data generation pipeline. Key steps include (a) Lidar preprocessing, (b) Stereo DSM, and (c) True disparity maps generation. We explain each step in an individual subsection.

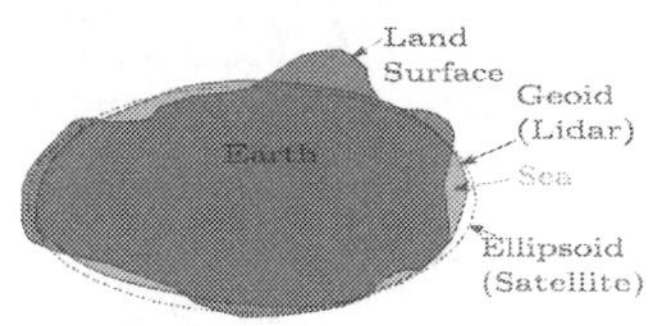

Figure 3. As UAVs that carry Lidars and satellites use different vertical datum, the two data sources must be calibrated to the same coordinate system.

3.1 Lidar Data Processing

Lidars measure a sparse point cloud $\{\mathbf{x}_i, i = 1, \cdots, N\}$ of the terrain. Each point $\mathbf{x}_i$ is a four-vector $\mathbf{x}_i = (x_i, y_i, z_i, c_i)$, where (x_i, y_i) indicates the Lambert conformal conic projection (LCC) coordinate of the point, z_i is the altitude of the point with respect to some zero altitude datum, and c_i is a semantic label of the point, e.g. building, bridge, water, etc., that is sometimes present in the data.

The first step of Lidar data processing is coordinate transformation. The Lidar point clouds follow NAVD88 geoid vertical datum, which uses the approximate mean sea level as the zero altitudes (Figure 3.) Meanwhile, the satellite imagery is collected under a different vertical datum, WGS84 ellipsoid. Therefore, the first step is to convert the altitude of each point z_i into the same vertical datum as the satellite imagery:

$$\tilde{z}_i = \text{Vertical Datum Conversion}(z_i).$$

This contribution has been peer-reviewed.
https://doi.org/10.5194/isprs-archives-XLVIII-M-1-2023-433-2023 | © Author(s) 2023. CC BY 4.0 License.

We achieve the conversion using a predefined transformation grid. We also transform the 2D location of each point from LCC coordinate (x_i, y_i) to Universal Transverse Mercator (UTM) coordinate $(\tilde{x}_i, \tilde{y}_i)$. We denote the point cloud after coordinate transformation as $\{\tilde{\mathbf{x}}_i = (\tilde{x}_i, \tilde{y}_i, \tilde{z}_i, c_i)\}$. For simplicity of notation, we omit the index range $i = 1, 2, \cdots$ hereafter.

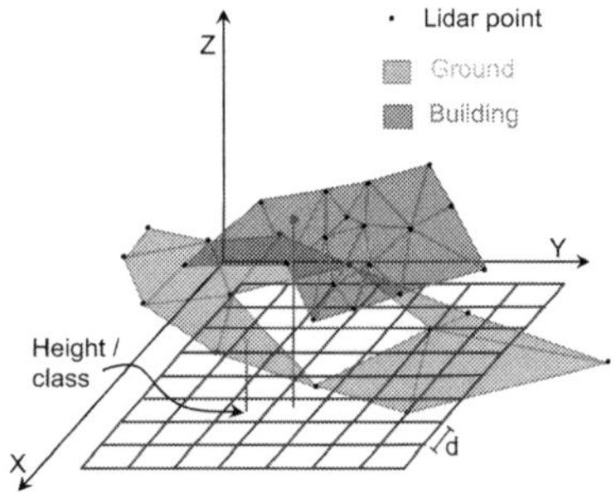

Figure 4. The process of generating dense 2D maps of altitude, i.e., DSM, and semantic classes from a point cloud. First, triangulate the point clouds and prune the redundant triangles to form a mesh. Then, project each pixel in the 2D map to the mesh and determine the value of the pixel, i.e., altitude or semantic class, by interpolating the vertices of the intersecting triangle. See detailed description in Section 3.1.1.

3.1.1 Lidar DSM Generation After the transformation, we rasterize the transformed point cloud $\{\tilde{\mathbf{x}}_i\}$ to a dense 2D altitude map, i.e., the DSM, and a semantic segmentation map. As in Figure 4, we first use Delaunay triangulation to obtain a mesh from the point cloud, $\mathcal{T} = \text{Delaunay}(\{(\tilde{x}_i, \tilde{y}_i, \tilde{z}_i)\})$. The mesh $\mathcal{T}$ contains a list of triangles t_j whose vertices are neighboring points in the point cloud, $\mathcal{T} = \{t_j\}$. Next, we prune the redundant triangles in the mesh $\mathcal{T}$ to form a new mesh $\mathcal{S}$ based on the following rule:

$\mathcal{S} = \{t_j \in \mathcal{T} \mid$ All edges of t_j are shorter than α
AND
The angle between t_j's normal vector
and all its neighboring triangles' normal vector
are below β
AND
The maximum altitude difference among t_j's vertices
is smaller than $\gamma\}$.

We use $\alpha = 10m$, $\beta = 20^\circ$, and $\gamma = 1m$. Then, as in Figure 4, we define a grid with a step size d and project each point in the grid along the z-axis to intersect with the mesh to find out its altitude. If the ray intersects a triangle t_j in the mesh, the height is estimated as the interpolated value of the three vertices of t_j. If a ray doesn't intersect any triangles, that point is marked as invalid. In our experiment, we define $d = 0.3m$ or $0.5m$.

Figure 5 shows a qualitative comparison between the Lidar DSMs in DFC (Bosch et al., 2019) and in Stellar for two regions included in both datasets. The average height difference between the two Lidar DSMs is less than 0.5 m. As highlighted using arrows in Figure 5, Stellar DSMs are more accurate along boundaries such as edges of buildings and include more fine details compared to DFC.

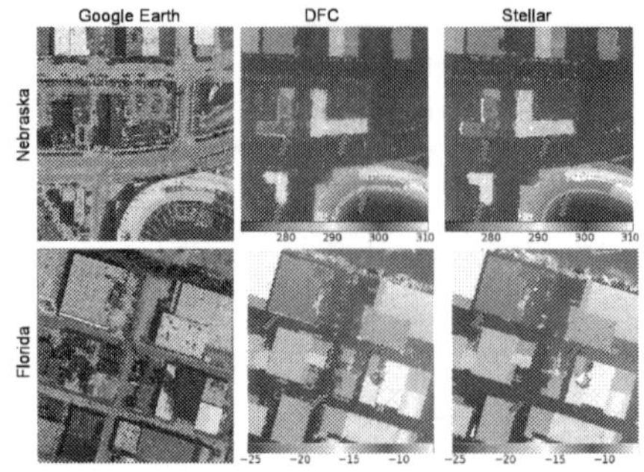

Figure 5. Qualitative comparison of the Lidar DSMs from DFC (Bosch et al., 2019) and Stellar. As highlighted using arrows, Stellar DSMs contain more fine details and are more accurate along building boundaries. (First column image source: ©Google Earth)

3.1.2 Semantic Segmentation As mentioned before, the semantic label may not be present in some point cloud data. In this case, we leverage the connected component analysis on the mesh $\mathcal{S}$ to cluster all triangles. Then we label the largest connected component as ground and the moderate size connected components, *i.e.*, components with at least 500 triangles, as buildings (Figure 4.) The semantic segmentation map is then generated using the same projection method described in Section 3.1.1. A sample semantic segmentation map is in Figure 1(e).

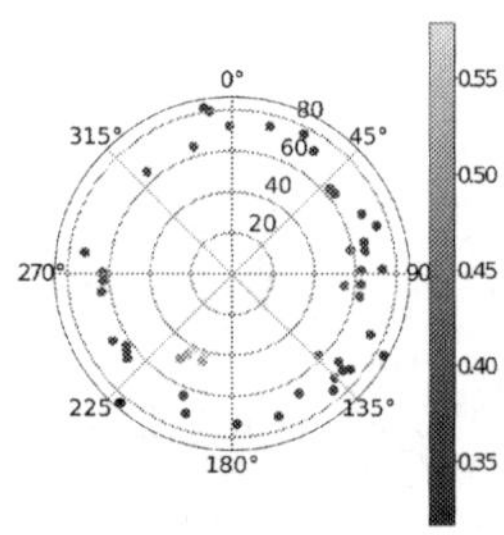

(a) Argentina (WV3)

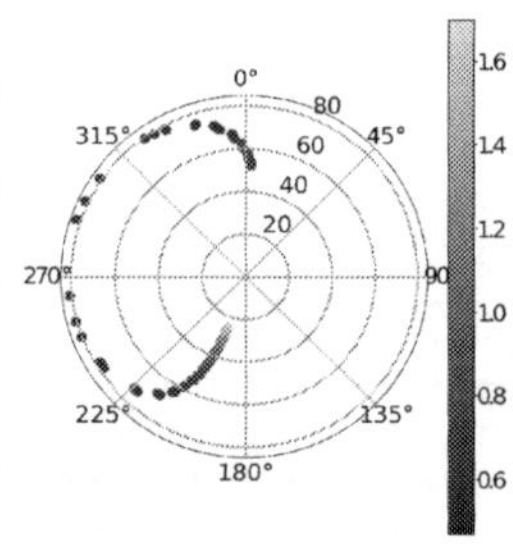

(b) Georgia (WV2)

Figure 6. Satellite positions per image for each region. Outer ring in each plot represents azimuth angle, the inner rings represent off-nadir angles and GSD (meters) for each image.

3.2 Satellite Stereo DSM

Each image $\mathbf{I}$ from the satellite comes with an approximate camera model $\mathcal{P}$ in rational polynomial coefficient (RPC) format, and metadata such as satellite (Figure 6), sun positions, timestamp, and ground sampling distance (GSD). The image we use is the panchromatic image that has a spatial resolution of 0.3-0.7m.

3.2.1 Data Alignment and Stereo Pair Selection The provided RPC camera model associated with each satellite image has a relative pointing error up to a few meters w.r.t. other images in the same region. Similar to the practice by Patil *et al.* (Patil et al., 2019a), we correct the relative point errors using bundle adjustment. This correction improves the stereo rectification accuracy. Bundle adjustment is the process that, given a

This contribution has been peer-reviewed.
https://doi.org/10.5194/isprs-archives-XLVIII-M-1-2023-433-2023 | © Author(s) 2023. CC BY 4.0 License.

set of images $\{\mathbf{I}_1, \cdots, \mathbf{I}_p\}$, determines the corresponding camera projection function of each image $\mathcal{P}_1, \cdots, \mathcal{P}_p$.

First, we use the SIFT/SURF to detect a set of corresponding key points in each image. We denote the ith key point in the jth image as $\mathbf{x}_i^j$. We also perform RANSAC to remove outliers to ensure correspondences among key points. Then, we unproject the ith key point for all images $\{\mathbf{x}_i^j, j = 1, \cdots, p\}$ using the estimated projection functions $\{\mathcal{P}_j, j = 1, \cdots, p\}$ to locate the corresponding 3D world point $\mathbf{X}_i$ via triangulation, and project $\mathbf{X}_i$ back to each view to calculate the reprojection error. We use this reprojection error as an objective function to search for the optimal camera projection functions:

$$\hat{\mathcal{P}}_{1,\cdots,p} = \underset{\mathcal{P}_{1,\cdots,p}}{\arg\min} \sum_{j \neq k} \sum_{i} ||\mathbf{x}_i^j - \mathcal{P}_j(\mathcal{P}_k^{-1}(\mathbf{x}_i^k))||^2. \quad (1)$$

In multi-date satellite stereo reconstruction, multiple factors such as view difference (Figure 6), sun angle difference, acquisition time difference, etc., influence the quality of DSM reconstruction. Given a set of n satellite images, there are ${}_n\mathrm{C}_2$ possible stereo pairs. We ignore stereo pairs with view angle differences less than $5°$ to avoid those with narrow baselines. All the rest possible stereo pairs are accessible to users in Stellar.

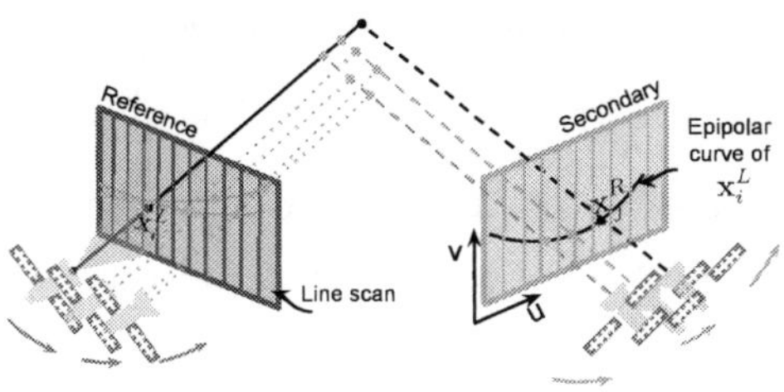

Figure 7. Satellite image sensors are pushbroom cameras, which have non-linear homography and non-conjugate epipolar curves. See detailed discussion in Section 3.2.2.

3.2.2 Epipolar Geometry of Satellite Imagers The linear pushbroom (LP) camera used in earth observation satellites has only one line of pixels in its photosensor. As the satellite moves along its orbit, the LP camera scans the earth (Figure 7.) During the scanning, the camera center moves at a constant speed. This has a significant impact on the epipolar geometry of two LP cameras. Meanwhile, cameras in our daily life are usually perspective, which has a unique camera center and a constant projection matrix for each captured image.

For a pair of images captured by LP cameras, there is a different fundamental matrix $\mathbf{F}_{ij}$ for every ith column of the left image and every jth column of the right image. As in Figure 7, for a point $\mathbf{x}_i^L$ on the ith column of the left image, it will correspond to an epipolar curve c_R on the right image (Gupta and Hartley, 1997). The epipolar curve c_R intersects the jth column at $\mathbf{x}_j^R$:

$$\mathbf{x}_j^R = (j, v), \quad (2)$$

where $(u', v', w') = \left(\mathbf{F}_{ij}\mathbf{x}_i^L\right) \times (1, 0, 0)$, and $v = v'/w'$.

Meanwhile, each point $\mathbf{x}_j^R$ corresponds to different epipolar curves on the left image, which means the epipolar curves are non-conjugate. Denote $\mathbf{C}_i^L$ and $\mathbf{C}_j^R$ of the camera center when capturing the ith column of the left image and the jth column of the right image. The baseline vector $\overrightarrow{\mathbf{C}_i^L\mathbf{C}_j^R}$ will vary its direction for different i, j in most situations (Habib et al., 2005). Thus, the disparity between the left and right images could become both positive and negative.

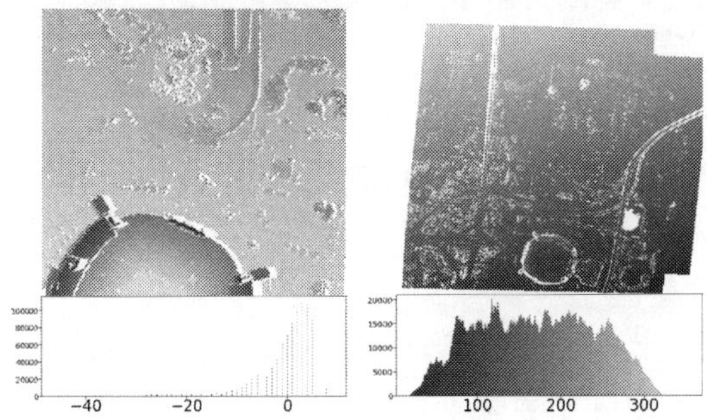

(a) DFC (b) Stellar

Figure 9. Polarity comparison of true disparity maps. The bottom plots are histograms of disparity. The disparity values of Stellar is always positive.

3.2.3 Stereo Rectification and Polarity Correction Rectifying a pair of satellite images is non-trivial because of the hyperbolic, non-conjugate epipolar geometry between the images. To deal with this, we divide the satellite images into tiles and assume each tile to follow the affine camera projective geometry. Thus, we can assume linear, conjugate epipolar geometry between each pair of tiles. De Franchis *et al.* also used similar tilewise affine assumptions (De Franchis et al., 2014), but we have several innovations compared to their practice. First, to efficiently identify corresponding tiles between the left and right images, we use the DEM sculpting approach (Patil et al., 2019b) to obtain a rough estimate of correspondences among tiles by using the low-resolution digital elevation model (DEM) of the earth. Second, we modify the rectification homography for the right image to apply additional translation to obtain unipolar disparities.

Given a set of images $\{\mathbf{I}_1, \cdots, \mathbf{I}_p\}$ with significant scene overlap and their corresponding bundle adjusted camera projection functions $\{\hat{\mathcal{P}}_1, \cdots, \hat{\mathcal{P}}_p\}$, we pick a stereo pair $(\mathbf{I}^L, \mathbf{I}^R) \in \{\mathbf{I}_1, \cdots, \mathbf{I}_p\}$ and their camera projection functions $(\mathcal{P}^L, \mathcal{P}^R) \in \{\hat{\mathcal{P}}_1, \cdots, \hat{\mathcal{P}}_p\}$. We divide each image in the selected stereo pair into n 500 pixel $\times$ 500 pixel tiles with 100-pixel overlap between neighboring tiles. This gives us a list of tiled stereo pairs $\{(\mathbf{I}_i^L, \mathbf{I}_i^R), i = 1, \cdots, n\}$.

We obtain a set of virtual correspondences in each tiled stereo pair using DEM-Sculpting (Patil et al., 2019b) as follows. The digital elevation model (DEM) is a two-dimensional altitude map of the terrain, similar to DSM but with a much lower spatial resolution. People already have access to the DEM of the entire earth. We uniformly sample points $\{\mathbf{x}_j^L, j = 1, \cdots, K_1\}$ on each left tile $\mathbf{I}_i^L$ and unproject these points onto the $30m$-resolution DEM to obtain the corresponding 3D world point $\mathbf{X}_j = (\mathcal{P}^L)^{-1}(\mathbf{x}_j^L)$. We then project each 3D point $\mathbf{X}_j$ onto the right tile to calculate the corresponding pixel location $\mathbf{x}_j^R = \mathcal{P}^R(\mathbf{X}_j)$. We compute another set of correspondences by creating another set of 3D world points $\tilde{\mathbf{X}}_j = \mathbf{X}_j + (0, 0, 100m)$ and project them onto $\mathbf{I}^L$ and $\mathbf{I}^R$. We ignore any point that falls outside the boundary of the tiles $(\mathbf{I}_i^L, \mathbf{I}_i^R)$. We denote the new set

https://doi.org/10.5194/isprs-archives-XLVIII-M-1-2023-433-2023 | © Author(s) 2023. CC BY 4.0 License.

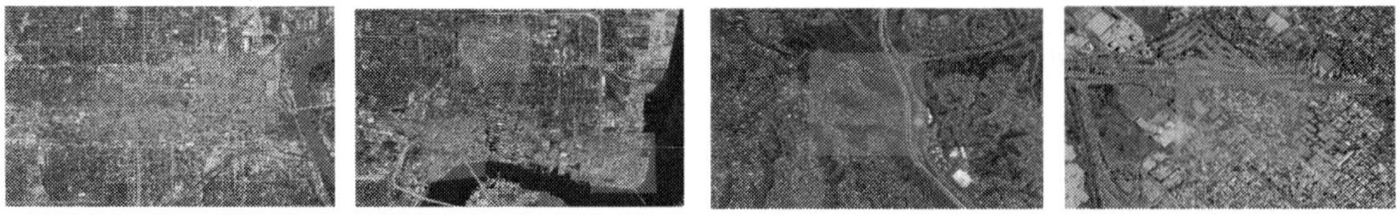

(a) Training sites in Nebraska. (b) Training sites in Florida. (c) Testing sites in California. (d) Testing sites in Argentina.

Figure 8. Training and testing sites.(Image source ©Google Earth)

of correspondences as $\{(\tilde{\mathbf{x}}_j^L, \tilde{\mathbf{x}}_j^R), j = 1, \cdots, K_2\}$. This new set of points creates an elevation envelope to account for the altitude of buildings, as they do not exist in the low-resolution DEM. In total, the process generates $K_1 + K_2$ virtual correspondences between the left and right tile $(\mathbf{I}_i^L, \mathbf{I}_i^R)$.

Then, we find the rectification homography $\mathbf{H}_i^L$ and $\mathbf{H}_i^R$ for the tile pair. Mathematically, the fundamental matrix $\mathbf{F}_i$ of the two tiles has the following form (Hartley and Zisserman, 2003).

$$\mathbf{F}_i = \begin{bmatrix} 0 & 0 & a \\ 0 & 0 & b \\ c & d & e \end{bmatrix},$$

as we assume affine camera model. Using the previously identified $K = K_1 + K_2$ virtual correspondences, denoted as $(\mathbf{x}_k^L, \mathbf{x}_k^R), k = 1, 2, \cdots, K$ for simplicity, we can estimate the five parameters of the fundamental matrix using the Gold Standard algorithm (Hartley and Zisserman, 2003). The epipoles for left and right images can be given as $\mathbf{e}^L = (-d, c, 0)^T$ and $\mathbf{e}^R = (-b, a, 0)^T$, respectively. We compute the stereo rectification homography as follows:

$$\mathbf{H}_i^L = \begin{bmatrix} s\mathbf{R}^L & \mathbf{t}^L \\ \mathbf{0}^{1\times 2} & 1 \end{bmatrix}, \quad \mathbf{H}_i^R = \mathbf{T}\tilde{\mathbf{H}}_i^R \tag{3}$$

where

$$\tilde{\mathbf{H}}_i^R = \begin{bmatrix} 1/s\mathbf{R}^R & \mathbf{t}^R \\ \mathbf{0}^{1\times 2} & 1 \end{bmatrix}, \mathbf{R}^L = \frac{1}{||\mathbf{e}||}\begin{bmatrix} -d & c \\ -c & -d \end{bmatrix},$$

$$\mathbf{R}^R = \frac{1}{||\mathbf{e}'||}\begin{bmatrix} -b & a \\ -a & -b \end{bmatrix}, \mathbf{T} = \begin{bmatrix} 0 & 0 & -t_x \\ 0 & 0 & 0 \\ 0 & 0 & 1 \end{bmatrix},$$

$s = \frac{||\mathbf{e}||}{||\mathbf{e}'||}$, $\mathbf{t}^L = (0, -t)^T$, $\mathbf{t}^R = (0, t)^T$ and $t = \frac{e}{2||\mathbf{e}||||\mathbf{e}'||}$. The matrix $\mathbf{T}$ shifts the right image horizontally by t_x so that the disparity is unipolar. The shift t_x is the maximum disparity of all the K rectified correspondences:

$$t_x = \max\{\tilde{\mathbf{H}}_i^R \mathbf{x}_k^R - \tilde{\mathbf{H}}_i^L \mathbf{x}_k^L, k = 1, 2, \cdots, K\}. \tag{4}$$

With the homography, we can obtain rectified tile pairs $\{(\tilde{\mathbf{I}}_i^L, \tilde{\mathbf{I}}_i^R)\}$. Then, we stitch all the rectified pairs together to form two rectified images $(\tilde{\mathbf{I}}^L, \tilde{\mathbf{I}}^R)$. In case of overlapping pixel value for multiple tiles, we pick the value from the tile with a smaller index i. For simplicity, we denote the entire rectification process as $\tilde{\mathbf{I}}^{L/R} = g^{L/R}(\mathbf{I}^{L/R})$, where $\tilde{\mathbf{I}}^{L/R}$ is the rectified left/right image and $g^{L/R}$ is the left/right rectification function.

Figure 9 shows a comparison of the disparity maps and their histograms for DFC and Stellar datasets. It is clear that Stellar contains all positive disparities.

3.2.4 DSM Generation from Disparity Maps In this section, we discuss an intermediate DSM generated from satellite images. This DSM is for aligning the Lidar data with the satellite data. Given a rectified stereo pair $(\tilde{\mathbf{I}}^L, \tilde{\mathbf{I}}^R)$, we use the modified tSGM algorithm (Patil et al., 2019b) to output two disparity maps $(\mathbf{D}^L, \mathbf{D}^R)$, where $\mathbf{D}^L$ is the left disparity and $\mathbf{D}^R$ is the right disparity. For a point $\mathbf{x}^L$ in the left image $\mathbf{I}^L$, we can use the following transformation to find the corresponding point on the right image $\mathbf{I}^R$:

$$\mathbf{x}^R = (g^R)^{-1}\left(\mathbf{D}^L(g^L(\mathbf{x}^L))\right).$$

By unprojecting points $\mathbf{x}^L$ and $\mathbf{x}^R$ using their corresponding camera projection functions and using triangulation, we can locate their corresponding 3D world point $\mathbf{X}$. We repeat this process for all points in the left and right images and can generate a 3D point cloud of the scene. Then we can use the same triangulation and interpolation process in Section 3.1.1 to produce a DSM.

3.3 Ground Truth Disparity Map Generation

The calibration of the Lidar DSM and the alignment DSM described in Section 3.2.4 follows the procedure in Nuth *et al.* (Nuth and Kääb, 2011). It minimizes the overall elevation differences between the two DSMs by performing a global translation and bias of the Lidar DSM.

After alignment, we project all points $\mathbf{X}$ from the Lidar DSM to the left and right rectified image via $\tilde{\mathbf{x}}^{L/R} = g^{L/R}(\mathcal{P}^{L/R}(\mathbf{X}))$ and calculate the true left disparity as the horizontal difference between the projected points: $\tilde{\mathbf{D}}^L(\tilde{\mathbf{x}}^L) = \tilde{\mathbf{x}}_u^L - \tilde{\mathbf{x}}_u^R$. We then treat $\tilde{\mathbf{D}}^L$ as a point cloud and use the triangulation and interpolation approach in Section 3.1.1 to obtain the ground truth of left disparity map $\mathbf{D}^L$. We repeat the same process to obtain $\mathbf{D}^R$. Then, we perform a left-right-right-left (LRRL) consistency check to remove pixels where both disparity values disagree, which could happen at occlusion. Mathematically, the LRRL check is:

$$\mathbf{D}^L(\mathbf{x}) = \begin{cases} \mathbf{D}^L(\mathbf{x}) & if \;\; |\mathbf{D}^L(\mathbf{x}) - \mathbf{D}^R(\mathbf{s})| \leq 1 \\ invalid & otherwise \end{cases} \tag{5}$$

where $\mathbf{s} = (\mathbf{x}_u - \mathbf{D}^L(\mathbf{x}), \mathbf{x}_v)$ and $\mathbf{x}$ is the position of a pixel in the left image.

4. PRELIMINARY EXPERIMENTAL RESULTS

This section describes our preliminary effort to analyze the effectiveness of Stellar. We select two pre-trained DL stereo architectures, fine-tune them on Stellar and DFC (Bosch et al., 2019) datasets, and analyze their performance. The information about the two DL architectures is as follows:

GANet (Zhang et al., 2019) follows the classic two-stage stereo matching pipeline. It first extracts dense features from the input image pair using a convolutional architecture to compute a

Method	California			Argentina		
	Good 3 (%) ↑	Avg (px) ↓	RMSE (px) ↓	Good 3 (%) ↑	Avg (px) ↓	RMSE (px) ↓
GANet (Sceneflow)	66.86	39.90	64.17	37.34	72.60	111.54
GANet (DFC)	70.36	71.32	103.61	28.27	141.18	174.24
GANet (Stellar)	71.96	25.41	36.05	48.74	37.9	46.32
RAFT-Stereo (Sceneflow)	84.48	16.76	26.32	48.61	47.27	62.83
RAFT-Stereo (DFC)	0.20	347.02	349.42	3.17	256.24	261.17
RAFT-Stereo (Stellar)	59.00	43.68	50.53	49.7	30.86	34.89

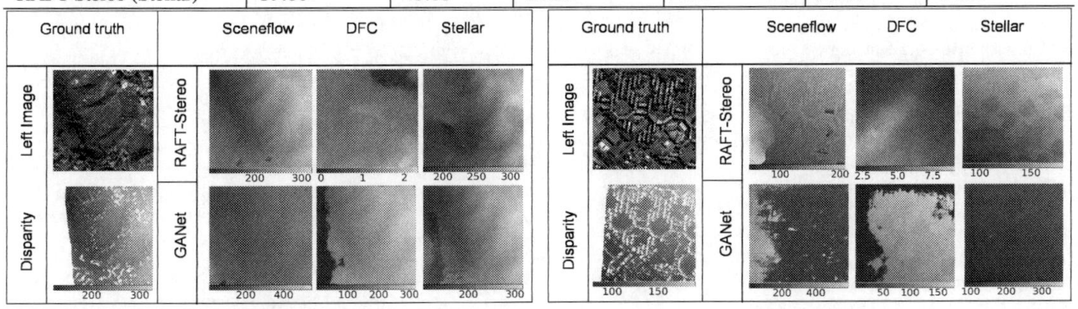

(a) California (b) Argentina

Figure 10. Preliminary quantitative (table) and qualitative (figure) analysis of two DL methods for satellite stereo. **Table.** Best and second best values are highlighted for each metric. Methods fine-tuned on Stellar generally perform the best on both testing regions, while RAFT-Stereo pretrained on Sceneflow performs the best on the California region. **Figures.** Visually, RAFT-Stereo pre-trained on Sceneflow preserves the best quality of fine details.

cost volume. Then, the network uses semi-global aggregation (SGA) layers and local guided aggregation (LGA) layers to perform an approximated semi-global matching procedure that is differentiable and computationally efficient. The whole architecture is differentiable for end-to-end training.

RAFT-Stereo (Lipson et al., 2021) is an end-to-end approach that directly estimates the disparity map from an image pair without generating the cost volume. The model first extracts feature maps from the image pair to build a multi-resolution correlation pyramid. Then it uses a recurrent architecture, the gated recurrent units (GRUs), to create and iteratively refine the output disparity map.

In our experiment, we use models pre-trained by the original authors on the Sceneflow dataset (Mayer et al., 2016) for both architectures. We fine-tune these pre-trained models on DFC and on a portion of Stellar separately, and test all models on a different portion of Stellar. Figure 8 shows the division of Stellar dataset for this experiment. We use image pairs of two cities as the training set and those of another two as the testing set. The testing images may contain different terrains than the training set. For example, the testing images of California in Figure 8c contain hills that are not present in the training images. Therefore, this experiment also analyzes the generalizability of the models.

The loss function for fine-tuning is:

$$\sum_i \mathcal{L}(\mathbf{D}_{gt}, f(\tilde{\mathbf{I}}_i^L, \tilde{\mathbf{I}}_i^R)), \tag{6}$$

where $\mathbf{D}_{gt}$ is the ground truth disparity map, $f = \{\text{GA-Net, RAFT-Stereo}\}$, $\{\tilde{\mathbf{I}}_i^L, \tilde{\mathbf{I}}_i^R\}$ are the ith rectified image pair in the training set. GANet use Huber loss and RAFT-Stereo use $\mathcal{L}_1$ loss as the loss function $\mathcal{L}$. The learning rate is 5×10^{-5} for RAFT-Stereo and 10^{-4} for GANet.

We adopt the following metrics for the quantitative evaluation of estimated disparity maps.

- **Good 3**: given as percentage of the valid pixels in $\mathbf{D}_{gt}$ such that $|\mathbf{D}_L(\mathbf{x}) - \mathbf{D}_{gt}(\mathbf{x})| < \delta$ for $\delta = 3$ where $\mathbf{x} = (u, v)^T$ is pixel position in a disparity map and $\mathbf{D}_L$ is an estimated left disparity map. For this metric, we consider all valid pixels in the ground truth disparity map, and it measures the density of estimated disparity maps.
- **Average Error**: Average error between estimated and ground truth disparity map considering valid pixels in both maps.
- **RMSE**: Root Mean Square Error (RMSE) between estimated and ground truth disparity map considering valid pixels in both maps.

Figure 10 shows the quantitative and qualitative evaluation of the two DL models. For GANet, the model fine-tuned with Stellar achieves the best performance in both testing areas. For RAFT-Stereo, the Stellar model performs the best on one testing region, while the pre-trained model achieves the highest accuracy on another. It is worth noticing that both models trained on DFC perform the worst for both regions, which is probably because the ground truth disparity of DFC is sparse and bipolar, and there is limited training data.

It is not crystal clear to us why the pre-trained RAFT-Stereo sometimes performs better than the same architecture fine-tuned on Stellar. A possible explanation is the training data for Stellar is still not sufficient for such sophisticated DL architectures. We envision more comprehensive analyses of DL satellite stereo methods using datasets such as Stellar in future studies.

5. ACKNOWLEDGEMENT

The authors thank Charles Brookshire for coming up with the name Stellar for our dataset.

REFERENCES

Albanwan, H., Qin, R., 2022. A Comparative Study on Deep-Learning Methods for Dense Image Matching of Multi-angle and Multi-date Remote Sensing Stereo Images. *arXiv preprint arXiv:2210.14031*.

Bosch, M., Foster, K., Christie, G., Wang, S., Hager, G. D., Brown, M., 2019. Semantic stereo for incidental satellite images. *2019 IEEE Winter Conference on Applications of Computer Vision (WACV)*, IEEE, 1524–1532.

Bosch, M., Kurtz, Z., Hagstrom, S., Brown, M., 2016. A multiple view stereo benchmark for satellite imagery. *2016 IEEE Applied Imagery Pattern Recognition Workshop (AIPR)*, IEEE, 1–9.

Brown, M., Goldberg, H., Foster, K., Leichtman, A., Wang, S., Hagstrom, S., Bosch, M., Almes, S., 2018. Large-scale public lidar and satellite image data set for urban semantic labeling. *Laser Radar Technology and Applications XXIII*, 10636, SPIE, 154–167.

Chang, J.-R., Chen, Y.-S., 2018. Pyramid stereo matching network. *Proceedings of the IEEE Conference on Computer Vision and Pattern Recognition*, 5410–5418.

Cheng, X., Zhong, Y., Harandi, M., Dai, Y., Chang, X., Li, H., Drummond, T., Ge, Z., 2020. Hierarchical neural architecture search for deep stereo matching. *Advances in Neural Information Processing Systems*, 33, 22158–22169.

De Franchis, C., Meinhardt-Llopis, E., Michel, J., Morel, J.-M., Facciolo, G., 2014. An automatic and modular stereo pipeline for pushbroom images. *ISPRS Annals of the Photogrammetry, Remote Sensing and Spatial Information Sciences*.

Gómez, A., Randall, G., Facciolo, G., von Gioi, R. G., 2022. An experimental comparison of multi-view stereo approaches on satellite images. *Proceedings of the IEEE/CVF Winter Conference on Applications of Computer Vision*, 844–853.

Gupta, R., Hartley, R. I., 1997. Linear pushbroom cameras. *IEEE Transactions on pattern analysis and machine intelligence*, 19(9), 963–975.

Habib, A. F., Morgan, M., Jeong, S., Kim, K.-O., 2005. Analysis of epipolar geometry in linear array scanner scenes. *The Photogrammetric Record*, 20(109), 27–47.

Hartley, R., Zisserman, A., 2003. *Multiple view geometry in computer vision*. Cambridge university press.

Hirschmuller, H., 2005. Accurate and efficient stereo processing by semi-global matching and mutual information. *2005 IEEE Computer Society Conference on Computer Vision and Pattern Recognition (CVPR'05)*, 2, IEEE, 807–814.

Kendall, A., Martirosyan, H., Dasgupta, S., Henry, P., Kennedy, R., Bachrach, A., Bry, A., 2017. End-to-end learning of geometry and context for deep stereo regression. *Proceedings of the IEEE international conference on computer vision*, 66–75.

Li, J., Wang, P., Xiong, P., Cai, T., Yan, Z., Yang, L., Liu, J., Fan, H., Liu, S., 2022a. Practical stereo matching via cascaded recurrent network with adaptive correlation. *Proceedings of the IEEE/CVF Conference on Computer Vision and Pattern Recognition*, 16263–16272.

Li, S., He, S., Jiang, S., Jiang, W., Zhang, L., 2022b. WHU-Stereo: A Challenging Benchmark for Stereo Matching of High-Resolution Satellite Images. *arXiv:2206.02342*. https://arxiv.org/abs/2206.02342.

Lipson, L., Teed, Z., Deng, J., 2021. Raft-stereo: Multilevel recurrent field transforms for stereo matching. *2021 International Conference on 3D Vision (3DV)*, IEEE, 218–227.

Mayer, N., Ilg, E., Häusser, P., Fischer, P., Cremers, D., Dosovitskiy, A., Brox, T., 2016. A large dataset to train convolutional networks for disparity, optical flow, and scene flow estimation. *IEEE International Conference on Computer Vision and Pattern Recognition (CVPR)*. arXiv:1512.02134.

Menze, M., Geiger, A., 2015. Object scene flow for autonomous vehicles. *Conference on Computer Vision and Pattern Recognition (CVPR)*.

Nuth, C., Kääb, A., 2011. Co-registration and bias corrections of satellite elevation data sets for quantifying glacier thickness change. *The Cryosphere*, 5(1), 271–290.

Patil, S., Comandur, B., Prakash, T., Kak, A. C., 2019a. A new stereo benchmarking dataset for satellite images. *arXiv preprint arXiv:1907.04404*.

Patil, S., Prakash, T., Comandur, B., Kak, A., 2019b. A comparative evaluation of SGM variants (including a new variant, tMGM) for dense stereo matching. *arXiv preprint arXiv:1911.09800*.

San Diego LiDAR Report, n.d. `https://noaa-nos-coastal-lidar-pds.s3.amazonaws.com/laz/geoid18/8614/supplemental/USGS_Eastern_San_Diego_County_LiDAR_ProjectReport.pdf`.

Scharstein, D., Hirschmüller, H., Kitajima, Y., Krathwohl, G., Nešić, N., Wang, X., Westling, P., 2014. High-resolution stereo datasets with subpixel-accurate ground truth. *German conference on pattern recognition*, Springer, 31–42.

Van Etten, A., Lindenbaum, D., Bacastow, T. M., 2018. Spacenet: A remote sensing dataset and challenge series. *arXiv preprint arXiv:1807.01232*.

Zabih, R., Woodfill, J., 1994. Non-parametric local transforms for computing visual correspondence. *Computer Vision-ECCV'94: Third European Conference on Computer Vision, Stockholm, Sweden, May 2-6, 1994. Proceedings*, 2, Springer Science & Business Media, 151.

Zbontar, J., LeCun, Y. et al., 2016. Stereo Matching by Training a Convolutional Neural Network to Compare Image Patches. *Journal of Machine Learning Research*, 17(1-32), 2.

Zhang, F., Prisacariu, V., Yang, R., Torr, P. H., 2019. Ga-net: Guided aggregation net for end-to-end stereo matching. *The IEEE Conference on Computer Vision and Pattern Recognition (CVPR)*.

Zhao, H., Zhou, H., Zhang, Y., Zhao, Y., Yang, Y., Ouyang, T., 2022. Eai-stereo: Error aware iterative network for stereo matching. *Proceedings of the Asian Conference on Computer Vision*, 315–332.

AUTHOR INDEX